The Elements

Name	Symbol	Atomic Number	Atomic Weight
actinium	Ac	89	[227]
aluminum	Al	13	26.981538 (2)
americium	Am	95	[243]
antimony	Sb	51	121.76 (1)
argon	Ar	18	39.948 (1)
arsenic	As	33	74.9216 (2)
astatine	At	85	[210]
barium	Ba	56	137.327 (7)
berkelium	Bk	97	[247]
beryllium	Be	4	9.012182 (3)
bismuth	Bi	83	208.98038 (2)
bohrium	Bh	107	[264]
boron	B	5	10.811 (7)
bromine	Br	35	79.904 (1)
cadmium	Cd	48	112.411 (8)
calcium	Ca	20	40.078 (4)
caesium	Cs	55	132.90545 (2)
californium	Cf	98	[251]
carbon	C	6	12.0107 (8)
cerium	Ce	58	140.116 (1)
chlorine	Cl	17	35.4527 (9)
chromium	Cr	24	51.9961 (6)
cobalt	Co	27	58.9332 (9)
copper	Cu	29	63.546 (3)
curium	Cm	96	[247]
dubnium	Db	105	[262]
dysprosium	Dy	66	162.5 (3)
einsteinium	Es	99	[252]
erbium	Er	68	167.26 (3)
europium	Eu	63	151.964 (1)
fermium	Fm	100	[257]
fluorine	F	9	18.9984032 (5)
francium	Fr	87	[223]
gadolinium	Gd	64	157.25 (3)
gallium	Ga	31	69.723 (1)
germanium	Ge	32	72.61 (2)
gold	Au	79	196.96655 (2)
hafnium	Hf	72	178.49 (2)
hassium	Hs	108	[269]
helium	He	2	4.002602 (2)
holmium	Ho	67	164.93032 (2)
hydrogen	H	1	1.00794 (7)
indium	In	49	114.818 (3)
iodine	I	53	126.90447 (3)
iridium	Ir	77	192.217 (3)
iron	Fe	26	55.845 (2)
krypton	Kr	36	83.8 (1)

(continued)

Name	Symbol	Atomic Number	Atomic Weight
lanthanum	La	57	138.9055 (2)
lawrencium	Lr	103	[262]
lead	Pb	82	207.2 (1)
lithium	Li	3	[6.941 (2)]
lutetium	Lu	71	174.967 (1)
magnesium	Mg	12	24.305 (6)
manganese	Mn	25	54.938049 (9)
meitnerium	Mt	109	[268]
mendelevium	Md	101	[258]
mercury	Hg	80	200.59 (2)
molybdenum	Mo	42	95.94 (1)
neodymium	Nd	60	144.24 (3)
neon	Ne	10	20.1797 (6)
neptunium	Np	93	[237]
nickel	Ni	28	58.6934 (2)
niobium	Nb	41	92.90638 (2)
nitrogen	N	7	14.00674 (7)
nobelium	No	102	[259]
osmium	Os	76	190.23 (3)
oxygen	O	8	15.9994 (3)
palladium	Pd	46	106.42 (1)
phosphorus	P	15	30.973762 (4)
platinum	Pt	78	195.078 (2)
plutonium	Pu	94	[244]
polonium	Po	84	[210]
potassium	K	19	39.0983 (1)
praseodymium	Pr	59	140.90765 (2)
promethium	Pm	61	[145]
protactinium	Pa	91	231.03588 (2)
radium	Ra	88	[226]
radon	Rn	86	[222]
rhenium	Re	75	186.207 (1)
rhodium	Rh	45	102.9055 (2)
rubidium	Rb	37	85.4678 (3)
ruthenium	Ru	44	101.07 (2)
rutherfordium	Rf	104	[261]
samarium	Sm	62	150.36 (3)
scandium	Sc	21	44.95591 (8)
seaborgium	Sg	106	[266]
selenium	Se	34	78.96 (3)
silicon	Si	14	28.0855 (3)
silver	Ag	47	107.8682 (2)
sodium	Na	11	22.98977 (2)
strontium	Sr	38	87.62 (1)
sulfur	S	16	32.066 (6)
tantalum	Ta	73	180.9479 (1)
technetium	Tc	43	[98]
tellurium	Te	52	127.6 (3)
terbium	Tb	65	158.92534 (2)

AN INTRODUCTION TO
MATERIALS ENGINEERING
AND SCIENCE

AN INTRODUCTION TO MATERIALS ENGINEERING AND SCIENCE

FOR CHEMICAL AND MATERIALS ENGINEERS

Brian S. Mitchell
Department of Chemical Engineering,
Tulane University

WILEY-INTERSCIENCE

A JOHN WILEY & SONS, INC., PUBLICATION

For general information on our other products and services please contact our Customer Care Department within the U.S. at 877-762-2974, outside the U.S. at 317-572-3993 or fax 317-572-4002.

Wiley also publishes its books in a variety of electronic formats. Some content that appears in print, however, may not be available in electronic format.

Library of Congress Cataloging-in-Publication Data:

Mitchell, Brian S., 1962-
    An introduction to materials engineering and science: for chemical and materials engineers
    Brian S. Mitchell
        p. cm.
    Includes bibliographical references and index.
    ISBN 978-0-471-43623-2
    1. Materials science. I. Title.

TA403.M685 2003
620.1'1—dc21   2003053451

To my parents; whose

Material was loam;

Engineering was labor;

Science was lore;

And greatest product was love.

CONTENTS

*Sections marked with an asterisk can be omitted in an introductory course.

This textbook is intended for use in a one- or two-semester undergraduate course in materials science that is primarily populated by chemical and materials engineering students. This is not to say that biomedical, mechanical, electrical, or civil engineering students will not be able to utilize this text, nor that the material or its presentation is unsuitable for these students. On the contrary, the breadth and depth of the material covered here is equivalent to most "traditional" metallurgy-based approaches to the subject that students in these disciplines may be more accustomed to. In fact, the treatment of biological materials on the same level as metals, ceramics, polymers, and composites may be of particular benefit to those students in the biologically related engineering disciplines. The key difference is simply the organization of the material, which is intended to benefit primarily the chemical and materials engineer.

This textbook is organized on two levels: by engineering subject area and by materials class, as illustrated in the accompanying table. In terms of topic coverage, this organization is transparent: By the end of the course, the student will have covered many of the same things that would be covered utilizing a different materials science textbook. To the student, however, the organization is intended to facilitate a deeper understanding of the subject material, since it is presented in the context of courses they have already had or are currently taking—for example, thermodynamics, kinetics, transport phenomena, and unit operations. To the instructor, this organization means that, in principle, the material can be presented either in the traditional subject-oriented sequence (i.e., in rows) or in a materials-oriented sequence (i.e., in columns). The latter approach is recommended for a two-semester course, with the first two columns covered in the first semester and the final three columns covered in the second semester. The instructor should immediately recognize that the vast majority of "traditional" materials science concepts are covered in the columns on metals and ceramics, and that if the course were limited to concepts on these two materials classes only, the student would receive instruction in many of the important topics covered in a "traditional" course on materials. Similarly, many of the more advanced topics are found in the sections on polymers, composites, and biological materials and are appropriate for a senior-level, or even introductory graduate-level, course in materials with appropriate supplementation and augmentation.

This textbook is further intended to provide a unique educational experience for the student. This is accomplished through the incorporation of instructional objectives, active-learning principles, design-oriented problems, and web-based information and visualization utilization. Instructional objectives are included at the beginning of each chapter to assist both the student and the instructor in determining the extent of topics and the depth of understanding required from each topic. This list should be used as a guide only: Instructors will require additional information they deem important or eliminate topics they deem inappropriate, and students will find additional topic coverage in their supplemental reading, which is encouraged through a list of references at the end

	Metals & Alloys	Ceramics & Glasses	Polymers	Composites	Biologics
Structure	Crystal structures, Point defects, Dislocations	Crystal structures, Defect reactions, The glassy state	Configuration, Conformation, Molecular Weight	Matrices, Reinforcements	Biochemistry, Tissue structure
Thermodynamics	Phase equilibria, Gibbs Rule Lever Rule	Ternary systems, Surface energy, Sintering	Phase separation, Polymer solutions, Polymer blends	Adhesion, Cohesion, Spreading	Cell Adhesion, Cell spreading
Kinetics	Transformations, Corrosion	Devitrification, Nucleation, Growth	Polymerization, Degradation	Deposition, Infiltration	Receptors, Ligand binding
Transport Properties	Inviscid systems, Heat capacity, Diffusion	Newtonian flow, Heat capacity, Diffusion	non-Newtonian flow, Heat capacity, Diffusion	Porous Flow, Heat capacity, Diffusion	Convection, Diffusion
Mechanical Properties	Stress-strain, Elasticity, Ductility	Fatigue, Fracture, Creep	Viscoelasticity, Elastomers	Laminates	Sutures, Bone, Teeth
Electrical, Magnetic & Optical Properties	Resistivity, Magnetism, Reflectance	Dielectrics, Ferrites, Absorbance	Ion conductors, Molecular magnets, LCDs	Dielectrics, Storage media	Biosensors, MRI
Processing	Casting, Rolling, Compaction	Pressing, CVD/CVI, Sol-Gel	Extrusion, Injection molding, Blow molding	Pultrusion, RTM, CVD/CVI	Surface modification
Case Studies	Compressed air tank	Ceramic piping	Polymeric packaging	Composite drive shaft	Tooth coatings

of each chapter. Active-learning principles are exercised through the presentation of example problems in the form of Cooperative Learning Exercises. To the student, this means that they can solve problems in class and can work through specific difficulties in the presence of the instructor. Cooperative learning has been shown to increase the level of subject understanding when properly utilized.* No class is too large to allow students to take 5–10 minutes to solve these problems. To the instructor, the Cooperative Learning Exercises are to be used only as a starting point, and the instructor is encouraged to supplement his or her lecture with more of these problems. Particularly difficult concepts or derivations are presented in the form of Example Problems that the instructor can solve in class for the students, but the student is encouraged to solve these problems during their own group or individual study time. Design-oriented problems are offered, primarily in the Level III problems at the end of each chapter,

*Smith, K. *Cooperative Learning and College Teaching*, **3**(2), 10–12 (1993).

that incorporate concepts from several chapters, that involve significant information retrieval or outside reading, or that require group activities. These problems may or may not have one "best" answer and are intended to promote a deeper level of understanding of the subject. Finally, there is much information on the properties of materials available on the Internet. This fact is utilized through the inclusion of appropriate web links. There are also many excellent visualization tools available on the Internet for concepts that are too difficult to comprehend in a static, two-dimensional environment, and links are provided to assist the student in their further study on these topics.

Finally, the ultimate test of the success of any textbook is whether or not it stays on your bookshelf. It is hoped that the extent of physical and mechanical property data, along with the depth with which the subjects are presented, will serve the student well as they transition to the world of the practicing engineer or continue with their studies in pursuit of an advanced degree.

BRIAN S. MITCHELL
Tulane University

◼◼◼ ACKNOWLEDGMENTS

The author wishes to thank the many people who have provided thoughtful input to the content and presentation of this book. In particular, the insightful criticisms and comments of Brian Grady and the anonymous reviewers are very much appreciated. Thanks also go to my students who have reviewed various iterations of this textbook, including Claudio De Castro, Shawn Haynes, Ryan Shexsnaydre, and Amanda Moster, as well as Dennis Johnson, Eric Hampsey, and Tom Fan. The support of my colleagues during the writing of this book, along with the support of the departmental staff, are gratefully acknowledged. Finally, the moral support of Bonnie, Britt, Rory, and Chelsie is what ultimately has led to the completion of this textbook—thank you.

BRIAN S. MITCHELL
Tulane University

CHAPTER 1

The Structure of Materials

1.0 INTRODUCTION AND OBJECTIVES

A wealth of information can be obtained by looking at the structure of a material. Though there are many levels of structure (e.g., atomic vs. macroscopic), many physical properties of a material can be related directly to the arrangement and types of bonds that make up that material. We will begin by reviewing some general chemical principles that will aid us in our description of material structure. Such topics as periodic structure, types of bonding, and potential energy diagrams will be reviewed. We will then use this information to look at the specific materials categories in more detail: metals, ceramics, polymers, composites, and biological materials (biologics). There will be topics that are specific to each material class, and there will also be some that are common to all types of materials. In subsequent chapters, we will explore not only how the building blocks of a material can significantly impact the properties a material possesses, but also how the material interacts with its environment and other materials surrounding it.

By the end of this chapter you should be able to:

- Identify trends in the periodic table for *IE, EA*, electronegativity, and atomic/ionic radii.
- Identify the type of bonding in a compound.
- Utilize the concepts of molecular orbital and hybridization theories to explain multiple bonds, bond angle, diamagnetism, and paramagnetism.
- Identify the seven crystal systems and 14 Bravais lattices.
- Calculate the volume of a unit cell from the lattice translation vectors.
- Calculate atomic density along directions, planes, and volumes in a unit cell.
- Calculate the density of a compound from its crystal structure and atomic mass.
- Locate and identify the interstitial sites in a crystal structure.
- Assign coordinates to a location, indices to a direction, and Miller indices to a plane in a unit cell.
- Use Bragg's Law to convert between diffraction angle and interplanar spacing.
- Read and interpret a simple X-ray diffraction pattern.
- Identify types of point and line defects in solids.

An Introduction to Materials Engineering and Science: For Chemical and Materials Engineers, by Brian S. Mitchell
ISBN 0-471-43623-2 Copyright © 2004 John Wiley & Sons, Inc.

- Calculate the concentration of point defects in solids.
- Draw a Burger's circuit and identify the direction of dislocation propagation.
- Use Pauling's rules to determine the stability of a compound.
- Predict the structure of a silicate from the Si/O ratio.
- Apply Zachariasen's rules to determine the glass forming ability of an oxide.
- Write balanced defect reaction equations using Kroger–Vink notation.
- Classify polymers according to structure or formability.
- Calculate the first three moments of a polymer molecular weight distribution.
- Apply principles of glass transition and polymer crystallinity to polymer classification.
- Identify nematic, smectic, and cholesteric structures in liquid crystalline polymers.
- Identify the components in a composite material.
- Approximate physical properties of a composite material based on component properties.
- Be conversant in terms that relate to the structure of biological materials, such as fibronectin and integrins.

1.0.1 The Elements

Elements are materials, too. Oftentimes, this fact is overlooked. Think about all the materials from our daily lives that are elements: gold and silver for our jewelry; aluminum for our soda cans; copper for our plumbing; carbon, both as a luminescent diamond and as a mundane pencil lead; mercury for our thermometers; and tungsten for our light bulb filaments. Most of these elements, however, are relatively scarce in the grand scheme of things. A table of the relative abundance of elements (Table 1.1) shows that most of our universe is made up of hydrogen and helium. A little closer to home, things are much different. A similar table of relative abundance (Table 1.2) shows that helium on earth is relatively scarce, while oxygen dominates the crust of our planet. Just think of how much molecular oxygen, water, and aluminosilicate rocks are contained in the earth's crust. But those are molecules—we are concentrating on atoms for the moment. Still, elements are of vital importance on earth, and the ones we use most often are primarily in the solid form.

Recall from your introductory chemistry course that the elements can be systematically arranged in a periodic table according to their electronic structure (see Table 1.3*). An overall look at the periodic table (Figure 1.1) shows that many elements are solids (white boxes) at room temperature. The fact that many of these elements remain solid well above ambient temperatures is important. As we heat to 1000°C, note that many of the IIIA–VA elements have melted (light shaded); also note how many of the alkali metals (IA) have vaporized (dark shaded), but how most of the transition elements are still in solid form. At 2000°C, the alkali earths are molten, and many of the transition elements have begun to melt, too. Note that the highest melting point element is carbon (Figure 1.1d). Keep in mind that this is in an inert atmosphere. What should happen to

*Note that the Lanthanide (atomic numbers 58–71) and Actinide (90–103) series elements, as well as the synthetic elements of atomic number greater than 87, are omitted from all the periodic tables in this text. With the possible exception of nuclear fuels such as uranium and plutonium, these elements are of little general engineering interest.

Table 1.1   Relative Abundance of Elements in the Universe

Element	Relative Abundance (Si = 1)
Hydrogen (H)	12,000
Helium (He)	2,800
Oxygen (O)	16
Nitrogen (N)	8
Carbon (C)	3
Iron (Fe)	2.6
Silicon (Si)	1
Magnesium (Mg)	0.89
Sulfur (S)	0.33
Nickel (Ni)	0.21
Aluminum (Al)	0.09
Calcium (Ca)	0.07
Sodium (Na)	0.045
Chlorine (Cl)	0.025

Table 1.2   Relative Abundance of Selected Elements in the Earth's Crust

Element	Relative Abundance (ppm)	Element	Relative Abundance (ppm)
Oxygen (O)	466,000	Fluorine (F)	300
Silicon (Si)	277,200	Strontium (Sr)	300
Aluminum (Al)	81,300	Barium (Ba)	250
Iron (Fe)	50,000	Zirconium (Zr)	220
Calcium (Ca)	36,300	Chromium (Cr)	200
Sodium (Na)	28,300	Vanadium (V)	150
Potassium (K)	25,900	Zinc (Zn)	132
Magnesium (Mg)	20,900	Nickel (Ni)	80
Titanium (Ti)	4,400	Molybdenum (Mo)	15
Hydrogen (H)	1,400	Uranium (U)	4
Phosphorus (P)	1,180	Mercury (Hg)	0.5
Manganese (Mn)	1,000	Silver (Ag)	0.1
Sulfur (S)	520	Platinum (Pt)	0.005
Carbon (C)	320	Gold (Au)	0.005
Chlorine (Cl)	314	Helium (He)	0.003

this element in the presence of oxygen? Such elements as tungsten, platinum, molybdenum, and tantalum have exceptional high-temperature properties. Later on we will investigate why this is so.

In addition, many elements are, in and of themselves, materials of construction. Aluminum and copper are just a few examples of elements that are used extensively for fabricating mechanical parts. Elements have special electrical characteristics, too. Silver and gold are used not just for jewelry, but also for a wide variety of electrical components. We will visit all of these topics in the course of this textbook.

1.0.2   Trends in the Periodic Table

A closer look at the periodic table points out some interesting trends. These trends not only help us predict how one element might perform relative to another, but also give us some insight into the important properties of atoms and ions that determine their performance. For example, examination of the melting points of the elements in Table 1.3 shows that there is a general trend to decrease melting point as we go down

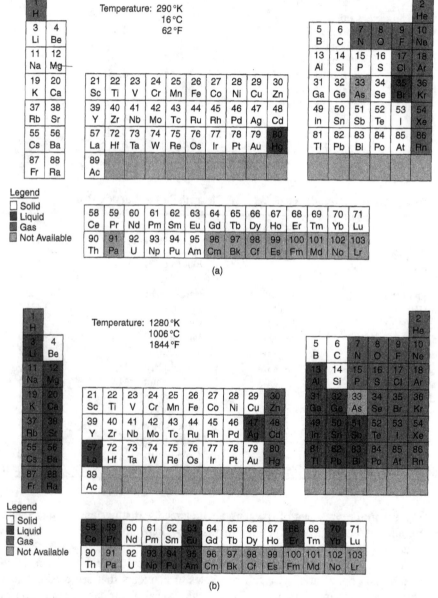

Figure 1.1   The periodic table of the elements at (a) room temperature, (b) 1000°C, (c) 2000°C, and (d) 3500°C.

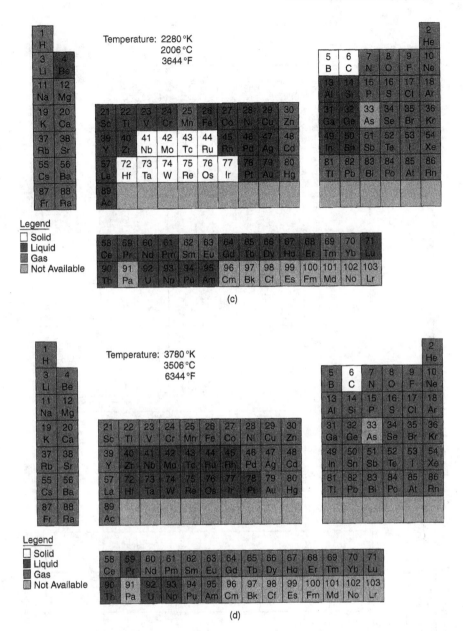

Figure 1.1 (*continued*).

a column for the alkali metals and alkali earth elements (columns IA and IIA), but that the column trend for the transition metals appears to be different. There are some trends that are more uniform, however, and are related to the electronic structure of the element.

1.0.2.1 *First Ionization Energy (IE)*.

The first *ionization energy*, abbreviated *IE*, is sometimes referred to as the "ionization potential." It is the energy required to remove

Table 1.3 Electronic Structure and Melting Points of the Elements

es = electronic structure
aw = atomic weight (average including isotopes)
mp = melting point, °C (sublimation temperatures enclosed in parentheses).

Z	Symbol	es	aw	mp
1	H	$1s^1$	1.008	−259
2	He	$1s^2$	4.003	—
3	Li	$He2s^1$	6.94	180.5
4	Be	$He2s^2$	9.012	1289
5	B	$Be2p^1$	10.81	~2103
6	C	$Be2p^2$	12.01	(3836)
7	N	$Be2p^3$	14.006	−210.0
8	O	$Be2p^4$	15.999	−218.8
9	F	$Be2p^5$	18.998	−219.7
10	Ne	$Be2p^6$	20.18	−249
11	Na	$Ne3s^1$	22.99	97.8
12	Mg	$Ne3s^2$	24.30	649
13	Al	$Mg3p^1$	26.98	660.4
14	Si	$Mg3p^2$	28.09	1414
15	P	$Mg3p^3$	30.974	44.1
16	S	$Mg3p^4$	32.06	112.8
17	Cl	$Mg3p^5$	35.45	−101.0
18	Ar	$Mg3p^6$	39.95	−189
19	K	$Ar4s^1$	39.10	63.2
20	Ca	$Ar4s^2$	40.08	840
21	Sc	$Ca3d^1$	44.96	1541
22	Ti	$Ca3d^2$	47.9	1672
23	V	$Ca3d^3$	50.94	1929
24	Cr	$K3d^5$	51.9	1863
25	Mn	$Ca3d^5$	54.93	1246
26	Fe	$Ca3d^6$	55.85	1538
27	Co	$Ca3d^7$	58.93	1494
28	Ni	$Ca3d^8$	58.71	1455
29	Cu	$K3d^{10}$	63.55	1084.5
30	Zn	$Ca3d^{10}$	65.37	419.6
31	Ga	$Ca4p^1$	69.72	29.8
32	Ge	$Ca4p^2$	72.59	938.3
33	As	$Ca4p^3$	74.92	(603)
34	Se	$Ca4p^4$	78.96	221
35	Br	$Ca4p^5$	79.90	−7.2
36	Kr	$Ca4p^6$	83.80	−157
37	Rb	$Kr5s^1$	85.47	39.5
38	Sr	$Kr5s^2$	87.62	769
39	Y	$Sr4d^1$	88.91	1528
40	Zr	$Sr4d^2$	91.22	1865
41	Nb	$Rb4d^4$	92.91	2471
42	Mo	$Rb4d^5$	95.94	2623
43	Tc	$Rb4d^6$	98.91	2204
44	Ru	$Rb4d^7$	101.7	2254
45	Rh	$Rb4d^8$	102.9	1963
46	Pd	$Kr4d^{10}$	106.4	1554
47	Ag	$Rb4d^{10}$	107.87	961.9
48	Cd	$Sr4d^{10}$	112.4	321.1
49	In	$Sr5p^1$	114.8	156.6
50	Sn	$Sr5p^2$	118.7	232.0
51	Sb	$Sr5p^3$	121.8	630.7
52	Te	$Sr5p^4$	127.6	449.6
53	I	$Sr5p^5$	126.9	113.6
54	Xe	$Sr5p^6$	131.3	−112
55	Cs	$Xe6s^1$	132.9	28.4
56	Ba	$Xe6s^2$	137.3	729
57	La	$Ba5d^1$	138.9	921
72	Hf	$Ba5d^2$	178.5	2231
73	Ta	$Ba5d^3$	180.9	3020
74	W	$Ba5d^4$	183.9	3387
75	Re	$Ba5d^5$	186.2	3186
76	Os	$Ba5d^6$	190.2	3033
77	Ir	$Xe5d^7$	192.2	2447
78	Pt	$Cs5d^9$	195.1	1772
79	Au	$Cs5d^{10}$	196.97	1064.4
80	Hg	$Ba5d^{10}$	200.6	−38.9
81	Tl	$Ba6p^1$	204.4	304
82	Pb	$Ba6p^2$	207.2	327.5
83	Bi	$Ba6p^3$	208.9	271.4
84	Po	$Ba6p^4$	210	254
85	At	$Ba6p^5$	210	—
86	Rn	$Ba6p^6$	222	−71

Source: Ralls, Courtney, Wulff, Introduction to Materials Science and Engineering, Wiley, 1976.

the most weakly bound (usually outermost) electron from an isolated gaseous atom

$$\text{atom (g)} + IE \longrightarrow \text{positive ion (g)} + e^- \tag{1.1}$$

and can be calculated using the energy of the outermost electron as given by the Bohr model and Schrödinger's equation (in eV):

$$IE = \frac{13.6Z^2}{n^2} \tag{1.2}$$

where Z is the effective nuclear charge and n is the principal quantum number.

As shown in Figure 1.2a, the general trend in the periodic table is for the ionization energy to increase from bottom to top and from left to right (why?). A quantity related to the IE is the *work function*. The work function is the energy necessary to remove an electron from the metal surface in thermoelectric or photoelectric emission. We will describe this in more detail in conjunction with electronic properties of materials in Chapter 6.

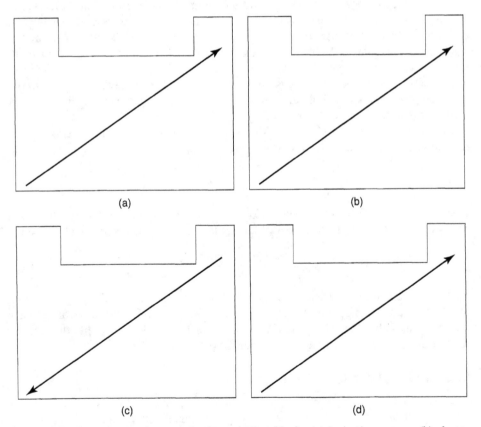

Figure 1.2   Some important trends in the periodic table for (a) ionization energy, (b) electron affinity, (c) atomic and ionic radii, and (d) electronegativity. Increasing values are in the direction of the arrow.

1.0.2.2 Electron Affinity (EA). *Electron affinity* is the reverse process to the ionization energy; it is the energy change (often expressed in eV) associated with an isolated gaseous atom accepting one electron:

$$\text{atom (g)} + e^- \longrightarrow \text{negative ion (g)} \tag{1.3}$$

Unlike the ionization energy, however, *EA* can have either a negative or positive value, so it is not included in Eq. (1.3). The *EA* is positive if energy is released upon formation of the negative ion. If energy is required, *EA* is negative. The general trend in the periodic table is again toward an increase in *EA* as we go from the bottom to top, and left to right (Figure 1.2b), though this trend is much less uniform than for the *IE*.

1.0.2.3 Atomic and Ionic Radii. In general, positive ions are smaller than neutral atoms, while negative ions are larger (why?). The trend in ionic and atomic radii is opposite to that of *IE* and *EA* (Figure 1.2c). In general, there is an increase in radius from top to bottom, right to left. In this case, the effective nuclear charge increases from left to right, the inner electrons cannot shield as effectively, and the outer electrons are drawn close to the nucleus, reducing the atomic radius. Note that the radii are only approximations because the orbitals, in theory, extend to infinity.

1.0.2.4 Electronegativity. The ionization energy and electron affinity are characteristics of isolated atoms; they say very little about how two atoms will interact with each other. It would be nice to have an independent measure of the attraction an atom has for electrons in a bond formed with another atom. *Electronegativity* is such a quantity. It is represented by the lowercase Greek letter "chi," χ. Values can be calculated using one of several methods discussed below. Values of χ are always relative to one another for a given method of calculation, and values from one method should not be used with values from another method.

Based upon a scale developed by Mulliken, electronegativity is the average of the ionization energy and the electron affinity:

$$\chi = \frac{IE + EA}{2} \tag{1.4}$$

There are other types of electronegativity scales as well, the most widely utilized of which is the one from the developer of the electronegativity concept, Linus Pauling:

$$\chi = \frac{0.31(n + 1 \pm c)}{r} + 0.5 \tag{1.5}$$

where n is the number of valence electrons, c is any formal valence charge on the atom and the sign corresponding to it, and r is the covalent radius. Typical electronegativity values, along with values of *IE* and *EA*, are listed in Table 1.4. We will use the concept of electronegativity to discuss chemical bonding.

1.0.3 Types of Bonds

Electronegativity is a very useful quantity to help categorize bonds, because it provides a measure of the *excess binding energy* between atoms A and B, Δ_{A-B} (in kJ/mol):

$$\Delta_{A-B} = 96.5(\chi_A - \chi_B)^2 \tag{1.6}$$

Table 1.4 Ionization Energies, Electron Affinities, and Electronegativities of the Elements[a]

Period 1

	1 H	2 He
IE	1310	2372
EA	67.4	−60.2
X	2.20	—

Period 2

	3 Li	4 Be	5 B	6 C	7 N	8 O	9 F	10 Ne
IE	519	900	799	1088	1406	1314	1682	2080
EA	77.0	−18.4	31.8	119.7	4.6	141.8	349.4	−54.8
X	0.98	1.57	2.04	2.55	3.04	3.44	3.98	—

Period 3

	11 Na	12 Mg	13 Al	14 Si	15 P	16 S	17 Cl	18 Ar
IE	498	736	577	787	1063	1000	1255	2372
EA	117.2	0	50.2	138.1	75.3	199.6	356.1	−60.2
X	0.93	1.31	1.61	1.90	2.19	2.58	3.16	—

Period 4

	19 K	20 Ca	21 Sc	22 Ti	23 V	24 Cr	25 Mn	26 Fe	27 Co	28 Ni	29 Cu	30 Zn	31 Ga	32 Ge	33 As	34 Se	35 Br	36 Kr
IE	419	590	632	661	653	653	715	761	757	736	745	904	577	782	966	941	1142	1351
EA	48.4	—	—	—	—	—	—	—	—	—	—	—	—	—	—	—	333.0	—
X	0.82	1.00	1.36	1.54	1.63	1.66	1.55	1.8	1.88	1.91	1.90	1.65	1.81	2.01	2.18	2.55	2.96	—

Period 5

	37 Rb	38 Sr	39 Y	40 Zr	41 Nb	42 Mo	43 Tc	44 Ru	45 Rh	46 Pd	47 Ag	48 Cd	49 In	50 Sn	51 Sb	52 Te	53 I	54 Xe
IE	402	548	636	669	653	695	699	724	745	803	732	866	556	707	833	870	1008	1172
EA	—	—	—	—	—	—	—	—	—	—	—	—	—	—	—	—	304.2	—
X	0.82	0.95	1.22	1.33	1.6	2.16	1.9	2.28	2.2	2.20	1.93	1.69	1.78	1.96	2.05	2.1	2.66	—

Period 6

	55 Cs	56 Ba	57 La	72 Hf	73 Ta	74 W	75 Re	76 Os	77 Ir	78 Pt	79 Au	80 Hg	81 Tl	82 Pb	83 Bi	84 Po	85 At	86 Rn
IE	377	502	540	531	577	770	761	841	887	866	891	1008	590	715	774	—	—	1038
EA	—	—	—	—	—	—	—	—	—	—	—	—	—	—	—	—	—	—
X	0.79	0.89	1.10	1.3	1.5	2.36	1.9	2.2	2.2	2.28	2.54	2.00	2.04	2.33	2.02	2.0	2.2	—

[a] Ionization energy (IE) and Electron affinities (EA) are expressed as kilojoules per mole. 1 eV = 96,490 J/mol.
Source: R. E. Dickerson, H. B. Gray, and G. P. Haight, *Chemical Principles*, 3rd ed., Pearson Education, Inc., 1979.

The excess binding energy, in turn, is related to a measurable quantity, namely the *bond dissociation energy* between two atoms, DE_{ij}:

$$\Delta_{A-B} = DE_{AB} - [(DE_{AA})(DE_{BB})]^{1/2} \tag{1.7}$$

The bond dissociation energy is the energy required to separate two bonded atoms (see Appendix 1 for typical values). The greater the electronegativity difference, the greater the excess binding energy. These quantities give us a method of characterizing bond types. More importantly, they relate to important physical properties, such as melting point (see Table 1.5). First, let us review the bond types and characteristics, then describe each in more detail.

1.0.3.1 Primary Bonds. *Primary bonds*, also known as "strong bonds," are created when there is direct interaction of electrons between two or more atoms, either through transfer or as a result of sharing. The more electrons per atom that take place in this process, the higher the bond "order" (e.g., single, double, or triple bond) and the stronger the connection between atoms. There are four general categories of primary bonds: *ionic, covalent, polar covalent,* and *metallic.* An ionic bond, also called a *heteropolar*

Table 1.5 Examples of Substances with Different Types of Interatomic Bonding

Type of Bond	Substance	Bond Energy, kJ/mol	Melting Point, (°C)	Characteristics
Ionic	CaCl	651	646	Low electrical conductivity, transparent, brittle, high melting point
	NaCl	768	801	
	LiF	1008	870	
	CuF_2	2591	1360	
	Al_2O_3	15,192	3500	
Covalent	Ge	315	958	Low electrical conductivity, very hard, very high melting point
	GaAs	~315	1238	
	Si	353	1420	
	SiC	1188	2600	
	Diamond	714	3550	
Metallic	Na	109	97.5	High electrical and thermal conductivity, easily deformable, opaque
	Al	311	660	
	Cu	340	1083	
	Fe	407	1535	
	W	844	3370	
van der Waals	Ne	2.5	−248.7	Weak binding, low melting and boiling points, very compressible
	Ar	7.6	−189.4	
	CH_4	10	−184	
	Kr	12	−157	
	Cl_2	31	−103	
Hydrogen bonding	HF	29	−92	Higher melting point than van der Waals bonding, tendency to form groups of many molecules
	H_2O	50	0	

bond, results when electrons are transferred from the more electropositive atom to the more electronegative atom, as in sodium chloride, NaCl. Ionic bonds usually result when the electronegativity difference between two atoms in a diatomic molecule is greater than about 2.0. Because of the large discrepancy in electronegativities, one atom will generally gain an electron, while the other atom in a diatomic molecule will lose an electron. Both atoms tend to be "satisfied" with this arrangement because they oftentimes end up with noble gas electron configurations—that is, full electronic orbitals. The classic example of an ionic bond is NaCl, but CaF_2 and MgO are also examples of molecules in which ionic bonding dominates.

A covalent bond, or *homopolar bond*, arises when electrons are shared between two atoms (e.g., H–H). This means that a binding electron in a covalent diatomic molecule such as H_2 has equal likelihood of being found around either hydrogen atom. Covalent bonds are typically found in homonuclear diatomics such as O_2 and N_2, though the atoms need not be the same to have similar electronegativities. Electronegativity differences of less than about 0.4 characterize covalent bonds. For two atoms with an electronegativity difference of between 0.4 and 2.0, a polar covalent bond is formed—one that is neither truly ionic nor totally covalent. An example of a polar covalent bond can be found in the molecule hydrogen fluoride, HF. Though there is significant sharing of the electrons, some charge distribution exists that results in a polar or *partial ionic character* to the bond. The percent ionic character of the bond can again be related to the electronegativities of the individual atoms:

$$\% \text{ ionic character} = 100\{1 - \exp[-0.25(\chi_A - \chi_B)^2]\} \tag{1.8}$$

Example Problem 1.1

What is the percent ionic character of H–F?

Answer: According to Table 1.4, the electronegativity of hydrogen is 2.20 and that of fluorine 3.98. Putting these values into Eq. (1.8) gives

$$\% \text{ ionic character of H–F} = 100[1 - \exp\{-0.25(2.20 - 3.98)^2\}] = 55\%$$

Adapted from Callister, W. D., *Materials Science and Engineering, An Introduction, 5th Ed.*, John Wiley & Sons, New York, 2000, page 28 (ionic separation distance changed).

The larger the electronegativity difference, the more ionic character the bond has. Of course, if the electronegativity difference is greater than about 2.0, we know that an ionic bond should result.

Finally, a special type of primary bond known as a metallic bond is found in an assembly of homonuclear atoms, such as copper or sodium. Here the bonding electrons become "decentralized" and are shared by the core of positive nuclei. Metallic bonds occur when elements of low electronegativity (usually found in the lower left region of the periodic table) bond with each other to form a class of materials we call *metals*. Metals tend to have common characteristics such as ductility, luster, and high thermal and electrical conductivity. All of these characteristics can to some degree be accounted for by the nature of the metallic bond. The model of a metallic bond, first proposed by Lorentz, consists of an assembly of positively charged ion cores surrounded by free electrons or an "electron gas." We will see later on, when we describe intermolecular forces and bonding, that the electron cloud does indeed have "structure" in the quantum mechanical sense, which accounts nicely for the observed electrical properties of these materials.

1.0.3.2 Secondary Bonds.

Secondary bonds, or *weak bonds*, occur due to indirect interaction of electrons in adjacent atoms or molecules. There are three main types of secondary bonding: *hydrogen bonding, dipole–dipole interactions*, and *van der Waals forces*. The latter, named after the famous Dutch physicist who first described them, arise due to momentary electric *dipoles* (regions of positive and negative charge) that can occur in all atoms and molecules due to statistical variations in the charge density. These intermolecular forces are common, but very weak, and are found in inert gases where other types of bonding do not exist.

Hydrogen bonding is the attraction between hydrogen in a highly polar molecule and the electronegative atom in another polar molecule. In the water molecule, oxygen draws much of the electron density around it, creating positively charged centers at the two hydrogen atoms. These positively charged hydrogen atoms can interact with the negative center around the oxygen in adjacent water molecules. Although this type of

---

HISTORICAL HIGHLIGHT

Dutch physicist *Johannes Diderik van der Waals* was born on November 23, 1837 in Leiden, the Netherlands. He was the eldest son of eight children. Initially, van der Waals was an elementary school teacher during the years 1856–1861. He continued studying to become headmaster and attended lectures on mathematics, physics, and astronomy at Leiden University. From 1866 onwards he taught physics and mathematics at a secondary school in The Hague. After a revision of the law, knowledge of Latin and Greek was no longer a prerequisite for an academic graduation, and in 1873 J. D. van der Waals graduated on the thesis: "Over de continuïteit van de gas—envloeistoftoestand" ("About the continuity of gaseous and liquid states"). In this thesis he published the well-known law:

$$\left(P + \frac{a}{V^2}\right)(V - b) = RT$$

This revision to the ideal gas law accounted for the specific volume of gas molecules and assumed a force between these molecules which are now known as "van der Waals forces." With this law, the existence of condensation and the critical temperature of gases could be predicted. In 1877 J. D. van der Waals became the first professor of physics at the University "Illustre" in Amsterdam. In 1880 he formulated his "law of corresponding states," in 1893 he devised a theory for capillary phenomena, and in 1891 he introduced his theory for the behavior of two mixtures of two materials. It was not possible to experimentally show the de-mixing of two gases into two separate gases under certain circumstances as predicted by this theory until 1941.

From 1875 to 1895 J.D. van der Waals was a member of the Dutch Royal Academy of Science. In 1908, at the age of 71, J. D. van der Waals resigned as a professor. During his life J. D. van der Waals was honored many times. He was one of only 12 foreign members of the "Academie des Sciences" in Paris. In 1910 he received the Nobel prize for Physics for the incredible work he had done on the equations of state for gases and fluids—only the fifth Dutch physicist to receive this honor. J. D. van der Waals died on March 8, 1923 at the age of 85.

Source: www.vdwaals.nl

bonding is of the same order of magnitude in strength as van der Waals bonding, it can have a profound influence on the properties of a material, such as boiling and melting points. In addition to having important chemical and physical implications, hydrogen bonding plays an important role in many biological and environmental phenomena. It is responsible for causing ice to be less dense than water (how many other substances do you know that are less dense in the solid state than in the liquid state?), an occurrence that allows fish to survive at the bottom of frozen lakes.

Finally, some molecules possess permanent charge separations, or dipoles, such as are found in water. The general case for the interaction of any positive dipole with a negative dipole is called dipole–dipole interaction. Hydrogen bonding can be thought of as a specific type of dipole–dipole interaction. A dipolar molecule like ammonia, NH_3, is able to dissolve other polar molecules, like water, due to dipole–dipole interactions. In the case of NaCl in water, the dipole–dipole interactions are so strong as to break the intermolecular forces within the molecular solid.

Now that the types of bonds have been reviewed, we will concentrate on the primary bond because it correlates more directly with physical properties in solids than do secondary bonds. Be aware that the secondary forces exist, though, and that they play a larger role in liquids and gases than in solids.

1.0.4 Intermolecular Forces and Bonding

We have described the different types of primary bonds, but how do these bonds form in the first place? What is it that causes a sodium ion and a chloride ion to form a compound, and what is it that prevents the nuclei from fusing together to form one element? These questions all lead us to the topics of intermolecular forces and bond formation. We know that atoms approach each other only to a certain distance, and then, if they form a compound, they will maintain some equilibrium separation distance known as the *bond length*. Hence, we expect that there is some *attractive energy* that brings them together, as well as some *repulsive energy* that keeps the atoms a certain distance apart.

Also known as *chemical affinity*, the attractive energy between atoms is what causes them to approach each other. This attraction is due to the electrostatic force between the nucleus and electron clouds of the separate atoms. It should make sense to you that the attractive energy (U_A) is inversely proportional to the separation distance, r; that is, the further the atoms are apart, the weaker the attraction:

$$U_A = -\frac{a}{r^m} \tag{1.9}$$

where a is a constant that we will describe in more detail in a moment, and m is a constant with a value of 1 for ions and 6 for molecules. Notice that there is a negative sign in Eq. (1.9). By convention, we will refer to the attractive energy as a "negative energy."

Once the atoms begin to approach each other, they can only come so close together due to the impenetrability of matter. The result is a repulsive energy, which we assign a positive value, again, by convention. The primary constituents of this repulsive energy are nucleus–nucleus and electron–electron repulsions. As with the attractive energy, the repulsive energy is inversely proportional to the separation distance; the closer the

Table 1.6   Values of the Repulsion Exponent

Noble Gas Ion Core	Outer Core Configuration	n
He	$1s^2$	5
Ne	$2s^2 2p^6$	7
Ar	$3s^2 3p^6$	9
Kr	$3d^{10} 4s^2 4p^6$	10
Xe	$4d^{10} 5s^2 5p^6$	12

atoms are, the more they repel each other:

$$U_R = \frac{b}{r^n} \tag{1.10}$$

where b and n are constants. The value of n, called the *repulsion exponent*, depends on the outer core configuration of the atom. Values of the repulsion exponent are given in Table 1.6.

The total, or potential energy of the system is then the sum of the attractive and repulsive components:

$$U = U_A + U_R = \frac{-a}{r^m} + \frac{b}{r^n} \tag{1.11}$$

The result is the *potential energy well* (see Figure 1.3). The well-known Lennard-Jones potential

$$U = -\frac{a}{r^6} + \frac{b}{r^{12}} \tag{1.12}$$

is a common potential energy function used in a number of models, including collision theory for kinetics. It is simply a special case of Eq. (1.11) with $n = 12$ (Xe configuration) and $m = 6$ (molecules).

It is oftentimes useful to know the forces involved in bonding, as well as the energy. Recall that energy and force, F, are related by

$$F = -\frac{dU}{dr} \tag{1.13}$$

We will see later on that we can use this expression to convert between force and energy for specific types of atoms and molecules (specific values of n and m). For now, this expression helps us find the *equilibrium bond distance*, r_0, which occurs when forces are equal (the sum of attractive and repulsive forces is zero) or at *minimum potential energy* (take the derivative and set it equal to zero):

$$F = -\frac{dU}{dr} = 0, \qquad \text{at } r = r_0 \tag{1.14}$$

This is not the same thing as the *maximum attractive force*, which we get by maximizing F:

$$F_{max} = \frac{dF}{dr} = -\frac{d^2U}{dr^2} = 0 \tag{1.15}$$

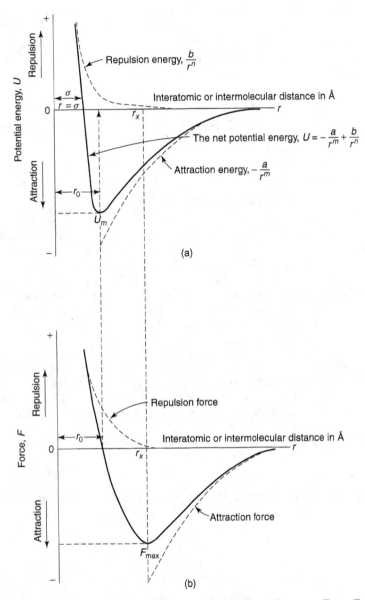

Figure 1.3 The interatomic (a) potential energy and (b) force diagrams. From Z. Jastrzebski, *The Nature and Properties of Engineering Materials*, 2nd ed. by Copyright © 1976 by John Wiley & Sons, Inc. This material is used by permission of John Wiley & Sons, Inc.

The forces are equal when the potential energy is a minimum and the separation distance is at the bond length, r_0. Differentiation of Eq. (1.11) and solving for r_0 in terms of a, b, n, and m gives

$$r_0 = \left(\frac{nb}{ma}\right)^{\frac{1}{n-m}} \tag{1.16}$$

The potential energy well concept is an important one, not only for the calculation of binding energies, as we will see in a moment, but for a number of important physical properties of materials. How tightly atoms are bound together in a compound has a direct impact on such properties as melting point, elastic modulus and thermal expansion coefficient. Figure 1.4 shows a qualitative comparison of a material that has a deep and narrow potential energy wells versus one in which the potential energy well is wide and shallow. The deeper well represents stronger interatomic attraction; hence it is more difficult to melt these substances, which have correspondingly large elastic moduli and low thermal expansion coefficients.

Cooperative Learning Exercise 1.1

Work with a neighbor. Consider the Lennard-Jones potential, as given by Eq. (1.12), for which $m = 6$ and $n = 12$. You wish to determine the separation distance, r, at which the maximum force occurs, F_{max}, in terms of the equilibrium bond distance, r_0.

Person 1: Use Eq. (1.16) with the values of m and n of the Lennard-Jones potential to solve for the constant a in terms of b and the equilibrium bond distance, r_0. Now perform the determination of F_{max} as given by Eq. (1.15): substitute this value of a back into Eq. (1.12), differentiate it twice with respect to r (remember that r_0 is a constant), and set this equal to zero (determine, then maximize the force function). Solve this equation for r in terms of r_0. The other constant should drop out.

Person 2: Use Eq. (1.16) with the values of m and n of the Lennard-Jones potential to solve for the constant b in terms of a and the equilibrium bond distance, r_0. Now perform the determination of F_{max} as given by Eq. (1.15); substitute this value of b back into Eq. (1.12), differentiate it twice with respect to r (remember that r_0 is a constant), and set this equal to zero (determine, then maximize the force function). Solve this equation for r in terms of r_0. The other constant should drop out.

Compare your answers. You should both get the same result.

Answer: $r = 1.1087 r_0$

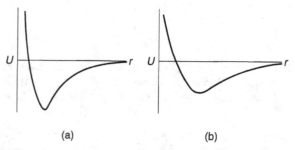

(a)   (b)

Figure 1.4 Schematic representation of the relationship between the shape of the potential energy well and selected physical properties. Materials with a deep well (a) have a high melting point, high elastic modulus, and low thermal expansion coefficient. Those with a shallow well (b) have a low melting point, low elastic modulus, and high thermal expansion coefficient. Adapted from C. R. Barrett, W. D. Nix, and A. S. Tetelman, *The Principles of Engineering Materials*. Copyright © 1973 by Prentice-Hall, Inc.

1.0.4.1 *The Ionic Bond.* To form an ionic bond, we must account for the complete transfer of electrons from one atom to the other. The easiest approach is to first transfer the electrons to form *ions*, then bring the two ions together to form a bond. Sodium chloride is a simple example that allows us to obtain both the bond energy and equilibrium bond distance using the potential energy approach. For this case, the potential energy, U, not only is the sum of the attractive and repulsive energies, U_A and U_R, respectively, but must also take into account the energy required to form ions from sodium and chlorine atoms, ΔE_{ions}. So, our energy expression looks like:

$$U = U_A + U_R + \Delta E_{ions} \tag{1.17}$$

which at the equilibrium bond distance gives the equilibrium potential energy, U_0:

$$U_0 = U_{A,0} + U_{R,0} + \Delta E_{ions} \tag{1.18}$$

Let us examine each of the three energies in Eq. (1.18) individually.

Energy is required to form an ion of Na^+ from elemental sodium. From Section 1.0.2.1, we already know that this process of removing an electron from an isolated atom is the ionization energy, *IE*, which for sodium is 498 kJ/mol. Similarly, energy is given off when Cl^- is formed by adding an electron to an isolated gaseous atom of chlorine. This is the electron affinity, *EA* (see Section 1.0.2.2), which for chlorine is -354 kJ/mol. So, the energy required to form ions is given by:

$$\Delta E_{ions} = IE_{Na} + EA_{Cl} = 498 - 354 = 144 \text{ kJ/mol} \tag{1.19}$$

For a diatomic molecule, the attraction between the two ions is strictly due to opposite charges, so the attractive force is given by *Coulomb's Law*:

$$F_A = (Z_1 e \times Z_2 e)/(4\pi\varepsilon_0 r^2) \tag{1.20}$$

where ε_o is a constant called the *electric permittivity* (8.854×10^{-12} $C^2/N \cdot m^2$), e is the charge of an electron (1.6×10^{-19} C), Z is respective *numbers of charge* of positive and negative ions ($Z_1 = Z_2 = 1$), and r is the separation distance between ions in meters. (We will learn more of the electric permittivity in Chapter 6.) Substituting the values of Z in for sodium and chloride ions gives

$$F_A = +e^2/(4\pi\varepsilon_0 r^2) \tag{1.21}$$

Energy is released by bringing the ions from infinite separation to their equilibrium separation distance, r_0. Recall that energy and force are related to one another by Eq. (1.13), so that the equilibrium attractive energy, $U_{A,0}$, can be found by integrating F_A:

$$U_{A,0} = -\int_{\infty}^{r_0} F_A \, dr = e^2/(4\pi\varepsilon_0 r_0) \tag{1.22}$$

Note the similarity in form of this expression for the attractive energy with that of Eq. (1.9). The exponent on r is 1, as it should be for ions ($m = 1$), and the other parameters can be grouped to form the constant, a. Recall that by definition the attractive energy is a negative energy, so we will end up inserting a minus sign in front of the expression in Eq. (1.22) to match the form shown in Eq. (1.9).

The repulsive energy can be derived simply by using the general expression given in Eq. (1.10) and solving for the constant b by minimizing the potential energy function [Eq. (1.11)] with a knowledge of the constant a from Eq. (1.22) and (1.9) (you should try this) to obtain:

$$U_{R,0} = e^2/(4\pi \varepsilon_0 n r_0) \tag{1.23}$$

where e and ε_0 are the same as for the attractive force, r_0 is again the equilibrium separation distance, and n is the repulsion exponent.

Inserting $U_{A,0}$ and $U_{R,0}$ into the main energy expression, Eq. (1.18), (recall that the attractive energy must be negative) gives us the equilibrium potential energy, U_0:

$$U_0 = \frac{-e^2}{4\pi \varepsilon_0 r_0} + \frac{e^2}{4\pi \varepsilon_0 n r_0} + \Delta E_{ions} \tag{1.24}$$

and simplifying gives:

$$U_0 = \left(1 - \frac{1}{n}\right)\left(\frac{-e^2}{4\pi \varepsilon_0 r_0}\right) + \Delta E_{ions} \tag{1.25}$$

We solved for the equilibrium bond distance, r_0, in Eq. (1.16), and the constants a and b have, in effect, just been evaluated. Inserting these values into Eq. (1.25), along with Eq. (1.19) and using $n = 8$ (why?), gives:

$$U_0 = \left(1 - \frac{1}{8}\right)\left[\frac{1(1.6 \times 10^{-19}\ \text{C})^2(6.02 \times 10^{23}\ \text{mol}^{-1})}{4\pi(8.854 \times 10^{-12}\ \text{C}^2/\text{N} \cdot \text{m}^2)(2.36 \times 10^{-10}\ \text{m})}\right] + 142$$

$$= -371\ \text{kJ/mol}$$

Similarly, we can calculate bond energies for any type of bond we wish to create. Refer to Appendix 1 for bond energy values.

When we have an ordered assembly of atoms called a *lattice*, there is more than one bond per atom, and we must take into account interactions with adjacent atoms that result in an increased interionic spacing compared to an isolated atom. We do this with the *Madelung constant*, α_M. This parameter depends on the structure of the ionic crystal, the charge on the ions, and the relative size of the ions. The Madelung constant fits directly into the energy expression (Eq. 1.25):

$$U_L = \left(\alpha_M - \frac{1}{n}\right)\left(\frac{-e^2}{4\pi \varepsilon_0 r_0}\right) + \Delta E_{ions} \tag{1.26}$$

For NaCl, $\alpha_M = 1.75$ so the *lattice energy*, U_L, is -811 kJ/mol. Typical values of the Madelung constant are given in Table 1.7 for different crystal structures (see Section 1.1.1). In general, the lattice energy increases (becomes more negative) with decreasing interionic distance for ions with the same charge. This increase in lattice energy translates directly into an increased melting point. For example, if we replace

Table 1.7 Typical Values for the Madelung Constant

Compound	Crystal Lattice (see Section 1.1.1)	α_M
NaCl	FCC	1.74756
CsCl	CsCl	1.76267
CaF_2	Cubic	2.51939
$CdCl_2$	Hexagonal	2.244
MgF_2	Tetragonal	2.381
ZnS (wurtzite)	Hexagonal	1.64132
TiO_2 (rutile)	Tetragonal	2.408
β-SiO_2	Hexagonal	2.2197

the chlorine in sodium chloride with other halogens, while retaining the cubic structure, the interionic spacing should change, as well as the melting point. We could also account for additional van der Waals interactions, but this effect is relatively small in lattices.

1.0.4.2 The Covalent Bond. Recall that covalent bonding results when electrons are "shared" by similar atoms. The simplest example is that of a hydrogen molecule, H_2. We begin by using *molecular orbital theory* to represent the bonding. Two atomic orbitals ($1s$) overlap to form two *molecular orbitals* (MOs), represented by σ: one *bonding orbital* ($\sigma 1s$), and one *antibonding orbital*, ($\sigma^* 1s$), where the asterisk superscript indicates antibonding. The antibonding orbitals are higher in energy than corresponding bonding orbitals.

The shapes of the electron cloud densities for various MOs are shown in Figure 1.5. The overlap of two s orbitals results in one σ-bonding orbital and one σ-antibonding orbital. When two p orbitals overlap in an end-to-end fashion, as in Figure 1.5b, they are interacting in a manner similar to $s-s$ overlap, so one σ-bonding orbital and one σ-antibonding orbital once again are the result. Note that all σ orbitals are symmetric about a plane between the two atoms. Side-to-side overlap of p orbitals results in one π-bonding orbital and one π-antibonding orbital. There are a total of four π orbitals: two for p_x and two for p_y. Note that there is one more *node* (region of zero electron density) in an antibonding orbital than in the corresponding bonding orbital. This is what makes them higher in energy.

As in the case of ionic bonding, we use a potential energy diagram to show how orbitals form as atoms approach each other, as shown in Figure 1.6. The electrons from the isolated atoms are then placed in the MOs from bottom to top. *As long as the number of bonding electrons is greater than the number of antibonding electrons, the molecule is stable.* For atoms with p and d orbitals, diagrams become more complex but the principles are the same. In all cases, there are the same number of molecular orbitals as atomic orbitals. Be aware that there is some change in the relative energies of the π and σ orbitals as we go down the periodic chart, particularly around O_2. As a result, you might see diagrams that have the $\pi 2p$ orbitals lower in energy than the $\sigma 2p$. Do not let this confuse you if you see some variation in the order of these orbitals in other texts or references. For our purposes, it will not affect whether the molecule is stable or not.

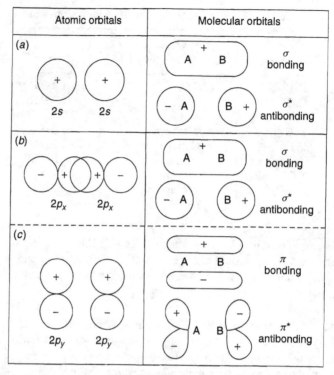

Figure 1.5   The shape of selected molecular orbitals formed from the overlap of two atomic orbitals. From K. M. Ralls, T. H. Courtney, and J. Wulff, *Introduction to Materials Science and Engineering*. Copyright © 1976 by John Wiley & Sons, Inc. This material is used by permission of John Wiley & Sons, Inc.

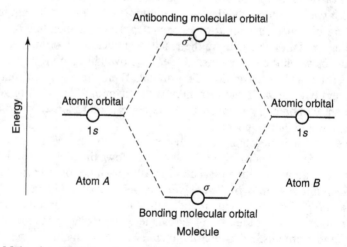

Figure 1.6   Molecular orbital diagram for the hydrogen molecule, H_2. Reprinted, by permission, from R. E. Dickerson, H. B. Gray, and G. P. Haight, Jr., *Chemical Principles*, 3rd ed., p. 446. Copyright © 1979 by Pearson Education, Inc.

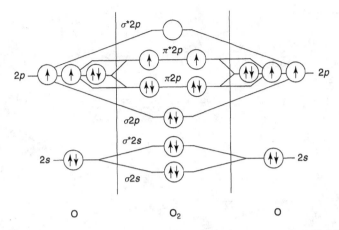

Figure 1.7 Molecular orbital diagram for molecular oxygen, O_2. From K. M. Ralls, T. H. Courtney, and J. Wulff, *Introduction to Materials Science and Engineering.* Copyright © 1976 by John Wiley & Sons, Inc. This material is used by permission of John Wiley & Sons, Inc.

Let us use molecular oxygen, O_2, as an example. As shown in Figure 1.7, each oxygen atom brings six outer-core electrons to the molecular orbitals. (Note that the $1s$ orbitals are not involved in bonding, and are thus not shown. They could be shown on the diagram, but would be at a very low relative energy at the bottom of the diagram.) The 12 total electrons in the molecule are placed in the MOs from bottom to top; according to *Hund's rule*, the last two electrons must be placed in separate π^*2p orbitals before they can be paired.

The pairing of electrons in the MOs can manifest itself in certain physical properties of the molecule. *Paramagnetism* results when there are unpaired electrons in the molecular orbitals. Paramagnetic molecules magnetize in magnetic fields due to the alignment of unpaired electrons. *Diamagnetism* occurs when there are all paired electrons in the MOs. We will revisit these properties in Chapter 6.

We can use molecular orbital theory to explain simple heteronuclear diatomic molecules, as well. A molecule such as hydrogen fluoride, HF, has molecular orbitals, but we must remember that the atomic orbitals of the isolated atoms have much different energies from each other to begin with. How do we know where these energies are relative to one another? Look back at the ionization energies in Table 1.4, and you see that the first ionization energy for hydrogen is 1310 kJ/mol, whereas for fluorine it is 1682 kJ/mol. This means that the outer-shell electrons have energies of −1310 ($1s$ electron) and −1682 kJ/mol ($2p$ electron), respectively. So, the electrons in fluorine are more stable (as we would expect for an atom with a much larger nucleus relative to hydrogen), and we can construct a relative molecular energy diagram for HF (see Figure 1.8) This is a case where the electronegativity of the atoms is useful. It qualitatively describes the relative energies of the atomic orbitals and the shape of the resulting MOs. The molecular energy level diagram for the general case of molecule AB where B is more electronegative than A is shown in Figure 1.9, and the corresponding molecular orbitals are shown in Figure 1.10. In Figure 1.9, note how the B atomic orbitals are lower in energy than those of atom A. In Figure 1.10, note how the number of nodes increases from bonding to antibonding orbitals, and also note how the electron probability is greatest around the more electronegative atom.

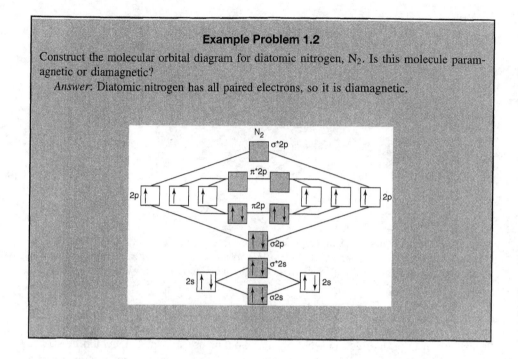

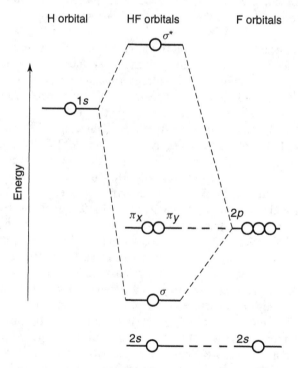

Figure 1.8 Molecular orbital diagram for HF. Reprinted, by permission, from R. E. Dickerson, H. B. Gray, and G. P. Haight, Jr., *Chemical Principles*, 3rd ed., p. 461. Copyright © 1979 by Pearson Education, Inc.

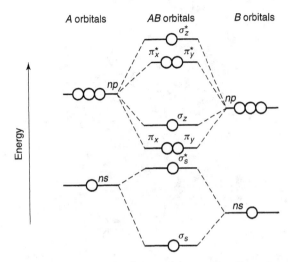

Figure 1.9   Molecular orbital diagram for the general case of a diatomic molecule AB, where B is more electronegative than A. Reprinted, by permission, from R. E. Dickerson, H. B. Gray, and G. P. Haight, Jr., *Chemical Principles*, 3rd ed., p. 464. Copyright © 1979 by Pearson Education, Inc.

Molecular orbitals don't explain everything and become increasingly more difficult to draw with more than two atoms. We use a model called *hybridization* to explain other effects, particularly in carbon compounds. Hybridization is a "mixing" of atomic orbitals to create new orbitals that have a geometry better suited to a particular type of bonding environment. For example, in the formation of the compound BeH_2, we would like to be able to explain why this molecule is linear; that is, the H–Be–H bond is 180°.

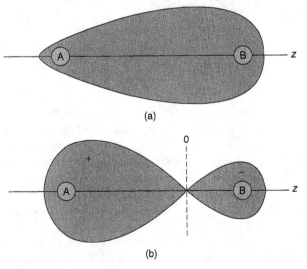

Figure 1.10   The shape of selected molecular orbitals for the diatomic molecule AB, where B is more electronegative than A: (a) σ, (b) σ^*, (c) π and (d) π^*.

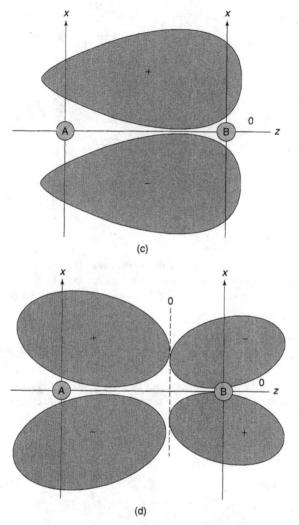

Figure 1.10 (*continued*).

The hydrogen atoms only have one electron each to donate, both in their respective $1s$ orbitals, but beryllium has two electrons in the $2s$ orbital, and because its principal quantum number is two, it also has $2p$ orbitals, even though they are empty.

The trick is to make two equivalent orbitals in Be out of the atomic orbitals so that each hydrogen will see essentially the same electronic environment. We can accomplish this by "mixing" the $2s$ orbital and one of the empty $2p$ orbitals (say, the $2p_z$) to form two equivalent orbitals we call "sp" hybrids, since they have both s and p characteristics. As with molecular orbital theory, we have to end up with the same number of orbitals we started with. The bonding lobes on the new sp_a and sp_b orbitals on Be are 180° apart, just as we need to form BeH$_2$. In this manner, we can mix any type of orbitals we wish to come up with specific bond angles and numbers of equivalent orbitals. The most common combinations are sp, sp^2, and sp^3 hybrids. In sp hybrids, one s and one p orbital are mixed to get two sp orbitals, both of which are 180° apart. A linear molecule results e.g., BeH_2, as shown in Figure 1.11. In sp^2 hybrids, one s and two p orbitals are mixed to obtain three sp^2 orbitals. Each orbital has 1/3 s and 2/3 p characteristic. A trigonal planar orbital arrangement results, with 120° bond angles. An example of a trigonal planar molecule is BF_3, as in Figure 1.12. Finally, in sp^3 hybrids, when one s and three p orbitals are mixed, four sp^3 orbitals result, each having 1/4 s and 3/4 p characteristic. The tetrahedral arrangement of orbitals creates a 109.5° bond angle, as is found in methane, CH_4 (Figure 1.13).

The concept of hybridization not only gives us a simple model for determining the correct geometry in simple molecules, but also provides us with a rationalization for multiple bonds. A *double bond* can result from sp^2 hybridization: one sp^2–sp^2 bond and one π bond that forms between the p orbitals not involved in hybridization. An example is in C_2H_4 (ethylene, Figure 1.14a), where each carbon undergoes sp^2 hybridization so that it can form an sp^2–$1s$ bond with two hydrogens and an sp^2–sp^2 bond with the other carbon. The remaining p orbitals on each carbon (say, p_z) share electrons, which form the C–C double bond. A *triple bond* can be explained in terms of sp hybridization. It is formed from one sp–sp bond and two π bonds which form

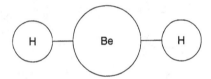

Figure 1.11   The linear structure of BeH_2.

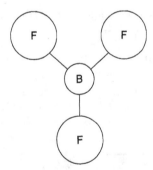

Figure 1.12   The trigonal planar structure of BF_3.

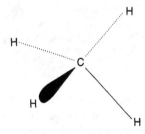

Figure 1.13   The tetrahedral structure of CH_4.

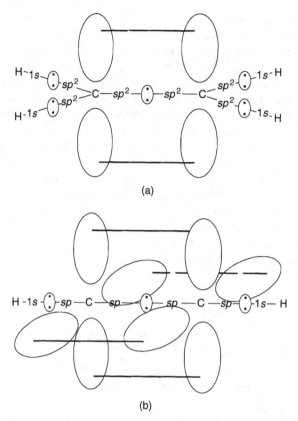

Figure 1.14   Hybridization resulting in (a) double bond and (b) triple bond.

between the two remaining p orbitals after hybridization. Acetylene (Figure 1.14b), C_2H_2, is such a compound in which both carbons undergo sp hybridization so that they can accommodate one bond with each other and one with hydrogen. Bonds can form between the remaining p orbitals, which in this case could be the p_z and p_y orbitals on each carbon, for a total of three bonds between the carbon atoms.

Example Problem 1.3

Consider the molecule NF_3. How can we explain the observation that the F–N–F bond angles in NF_3 are 107.3° and not 120°, as we might predict?

  Answer: Nitrogen undergoes sp^3 hybridization, not sp^2, so it is tetrahedral. The additional sp^3 orbital is occupied by a lone pair of electrons from the nitrogen. This lone pair results in electron–electron repulsion that causes the other sp^3 orbitals bonded to fluorines to be closer together than the normal 109° tetrahedral bond angle, hence the 107.3 F–N–F bond angle.

1.0.4.3 The Metallic Bond. Some elements, namely those in the first two columns of the periodic table (IA and IIA) and the transition metals (e.g., Co, Mn, Ni), not only have a propensity for two atoms to share electrons such as in a covalent bond, but also have a tendency for large groups of atoms to come together and share valence electrons in what is called a *metallic bond*. In this case, there are now N atoms in the lattice, where N is a large number. There are N atomic orbitals, so there must be N MOs, many of which are *degenerate*, or of the same energy. This leads to *bands* of electrons, as illustrated in Figure 1.15 for sodium. The characteristics of the metallic bond are that the valence electrons are not associated with any particular atom in the lattice, so they form what is figuratively referred to as an *electron gas* around the solid core of metallic nuclei. As a result, the bonds in metals are nondirectional, unlike covalent or ionic bonds in which the electrons have a finite probability of being around a particular atom

The electrons not involved in bonding remain in what is called the *core band*, whereas the valence electrons that form the electron gas enter into the *valence band*.

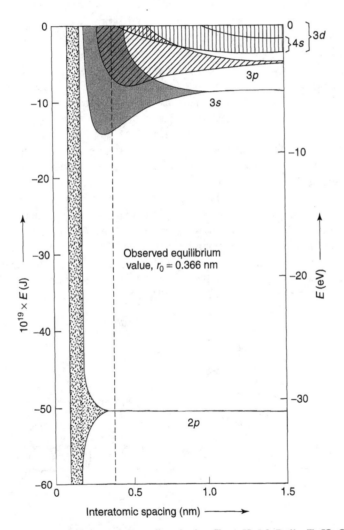

Figure 1.15   Energy band diagram for a sodium lattice. From K. M. Ralls, T. H. Courtney, and J. Wulff, *Introduction to Materials Science and Engineering*. Copyright © 1976 by John Wiley & Sons, Inc. This material is used by permission of John Wiley & Sons, Inc. After J. C. Slater, *Phys. Rev.*, **45**, 794 (1934).

The remaining unfilled orbitals form higher-energy bands, called the *conduction band.* Keep in mind that even though the *d* and *f* orbitals may not be filled with electrons, they still exist for many of the heavier elements, so they must be included in the molecular orbital diagram. We will see later in Chapter 6 that the conduction band plays a very important role in the electrical, thermal, and optical properties of metals.

1.1 STRUCTURE OF METALS AND ALLOYS

Since the electrons in a metallic lattice are in a "gas," we must use the core electrons and nuclei to determine the structure in metals. This will be true of most solids we will describe, regardless of the type of bonding, since the electrons occupy such a small volume compared to the nucleus. For ease of visualization, we consider the atomic cores to be hard spheres. Because the electrons are delocalized, there is little in the way of electronic hindrance to restrict the number of neighbors a metallic atom may have. As a result, the atoms tend to pack in a *close-packed* arrangement, or one in which the maximum number of *nearest neighbors* (atoms directly in contact) is satisfied.

Refer to Figure 1.16. The most hard spheres one can place in the plane around a central sphere is six, regardless of the size of the spheres (remember that all of the spheres are the same size). You can then place three spheres in contact with the central sphere both above and below the plane containing the central sphere. This results in a total of 12 nearest-neighbor spheres in contact with the central sphere in the close-packed structure.

Closer inspection of Figure 1.16a shows that there are two different ways to place the three nearest neighbors above the original plane of hard spheres. They can be directly aligned with the layer below in an ABA type of structure, or they can be rotated so that the top layer does not align core centers with the bottom layer, resulting in an ABC structure. This leads to two different types of close-packed structures. The ABAB... structure (Figure 1.16b) is called *hexagonal close-packed* (HCP) and the ABCABC... structure is called *face-centered cubic* (FCC). Remember that both

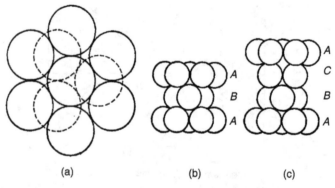

(a)   (b)   (c)

Figure 1.16   Close-packing of spheres. (a) Top view, (b) side view of ABA structure, (c) side view of ABC structure. From Z. Jastrzebski, *The Nature and Properties of Engineering Materials,* 2nd ed. Copyright © 1976 by John Wiley & Sons, Inc. This material is used by permission of John Wiley & Sons, Inc.

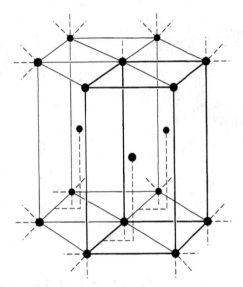

Figure 1.17 The extended unit cell of the hexagonal close-packed (HCP) structure.

of these close-packed arrangements have a *coordination number* (number of nearest neighbors surrounding an atom) of 12: 6 in plane, 3 above, and 3 below.*

Keep in mind that for close-packed structures, the atoms touch each other in all directions, and all nearest neighbors are equivalent. Let us first examine the HCP structure. Figure 1.17 is a section of the HCP lattice, from which you should be able to see both hexagons formed at the top and bottom of what is called the *unit cell.* You should also be able to identify the ABA layered structure in the HCP unit cell of Figure 1.17 through comparison with Figure 1.16. Let us count the number of atoms in the HCP unit cell. The three atoms in the center of the cell are completely enclosed. The atoms on the faces, however, are shared with adjacent cells in the lattice, which extends to infinity. The center atoms on each face are shared with one other HCP unit cell, either above (for the top face) or below (for the bottom face), so they contribute only half of an atom each to the HCP unit cell under consideration. This leaves the six corner atoms on each face (12 total) unaccounted for. These corner atoms are at the intersection of a total of *six* HCP unit cells (you should convince yourself of this!), so each corner atom contributes only one-sixth of an atom to our isolated HCP unit cell. So, the total number of whole atoms in the HCP unit cell is

$$3 \times 1 \quad = 3 \text{ center atoms}$$

$$2 \times (1/2) = 1 \text{ face atom}$$

$$12 \times (1/6) = 2 \text{ corner atoms}$$

<hr>

6 total atoms

*At this point, you may find it useful to get some styrofoam spheres or hard balls to help you visualize some of the structures we will describe. We will use a number of perspectives, views, and diagrams to build these structures. Some will treat the atoms as solids spheres that can touch each other, some will use dots at the center of the atoms to help you visualize the larger structure. Not all descriptions will help you—find those perspectives that work best for you.

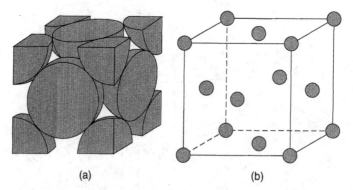

Figure 1.18 The face-centered cubic (FCC) structure showing (a) atoms touching and (b) atoms as small spheres. Reprinted, by permission, from W. Callister, *Materials Science and Engineering: An Introduction*, 5th ed., p. 32. Copyright © 2000 by John Wiley & Sons, Inc.

Counting the atoms in the FCC structure is performed in a similar manner, except that visualizing the FCC structure takes a little bit of imagination and is virtually impossible to show on a two-dimensional page. Take the ABC close-packed structure shown in Figure 1.16c, and pick three atoms along a diagonal. These three atoms form the diagonal on the face of the FCC unit cell, which is shown in Figure 1.18. There is a trade-off in doing this: It is now difficult to see the close-packed layers in the FCC structure, but it is much easier to see the cubic structure (note that all the edges of the faces have the same length), and it is easier to count the total number of atoms in the FCC cell. In a manner similar to counting atoms in the HCP cell, we see that there are *zero* atoms completely enclosed by the FCC unit cell, six face atoms that are each shared with an adjacent unit cell, and eight corner atoms at the intersection of eight unit cells to give

$$6 \times (1/2) = 3 \text{ face atoms}$$

$$8 \times (1/8) = 1 \text{ corner atom}$$

$$\overline{ 4 \text{ total atoms} }$$

Remember that both HCP and FCC are close-packed structures and that each has a coordination number of 12, but that their respective unit cells contain 6 and 4 total atoms. We will now see how these two special close-packed structures fit into a larger assembly of crystal systems.

1.1.1 Crystal Structures

Our description of atomic packing leads naturally into *crystal structures*. While some of the simpler structures are used by metals, these structures can be employed by heteronuclear structures, as well. We have already discussed FCC and HCP, but there are 12 other types of crystal structures, for a total of 14 *space lattices* or *Bravais lattices*. These 14 space lattices belong to more general classifications called *crystal systems*, of which there are seven.

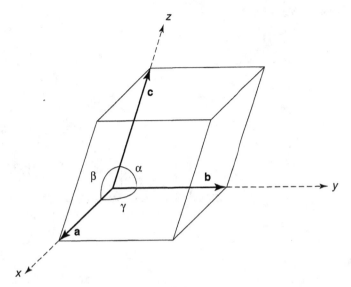

Figure 1.19 Definition of a coordinate system for crystal structures.

Before describing each of the space lattices, we need to define a coordinate system. The easiest coordinate system to use depends upon which crystal system we are looking at. In other words, the coordinate axes are not necessarily orthogonal and are defined by the unit cell. This may seem a bit confusing, but it simplifies the description of cell parameters for those systems that do not have crystal faces at right angles to one another. Refer to Figure 1.19. For each crystal system, we will define the space lattice in terms of three axes, x, y, and z, with interaxial angles α, β, γ. Note that the interaxial angle α is defined by the angle formed between axes z and y, and also note that angles β and γ are defined similarly. Only in special circumstances are α, β, γ equal to 90°. The distance along the y axis from the origin to the edge of the unit cell is called the *lattice translation vector*, **b**. Lattice translation vectors **a** and **c** are defined similarly along the axes x and z, respectively. The magnitudes (lengths) of the lattice translation vectors are called the *lattice parameters*, a, b, and c. We will now examine each of the seven crystal systems in detail.

1.1.1.1 *Crystal Systems.* The cubic crystal system is composed of three space lattices, or unit cells, one of which we have already studied: *simple cubic* (SC), *body-centered cubic* (BCC), and *face-centered cubic* (FCC). The conditions for a crystal to be considered part of the cubic system are that the lattice parameters be the same (so there is really only one lattice parameter, a) and that the interaxial angles all be 90°.

The simple cubic structure, which is the basic structural unit of many primitive cells, including NaCl, is not a close-packed structure (see Figure 1.20). In fact, it contains about 48% void space; and as a result, it is not a very dense structure. The large space in the center of the SC structure is called an *interstitial site*, which is a vacant position between atoms that can be occupied by a small impurity atom or alloying element. In this case, the interstitial site is surrounded by eight atoms. All eight atoms in SC are equivalent and are located at the intersection of eight adjacent unit cells, so that there are $8 \times (1/8) = 1$ total atoms in the SC unit cell. Notice that

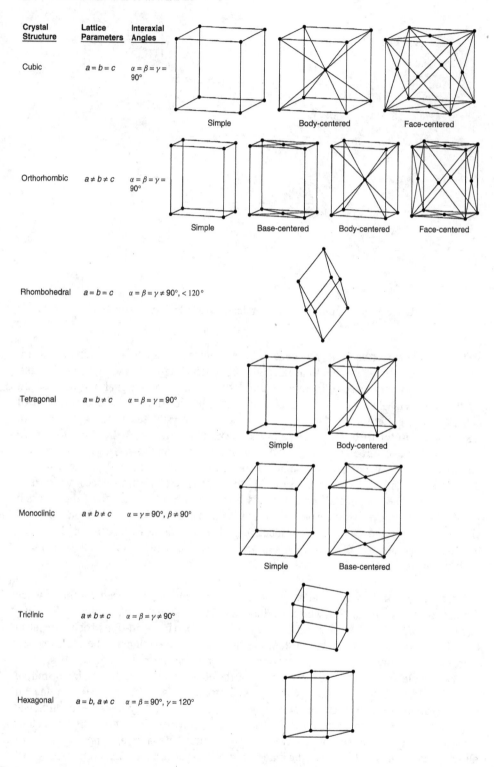

Crystal Structure	Lattice Parameters	Interaxial Angles
Cubic	$a = b = c$	$\alpha = \beta = \gamma = 90°$
Orthorhombic	$a \neq b \neq c$	$\alpha = \beta = \gamma = 90°$
Rhombohedral	$a = b = c$	$\alpha = \beta = \gamma \neq 90°, < 120°$
Tetragonal	$a = b \neq c$	$\alpha = \beta = \gamma = 90°$
Monoclinic	$a \neq b \neq c$	$\alpha = \gamma = 90°, \beta \neq 90°$
Triclinic	$a \neq b \neq c$	$\alpha = \beta = \gamma \neq 90°$
Hexagonal	$a = b, a \neq c$	$\alpha = \beta = 90°, \gamma = 120°$

Figure 1.20 Summary of the 14 Bravais space lattices.

Figure 1.20 shows the atoms as points, but the atoms actually occupy a larger space than that. In fact, for SC, the atoms touch along the edge of the crystal.

Body-centered cubic (BCC) is the unit cell of many metals and, like SC, is not a close-packed structure. The number of atoms in the BCC unit cell are calculated as follows:

$$1 \times 1 \quad = 1 \text{ center atom}$$

$$8 \times (1/8) = 1 \text{ corner atom}$$

$$\overline{\phantom{2 \text{ total atoms}}}$$

$$2 \text{ total atoms}$$

Finally, face-centered cubic (FCC) has already been described (Figure 1.18). Even though FCC is a close-packed structure, there are interstitial sites, just as in SC. There are actually two different types of interstitial sites in FCC, depending on how many atoms surround the interstitial site. A group of four atoms forms a *tetrahedral interstice*, as shown in Figure 1.21. A group of six atoms arranged in an octahedron (an eight-sided geometric figure), creates an *octahedral interstice* (Figure 1.22). Figure 1.23 shows the locations of these interstitial sites within the FCC lattice. Note that there are eight *total* tetrahedral interstitial sites in FCC and there are four *total* octahedral interstitial sites in FCC (prove it!), which are counted in much the same way as we previously counted the total number of atoms in a unit cell. We will see later on that these interstitial sites play an important role in determining solubility of impurities and phase stability of alloys.

Interstitial sites are the result of packing of the spheres. Recall from Figure 1.18 that the spheres touch along the face diagonal in FCC. Similarly, the spheres touch along the body diagonal in BCC and along an edge in SC. We should, then, be able to calculate the lattice parameter, a, or the length of a face edge, from a knowledge of the sphere radius. In SC, it should be evident that the side of a unit cell is simply $2r$. Application of a little geometry should prove to you that in FCC, $a = 4r/\sqrt{2}$. The relationship between a and r for BCC is derived in Example Problem 1.4; other geometric relationships, including cell volume for cubic structures, are listed in Table 1.8. Finally, atomic radii for the elements can be found in Table 1.9. The radius of an atom is not an exactly defined quantity, and it can vary depending upon the bonding environment in which

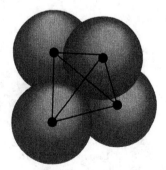

Figure 1.21   A tetrahedral interstice. From W. D. Kingery, H. K. Bowen, and D. R. Uhlmann, *Introduction to Ceramics*. Copyright © 1976 by John Wiley & Sons, Inc. This material is used by permission of John Wiley & Sons, Inc.

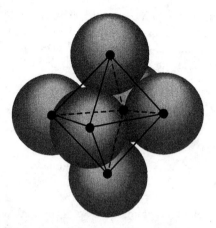

Figure 1.22 An octahedral interstice. From W. D. Kingery, H. K. Bowen, and D. R. Uhlmann, *Introduction to Ceramics*. Copyright © 1976 by John Wiley & Sons, Inc. This material is used by permission of John Wiley & Sons, Inc.

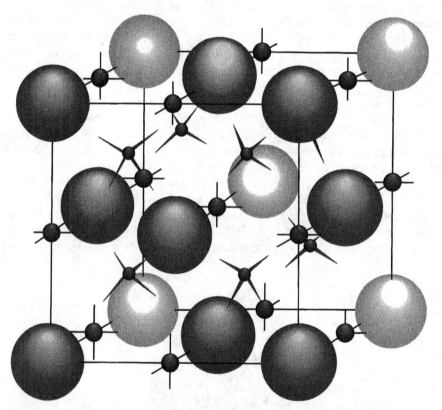

Figure 1.23 Location of interstitial sites in FCC. From W. D. Kingery, H. K. Bowen, and D. R. Uhlmann, *Introduction to Ceramics*. Copyright © 1976 by John Wiley & Sons, Inc. This material is used by permission of John Wiley & Sons, Inc.

Example Problem 1.4

Molybdenum has a BCC structure with an atomic radius of 1.36 Å. Calculate the lattice parameter for BCC Mo.

Answer: We know that the molybdenum atoms touch along the body diagonal in BCC, as shown in the projection at right. The length of the body diagonal, then, is $4r$, and is related to the lattice parameter, a (which is the length of the cube edge, not the length of the face diagonal, which is $a\sqrt{2}$) by application of the Pythagorean theorem:

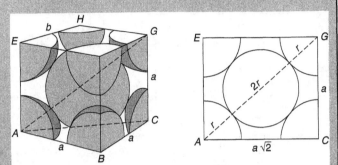

Reprinted, by permission, from Z. Jastrzebski, *The Nature and Properties of Engineering Materials*, p. 47, 2nd ed. Copyright © 1976, John Wiley & Sons, Inc.

$$(4r)^2 = a^2 + (a\sqrt{2})^2 = 3a^2$$

$$a = 4r/\sqrt{3} = 4(1.36)/\sqrt{3} = 3.14$$

The lattice parameter for BCC Mo is 3.14 Å, which is consistent with the value in Table 1.11.

Table 1.8   Summary of Important Parameters in the Cubic Space Lattices

	Simple Cubic	Face-Centered Cubic	Body-Centered Cubic
Unit cell side, a	$2r$	$4r/\sqrt{2}$	$4r/\sqrt{3}$
Face diagonal	$\sqrt{2}(2r)$	$4r$	$\sqrt{2/3}(4r)$
Body diagonal	$\sqrt{3}(2r)$	$\sqrt{3/2}(4r)$	$4r$
Number of atoms	1	4	2
Cell volume	$8r^3$	$\dfrac{32r^3}{\sqrt{2}}$	$\dfrac{64r^3}{3\sqrt{3}}$

r = atomic radius.

it finds itself. As a result, three types of radii are listed for each element in Table 1.9: an *atomic radius* of an isolated atom, an *ionic radius*, and a *metallic radius*. Just as in Figure 1.2 for electronic structure, there are some important trends in the atomic radii. The atomic radius tends to increase as one goes down the column in a series. This is due to the addition of energy levels and more electron density. Radii tend to decrease as we move across a row, because there is less shielding from inner electrons

Table 1.9   Atomic, Ionic, and Metallic Radii of the Elements

Many elements have multiple valence states. Additional ionic radii are listed below.

Ion	radius	Ion	radius	Ion	radius
Pb^{4-}	2.15	Mn^{3+}	0.70	Fe^{2+}	0.87
Pb^{2+}	1.32	Co^{2+}	0.82	W^{6+}	0.65
S^{6+}	0.34	Mo^{4+}	0.68	Cr^{3+}	0.64
Te^{4+}	0.89	V^{4+}	0.61	Sn^{4-}	2.15
Si^{4-}	1.98	Tl^{+}	1.49		
Ti^{3+}	0.69	Se^{6+}	0.30–0.04		

Main table (Atomic / Ionic / Metallic radii). The group header indicates the ion for which the ionic radius is given.

Z	Element	Ion	Atomic	Ionic	Metallic
1	H		0.46	1.54	—
2	He		—	—	—
3	Li		1.52	0.78	1.23
4	Be	2+	1.14	0.54	0.89
5	B	3+	0.97	0.2	0.81
6	C	4+	0.77	<0.2	—
7	N	5+	0.71	0.1	—
8	O	2−	0.6	1.32	—
9	F	1−	—	1.33	—
10	Ne		—	1.60	—
11	Na		1.86	0.98	1.57
12	Mg	2+	1.60	0.78	1.36
13	Al	3+	1.43	0.57	1.25
14	Si	4+	1.17	0.39	1.17
15	P	5+	1.09	0.3	—
16	S	2−	1.06	1.74	—
17	Cl	1−	1.07	1.81	—
18	Ar		1.92	—	—
19	K		2.31	1.33	2.03
20	Ca	2+	1.97	1.06	1.74
21	Sc	3+	1.60	0.8^{2+}	1.44
22	Ti	4+	1.47	0.64	1.32
23	V	5+	1.32	0.4^{5+}	1.22
24	Cr	6+	1.25	0.3	1.18
25	Mn	4+	1.12	0.52	1.17
26	Fe	4+	1.24	0.67^{3+}	1.17
27	Co	3+	1.25	0.65	1.16
28	Ni	2+	1.25	0.78	1.15
29	Cu	1+	1.28	0.96	1.17
30	Zn	2+	1.33	0.83	1.25
31	Ga	3+	1.35	0.62	1.25
32	Ge	4+	1.22	0.44	1.22
33	As	5+	1.25	0.69	—
34	Se	2−	1.16	1.91	—
35	Br	1−	1.19	1.96	—
36	Kr		1.97	—	—
37	Rb		2.51	1.49	2.16
38	Sr	2+	2.15	1.27	1.91
39	Y	3+	1.81	1.06	1.62
40	Zr	4+	1.58	0.89	1.45
41	Nb	5+	1.43	0.74	1.34
42	Mo	6+	1.36	0.65	1.30
43	Tc	4+	—	—	1.27
44	Ru	4+	1.34	0.65	1.25
45	Rh	3+	1.34	0.68	1.25
46	Pd	2+	1.37	0.50	1.28
47	Ag	1+	1.44	1.13	1.34
48	Cd	2+	1.50	1.03	1.41
49	In	3+	1.57	0.91	1.50
50	Sn	4+	1.58	0.74	1.40
51	Sb	5+	1.61	0.9^{3+}	—
52	Te	2−	1.43	0.89	—
53	I	1−	1.36	2.20	—
54	Xe		2.18	—	—
55	Cs		2.65	1.65	2.35
56	Ba	2+	2.17	1.3	1.98
57	La	3+	1.87	1.22	1.69
72	Hf	4+	1.59	0.84	1.44
73	Ta	5+	1.47	0.68	1.34
74	W	6+	1.37	0.68	1.30
75	Re	4+	1.38	0.72	1.28
76	Os	4+	1.35	0.67	1.26
77	Ir	3+	1.35	0.66^{4+}	1.27
78	Pt	2+	1.38	0.52	1.30
79	Au	1+	1.44	1.37	1.34
80	Hg	2+	1.50	1.12	1.44
81	Tl	3+	1.71	1.49	1.55
82	Pb	4+	1.75	0.84	1.54
83	Bi	5+	1.82	1.2^{3+}	—
84	Po	2−	1.40	0.76^{6+}	—
85	At	1−	—	0.6^{7+}	—
86	Rn		—	—	—

Source: Materials Science and Engineering Handbook; Pauling, *Nature of the Chemical Bond.* All values in angstroms, Å (1Å = 0.1 nm).

and the outer-core electrons are drawn more tightly toward the nucleus. There are some notable exceptions, or "jumps," in the row trend (why?). In general, the ionic radius is much smaller for positive ions and much larger for negative ions than the corresponding isolated atom (why?), and it follows the same general trend as for the isolated atoms. For the discussion of elemental, crystalline solids, the metallic radius is most appropriate. We will find later that the ionic values will be equally important for heteronuclear structures. There are other types of radii, such as covalent radii and van der Waals radii. The former is highly dependent upon the type of covalent bond. For example, a carbon atom in a carbon–carbon single bond has a covalent radius of 1.54 Å, whereas the same atom in a carbon–carbon triple bond is only 1.35 Å.

Continuing with our survey of the seven crystal systems, we see that the *tetragonal* crystal system is similar to the cubic system in that all the interaxial angles are 90°. However, the cell height, characterized by the lattice parameter, c, is not equal to the base, which is square ($a = b$). There are two types of tetragonal space lattices: *simple tetragonal*, with atoms only at the corners of the unit cell, and *body-centered tetragonal*, with an additional atom at the center of the unit cell.

Orthorhombic crystals are similar to both tetragonal and cubic crystals because their coordinate axes are still orthogonal, but now all the lattice parameters are unequal. There are four types of orthorhombic space lattices: *simple orthorhombic, face-centered orthorhombic, body-centered orthorhombic*, and a type we have not yet encountered, *base-centered orthorhombic*. The first three types are similar to those we have seen for the cubic and tetragonal systems. The base-centered orthorhombic space lattice has a lattice point (atom) at each corner, as well as a lattice point only on the top and bottom faces (called basal faces). All four orthorhombic space lattices are shown in Figure 1.20.

There is only one space lattice in the *rhombohedral* crystal system. This crystal is sometimes called *hexagonal R* or *trigonal R*, so don't confuse it with the other two similarly-named crystal systems. The rhombohedral crystal has uniform lattice parameters in all directions and has equivalent interaxial angles, but the angles are nonorthogonal and are less than 120°.

The crystal descriptions become increasingly more complex as we move to the *monoclinic* system. Here all lattice parameters are different, and only two of the interaxial angles are orthogonal. The third angle is not 90°. There are two types of monoclinic space lattices: *simple monoclinic* and *base-centered monoclinic*. The *triclinic* crystal, of which there is only one type, has three different lattice parameters, and none of its interaxial angles are orthogonal, though they are all equal.

Finally, we revisit the hexagonal system in order to provide some additional details. The lattice parameter and interaxial angle conditions shown in Figure 1.20 for the hexagonal cell refer to what is called the *primitive cell* for the hexagonal crystal, which can be seen in the front quadrant of the extended cell in Figure 1.17. The primitive hexagonal cell has lattice points only at its corners and has one atom in the center of the primitive cell, for a *basis* of two atoms. A basis is a unit assembly of atoms identical in composition, arrangement, and orientation that is placed in a regular manner on the lattice to form a space lattice. You should be able to recognize that there are three equivalent primitive cells in the extended HCP structure. The HCP *extended cell*, which is more often used to represent the hexagonal structure, contains a total of six atoms, as we calculated earlier. In the extended structure, the ratio of the height of

Table 1.10   Axial Ratios for Some HCP Metals

Metal	c/a
Be, Y	1.57
Hf, Os, Ru, Ti	1.58
Sc, Zr	1.59
Tc, Tl	1.60
La	1.61
Co, Re	1.62
Mg	1.63
Zn	1.85
Cd	1.89
Ideal (sphere packing)	1.633

the cell to its base, c/a, is called the *axial ratio*. Table 1.10 lists typical values of the axial ratio for some common HCP crystals.

A table of crystal structures for the elements can be found in Table 1.11 (excluding the Lanthanide and Actinide series). Some elements can have multiple crystal structures, depending on temperature and pressure. This phenomenon is called *allotropy* and is very common in elemental metals (see Table 1.12). It is not unusual for close-packed crystals to transform from one stacking sequence to the other, simply through a shift in one of the layers of atoms. Other common allotropes include carbon (graphite at ambient conditions, diamond at high pressures and temperature), pure iron (BCC at room temperature, FCC at 912°C and back to BCC at 1394°C), and titanium (HCP to BCC at 882°C).

1.1.1.2 Crystal Locations, Planes, and Directions. In order to calculate such important quantities as cell volumes and densities, we need to be able to specify locations and directions within the crystal. *Cell coordinates* specify a position in the lattice and are indicated by the variables u, v, w, separated by commas with no brackets:

u distance along the lattice translation vector **a**
v distance along the lattice translation vector **b**
w distance along the lattice translation vector **c**

HISTORICAL HIGHLIGHT

On warming, gray (or α) tin, with a cubic structure changes at 13.2°C into white (or β) tin, the ordinary form of the metal, which has a tetragonal structure. When tin is cooled below 13.2°C, it changes slowly from white to gray. This change is affected by impurities such as aluminum and zinc and can be prevented by small additions of antimony or bismuth. The conversion was first noted as growths on organ pipes in European cathedrals, where it was thought to be the devils work. This conversion was also speculated to be caused by microorganisms and was called "tin plague" or "tin disease."

Source: www.webelements.com/ webelements/elements/text/key/Sn.html

Table 1.11 Common Crystal Structures, Densities, and Lattice Parameters of the Elements

	1 H	2 He
Struct.	hcp	fcc
ρ, g/cc	n/a	n/a
a, Å	4.70	4.24

	3 Li	4 Be	5 B	6 C	7 N	8 O	9 F	10 Ne
Struct.	bcc	hcp	rhom	diam	hcp	mon	mon	fcc
ρ, g/cc	0.533	1.85	2.47	3.51	n/a	n/a	n/a	n/a
a, Å	3.51	2.29	5.06	3.56	3.86	5.40	5.50	4.42

	11 Na	12 Mg	13 Al	14 Si	15 P	16 S	17 Cl	18 Ar
Struct.	bcc	hcp	fcc	diam	tricl	orth	orth	fcc
ρ, g/cc	0.966	1.74	2.70	2.33	1.82	2.09	n/a	n/a
a, Å	4.29	3.21	4.05	5.43	11.5	10.4	6.22	5.26

	19 K	20 Ca	21 Sc	22 Ti	23 V	24 Cr	25 Mn	26 Fe	27 Co	28 Ni	29 Cu	30 Zn	31 Ga	32 Ge	33 As	34 Se	35 Br	36 Kr
Struct.	bcc	fcc	hcp	hcp	bcc	bcc	cubic	bcc	hcp	fcc	fcc	hcp	orth	diam	rhom	mon	orth	fcc
ρ, g/cc	0.862	1.53	2.99	4.51	6.09	7.19	7.47	7.87	8.8	8.91	8.93	7.13	5.91	5.32	5.78	4.81	n/a	n/a
a, Å	5.33	5.56	3.31	2.95	3.03	2.88	8.91	2.86	2.51	3.52	3.61	2.66	4.52	5.65	3.76	9.05	6.72	5.71

	37 Rb	38 Sr	39 Y	40 Zr	41 Nb	42 Mo	43 Tc	44 Ru	45 Rh	46 Pd	47 Ag	48 Cd	49 In	50 Sn	51 Sb	52 Te	53 I	54 Xe
Struct.	bcc	fcc	hcp	hcp	bcc	bcc	hcp	hcp	fcc	fcc	fcc	hcp	tetrag	tetrag	trig	trig	orth	fcc
ρ, g/cc	1.53	2.58	4.48	6.51	8.58	10.22	11.5	12.36	12.42	12.0	10.5	8.65	7.29	7.29	6.69	6.25	4.95	n/a
a, Å	5.59	6.08	3.65	3.61	3.30	3.14	2.74	2.71	3.80	3.88	4.08	2.98	3.25	5.83	4.31	4.46	7.18	6.20

	55 Cs	56 Ba	57 La	72 Hf	73 Ta	74 W	75 Re	76 Os	77 Ir	78 Pt	79 Au	80 Hg	81 Tl	82 Pb	83 Bi	84 Po	85 At	86 Rn
Struct.	bcc	bcc	hcp	hcp	bcc	bcc	hcp	hcp	fcc	fcc	fcc	rhom	hcp	fcc	rhom	cub	n/a	fcc
ρ, g/cc	1.91	3.59	6.17	13.3	16.7	19.3	21.0	22.58	22.55	21.4	19.28	n/a	11.87	11.34	9.80	9.2	n/a	n/a
a, Å	6.14	5.01	3.77	3.20	3.30	3.17	2.76	2.73	3.84	3.92	4.07	3.01	3.46	4.95	4.74	3.36	n/a	n/a

Note that many elements have allotropes. Cited structures are generally the most stable.

Source: *Materials Science and Engineering Handbook*; Kittel, *Solid State Physics*; http://www.webelements.com/

39

Table 1.12 Some Metal Allotropes

Metal	R.T. Crystal Structure	Structure at Other Temperatures
Ca	FCC	BCC (>447°C)
Co	HCP	FCC (>427°C)
Hf	HCP	BCC (>1742°C)
Fe	BCC	FCC (>912°C)
		BCC (>1394°C)
Li	BCC	BCC (< −193°C)
Na	BCC	BCC (< −233°C)
Sn	BCT	Cubic (<13°C)
Tl	HCP	BCC (>234°C)
Ti	HCP	BCC (>883°C)
Y	HCP	BCC (>1481°C)
Zr	HCP	BCC (>872°C)

For example, the center atom in the BCC space lattice (see Figure 1.20) has cell coordinates of 1/2, 1/2, 1/2. Any two points are equivalent if the fractional portions of their coordinates are equal:

$$1/2, 1/2, 1/2 \equiv -1/2, -1/2, -1/2 \text{ (center)}$$

$$0, 0, 0 \equiv 1, 0, 1 \text{ (corner)}$$

A *cell direction* is designated by the vector **r**, which is a combination of the lattice translation vectors **a**, **b**, and **c**:

$$\mathbf{r} = u\mathbf{a} + v\mathbf{b} + w\mathbf{c} \tag{1.27}$$

A direction can also be specified with the cell coordinates in square brackets, with commas and fractions removed:

$$[1\ 1\ 1] \equiv [1/2\ 1/2\ 1/2] \equiv 1/2\mathbf{a} + 1/2\mathbf{b} + 1/2\mathbf{c}$$

Negative directions are indicated by an overbar [1$\bar{1}$1], and are called the "one negative one one" direction. All directions are relative to the origin where the three lattice translation vectors originate (see Figure 1.19).

The *cell volume*, V, can be calculated using the lattice translation vectors:

$$V = |\mathbf{a} \times \mathbf{b} \cdot \mathbf{c}| \tag{1.28}$$

Mathematically, this is a triple scalar product and can be used to calculate the volume of any cell, with only a knowledge of the lattice translation vectors. If the lattice parameters and interaxial angles are known, the following expression for V can be derived from the vector expression:

$$V = abc(1 - \cos^2 \alpha - \cos^2 \beta - \cos^2 \gamma + 2\cos \alpha \cos \beta \cos \gamma)^{1/2} \tag{1.29}$$

This looks complicated, but for orthogonal systems ($\alpha, \beta, \gamma = 90°$) it reduces quite nicely to the expected expression:

$$V = abc \tag{1.29a}$$

Now that we know how to find the cell volume, we can use some previous information to calculate an important property of a material, namely, its *density*, which we represent with the lowercase Greek letter rho, ρ. For example, aluminum has an FCC space lattice. Recall that there are four atoms in the FCC unit cell. We know that each aluminum atom has an atomic weight of 27 g/mol. From Table 1.11, the cubic lattice parameter for aluminum is 4.05 Å, or 0.405 nm (4.05×10^{-8} cm). This gives us a volume of $a^3 = 6.64 \times 10^{-23}$ cm^3. You should confirm that the theoretical density for aluminum is then:

$$\rho = \frac{(4 \text{ atoms/unit cell})(27 \text{ g/mol})}{(6.02 \times 10^{23} \text{ atoms/mol})(6.64 \times 10^{-23} \text{ cm}^3)} = 2.70 \text{ g/cm}^3 \text{ per unit cell}$$

Cooperative Learning Exercise 1.2

The actinide-series element protactinium (Pa, AW = 231.04) has a body-centered tetragonal structure with cell dimensions $a = 0.3925$ nm, $c = 0.3238$ nm.

Person 1: Determine the weight, in grams, of a single unit cell of Pa.
Person 2: Calculate the volume of a single Pa unit cell.
Combine your answers appropriately to arrive at the density.

Answer: 15.38 g/cm^3

In addition to cell coordinates and directions, *crystal planes* are very important for the determination and analysis of structure. We begin with the cell's coordinate system, with axes $x, y,$ and z. Recall that the axes are not necessarily orthogonal and that a, b, and c are the lattice parameters. Look at Figure 1.24. The equation of an arbitrary plane with intercepts A, B, and C, relative to the lattice parameters is given by

$$\frac{1}{A}\frac{x}{a} + \frac{1}{B}\frac{y}{b} + \frac{1}{C}\frac{z}{c} = 1 \tag{1.30}$$

We designate a plane by *Miller indices*, $h, k,$ and l, which are simply the reciprocals of the intercepts, A, B, and C:

$$\left(\frac{1}{A}\frac{1}{B}\frac{1}{C}\right) = (hkl) \tag{1.31}$$

Note that the Miller indices are enclosed in parentheses and **not** separated by commas. Miller indices are determined as follows:

- Remove all indeterminacy; that is, the planes should have nonzero intercepts.
- Find intercepts along three axes of the crystal system.
- Take the reciprocals of the intercepts in terms of a, b, and c.

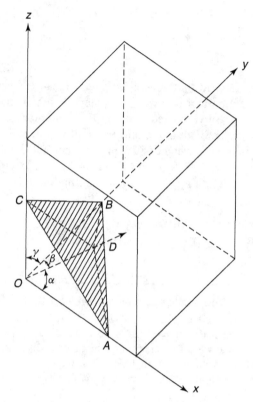

Figure 1.24 Definition of Miller indices for an arbitrary plane (shaded area). From Z. Jastrzebski, *The Nature and Properties of Engineering Materials*, 2nd ed. Copyright © 1976 by John Wiley & Sons, Inc. This material is used by permission of John Wiley & Sons, Inc.

- Multiply reciprocals by the smallest integer necessary to convert them into a set of integers.
- Enclose resulting integers in parentheses, (hkl), without commas.

Any planes that have common factors are parallel. For example, a (222) and a (111) plane are parallel, as are (442) and (221) planes. As with cell directions, a minus sign (in this case, indicating a negative intercept) is designated by an overbar. The ($2\bar{2}1$) plane has intercepts at 1/2, −1/2, and 1 along the x, y, and z axes, respectively. Some important planes in the cubic crystal system are shown in Figure 1.25.

In a manner similar to that used to calculate the density of a unit cell, we can calculate the density of atoms on a plane, or *planar density*. The perpendicular intersection of a plane and sphere is a circle, so the radius of the atoms will be helpful in calculating the area they occupy on the plane. Refer back to Example Problem 1.4 when we calculated the lattice parameter for a BCC metal. The section shown along the body diagonal is actually the (110) plane. The body-centered atom is entirely enclosed by this plane, and the corner atoms are located at the confluence of four adjacent planes, so each contributes 1/4 of an atom to the (110) plane. So, there are a total of two atoms on the (110) plane. If we know the lattice parameter or atomic radius, we can calculate the *area of the plane*, A_p, the *area occupied by the atoms*, A_c, and the corresponding

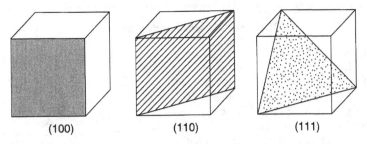

Figure 1.25   Some important planes in the cubic space lattice.

planar density, $PD = A_c/A_p$. We will see in Chapter 5 that an important property of metals, ductility, is affected by planar density.

1.1.1.3 *Interplanar Spacings.* To this point, we have concentrated on planes in an isolated cell. A crystal lattice, of course, is composed of many individual unit cells, with the planes extending in all directions. So, a real crystal lattice has many (111) planes, for example, all of which are parallel to one another. There is a uniform distance between like planes in a lattice, which we call the *interplanar spacing* and designate with d, the perpendicular distance between adjacent planes in a set. Note that even though the (111) and (222) planes are parallel to one another, they are **not** the same plane, since their planar densities may be much different depending on the lattice (for example, compare these two planes in simple cubic). What we are calculating here is the perpendicular distance between the same plane in adjacent cells.

The general expression for d in terms of lattice parameters and interaxial angles is somewhat complicated

$$d = V \left[\begin{array}{l} h^2b^2c^2\sin^2\alpha + k^2a^2c^2\sin^2\beta + l^2a^2b^2\sin^2\gamma \\ + 2hlab^2c(\cos\alpha\cos\gamma - \cos\beta) + 2hkabc^2(\cos\alpha\cos\beta - \cos\gamma) \\ + 2kla^2bc(\cos\beta\cos\gamma - \cos\alpha) \end{array} \right]^{-1/2}$$

(1.32)

where V is the cell volume as defined by Eq. (1.28). Note that the Miller indices of the plane under consideration are included here, to distinguish the distance between different planes in the same space lattice. For orthogonal systems (orthorhombic, tetragonal, and cubic), $\alpha = \beta = \gamma = 90°$

$$\frac{1}{d} = \left(\frac{h^2}{a^2} + \frac{k^2}{b^2} + \frac{l^2}{c^2} \right)^{1/2}$$

(1.33)

For a cubic system ($a = b = c$), this expression simplifies even further to

$$d = \frac{a}{\sqrt{h^2 + k^2 + l^2}}$$

(1.34)

We will see that the interplanar spacing is an important parameter for characterizing many types of materials.

Cooperative Learning Exercise 1.3

The (111) plane in FCC is shown at right.

Person 1: Calculate the area of the (111) plane, A_p, in the FCC cell in terms of the atomic radius, r.

Person 2: Calculate the area occupied by the atoms, A_c, in the (111) plane of the FCC cell in terms of r.

Combine your answers appropriately to arrive at the planar density, *PD*. If $r = 1.36$ Å, what is the *number density* (number of atoms per unit area) for the (111) plane in FCC?

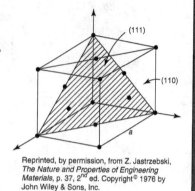

Reprinted, by permission, from Z. Jastrzebski, *The Nature and Properties of Engineering Materials*, p. 37, 2nd ed. Copyright© 1976 by John Wiley & Sons, Inc.

Answer: $PD = 0.91$; number density $= 1.72 \times 10^{15}$ atoms/cm^2

1.1.2 X-Ray Diffraction

The interplanar spacing between adjacent planes allows us to use a very powerful tool for structural determination called *X-ray diffraction (XRD)*. If we bombard a crystal with X rays of a certain wavelength, λ, at an incident angle, θ, the X rays can either pass directly through the crystal or, depending on the angle of incidence, interact with certain atoms in the lattice. This interaction is best visualized as a "shooting gallery" in which the incident X rays "bounce off" the hard-core spheres in the lattice (see Figure 1.26). In reality, the X-ray photons are interacting with the electron density around the atoms, leading to diffraction. The path difference for rays reflected from adjacent planes is

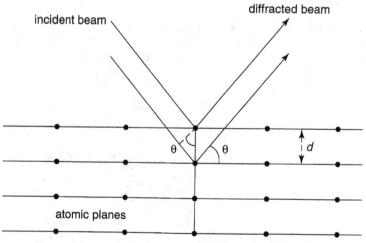

Figure 1.26   Schematic illustration of incident radiation diffraction by a crystal lattice.

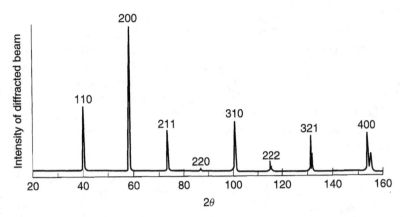

Figure 1.27   X-Ray diffraction pattern for tungsten. Adapted from A. G. Guy, and J. J. Hren, *Elements of Physical Metallurgy*, p. 208, 3rd ed. Copyright © 1974 by Addison-Wesley.

$2d(\sin \theta)$, and *constructive interference* occurs when the path difference is an integer, n, of the wavelength:

$$n\lambda = 2d \sin \theta \qquad (1.35)$$

Equation (1.35) is called *Bragg's law* and is a very important result. It says that if we bombard a crystal lattice with X rays of a known wavelength and at a known angle, we will be able to detect diffracted X rays of various intensities that represent a specific interplanar spacing in the lattice. The result of scanning through different angles at a fixed wavelength and counting the number (intensity) of diffracted X rays is a diffraction pattern, an example of which is shown in Figure 1.27. Note how the sum of the Miller indices, or *order*, for each plane increases from left to right in the pattern, and also note how different planes have different intensities. Also, not all planes result in diffracted X rays. From the abscissa (which is by convention listed as 2θ for instrumentation reasons), the interplanar spacing, d, can be calculated using Bragg's law.

The study of crystals and their X-ray patterns is both fascinating and complex. From a known crystal structure, a theoretical diffraction pattern can be constructed using vector algebra, including such considerations as *structure factors* which account for disallowed interactions with the crystal lattice and determine the relative intensity of the diffracted beams. For our purposes, it is sufficient to know that a diffraction pattern can be obtained experimentally in a relatively routine fashion, and the resulting pattern is characteristic of a specific crystalline material.

1.1.3 Point Defects

Now that the most important aspects of perfect crystals have been described, it is time to recognize that things are not always perfect, even in the world of space lattices. This is not necessarily a bad thing. As we will see, many important materials phenomena that are based on defective structures can be exploited for very important uses. These *defects*, also known as *imperfections*, are grouped according to spatial extent.

Point defects have zero dimension; *line defects*, also known as *dislocations*, are one-dimensional; and *planar defects* such as *surface defects* and *grain boundary defects* have two dimensions. These defects may occur individually or in combination.

Let us first examine what happens to a crystal when we remove, add, or displace an atom in the lattice. We will then describe how a different atom, called an *impurity* (regardless of whether or not it is beneficial), can fit into an established lattice. As shown by Eq. (1.36), point defects have equilibrium concentrations that are determined by temperature, pressure, and composition. This is not true of all types of dimensional defects that we will study.

$$N_d = N \exp\left(-E_d \Big/ k_\mathrm{B} T\right) \tag{1.36}$$

In Eq. (1.36), N_d is the equilibrium number of point defects, N is the total number of atomic sites per volume or mole, E_d is the activation energy for formation of the defect, k_B is Boltzmann's constant (1.38×10^{-23} J/atom $\cdot$ K), and T is absolute temperature. Equation (1.36) is an *Arrhenius-type expression* of which we will see a great deal in subsequent chapters. Many of these Arrhenius expressions can be derived from the Gibbs free energy, ΔG.

When an atom is missing from a lattice, the resulting space is called a *vacancy* (not to be confused with a "hole," which has an electronic connotation), as in Figure 1.28. In this case, the activation energy, E_d, is the energy required to remove an atom from the lattice and place it on the surface. The activation energy for the formation of vacancies in some representative elements is given in Table 1.13, as well as the corresponding vacancy concentration at various temperatures. Note that the vacancy concentration decreases at lower temperatures. In a nonequilibrium situation, such as rapid cooling from the melt, we would not expect the equilibrium concentration to be attained. This

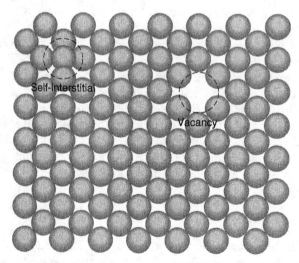

Figure 1.28 Representation of a vacancy and self-interstitial in a crystalline solid. From K. M. Ralls, T. H. Courtney, and J. Wulff, *Introduction to Materials Science and Engineering.* Copyright © 1976 by John Wiley & Sons, Inc. This material is used by permission of John Wiley & Sons, Inc.

Table 1.13 **Formation Energy of Vacancies for Selected Elements and Equilibrium Concentrations at Various Temperatures**

Element	E_d (kJ/mol)	Melting Point, T_m (°C)	N_d (vacancies/cm³)			
			25°C	300°C	600°C	T_m
Ag	106.1	960	1.5×10^4	1.5×10^{13}	3.0×10^{16}	7.8×10^{17}
Al	73.3	660	1.0×10^{10}	1.2×10^{16}	2.4×10^{18}	5.0×10^{18}
Au	94.5	1063	1.5×10^6	1.5×10^{14}	1.5×10^{17}	1.2×10^{19}
Cu	96.4	1083	1.1×10^6	1.4×10^{14}	1.4×10^{17}	9.0×10^{18}
Ge	192.9	958	<1	1.3×10^5	1.3×10^{11}	8.2×10^{13}
K	38.6	63	2.1×10^{15}	—	—	1.3×10^{16}
Li	39.5	186	4.7×10^{15}	—	—	1.4×10^{18}
Mg	85.8	650	4.4×10^7	6.4×10^{14}	3.5×10^{17}	5.7×10^{17}
Na	38.6	98	4.0×10^{15}	—	—	1.0×10^{17}
Pt	125.4	1769	8.7	2.7×10^{11}	2.0×10^{15}	4.2×10^{19}
Si	221.8	1412	<1	3.1×10^2	2.5×10^9	8.0×10^{15}

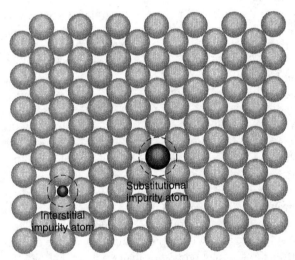

Figure 1.29 Representation of interstitial and substitutional impurity atoms in a crystalline solid. From K. M. Ralls, T. H. Courtney, and J. Wulff, *Introduction to Materials Science and Engineering*. Copyright © 1976 by John Wiley & Sons, Inc. This material is used by permission of John Wiley & Sons, Inc.

is indeed the case, and vacancy concentrations in rapidly quenched metals are much closer to the liquid concentration than they are to the equilibrium solid concentration.

The second type of point defect is called an impurity. Impurities can occur in two ways: as an interstitial impurity, in which an atom occupies an interstitial site (see Figures 1.21, 1.22, and 1.29); or when an impurity atom replaces an atom in the perfect lattice (see Figure 1.29). In the first instance, either the same atom as in the lattice, or an impurity atom, can occupy an interstitial site, causing considerable lattice strain as the atomic planes distort slightly to accommodate the misplaced atom. The amount of strain created depends on how large the atom is relative to lattice atoms. It is also possible for a lattice atom to move off of a lattice site and occupy an interstitial site. In this case, both of the defects shown in Figure 1.28 occur simultaneously, and a defect pair known as a *Frenkel defect* (or Frenkel disorder) occurs. In a pure Frenkel defect, there are always equal concentrations of interstitial impurities and vacancies.

The second type of impurity, substitution of a lattice atom with an impurity atom, allows us to enter the world of alloys and intermetallics. Let us diverge slightly for a moment to discuss how control of substitutional impurities can lead to some useful materials, and then we will conclude our description of point defects. An *alloy*, by definition, is a metallic solid or liquid formed from an intimate combination of two or more elements. By "intimate combination," we mean either a liquid or solid solution. In the instance where the solid is crystalline, some of the impurity atoms, usually defined as the minority constituent, occupy sites in the lattice that would normally be occupied by the majority constituent. Alloys need not be crystalline, however. If a liquid alloy is quenched rapidly enough, an amorphous metal can result. The solid material is still an alloy, since the elements are in "intimate combination," but there is no crystalline order and hence no substitutional impurities. To aid in our description of substitutional impurities, we will limit the current description to crystalline alloys, but keep in mind that amorphous alloys exist as well.

The extent to which a lattice will allow substitutional impurity atoms depends on a number of things. The factors affecting the solubility of one element in another are summarized in a set of guidelines called the *Hume–Rothery rules*, though they are really not rules at all. As you can imagine, atomic size plays an important role in determining solubility. The first Hume–Rothery "rule" states that if the atomic size of the host lattice and impurity atom differ by more than about 14%, the solubility of the impurity in the lattice will be small. Refer to Table 1.9 for values of atomic size. The second rule involves electronegativity. We mentioned earlier in this chapter that electronegativity is an important concept, and it plays an important role in determining not only how soluble an impurity is, but also what type of bond will result. In general, the larger the electronegativity difference, $\Delta\chi$, between the host atom and the impurity, the greater the tendency to form compounds and the less solubility there is. So, elements with similar electronegativities (refer to Table 1.4) tend to alloy, whereas elements with large $\Delta\chi$ tend to have more ionic bonds (see Section 1.0.3) and form *intermetallics*. Intermetallics are similar to alloys, but the bonding between the different types of atoms is partly ionic, leading to different properties than traditional alloys. The third rule deals with crystal structures. One would expect like crystal structures to be more compatible, and this is generally the case. Refer to Table 1.11 for typical crystal structures, but keep in mind that the elements can have multiple structures depending on temperature, and remember that this can affect the stability of the alloy. Finally, all other things being equal, the fourth Hume–Rothery rule states that a metal of lower valency is more likely to dissolve one of higher valency than vice versa. Common valences of the elements are listed in Table 1.9. Again, elements can have multiple oxidation states. An interesting corollary to the fourth rule is that the total number of valence electrons per atom can be used as a guideline in determining the crystal structure of the alloy. As summarized in Table 1.14, by summing the valence electrons of the elements in the alloy and dividing by the number of types of atoms (binary $= 2$, ternary $= 3$, etc.), it is sometimes possible to predict the crystal structure of an alloy. The "complex cubic" structures include cubic structures other than SC, BCC, and FCC, which we have not yet described, such as the diamond structure. As an example of this corollary, the binary

Table 1.14 **Common Crystal Structures of Alloys Based on Valences of Components**

Valence Electrons/Atom	Structure
3/2	BCC, complex cubic, HCP
21/13	Complex cubic
7/4	HCP

alloy formed between Cu (+1 valence) and Be (+2 valence) has $(1 + 2)/2 = 3/2$ valence electrons/atom, and it turns out to have the BCC structure, which is different than either of the two component structures.

Cooperative Learning Exercise 1.4

Person 1: Do Cu and Ni satisfy the first and second Hume–Rothery rules for complete solid solubility?

Person 2: Do Cu and Ni satisfy the third and fourth Hume–Rothery rules for complete solid solubility?

Compare your answers. Would you predict that Cu and Ni have complete, partial, or no solid solubility in each other?

Answer: $\Delta r\% = 2.3\%$ (<14%); $\Delta \chi = 0.01$ (small); both are FCC; Cu is lower valence than Ni. All four are satisfied; Cu and Ni are completely soluble.

This concludes our diversion into alloys for the time being. From this point on, we will often describe metals and alloys in similar terms, and we will make distinctions between the two classes of materials only when there are substantial dissimilarities between them. Returning now to our description of point defects, we have but one type of point defect pair left to describe. Similar to a Frenkel defect in which both a vacancy and interstitial impurity must occur simultaneously, a *Schottky defect* (a.k.a. Schottky disorder or imperfection) arises in ionic solids when a cation–anion vacancy pair is formed. Recall that ionic compounds occur when there is a large electronegativity difference between the components, so that a Schottky defect normally occurs in binary ionic compounds such as sodium chloride. Though the ionic compounds we will use as illustrations here are not technically metals or alloys, keep in mind that metallic solids such as intermetallics can have ionic bonding. In sodium chloride, removal of one sodium ion and one chloride ion from the lattice results in a Schottky defect (see Figure 1.30). In ionic solids where the cation (positively charged ion) and anion (negatively charged ion) have the same absolute charge (e.g., $|Na^+| = 1$, $|Cl^-| = 1$), a Schottky defect arises from the same number of vacancies in both ions. For ionic solids in which the anion and cation have different absolute valencies (e.g., CaF_2), a nonstoichiometric compound must be formed in order to maintain charge neutrality in the lattice; that is, two fluorine ions (F^-) must leave for every calcium ion (Ca^{2+}) that is removed from the lattice. Because atoms must leave the ionic lattice, Schottky

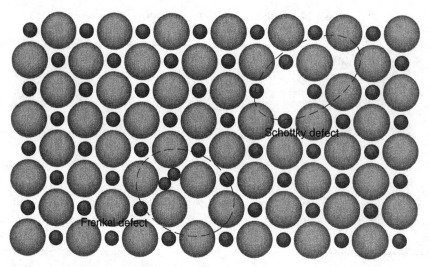

Figure 1.30   Representation of Frenkel and Schottky defects in a crystalline solid. Adapted from W. G. Moffatt, G. W. Pearsall, and J. Wulff, *The Structure and Properties of Materials*, Vol. 1. Copyright © 1964 by John Wiley & Sons, Inc.

defects normally occur only at lattice perturbations such as grain boundaries or surfaces so that the removed atoms have someplace to go. We will describe Schottky defects in more detail when we come to inorganic materials, such as oxides, where binary compounds are more prevalent.

1.1.4   Line Defects and Dislocations

We now move on to defects that have some spacial extent, even if only in one dimension. As we continue to increase the geometric complexity of these defects, you may find it more difficult to visualize them. As with crystal structures, three-dimensional models may help you with visualization, and do not limit yourself to one representation of a specific defect—look for multiple views of the same thing.

The first type of one-dimensional defect, or *line defect*, is called a *dislocation*. A dislocation is a linear disturbance of the atomic arrangement in a crystal caused by the displacement of one group of atoms from an adjacent group. There are three types of dislocations: *edge dislocations, screw dislocations*, and a combination of these two, termed *mixed dislocations*. An edge dislocation occurs when a single atomic plane does not extend completely through the lattice. The termination of this half-plane of atoms creates a defect line (dislocation line) in the lattice (line *DC* in Figure 1.31). The edge dislocation is designated by a perpendicular sign, either ⊥ if the plane is above the dislocation line or ⊤ if the plane is below the dislocation line. Edge dislocations can be quantified using a vector called the *Burger's vector*, **b**, which represents the relative atomic displacement in the lattice due to the dislocation (see Figure 1.32). The Burger's vector is determined as follows:

- Define a positive direction along the dislocation line. This is usually done *into* the crystal.

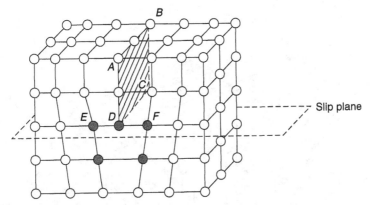

Figure 1.31 Edge dislocation *DC* results from introducing an extra half-plane of atoms *ABCD*, where *DC* is considered a positive dislocation designated by ⊥. From Z. Jastrzebski, *The Nature and Properties of Engineering Materials*, 2nd ed. Copyright © 1976 by John Wiley & Sons, Inc. This material is used by permission of John Wiley & Sons, Inc.

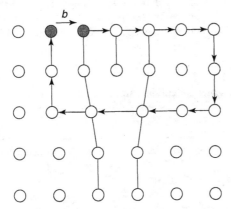

Figure 1.32 Burger's circuit around an edge dislocation. From Z. Jastrzebski, *The Nature and Properties of Engineering Materials*, 2nd ed. Copyright © 1976 by John Wiley & Sons, Inc. This material is used by permission of John Wiley & Sons, Inc.

- Construct a plane of atoms perpendicular to the dislocation line.
- Trace out a *clockwise* path around the dislocation line moving n lattice vectors in each of the four mutually perpendicular directions.
- The Burger's vector is drawn from the *finish* to the *start* of the path.

You can start the clockwise path, called the *Burger's circuit*, anywhere, as long as you entirely enclose the dislocation with the circuit.

The second type of line defect, the screw dislocation, occurs when the Burger's vector is parallel to the dislocation line (*OC* in Figure 1.33). This type of defect is called a screw dislocation because the atomic structure that results is similar to a screw. The Burger's vector for a screw dislocation is constructed in the same fashion as with the edge dislocation. When a line defect has both an edge and screw dislocation

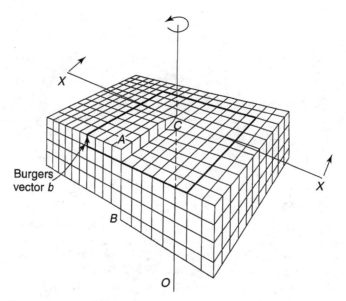

Figure 1.33 Representation of defect line (*OC*), Burger's circuit and Burger's vector in a screw dislocation. From Z. Jastrzebski, *The Nature and Properties of Engineering Materials*, 2nd ed. Copyright © 1976 by John Wiley & Sons, Inc. This material is used by permission of John Wiley & Sons, Inc.

Table 1.15 Comparison of Dislocation Characteristics

Dislocation Type	Burger's Vector	Propagation Direction
Edge	⊥ to dislocation line	⊥ to dislocation line, ∥ to Burger's vector
Screw	∥ to dislocation line	⊥ to dislocation line, ⊥ to Burger's vector
Mixed	Neither ∥ nor ⊥ to dislocation line	Neither ∥ nor ⊥ to dislocation line, Burger's vector

component, a mixed dislocation results. In this case, the Burger's vector is neither parallel nor perpendicular to the dislocation line, but can be resolved into edge and screw components. A comparison of the three types of dislocations is summarized in Table 1.15.

Dislocations can move through a crystal lattice. This is most easily visualized in the edge dislocation, where the half-plane of atoms simply moves perpendicular to the defect line (line *DC* in Figure 1.31). The lattice atoms below the half-plane incorporate the atoms above them into their plane, but must create a new half-plane of atoms to do so. This is how the half-plane propagates. The dislocation movement can be very rapid—approaching the speed of sound. Screw and mixed dislocations move in more complex, but similar, fashions. The propagation of all three dislocation types is summarized in Figure 1.34. We will see that dislocation propagation plays an important role in ductility and crystal slip.

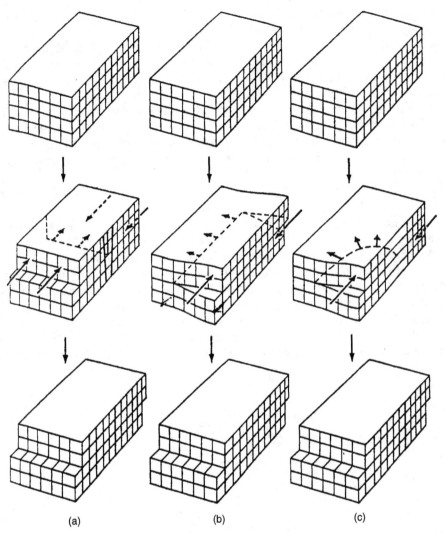

Figure 1.34 Representation of propagation modes for (a) edge, (b) screw, and (c) mixed dislocations.

1.1.5 Planar Defects

Finally, two-dimensional defects can occur in crystals. There are two categories of planar defects: *stacking faults* and *grain boundaries*.

Stacking faults arise due to imperfections in the stacking sequence of atomic planes. Recall that the FCC structure is the result of ABCABC... stacking of close-packed planes. If a plane of atoms is missing from this sequence, as in ABCAB|ABC, an *intrinsic stacking fault* results. If an additional plane is inserted into the sequence (e.g., an extra C plane in the sequence ABCA|C|BCABC), an *extrinsic stacking fault* results. Finally, a *twinning fault* occurs when the sequence reverses itself about a *mirror plane*, as in ABCABCBACBA, where C is the mirror plane and CBA is the reverse sequence to ABC.

To this point, we have concentrated on single crystals. Most crystalline materials are *polycrystalline*—that is, composed of many small crystals, or *grains*, that usually have random crystallographic orientation relative to each other. Unless the grains are at a surface, they are adjacent to other grains, not necessarily of the same orientation. The region where they intersect is called a *grain boundary*. There are two general types of grain boundaries: tilt and twist.

A *tilt grain boundary* is actually a set of edge dislocations (see Figure 1.35). The *angle of misorientation*, θ, characterizes the tilt grain boundary and is defined as the angle between the same directions in adjacent crystals. The angle of misorientation can be calculated from the Burger's vector and the vertical separation between edge dislocations, h:

$$\tan \theta = b/h \tag{1.37}$$

A high-angle tilt grain boundary results when $\theta > 15°$. For $\theta < 10°$, a low-angle tilt grain boundary results, and Eq. (1.37) can be simplified to $\tan \theta \approx \theta = b/h$. In a similar manner, a *twist grain boundary* is a set of screw dislocations.

We will have much more to say about planar defects, particularly their effect in determining the physical properties of bulk crystalline solids. In this regard, not only will defect structure play an important role, but so will crystallite size. Generally, polycrystalline materials are termed *coarse-grained, fine-grained*; and *ultra-fine-grained* in order of decreasing crystallite size. There are additional methods for classifying the grain size of materials, such as that used for metals by the American Society for

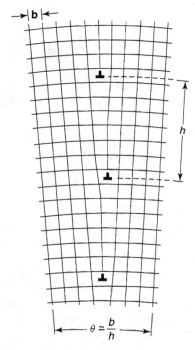

Figure 1.35   Representation of a tilt grain boundary. From K. M. Ralls, T. H. Courtney, and J. Wulff, *Introduction to Materials Science and Engineering*. Copyright © 1976 by John Wiley & Sons, Inc. This material is used by permission of John Wiley & Sons, Inc.

Testing Materials (ASTM). A more recent development in this terminology has been the addition of the term *nanocrystalline* to describe materials with crystallite sizes of 10–100 nm. Though we will not spend a great deal of time on the emerging area of *nanostructured materials* (at least not as a separate topic), this is an appropriate place to introduce the concept. A nanostructured material is one in which at least one critical dimension of the material is less than 100 nm in size. Nanocrystalline materials qualify for this designation since their grain sizes fall below the 100-nm range. We will see that a number of materials and phenomena meet this requirement, whether or not we ultimately refer to them as nanostructured materials.

1.2 STRUCTURE OF CERAMICS AND GLASSES

Inorganic materials constitute the largest class of solids in the world. We have already described metals; and while they are not organic (they contain no biological carbon), they are also not inorganic in the strict sense of the word—they are metals due to the unique characteristics of their valence electronic structure. Inorganic materials are typically compounds, such as metal oxides, carbides, or nitrides. They possess many interesting properties that we will only begin to describe at this point. They can also differ structurally from other types of materials like metals and polymers. Let us begin by describing the structure of inorganic materials.

1.2.1 Pauling's Rules

Recall that the structure of a crystal is determined mostly by how the atoms pack together. The same is true of binary compounds such as alloys, and of binary compounds that contain noncovalent bonds, such as ionic compounds. In addition to the concept of electronegativity, Linus Pauling also produced a set of generalizations that are used to describe the majority of ionic crystal structures. Pauling's first rule states that *coordination polyhedra* are formed. Coordination polyhedra are three-dimensional geometric constructions such as tetrahedra and octahedra. Which polyhedron will form is related to the radii of the anions and cations in the compound. Consider the two-dimensional representation of a binary ionic compound shown in Figure 1.36. The anions (open circles) are larger than the cations (why?), and a central cation cannot remain in contact with the surrounding anions if the anion radius is larger than a certain value. Thus, the structure in Figure 1.36c is unstable, whereas the structures in Figures 1.36a and 1.36b are both stable. Note that the cation–anion distance is simply

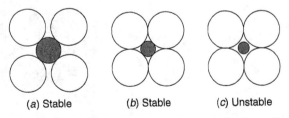

(a) Stable   (b) Stable   (c) Unstable

Figure 1.36   (a, b) Stable and (c) unstable coordination configurations. From W. D. Kingery, H. K. Bowen, and D. R. Uhlmann, *Introduction to Ceramics*. Copyright © 1976 by John Wiley & Sons, Inc. This material is used by permission of John Wiley & Sons, Inc.

the sum of the cation–anion radii for the stable structures. It is also true that the coordination number is determined by the *radius ratio* of the two ions (R_{anion}/R_{cation}). The larger the central cation, the more anions that can be packed around it. It is now evident why our description of ionic radii in Table 1.9 is so important. For each coordination number, there is some critical value of the radius ratio above which the structure will not be stable. These limits are summarized in Figure 1.37. Pauling's second rule on the packing of ions states that local electrical neutrality is maintained. We use a quantity called *bond strength* to assure electrical neutrality, where the bond strength is the ratio of formal charge on a cation to its coordination number. For example, silicon has a formal charge of +4 and a coordination number of 4, so that its strength

Coordination Number	Anion-Cation Radius Ratio	Coordination Geometry
2	> 6.45	
3	4.45–6.45	
4	2.42–4.45	
6	1.37–2.42	
8	1.0–1.37	

Figure 1.37 Critical radius ratios for various coordination numbers. Adapted, from W. Callister, *Materials Science and Engineering: An Introduction*, 5th ed., p. 384. Copyright © 2000 by John Wiley & Sons, Inc.

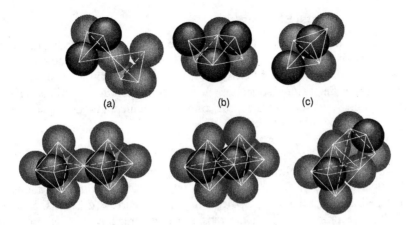

Figure 1.38 Tetrahedra (*top*) and octahedra (*bottom*) linked by sharing (a) corners, (b) edges, and (c) faces. From W. D. Kingery, H. K. Bowen, and D. R. Uhlmann, *Introduction to Ceramics*. Copyright © 1976 by John Wiley & Sons, Inc. This material is used by permission of John Wiley & Sons, Inc.

is $4/4 = 1$. For aluminum, a formal charge of $+3$ and a coordination number of six gives a bond strength of 1/2. So, the total strength of the bonds reaching an anion from all surrounding cations is equal to the charge of the anion. Pauling's third rule tells us how to link these polyhedra together (see Figure 1.38). Corners, rather than faces or edges, tend to be shared in stable structures. This is due to the fact that the cation separation between adjacent polyhedra decreases as edges and faces are shared, increasing repulsion and leading to unstable structures. Pauling's fourth rule is similar to the third, stating that polyhedra formed about cations of low coordination number and high charge tend to be linked by corners. The fifth and final rule states that the number of different constituents in a structure tends to be small; that is, it is difficult to efficiently pack different-sized polyhedra into a single structure.

Cooperative Learning Exercise 1.5

Magnesium silicate, Mg_2SiO_4, consists of a HCP arrangement of O^{2-} ions with the cations occupying interstitial spaces. Use the ionic radii of the elements given in Table 1.9 to answer the following questions.

    Person 1: What is the coordination number for the Mg^{2+} ions and what type of interstitial site will they occupy?

    Person 2: What is the coordination number for the Si^{4+} ions and what type of interstitial site will they occupy?

    Combine your information. According to Table 1.16, what is the fraction of the total type of interstitial site occupied by each cation, and what is the structure name of this compound?

Answer: $R_O/R_{Mg} = 1.32/0.78 = 1.69$ (octahedral); $R_O/R_{Si} = 1.32/0.39 = 3.39$ (tetrahedral); 1/2 of octahedral sites are occupied; $\frac{1}{8}$ of tetrahedral sites are occupied; this is the olivine structure.

1.2.2 Ceramic Crystal Structures

Most crystalline inorganic compounds are based on nearly close-packing of the anions (generically referred to as O or X, though oxygen is the most common anion) with metal atom cations (generically called M or A) placed interstitially within the anion lattice. A summary of some simple ionic structures and their corresponding coordination numbers is given in Table 1.16. Some of the most common ceramic and ionic crystal structures are described in the following sections.

1.2.2.1 Rock Salt Structure. The compounds NaCl, MgO, MnS, LiF, FeO, and many other oxides such as NiO and CaO have the so-called *rock salt structure*, in which the anions, such as Cl or O, are arranged in an FCC array, with the cations placed in the octahedral interstitial sites, also creating an FCC array of cations, as shown in Figure 1.39. The primitive cell of the rock salt structure, however, is simple cubic, with anions at four corners, and cations at alternating corners of the cube, as given by one of the quadrants in Figure 1.39. As indicated in Table 1.16, the coordination number of both anions and cations is 6, so the cation–anion radius ratio is between 1.37 and 2.42, as indicated in Figure 1.37.

1.2.2.2 Diamond Structure. Another common ceramic structure arises when the tetrahedral sites in an FCC array of anions are occupied. For example, an FCC array

Table 1.16   Table of Some Simple Ionic Structures and Their Corresponding Coordination Numbers

Anion Packing	Coordination Number of M and O	Sites by Cations	Structure Name	Examples
Cubic close-packed	6:6 MO	All oct.	Rock salt	NaCl, KCl, LiF, KBr, MgO, CaO, SrO, BaO, CdO, VO, MnO, FeO, CoO, NiO
Cubic close-packed	4:4 MO	1/2 tet.	Zinc blende	ZnS, BeO, SiC
Cubic close-packed	4:8 M_2O	All tet.	Antifluorite	Li_2O, Na_2O, K_2O, Rb_2O, sulfides
Distorted cubic close-packed	6:3 MO_2	1/2 oct.	Rutile	TiO_2, GeO_2, SnO_2, PbO_2, VO_2, NbO_2, TeO_2, MnO_2, RuO_2, OsO_2, IrO_2
Cubic close-packed	12:6:6 ABO_3	1/4 oct. (B)	Perovskite	$CoTiO_3$, $SrTiO_3$, $SrSnO_3$, $SrZrO_3$, $SrHfO_3$, $BaTiO_3$
Cubic close-packed	4:6:4 AB_2O_4	1/8 tet. (A) 1/2 oct. (B)	Spinel	$FeAl_2O_4$, $ZnAl_2O_4$, $MgAl_2O_4$
Cubic close-packed	4:6:4 $B(AB)O_4$	1/8 tet. (B) 1/2 oct. (A, B)	Spinel (inverse)	$FeMgFeO_4$, $MgTiMgO_4$
Hexagonal close-packed	4:4 MO	1/2 tet.	Wurtzite	ZnS, ZnO, SiC
Hexagonal close-packed	6:6 MO	All oct.	Nickel arsenide	NiAs, FeS, FeSe, CoSe
Hexagonal close-packed	6:4 M_2O_3	2/3 oct.	Corundum	Al_2O_3, Fe_2O_3, Cr_2O_3, Ti_2O_3, V_2O_3, Ga_2O_3, Rh_2O_3
Hexagonal close-packed	6:6:4 ABO_3	2/3 oct. (A, B)	Ilmenite	$FeTiO_3$, $NiTiO_3$, $CoTiO_3$
Hexagonal close-packed	6:4:4 A_2BO_4	1/2 oct. (A) 1/8 tet. (B)	Olivine	Mg_2SiO_4, Fe_2SiO_4
Simple cubic	8:8 MO	All cubic	CsCl	CsCl, CsBr, CsI
Simple cubic	8:4 MO_2	1/2 cubic	Fluorite	ThO_2, CeO_2, PrO_2, UO_2, ZrO_2, HfO_2, NpO_2, PuO_2, AmO_2
Connected tetrahedra	4:2 MO_2	—	Silica types	SiO_2, GeO_2

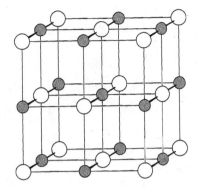

Figure 1.39   The rock salt crystal structure.

of sulfur anions with Zn ions in the tetrahedral positions gives the compound ZnS and results in the *zinc blende structure* (see Figure 1.40). The atoms can be all alike also, for example carbon, in which case diamond results. In either case, the atoms in the tetrahedral sites have a coordination number of four (by definition). Many important compounds have the zinc blende or diamond structure, including SiC.

1.2.2.3 Spinel Structure. Many compounds are formed when there is more than one metal cation in the lattice. Such is the case with the *spinel* structure, which has the general formula AB_2O_4, where A and B are different metal cations, such as in magnesium aluminate, $MgAl_2O_4$ (see Figure 1.41). This structure can be viewed as a combination of the rock salt and zinc blende structures. The anions, usually oxygen, are again placed in an FCC array. In a normal spinel, the divalent A ions are on tetrahedral sites, and the trivalent B atoms are on octahedral sites. In an inverse spinel, divalent A atoms and half of the trivalent B atoms are on octahedral sites, with the other half of the B^{3+} atoms on tetrahedral sites. Many of the ferrites, such as Fe_3O_4 (in which iron has two different coordination states), have the inverse spinel structure.

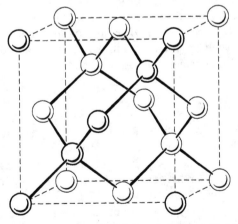

Figure 1.40   The diamond (zinc blende) crystal structure. From K. M. Ralls, T. H. Courtney, and J. Wulff, *Introduction to Materials Science and Engineering*. Copyright © 1976 by John Wiley & Sons, Inc. This material is used by permission of John Wiley & Sons, Inc.

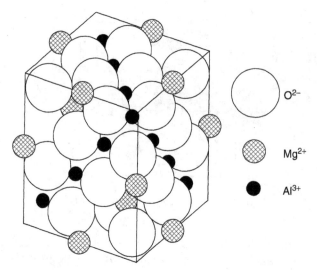

Figure 1.41   The spinel crystal structure of $MgAl_2O_4$. Reprinted, by permission, from C. Kittel, *Introduction to Solid State Physics*, p. 447. Copyright © 1957 by John Wiley & Sons, Inc.

1.2.2.4   *Other Important Ceramic Structures.*

There are many additional crystal structures that arise due to compounds of both simple and complex stoichiometry. These structures generally have specific names associated with them that have developed out of geology and crystallography over many years. For example, the *corundum* structure is common to Al_2O_3, the *rutile* structure comes from one of the forms of TiO_2, and a number of important ceramics, such as CaF_2, have the *fluorite* structure. One structure with current technological importance is the *perovskite* structure (see the $CaTiO_3$ perovskite structure in Figure 1.42). Many important ceramics with unique electrical and dielectric properties have the perovskite structure, including barium titanate, $BaTiO_3$, and *high-temperature superconductors* (HTS). The perovskites have the general formula of ABO_3, but in the case of most superconductors, the A cation consists of more than one type of atom, such as in $Y_1Ba_2Cu_3O_{7-x}$, or the so-called "1–2–3" superconductor, in which the perovskite structure is tripled, and one ytrrium atom is replaced for every third barium atom; there is usually less than the stoichiometric nine oxygen atoms required in this structure in order for enough oxygen vacancies to form and superconductivity to result.

1.2.3   Silicate Structures*

The *silicates*, made up of base units of silicon and oxygen, are an important class of ceramic compounds that can take on many structures, including some of those we have already described. They are complex structures that can contain several additional atoms such as Mg, Na, K. What makes the silicates so important is that they can be either crystalline or *amorphous* (glassy) and provide an excellent opportunity to compare these two disparate types of structure. Let us first examine the crystalline state, which will lead us into the amorphous state.

The structural unit for the simplest silicate, SiO_2, also known as *silica*, is the tetrahedron (see Figure 1.43). This is the result of applying Pauling's principles (Section

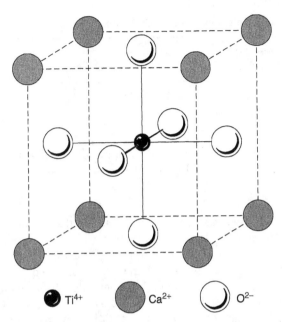

Figure 1.42 The perovskite crystal structure of CaTiO$_3$. From W. D. Kingery, H. K. Bowen, and D. R. Uhlmann, *Introduction to Ceramics.* Copyright © 1976 by John Wiley & Sons, Inc. This material is used by permission of John Wiley & Sons, Inc.

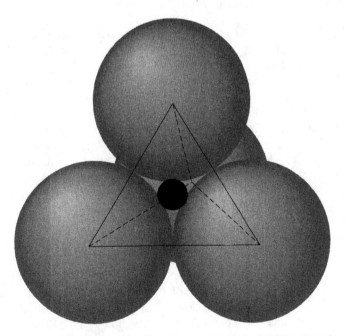

Figure 1.43 The $(SiO_4)^{4-}$ tetrahedron. The silicon atom is the solid circle at the center of the tetrahedron; large open circles are oxygens.

Table 1.17   Structural Units Observed in Crystalline Silicates

O/Si Ratio	Silicon–Oxygen Groups	Structural Units	Examples
2	SiO_2	Three-dimensional network	Quartz
2.5	Si_4O_{10}	Sheets	Talc
2.75	Si_4O_{11}	Chains	Amphiboles
3.0	SiO_3	Chains, rings	Pyroxenes, beryl
3.5	Si_2O_7	Tetrahedra sharing one oxygen ion	Pyrosilicates
4.0	SiO_4	Isolated orthosilicate tetrahedra	Orthosilicates

1.2.1) to a compound between silicon and oxygen. The data in Table 1.9 indicate that the anion/cation ratio in SiO_2 is $R_O/R_{Si} = (1.32)/(0.39) = 3.3$, which, according to Figure 1.37, dictates the tetrahedron as the base structural unit. Note that the SiO_4 tetrahedron has a formal charge of -4, which must be neutralized with cations, such as other Si atoms, in real compounds. Pauling's second rule tells us that the bond strength in silicon is 1, and the third and fourth rules tell us that corners of the tetrahedra are generally shared. This is not always the case, and different macroscopic silicate structures result depending on how the tetrahedra are combined. Corners, edges, or faces of tetrahedra can be shared. As the nature of combination of the tetrahedra changes, so must the O/Si ratio, and charge neutrality is maintained through the addition of cations. These structures are summarized in Table 1.17 and will be described separately.

1.2.3.1   Crystalline Silicate Network. When all four corners of the SiO_4 tetrahedra are shared, a highly ordered array of networked tetrahedra results, as shown in Figure 1.44. This is the structure of *quartz*, one of the crystalline forms of SiO_2. Notice that even though the O/Si ratio is exactly 2.0, the structure is still composed of isolated $(SiO_4)^{4-}$ tetrahedra. Each oxygen on a corner is shared with one other tetrahedron, however, so there are in reality only two full oxygen atoms per tetrahedron. There are actually several structures, or *polymorphs*, of crystalline silica, depending on the temperature. Quartz, with a density of 2.655 g/cm^3, is stable up to about 870°C, at which point it transforms into *tridymite*, with a density of 2.27 g/cm^3. At 1470°C, tridymite transforms to *cristobalite* (density = 2.30 g/cm^3), which melts at around 1710°C. There are "high" and "low" forms of each of these structures, which result from slight, albeit rapid, rotation of the silicon tetrahedra relative to one another.

1.2.3.2   Silicate Sheets. If three of the four corners of the $(SiO_4)^{4-}$ tetrahedron are shared, repeat units of $(Si_2O_5)^{2-}$ or $(Si_4O_{10})^{4-}$ result, with a corresponding O/Si of 2.5. Table 1.17 tells us, and Figure 1.45 shows us, that sheet structures are the result of sharing three corners. In these structures, additional cations or network modifiers, such as Al^{3+}, K^+, and Na^+, preserve charge neutrality. Through simple substitution of selected silicon atoms with aluminum atoms, and some hydroxide ions (OH^-) for oxygen atoms, complex and amazing sheet structures can result. One such common example is *muscovite*, $K_2Al_4(Si_6Al_2)O_{20}(OH)_4$, more commonly known as *mica* (Figure 1.46). The large potassium ions between layers create planes that are easily cleaved, leading to the well-known thin sheets of mica that can be made thinner and thinner in a seemingly endless fashion. It is, in fact, possible to obtain atomically smooth surfaces of mica.

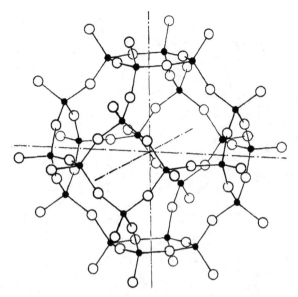

Figure 1.44 The structure of quartz, showing the three-dimensional network of SiO_4 tetrahedra. Reprinted, by permission, from L. G. Berry, B. Mason, and R. V. Dietrich, *Mineralogy: Concepts, Descriptions, Determinations*, p. 388, 2nd ed. Copyright © 1983, Freeman Publishing, Inc.

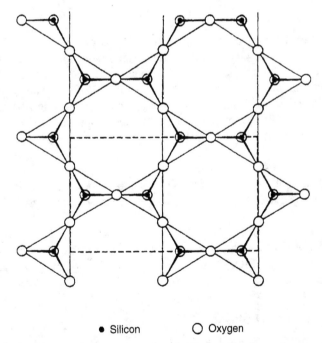

● Silicon   ○ Oxygen

Figure 1.45 Top view of a silicate sheet structure resulting from sharing three corners of the SiO_4 tetrahedra. From W. D. Kingery, H. K. Bowen, and D. R. Uhlmann, *Introduction to Ceramics*. Copyright © 1976 by John Wiley & Sons, Inc. This material is used by permission of John Wiley & Sons, Inc.

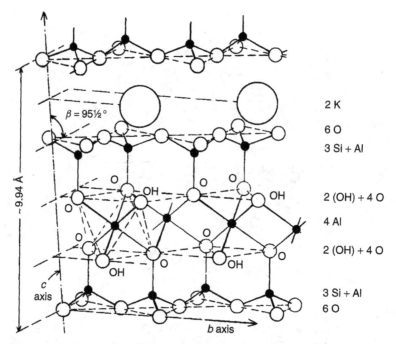

Figure 1.46   The structure of muscovite (mica), a sheet silicate. Reprinted, by permission, from L. G. Berry, B. Mason, and R. V. Dietrich, *Mineralogy: concepts, descriptions, determinations*, p. 431, 2nd ed. Copyright © 1983 by Freeman Publishing, Inc.

1.2.3.3 Silicate Chains and Rings. Sharing two out of the four corners of the SiO_4 tetrahedra results in chains. The angle formed between adjacent tetrahedra can vary widely, resulting in unique structures such as rings (see Figure 1.47). In all cases, when only two corners are shared, the repeat unit is $(SiO_3)^{2-}$, and the O/Si is 3.0. Slight variations in the O/Si ratio can also take place, and result in partially networked structures such as double chains, in which two silicate chains are connected periodically by a bridging oxygen. *Asbestos* is such a double chain, with O/Si = 2.75.

1.2.3.4 Pyrosilicates. One corner of the SiO_4 tetrahedron shared results in a $(Si_2O_7)^{6-}$ repeat unit and a class of compounds called the *pyrosilicates*. Again, counterions are necessary to maintain charge neutrality. The pyrosilicates are non-networked and have an O/Si of 3.5.

1.2.3.5 Orthosilicates. Finally, no tetrahedral corners shared gives an O/Si of 4.0, and it results in isolated $(SiO_4)^{4-}$ tetrahedra. These class of materials are referred to as the *orthosilicates*.

1.2.4 The Structure of Glasses*

The SiO_4 tetrahedra of silicates need not be arranged regularly. A perfect, three-dimensional SiO_2 lattice like quartz can be "disrupted" by introducing other cations, such as Na^+. This leads to a more random, yet still networked, structure called a *glass*. Such a random network glass is illustrated in Figure 1.48.

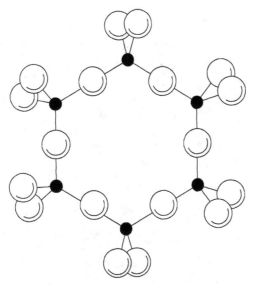

Figure 1.47 A silicate ring, beryl, with two corners of the SiO_4 tetrahedra shared. From K. M. Ralls, T. H. Courtney, and J. Wulff, *Introduction to Materials Science and Engineering.* Copyright © 1976 by John Wiley & Sons, Inc. This material is used by permission of John Wiley & Sons, Inc.

Before proceeding further with a description of the structural aspects of glasses, it is advisable to be aware of some pitfalls in nomenclature that abound in the area of glass science. Technically, a glass is a type of *noncrystalline solid* that is formed from the melt. Thus, a glass need not contain silicon or oxygen at all, but it does need to be obtained by cooling a substance from the molten state. The distinction in processing condition is necessary to distinguish glasses from other types of amorphous materials that also do not contain a regular, repeating structure, but that are formed through other processing routes, such as from the vapor phase, in which case they are called *amorphous solids*, or by dehydrating a *sol* to form a *gel*. These distinctions are summarized in Figure 1.49. We will describe some of these processing techniques in later chapters, but for now we simply note that glasses must technically be formed from the melt and that there are no restrictions on the chemical constituents of a glass. Additional characteristics of glasses that are sometimes described in the glass literature include a rigid material, a glass transition, T_g, (see Section 1.3.7), and a viscosity greater than about 10^{15} poise. The viscosity distinction is an important one, since some consider a glass to be a liquid of high viscosity. Finally, the term *vitreous* is sometimes used in connection with glasses. This term is usually reserved for glassy materials that can be crystallized through proper heat treatment in a process known as *devitrification*.

The distinction between an amorphous material and a glass is an important one. For example, an SiO_2 glass prepared from the melt has a noticeably different X-ray diffraction pattern than a solid SiO_2 gel derived from dehydration of a solution (see Figure 1.50). In both cases, the glass and gel have no *long-range order* in comparison to cristobalite, which is highly crystalline and exhibits distinct X-ray diffraction lines. The increase in the intensity of the gel pattern at small angles

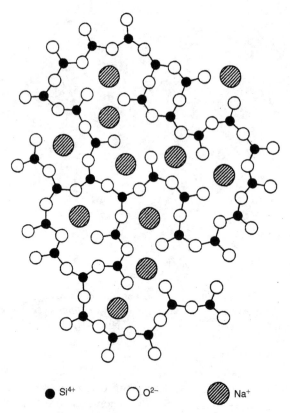

● Si⁴⁺   ○ O²⁻   ◍ Na⁺

Figure 1.48 Schematic representation of a random network sodium silicate glass. From W. D. Kingery, H. K. Bowen, and D. R. Uhlmann, *Introduction to Ceramics*. Copyright © 1976 by John Wiley & Sons, Inc. This material is used by permission of John Wiley & Sons, Inc.

HISTORICAL HIGHLIGHT

Some historians credit the added illumination made possible by glass with heightening interest in cleanliness and hygiene. Windows made dirt more visible. And thanks to superior mirrors—made with transparent glass that reflected properly from the thin mental foil on one side—people could see and understand themselves and their conditions more accurately than ever before; glass, a miraculous substance that is at once as solid as a rock and as invisible as air, shed as much light on people's minds as on their surroundings. Moreover, the magnifying powers of glass eventually enlightened scientist as well, enabling them to understand what it is inside of materials that makes the stuff of the world the way it is.

Source: Reprinted, by permission, from I. Amato, *Stuff*, p. 32. Copyright © 1997 by Harper Collins Publishers, Inc.

---

is due to microporous structures that result from the removal of water during drying—inhomogeneities that are not present in the silica glass. Both the glass and the gel do possess *short-range order*, however, as indicated by the broad peak centered at a *d*-spacing of about 0.12 nm. This short-range order is attributed chiefly to the SiO_4 tetrahedral structural unit present in all silicates.

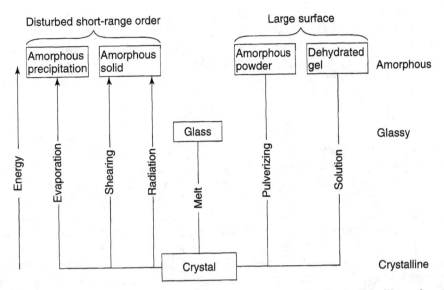

Figure 1.49 Comparison of preparation procedures of noncrystalline solids illustrating the difference between glassy and amorphous solids. Reprinted, by permission, from H. Scholze, *Glass*, p. 123. Copyright © 1991 by Springer-Verlag.

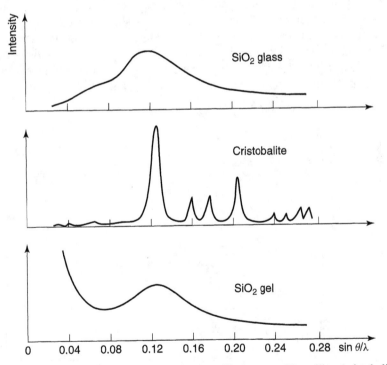

Figure 1.50 X-ray diffraction patterns of vitreous silicon, crystalline silica (cristobalite), and sol–gel-derived silica. Reprinted, by permission from H. Scholze, *Glass*, p. 97. Copyright © 1991 by Springer-Verlag.

Though it may seem nonsensical to obtain X-ray diffraction patterns from materials that have no long-range structure (i.e., crystallinity) since there is no spacing between crystal planes, the X-ray pattern provides additional information that is useful in analyzing the structure of glasses. Even though the peaks in an amorphous pattern are broad, we can extract additional information using something called the *radial distribution function* (rdf):

$$\text{rdf} = 4\pi r^2 n_0 g(r) \tag{1.38}$$

where $g(r)$ is the *pair distribution function* between adjacent atoms—that is, the probability of finding another atom a distance r from the reference atom located at $r = 0$. The pair distribution function is determined from various diffraction experiments (electron, neutron, X ray). The quantity n_0 in Eq. (1.38) is the average number density $= N/V$. If we plot rdf versus r, we obtain a curve similar to the one shown in Figure 1.51. The dotted line represents the parabola $4\pi r^2$, and deviations from the dotted line indicate regions of greater probability for finding an atom. The "peaks," then, correspond to likely bond distances as indicated: Si–O, O–O, and Si–Si. The radial distribution function is useful for characterizing not only glasses, but liquids and polymers as well.

So, we have seen that glasses have short-range structure, but no long-range structure, at least relative to the wavelength of the probing X rays. But can we predict, or at least rationalize, which compounds will form glasses readily, and which ones will not? The answer is, "Yes." Once again, there are several sets of "rules," or guidelines, for describing the ability of certain cation/anion pairs to form glassy compounds. We will look at three such sets of guidelines, and you should recognize some of their components from earlier guidelines, such as Pauling's rules and the Hume–Rothery rules. Although we know that glasses by definition may consist of any types of cations and anions, oxide glasses are by far the most common and industrially most important. We will limit our discussion to oxide glasses for the moment.

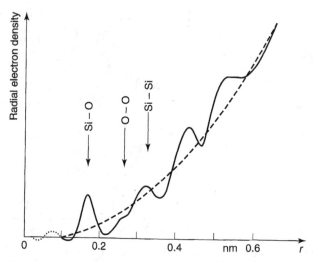

Figure 1.51   Radial distribution of electron densities of vitreous silica from X-ray exposures (Scholze). Reprinted, by permission, from Scholze, H., *Glass*, p. 98. Copyright © 1991 by Springer-Verlag.

1.2.4.1 *Zachariasen Rules.* In 1932, W. H. Zachariasen considered the conditions for constructing a random network like the one shown in Figure 1.48 and proposed four rules for the formation of oxide glasses:

- An anion (oxygen atom) is linked to not more than two glass-forming cations (metal atoms).
- The coordination numbers of the glass-forming atoms (cations) is small, four or less.
- The oxygen polyhedra (structural units) share corners with each other, not edges or faces.
- The polyhedra are linked in a 3-D network (at least three corners of each polyhedra should be shared).

The "structural polyhedra" are those that we have already been using: triangles, tetrahedra, and octahedra. Zachariasen's rules, as supported and modified by Warren, came to be known as the *random network theory* and, despite its limitations, is still widely used.

1.2.4.2 *Stanworth Rules.* In the late 1940s and early 1950s, Stanworth proposed a set of much simpler guidelines that did not rely on the formation of polyhedra. He suggested that the primary criteria for glass formation in metal oxide glasses were

- A cation valence ≥ 3
- An increasing tendency for glass formation with decreasing cation size
- A cation electronegativity between 1.5 and 2.1

Based on what we already know, there is circumstantial evidence to support these guidelines. For example, using the electronegativity values in Table 1.4, we see that two well-known glass-forming metal oxides B_2O_3 and SiO_2 meet the electronegativity criterion ($\chi_B = 2.04$, $\chi_{Si} = 1.90$), whereas Na_2O does not ($\chi_{Na} = 0.93$). Similarly, both B^{3+} and Si^{4+} have valencies greater than or equal to three, and they have relatively small cation sizes (0.2 and 0.39 Å, respectively) in comparison to other 3+ valence, non-network-forming cations like Co and Fe (0.65 and 0.67 Å, respectively).

1.2.4.3 *Oxide Glass Cations.* Perhaps the single most useful guideline is a table (Table 1.18) that classifies cations into three categories: *glass formers, intermediates,* and *modifiers*. These classifications are actually an extension of the random network model, but also include some of Stanworth's guidelines. Glass formers are cations with a valence greater than or equal to three (this is a Stanworth rule) and a coordination number less than four (a Zachariasen rule). Note that there are many more glass-forming cations than just silicon. Intermediates are cations of lower valence and higher coordination number that can sometimes act as glass formers (such as aluminum), but can also act as network modifiers. A network modifier is a cation that serves to interrupt the random, glass network, partly by being of high enough valence to provide additional oxygen to the network, thereby increasing the oxygen-to-metal atom ratio and destroying the network (see Table 1.17). Note that some cations can be in several categories, such as Pb, which can have multiple oxidation states. We know that leaded-glass exists for such important applications as television screens, but the role of lead can be that of either an intermediate or a network modifier. Sodium is a well-known

Table 1.18 Coordination Number and Bond Strength of Oxides

	M in MO_x	Valence	Dissociation Energy per MO_x (kcal/g-atom)*	Coordination Number	Single-Bond Strength (kcal/g-atom)*
Glass formers	B	3	356	3	119
	Si	4	424	4	106
	Ge	4	431	4	108
	Al	3	402–317	4	101–79
	B	3	356	4	89
	P	5	442	4	111–88
	V	5	449	4	112–90
	As	5	349	4	87–70
	Sb	5	339	4	85–68
	Zr	4	485	6	81
Intermediates	Ti	4	435	6	73
	Zn	2	144	2	72
	Pb	2	145	2	73
	Al	3	317–402	6	53–67
	Th	4	516	8	64
	Be	2	250	4	63
	Zr	4	485	8	61
	Cd	2	119	2	60
Modifiers	Sc	3	362	6	60
	La	3	406	7	58
	Y	3	399	8	50
	Sn	4	278	6	46
	Ga	3	267	6	45
	In	3	259	6	43
	Th	4	516	12	43
	Pb	4	232	6	39
	Mg	2	222	6	37
	Li	1	144	4	36
	Pb	2	145	4	36
	Zn	2	144	4	36
	Ba	2	260	8	33
	Ca	2	257	8	32
	Sr	2	256	8	32
	Cd	2	119	4	30
	Na	1	120	6	20
	Cd	2	119	6	20
	K	1	115	9	13
	Rb	1	115	10	12
	Hg	2	68	6	11
	Cs	1	114	12	10

*Multiply by 4.184 to obtain units of kJ/mol.
Source: W. D. Kingery, H. K. Bowen, and D. R. Uhlmann; *Introduction to Ceramics*. Copyright © 1976 by John Wiley & Sons, Inc.

network modifier and is added to sand (quartz) in the form of Na_2O to form sodium silicates, which constitute a large class of glasses and, in the aqueous solution form, a large class of adhesives.

1.2.5 Glass Ceramics

There are materials that are hybrids between glasses and ceramics. *Glass ceramics* are a family of fine-grained crystalline materials achieved through controlled crystallization

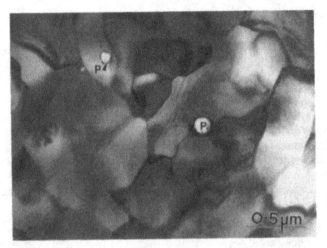

Figure 1.52 Microstructure of a typical glass ceramic with a fine grain size, minimal residual glass, and little porosity (P). From W. E. Lee and W. M. Rainforth, *Ceramic Microstructures*. p. 59; Copyright © 1994 by William E. Lee and W. Mark Rainforth, with kind permission of Kluwer Academic Publishers.

of glasses. They are nonporous and are either opaque or transparent. Their optical transparency arises from the fact that the crystals are smaller than the wavelength of visible light such that the light waves are transmitted without significant scattering (see Figure 1.52). Like glasses, glass ceramics have excellent corrosion and oxidation resistance, but have improved impact strength and dimensional stability over both glasses and ceramics. Glass ceramics are commonly made of compositions in the $MgO-Al_2O_3-SiO_2$, and $Li_2O-Al_2O_3-SiO_2$ systems, with the main crystal phases being *cordierite* ($Mg_2Al_4Si_5O_{18}$) and β-quartz and β-*spudomene* ($Li_2O \cdot Al_2O_3 \cdot 4SiO_2$) solid solutions. The residual glass phase is usually $1-10\%$ by volume.

The desired crystalline microstructure is achieved through the addition of *nucleating agents*, often titanium dioxide, to the glass prior to heat treating. The glass is heated to an appropriate nucleating temperature that allows the maximum number of crystal nuclei to form, followed by heat treatment at a higher temperature to grow the crystals (see chapter 3). The result is a highly crystalline material with tiny crystallites, on the order of a few nanometers. The processing of glass ceramics will be described in detail in a later chapter. Glass ceramics are used as ovenware and tableware, electrical insulators, substrates for circuit boards, architectural cladding, and for heat exchangers and regenerators.

1.2.6 Defect Reactions*

As with metals, ceramic crystals are not perfect. They can contain all of the same types of defects previously described in Sections 1.1.3–1.1.5. What is unique about ceramic crystals, particularly oxide ceramics, is that the concentration of point defects, such as vacancies and interstitials, is not only determined not only by temperature, pressure, and composition, but can be influenced greatly by the concentration of gaseous species in which they come in contact (e.g., gaseous oxygen). The concentration of gaseous species affects the crystal structure, which in turn can affect physical properties such as conductivity and thermal expansion. This opens up a multitude of applications for ceramic materials, ranging from sensors and actuators to nanoscale reactors. In order to describe some of these phenomena, we will need to accurately describe the reactions that are taking place, and quantify their effect on point defect concentrations. This is done through the development of *defect reactions*, in which defect concentrations are treated like any other chemical species.

1.2.6.1 Kroger–Vink Notation. We must slightly modify the notation we use for chemical reactions to avoid confusion between vacancies, which have a charge associated with them, and formal valence charges on isolated ions, like Ca^{2+}. To do this, a system of notation has been developed called *Kroger–Vink notation*. Consider the generic binary crystalline compound MX. Recall that a vacancy occurs when an atom is removed from a lattice position. In the binary compound, there can be two types of vacancies: one created by removing an M atom, designated V_M, the other from a missing X atom, designated V_X. As in all Kroger–Vink notation, the primary symbol, in this case "V" for vacancy, indicates the type of species, and the subscript, in this case "M" or "X," designates the lattice site. Similarly, interstitial sites are designated with a subscript "i," and the atom occupying the interstitial position is indicated by either an M or an X. Thus, M_i represents a metal atom interstitial and X_i represents a counterion interstitial. The subscript does not tell what type of interstitial site is being occupied—for example, tetrahedral versus octahedral.

It is theoretically possible for cations to occupy anion sites, and vice versa. Kroger–Vink notation, then, dictates that an M atom on an X site be designated as M_X and that an X atom on an M site be designated as X_M. Recall that we can have defect clusters, such as a Frenkel defect. Defect clusters are enclosed in parentheses—for example, $(V_M V_X)$ or $(X_i X_M)$—to indicate that the individual defects are associated with one another. Impurity atoms are also coded as to lattice position. If we introduce a metal impurity atom L into our compound MX, it might occupy a metal cation site, and is thus designated as L_M. Similarly, S_i is an S impurity atom on an interstitial site.

Two species we have not yet discussed are *free electrons* and *free holes*. We will use these species extensively in describing electronic properties of materials, but for now we simply note that they are dealt with like any other species. A free electron is indicated with "e" and has a charge associated with it, which is designated with a superscript prime, e', to differentiate it from a formal valance charge $(-)$. A free electron is not localized and is free to move about the lattice. As a result, it does not occupy a specific lattice site and carries no subscript. An electron hole, which carries a positive charge, is also delocalized and is designated by $h^\bullet$. Here, the superscript dot indicates a positive charge in Kroger–Vink notation. We also use superscripts to indicate charges on atoms and vacancies. If, for example, we remove the ion M^+ from our MX lattice, the remaining vacancy has a negative charge associated with it since the original lattice was charge-neutral. The symbol V'_M represents a negatively charged metal vacancy. Similarly, if we remove X^- from the lattice, a positive charge is left on the vacancy. The symbol $V_X^\bullet$ is used for a positively charged anion vacancy. For a specific compound, the symbols M and X are replaced with the actual atomic symbol. For example, $Zn_i^{\bullet\bullet}$ represents a Zn ion on interstitial site with a resulting 2^+ charge. The Kroger–Vink notation is summarized in Table 1.19.

It should come as no surprise that defects have concentrations—for example, $[Zn_i^{\bullet\bullet}]$—and we can write reactions with these defects. As with balancing equations, which

Table 1.19   Summary of Kroger–Vink Notation

The Notation...	Represents a(n)...
V	Lattice vacancy
h	Free hole
e	Free electron
M (e.g., Ca, Al...)	Cation atom
X (e.g., O, Cl...)	Anion atom

Subscripts

i	Interstitial lattice position
M	Cation lattice position
X	Anion lattice position

Superscripts

•	Positive charge
′	Negative charge

you learned how to do in general chemistry, there are no set rules—there is a bit of guesswork and art involved. There are a few general guidelines that should be followed in balancing defect reactions, however. The first guideline involves site relation. The number of M sites must be in correct proportion to X sites as dictated by the compound stoichiometry. For example, the ratio M:X is 1:1 in MgO and 1:2 in UO_2.

The second guideline deals with site creation, and it states that defect changes that alter the number of lattice sites must not change the overall site relation (guideline 1). Site creation is easily recognized from the subscripts: Species such as V_M, V_X, M_M, X_M, and so on, create sites, whereas the species e', $h^\bullet$, M_i, L_i, and so on, do not create sites. As with regular reactions, the third guideline states that mass balance must be maintained; that is, any species appearing on the left side must appear on the right side of the equation. Remember that subscript symbols only indicate sites and are not involved in the mass balance. The fourth guideline in balancing defect reaction equations simply says that electrical neutrality must be maintained. Both sides of the defect equation should have the same total effective charge, but that charge need not necessarily be zero. Finally, there is no special distinction for surface sites. Lattice positions at the surface are treated like every other position in the lattice. See example problem 1.5 for details on balancing defect reaction equations.

Now that we know how to write defect equations, let's look at Frenkel and Schottky defects in more detail.

1.2.6.2   Defect Reaction Equilibrium Constants.   Recall that a Frenkel disorder is a self interstitial–vacancy pair. In terms of defect concentrations, there should be equal concentrations of vacancies and interstitials. Frenkel defects can occur with metal atoms, as in AgBr:

$$V_i + Ag_{Ag} \iff Ag_i^\bullet + V'_{Ag} \tag{1.39}$$

where $Ag_i^{\bullet}$ is a silver atom on an interstitial site with a $+1$ charge, and V'_{Ag} is a silver vacancy with a -1 charge; or with anions, such as oxygen in Y_2O_3:

$$O_O + V_i \rightleftarrows O''_i + V_O^{\bullet\bullet} \tag{1.40}$$

Example Problem 1.5

Write a defect reaction equation for the substitution of a $CaCl_2$ molecule into a KCl lattice.

Solution: There are actually two ways that $CaCl_2$ can be placed in the KCl lattice: substitutionally and interstitially. The defect reaction equation for substitution is

$$CaCl_2(s) + 2K_K + 2Cl_{Cl} \longrightarrow Ca_K^{\bullet} + V'_K + 2Cl_{Cl} + 2KCl(g)$$

Again, there are no set "rules" for balancing this equation, but we can describe some of the guidelines as they relate to this example.

(a) *Site relation*

- KCl sites must be 1:1
- Two K sites are used, so two Cl sites must be used. Notice that the chlorines are all equivalent, and that the Cl brought in by the $CaCl_2$ simply occupies existing Cl sites, with the removal the previous chlorine with gaseous KCl. A legitimate simplification would be to remove $2Cl_{Cl}$ from both sides of the defect reaction equation.

(b) *Site creation*

- vacancy creation doesn't change site balance.

(c) *Mass balance*

- KCl is given off as gas. This is a common way of "getting rid" of solid species. Don't be concerned that a solid is turning into a gas—it is definitely possible.

(d) *Electrical neutrality*

- Keep in mind that we have strongly ionic species in this example; charges are involved.
- Placing a Ca^{2+} ion on a K^+ site gives a net $+1$ charge on the site, Ca_K.
- A vacancy must be created in order to preserve charge neutrality and maintain site relation. This is a "trick" that you will have to learn.

For interstitial substitution of $CaCl_2$ in KCl, the defect reaction equation is

$$CaCl_2(s) + 2K_K + 2Cl_{Cl} \longrightarrow Ca_i^{\bullet\bullet} + 2V'_K + 2Cl_{Cl} + 2KCl(g)$$

The details of balancing this reaction are left to the reader.

As with all reactions, defect reactions are subject to the law of mass action [see Eq. (3.4) for more details), so an equilibrium constant, K_F, can be written:

$$K_F = \frac{[O_i''][V_O^{\bullet\bullet}]}{[O_O][V_i]} \tag{1.41}$$

We can simplify this expression by noting that defect concentrations are usually small; that is, $[V_i] \approx [O_O] \approx 1$, so Eq. (1.41) becomes:

$$K_F = [O_i''][V_O^{\bullet\bullet}] \tag{1.42}$$

Frenkel defects form interstitial–vacancy pairs, so that $[O_i''] = [V_O^{\bullet\bullet}]$, and Equation (1.42) reduces further to

$$\sqrt{K_F} = [O_i''] = [V_O^{\bullet\bullet}] \tag{1.43}$$

This is the general expression for the equilibrium constant of oxygen interstitials in Y_2O_3.

The defect concentration comes from thermodynamics. While we will discuss thermodynamics of solids in more detail in Chapter 2, it is useful to introduce some of the concepts here to help us determine the defect concentrations in Eq. (1.43). The *free energy* of the disordered crystal, ΔG, can be written as the free energy of the perfect crystal, ΔG_0, plus the free energy change necessary to create n interstitials and vacancies ($n_i = n_v = n$), Δg, less the entropy increase in creating the interstitials; ΔS_c at a temperature T:

$$\Delta G = \Delta G_0 + n\Delta g - T\Delta S_c \tag{1.44}$$

Equation (1.44) states that the structural energy increases associated with the creation of defects are offset by entropy increases. The *entropy* is the number of ways the defects (both interstitials and vacancies) can be arranged within the perfect lattice, and it can be approximated using statistical thermodynamics as

$$\Delta S_c = k_B \ln\left[\frac{N!}{(N-n_i)!n_i!}\right]\left[\frac{N!}{(N-n_v)!n_v!}\right] \tag{1.45}$$

where k_B is Boltzmann's constant and N is the total number of lattice sites. Use of Stirling's approximation ($\ln N! = N \cdot \ln N - N$) and the fact that $n_i = n_v = n$ gives

$$S_c = 2k_B[N \ln N - (N-n)\ln(N-n) - n\ln n] \tag{1.46}$$

The free energy is then

$$\Delta G = \Delta G_0 + n\Delta g - 2k_B T\left[N \ln\left(\frac{N}{N-n}\right) + n\ln\left(\frac{N-n}{n}\right)\right] \tag{1.47}$$

At equilibrium, the free energy change with respect to the number of defects is a minimum, so we can obtain a relationship for the concentration of defects, n/N (assuming $N - n \approx N$):

$$\frac{n}{N} = \exp\left(\frac{-\Delta g}{2k_B T}\right) \tag{1.48}$$

The free energy change is usually approximated by the *enthalpy* change (additional entropy changes are small). Refer back to Table 1.13 for typical defect concentrations at various temperatures, and note that the defect concentrations are orders of magnitude smaller at 100°C, especially for large enthalpies of formation (approximated by E_d in Table 1.13).

The equilibrium concentration for a Schottky disorder can be found in a similar manner. Recall that a Schottky defect is a cation–anion defect pair. For example, the migration of an MgO molecule to the surface in an MgO crystal can be described as follows:

$$Mg_{Mg} + O_O \iff V''_{Mg} + V^{\bullet}_O + Mg_{surf} + O_{surf} \qquad (1.49)$$

Recall that surface sites are indistinguishable from lattice sites, so we usually write

$$null \longrightarrow V''_{Mg} + V^{\bullet\bullet}_O \qquad (1.50)$$

where the term "null" simply means that the vacancies form from the perfect lattice, and that all cations and anions are equivalent, so that any could be used in this equilibrium expression. The Schottky equilibrium constants K_s is then

$$K_s = [V''_{Mg}][V^{\bullet\bullet}_O] \qquad (1.51)$$

and since the concentration of both types of vacancies must, by definition of the Schottky defect, be equivalent, the equilibrium constant simplifies to

$$\sqrt{K_s} = [V''_{Mg}] = [V^{\bullet\bullet}_O] \qquad (1.52)$$

We will see in subsequent chapters how defect reactions can be used to quantitatively describe important defect-driven phenomena, particularly in ceramics.

1.3 STRUCTURE OF POLYMERS

The term *polymer* comes from "poly," meaning many, and "mer," meaning units. Hence, polymers are composed of many units—in this case, structural units called *monomers*. A monomer is any unit that can be converted into a polymer. Similarly, a *dimer* is a combination of two monomers, a *trimer* is a combination of three monomers, and so on. Before describing the chemical composition of typical monomers and how they are put together to form polymers, it is useful to have a brief organic chemistry review. You may wish to refer to an organic chemistry text for more detailed information. We will reserve discussion of how these organic molecules are brought together to form polymers until Chapter 3 when the kinetics of polymerization are described. For the remainder of the description of polymer structure, it is sufficient to know that polymer chains are formed from the reaction of many monomers to form long-chain hydrocarbons, sometimes called *macromolecules*, but more commonly referred to as polymers.

1.3.1 Review of Organic Molecules

Alkanes, also called *paraffins*, are composed of all C–C, saturated bonds and have the general formula C_nH_{2n+2}. The naming conventions and typical properties of the first

Table 1.20   The Alkane Series

Substance	Molecular Weight	T_m (°C)	T_v (°C)
Methane, CH_4	16.04	−182.5	−164
Ethane, C_2H_6	30.07	−183.3	−88.6
Propane, C_3H_8	44.10	−189.7	−42.1
Butane, C_4H_{10}	58.13	−138.4	−0.5
Pentane, C_5H_{12}	72.15	−129.7	36.1
Hexane, C_6H_{14}	86.18	−94.3	69.0
Heptane, C_7H_{16}	100.21	−90.6	98.4
Octane, C_8H_{18}	114.23	−56.8	125.7
Nonane, C_9H_{20}	128.26	−53.7	150.8
Decane, $C_{10}H_{22}$	142.29	−29.7	174.1

Table 1.21   Summary of Chain and Cyclic Hydrocarbons

Hydrocarbons	Formula	Characteristics
Chain		
Alkane series	C_nH_{2n+2}	All single C–C bonds
Alkene series	C_nH_{2n}	One double C=C bond
Alkadiene series	C_nH_{2n-2}	Two double C=C bonds
Alkyne series	C_nH_{2n-2}	One triple C ≡ C bond
Alkadiyne series	C_nH_{2n-6}	Two triple C ≡ C bonds
Cyclic		
Cycloalkane series	C_nH_{2n}	All single C–C bonds, cyclic
Cycloalkene series	C_nH_{2n-2}	One double C=C bond, cyclic
Aromatic	Various	Ring structures, based on the benzene ring, in which single and double carbon bonds alternate

10 alkanes are summarized in Table 1.20. The *alkenes* have one C=C double bond, and *alkynes* have a carbon–carbon triple bond. The formulae of these higher-order hydrocarbons are summarized in Table 1.21.

Many chemical reactions involving organic molecules, particularly polymerization reactions, involve *functional groups*. A functional group is an atom or groups of atoms that show a relative constancy of properties when attached to different carbon chains. For example, the attachment of a hydroxyl group, –OH, to hydrocarbons, leads to a group of compounds called *alcohols*, which, taken as a class, have similar properties, such as a more polar nature than the parent hydrocarbon. Some of the more important functional groups in polymer chemistry are listed in Table 1.22.

Another important concept from organic chemistry that has an impact on polymer structure is that of *isomerism*. Recall that *structural isomers* are molecules that have the same chemical formula, but different molecular architectures. For example, there are two different types of propyl alcohols, both with the same formula, depending on where the —OH functional group is placed on the carbon backbone (see Figure 1.53). Another type of isomerism results in *stereoisomers*, in which the functional groups are

Table 1.22   Some of the More Important Functional Groups in Polymer Chemistry

Group	General Formula	Compound Class	Example
$-Cl$	$R-Cl$	Chlorides	Methyl chloride, CH_3Cl
$-OH$	$R-OH$	Alcohols	Ethyl alcohol, CH_3CH_2OH
$-C\overset{O}{\underset{OH}{}}$	$R-C\overset{O}{\underset{OH}{}}$	Acids (or carboxylic acids)	Acetic acid, CH_3COOH
$\overset{}{\underset{}{}}C=O$	$\overset{R}{\underset{R}{}}C=O$	Ketones	Acetone, CH_3COCH_3
$\overset{H}{\underset{}{}}C=O$	$\overset{H}{\underset{R}{}}C=O$	Aldehydes	Propionaldehyde, CH_3CH_2CHO
$-O-$	$R-O-R$	Ethers	Ethyl ether, $CH_3CH_2OCH_2CH_3$
$-NH_2$	$R-NH_2$	Amines	Propylamine, $CH_3(CH_2)_2\,NH_2$
$-C\overset{O}{\underset{NH_2}{}}$	$R-C\overset{O}{\underset{NH_2}{}}$	Amides	Acetamide, CH_3CONH_2
$-C\overset{O}{\underset{O-R'}{}}$	$R-C\overset{O}{\underset{O-R'}{}}$	Esters	Ethyl acetate, $CH_3COOC_2H_6$

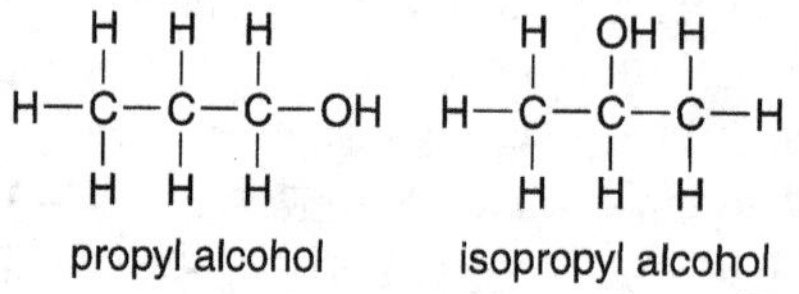

propyl alcohol   isopropyl alcohol

Figure 1.53   An example of polymer structural isomers.

in the same position on the chain but occupy different geometric positions. For example, the *cis* and *trans* forms of butene are stereoisomers (see Figure 1.54). An asymmetric carbon is required for stereoisomers to form. We will see that asymmetric carbons play an important role in determining the tacticity of polymers. Finally, two terms related to the structure and chemistry of the repeat units are *conformation* and *configuration*. These terms are sometimes incorrectly used interchangeably, but they are quite different. The conformation of a molecule refers to how the bonds, such as those found in the

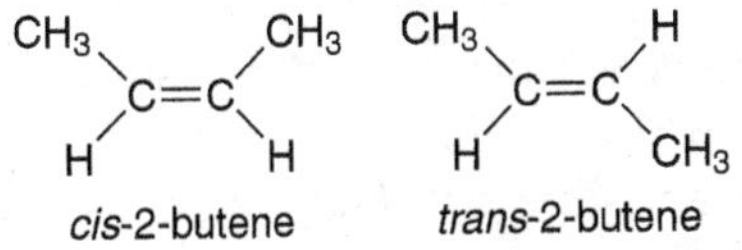

CH₃, ,CH₃ CH₃, ,H
‾C═C‾   ‾C═C‾
H′   ‵H   H′   ‵CH₃

cis-2-butene   trans-2-butene

Figure 1.54 An example of polymer stereoisomers.

backbone of a polymer, are rotated to cause the chain to fold or straighten out. Changes in conformation involve bond rotation only. For example, some stereoisomers can convert between *cis* and *trans* conformation through bond rotation (except for those shown in Figure 1.54, since bond rotation about a double bond is not allowed). Changes in configuration, however, require that bonds be broken and reformed. For example, to change between the isopropyl and propyl alcohol configurations in Figure 1.53, the hydroxyl functional group would have to be moved from the center to an end carbon.

1.3.2 Polymer Classification

It is useful to classify polymers in order to make generalizations regarding physical properties, formability, and reactivity. The appropriate classification scheme can change, however, because there are several different ways in which to classify polymers. The first scheme groups polymers according to their chain chemistry. *Carbon-chain polymers* have a backbone composed entirely of carbon atoms. In contrast, *heterochain polymers* have other elements in the backbone, such as oxygen in a polyether, –C–O–C–. We can also classify polymers according to their macroscopic structure—that is, independent of the chemistry of the chain or functional groups. There are three categories of polymers according to this scheme: *linear, branched,* and *networked* (crosslinked) polymers. Refer to Figure 1.55 for a schematic representation of this classification scheme. Finally, polymers can be classified according to their

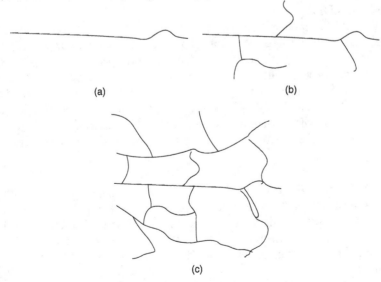

(a)   (b)

(c)

Figure 1.55 Classification of polymers according to macroscopic structure: (a) linear, (b) branched and (c) networked.

formability. Polymers that can be repeatedly shaped and reshaped are called *thermo-plastics*, whereas those polymers that cannot be reshaped at any temperature once they are set are termed *thermosets*. Often times network polymers are thermosets, and linear and branched polymers are thermoplastics. Hence, the thermoplastic/thermoset distinction is worthy of some elaboration.

1.3.2.1 Thermoplastic Polymers.

1.3.2.1 Thermoplastic Polymers. Most thermoplastic polymers are used in high-volume, widely recognized applications, so they are often referred to as *commodity plastics*. (We will elaborate upon the distinction between a polymer and a plastic in Chapter 7, but for now we simply note that a plastic is a polymer that contains other additives and is usually identified by a variety of commercial trade names. There are numerous databases, both in books [1] and on the Internet [2], that can be used to iden-tify the primary polymer components of most plastics. With a few notable exceptions, we will refer to most polymers by their generic chemical name.) The most common commodity thermoplastics are polyethylene (PE), polypropylene (PP), polyvinyl chlo-ride (PVC) and polystyrene (PS). These thermoplastics all have in common the general repeat unit $-(CHX-CH_2)-$, where $-X$ is $-H$ for PE, $-CH_3$ for PP, $-Cl$ for PVC, and a benzene ring for PS. When we discuss polymerization reactions in Chapter 3, we will see that all of these thermoplastics can be produced by the same type of reaction.

In their simplest forms, the thermoplastics are linear, carbon chain polymers. There are methods for creating branches, especially in polyethylene, while still maintaining thermoplasticity. Increased branching tends to decrease the density, melting point, and certain mechanical properties of the polymer, but increases transparency and impact toughness. Thus, branched PE is important for many packaging applications. Other special types of PE include low-density PE (LDPE), high-density PE (HDPE), and linear low-density PE (LLDPE). HDPE is used when greater stiffness is required, such as in milk, water and detergent bottles. LDPE is used for many standard piping applications. LLDPE has a low density like LDPE, but a linear structure much like HDPE. It is less expensive to produce than LDPE, and it generally has better mechanical properties than LDPE.

PP, PVC, and PS have in common an asymmetric carbon atom in their backbone, a fact that leads to interesting structural properties and is elaborated upon in the next section. They all have specific advantages leading to a variety of applications, which will also be discussed in subsequent chapters. For the moment, it is important to recognize that they are linear thermoplastics, whose properties can be tailored through blending, branching, and additives. It is important to note at this point that the ability of thermoplastics to soften when heated and harden upon cooling is what leads to the principle of *recycling*. The structure of some of these common thermoplastics can be found in Appendix 2.

1.3.2.2 Thermoset Polymers.

1.3.2.2 Thermoset Polymers. There are many important examples of thermoset polymers, a subset of which are sometimes referred to as *resins*. The thermoplastic PE can be treated by electron radiation or chemical means to form chemical bonds between adjacent chains called *crosslinks*. In this process, some of the carbon–hydrogen or carbon–carbon bonds in the linear chain are broken, creating free radicals, which react with free radicals on other chains to form bridges between the chains. When the crosslinking is brought about by chemical means, the term *curing* is often employed. The result is that unlike linear PE, crosslinked PE has no distinct melting point due to limited chain mobility and will eventually degrade upon heating. This is a common characteristic of thermoset polymers. Due to their three-dimensional structure, they cannot be reshaped upon heating. Instead, they tend to remain rigid, or soften only slightly, until the backbone begins to break, leading to *polymer degradation*. While this may not seem like a desirable quality, it means that thermoset polymers are generally stiffer than thermoplastics. In fact, the crosslinks tend to increase the long-term thermal and dimensional stability of the polymer, such that thermosets find wide use in very large and complex parts where thermal stability is important.

The chemical structures of thermosets are generally much more diverse than the commodity thermoplastics. The most common types of thermosets are the phenol-formaldehydes (PF), urea-formaldehydes (UF), melamine-formaldehydes (MF), epoxies (EP), polyurethanes (PU), and polyimides (PI). Appendix 2 shows the chemical structure of these important thermosetting polymers.

A related, yet distinctly different, class of crosslinked polymers are the *elastomers*. Though they are structurally different than the thermosets, we will include them here since they tend to decompose when heated, rather than flow. The presence of crosslinks in some polymers allows them to be stretched, or elongated, by large amounts. Polymers that have more than 200% elastic elongation (three times the original length) and can be returned to their original form are termed elastomers. They are like thermoplastics in that they readily elongate, but the presence of crosslinks limits the elongation prior to breakage and allows the polymer to return to its original shape. Natural rubber is an important elastomer. Crosslinks are added to an emulsion of rubber, called *latex*, through the addition of heat and sulfur. The sulfur creates chemical bonds between the rubber chains in a process known as *vulcanization*. Other common elastomers are polyisoprene, butadiene rubber (BR), styrene butadiene rubber (SBR), silicones, and fluoroelastomers.

1.3.3 Tacticity

An important phenomenon in some, but not all, industrially significant polymers arises from the ability of organic monomers with asymmetric carbons to form stereoisomers. When the repeat units (monomer units) are placed together in a long chain, monomers such as the one shown in Figure 1.56 can add to the growing chain in one of two ways: with the R group (which may be a chlorine, methyl group, or some other functional group) sticking out of the plane, or with it pointing back into the plane of the paper. The variation in how the functional groups are arranged leads to differences in *tacticity*. Tacticity is variations in the configuration (not conformation!) of polymer chains as a result of the sequence of asymmetric carbon centers in the repeat units. There are three types of tacticity in polymers: isotactic, syndiotactic, and atactic. In *isotactic*

Figure 1.56 An example of an asymmetric carbon atom.

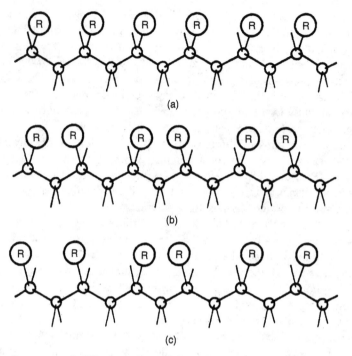

Figure 1.57 Summary of different types of tacticity, where R represents an atom other than hydrogen: (a) isotactic; (b) syndiotactic; and (c) atactic. Reprinted, by permission, from P. Hiemenz, *Polymer Chemistry: The Basic Concepts*, p. 27. Copyright © 1984 by Marcel Dekker, Inc.

polymers, the functional groups are all arranged on the same side of the asymmetric carbon, as in Figure 1.57a. Figure 1.57b shows the *syndiotactic* arrangement, in which the R groups perfectly alternate between sticking in and out of the plane of the carbon backbone. Finally, it is possible (and common) to have a perfectly random arrangement of functional groups, as in Figure 1.57c, in which case an *atactic* polymer results. Notable examples of polymers that exhibit tacticity are polyvinylchloride (PVC, R = Cl), polystyrene (PS, R = Ph, phenyl group), and polypropylene (PP, R = CH$_3$).

1.3.4 Copolymers

Polymers can be made that contain more than one type of repeat unit. For example, the R group on the asymmetric carbon in Figure 1.56 could be chlorine in some of the monomer units and fluorine in the rest. Such polymers are called *copolymers*. The ratio of the two types of monomers can vary from 0 to 1, and there can be more than two types of monomers in a copolymer. The presence of more than one type of repeat unit opens up many possibilities for variation in the structure of the polymer, or *chain architecture* as it is sometimes called. We will not describe the myriad of possible variations and the important consequences in terms of polymer physical properties, but here merely categorize copolymers in some broad, structural terms.

If the two monomers of a bi-component copolymer are perfectly alternating, an *alternating copolymer* results, as shown in Figure 1.58a. If the monomer units alternate in a

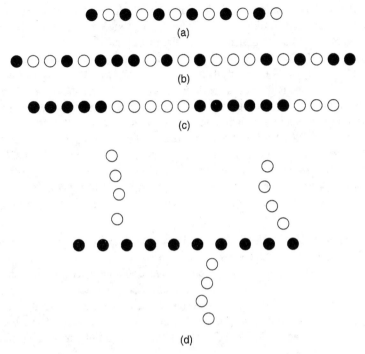

Figure 1.58 Summary of copolymer classifications: (a) alternating, (b) random, (c) block, and (d) graft.

random fashion, a *random copolymer* results, as in Figure 1.58b. Figure 1.58c shows a *block copolymer*, in which the different monomers prefer to add to each other, resulting in large segments of one type of monomer, followed by large segments of the other type of monomer. Finally, a second type of monomer can be added as a branch to a polymer backbone, as shown in Figure 1.58d, resulting in a *graft copolymer*. These different types of copolymers can have markedly different physical properties and are the basis for many important materials in the healthcare and consumer products industries.

1.3.5 Molecular Weight

We have described some of the general characteristics of polymers, and how they can be grouped according to structure, but we have not addressed any of the more quantitative aspects of polymer structures. For instance, we have stated that a polymer is made up of many monomer (repeat) units, but how many of these repeat units do we typically find in a polymer? Do all polymer chains have the same number of repeat units? These topics are addressed in this section on polymer molecular weight. Again, the kinetics of polymer formation are not discussed until Chapter 3—we merely assume here that the polymer chains have been formed and that we can count the number of repeat units in each chain.

1.3.5.1 Degree of Polymerization. The number of repeat units in an isolated polymer chain is called the *degree of polymerization*, x_n. If the monomer from which our polymer is formed has an initial molecular weight of M_0, then the molecular weight of an isolated polymer chain is $x_n \cdot M_0$. For example, polyethylene, $-(H_2C–CH_2)-$, is made up of ethylene repeat units, each of which has a molecular weight $M_0 = 2(12) + 4(1) = 28$ g/mol. Let us assume that an isolated polyethylene chain has 1000 repeat units, so $x_n = 1000$. The molecular weight of this isolated polyethylene chain would be $1000 \cdot (28$ g/mol$) = 28,000$ g/mol. Notice that we are neglecting the contribution of the terminal hydrogens located at both ends of the polyethylene chain. Since typical molecular weights are of the order 10^6 and larger, terminal groups are not a significant contribution to the molecular weight, and neglecting them does not introduce substantial error in calculations. We will also see that the molecular weight of the repeat unit in the polymer chain is not necessarily exactly the same as the molecular weight of the monomer. Small molecules can sometimes be produced during polymerization reactions, such as condensation reactions in which water is formed as a by-product. Be sure to use the molecular weight of the repeat unit in your calculations.

1.3.5.2 *Average Molecular Weight.*

In a typical polymer, not all chains have the same length. Consequently, there is a distribution of x_n values, and the molecular weight, M_i, varies from chain to chain. We could construct a histogram of the distribution of the number of chains, N_i, with molecular weight M_i, as shown in Figure 1.59. Molecular weights have discrete values, of course, and it is possible to find some mathematical function, $f(M_i)$ that appropriately describes the curve represented by the dotted line in Figure 1.59. For molecular weights, this function is simply the *number fraction*, n_i:

$$n_i = \frac{N_i}{\sum N_i} \tag{1.53}$$

Recall from calculus (or physics or statistics) that distributions can have k different moments about the origin, μ_k', which for a discrete random variable, M_i, are given by

$$\mu_k' = \sum_i n_i (M_i)^k \tag{1.54}$$

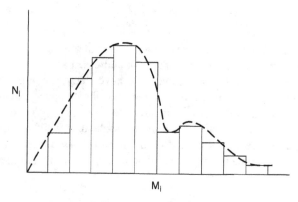

Figure 1.59 Histogram showing the number of molecules, N_i, having the molecular weight M_i. Reprinted, by permission, from P. Heimenz, *Polymer Chemistry: The Basic Concepts*, p. 35. Copyright © 1984 by Marcel Dekker, Inc.

The first moment of the distribution ($k = 1$) is simply the average molecular weight, called the *number-average molecular weight*, $\overline{M}_n$,

$$\mu_1' = \overline{M}_n = \sum n_i M_i = \frac{\sum N_i M_i}{\sum N_i} \tag{1.55}$$

or, in terms of the *average degree of polymerization*, $\overline{x}_n$,

$$\overline{M}_n = \overline{x}_n M_0 \tag{1.56}$$

where M_0 is the molecular weight of a repeat unit.

The second moment of the molecular weight distribution is then

$$\mu_2' = \sum n_i M_i^2 \tag{1.57}$$

A ratio of the second moment to the first moment (number-average molecular weight) is called the *weight-average molecular weight*, $\overline{M}_w$, which is the average based on the *weight* of each polymer chain:

$$\overline{M}_w = \frac{\sum N_i M_i^2}{\sum N_i M_i} = \frac{\sum n_i M_i^2}{\sum n_i M_i} \tag{1.58}$$

or, in terms of the mass (in grams) of species with molecular weight M_i, $m_i = N_i M_i$

$$\overline{M}_w = \frac{\sum m_i M_i}{\sum m_i} = \sum w_i M_i \tag{1.59}$$

where w_i is the *weight fraction*, $m_i / \Sigma m_i$.

There are other types of molecular weights based on higher-order moments to the distribution (see Table 1.23), but $\overline{M}_n$ and $\overline{M}_w$ are the most widely used quantities. In

Table 1.23   Summary of the Most Common Molecular Weight Averages

Average	Definition	Methods of Determination
$\overline{M}_n$	$\dfrac{\sum N_i M_i}{\sum N_i}$	Osmotic pressure, colligative properties, end group analysis
$\overline{M}_w$	$\dfrac{\sum N_i M_i^2}{\sum N_i M_i}$	Light scattering, sedimentation velocity
$\overline{M}_z$	$\dfrac{\sum N_i M_i^3}{\sum N_i M_i^2}$	Sedimentation equilibrium
$\overline{M}_v$	$\left(\dfrac{\sum N_i M_i^{1+a}}{\sum N_i M_i}\right)^{1/a}$	Intrinsic viscosity (a is characteristic of a system and lies between 0.5 and 1.0)

general, $\overline{M}_n \neq \overline{M}_w$, but taken together as a ratio, they provide a measure of the breadth of the molecular weight distribution. This ratio is called the *polydispersity index*, and it takes on values greater than or equal to 1.0:

$$\frac{\overline{M}_w}{\overline{M}_n} \geq 1.0 \tag{1.60}$$

If all chains have exactly the same weight and number of repeat units, then the system is termed *monodispersed*, and the polydispersity index is exactly 1.0. Most real polymers have rather large polydispersity indexes, but some standards used for chromatography calibration, such as polystyrene, can have values approaching unity. Calculating molecular weights and determining which form of Eqs. (1.55), (1.58), and (1.59) to use requires some practice and patience. A Cooperative Learning Exercise is provided below, but you are encouraged to consult any of the excellent textbooks on polymer science listed at the end of this chapter for further information on molecular weight calculations and determination.

Cooperative Learning Exercise 1.6

Consider the following collection of polymer chains:

10 molecules	2800 MW
5 molecules	3000 MW
4 molecules	1200 MW
2 molecules	3600 MW
1 molecule	1000 MW

Person 1: Calculate the number average molecular weight.
Person 2: Calculate the weight average molecular weight.
Combine your answers to determine the polydispersity index.

Answer:

$$\overline{M}_w = \frac{28,000(2800) + 15,000(3000) + 4800(1200) + 7200(3600) + 1000(1000)}{56,000}$$

$$= \frac{1.56 \times 10^8}{56,000} = 2787$$

$$\overline{M}_n = \frac{10(2800) + 5(3000) + 4(1200) + 2(3600) + 1(1000)}{22}$$

$$= \frac{56,000}{22} = 2545$$

$$\overline{M}_w/\overline{M}_n = 2787/2545 = 1.095$$

1.3.6 Polymer Crystallinity

An interesting and important structural characteristic of many polymers is that they are easily transformed from the amorphous to the crystalline state. Unlike inorganic glasses, which generally need a great deal of energy to devitrify (crystallize), or metals that require nonequilibrium conditions to form amorphous structures, many polymers have amorphous to crystalline transition temperatures that are near room temperature. Not all polymers crystallize readily, however. Intuitively, we would expect that a jumbled mass of spaghetti-like strands that make up a polymer solution or melt would tend to be amorphous, and this is indeed the case. But there are a number of structural factors that contribute to the ability of the amorphous polymer chains to rearrange themselves into an ordered structure. These factors include chain architecture (the chemical constituents and bond angles of the backbone and side groups), order and regularity (e.g., tacticity), intermolecular forces (both within an individual chain and between adjacent chains, such as hydrogen bonding), and steric effects (the size of side groups and branches).

1.3.6.1 Types of Bulk Polymer Crystallinity. Polymer crystallinity is a complicated subject, to which numerous books and symposia are devoted; but for our purposes, we can classify crystallinity in bulk polymers into two general categories: extended chain and folded chain. *Extended chain crystallinity* arises in many polymers with highly regular structures, such as polyethylene, poly(vinyl alcohol), syndiotactic polymers of poly(vinyl chloride) and poly(1,2-butadiene), most polyamides, and cellulose. In these molecules, the so-called "planar zigzag" structure shown in Figure 1.60a possesses the minimum energy for an isolated section of the chain and is therefore the thermodynamically favored conformation. Side groups, if they are small enough and arranged in a regular fashion, as in the syndiotactic structure, need not prevent crystallinity (see Figure 1.60b), but as the bulkiness and irregularity of the side groups grow, crystallization becomes more and more difficult. As a result, highly branched molecules, such as branched polyethylene, do not crystallize, even though polyethylene itself is easily crystallized. Similarly, networked polymers do not have the freedom to move in a way such that extended chain crystallinity can occur.

As with the other classes of materials, polymers can be either single crystals or polycrystalline. Polycrystalline polymers are more appropriately termed *semicrystalline polymers*, since the region between crystalline domains in polymers can be quite large and result in a significant amorphous component to the polymer. The crystalline regions in semicrystalline polymers are called *crystallites*, which have dimensions of several hundred angstroms, but the length of polymer chains is generally much larger than this. For example, a polyethylene chain with the extended chain structure shown in Figure 1.60a with molecular weight 50,000 has an end-to-end length of about 4500 Å. How can this be? The second type of crystalline structure in polymers chain folding gives as an explanation. Polymer chains can fold in a regular fashion to form plate-like crystallites called *lamellae*, as shown schematically in the insert of Figure 1.61. Notice that the polymer chains not only fold, but can extend from one lamella to another to form amorphous regions. In polymers crystallized from the melt, these lamellae often radiate from a central nucleation site, forming three-dimensional spherical structures called *spherulites* (see Figures 1.61 and 1.62). In cross-polarized light, the spherulites form a characteristic *Maltese cross* pattern due to birefringent effects associated with the lamellar structures.

A polymer crystal structure related to chain folding is called the *fringed micelle model*, in which the polymer chains do not fold in a regular fashion but extend from one crystalline region to another, again forming amorphous regions between the crystallites (see Figure 1.63). While the fringed micelle model is no longer the preferred one for

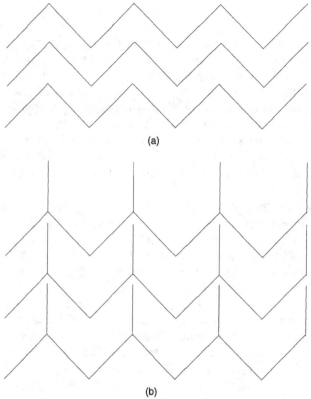

Figure 1.60   Schematic illustration of extended chain crystallinity in polymers (a) polyethylene and (b) polypropylene.

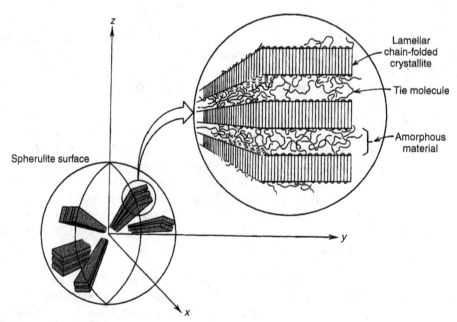

Figure 1.61   Schematic illustration of chain folding leading to lamellar crystallites (*inset*) and lamellar stacking to form spherulites.

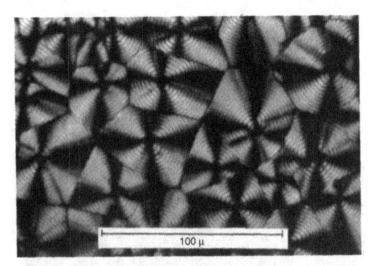

Figure 1.62   Cross-polarized micrograph of polyethylene showing spherulitic structure. From K. M. Ralls, T. H. Courtney, and J. Wulff, *Introduction to Materials Science and Engineering.* Copyright © 1976 by John Wiley & Sons, Inc. This material is used by permission of John Wiley & Sons, Inc.

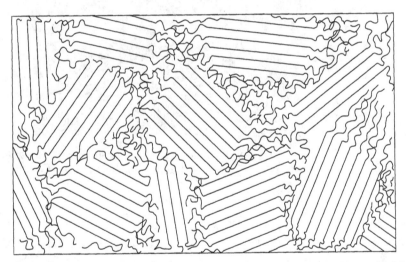

Figure 1.63   The fringed-micelle model of polymer crystallinity. From K. M. Ralls, T. H. Courtney, and J. Wulff, *Introduction to Materials Science and Engineering.* Copyright © 1976 by John Wiley & Sons, Inc. This material is used by permission of John Wiley & Sons, Inc.

describing crystallinity in most polymers, it is still used to describe the structure of highly oriented polymers, such as occurs in the stretching of rubber, viscose rayon, and Kevlar®.

The amorphous and crystalline regions each have different densities, with the crystalline density ρ_c being higher than the amorphous density ρ_a due to a more compact structure. The percent crystallinity in a semicrystalline polymer with bulk density ρ_s

can then be calculated from the respective crystalline and amorphous densities:

$$\% \text{ crystallinity} = \frac{\rho_c(\rho_s - \rho_a)}{\rho_s(\rho_c - \rho_a)} \times 100 \tag{1.61}$$

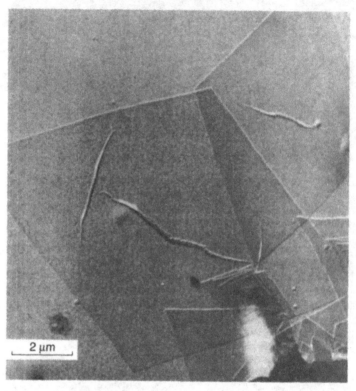

Figure 1.64 Polyethylene single crystals. Reprinted, by permission, from P. Heimenz, *Polymer Chemistry: The Basic Concepts*, p. 239. Copyright © 1984 by Marcel Dekker, Inc.

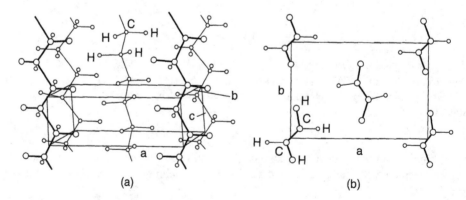

Figure 1.65 (a) Crystal structure of polyethylene unit cell shown in relation to chains. (b) View of unit cell perpendicular to chain axis. Reprinted, by permission, from Heimenz, P., *Polymer Chemistry: The Basic Concepts*, p. 236. Copyright © 1984 by Marcel Dekker, Inc.

Polymer single crystals possess the density of the crystal, ρ_c. Though polymer single crystals do not usually form in the bulk, but rather from more carefully controlled formation techniques such as vapor deposition, we will describe them here since they generally have the folded-chain crystal structure. An example of a polymer single crystal is shown in Figure 1.64. One would expect that the regular array of folded chains in a single crystal, or in a semicrystalline polymer for that matter, leads to diffraction phenomena when bombarded with X rays, in much the same way that lower-molecular-weight materials like sodium chloride or aluminum do. This is indeed the case, and the study of X-ray diffraction phenomena in polymers is a large field of interest. The primary difference in the diffraction patterns of polymer crystals is due to the fact that molecules, in this case polymer chains, rather than atoms, make up the lattice points of the unit cells. For example, the unit cell of polyethylene is orthorhombic, with polyethylene chains forming the lattice points, as illustrated in Figure 1.65. As a result, the dimensions of the unit cells in polymer crystals tend to be much bigger than those for ceramics and metals. The orthorhombic unit cell for polyethylene in Figure 1.65 has dimensions $a = 7.42$ Å, $b = 4.95$ Å, $c = 2.55$ Å, with two chains per unit cell: one in the center and $4 \times (1/4)$ at each corner where four unit cells come together. The larger lattice parameters and increased interplanar spacings mean that the diffraction angles for polymers are generally much smaller than for ceramics and metals [recall that $\theta \propto 1/d$, see Eq. (1.35)]. Hence, *small-angle X-ray scattering* (SAXS) is often used for polymer structural characterization instead of the traditional *wide-angle X-ray scattering* (WAXS).

As with ceramics and metals, polymer crystals can have multiple crystal forms. Polyethylene has a metastable monoclinic form and a orthohexagonal high pressure form. A list of some of the more common polymers and their corresponding crystal structures is given in Table 1.24. Finally, X-ray diffraction can be used to determine the amorphous to crystalline ratio in semicrystalline polymers in much the same way that Eq. (1.61) can be used. Figure 1.66 shows a schematic illustration of the X-ray diffraction patterns for semicrystalline and amorphous polyethylene. The estimation of crystalline content is based upon a ratio of the peak areas in the two samples.

More so than in metals, glasses, and ceramics, the microstructure in polymers is easily altered and within the operating temperature and pressure of many industrial and biological processes, transitions between the amorphous and crystalline state, and the ratio between the amorphous and crystalline components, can easily take place. As we mentioned in the previous section, the amorphous component of a polymer is less dense than the crystalline component. Conversely, the *specific volume*, or volume per mole in cubic centimeters, is lower for polymer crystals than it is for amorphous polymers. This distinction is best understood by observing the volume change of a polymer melt as it cools. Consider a molten polymer at point A in Figure 1.67. If we cool the polymer melt slowly, as in path ABG, the polymer chains have sufficient time to rearrange, fold, and form lamellar structures, resulting in a crystalline polymer, provided, of course, that they have the propensity to fold in the first place, and are not prevented from doing so by such factors as steric effects. The point at which the melt solidifies in the form of crystals is called the *crystalline melting point*, T_m, and is characterized by a sharp decrease in the specific volume and an increase in the density. If the same polymer melt is cooled rapidly, as in the path $ABCD$, a supercooled liquid is first obtained at point C, followed by an amorphous solid, or *glassy polymer*, at point D, due to insufficient time for the large molecules to arrange themselves in an ordered structure. The point

Table 1.24 The Crystal Structures of Selected Polymers

Polymer	Crystal System	Lattice Constants, Å			Crystal Density (g/cm³)
		a	b	c	
Polyethylene	Orthorhombic	7.417	4.945	2.547	1.00
Polytetrafluoroethylene	Trigonal (>19°C)	5.66	—	19.50	2.30
Isotactic	α-Monoclinic	6.65	20.96	6.50	0.936
polypropylene	β-Hexagonal	19.08	—	6.49	0.922
Syndiotactic	Orthorhombic	14.50	5.60	7.40	0.93
polypropylene					
Isotactic polystyrene	Trigonal	21.90	—	6.65	1.13
Poly(vinyl chloride)	Orthorhombic	10.6	5.4	5.1	1.42
Poly(vinyl alcohol)	Monoclinic	7.81	2.25	5.51	1.35
Poly(vinyl fluoride)	Orthorhombic	8.57	4.95	2.52	1.430
Poly(vinylidine	α-Monoclinic	4.96	9.64	4.62	1.925
fluoride)	β-Orthorhombic	8.58	4.91	2.56	1.973
Isotactic poly(methyl	Orthorhombic	20.98	12.06	10.40	1.26
methacrylate)					
trans-1,4-Polybutadiene	Monoclinic	8.63	9.11	4.83	1.04
cis-1,4-Polybutadiene	Monoclinic	4.60	9.50	8.60	1.01
Poly(ethylene oxide)	Monoclinic	8.05	13.04	19.48	1.228
	triclinic	4.71	4.44	7.12	1.197
Isotactic	Orthorhombic	10.46	4.66	7.03	1.126
poly(propylene					
oxide)					
Nylon 66	α-Triclinic	4.9	5.4	17.2	1.24
	β-Triclinic	4.9	8.0	17.2	1.248
2,6-Polyurethane	Triclinic	4.93	4.58	16.8	1.27
	Triclinic	4.59	5.14	13.9	1.33
3,6-Polyurethane	Monoclinic	4.70	8.66	33.9	1.34
Polyketone	Orthorhombic	7.97	4.76	7.57	1.296
Poly(ethylene sulfide)	Orthorhombic	8.50	4.95	6.70	1.416
Polyisobutylene	Orthorhombic	6.88	11.91	18.60	0.972
Poly(isobutylene oxide)	Orthorhombic	10.76	5.76	7.00	1.10
Poly(ethylene sulfide)	Orthorhombic	8.50	4.95	6.70	1.60
Isotactic poly(vinyl	Trigonal	16.25	—	6.50	1.168
methyl ether)					

Source: Tadokoro, *Structure of Crystalline Polymers.*

at which the slope of the specific volume with temperature curve decreases (between C and D), representing solidification, is called the *glass transition temperature*, T_g. The reasoning behind the term "glass transition temperature" becomes more apparent if we turn around and begin slowly heating the glassy polymer. At some point, there is sufficient mobility in the polymer chains for them to begin to align themselves in a regular array and form crystallites. The polymer is not yet molten—there is simply short range chain movement that results in an amorphous to crystalline transformation in the solid state. This point is also T_g. As we continue to heat the sample, the now-crystalline polymer eventually reaches T_m and melts. Most polymers are a combination

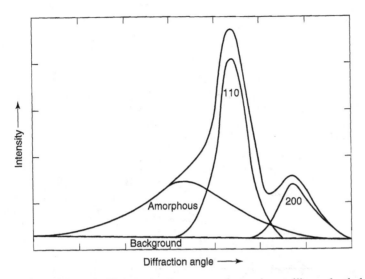

Figure 1.66 Resolution of the X-ray scattering curve of a semicrystalline polyethylene sample into contributions from crystalline (110 and 200 planes) and amorphous components. From F. W. Billmeyer, *Textbook of Polymer Science*, 3rd ed. Copyright © 1984 by John Wiley & Sons, Inc. This material is used by permission of John Wiley & Sons, Inc.

of amorphous and crystalline components and are the result of intermediate cooling paths such as *ABEF* in Figure 1.67. Keep in mind that some polymers, no matter how slowly we cool them, cannot crystallize, and they follow the glassy path *ABCD* all the time. We will have more to say about the glass transition temperature in Section 1.3.7.

1.3.6.2 *Liquid Crystalline Polymers.* One class of polymers that requires some special attention from a structural standpoint is *liquid crystalline polymers*, or LCPs. Liquid crystalline polymers are nonisotropic materials that are composed of long molecules parallel to each other in large clusters and that have properties intermediate between those of crystalline solids and liquids. Because they are neither completely liquids nor solids, LCPs are called *mesophase* (intermediate phase) materials. These mesophase materials have liquid-like properties, so that they can flow; but under certain conditions, they also have long-range order and crystal structures. Because they are liquid-like, LCPs have a translational degree of freedom that most solid crystals we have described so far do not have. That is, crystals have three-dimensional order, whereas LCPs have only one- or two-dimensional order. Nevertheless, they are called "crystals," and we shall treat them as such in this section.

In many cases, these polymer chains take on a rod-like (*calamitic* LCPs) or even disc-like (*discotic* LCPs) conformation, but this does not affect the overall structural classification scheme. There are many organic compounds, though not polymeric in nature, that exhibit liquid crystallinity and play important roles in biological processes. For example, arteriosclerosis is possibly caused by the formation of a cholesterol containing liquid crystal in the arteries of the heart. Similarly, cell wall membranes are generally considered to have liquid crystalline properties. As interesting as these examples of liquid crystallinity in small, organic compounds are, we must limit the current discussion to polymers only.

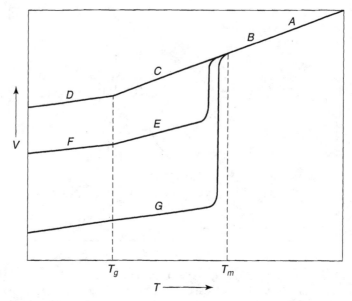

Figure 1.67   Specific volume as a function of temperature on cooling from the melt for a polymer that tends to crystallize. Region A is liquid, B liquid with elastic response, C supercooled liquid, D glass, E crystallites in a supercooled liquid matrix, F crystallites in a glassy matrix, and G completely crystalline. Paths $ABCD$, $ABEF$, and ABG represent fast, intermediate, and very slow cooling rates, respectively. From K. M. Ralls, T. H. Courtney, and J. Wulff, *Introduction to Materials Science and Engineering*. Copyright © 1976 by John Wiley & Sons, Inc. This material is used by permission of John Wiley & Sons, Inc.

There are three categories of LCPs, grouped according to the arrangement of the molecules: smectic, nematic, and cholesteric. *Nematic* (from the Greek term meaning "thread-like") LCPs have their molecules aligned along the chain axis, as shown schematically in Figure 1.68. Nematic liquids have low viscosity, and tend to be turbid, or "cloudy." *Smectic* (from the Greek term for "soap") LCPs have an additional level of structure, in that the polymer chains are also aligned along the chain axis, but they also segregate into layers. Smectic liquids are also turbid, but tend to be highly viscous. Finally, *cholesteric* (from the Greek term for "bile") LCPs have layered structures, but the aligned polymer chains in one layer are rotated from the aligned polymer chains in adjacent layers. Cholesteric LCPs are also highly viscous, but often possess novel photochromic, optical, thermochromic, and electro-optical properties.

Clearly, not all polymeric molecules possess the ability to form LCPs. Generally, LCPs have a molecular structure in which there are two regions with dissimilar chemical properties. For example, chains that consist of aliphatic–aromatic, dipolar–nonpolar, hydrophobic–hydrophilic, flexible–rigid, or hydrocarbon–fluorocarbon combinations of substantial size have a propensity to form LCPs. The two different portions of the chains can interact locally with similar regions of adjacent chains, leading to ordering. Liquid crystalline polymers that rely on structural units in the backbone, or main chain, to impart their crystallinity are called *main-chain LCPs*. The building blocks of typical main-chain LCPs are shown in Figure 1.69. Side-chain LCPs crystallize due to interaction of the side chains, or branches, as shown in Figure 1.69. Side

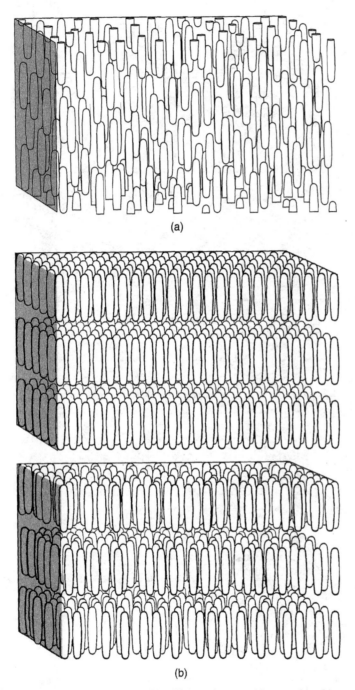

(a)

(b)

Figure 1.68   The structure of liquid crystalline polymers (a) nematic, (b) smectic and (c) cholesteric. Reprinted, by permission, from J. L. Fergason, *Scientific American*, **211**(2), pp. 78, 80. Copyright © 1964 by Scientific American, Inc.

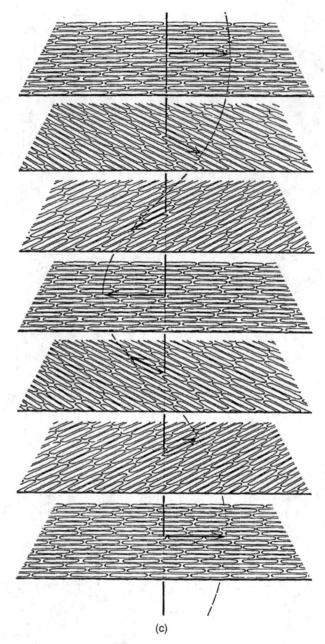

(c)

Figure 1.68   *(continued)*.

chains play an important role in determining not only the liquid crystalline activity of polymers, but the type of structure that will form as well. For example, in Table 1.25, we see that the number of hydrocarbon units in the side chain (not the backbone repeat unit) affects whether the resulting LCP is nematic or smectic.

We have used the general term "liquid" to describe this special class of polymers, but we know that a liquid can be either a melt or a solution. In the case of LCPs, both

Table 1.25   Effect of the Flexible Tail on the Structure of a Side-Chain LCP

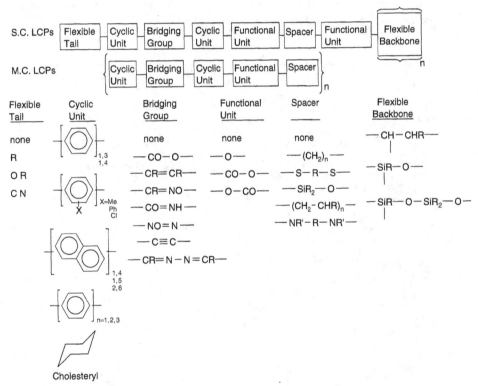

Number	n	R	Structure
1	2	OCH_3	Nematic
2	2	OC_3H_7	Smectic
3	6	OCH_3	Nematic
4	6	OC_6H_{13}	Smectic

Figure 1.69   General structure of main-chain (M.C.) and side-chain (S.C.) LCPs. Adapted from T. S. Chung, The recent developments of thermotropic liquid crystalline polymers, *Polymer Engineering and Science*, **26**(13), p. 903. Copyright © 1986, Society of Plastics Engineers.

types of liquids can occur. A polymer that exhibits crystallinity in the melt and that undergoes an ordered–disordered transformation as a result of thermal effects is called a *thermotropic* LCP. A polymer requiring a small molecule solvent in order to exhibit crystallinity is termed a *lyotropic* LCP. All three types of LCP structures can occur in either thermotropic or lyotropic polymers, and both are industrially relevant materials.

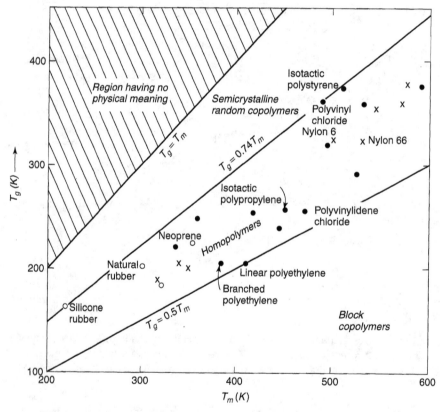

Figure 1.70   The repeat structure of Kevlar®.

Perhaps the most widely utilized (and studied) lyotropic LCP is poly *p*-phenylene terephthalamide (PPTA), more commonly known as Kevlar® (see Figure 1.70). Kevlar® belongs to the class of aramids that are well known for their LCP properties. Because these polymers are crystalline in solution, they are often spun into filaments, from which the solvent is subsequently removed in order to retain the aligned polymer structure. The result is a highly oriented, strong filament that can be used for a wide variety of structural applications. Most thermotropic LCPs are polyesters or copolymers that can be melted and molded into strong, durable objects.

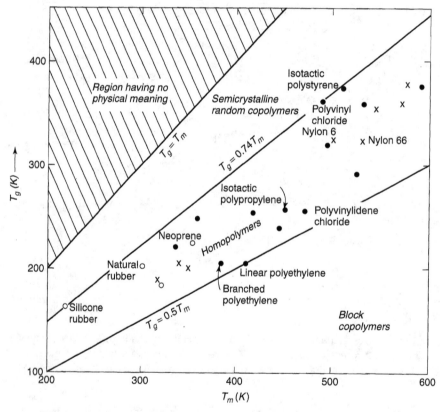

Figure 1.71   The glass transition temperature, T_g, as a function of crystalline melting point, T_m, for homopolymers. Filled circles are addition homopolymers, open circles are elastomers, and crosses are condensation homopolymers. From K. M. Ralls, T. H. Courtney, and J. Wulff, *Introduction to Materials Science and Engineering*. Copyright © 1976 by John Wiley & Sons, Inc. This material is used by permission of John Wiley & Sons, Inc.

Table 1.26 Predominant Properties of Crystalline Polymers

Temperature Range	Degree of Crystallinity		
	Low (5–10%)	Intermediate (20–60%)	High (70–90%)
Above T_g	Rubbery	Leathery, tough	Stiff, hard, brittle
Below T_g	Glassy, brittle	Hornlike, tough	Stiff, hard, brittle

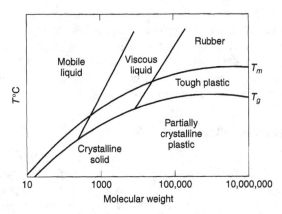

Figure 1.72 Approximate relations among molecular weight, T_g, T_m and polymer properties. From F. W. Billmeyer, *Textbook of Polymer Science*, 3rd ed. Copyright © 1984 by John Wiley & Sons, Inc. This material is used by permission of John Wiley & Sons, Inc.

1.3.7 The Glass Transition

The glass transition is an important parameter, since many physical properties change when polymer chains gain mobility. This is an important temperature because it marks the point where the polymer becomes more easily deformed, or more ductile. As summarized in Table 1.26, totally non crystalline polymers are very brittle below the glass transition, whereas polymers with intermediate crystallinity are very tough above the glass transition. Most homopolymers have glass transition temperatures that are $0.5–0.75T_m$, even though the crystalline melting point can vary by several hundred degrees between polymer types (see Figure 1.71). Most plastics are designed to be used between T_m and T_g, as shown in Figure 1.72.

To conclude this section on polymers, we should note that we have used the term polymer almost exclusively to refer to organic macromolecules. The term *plastic* refers not only to organic substances of high molecular weight, but also to such substances that at some point in their manufacture have been shaped by flow. Thus, the term plastic is more specific than the term polymer, and this term carries with it an indication of its processing history. As we will see in Chapter 7, there are many materials that can be considered polymers, yet are formed by routes other than melt processing.

1.4 STRUCTURE OF COMPOSITES

The first three sections of this chapter have described the three traditional primary classifications of materials: metals, ceramics, and polymers. There is an increasing emphasis on combining materials from these different categories, or even different materials within each category, in such a way as to achieve properties and performance that are unique. Such materials are called *composites*. A composite can be different things, depending on the level of definition we use. In the most basic sense, all materials except elements are composites. For example, a binary mixture of two elements, like an alloy, can be considered a composite structure on an atomic scale. In terms of microstructure, which is a larger scale than the atomic level definition, composites are composed of crystals, phases and compounds. With this definition, steel, which is a suspension of carbon in iron, is a composite, but brass, a single-phase alloy, is not a composite. If we move up one more level on the size scale, we find that there are macrostructural composites: materials composed of fibers, matrices, and particulates—they are *materials systems*. This highest level of structural classification is the one we will use, so our definition of a composite is this: *a material brought about by combining materials differing in composition or form on a <u>macroscale</u> for the purpose of obtaining specific characteristics and properties.*

1.4.1 Composite Constituents

The constituents in a composite retain their identity such that they can be physically identified and they exhibit an interface between one another. This concept is graphically summarized in Figure 1.73. The body constituent gives the composite its bulk form, and it is called the *matrix*. The other component is a *structural constituent*, sometimes called the *reinforcement*, which determines the internal structure of the composite. Though the structural component in Figure 1.73 is a fiber, there are other geometries that the structural component can take on, as we will discuss in a subsequent section. The region between the body and structural constituents is called the *interphase*. It is quite common (even in the technical literature), but incorrect, to use the term *interface* to describe this region. An interface is a two-dimensional construction—an area having a common boundary between the constituents—whereas an interphase is a three-dimensional phase between the constituents and, as such, has its own properties. It turns out that these interphase properties play a very important role in determining the ultimate properties of the bulk composite. For instance, the interphase is where

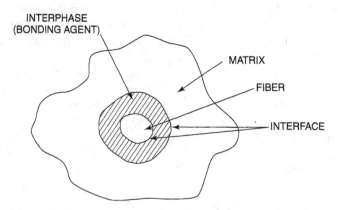

Figure 1.73 Schematic illustration of composite constituents. Reprinted, by permission, from M. Schwartz, *Composite Materials Handbook*, 2nd ed., p. 1.4. Copyright © 1992 by McGraw-Hill.

mechanical stresses are transferred between the matrix and the reinforcement. The interphase is also critical to the long-term stability of a composite. It will be assumed that there is always an interphase present in a composite, even though it may have a thickness of only an atomic dimension.

The chemical composition of the composite constituents and the interphase is not limited to any particular material class. There are metal–matrix, ceramic–matrix, and polymer–matrix composites, all of which find industrially relevant applications. Similarly, reinforcements in important commercial composites are made of such materials as steel, E-glass, and Kevlar®. Many times a bonding agent is added to the fibers prior to compounding to create an interphase of a specified chemistry. We will describe specific component chemistries in subsequent sections.

1.4.2 Composite Classification

There are many ways to classify composites, including schemes based upon (1) materials combinations, such as metal–matrix, or glass-fiber-reinforced composites; (2) bulk-form characteristics, such as laminar composites or matrix composites; (3) distribution of constituents, such as continuous or discontinuous; or (4) function, like structural or electrical composites. Scheme (2) is the most general, so we will utilize it here. We will see that other classification schemes will be useful in later sections of this chapter.

As shown in Figure 1.74, there are five general types of composites when categorized by bulk form. *Fiber composites* consist of fibers; with or without a matrix. By definition, a fiber is a particle longer than 100 µm with a length-to-diameter ratio (*aspect ratio*) greater than 10:1. *Flake composites* consist of flakes, with or without a matrix. A flake is a flat, plate-like material. *Particulate composites* can also have either a matrix or no matrix along with the particulate reinforcement. Particulates are roughly

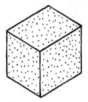

FIBER COMPOSITE   PARTICULATE COMPOSITE

LAMINAR COMPOSITE

FLAKE COMPOSITE   FILLED COMPOSITE

Figure 1.74 Classes of composites. Reprinted, by permission, from Schwartz, M., *Composite Materials Handbook*, 2nd ed., p. 1.7. Copyright © 1992 by McGraw-Hill.

spherical in shape in comparison to fibers or flakes. In a *filled composite*, the reinforcement, which may be a three-dimensional fibrous or porous structure, is continuous and often considered the primary phase, with a second material added through such processes as *chemical vapor infiltration* (CVI). Finally, *laminar composites* are composed of distinct layers. The layers may be of different materials, or the same material with different orientation. There are many variations on these classifications, of course, and we will see that the components in fiber, flake, and particulate composites need not be distributed uniformly and may even be arranged with specific orientations.

Cooperative Learning Exercise 1.7

We will see in Section 5.4.2 that the elastic modulus of a unidirectional, continuous-fiber-reinforced composite depends on whether the composite is tested along the direction of fiber orientation (parallel) or normal to the fiber direction (transverse). In fact, the elastic modulus parallel to the fibers, E_1, is given by Eq. (1.62), whereas the transverse modulus, E_2, is given by Eq. (1.63). Consider a composite material that consists of 40% (by volume) continuous, uniaxially aligned, glass fibers ($E_f = 76$ GPa) in a polyester matrix ($E_m = 3$ GPa).

Person 1: Calculate the elastic modulus parallel to the fiber direction, E_1.

Person 2: Calculate the elastic modulus transverse to the fiber direction, E_2.

Compare your answers. What would you expect the composite modulus to be if the same volume fraction of fibers were randomly oriented instead of uniaxially aligned?

Answer: $E_1 = (0.4)(76) + (0.6)(3) = 32.2$ GPa; $E_2 = (76)(3)/[(76)(0.6) + (0.4)(3)] = 4.87$ GPa

---

1.4.3 Combination Effects in Composites

There are three ways that a composite can offer improved properties over the individual components, collectively called *combination effects*. A *summation effect* arises when the contribution of each constituent is independent of the others. For example, the density of a composite is, to a first approximation, simply the weighted average of the densities of its constituents. The density of each component is independent of the other components. Elastic modulus is also a summation effect, with the upper limit $E_c(u)$ given by

$$E_c(u) = V_p E_p + V_m E_m \tag{1.62}$$

and the lower limit, $E_c(l)$ given by

$$E_c(l) = \frac{E_m E_p}{V_m E_p + V_p E_m} \tag{1.63}$$

where E_m and E_p are the elastic moduli of the matrix and particulate, respectively; and V_m and V_p are their respective volume fractions. We will discuss mechanical properties in more detail in Chapter 5, but the point of Eqs. (1.62) and (1.63) is that summation properties can be added appropriately to give an estimate of the composite properties. A *complementation effect* occurs when each constituent contributes separate properties. For example, laminar composites are sandwich-type composites composed of several layers of materials. Sometimes the outer layer is simply a protective coating, such as a polymeric film, that imparts corrosion resistance to the composite. This outer layer serves no structural purpose, and contributes only a specific property to the overall composite—in this case, corrosion resistance. Finally, some constituent properties are not independent of each other, and an *interaction effect* occurs. In this case, the composite property may be higher than either of the components, and the effect may be synergistic rather than additive. For example, it has been observed that the strength of some glass-fiber-reinforced plastic composites is greater than either the matrix or the reinforcement component by itself. In the subsequent sections, we look at the individual components of a composite and see what each can contribute, and how it helps to improved properties of the composite.

1.4.4 The Composite Matrix

The matrix serves two primary functions: to hold the fibrous phase in place and to deform and distribute the stress under load to the reinforcement phase. In most cases, the matrix material for a fiber composite has an elongation at break greater than the fiber; that is, it must deform more before breaking. It is also beneficial to have a matrix that encapsulates the reinforcement phase without excessive shrinkage during processing. A secondary function of the matrix is to protect the surface of the reinforcement. Many reinforcements tend to be brittle, and the matrix protects them from abrasion and scratching, which can degrade their mechanical properties. The matrix can also protect the reinforcement component from oxidation or corrosion. In this way, many fibers with excellent mechanical properties, such as graphite fibers, can be used in oxidizing environments at elevated temperatures due to protection by the matrix constituent.

Most *fiber-matrix composites* (FMCs) are named according to the type of matrix involved. *Metal-matrix composites* (MMCs), *ceramic-matrix composites* (CMCs), and *polymer-matrix composites* (PMCs) have completely different structures and completely different applications. Oftentimes the temperature at which the composite must operate dictates which type of matrix material is to be used. The maximum operating temperatures of the three types of FMCs are listed in Table 1.27.

Most structural PMCs consist of a relatively soft matrix, such as a thermosetting plastic of polyester, phenolic, or epoxy, sometimes referred to as *resin-matrix composites*. Some typical polymers used as matrices in PMCs are listed in Table 1.28. The list of metals used in MMCs is much shorter. Aluminum, magnesium, titanium, and iron- and nickel-based alloys are the most common (see Table 1.29). These metals are typically utilized due to their combination of low density and good mechanical properties. Matrix materials for CMCs generally fall into four categories: glass ceramics like lithium aluminosilicate; oxide ceramics like aluminum oxide (alumina) and mullite; nitride ceramics such as silicon nitride; and carbide ceramics such as silicon carbide.

Table 1.27 Approximate Upper Temperature Limits for Continuous Operation of Composites

Composite	Maximum Operating Temperature ($^\circ$C)
Polymer matrix	400
Metal matrix	1000
Ceramic matrix	1300

Table 1.28   Some Typical Polymers Used as Matrices in PMCs

Matrix Material	Process Temperature (°C)	Upper Use Temperature (°C)
Thermosetting		
Polyester (phthalic/maleic type)	RT	70
Vinyl ester	RT	125
Epoxy	150	125
Epoxy	200	175
Phenolic	250	200
Cyanates (triazines)	250	200
Bismaleimides	250	225
Nadic end-capped polyimides (e.g., PMR-15)	316	316
Thermoplastic		
Polysulfone	325	180
Polyamide	250	100
Polycarbonate	280	100
Polyphenylene oxide (PPO)	280	100
Polysulfides (PPS)	300	150
Polyether ether ketone (PEEK)	370	175
Polyether sulfone (PES)	350	175
Polyamide-imides	325	200
Polyetherimide	400	275
Polyimide	370	316
Polyarylate	400	300
Polyester (liquid crystalline)	300	150

Table 1.29   Common Metals for Metal-Matrix Composites

Matrix Material	Fabrication Method	Typical Composite Density (g/cm^3)	Use Temperature (°C)
Aluminum	Diffusion bonding		350
	Hot molding	2.62–3.45	
	Powder metallurgy		
	Liquid processing		
Magnesium	Liquid processing	1.82–2.80	300
	Diffusion bonding		
Titanium	Diffusion bonding	3.76–4.00	650
Iron-, nickel-based alloys	Diffusion bonding	5.41–11.7	800–1150

The processing techniques used for CMCs can be quite exotic (and expensive), such as chemical vapor infiltration (CVI), or through pyrolysis of polymeric precursors. Their maximum use temperatures are theoretically much higher than most MMCs or PMCs, exceeding 1800°C, although the practical use temperature is often much lower due to creep deformation (see Table 1.30). The attraction of CMCs, then, is for use in high-temperature structural applications, such as in combustion engines.

1.4.5   The Composite Reinforcement

In Section 1.4.2, we described several classification schemes for composites, including one that is based upon the distribution of the constituents. For reinforced composites, this scheme is quite useful, as shown in Figure 1.75. In reinforced composites, the reinforcement is the structural constituent and determines the internal structure of the composite. The reinforcement may take on the form of particulates, flakes, lamina, or fibers or may be generally referred to as "filler." Fibers are the most common type of reinforcement, resulting in fiber-matrix composites (FMCs). Let us examine some of these reinforcement constituents in more detail.

1.4.5.1   Fiber-Matrix Composites.   As shown in Figure 1.75, there are two main classifications of FMCs: those with continuous fiber reinforcement and those with discontinuous fiber reinforcement. *Continuous-fiber-reinforced composites* are made from fiber *rovings* (bundles of twisted filaments) that have been woven into two-dimensional sheets resembling a cloth fabric. These sheets can be cut and formed to a desired shape, or *preform*, that is then incorporated into a composite matrix, typically a thermosetting resin such as epoxy. Metallic, ceramic, and polymeric fibers of specific compositions can all be produced in continuous fashions, and the properties of the

Table 1.30   Common Ceramics for Ceramic-Matrix Composites

Matrix Material	Density (g/cm^3)	Use Temperature (°C)
Alumina, Al$_2$O$_3$	4.0	~1000
Glass ceramics	2.7	900
Si$_3$N$_4$	3.1	~1300
SiC	3.2	~1300

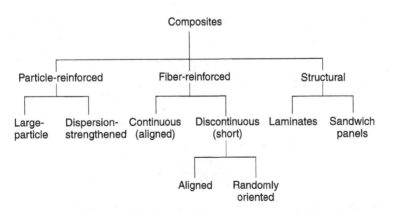

Figure 1.75   Types of reinforced composites.

resulting composite are highly dependent not only on the type of fiber and matrix used, but also on the processing techniques with which they are formed. We will discuss continuous-fiber-reinforced composites in more detail in Sections 5.4 and 7.4.

Discontinuous-fiber-reinforced composites are much more widely used, and there are some general underlying principles that affect their overall properties. Four main factors contribute to the performance level of a fiber in discontinuously reinforced FMCs. The first factor is fiber orientation. As shown in Figure 1.76, there are several ways that short fibers can be oriented within the matrix. *One-dimensional reinforcement* occurs when the fibers are oriented along their primary axis. This offers maximum mechanical strength along the orientation axis, but results in *anisotropic composites*; that is, the mechanical and physical properties are not the same in all directions. *Planar reinforcement* occurs with two-dimensional orienting of the fibers, as often occurs with woven fabrics. The fabric, as is common in woven glass fibers, is produced in sheets, and it is laid down (much like a laminate) to produce a *two-dimensional reinforcement* structure. *Three-dimensional reinforcement* results from the random orientation of the fibers. This creates an isotropic composite, in which the properties are the same in all directions, but the reinforcing value is often decreased compared to the aligned fibers.

The second factor that affects performance in discontinuously reinforced FMCs is fiber length. This has an effect primarily on the ease with which the composite can be manufactured. Very long fibers can create difficulties with methods used to create discontinuously reinforced FMCs and can result in nonuniform mechanical properties. The third factor is also related to fiber geometry, namely, the fiber shape. Recall that the

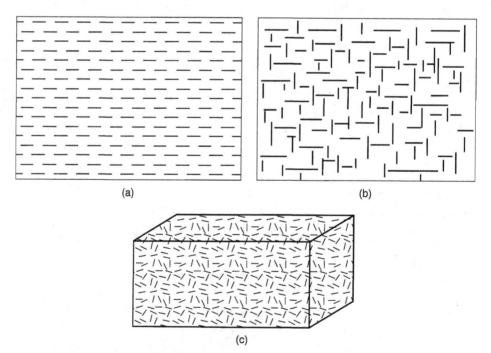

(a)

(b)

(c)

Figure 1.76 Types of fiber reinforcement orientation (a) one-dimensional, (b) two-dimensional, and (c) three-dimensional.

definition of a fiber is a particulate with length greater than 100 μm and an aspect ratio greater than 10:1. This definition allows for a great deal of flexibility in the geometry of the fiber. For example, aspect ratio can vary widely; many reinforcement filaments have aspect ratios much larger than 10:1. It is also not necessary that the fiber cross section be exactly circular. Hexagonal, ellipsoidal, and even annular (hollow fiber) cross sections are quite common. Finally, it is also not necessary that the fiber, even those with circular cross sections, be exactly cylindrical. "Dumbbell"-shaped fibers are very common, and even preferable, since mechanical stresses tend to concentrate at the fiber ends. The dumbbell shape helps distribute these stresses.

The final factor affecting the reinforcement performance is its composition. Chemistry affects properties, and strength is usually the most important property of a reinforcing fiber. Though we will concentrate on mechanical properties of materials in Chapter 5, it is useful at this point to familiarize yourself with some of the common fiber reinforcements, as summarized in Table 1.31. It is worth noting that oftentimes the important design consideration for reinforcement materials (or the matrix, for that matter) is not the absolute value of a particular design criterion, such as tensile strength

Table 1.31   Selected Properties of Some Common Reinforcing Fibers

Fiber	Density (g/cm^3)	Melting Point (°C)	Specific Modulus (GPa · cm^3/g)	Specific Strength (MPa · cm^3/g)
Aluminum	2.70	660	27	230
Steel	7.87	1621	25	530
Tantalum	16.6	2996	12	37
Titanium	4.72	1668	24	410
Tungsten	19.3	3410	21	220
Boron	2.30	2100	192	1500
Beryllium	1.84	1284	165	710
Molybdenum	10.2	2610	35	140
Aluminum oxide	3.97	2082	132	170
Aluminosilicate	3.90	1816	26	1060
Asbestos	2.50	1521	60	550
Beryllium carbide	2.44	2093	127	420
Beryllium oxide	3.03	2566	116	170
Carbon	1.76	3700	114	1570
Graphite	1.50	3650	230	1840
E-glass	2.54	1316	28	1360
S-glass	2.49	1650	34	1940
Quartz	2.20	1927	32	407
SiC (on tungsten)	3.21	2316	140	1000
Si$_3$N$_4$	2.50	1900	100	1344
BN	1.91	3000 (decomp.)	47	722
ZrO$_2$	4.84	2760	71	427
Wood	0.4–0.8	—	17	—
Polyamide (Kevlar)	1.14	249	2.5	730
Polyester	1.40	249	2.9	490
Polypropylene	0.9	154	1.8	77

or modulus, but the value per unit weight, such as *specific strength* or *specific modulus*. These values are listed in Table 1.31 rather than absolute values to illustrate this point. This fact is extremely important for many applications, such as automotive and aerospace composites, for which weight savings is paramount. Note also that reinforcing fibers come from all materials classes. Let us examine some of the more common fibers in more detail.

Organic fibers generally have very low specific gravities, so they are attractive for applications where strength/weight ratio is important. They are also very common for textile applications, since some organic fibers are easily woven. Carbon fibers offer excellent thermal shock resistance and a very high strength-to-weight ratio. There are two general types of carbon fibers, PAN-based carbon fibers and pitch-based carbon fibers. PAN-based carbon fibers are manufactured by the pyrolysis of *polyacrylonitrile* (PAN), as illustrated in Figure 1.77. The PAN is polymerized and fibers spun from the pre-polymer. The PAN fibers are then pyrolyzed to remove the hydrogens and form benzene-ring structures. *Pitch-based fibers* are produced in a similar pyrolysis process of a precursor fiber, except that the precursor in this case is *pitch*. Pitch is actually a liquid crystalline material, often called *mesophase pitch*, and is composed of a complex mixture of thousands of different species of hydrocarbons and heterocyclic molecules. It is the residual product of petroleum refining operations.

Wood fibers are technically organic, and though we do not discuss wood as a separate materials class in this text, it is an important structural material. Wood fibers, often in the form of wood flour, possess a variety of properties depending upon the type of tree from which they are derived, but are used extensively in low-cost composites. Wood fiber has a good strength/weight ratio and provides a use for recycled paper products. Turning to synthetic organic fibers, we see that polyamide fibers, such as Kevlar®, offer excellent specific mechanical properties. Kevlar® is used in many applications where high toughness is required, such as ropes and ballistic cloths. In addition to polyamides, polyesters, and polypropylene fibers listed in Table 1.31, nylon and polyethylene are other common polymeric fibers used for composites. In all cases, an added attraction of synthetic fibers is their chemical inertness in most matrix materials.

Glass fibers are the most common reinforcing fiber due to their excellent combination of mechanical properties, dielectric properties, thermal stability and relatively low cost. As a result, there are many different types of silicate glass fibers, all with varying properties designed for various applications (see Table 1.32). The majority

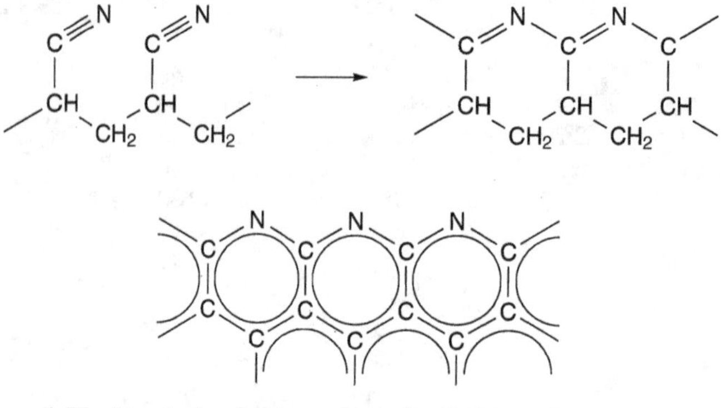

Figure 1.77   Pyrolysis of polyacrylonitrile (PAN) to form carbon fibers.

Table 1.32   Composition of Commercial Glass Fibers

	Composition (wt%)										
	SiO_2	Al_2O_3	Fe_2O_3	B_2O_3	ZrO_2	MgO	CaO	Na_2O	K_2O	Li_2O	TiO_2	F_2
A-glass (typical)	73	1	0.1			4	8	13	0.5			
E-glass (range)	52–56	12–16	0–0.5	8–13		0–6	16–25	← <1 total →			0	0–1.5
AR glass (range)	60–70	0–5			15–20		0–10	10–15			0–5	
C-glass (range)	59–64	3.5–5.5	0.1–0.3	6.5–7		2.5–3.5	13.5–14.5	8.5–10.5	0.4–0.7			
S and R glasses (range)	50–85	10–35				← 4–25 total →		0				

Table 1.33   Some Properties of Commercial Glass Fibers

	Liquidus Temperature (°C)	Working Temperature ($\eta = 100$ Pa s) (°C)	Density (g cm^{-3})	Coefficient of Thermal Expansion (°C^{-1})	Refractive Index	Young's Modulus (GN · m^{-2})	Strength (MN · m^{-2})	
							Undamaged Filament	Strand from Roving
A-glass	1140	1220	2.46	7.8×10^{-6}	1.52	72	3500	
E-glass	1400	1210	2.54	4.9×10^{-6}	1.55	72	3600	1700–2700
AR glass	1180–1200	1280–1320	2.7	7.5×10^{-6}	1.56	70–75	3600	1500–1900
C-glass			~2.5					
S- and R-glasses			~2.5			~85	~4500	2000–3000

component in all of these glass fibers is SiO_2, with various amounts of intermediates and modifiers added to improve strength, chemical resistance and temperature resistance. Some properties of these glass fibers are listed in Table 1.33. The two most common types of silicate glass fibers are *E-glass* (for "electronic" glass) and *S-glass* (for "strength" glass). S-glass was developed to provide improved strength in comparison to E-glass, while maintaining most of the same properties as E-glass. Both glass filaments are widely used in polymer–matrix composites, especially with epoxy-based matrixes. The result is a *glass-fiber-reinforced (GFR) composite* that is used extensively in automotive, aerospace, marine, electronics, and consumer product industries.

As a class, ceramic fibers offer better thermal resistance than glass fibers, and they are the preferred reinforcement in high-temperature structural composites. There are a number of commercial oxide-based fibers available, such as Saffil (SiO_2/Al_2O_3), Nextel ($SiO_2/B_2O_3/Al_2O_3$), Fiberfrax (SiO_2/Al_2O_3), and Kaowool (SiO_2/Al_2O_3), as well as nearly pure single-component metal oxide fibers of Al_2O_3 and ZrO_2. There are also a number of slag-based fibers of varying composition, based upon SiO_2, Al_2O_3, MgO, and CaO, which are recovered from smelting operations. The more refractory fibers consist of nitrides and carbides, such as Si_3N_4 and SiC. These fibers are usually produced by more exotic techniques. For example, Si_3N_4 fibers are produced from polymeric precursors such as polysilazanes (Si–N), polycarbosilanes (Si–CH$_2$), or polysilanes (Si–Si). Silicon carbide fibers are currently substrate-based and are formed by the deposition of SiC on a metallic filament such as tungsten or carbon. Silicon carbide whiskers can also be produced by the pyrolysis of polycarbosilane precursors.

Finally, metallic fibers find some limited applications as reinforcement in composites. They are generally not desirable due to their inherently high densities and because they present difficulties in coupling to the matrix. Nonetheless, tungsten fibers are used in metal-matrix composites, as are steel fibers in cement composites. There is increasing interest in *shape memory alloy* filaments, such as Ti–Ni (Nitanol) for use in piezoelectric composites. We will discuss shape-memory alloys and nonstructural composites in later chapters of the text.

1.4.5.2 Particulate Composites. Particulate composites encompass a wide range of materials, from cement reinforced with rock aggregates (concrete) to mixtures of ceramic particles in metals, called *cermets*. In all cases, however, the particulate composite consists of a reinforcement that has similar dimensions in all directions (roughly spherical), and all phases in the composite bear a proportion of an applied load. The percentage of particulates in this class of composites range from a few percent to 70%.

One important class of particulate composites is *dispersion-hardened alloys*. These composites consist of a hard particle constituent in a softer metal matrix. The particle constituent seldom exceeds 3% by volume, and the particles are very small, below micrometer sizes. The characteristics of the particles largely control the property of the alloy, and a spacing of 0.2–0.3 μm between particles usually helps optimize properties. As particle size increases, less material is required to achieve the desired interparticle spacing. Refractory oxide particles are often used, although intermetallics such as $AlFe_3$ also find use. Dispersion-hardened composites are formed in several ways, including surface oxidation of ultrafine metal powders, resulting in trapped metal oxide particles within the metal matrix. Metals of commercial interest for dispersion-hardened alloys include aluminum, nickel, and tungsten.

A *cermet* is a particulate composite similar to a dispersion-hardened alloy, but consists of larger ceramic grains (*cer-*) held in a metal matrix (*-met*) (see Figure 1.78). The refractory particulates can be from the oxide category, such as alumina (Al_2O_3),

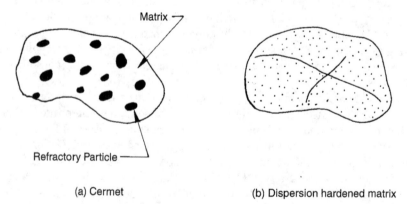

(a) Cermet   (b) Dispersion hardened matrix

Figure 1.78 Comparison of (a) cermet and (b) dispersion-hardened alloy. Reprinted, by permission, from M. Schwartz, *Composite Materials Handbook*, 2nd ed., p. 1.32. Copyright © 1992 by McGraw-Hill.

magnesia (MgO), or thoria (ThO$_2$), or from the carbide category, like tungsten carbide (WC), chromium carbide (Cr$_3$C$_2$) or titanium carbide (TiC). Cermets are formed by traditional powder-metallurgical techniques, although bonding agents must sometimes be added to improve bonding between the ceramic particulate and the metallic matrix. Cermets are used in a number of applications, including (a) fuel elements and control rods in the nuclear industry, (b) pulse magnetrons, and (c) cutting tools.

Finally, metal- and resin-bonded composites are also classified as particulate composites. *Metal-bonded composites* included structural parts, electrical contact materials, metal-cutting tools, and magnet materials and are formed by incorporating metallic or ceramic particulates such as WC, TiC, W, or Mo in metal matrixes through traditional powder metallurgical or casting techniques. *Resin-bonded composites* are composed of particulate fillers such as silica flour, wood flour, mica, or glass spheres in phenol-formaldehyde (Bakelite), epoxy, polyester, or thermoplastic matrixes.

1.4.6 The Composite Interphase

As mentioned in the introduction to the section on composites, the interphase, or the region between the matrix and the reinforcement, is often the primary determinant of mechanical properties of the composite. It can have an effect on other properties as well, but since it serves to transfer loads from the matrix to the reinforcement, its primary impact is on the strength of the composite.

There are several factors that affect the composition and spatial extent of the interphase. The first is *wettability*, or the ease with which the liquid matrix wets the reinforcing constituent prior to solidification. (There are other ways to form composites, of course, but incorporation of a solid particle into a liquid matrix is common and is germane to the description of wetting.) The ability of a liquid to wet a solid is measured by an interfacial *contact angle*, as shown in Figure 1.79. Without going into the thermodynamics of wetting at this point, it is important to the current discussion only to know that low contact angles are representative of "good" wetting (i.e., favorable liquid–solid interactions), whereas high contact angles (greater than about 90°) are indicative of "poor" wetting and unfavorable liquid–solid interactions. Obviously, a favorable liquid–solid interaction is desirable in order to obtain good matrix–fiber interactions. Sometimes this is not possible due to the disparate chemistries of the

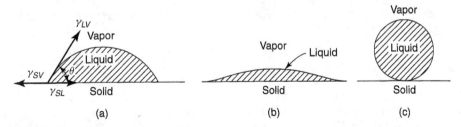

Figure 1.79 Interaction of a liquid droplet with a solid substrate: (a) Partially wetting (0° < θ < 180°), (b) completely wetting ($\theta \approx$ 0°), and (c) completely nonwetting ($\theta = $ 180°). Interfacial energies between the solid–liquid, solid–vapor, and vapor–liquid are represented by γ_{SL}, γ_{SV}, and γ_{LV}, respectively. From K. M. Ralls, T. H. Courtney, and J. Wulff, *Introduction to Materials Science and Engineering.* Copyright © 1976 by John Wiley & Sons, Inc. This material is used by permission of John Wiley & Sons, Inc.

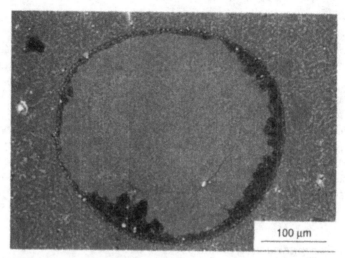

Figure 1.80   SEM photomicrograph of a CaO–Al$_2$O$_3$ fiber embedded in 4032 aluminum alloy [3]. Reprinted, by permission, copyright © 1994 by Kluwer Academic Publishers.

constituents, such as is found between silicate glass fibers and polymeric matrixes. In such cases, *wetting agents* are sometimes added to improve wettability. Another important factor in determining the makeup of the interphase is the inherent reactivity between the components. It is possible for the reinforcement and matrix to react with one another, either due to direct chemical reaction or through longer-term diffusion of components from one phase to another. For example, Figure 1.80 shows a CaO–Al$_2$O$_3$ fiber embedded in an aluminum matrix with traces of silicon [3]. Elemental analysis of the interphase (dark region between the fiber and the matrix) shows it contains silicon, which has diffused into the fiber surface during melt processing of the aluminum matrix to form an aluminum–silicon oxide compound.

Despite the emphasis on favorable interactions between the matrix and reinforcement and compound formation between them, it may be beneficial in certain circumstances for the interaction between the two primary constituents to be relatively weak. This is especially true in ceramic–ceramic composites, where both constituents are brittle, and the only way to impart some ductility on the composite is for the interphase to fail "gracefully"—that is, for the fibers to actually "pull out" of the matrix in a controlled manner. Optimization of the interphase properties in advanced composites is currently the focus of much research.

1.4.7   Functionally Graded Materials

Functionally graded materials (FGMs) feature gradual transitions in microstructure and composition that impart functional performance to the component. This term was first developed in the 1950s, but came into common use in the 1980s as new materials for the Space Plane were being conceived and developed. It was recognized early on that the material requirements would be extreme for this application. For the surface that contacts high-temperature gases at thousands of degrees, ceramics would need to be used, but for the surface that provides cooling, metallic materials would be required to provide sufficient thermal conductivity and mechanical strength. An example of

Relaxation function for thermal stress

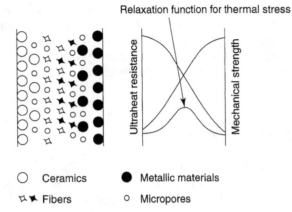

| ○ | Ceramics | ● | Metallic materials |
| ❑★ | Fibers | ○ | Micropores |

Figure 1.81 Conceptual diagram of functionally graded materials (FGMs). From M. Koizumi and M. Niino, Overview of FGM research in Japan, *MRS Bulletin*, **XX**(1), 19 (1995). Reproduced by permission of MRS Bulletin.

an FGM for this application is shown in Figure 1.81. The constituents are graded along the thickness of this composite, as are the resulting physical and mechanical properties. Another place where FGMs are common are on coatings. *Thermal barrier coatings* (TBCs) and wear-resistant coatings are often formed in a gradient fashion on the surface of bulk components. The gradient is primarily a result of the processing technique. *Plasma vapor deposition* (PVD), *chemical vapor deposition* (CVD), *thermal spray processing*, and *plasma spray processing* are but a few of the techniques used to form FGMs.

Though the term "functionally graded materials" is relatively young, the concept is not. Nature has been producing FGMs for a long time. Figure 1.82 shows a cross

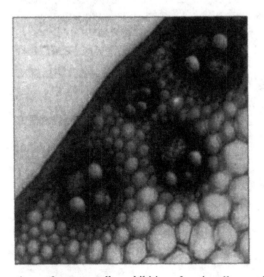

Figure 1.82 Cross-section of corn stalk exhibiting functionally graded structure. From Shigeyasu Amada, Hierarchical functionally gradient structures of bamboo, barley, and corn, *MRS Bulletin*, Vol. **XX**(1), 35 (1995). Reproduced by permission of MRS Bulletin.

section of a corn plant. The bundle sheaths on the outside of the stalk provide structural integrity to the plant. The fiber strength in the bundles is about 10 times that of the matrix. But the fiber distribution is dense in the outer region and sparse in the inner region. Materials developed as a result of adaptation of structural elements found in nature are called *biomimetic structures*—an area that is receiving considerable attention in the research community today.

1.5 STRUCTURE OF BIOLOGICS

The boundary between the disciplines of biology and materials engineering is becoming increasingly blurred. Molecules that were once thought of as being biological "macro-molecules" such as DNA are now being used as templates to produce molecules with desired gene sequences, and live tissue cultures are being used to reconstruct human body parts much as an auto body is repaired with fiberglass and epoxy. More impor-tantly, however, the scientific and engineering principles behind the application and development of biological materials are becoming more well understood, so it is useful to provide some information on the structure of these materials in the context of their use as materials of (human) construction. We begin with some simple biochemistry, then use this to describe how nature uses these building blocks to create complex composite structures of tissue, both hard and soft.

1.5.1 Review of Biological Molecules

Just as we began our description of polymer structure with an organic chemistry review, let us begin our introduction to biological materials with some simple biochemistry.

1.5.1.1 Amino Acids and Proteins. *Proteins* are the molecules that perform the functions of life. They can be enzymes that catalyze biological reactions, or they can be the receptor site on a membrane that binds a specific substance. Proteins are important parts of both bones—the so-called *hard biologics*—and the *soft biologics* such as muscle and skin. Any discussion of the structure of living organisms must begin with the structure of proteins.

   Proteins are composed of *amino acids*. As shown in Figure 1.83, an amino acid has a carboxyl group, an amino group (refer to Table 1.22 for a summary of functional

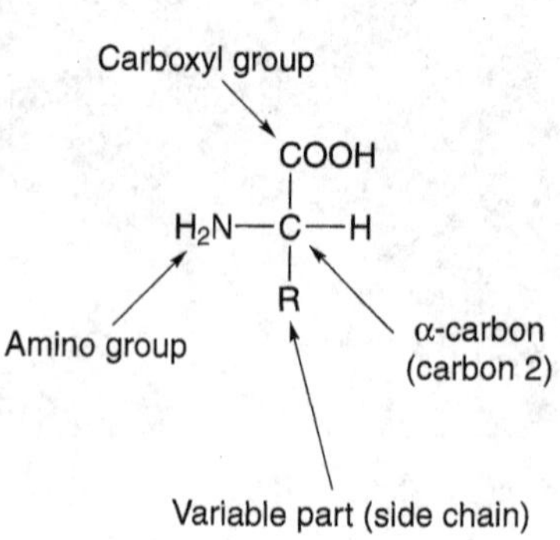

Figure 1.83   General structure of an amino acid.

groups), a central atom identified as the alpha (α) carbon, and a variable part know as the side chain (R). Normally, the amino acid is in its dissociated state, such that the terminal hydrogen on the carboxyl group moves to the amino group, thereby creating a carboxylate group (COO^-) and an ammonium group (NH_3^+) known as a *zwitterion*. The molecule remains neutral overall, and there may be several stable zwitterion forms, depending on the amino acid. Note also that the α-carbon is an asymmetric carbon, resulting in a chiral molecule for all but one of the naturally occurring amino acids. All the amino acids are found in only one of the stereoisomer configurations (the L-configuration).

There are only 20 amino acids that make up proteins (technically, 19 amino acids and one imino acid), differing only in the type of R group they contain (see Table 1.34). Each amino acid has a specific name and three-letter designation. Most amino acids decompose instead of melting due to the strong intermolecular electrostatic attractions, with decomposition temperatures ranging from 185–315°C, and they are only sparingly soluble in water, with the exceptions of glycine, alanine, proline, lysine, and arginine.

To form proteins, the carboxylic acid group on one amino acid reacts with the amine group on another molecule in a condensation reaction that forms one water molecule and a –CO–NH–CHR– linkage known as a *peptide bond*. A molecule containing more than about 100 amino acid sequences is called a *polypeptide*, and a protein is composed of one (or more) polypeptide chains. Thus, the number of possible proteins from the 20 amino acids is enormous ($\sim 20^{100}$). Replacing even one amino acid in the sequence of a protein can change its function completely. Sickle cell anemia is the result of replacing only one valine amino acid with a glutamic acid unit in one protein chain of the hemoglobin molecule. The peptide sequence is named by starting at the N(-amino) terminus of the polypeptide.

As with the polymers we have already described, peptides and proteins can possess complex conformations by rotation of bonds in the backbone and interaction between side chains. The stability of these structures is strongly dependent upon the R groups and, hence, the specific amino acid sequence, as well as the environment in which the protein finds itself. Figure 1.84 illustrates a few of the common conformations found in proteins. The α-*helix* occurs when the chain coils like a right-hand screw to form a cylinder, and it is the result of hydrogen bonding between the C=O and N–H in adjacent turns of the helix. Only the right-handed α-helix occurs in nature, and its presence results in an electric dipole with excess positive charge at one end and excess negative charge at the other. In the β-sheet, the peptide chain is much more extended, with 0.35 nm between adjacent peptide groups, in comparison to 0.15 nm for the α-helix. These sheet structures are also the result of hydrogen bonding. The α-helix and the β-sheet are examples of *secondary structure* in proteins. Tertiary and quaternary structures also exist, but are beyond the scope of this text. Depending on the nature of the side chains, there can also be hydrophobic interactions within the chain, leading to chain extension. Finally, disulfide bonds can occur when two cysteine residues react to form a covalent –S–S– bond. The breaking of disulfide bonds, or any action that leads to an alteration in the structure of a protein as to render it inactive, leads to *denaturation*. Denatured proteins also tend to have decreased solubility.

1.5.1.2 DNA and RNA. Like proteins, *deoxyribonucleic acid* (DNA) and *ribonucleic acid* (RNA) are polymers, but instead of amino acids as repeat units, they are

Table 1.34 Common Amino Acids

Structure	Name	Abbreviation
$\begin{array}{c}NH_2\\|\\R\!-\!CHCOOH\end{array}$	Name	Abbreviation
$\begin{array}{c}NH_2\\|\\H\!-\!CHCOOH\end{array}$	Glycine	Gly
$\begin{array}{c}NH_2\\|\\CH_3\!-\!CHCOOH\end{array}$	Alanine	Ala
$\begin{array}{c}CH_3\ \ NH_2\\|\ \ \ \ |\\CH_3CH\!-\!CHCOOH\end{array}$	Valine	Val
$\begin{array}{c}CH_3\ \ \ \ \ \ NH_2\\|\ \ \ \ \ \ \ \ \ |\\CH_3CHCH_2\!-\!CHCOOH\end{array}$	Leucine	Leu
$\begin{array}{c}CH_3\ \ NH_2\\|\ \ \ \ \ |\\CH_3CH_2CH\!-\!CHCOOH\end{array}$	Isoleucine	Ile
$\begin{array}{c}NH_2\\|\\CH_3SCH_2CH_2\!-\!CHCOOH\end{array}$	Methionine	Met
$\begin{array}{c}CH_2\diagdown\\ \diagup\ \ \ \ NH\\CH_2\ \ \ \ |\\ \diagdown\ \ \ \ CHCOOH\\CH_2\diagup\end{array}$	Proline	Pro
Phenyl–$CH_2\!-\!\overset{NH_2}{\overset{|}{CHCOOH}}$	Phenylalanine	Phe
Indolyl–$CH_2\!-\!\overset{NH_2}{\overset{|}{CHCOOH}}$	Tryptophan	Trp
$\begin{array}{c}NH_2\\|\\HOCH_2\!-\!CHCOOH\end{array}$	Serine	Ser
$\begin{array}{c}OH\ \ NH_2\\|\ \ \ \ |\\CH_3CH\!-\!CHCOOH\end{array}$	Threonine	Thr
$\begin{array}{c}NH_2\\|\\HSCH_2\!-\!CHCOOH\end{array}$	Cysteine	Cys
HO–C$_6$H$_4$–$CH_2\!-\!\overset{NH_2}{\overset{|}{CHCOOH}}$	Tyrosine	Tyr

Table 1.34   (*continued*)

$\begin{array}{c} NH_2 \\ | \\ R-CHCOOH \end{array}$	Name	Abbreviation
$\begin{array}{cc} O & NH_2 \\ || & | \\ H_2NCCH_2-CHCOOH \end{array}$	Asparagine	Asn
$\begin{array}{cc} O & NH_2 \\ || & | \\ H_2NCCH_2CH_2-CHCOOH \end{array}$	Glutamine	Gln
$\begin{array}{cc} O & NH_2 \\ || & | \\ HOCCH_2-CHCOOH \end{array}$	Aspartic acid	Asp
$\begin{array}{cc} O & NH_2 \\ || & | \\ HOCCH_2CH_2-CHCOOH \end{array}$	Glutamic acid	Glu
$\begin{array}{c} NH_2 \\ | \\ H_2NCH_2CH_2CH_2CH_2-CHCOOH \end{array}$	Lysine	Lys
$\begin{array}{cc} NH & NH_2 \\ || & | \\ H_2NCNHCH_2CH_2CH_2-CHCOOH \end{array}$	Arginine	Arg
Histidine imidazole $-CH_2-\overset{NH_2}{\underset{|}{C}}HCOOH$	Histidine	His

composed of a chain of nucleotides. Each nucleotide is composed of three basic structural units: a *base*, a *sugar*, and a *phosphate group* (see Figure 1.85). One base with its sugar (and without the phosphate group) is called a *nucleoside*. The sugar in RNA is called *ribose*, which is reduced in DNA by a loss of oxygen at the 2′ carbon to form *deoxyribose* (see Figure 1.86). There are only five primary bases found in polynucleotides: two *purines* represented by *adenine* (A) and *guanine* (G); and three *pyrimidines* represented by *cytosine* (C), *thymine* (T), and *uracil* (U). Thymine is found only in DNA nucleotides, and uracil only in RNA nucleotides, which results in four DNA nucleotides (see Figure 1.87) and four RNA nucleotides. The four DNA nucleotides are 2′-deoxyadenosine monophosphate (dAMP), 2′-deoxyguanosine monophosphate (dGMP), 2′-deoxycytidine monophosphate (dCMP), and thymidine monophosphate (TMP), the latter of which is already assumed to have 2′-deoxyribose as the sugar since it occurs only in DNA and not in RNA.

The nucleic acid polymer is formed when the nucleotides attach to one another through phosphodiester bonds, which connect the 3′-OH group of one nucleotide to the 5′-OH group of another nucleotide through the phosphate group. The order of the nucleotides in the chain is the primary structure of the DNA or RNA molecule, and it can be represented in short-hand notation with only the base pair designation

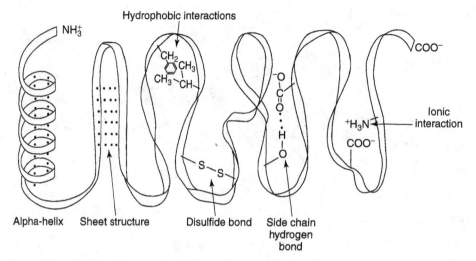

Figure 1.84 Some common structural elements in a hypothetical protein molecule. Reprinted, by permission, from M. E. Houston, *Biochemistry Primer for Exercise Science*, 2nd ed., p. 9. Copyright © 2001 by Michael E. Houston.

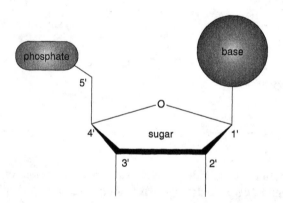

Figure 1.85 The chemical structure of DNA and RNA nucleotides. From H. R. Matthews, R. Freedland, and R. L. Miesfeld, *Biochemistry: A Short Course*. Copyright © 1997 by John Wiley & Sons, Inc. This material is used by permission of John Wiley & Sons, Inc.

(A,G,C,T, or U). Some sequences are highly repetitive, such as the Alu sequence, which is a sequence of 123 base pairs that occurs millions of times in the chain. Some sequences are specific binding sites for certain proteins, as in the TATAA sequence found near the start of many DNA molecules.

As with proteins, the nucleic acid polymers can denature, and they have secondary structure. In DNA, two nucleic acid polymer chains are twisted together with their bases facing inward to form a *double helix*. In doing so, the bases shield their hydrophobic components from the solvent, and they form hydrogen bonds in one of only two specific patterns, called *base pairs*. Adenine hydrogen bonds only with thymine (or uracil in RNA), and guanine pairs only with cytosine. Essentially every base is part of a base pair in DNA, but only some of the bases in RNA are paired. The double-helix structure

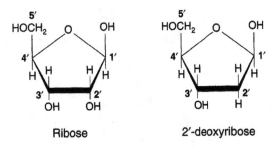

Figure 1.86 The chemical structure of the sugars in RNA (ribose) and DNA (deoxyribose). Reprinted, by permission, from M. E. Houston, *Biochemistry Primer for Exercise Science*, p. 30, 2nd ed. Copyright © 2001 by Michael E. Houston.

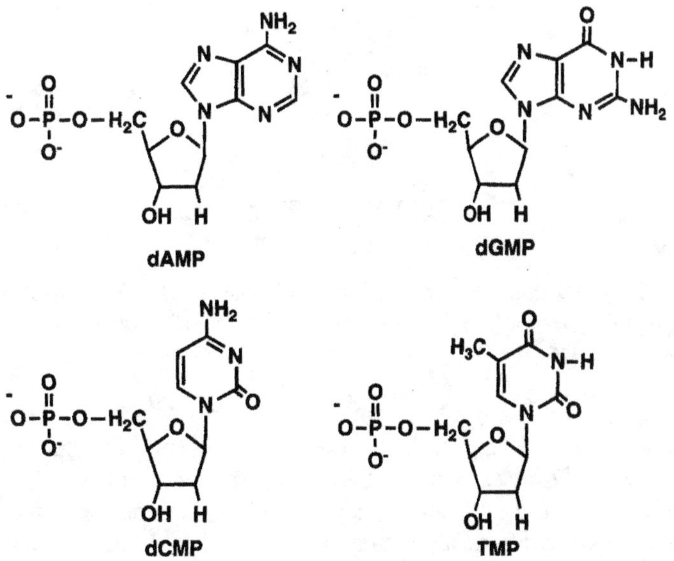

Figure 1.87 The four DNA nucleotides. Reprinted, by permission, from D. E. Schumm, *Essentials of Biochemistry*, 2nd ed., p. 17. Copyright © 1995 by Little, Brown and Company, Inc.

formed by DNA is comprised not only of two nucleic acid polymer chains, but the chains have complementary sequences such that when they are wound around each other in an antiparallel fashion, each base is opposite its appropriate partner and a base pair is automatically formed. As with proteins, this double helix is also right-handed, with the phosphate groups on the outside of the structure where they can interact with solvent ions. Thus, there are two "grooves" formed between the phosphate chains (see Figure 1.88): a *major groove* and a *minor groove*. The edges of the base pairs are accessible to the solvent in the grooves and provide regions where specific protein binding can occur. The double helix undergoes further conformational changes, called *supercoiling*, which allows a single molecule of human DNA, which is nearly one meter long if stretched out, to fit into the nucleus of a cell.

DNA is purely a molecular code: The molecule itself executes no function. A specific section of a DNA molecule, known as a *gene*, is used only as a blueprint

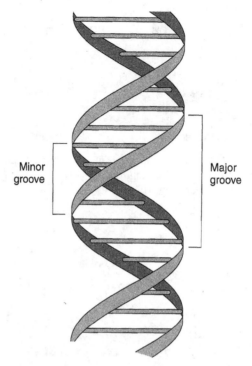

Minor groove

Major groove

Figure 1.88   Major and minor grooves in the DNA double helix. Reprinted, by permission, from D. E. Schumm, *Essentials of Biochemistry*, 2nd ed., p. 18. Copyright © 1995 by Little, Brown and Company, Inc.

to produce a nucleic acid chain called *messenger RNA* (mRNA) that carries out the synthesis of proteins. From a sequence of four bases sequence in DNA, mRNA specifies the correct sequence of the 20 amino acids required to produce a specific protein. Interestingly, only about 5% of all human DNA is used to produce protein molecules. Nonetheless, this is still roughly 100,000 gene sequences. There is currently an effort underway to map all of the human genes, called the Human Genome Project, which you can learn more about by visiting their website at the National Institutes of Health: *www.nhgri.nih.gov/*.

1.5.1.3 Cells. We finally come to what are the direct building blocks of biological materials: *cells*. Cells are assemblies of molecules enclosed within a *plasma membrane* that carry out specific functions. The human body contains over 10^{14} cells, all of which take in nutrients, oxidize fuels, and excrete waste products. Despite their varied functions, all cells have a similar internal organization. We will concentrate on this internal organization for now and will leave the topics of cell reproduction, energy production, and related concepts to the molecular biologist.

Surrounding the outside of all cells is the plasma membrane (see Figure 1.89). It is composed primarily of *lipids* and is selectively permeable, limiting the exchange of molecules between the inside and outside of the cell. The outside of the plasma membrane contains all the carbohydrates and receptor sites. The *cytoplasm* includes everything inside the plasma membrane except for the nucleus. Energy is generated

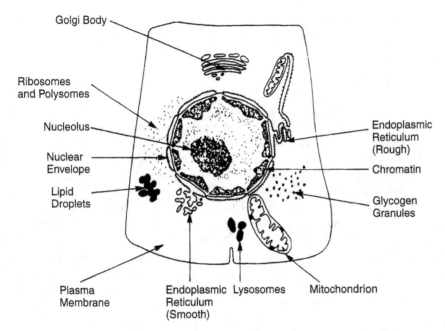

Golgi Body

Ribosomes
and Polysomes

Nucleolus

Nuclear
Envelope

Lipid
Droplets

Endoplasmic
Reticulum
(Rough)

Chromatin

Glycogen
Granules

Plasma
Membrane

Endoplasmic
Reticulum
(Smooth)

Lysosomes

Mitochondrion

Figure 1.89   Diagram of a typical human cell showing some of the subcellular structures. Reprinted, by permission, from D. E. Schumm, *Essentials of Biochemistry*, 2nd ed., p. 4. Copyright © 1995 by Little, Brown and Company, Inc.

in the cell by the *mitochondria*, where cellular fuel is oxidized. There may be several mitochondria per cell. The *Golgi complex* is composed of parallel membrane sacs and is used to secrete proteins. The *endoplasmic reticulum* (ER) is a system of membranes that store, segregate, and transport substances within the cell. The ER is continuous with the Golgi complex and the *nuclear membrane*. The nuclear membrane, or envelope, surrounds the nucleolus, which contains RNA, as well as DNA and other proteins. Normally, the DNA is a diffuse tangle of fine threads called *chromatin*. The rest of the cytoplasm is composed of *lysosomes* which degrade nucleic acids, proteins, and complex carbohydrates; *peroxisomes*, which contain a variety of oxidation enzymes; *polysomes*, which are engaged in protein synthesis using the mRNA; and *glycogen*, which is a polymer of glucose, used for energy.

Cells with similar structure and function group together to form *tissue*. Despite the astounding diversity of cell types and functions, there are really only four major types of tissue: *epithelial, connective, muscular*, and *nervous*. Epithelial tissues are usually delicate cells that form linings of internal structures and organs. They also form the outer covering of our bodies, called *skin*. Connective tissues are found in the walls of organs where they provide structural support. *Bone, cartilage, ligaments*, and *tendons* are all types of connective tissue. Muscle tissue enables the body to move, and it is characterized by its ability to contract. Nervous tissue is composed of highly specialized cells called *neurons*, and it is characterized by its ability to translate stimulation into electrochemical nervous impulses. The region between cells in a tissue is equally important. It is often termed the *extracellular matrix*, and it contains proteins and ions that perform vital functions. We will see that the extracellular matrix components have a profound effect on how well foreign materials are accepted (or rejected) by the body. For the time being, however, we will concentrate on the tissue itself by further generalizing biological materials as either "hard" or "soft" materials.

1.5.2 Hard Biologics

Though most (by volume) of the human body is composed of soft materials such as skin and muscle, we would have limited abilities to eat, move, and protect ourselves were it not for hard biologics such as bone and teeth. From an industrial standpoint, hard tissue repair accounts for roughly 20% of the multi-billion dollar biomaterials industry worldwide [4], with a projected growth rate of 7–12% annually.

The delineation between hard and soft biologics is not always clear, such as in the case of the connecting tissue between bone and muscle called *tendons*, but for our purposes, hard biologics contain some significant fraction of inorganic components. For both bone and teeth, the primary inorganic constituent belongs to the *calcium phosphate* family, $CaO–P_2O_5$. Actually, most calcium phosphates found in biological environments have water associated with them, so that the calcium phosphates we will be discussing belong to the ternary system $CaO–P_2O_5–H_2O$. The specific calcium phosphate phase depends on such factors as temperature and pH of the environment, as well as the chemical reactivity of that phase with its surroundings. The most prevalent form of calcium phosphate in the human body is *hydroxyapatite* (sometimes called *hydroxyl*apatite and abbreviated HA or HAp), $Ca_{10}(PO_4)_6(OH_2)$. Hydroxyapatite can occur both naturally and synthetically and has a hexagonal crystal structure (see Figure 1.90), with Ca^{2+} and $(PO_4)^{3-}$ ions arranged about columns of $(OH)^-$ ions [5–9]. Other biologically important calcium phosphates include dicalcium phosphate, $CaHPO_4$, and tricalcium phosphate (TCP), $Ca_3(PO_4)_2$. The remaining chemical constituents in teeth and bone are listed in Table 1.35.

Back in the 1920s, the basic structural features of apatite lattices were first worked out independently by de Jong [5], Mehmen [6], and st. Naray-Szabo [7]. de Jong showed for the first time, with the then relatively new technique of X-ray diffraction, that the mineral in bone bore a close resemblance to naturally occurring hydroxyapatite (HA). In this pioneering X-ray diffraction study, de Jong also observed that the apatite crystals in bone were extremely minute and ill-defined. However, a detailed spatial arrangement of the constituent ions in the apatite structure was not firmly established. It was only after 25 years that Posner et al. [8] could arrive at these structural aspects from X-ray diffraction studies on synthetically prepared singled crystals of HA. That animal bones and teeth contain HA as an ingredient was proved by identification of XRD patterns of the former with those of naturally occurring HA, as shown at right. Pattern (a) is from the synthetic crystalline sample, pattern (b) is from the synthetic amorphous sample, and pattern (c) is from the bone sample.

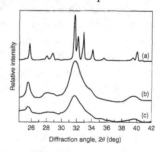

Source: Narasaraju and Phebe [9].

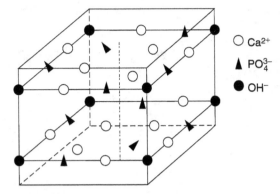

Figure 1.90   The unit cell perspective of hydroxyapatite. Reprinted, by permission, from T. S. B. Narasaraju, and D. E. Phebe, *Journal of Materials Science*, **1**, 1, Copyright © 1996, Chapman & Hall.

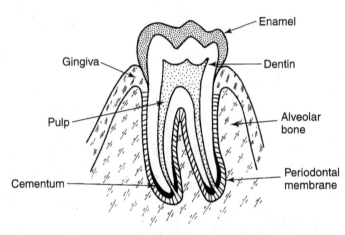

Figure 1.91   Schematic diagram of a human tooth. Reprinted, by permission, from G. Lewis, *Selection of Engineering Materials*, p. 401. Copyright © 1990 by Prentice-Hall, Inc.

The primary components of a human tooth are shown in Figure 1.91. The outer surface of the tooth is composed entirely of *enamel*. Enamel is a smooth, white, semi-transparent material that varies in thickness from 0.01 mm to about 2.5 mm over the grinding surfaces of molars. It is extremely hard and brittle and not very strong. The main constituent inorganic phase (95 wt%) is HA, with an organic phase (1 wt%) made up of protein, carbohydrates, lipids, and other matter. The remaining phase (about 4 wt%) is water, mostly present as a shell surrounding the HA crystallites. *Dentin*, the most important part of the tooth structure, is covered by the enamel in the crown and the *cementum* in the root. Dentin is a calcified mesh of collagen fibrils surrounding the cellular pathways of the dentin-forming cells. The resultant tubular structure contains about 50% HA (but the crystal size is smaller and the orientation different from that in enamel), 32% collagen, 8% mucopolysaccharides, and 10% water.

Bone is also a composite material composed of about 70% HA, as well as collagen (see Section 1.5.3), but in contrast to enamel, bones also contain blood vessels that help

Table 1.35 Composition and Selected Properties of Inorganic Phases in Adult Human Enamel, Dentine, and Bone

	Enamel	Dentine	Bone
Composition[a]			
Calcium, Ca^{2+} [b]	36.5	35.1	34.8
Phosphorus, as P	17.7	16.9	15.2
(Ca/P) molar[b]	1.63	1.61	1.71
Sodium, Na^+ [b]	0.5	0.6	0.9
Magnesium, Mg^{2+} [b]	0.44	1.23	0.72
Potassium, K^+ [b]	0.08	0.05	0.03
Carbonate, as CO_3^{2-} [c]	3.5	5.6	7.4
Fluoride, F^- [b]	0.01	0.06	0.03
Chloride, Cl^- [b]	0.30	0.01	0.13
Pyrophosphate, $P_2O_7^{4-}$	0.022	0.10	0.07
Total inorganic (mineral)	97.0	70.0	65.0
Total organic[d]	1.5	20.0	25.0
Absorbed H_2O	1.5	10.0	10.0
Trace elements: Sr^{2+}, Pb^{2+}, Zn^{2+}, Cu^{2+}, Fe^{3+}, etc.			
Crystallographic properties			
Lattice parameters (± 0.003 Å)			
a axis	9.441	9.42	9.41
c axis	6.880	6.88	6.89
Crystallinity index[e]	70–75	33–37	33–37
Crystallite size (average), Å	1300×300	200×40	250×30
Ignition products (800°C)	β-TCP + HAp	β-TCP + HAp	HAp + CaO

[a] Weight %.
[b] Ashed sample.
[c] Unashed sample, IR method.
[d] Principal organic component: enamel, noncollagenous; dentine and bone, collagenous.
[e] Calculated from ratio of coherent/incoherent scattering, mineral, HAp = 100.
Source: Suchanek, W. and M. Yoshimura, *J. Mater. Res.*, **13**(1), 99 (1998).

the bone regenerate itself (see Figure 1.92). The collagen fibers are a bundled array of crosslinked helical peptides. Bone tissue replaces itself through the action of cells called *osteoclasts*. Osteoclasts produce acids that resorb HA and break down collagen. The resulting calcium and proteins cause other cells called *osteoblasts* to lay down new matrix that mineralizes and forms HA and collagen. Bone cells produce growth factors, such as bone *morphogenetic proteins*, to increase or decrease bone regrowth. The blood vessels navigate through the porous bone structure, carrying important compounds such as *calcitonin* (a thyroid hormone that prevents bone resorption), *parathyroid hormone* and *calcitrol* (which regulate calcium and phosphate metabolism), and *prostaglandins* (fatty acids that perform hormone-like functions).

As you can see, bone is a very complex composite material and, as such, is difficult to artificially replicate. Nonetheless, there are a number of materials of all types that are being used as bone replacements. Polymethylmethacrylate (PMMA), titanium, graphite/polyethyletherketone (PEEK) composites, and tricalcium phosphate, among

---

Cooperative Learning Exercise 1.8

The crystal structure of hydroxyapatite, shown in Figure 1.90, is a hexagonal unit cell with $a = 9.42$ Å and $c = 6.88$ Å. The relationship between interplanar diffraction spacing, d, and the lattice parameter for the HCP structure, analogous to Eqs. (1.33) and (1.34), is

$$d = \frac{1}{\sqrt{\frac{4}{3}\left(\frac{h^2 + hk + k^2}{a^2} + \frac{l^2}{c^2}\right)}}$$

The diffraction pattern for various forms of hydroxyapatite is shown in the Historical Highlight on page 122. Use this information to calculate the following.

Person 1: Use Bragg's Law [Equation (1.35)] to calculate the d-spacing (in nm) for the first diffraction peak in hydroxyapatite. Assume a first-order diffraction and an X-ray source of $\lambda = 0.1537$ nm.

Person 2: Derive a relationship in simplest terms for the d-spacing of hydroxyapatite in terms of the Miller indices only (h, k, and l). Use the cell parameters in nm.

Combine your information to determine the Miller indices of the first diffraction peak for hydroxyapatite.

Answer: $d = 0.355$ nm; $\frac{d}{1} = 1.51(h^2 + hk + k^2) + 2.821l^2$; $(hkl) = (111)$ for $2\theta = 25°$.

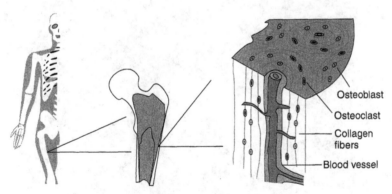

Figure 1.92   The structure of human bone. Reprinted with permission from S. K. Ritter, *Chemical & Engineering News*, p. 27, August 25, 1997. Copyright © 1997 American Chemical Society.

numerous others, have all been used to repair or replace bone in humans. A review of the structure and effectiveness of these different materials is beyond the scope of this text, but the reader should be aware that this is one of the developing areas of materials engineering. Refer to some of the more recent review articles available on this topic [10–14], and keep an eye out for new developments as they come along.

1.5.3 Soft Biologics

Unlike hard biological materials that are composed primarily of one compound (calcium phosphate) and perform a limited number of functions (primarily structural), soft biological materials are highly diversified, performing a myriad of highly specialized and/or combinatorial functions. Earlier, we classified tissue into four general categories (muscle, nervous, epithelial, and connective). Although connective tissue is primarily a hard material, there are a number of connective tissues that are more like the other soft tissues, inasmuch as they do not have a significant fraction of the calcium phosphate-based inorganic phase. Instead of taking the traditional biology-based approach to classifying and characterizing soft materials, let us instead concentrate on four important proteins found in the extracellular matrix that allow these tissue types to execute their intended functions, as well as control our ability to introduce foreign objects—biomaterials—into the human body. These proteins are *collagen, elastin, fibronectin,* and *laminin.*

Collagen is one of the most important and abundant substances in the human body. It is not a single protein, but rather a group of at least five different proteins that have a similar structure. Collagen contains 30% glycine, 20% proline and hydroxyproline, and a modified version of hydroxylysine. The secondary structure of collagen is a triple helix (see Figure 1.93), but not an α-helix, because the high proline content prevents the formation of the α-helix. The three chains in the helix may be identical or different. There are at least 10 different types of collagen found in connective tissue (see Table 1.36), with types I–III having the ability to form fibers called *fibrils*. Type I collagen is the principal structural component of most tissue. Type II and III collagens

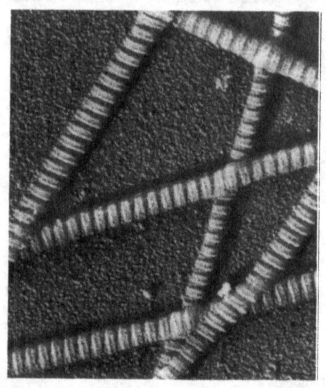

Figure 1.93 Electron photomicrograph of collagen. Reprinted, by permission, from *Chemistry of Advanced Materials*, L. V. Interrante and M. J. Hampden-Smith, editors, p. 507. Copyright © 1998 by Wiley-VCH, New York.

Table 1.36 Types of Collagen Found in Tissue[a]

Collagen Type	Tissue or Organ	Location
I	Tendon, skin, bone and fascia	Thick extracellular fibrils and fibers
II	Cartilage	Thin fibrils around cartilage cells
III	Cardiovascular tissue	Intermediate-size extracellular fibrils
IV	Basement membranes	Network-forming component
V	Tendon, skin and cardiovascular tissue	Pericellular matrix around cells
VI	Cardiovascular tissue, placenta, uterus, liver, kidney, skin, ligament and cornea	Extracellular matrix
VII	Skin	Anchoring fibrils
VIII	Cardiovascular tissue	Around endothelial cells
IX	Cartilage	Extracellular matrix
X	Cartilage	Extracellular matrix
$1\alpha, 2\alpha, 3\alpha$	Cartilage	Extracellular matrix

[a] See G. R. Martin, R. Timpl, R. K. Muller, and K. Kuhn, *Trends Biochem. Sci.*, **9**, 285 (1985).

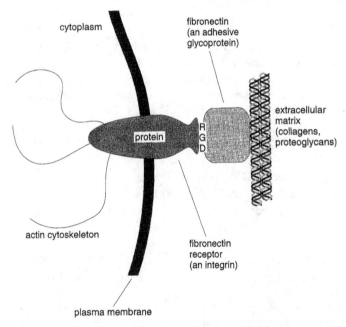

Figure 1.94 The function of integrin. From H. R. Matthews, R. Freedland, and R. L. Miesfeld, *Biochemistry: A Short Course*. Copyright © 1997 by John Wiley & Sons, Inc. This material is used by permission of John Wiley & Sons, Inc.

are found in cartilage (II) and cardiovascular tissue (III), among other places. Type IV collagen is found in basement membranes, which are sheet-like structures found beneath epithelial cells or blood vessel linings (*endothelial cells*).

Elastin is a protein also found in connective tissue that imparts an ability for these tissues to undergo large shape and size changes without permanent damage to the tissue. Elastin is the amorphous component (up to 90%) of elastic fibers that are found in the extracellular matrix of most tissue. For example, the aorta has an elastin content as high as 30–60% of the dry weight—higher than that of any other tissue. Skin, in contrast, contains less than 5% elastin by weight. Elastin is similar to collagen in that it is composed of about 33% glycine and 13% proline, but it contains no hydroxyproline, and contains 10–14% valine and 21–24% alanine. The later two amino acids are nonpolar and do not form hydrogen bonds with water molecules. There are a number of crosslinks between the elastin chains, as one would expect for an elastic substance.

Fibronectin and *laminin* are adhesion proteins responsible for linking the outer surface of cells to collagens and other components in the extracellular matrix. Both fibronectin and laminin are *glycoproteins*—proteins that contain polysaccharide residues. Fibronectin binds to cells through a *tripeptide sequence* (–Arg–Gly–Glu–) called *RGD*, which binds to the *fibronectin receptor*, one of a family of cellular transmembrane proteins called *integrins* (see Figure 1.94). Integrins link the extracellular matrix to the cytoskeleton. We will discuss integrins in more detail in Chapter 3.

REFERENCES

Cited References

1. *Index of Polymer Trade Names*, Fachinformationszentrum Chemie GmbH, Berlin, 1987.
2. www.matweb.com/search/SearchTradeName.asp
3. Sung, Y. M., K.-Y. Yon, S. A. Dunn, and J. A. Koutsky, Wetting behavior and mullite formation at the interface of inviscid melt-spun CaO–Al$_2$O$_3$ fibre-reinforced Al–Si (4032) composite, *J. Mater. Sci.*, **29**, 5583–5588 (1994).
4. Suchanek, W., and M. Yoshimura, Processing and properties of hydroxyapatite-based biomaterials for use as hard tissue replacement implants. *J. Mater. Res.*, **13**(1), 94 (1998).
5. de Jong, W. F., *Rec. Trav. Chem. Pays-Bas*, **45**, 445 (1926).
6. Mehmen, M., *Z. Kristallogr.*, **75**, 323 (1930).
7. St. Naray-Szabo, *Z. Kristallogr.*, **75**, 387 (1930).
8. Posner, A., A. Perloff, and A. F. Diorio, *Acta Crystallogr.*, **11**, 308 (1958).
9. Narasaraju, T. S., and D. E. Phebe, Some physico-chemical aspects of hydroxyapatite, *J. Mater. Sci.*, **31**, 1 (1996).
10. Ritter, S. K., Boning up, *Chem. Eng. News*, August 25, 1997, p. 27.
11. Lavernia, C., and J. M. Schoenung, Calcium phosphate ceramics as bone substitutes, *Ceram. Bull.*, **70**(1), 95 (1991).
12. Kelsey, D. J., G. S. Springer, and S. B. Goodman, Composite implant for bone replacement, *J. Compos. Mater.*, **31**(16), 1593 (1997).
13. Dee, K. C., and R. Bizios, Proactive biomaterials and bone tissue engineering, *Biotech. Bioeng.*, **50**, 438 (1996).
14. Mansur, C., M. Pope, M. R. Pascucci, and S. Shivkumar, Zirconia-calcium phosphate composite for bone replacement, *Ceram. Int.*, **24**, 77 (1998).

General

Wyckoff, R. W. G., *Crystal Structures*, 2nd ed., Interscience, New York, 1963.

Ralls, Kenneth M., T. H. Courtney, and J. Wulff, *Introduction to Materials Science and Engineering*, John Wiley & Sons, New York, 1976.

Jastrzebski, Z., *The Nature and Properties of Engineering Materials*, 2nd ed., John Wiley & Sons, New York, 1976.

Barrett, Craig R., W. D. Nix, and A. S. Tetelman, *The Principles of Engineering Materials*, Prentice-Hall, New York, 1973.

Callister, William D., *Materials Science and Engineering, An Introduction*, 5th ed., John Wiley & Sons, New York, 2000.

Handbook of Industrial Materials, 2nd ed., Elsevier, Oxford, 1992.

Encyclopedia of Chemical Technology, H. Mark et al., eds., John Wiley & Sons, New York, 1972.

Treatise on Materials Science and Technology, Vols, 1–19, Academic Press, New York.

Taylor, G. D., *Construction Materials*, Longman Scientific, Essex, 1991.

Materials Chemistry, L. V. Interrante, L. A. Casper, and A. B. Ellis, eds., ACS Advances in Chemistry Series, Volume 245, American Chemical Society, Washington, D.C., 1995.

Materials Handbook, G. S. Brady and H. R. Clauser, eds., 13th ed., McGraw-Hill, New York, 1991.

Amato, I., *Stuff—The Materials The World Is Made Of*, Basic Books, New York, 1997.

Metals

Darkin, L. S. and R. W. Gurry, *Physical Chemistry of Metals*, McGraw-Hill, New York, 1953.

Metals Handbook, 9th ed., ASM Handbook Committee, W. H. Cubberly, director, American Society for Metals, Metals Park, OH, 1978.

Ceramics

Kingery, W. D., H. K. Bowen, and D. R. Uhlmann, *Introduction to Ceramics*, 3rd ed., John Wiley & Sons, New York, 1993.

Somiya, S. *Advanced Technical Ceramics*, Academic Press, New York, 1989.

Yanagida, H, K. Koumoto, and M. Miyayama, *The Chemistry of Ceramics*, John Wiley & Sons, New York, 1996.

Glass

Doremus, R. H., *Glass Science*, John Wiley & Sons, New York, 1973.

Scholze, H., *Glass—Nature, Structure and Properties*, Springer-Verlag, New York, 1991.

Morey, G. W., *The Properties of Glass*, 2nd ed., Reinhold, New York, 1954.

Bansal, N. P., and R. H. Doremus, *Handbook of Glass Properties*, Academic Press, Orlando, FL, 1986.

Experimental Techniques of Glass Science, C. J. Simmons and O. H. El-Bayoumi, eds., American Ceramic Society, Westerville, OH, 1993.

Polymers

Billmeyer, F. W., *Textbook of Polymer Science*, 3rd ed., John Wiley & Sons, New York, 1984.

Encyclopedia of Polymer Science and Engineering, Herman F. Mark, et al., eds., John Wiley & Sons, New York, 1985.

Hiemenz, P., *Polymer Chemistry*, Marcel Dekker, New York, 1984.

Rodriguez, F., *Principles of Polymer Systems*, 2nd ed., McGraw-Hill, New York, 1982.

Tadokoro, H., *Structure of Crystalline Polymers*, Krieger, Malabar, FL, 1990.

Liquid Crystalline Polymers

Cser, F., Relationship between chemistry and properties of liquid crystalline polymers, *Mater. Forum*, **14**, 81–91 (1990).

Chung, T.-S, The recent developments of Thermotropic Liquid Crystalline Polymers, *Polym. Eng. Sci.*, **26**(13), 901–919 (1986).

Goodby, J. W., Melting phenomena and liquid-crystalline behavior, *Chemlog Highlights*, **11**, 3–7 (1987).

Fergason, J. L., Liquid Crystals, *Sci. Am.*, **211**(2), 76 (1964).

Composites

Composite Materials Handbook, M. Schwartz, ed., 2nd ed., McGraw-Hill, New York, 1984.

Concise Encyclopedia of Composite Materials, A. Kelly, ed., Pergamon, Elmsford, New York, 1994.

Suresh, S., and A. Mortensen, *Fundamentals of Functionally Graded Materials: Processing and Thermomechanical Behavior of Graded Metals and Metal–Ceramic Composites*, Ashgate Publishing Co., Brookfield, VT (1999).

Biologics

Matthews, H. R., R. Freedland, and R. L. Miesfeld, *Biochemistry: A Short Course*, John Wiley & Sons, New York, 1997.

Schumm, D. E., *Essentials of Biochemistry*, 2nd ed., Little, Brown & Co., Boston, 1995.

Houston, M. E., *Biochemistry Primer for Exercise Science, Human Kinetics*, Champaign, IL, 1995.

Silver, F. H., *Biological Materials: Structure, Mechanical Properties, and Modeling of Soft Tissues*, New York University Press, New York, 1987.

DeCoursey, R. M., and J. L. Renfro, *The Human Body*, 5th ed., McGraw-Hill, New York, 1980.

PROBLEMS

Level I

1.I.1 Calculate the force of attraction between a K^+ and O^{2-} ion whose centers are separated by a distance of 2.0 nm. Adapted from Callister problem 2.12, page 28 (ionic separation distance changed).

1.I.2 Estimate the % ionic character of the interatomic bonds in the following compounds: TiO_2, ZnTe, CsCl, InSb, and $MgCl_2$. Adapted from Callister problem 2.19, p. 29 (wording changed, but calculations the same).

1.I.3 An amino acid has three ionizable groups, the α-amino and α-carbonyl groups and a side chain that can be positively charged. The pH values are 3, 9 and 11, respectively. Which of the following pH values is nearest to the isoelectric point (the point at which the overall net charge is zero) for this amino acid: 1.1, 5.3, 12.2? Explain your answer.

1.I.4 Calculate the radius of a palladium atom, given that Pd has an FCC crystal structure, a density of 12.0 g/cm^3, and an atomic weight of 106.4 g/mol. Adapted from Callister, problem 3.9 and 3.10, p. 60 (different element and properties).

1.I.5 Cite the indices of the direction that results from the intersection of each of the following pair of planes within a cubic crystal: (a) (110) and (111) planes;

(b) (110) and ($1\bar{1}0$) planes; (c) ($10\bar{1}$) and (001) planes. Callister problem 3.41, p. 64.

1.I.6 (a) Can fully cured Bakelite be ground up and reused? Explain. (b) Can polyethylene be ground up and reused? Explain.

1.I.7 (a) Calculate the molecular weight of polystyrene having $\bar{x} = 100,000$. (b) Calculate the approximate extended chain length of one of the molecules.

1.I.8 In the formaldehyde molecule, H_2CO, a double bond exists between the carbon and oxygen atoms. (a) What type of hybridization is involved? (b) The molecule is found to be planar; one bond between the C and O atoms is a σ bond, and the other is a π bond. With a simple sketch, show the atomic orbital overlap that is responsible for the π bond.

1.I.9 Calculate the energy of vacancy formation in aluminum, given that the equilibrium number of vacancies at $500°C$ is 7.57×10^{23} m^{-3}. State your assumptions. Adapted from Callister, problem 4.3, p. 89 (changed substantially).

1.I.10 Draw an orthorhombic cell, and within that cell draw a $[\bar{2}11]$ direction and a ($0\bar{2}1$) plane.

1.I.11 Which of the following molecules is (are) paramagnetic: $O_2{}^{2+}$; $Be_2{}^{2+}$; $F_2{}^{2+}$?

1.I.12 Estimate the coordination number for the cation in each of these ceramic oxides: Al_2O_3, B_2O_3, CaO, MgO, SiO_2, and TiO_2.

1.I.13 Which ions or atoms of the following pairs have the greatest radius: K/K^+; O/O^{2-}; H/He; Co/Ni; Li/Cl; Li^+/Cl^-; Co^{2+}/Ni^{2+}?

1.I.14 Draw structural formulas comparing starch with cellulose.

1.I.15 Show the centers of positive and negative charge in (i) CCl_4, (ii) $C_2H_2Cl_2$, and (iii) CH_3Cl. Which of these molecules can have two forms?

1.I.16 Which of the following substitutions in an α-helical part of a protein is most likely to affect the function of the protein: Glu→Asp; Lys→Arg; Val→Phe; Ser→Cys; or Gln→Pro?

Level II

1.II.1 A somewhat inaccurate, but geometrically convenient way of visualizing carbon bonding is to consider the carbon nucleus at the center of a tetrahedron with four valence electron clouds extending to corners of the tetrahedron. In this scheme, a carbon–carbon single bond represents tetrahedra joined tip-to-tip, a double bond represents tetrahedra joined edge-to-edge, and a triple bond represents tetrahedra joined face-to-face. Calculate the expected ratio of single, double, and triple bond lengths according to this geometrical interpretation and compare with the measured bond lengths shown below. Comment on your results.

Bond Type	Bond Length (nm)
C–C single bond	0.154
C–C double bond	0.134
C–C triple bond	0.120

1.II.2 A recent article [James, K. and J. Kohn, New biomaterials for tissue engineering, *MRS Bull.*, **21**(11), 22–26 (1996)] describes the use of tyrosine-derived polycarbonates for tissue engineering, specifically as a resorbable substrate for small bone fixation. Three similar polycarbonates (DTE, DTH, and DTO) were considered, all having the general structure shown below where the size of the "pendant" chain (side chain in the circle) can be varied during synthesis. If $x = 2$, the pendant chain contains an ethyl group, and the polymer is called DTE ("E" for ethyl). Similarly, $x = 6$ for DTH ("hexyl") and $x = 8$ for DTO ("octyl"). If the weight average molecular weight for DTH is 350,000, what is its number average degree of polymerization, assuming that it is monodispersed?

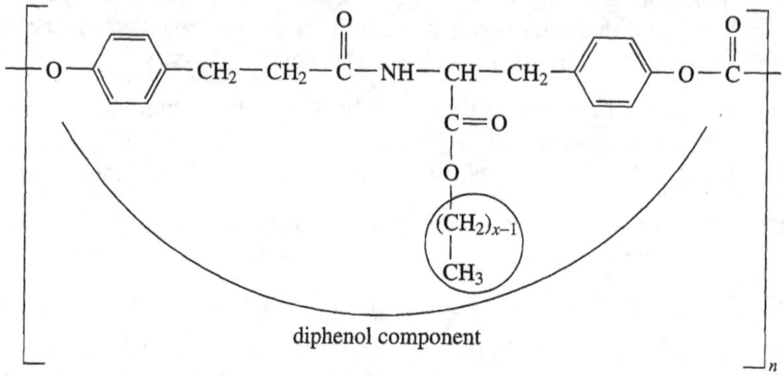

diphenol component

1.II.3 For both FCC and BCC crystal structures, the Burger's vector **b** may be expressed as

$$\mathbf{b} = \tfrac{1}{2}a[hkl]$$

where a is the unit cell length and $[hkl]$ is the crystallographic direction having the greatest linear atomic density. (a) What are the Burger's vector representations for FCC, BCC, and SC structures? (b) If the magnitude of the Burger's vector $|\mathbf{b}|$ is

$$|\mathbf{b}| = \tfrac{1}{2}a(h^2 + k^2 + l^2)^{\frac{1}{2}}$$

determine the values of $|\mathbf{b}|$ for aluminum and tungsten. Callister problem 4.25, p. 90.

1.II.4 Bragg's Law [Eq. (1.35)] is a necessary but not sufficient condition for diffraction by real crystals. It specifies when diffraction will occur for unit cells having atoms positioned only at cell corners. However, atoms situated at other sites (e.g., face and interior positions in FCC or BCC) act as extra scattering centers, which can produce out-of-phase scattering at certain Bragg angles. The net result is the absence of some diffracted beams that, according to Eq. (1.35), should be present. For example, for the BCC crystal structure, $h + k + l$ must be even if diffraction is to occur, whereas for FCC, h, k, and l must all be either odd or even. Use this information to determine the Miller indices for the first five reflections that are present for a single atom BCC and FCC unit cell. The first reflection is defined to be the one closest to $2\theta = 0$. (Contributed by Brian Grady)

1.II.5 Indicate which of the following pairs of metals would not be likely to form a continuous series of solid solutions: Ta–W; Pt–Pb, Co–Ni, Co–Zn, and Ti–Ta. Check your predictions in the *Metals Handbook*.

1.II.6 An article related to acrylic bone cements [Abboud, M. et al., PMMA-based composite materials with reactive ceramic fillers: IV. Radiopacifying particles embedded in PMMA beads for acrylic bone cements, *J. Biomed. Mater. Res.*, **53**(6), 728 (2000)] provides the following information on the PMMA matrix used in these cements: $\overline{M}_w = 295{,}000$; $\overline{M}_w/\overline{M}_n = 2.2$. Calculate the number average degree of polymerization for the PMMA used in this study.

Level III

1.III.1 Al_2O_3 will form a limited solid solution in MgO. At a specific temperature called the "eutectic temperature" (1995°C), approximately 18 wt% of Al_2O_3 is soluble in MgO. Predict the change in density on the basis of (a) interstitial Al^{3+} ions and (b) substitutional Al^{3+} ions.

1.III.2 The three materials listed in the table below are available in either fiber or sheet form. Each material may also be used as a matrix. The individual physical and chemical characteristics listed in the table are independent of geometry.

	Strength (kpsi)	Density (g/cm³)	Oxidation Resistance
Polymer	1	1	Poor
Metal	97	7	Poor
Ceramic	21	3	Excellent

Design a composite that has good oxidation resistance, a density of less than 3.0 g/cm³ and an *isotropic* strength of at least 30 kpsi. You need not use all three materials in your design.

Assume:
- Density is a summation effect; the total density is a weight average of the components.
- Oxidation resistance is a complementary effect.
- Strength is either an interactive or a summation effect, depending on the form of the material. The total strength of the composite is three times the strength of the matrix for one-dimensional fiber orientation in a fiber-matrix composite (FMC). The total strength is two times the strength of the matrix for two-dimensional fiber orientation in an FMC. Three-dimensional (random) fiber orientation, or a nonfibrous composite causes the total strength to be a weight average of all the components.

Describe the form (e.g., fiber, matrix, layer, etc.) of each material in your composite and the weight fraction of each component. Also indicate the composite density and strength. Make a diagram of your composite, indicating the different components and any important features.

1.III.3* As an oxide modifier (such as Na_2O) is added to silica glass, the oxygen-to-silicon ratio increases, and it is empirically observed that the limit of glass formation is reached when O/Si is about 2.5 to 3. Explain, in terms of structure, why a soda–silica mixture such that 2 < O/Si < 2.5 will form a glass, whereas a soda–silica mixture such that O/Si = 3 will crystallize rather than forming a glass.

1.III.4   In an article [M. S. Dresselhaus et al., Hydrogen adsorption in carbon materials, *MRS Bull.*, **24**(11), 45 (1999)] on the storage of molecular hydrogen,
the use of carbon as an economical, safe, hydrogen storage medium for
a hydrogen-fueled transportation system is discussed. Use the following
excerpts from this article to provide answers to the following questions.

"To gain insight into the hydrogen adsorption problem, it is first necessary
to review a few basic facts about hydrogen molecules and the surfaces to
which they might bind. In the ground state, the hydrogen molecule is nearly
spherical ... and the intermolecular interaction between H_2 molecules is
weak. Experimentally, solid hydrogen at 4.2 K forms a hexagonal close-
packed structure, with lattice parameters $a = 3.76$ Å and $c = 6.14$ Å."
(a) What is the axial ratio for the hexagonal cell of solid hydrogen molecules?
(b) What is the theoretical axial ratio for a hexagonal cell? (c) Compare your
answers to parts (a) and (b). What does the difference between them, if any,
mean physically?

The article continues:

"Using purely geometric arguments, we can thus gain a simple geometric estimate for the close-packing capacity of hydrogen molecules above a
plane of graphite. Graphite has a honeycomb structure, with an in-plane lattice parameter, $a_g = 2.46$ Å and an interplanar separation of 3.35 Å. Since
the value of the ... diameter for the hydrogen molecule is greater than a_g,
the closest packing of hydrogen molecules would have to be incommensurate with the (graphite surface). Commensurate H_2 adsorption on a two-
dimensional ... superlattice would yield a lattice constant of $a = 4.26$ Å."
(See figure below.)

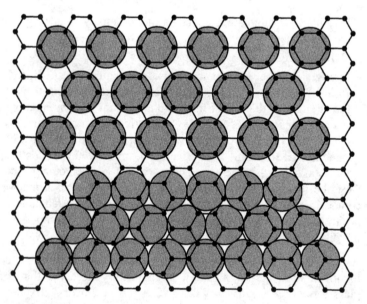

Relative density of a $\sqrt{3} \times \sqrt{3}$ commensurate (top) and an incommensurate (bottom) monolayer
of H_2 on a graphite surface. Reprinted, by permission, from M. S. Dresselhaus, K. A. Williams,
and P. C. Eklund, *MRS Bulletin*, **24**(11), p. 47. Copyright © 1999 by Materials Research Society.

(d) What is the number of nearest neighbors that a hydrogen molecule contained in a single, incommensurate, adsorbed layer on a surface will have? (e) What is the ratio of the number of hydrogen atoms to carbon atoms (H:C) for commensurate packing? (f) What is the weight percentage (g H/g C $\times$ 100) of atomic hydrogen that could be adsorbed onto a graphite surface using commensurate adsorption? (g) At what diffraction angle (2θ) would you expect to see a peak due to the interplanar separation of graphite layers? Assume you are using Cu Kα radiation with wavelength $\lambda = 1.54$ Å. (h) Give at least two reasons why carbon (graphite) would be desirable as a hydrogen storage medium in fuel cells.

CHAPTER 2

Thermodynamics of Condensed Phases

2.0 INTRODUCTION AND OBJECTIVES

Thermodynamics is the foundation of all the engineering and many of the natural science disciplines, and materials engineering and science is no exception. A certain familiarity with the tenants of thermodynamics—First and Second Laws, enthalpy and entropy—will be assumed in this chapter, since you have probably been exposed to them already in chemistry and physics and possibly in an introductory thermodynamics course. A review of these concepts is certainly in order—not only to reacquaint you with them, but to establish certain conventions (sign conventions, which letters represent which quantities) that can vary between disciplines, instructors, and books. As we traverse through this review and begin applying thermodynamic principles to real material systems, you should be delighted to notice that some of the concepts you may have found confusing in other courses (e.g., Carnot cycles, compressibility factors, and fugacities) are not of concern here. There is even little need to distinguish between internal energy and enthalpy. This stems from the fact that we are dealing (almost) exclusively with *condensed systems*—that is, liquids and solids. Though gases can certainly behave as material systems, we will not find much utility in treating them as such.

By the end of this chapter, you should be able to:

- Identify the number of components present, the number of phases present, the composition of each phase, and the quantity of each phase from unary, binary, and ternary phase diagrams—that is, apply the Gibbs Phase Rule.
- Apply the Lever Rule to a two-phase field in a binary phase diagram.
- Identify three-phase reactions in binary component systems.
- Calculate the free energy and heat of mixing for a simple binary mixture.
- Define surface energy, and relate it to thermodynamic quantities such as free energy.
- Apply the Laplace equation to determine the pressure across a curved surface.
- Apply the Young equation to relate contact angle with surface energies.
- Identify the three stages of sintering, and describe how surface energy drives each process.
- Differentiate between binodal and spinodal decomposition in polymer mixtures.

An Introduction to Materials Engineering and Science: For Chemical and Materials Engineers, by Brian S. Mitchell
ISBN 0-471-43623-2 Copyright © 2004 John Wiley & Sons, Inc.

- Differentiate between adhesion, cohesion, and spreading, and calculate the work associated with all three.
- Describe cell–cell adhesion in terms of free energy concepts.

2.0.1 Internal Energy

The letter U will be used to represent the *internal energy* of a system. Recall that the internal energy of a system is the sum of the microscopic kinetic and potential energies of the particles. *Kinetic energy* is the energy due to motion of the particles, including translation, rotation, and vibration. The *potential energy* is the energy due to composition. We saw in Chapter 1 how there is potential energy stored in chemical bonds, and how the potential energy between two atoms or ions changes as a function of separation distance. The quantity ΔU represents the change of internal energy for the system from some initial state to some final state.

The letter E will be used to represent the *total energy* of a system. The quantity ΔE represents the change of total energy from an initial state to a final state. The total energy is comprised of the internal energy, U, the kinetic energy of the system, and the potential energy of the system. Do not confuse the *macroscopic* kinetic and potential energies, ΔE_k, and ΔE_p, respectively, with the *microscopic* kinetic and potential energy contributions to the internal energy just described above. By way of analogy, the former has to do with a rock being pushed off a 10-meter-high ledge, while the latter has to do with the bonds between molecules in the rock and the movement of the individual atoms within the rock.

Already, you should be thinking to yourself "But the particles in solids really don't move that much!" and you are certainly correct. They do move or "translate" in the liquid state of that same solid, however, and don't forget about rotation and vibration, which we will see in subsequent chapters can be very important in solids. But along this line of thinking, we can simplify the First Law of Thermodynamics, which in general terms can be written for a closed system (no transfer of matter between the system and surroundings) as

$$\Delta U + \Delta E_k + \Delta E_p = \pm Q \pm W \tag{2.1}$$

where Q is the *heat* transferred between the surroundings and the system, and W is the *work* performed. The signs used in front of Q and W are a matter of convention. Each is taken as positive when the exchange is from the surroundings to the system—that is, *by* the surroundings *to* the system. This is the convention we will use. This convention is universally accepted for Q, but not so for W. In this way, *exothermic processes* in which heat is transferred from the system to the surroundings have a negative sign in front of Q; and *endothermic processes*, in which heat is transferred from the surroundings to our system, have a positive sign. The sign convention for W is less problematic. We will see that in condensed, closed systems, the work term will not be of interest to us and will be ignored, since it mostly arises from pressure–volume (PV) work. Similarly, the macroscopic kinetic and potential energies, ΔE_k, and ΔE_p, respectively, are not of importance, so that the functional form of the First Law is

$$\Delta U = Q \tag{2.2}$$

or, in differential form for infinitesimal changes of state,

$$dU = dQ \tag{2.3}$$

Both U and Q have units of joules, J, in the SI system, dynes (dyn) in cgs, and calorie (cal) in American engineering units.

2.0.2 Enthalpy

A closely related quantity to the internal energy is the *enthalpy*, H. It, too, has SI units of joules and is defined as the internal energy plus the pressure–volume product, PV. As in most cases, we are concerned with changes in internal energy and enthalpy from one state to another, so that the definition of enthalpy for infinitesimal changes in state is

$$dH = dU + d(PV) \tag{2.4}$$

Substitution of Eq. (2.3) into (2.4) and recognition that for condensed systems the $d(PV)$ term is negligible leads to

$$dH = dQ \tag{2.5}$$

which in integrated form is

$$\Delta H = Q \tag{2.6}$$

2.0.3 Entropy

Entropy will be represented by the letter S. Entropy is a measure of randomness or disorder in a system and has SI units of J/K. Recall that the Second Law of Thermodynamics states that *the entropy change of all processes must be positive*. We will see that the origins of entropy are best described from statistical thermodynamics, but for now let us concentrate on how we can use entropy to describe real material systems.

The differential change in entropy for a closed system from one state to another is, by definition, directly proportional to the change in reversible heat, dQ_{rev}, and inversely proportional to the absolute temperature, T:

$$dS = \frac{dQ_{rev}}{T} \tag{2.7}$$

Primarily reversible processes will be studied in this chapter, so that it is not necessary to retain the subscript on Q. Also, equilibrium transformations will be the primary focus of this chapter, such that we will be concerned mostly with constant-temperature processes. These facts allow us to simplify Eq. (2.7) by integration

$$\Delta S = \frac{Q}{T} \tag{2.8}$$

Substitution of Eq. (2.6) into (2.8) gives as a useful relationship between the entropy and enthalpy for constant pressure and temperature processes:

$$\Delta S = \frac{\Delta H}{T} \tag{2.9}$$

Entropy also plays a role in the Third Law of Thermodynamics, which states that *the entropy of a perfect crystal is zero at zero absolute temperature.*

2.0.4 Free Energy

The most useful quantity for this chapter is the *Gibbs free energy, G*. The Gibbs free energy for a closed system is defined in terms of the enthalpy and entropy as

$$G = H - TS \tag{2.10}$$

which in differential form at constant pressure and temperature is written

$$dG = dH - TdS \tag{2.11}$$

or in the integrated form

$$\Delta G = \Delta H - T\Delta S \tag{2.12}$$

Recall that the important function of the free energy change from one state to another is to determine whether or not the process is *spontaneous*—that is, thermodynamically favored. The conditions under which a process is considered spontaneous are summarized in Table 2.1. The "processes" described here in a generic sense are the topic of this chapter, as are the implications of "equilibrium" to material systems. The issue of "spontaneity" and what this means (or does not mean) to the rate at which a process occurs is the subject of Chapter 3.

HISTORICAL HIGHLIGHT

Josiah Willard Gibbs was born in New Haven, CT on February 11, 1839 and died in the same city on April 28, 1903. He graduated from Yale College in 1858, received the degree of doctor of philosophy in 1863 and was appointed a tutor in the college for a term of three years. After his term as tutor he went to Paris (winter 1866/67) and to Berlin (1867), where he heard the lectures of Magnus and other teachers of physics and mathematics. In 1868 he went to Heidelberg where Kirchhoff and Ostwald were then stationed before returning to New Haven in June 1869. Two years later he was appointed Professor of mathematical physics in Yale College, a position he held until the time of his death. In 1876 and 1878 he published the two parts of the paper "On the Equilibrium of Heterogeneous Substances," which is generally considered his most important contribution to physical sciences. It was translated into German in 1881 by Ostwald and into French in 1889 by Le Chatelier.

Outside his scientific activities, J. W. Gibbs's life was uneventful; he made but one visit to Europe, and with the exception of those three years and of summer vacations in the mountains, his whole life was spent in New Haven. His modesty with regard to his work was proverbial among all who knew him; there was never any tendency to make the importance of his work an excuse for neglecting even the most trivial of his duties, and he was never too busy to devote as much time and energy as might be necessary to any of his students who sought his assistance.

Source: www.swissgeoweb.ch/minpet/groups/thermodict/notes/gibbspaper.html

Table 2.1 **Summary of Free Energy Effects on Process Spontaneity**

$\Delta G < 0$	Process proceeds spontaneously
$\Delta G > 0$	Process not spontaneous
$\Delta G = 0$	Process at equilibrium

2.0.5 Chemical Potential

The final thermodynamic quantity for review is the *chemical potential*, which is represented with the Greek letter mu, μ. The chemical potential can be defined in terms of the partial derivative of any of the previous thermodynamic quantities with respect to the number of moles of species i, n_i, at constant n_j (where j indicates all species other than i) and thermodynamic quantities as indicated:

$$\mu_i = \left(\frac{\partial U}{\partial n_i}\right)_{S,V,n_j} = \left(\frac{\partial H}{\partial n_i}\right)_{P,S,n_j} = \left(\frac{\partial G}{\partial n_i}\right)_{T,P,n_j} \tag{2.13}$$

The advantage of the chemical potential over the other thermodynamic quantities, U, H, and G, is that it is an *intensive quantity*—that is, is independent of the number of moles or quantity of species present. Internal energy, enthalpy, free energy, and entropy are all *extensive variables*. Their values depend on the extent of the system—that is, how much there is. We will see in the next section that intensive variables such as μ, T, and P are useful in defining equilibrium.

2.1 THERMODYNAMICS OF METALS AND ALLOYS

In Chapter 1, the assertion was made that "Elements are materials, too!" We will use this fact to begin our description of equilibrium transformation processes in materials. Not all elements are metals, of course, but most of the elements that are of interest from a materials application standpoint are.

2.1.1 Phase Equilibria in Single-Component Systems

2.1.1.1 Gibbs Phase Rule. The goal of this section section is to predict what will happen to our element when it is subjected to changes in those variables that we can manipulate, usually temperature and pressure. For example, what happens when we heat a sample of pure sulfur? It will probably melt at some point. What happens when we subject carbon to very high pressures? We predict that diamond will form. We seek quantitative explanations of these phenomena and an ability to predict under what conditions they will occur.

Consider a single *component*, A. A component is a chemical constituent (element or compound) that has a specified composition. For simplicity, we will assume that component A is an element, but we will see in subsequent sections that it can be anything that does not undergo a chemical change, including compounds. But for now, we have an element, A, that can exist in two *phases*, which we will designate α and β. A phase is defined as a homogeneous portion of a system that has uniform physical and chemical characteristics. It need not be continuous. For example, a carbonated beverage consists of two phases: the liquid phase, which is continuous, and the gas phase, which is dispersed in the liquid phase as discrete bubbles. For the current discussion, it is easiest to visualize the two phases as a solid and a liquid, respectively, since we are pretty familiar with the processes of melting and solidification, but these phases could be two solids, two liquids, a liquid and a gas, or even a solid and a gas.

Phase α and phase β are in *equilibrium* with one another. What does this mean? First of all, it means that although there is probably an exchange of atoms between the two phases; (i.e., some of the solid phase α is melting to form β, and some of the liquid phase β is solidifying to form α) these processes are occurring at essentially equal rates such that the relative amounts of each phase are unchanged. This is known as a *dynamic equilibrium*. In terms of intensive variables, equilibrium means that

$$T_\alpha = T_\beta \tag{2.14}$$

$$P_\alpha = P_\beta \tag{2.15}$$

$$\mu_\alpha = \mu_\beta \tag{2.16}$$

where T_α and P_α are the temperature and pressure of the solid phase α; T_β and P_β are the temperature and pressure of the liquid phase β, and μ_α and μ_β are chemical potentials of each phase. Thus, we have six intensive quantities that establish equilibrium between the two phases. However, these six variables are not all independent. Changing any one of them can affect the others. This can be shown mathematically by assuming that the chemical potential is a function of temperature and pressure:

$$\mu_\alpha = \mu_\alpha(T_\alpha, P_\alpha) \tag{2.17}$$

$$\mu_\beta = \mu_\beta(T_\beta, P_\beta) \tag{2.18}$$

So, for a one-component system containing two phases in equilibrium, we have three thermodynamic conditions of equilibrium [Eqs. (2.14)–(2.16)] and four unknown parameters, T_α, P_α, T_β, and P_β. If we arbitrarily assign a value to one of the parameters, we can solve for the other three (three equations, three unknowns).

Let us extend this analysis to the general case of C independent, nonreacting components, so that we might arrive at a very useful, general conclusion. Instead of one component, we now have C; and instead of two phases (liquid and solid), we now have an arbitrary number, ϕ. The conditions of equilibrium are now analogous to Eqs. (2.14)–(2.16):

$$T_\alpha = T_\beta = T_\gamma = \ldots T_\phi \tag{2.19}$$

$$P_\alpha = P_\beta = P_\gamma = \ldots P_\phi \tag{2.20}$$

$$\mu_{1\alpha} = \mu_{1\beta} = \mu_{1\gamma} = \ldots \mu_{1\phi}$$

$$\mu_{2\alpha} = \mu_{2\beta} = \mu_{2\gamma} = \ldots \mu_{2\phi} \tag{2.21}$$

$$\vdots$$

$$\mu_{C\alpha} = \mu_{C\beta} = \mu_{C\gamma} = \ldots \mu_{C\phi}$$

There are a total of $(\phi - 1)$ equalities for temperature in Eq. (2.19), $(\phi - 1)$ equalities for pressure in Eq. (2.20), and $(\phi - 1)$ equalities for chemical potential of each of the components, C in Eq. (2.21). There are C equations in (2.21), so the total number of equations is $(C + 2)$. Thus, we have $(\phi - 1)(C + 2)$ total equalities, or restrictions, to be satisfied for equilibrium. The total number of intensive variables (excluding composition) in the system is $\phi(C + 2)$. The difference between the number of intensive variables and the number of independent restrictions is known as the *degrees of freedom* of the system, F. The degrees of freedom are the number of variables (including composition) that must be specified in order for the system to be defined in a strict, mathematical sense. So,

$$\text{degrees of freedom} = \text{number of intensive variables}$$
$$- \text{number of independent restrictions}$$

$$F = \phi(C + 1) - (\phi - 1)(C + 2) \tag{2.22}$$
$$F = C - \phi + 2 \tag{2.23}$$

Equation (2.23) is a very important result. It is known as the *Gibbs Phase Rule*, or simply the "phase rule," and relates the number of components and phases to the number of degrees of freedom in a system. It is a more specific case of the general case for N independent, noncompositional variables

$$F = C - \phi + N \tag{2.24}$$

We utilized only temperature and pressure as independent, noncompositional variables in our derivation ($N = 2$), which are of the most practical importance.

Return now to the case of a single component and two phases, $C = 1$ and $\phi = 2$, so that the phase rule for our element is

$$F = 1 - 2 + 2 = 1 \tag{2.25}$$

This means that when phases α and β coexist at equilibrium, only one variable may be changed independently. For example, if temperature is changed, pressure cannot be changed simultaneously without affecting the balance of equilibrium. In those cases where $F = 2$, both temperature and pressure can be changed without affecting the balance of equilibrium (number of phases present); and when $F = 0$, none of the intensive variables can be changed without altering equilibrium between phases.

2.1.1.2   Unary Phase Diagrams.
The phase rule provides us with a powerful tool for connecting the thermodynamics of phase equilibria with graphical representations of the phases known as *phase diagrams*. Phase diagrams tell us many things, but at a minimum, describe the number of components present, the number of phases present, the composition of each phase, and the quantity of each phase. There are several ways in which this information can be presented. We start with the free energy.

Take, for example, the plot of G versus temperature for elemental sulfur, represented by the bottom diagram in Figure 2.1. We know from experiments and observation that there are four phases we have to consider for sulfur: two solid forms (a low-temperature orthorhombic form, R, and high-temperature monoclinic form, M), liquid (L), and vapor (or gas, V). The lines of G versus T for each phase, which are partially solid and continue on as dashed lines, are constructed at constant pressure using Eq. (2.10),

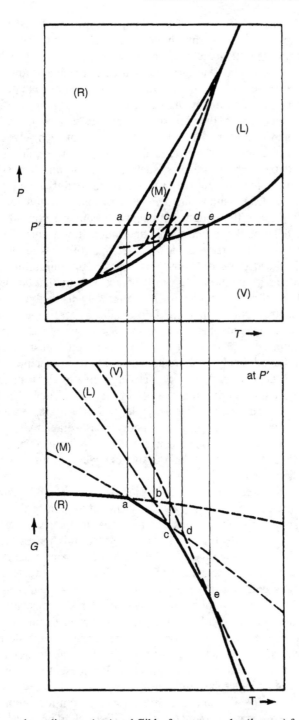

Figure 2.1   Unary phase diagram (*top*) and Gibbs free energy plot (*bottom*) for elemental sulfur. Reprinted, by permission, from D. R. Gaskell, *Introduction to Metallurgical Thermodynamics*, 2nd ed., p. 178, Copyright © 1981 by Hemisphere Publishing Corporation.

where each phase has a different enthalpy and entropy associated with it. The solid line in this figure represents the minimum free energy, regardless of which phase it represents, at a specified temperature. Hence, at this pressure (which is unspecified, but fixed at P'), the orthorhombic form, R, is the most stable at low temperatures. As temperature is increased, monoclinic (M) becomes more stable at point a, followed by the liquid, L, at point c, and finally the vapor phase (V) at point e.

Plots of free energy as a function of temperature can be made at any pressure (in theory), but they are not particularly useful in and of themselves. Free energy is something we cannot generally measure directly, but temperature and pressure are. Hence, plots of two intensive variables, such as pressure and temperature, are much more practical. Such plots are produced by simply translating the information in the G versus T plot at various pressures onto a P versus T plot, as shown in the top diagram of Figure 2.1. Point a in the bottom figure now becomes a point on the solid–solid equilibrium line between the orthorhombic and monoclinic forms of sulfur. Point b is distinctly in the monoclinic phase field, M, since this form has the lowest free energy at this temperature and pressure. Point c is also on an equilibrium phase boundary between the monoclinic and liquid, L, forms of sulfur. Note that this is the melting (or solidification) point of sulfur at pressure P'. Point d is distinctly in the liquid region, and point e is on the liquid–vapor equilibrium line. These plots of pressure versus temperature for a single-component system are called *unary phase diagrams*. There is no limit to the number of phases that may be present on a unary diagram. The only constraint is that there be only one component—that is, no chemical transformations are taking place. Although our discussion here is limited to elements and metals, keep in mind that we are not limiting the number of atoms that make up the component, only the number of components. For example, water is a single component that is composed of hydrogen and oxygen. In a unary phase diagram for water, only the phase transformations of that component are shown—for example, ice, water, and steam. Two hydrogens are always bound to one oxygen. For now, we will continue to limit our discussion to metals.

The power of the phase rule is illustrated in a second example. Consider the $T–P$ phase diagram for carbon shown in Figure 2.2. First of all, notice that pressure is plotted as the independent variable, instead of the usual temperature variable, but given what we know about the effects of pressure on carbon, this makes sense. Let us first examine the equilibrium between two solid forms of carbon: graphite and diamond. This equilibrium is shown graphically by the line A–B in Figure 2.2. Application of Eq. (2.23) to any point on this line, or *phase boundary*, results in $F = 1$ ($C = 1$ component, carbon; $\phi = 2$, graphite and diamond). This means that any change in temperature requires a corresponding change in pressure to maintain this equilibrium. Temperature and pressure are not independent. Similarly, the phase boundaries represented by curves B–C (diamond–liquid), C–D (diamond–metallic carbon), C–E (metallic carbon–liquid), and J–B (graphite–liquid) represent lines of $F = 1$.

Any point within a single-phase region ($\phi = 1$) results in $F = 2$. Point K is such a point. It is located in the diamond *phase field*, where both temperature and pressure can be changed independently without creating or destroying the phase. No equilibrium exists here—there is only one phase.

Since $F = 2$ results in an area (phase field), and $F = 1$ results in a line (phase boundary), we can predict that $F = 0$ should occur at a point, and indeed it does. Point B is such a point ($\phi = 3$; graphite, diamond, and liquid; $C = 1$, carbon), as is

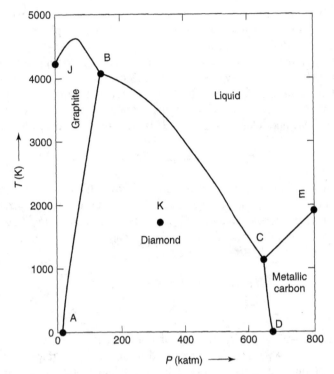

Figure 2.2 Temperature–Pressure unary phase diagram for carbon. From K. M. Ralls, T. H. Courtney, and J. Wulff, *Introduction to Materials Science and Engineering.* Copyright © 1976 by John Wiley & Sons, Inc. This material is used by permission of John Wiley & Sons, Inc.

point C. There are no degrees of freedom at this point—any variation in temperature or pressure will result in movement into a distinctly separate phase field, and at least one of the phases must necessarily be lost. In the cases where the three phases in coexistence are solid, liquid, and vapor, this *invariant point* is known as the *triple point.*

2.1.2 Phase Equilibria in Binary-Component Systems

Most metals of practical importance are actually mixtures of two or more metals. Recall from Section 1.1.3 that these "intimate mixtures" of metals are called alloys, and when the bond between the metals is partially ionic, they are termed *intermetallics.* For the purposes of this chapter, and especially this section, we will not need to distinguish between an intermetallic and an alloy, except to note that when a compound is indicated on a phase diagram (e.g., $CuAl_2$), it indicates an intermetallic compound. We are concerned only with the thermodynamics that describe the intimate mixing of two species under equilibrium conditions. The factors affecting how two metal atoms mix has already been described in Section 1.1.3. Recall that the solubility of one element in another depends on the relative atomic radii, the electronegativity difference between the two elements, the similarity in crystal structures, and the valencies of the two elements. Thermodynamics does not yet allow us to translate these properties of atoms directly into free energies, but these factors are what contribute to the free energy of the metallic mixture and determine whether the species are soluble in one another, whether they form intermetallic compounds, or whether they phase separate into two distinct phases of different composition.

The conditions for equilibrium have not changed, and application of the phase rule is conducted as in the previous section. The difference now is that composition can be counted as an intensive variable. Composition is accounted for through direct introduction into the thermodynamic quantities of enthalpy and entropy. The free energy of a mixture of two pure elements, A and B, is still given by the definition

$$G = H - TS \tag{2.10}$$

but the enthalpy and entropy for the mixture are now taken as a combination of the enthalpies and entropies of components A and B, plus parameters that accounts for the interaction of the two species due to differences in crystal structure, electronegativity, valence, and atom radii, known as the *enthalpy* and *entropy of mixing*, ΔH_{mix} and ΔS_{mix}, respectively:

$$H = X_A H_A^\circ + X_B H_B^\circ + \Delta H_{mix} \tag{2.26}$$
$$S = X_A S_A^\circ + X_B S_B^\circ + \Delta S_{mix} \tag{2.27}$$

Here X_A and X_B are the mole fractions of component A and B, respectively, and are related by $(X_A + X_B) = 1$. We have used a superscript circle on the enthalpies and entropies of pure components A and B to indicate that these are *standard state* enthalpies and entropies of the pure components. The standard state of a component in a condensed system is its stable state at the particular temperature and pressure of interest. So, depending on the temperature and pressure of the system, the standard state could be either a liquid or a solid for either of components A and B.

Substitution of Eqs. (2.26) and (2.27) into Eq. (2.10), and using a similar definition for standard state free energies, leads to the following relationship for the free energy of the mixture

$$G = X_A(H_A^\circ - T S_A^\circ) + X_B(H_B^\circ - T S_B^\circ) + \Delta H_{mix} - T \Delta S_{mix}$$
$$G = X_A G_A^\circ + X_B G_B^\circ + \Delta G_{mix} \tag{2.28}$$

where

$$\Delta G_{mix} = \Delta H_{mix} - T \Delta S_{mix} \tag{2.29}$$

Though Eq. (2.28) is rigorously correct, it contains a rather ill-defined parameter that has been introduced specifically to account for the mixing of the two species, namely, ΔG_{mix}. In order to proceed further, we must explore this quantity and its definition [given in Eq. (2.29)] more thoroughly.

The entropy of mixing can be thought of as a measure of the increase in the number of spatial configurations that become available to the system as a result of the mixing process, ΔS_{conf}, which can be shown with statistical thermodynamic arguments to be

$$\Delta S_{conf} = \Delta S_{mix} = -R(X_A \ln X_A + X_B \ln X_B) \tag{2.30}$$

where R is the gas constant.

Equation (2.30) shows that the entropy of mixing in a binary component solution is dependent only on the composition (relative number of moles of components) in the solution and is independent of temperature.

Similarly, it can be shown that the enthalpy of mixing can be approximated by the internal energy of mixing [see Eq. (2.4)] for condensed phases, which in turn can be related to the mole fractions of the two components, and an *interaction energy*, α, which also has units of joules:

$$\Delta H_{mix} = \alpha X_A X_B \tag{2.31}$$

If $\alpha < 0$, *exothermic mixing* occurs ($\Delta H_{mix} < 0$) due to the fact that an A–B bond is stronger (more negative) than either the A–A or B–B bonds and unlike nearest neighbors are favored in the system. *Endothermic mixing* ($\Delta H_{mix} > 0$) occurs when like nearest neighbors are favored, and A–A and B–B bonds are stronger (more negative) than A–B bonds, resulting in $\alpha > 0$. For an ideal solution, the bond energies of A–A, B–B, and A–B bonds are identical, there is no preference for either atom as a nearest neighbor in solution, and both the interaction energy and the heat of mixing are zero:

$$\alpha^{ideal} = \Delta H_{mix}^{ideal} = 0 \tag{2.32}$$

The free energy of mixing is found by substituting (2.31) and (2.30) into (2.29) to obtain

$$\Delta G_{mix} = \alpha X_A X_B + RT(X_A \ln X_A + X_B \ln X_B) \tag{2.33}$$

which is then substituted into (2.28) for ΔG_{mix} to obtain

$$G^{reg} = X_A G_A^\circ + X_B G_B^\circ + \alpha X_A X_B + RT(X_A \ln X_A + X_B \ln X_B) \tag{2.34}$$

Solutions whose free energies follow Eq. (2.34) are said to be *regular*, to distinguish them from irregular solutions, wherein the entropy of mixing is governed by a relationship other than Eq. (2.30). The free energy for a mixture that behaves as an *ideal solution* ($\alpha = 0$) reduces to

$$G^{ideal} = X_A G_A^\circ + X_B G_B^\circ + RT(X_A \ln X_A + X_B \ln X_B) \tag{2.35}$$

Keep in mind that at this point, the mixtures can be either solid or liquid, depending on the temperature, and that both solid and liquid mixtures may coexist at certain temperatures and compositions. Thus, when performing actual free energy calculations using Eqs. (2.34) or (2.35), the standard state free energies for both components must be carefully selected. This process is illustrated in Example Problem 2.1.

The conditions for equilibrium between two phases in coexistence are still the same as for the case of a single component; that is, Eqs. (2.19)–(2.21) still apply. The phase rule is also still applicable. Before applying these principles to a binary system, let us first examine how a binary phase diagram can be constructed from the free energy equations.

2.1.2.1 Binary Phase Diagrams.
Just as in the case of the unary phase diagram in the previous section, we can construct graphical representations of the equilibrium phases present in a binary component system by first plotting, with the assistance of Eq. (2.33), free energy of mixing versus composition for a system consisting of two

Example Problem 2.1

Cesium, Cs, and rubidium, Rb, form ideal solutions in the liquid phase, and regular solutions in the solid phase. Their standard state Gibbs free energy changes of melting as a function of temperature, $\Delta G^\circ_{m,Cs} = (G^\circ_{Cs,L} - G^\circ_{Cs,S})$ and $\Delta^\circ_{m,Rb} = (G^\circ_{Rb,L} - G^\circ_{Rb,S})$ respectively, are given by the following empirical relationships (in J)

$$\Delta G^\circ_{m,Cs} = 2100 - 6.95T \qquad \Delta G^\circ_{m,Rb} = 2200 - 76.05T$$

1. Determine the melting temperatures for Cs and Rb.
2. Determine the relationship for Gibbs free energy of mixing as a function of composition at 9.7°C (282.7 K) for both liquid and solid solutions.

Solution

1. The melting temperatures are found from the free energy expressions by setting them equal to zero, since for each pure component, the liquid and solid free energies must be equal for both phases to co-exist at the melting temperature. For Cs

$$\Delta G^\circ_{m,Cs} = 2100 - 6.95T_{m,Cs} = 0$$

$$T_{m,Cs} = 302 \text{ K} = 29.0°\text{C}$$

Similarly, $T_{m,Rb} = 312$ K (38.9°C).

2. First, we note that the temperature under consideration (282.7 K) is below the melting point of both elements, so that the solid-state free energies of both components should be selected as the reference state for our free energy calculations. For the liquid solution, which is ideal, the free energy change of mixing relative to the solid components is found by subtracting the weighted average (in terms of mole fraction) of the standard state free energies of the components from the free energy of the ideal mixture, as given by Eq. (2.35)

$$\Delta G^{ideal}_L = [X_{Cs}G^\circ_{Cs,L} - X_{Rb}G^\circ_{Rb,L} + RT(X_{Cs}\ln X_{Cs} + X_{Rb}\ln X_{Rb})]$$
$$- [X_{Cs}G^\circ_{Cs,S} - X_{Rb}G^\circ_{Rb,S}]$$

which, upon combining terms and using the definitions of free energy changes of melting, simplifies to

$$\Delta G^{ideal}_L = X_{Cs}\Delta G^\circ_{m,Cs} + X_{Rb}\Delta G^\circ_{m,Rb} + RT(X_{Cs}\ln X_{Cs} + X_{Rb}\ln X_{Rb})$$

Substitution of the empirical free energy relationships provided for each component, simplification using $X_{Cs} + X_{Rb} = 1$, and evaluation at $T = 282.7$ K gives

$$\Delta G^{ideal}_L = 205X_{Rb} + 133(1 - X_{Rb}) + (8.314)(282.7)$$
$$[(1 - X_{Rb})\ln(1 - X_{Rb}) + X_{Rb}\ln X_{Rb}]$$

This is the free energy change of mixing for the liquid as a function of composition at 282.7 K.

A similar evaluation for the solid solution starts with Eq. (2.34), since the solid solution is regular and not ideal. This equation simplifies rapidly, since the standard state free energies of the components are in the solid phase in both the separate and mixed states, such that

$$\Delta G_S^{reg} = \alpha X_A X_B + RT(X_A \ln X_A + X_B \ln X_B)$$

which, when evaluated at 282.7 K, becomes

$$\Delta G_S^{reg} = \alpha X_{Rb}(1 - X_{Rb}) + (8.314)(282.7)[X_{Rb} \ln X_{Rb} + (1 - X_{Rb}) \ln(1 - X_{Rb})]$$

Cooperative Learning Exercise 2.1

Using the data in Example Problem 2.1 and assuming that $\alpha = 668$ J, determine the composition of a mixture of Cs and Rb that will melt *congruently* at 282.7 K—that is, will go from a solid solution to a liquid solution without changing composition (going through phase separation).

Person 1: Calculate the Gibbs free energy of mixing for the liquid solution from $X_{Rb} = 0$ to 1, in increments of 0.1. (*Note*: You will have to use approximations for $X_{Rb} = 0.0$ and 1.0. Why?)

Person 2: Calculate the Gibbs free energy of mixing for the solid solution from $X_{Rb} = 0$ to 1, in increments of 0.1. (*Note*: You will have to use approximations for $X_{Rb} = 0.0$ and 1.0. Why?)

Plot both of your values on one graph, and estimate the value of X_{Rb} at which the free energies of the liquid and solution solutions are equivalent.

Answer: $X_{Rb} \approx 0.45$. This is also easily accomplished using a spreadsheet and equation solver.

---

pure elements A and B. Such a plot is shown in Figure 2.3. It should be noted that the free energy of mixing is selected as the ordinate in these plots, since we are concerned only with the change in free energy that results when the two components are taken from separate, standard states into the mixture. The mixture may be either a liquid or a solid.

Figure 2.3 requires careful consideration. Figures 2.3a through 2.3e are free energy of mixing curves as a function of concentration (here plotted as X_B, where $X_B = 1.0$ at pure B and $X_B = 0$ at pure A) at decreasing temperatures ($T_5 > T_4 > T_3 > T_2 > T_1$). Temperatures T_1 and T_2 are selected such that they are at or below the melting point (T_B) of the lowest melting point species, B. Temperatures T_4 and T_5 are selected such that they are at or above the melting point (T_A) of the higher melting species, A. Temperature T_3 is between the melting points of the individual components. Figure 2.3f results from application of the equilibrium criteria at all temperatures, including T_1 through T_5. Notice that the binary-component phase diagram is a plot of temperature versus composition, and not pressure versus temperature as was the case for a unary phase diagram. This has to do with the relative importance of our intensive variables.

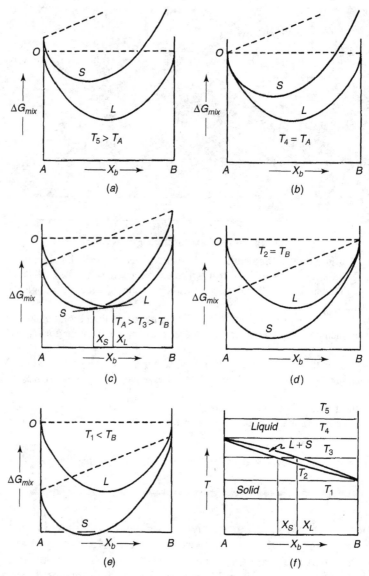

Figure 2.3 Free energy of mixing curves for solid and liquid phases at various temperatures (a–e) and resulting temperature–composition phase diagram for a completely soluble binary component system (f). From O. F. Devereux, *Topics in Metallurgical Thermodynamics*. Copyright © 1983 by John Wiley & Sons, Inc. This material is used by permission of John Wiley & Sons, Inc.

For the unary diagram, we only had one component, so that composition was fixed. For the binary diagram, we have three intensive variables (temperature, pressure, and composition), so to make an $x-y$ diagram, we must fix one of the variables. Pressure is normally selected as the fixed variable. Moreover, pressure is typically fixed at 1 atm. This allows us to plot the most commonly manipulated variables in a binary component system: temperature and composition.

In Figure 2.3a, at a temperature T_5, the liquid phase free energy of mixing is lower than the solid phase free energy of mixing at all compositions. This means that the liquid is more stable at this temperature for all compositions, and indeed this is the case in Figure 2.3f. Any composition at temperature T_5 will be located in the single-phase, liquid region of the phase diagram. A similar situation applies to temperature T_4, shown in Figure 2.3b. Any changes in composition of the system at these temperatures e.g., adding more B to the mixture) will be reflected in a commensurate change in composition of the liquid phase. At temperatures T_2 and T_1, Figures 2.3d and 2.3e, respectively, the solid solution is the most stable at all compositions. Since A and B mix completely at all compositions, any change in composition at these temperatures e.g., adding more A to the mixture) will result in a corresponding change in the solid alloy.

The situation shown in Figure 2.3c is a little more complex. Here, the solid phase has the lowest free energy at A-rich compositions ($X_B < X_S$), and the liquid phase has the lowest free energy at B-rich compositions ($X_B > X_L$). In between the two minima of the solid and liquid free energy curves, $X_S < X_B < X_L$, both the liquid and solid phases can coexist. Why is this? Recall that the stability criteria are as follows for a two-phase system:

$$T_\alpha = T_\beta \tag{2.14}$$

$$P_\alpha = P_\beta \tag{2.15}$$

$$\mu_\alpha = \mu_\beta \tag{2.16}$$

Pressure has already been fixed, as has temperature for the diagram in Figure 2.3c. The only criterion of concern here is that the chemical potentials of the two phases be equal. Recall that the chemical potential is defined as the derivative of the free energy with respect to composition

$$\mu_i = \left(\frac{\partial G}{\partial n_i}\right)_{T,P,n_j} \tag{2.13}$$

or, for the two phases under consideration here,

$$\mu_S = \left(\frac{\partial \Delta G_{mix,S}}{\partial X_B}\right)_{T,P,n_L} = \mu_L = \left(\frac{\partial \Delta G_{mix,L}}{\partial X_B}\right)_{T,P,n_S} \tag{2.36}$$

Graphically, this condition can only occur when the two free energy of mixing curves share the same tangent, which represents the derivative of the curve at that point. This tangent is shown in Figure 2.3c. This is the situation that exists at all temperatures between T_4 and T_2, so there exists a two-phase region at all compositions between these two temperatures, as reflected in Figure 2.3f. Recall that the minima of the two free energy of mixing curves at any given temperature are located by taking the second derivative and setting it equal to zero:

$$\left(\frac{\partial^2 \Delta G_{mix,S}}{\partial X_B^2}\right) = 0, \qquad \left(\frac{\partial^2 \Delta G_{mix,L}}{\partial X_B^2}\right) = 0 \tag{2.37}$$

The locus of points generated by these conditions at various temperatures constructs the phase boundary lines between the solid and liquid phases shown in Figure 2.3f.

That is, there is a different solid and liquid composition, X_S and X_L, resulting from the tangent between the free energy of mixing curves, at all temperatures between T_2 and T_4, as illustrated in Figure 2.3c. We will return to the conditions stipulated by Eq. (2.37) in subsequent sections.

So, for this binary solution of components A and B, which mix perfectly at all compositions, there is a two-phase region at which both solid and liquid phases can coexist. The uppermost boundary between the liquid and liquid + solid phase regions in Figure 2.3f is known as the *liquidus*, or the point at which solid first begins to form when a melt of constant composition is cooled under equilibrium conditions. Similarly, the lower phase boundary between the solid and liquid + solid phase regions is known as the *solidus*, or the point at which solidification is complete upon further equilibrium cooling at a fixed composition.

Let us examine a real alloy phase diagram. Figure 2.4 is the binary phase diagram for the system Cu–Ni. First of all, note that the composition is indicated in terms of weight fraction of one of the components, in this case Ni, and not mole fraction. The conversion between the two is straightforward if the molecular weights are known, and weight fraction is used only because it is more useful experimentally than mole fraction.

In terms of the phase rule, we must utilize Eq. (2.24) instead of (2.23), since we have only one noncompositional intensive variable, T (pressure is fixed), so $N = 1$. Application of Eq. (2.24) to points A and C in Figure 2.4 indicates that the liquid and solid regions ($\phi = 1; C = 2$) correspond to $F = 2$, which means that both temperature and composition can be altered independently in these regions. Point B is in the liquid + solid two-phase region ($\phi = 2; C = 2$), so that $F = 1$. Temperature and composition are not independent in this region; any change in temperature necessarily results in a

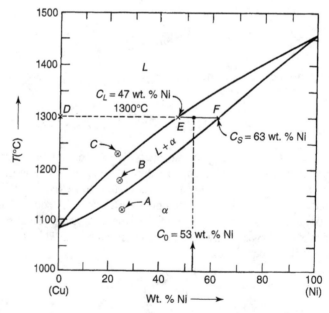

Figure 2.4 The Cu–Ni phase diagram, illustrating the use of the lever rule. From K. M. Ralls, T. H. Courtney, and J. Wulff, *Introduction to Materials Science and Engineering*. Copyright © 1976 by John Wiley & Sons, Inc. This material is used by permission of John Wiley & Sons, Inc.

change in composition of both the liquid and the solid phases. The liquidus and solidus phase boundaries are included in this two-phase region.

More involved phase diagrams result when the two elements, A and B, are only slightly soluble in each other, or when more than one solid alloy phase can be formed from the two components. The principles of generating the phase diagram from free energy of mixing versus composition diagrams still apply, however, as do the application of the stability requirements represented by Eq. (2.36). For example, when A and B are insoluble in the solid phase, a phase diagram such as that shown in Figure 2.5f can result. As in the previous example, the phase boundaries are constructed from the free energy of mixing curves as a function of composition at various temperatures. The primary difference here is that the solid free energy curve is a straight line instead of a curve with a minimum, such that at intermediate temperatures, as in Figures 2.3c and 2.3d, tangents can be drawn between the free energy curves of the pure solids (which occur at the pure component free energies) and the liquid solution, sometimes involving both pure components, as in Figure 2.3d. When the free energy of the pure solids segment is the tangent to the liquids solution (located exactly at the minimum of the liquid solution free energy curve), an interesting phenomenon occurs. It is at this temperature–composition combination, called the *eutectic point*, at which the liquid, solid A, and solid B coexist. According to the phase rule, there should be zero degrees of freedom, and indeed this is the case ($\phi = 3, C = 2, N = 1$). We will learn more about the eutectic and other three-phase transformations in section 2.1.2.3.

Another typical binary phase diagram that results when two different solid solutions of A and B can form, here labeled α and β, is shown in Figure 2.6f. Again, the phase diagram is constructed from the free energy of mixing as a function of concentration curves. In this case, it is always possible to draw a tangent between the free energy curves of the α and β phases at low temperatures, and in some cases, one tangent will be sufficient to touch all three curves for the α, β and liquid phases, as in Figures 2.6d and 2.6e. Note that unlike the binary component system in Figure 2.5 in which A and B were never soluble in the solid phase, in this system, A and B are always soluble in each other to some extent. Thus, the composition of the solid α phase will vary, as will that of the β phase. This can be thought of as each major component having an ability to dissolve small amounts of the impurity phase, either as interstitial or substitutional impurities (see Section 1.1.3). This phase diagram also has a eutectic point, found at the intersection of the X_L composition line and the T_1 isotherm.

Close examination of a variety of binary component phase diagrams allows us to draw a number of generalizations regarding the spatial relationship between phases in a diagram. These are summarized in Table 2.2.

2.1.2.2 The Lever Rule.

The phase rule tells us that there is only one degree of freedom in a two-phase region of a binary component phase diagram—for example, the solid + liquid region of the Cu–Ni phase diagram in Figure 2.4. If we change temperature in this region, the compositions of the two phases must also change. This means that the relative amounts of the two phases must, in turn, be adjusted. If there is an equal amount (in terms of weight) of the solid and liquid phase, regardless of their compositions, removing some of component B from the liquid and moving it to the solid reduces the weight of the liquid phase and increases the weight of the solid phase. The relationship between composition of the two phases and the relative amount

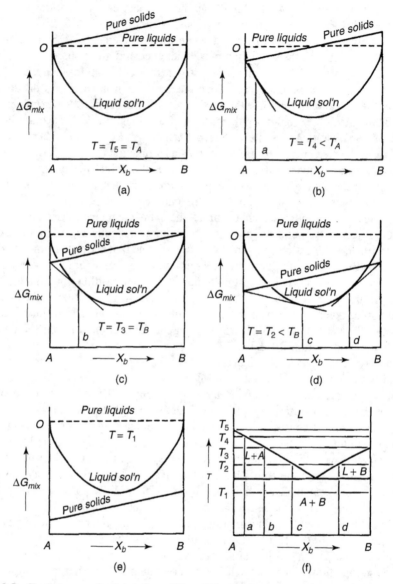

Figure 2.5 Free energy of mixing for solid and liquid phases at various temperatures (a–e) and resulting temperature–composition phase diagram for a slightly soluble eutectic binary component system (f). From O. F. Devereux, *Topics in Metallurgical Thermodynamics*. Copyright © 1983 by John Wiley & Sons, Inc. This material is used by permission of John Wiley & Sons, Inc.

of each phase in a phase diagram is given by the *lever rule*. We will develop the lever rule by way of example.

Consider the initial alloy composition $C_0 = 53$ wt% Ni, 47 wt% Cu indicated by the vertical arrow and dashed line in Figure 2.4. At 1300°C, this composition will exist as both liquid and solid (called the α phase), according to the two-phase region in the diagram. Let W_α and W_L be the fractional amounts by weight of solid and liquid,

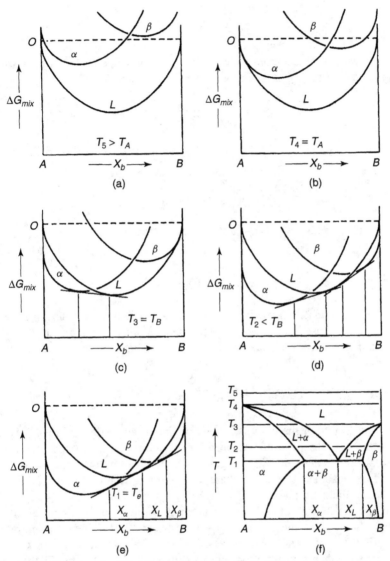

Figure 2.6   Free energy of mixing for solid and liquid phases at various temperatures (a–e) and resulting temperature–composition phase diagram for a mostly insoluble binary component system (f). From O. F. Devereux, *Topics in Metallurgical Thermodynamics.* Copyright © 1983 by John Wiley & Sons, Inc. This material is used by permission of John Wiley & Sons, Inc.

Table 2.2   Summary of Phase Relationships in a Phase Diagram

- One-phase regions may touch each other only at single points, never along boundary lines.
- Adjacent one-phase regions are separated from each other by two-phase regions involving the same two phases.
- Three two-phase regions must originate upon every three-phase isotherm; that is, six boundary lines must radiate from each three phase reaction horizontal.
- Two three-phase isotherms may be connected by a two-phase region provided that there are two phases which are common to both of the three-phase equilibria.

respectively, such that

$$W_\alpha + W_\beta = 1 \tag{2.38}$$

Also, the mass of each component, in this case Ni and Cu, distributed between the two phases must equal their masses in the original composition:

$$W_\alpha C_S + W_\beta C_L = C_0 \tag{2.39}$$

Solution of Eqs. (2.38) and (2.39) gives the weight fraction of liquid in terms of the compositions

$$W_\beta = \frac{C_S - C_0}{C_S - C_L} \tag{2.40}$$

and the weight fraction of solid in terms of compositions

$$W_\alpha = \frac{C_0 - C_L}{C_S - C_L} \tag{2.41}$$

Equations (2.40) and (2.41) are the lever rule and can be used to determine the relative amounts of each phase in any two-phase region of a binary component phase diagram. For the example under consideration, the amount of liquid present turns out to be

$$W_\beta = \frac{63 - 53}{63 - 47} = 0.625 \tag{2.42}$$

So 62.5% by weight of the alloy is present as liquid, and 37.5% is solid. The compositions of the two phases are determined by extending the horizontal line, or *tie line*, to the phase boundaries (line EF in Figure 2.4), then reading the compositions of these phases off the abscissa. The liquid phase has a composition of 47 wt% Ni/53 wt% Cu (point E), and the solid phase is 63 wt% Ni/37% Cu (point F).

There are a few important points regarding application of the lever rule. First, it can only be used in the form of Eqs. (2.40) and (2.41) when concentrations are expressed in terms of weight percentages. Mole percentages must first be converted to weight percents when concentrations are so expressed. Second, Eqs. (2.40) and (2.41) can be used by simply measuring the distances of the corresponding line segments directly off the diagram. For example, the line EF in Figure 2.4 represents the quantity $C_S - C_L$. If it is measured using a ruler in, say, millimeters, then $C_0 - C_L$ could be measured similarly, and W_α determined from a ratio of these two measurements. In fact, this method is much more accurate than extrapolating lines from the phase boundaries down to the abscissa (which can introduce error) and then interpolating the composition from markings that may, in some cases, only give markers every 20 wt% or so. Finally, as you begin to work problems with the lever rule, you may find that the ratio of line segments given by Eqs. 2.40 and 2.41 is a little bit counterintuitive. That is, the line segment in the numerator for the weight fraction of solids [Eq. (2.41)] is that *opposite* the solid-liquid + solid phase boundary, and the line segment in the numerator for the liquid fraction [Eq. (2.42)] is opposite the liquid-solid + liquid phase boundary. A little practice will eliminate this confusion.

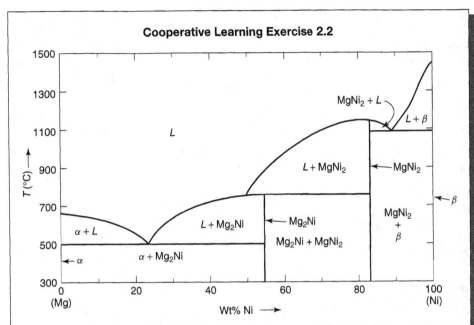

Cooperative Learning Exercise 2.2

Reprinted, by permission, from K. M. Ralls, T. H. Courtney, and J. Wulff, *Introduction to Materials Science and Engineering*, p. 333. Copyright © 1976 by John Wiley & Sons, Inc.

Consider the Mg–Ni binary phase diagram shown above. Note that in this system, intermetallic compound formation is common; for example, Mg_2Ni, $MgNi_2$ are stoichiometric compounds.

Person 1: Determine the phases present, the relative amounts of each phase, and the composition of each phase when the only *incongruently melting* intermetallic compound in this system is heated to 800°C.

Person 2: Determine the phases present, the relative amounts of each phase, and the composition of each phase when the liquid phase from Person 1's problem is cooled to 600°C.

Answer:
Person 1: 88% L (51 wt% Ni, 49 wt% Mg), 12% $MgNi_2$ (83 wt% Ni, 17 wt% Mg).
Person 2: 14% L (27 wt% Ni, 73 wt% Mg); 86% Mg_2Ni (55 wt% Ni, 45 wt% Mg).

2.1.2.3 *Three-Phase Transformations in Binary Systems.* Although this chapter focuses on the equilibrium between phases in binary component systems, we have already seen that in the case of a eutectic point, phase transformations that occur over minute temperature fluctuations can be represented on phase diagrams as well. These transformations are known as *three-phase transformations*, because they involve three distinct phases that coexist at the transformation temperature. Their characteristic shapes as they occur in binary component phase diagrams are summarized in Table 2.3. Here, the Greek letters α, β, γ, and so on, designate solid phases, and L designates the liquid phase. Subscripts differentiate between immiscible phases of different compositions. For example, L_I and L_{II} are immiscible liquids, and α_I and α_{II} are allotropic solid phases (different crystal structures).

Table 2.3   Common Three-phase Transformations in Condensed Binary Systems

Name of Reaction	Equation	Phase Diagram Characteristic
Monotectic	$L_I \xrightarrow{\text{cooling}} \alpha + L_{II}$	
Monotectoid	$\alpha_1 \xrightarrow{\text{cooling}} \alpha_2 + \beta$	
Eutectic	$L \xrightarrow{\text{cooling}} \alpha + \beta$	
Eutectoid	$\alpha \xrightarrow{\text{cooling}} \beta + \gamma$	
Syntectic	$L_I + L_{II} \xrightarrow{\text{cooling}} \alpha$	
Peritectic	$\alpha + L \xrightarrow{\text{cooling}} \beta$	
Peritectoid	$\alpha + \beta \xrightarrow{\text{cooling}} \gamma$	

Source: K. M. Ralls, T. H. Courtney and J. Wulff, *Introduction to Materials Science and Engineering*, p. 331. Copyright © 1976 by John Wiley & Sons, Inc.

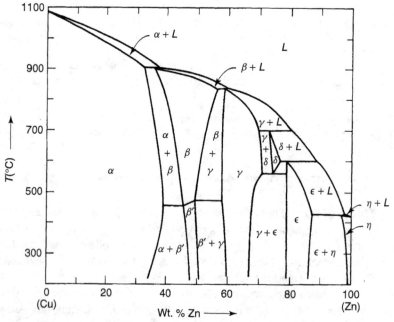

Figure 2.7 Cu–Zn phase diagram, illustrating a number of three-phase reactions. From K. M. Ralls, T. H. Courtney, and J. Wulff, *Introduction to Materials Science and Engineering*. Copyright © 1976 by John Wiley & Sons, Inc. This material is used by permission of John Wiley & Sons, Inc.

The Cu–Zn system (see Figure 2.7) displays a number of *intermediate solid solutions* that arise due to limited solubility between the two elements. For example, at low wt% Zn, which incidently is the composition of alloys known as *brass*, the relatively pure copper α phase is able to accommodate small amounts of Zn as an impurity in the crystal structure. This is known as a *terminal solid phase*, and the solubility limit where intermediate solid solutions (such as $\alpha + \beta$) begin to occur is called the *solvus* line. Some of the three-phase transformations that are found in this diagram include a peritectic ($\delta + L \rightarrow \varepsilon$) and a eutectoid ($\delta \rightarrow \gamma + \varepsilon$). Remember that these three-phase transformations are defined for equilibrium cooling processes, not heating or nonequilibrium conditions.

2.1.3 The Iron–Carbon Phase Diagram*

Iron and its alloys continue to be the most widely utilized metallic systems for construction. The process by which *iron-based alloys*, including *steels*, are produced from iron

---

HISTORICAL HIGHLIGHT

Iron was the first metal for the masses. As far back as five thousand years ago, metalworkers must have somehow gotten their furnaces especially hot on a day when they had charged the furnace with some iron-rich rock, maybe blood-red hematite. Perhaps it was a consistent blast from the sea that made the difference. From this they obtained a silver-gray metal that they had never seen before. It was iron. One of the oldest known iron implements is a dagger blade from 1350 B.C.

The downside of iron is that wrenching it from ore requires a temperature hundreds of degrees higher than that needed to extract copper or tin. In fact, the melting temperature was too high for ancient furnaces, so early iron-workers got their metal by beating it out of a solid "bloom," a conglomeration of metal and rocky slag that forms when the ore is heated to temperatures high enough to loosen things up in the ore but not high enough to melt iron.

The earliest iron turned out to be no better than bronze. Often it was worse, especially for weaponry. Compared to copper and bronze, it was a slightly crazy metal, with unpredictably varying qualities, so a smith never quite knew what to expect from a finished iron implement or weapon—often they came out too soft or too brittle.

The reason for this fickleness is the special role that carbon atoms play in the properties of iron-based metals, including steel. No one knew in the early days of iron that carbon atoms from the burning fuel in their bloomery and smithing furnaces were actually becoming part of the metal's internal anatomy and changing its properties. In the smelting of copper or tin, carbon's role is to remove oxygen without remaining in the metal. The iron that smiths wrought from the blooms and then formed into weapons and implements absorbed varying amounts of carbon. Too little carbon resulted in a consistency closer to that of pure iron, which is soft and malleable. A man wielding a pure iron sword would be no match for an opponent swinging a bronze sword. Too much carbon in the iron, however, yielded a brittle metal that could shatter like pottery.

Most often, smiths ended up with iron containing somewhere around 1 to 3 percent carbon by weight. When lucky, they would get somewhere between .1 and 1 percent carbon. When that happened, the metal underwent a dramatic personality change. It became much harder and tougher, and it could hold a much sharper edge. When a smith managed to keep the carbon content within this range, he made steel.

Source: *Stuff*, I. Amato, pp. 28–29.

ore involves the reduction of iron oxide by means of carbon to form a product known as *pig iron*, which contains approximately 4% carbon. As a result, the iron–carbon phase diagram is an industrially important one. The compounds formed between iron and carbon are sufficiently complex in their microstructures and properties to warrant intensive investigation and more detailed description. However, the iron–carbon phase diagram is really just another binary phase diagram, and, as such, we have all the tools we need to fully understand it.

The iron–carbon phase diagram at low weight percentages of carbon is shown in Figure 2.8. Actually, the phase boundary at 6.69 wt% carbon represents the compound iron carbide, Fe_3C, known as *cementite*, so that the phase diagram in Figure 2.8 is more appropriately that of Fe_3C–C.

The allotropy of elemental iron plays an important role in the formation of iron alloys. Upon solidification from the melt, iron undergoes two allotropic transformations (see Figure 2.9). At 1539°C, iron assumes a BCC structure, called *delta-iron* (δ-Fe). Upon further cooling, this structure transforms to the FCC structure at 1400°C, resulting in *gamma-iron* (γ-Fe). The FCC structure is stable down to 910°C, where it transforms back into a low-temperature BCC structure, *alpha-iron* (α-Fe). Thus, δ-Fe and α-Fe are actually the same form of iron, but are treated as distinct forms due to their two different temperature ranges of stability.

Carbon is soluble to varying degrees in each of these allotropic forms of iron. The solid solutions of carbon in α-Fe, γ-Fe, and δ-Fe are called, respectively, *ferrite*, *austenite*, and *δ-ferrite*. So, for example, the single-phase region labeled as γ in

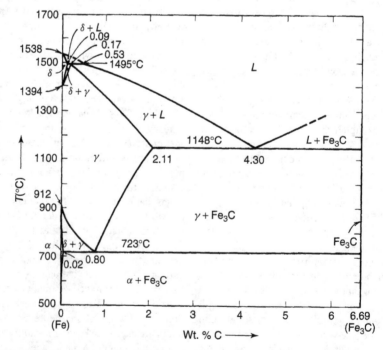

Figure 2.8 The Fe–C phase diagram (low weight % C). From K. M. Ralls, T. H. Courtney, and J. Wulff, *Introduction to Materials Science and Engineering.* Copyright © 1976 by John Wiley & Sons, Inc. This material is used by permission of John Wiley & Sons, Inc.

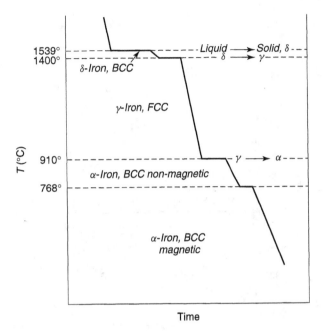

Figure 2.9 Cooling curve for pure iron. Reprinted, by permission, from Committee on Metallurgy, *Engineering Metallurgy*, p. 245. Copyright © 1957 by Pitman Publishing.

Figure 2.8 could be (and is in many diagrams) labeled as the austenite phase region. Similarly, the α and δ phase can be called the ferrite and δ-ferrite regions. Keep in mind, however, that these single-phase regions are simply a solid solution between one of the allotropic forms of elemental iron and carbon. Note that the ferrite region is particularly limited in both temperature and composition, as is δ-ferrite, and that austenite has a particularly large range of stability at higher temperatures. All alloys with carbon contents between 0.5 and 2 wt% C solidify as austenite. Alloys containing more than 2% carbon are subject to a eutectic transformation that forms austenite and cementite. At 0.8% carbon, what has solidified as austenite undergoes a eutectoid transformation at 723°C to an intimate mixture of ferrite and cementite. This mixture is given a special name, *pearlite*. Recall that phases need not be continuous, and such is the case in pearlite where the cementite (Fe_3C) forms lamellar structures within the austenite phase, which can also have some carbon dissolved in it (see Figure 2.10).

The eutectoid transformation is an important one not only for this specific carbon composition, but for classifying all types of steels. *Carbon steels* have carbon contents between 0.1 and 1.5 wt%. Those with carbon contents less than 0.8% are termed *hypoeutectoid steels*, and those with greater than 0.8% C are called *hypereutectoid steels*. Further classifications of steels are given in Table 2.4.

When austenite is cooled under more rapid conditions, a compound called *bainite* is produced. Bainite is a nonequilibrium product that is similar to pearlite, but consists of a dispersion of very small Fe_3C particles between the ferrite plates. Bainite formation is favored at a high degree of supercooling from the austenite phase, whereas pearlite forms at low degrees of supercooling, or more equilibrium cooling.

Finally, if an austenite steel is rapidly quenched, *martensite* steel can form. Steel martensite possesses a body-centered tetragonal structure (see Figure 2.11), which is

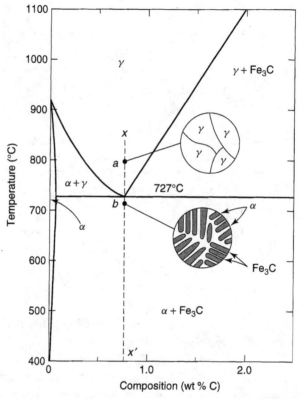

Figure 2.10 Schematic representation of the microstructures for a eutectoid transformation in the Fe–C system. Reprinted, by permission, from W. Callister, *Materials Science and Engineering: An Introduction*, 5th ed., p. 277. Copyright © 2000 by John Wiley & Sons, Inc.

Table 2.4   Classification of Ferrous Alloys

Component (wt%)	Classification	Examples
C < 0.01	Iron	
0.1 < C < 1.5	Carbon steel	1000 series—ferrite/pearlite
		2000 series—ferrite/pearlite or bainite
		3000 series—martensitic 4000 series
2 < C < 4.5	Cast iron	Gray iron
		White iron
Various	Alloy steels	Ni, Mn, Cu, Co, C, N—austenite stabilizers
		Cr, Mo, Si, W, V, Sn, Cb, Ti—ferrite stabilizers
Cr > 12	Stainless steels	200/300—austenitic
		400—ferritic
		400/500—martensitic

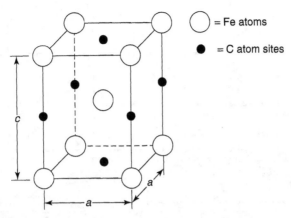

Figure 2.11   The body-centered tetragonal unit cell of steel martensite. From K. M. Ralls, T. H. Courtney, and J. Wulff, *Introduction to Materials Science and Engineering.* Copyright © 1976 by John Wiley & Sons, Inc. This material is used by permission of John Wiley & Sons, Inc.

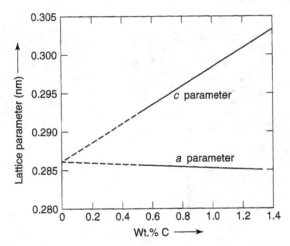

Figure 2.12   Variation of tetragonality in steel martensite with axial ratio. From K. M. Ralls, T. H. Courtney, and J. Wulff, *Introduction to Materials Science and Engineering.* Copyright © 1976 by John Wiley & Sons, Inc. This material is used by permission of John Wiley & Sons, Inc.

only slightly distorted from the BCC structure of austenite. The distortion arises from interstitial carbon atoms that occupy one set of preferred octahedral interstitial sites. During the martensitic transformation upon rapid cooling, carbon atoms residing at equivalent octahedral sites in BCC change position slightly to only one (along the c-axis) of the three possible sets of octahedral sites in the body-centered tetragonal structure, thus creating the distortion. The extent of tetragonal distortion, as measured by the axial ratio, increases with increasing carbon content (see Figure 2.12). The transformation process from austenite to martensite is very rapid, because it involves only a distortion of the BCC unit cell, as opposed to the BCC–FCC transition to pearlite or bainite, which is a diffusion-controlled process. This type of diffusionless transformation is termed a *martensitic transformation* and can occur in alloys other than steel.

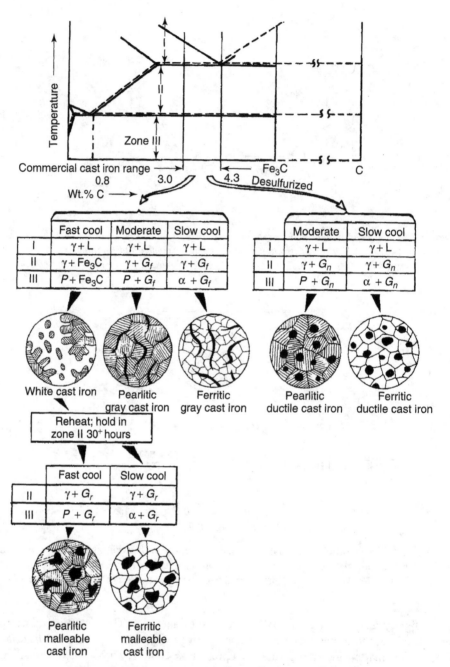

Figure 2.13 Microstructures obtained by varying thermal treatments in cast irons (G_f = graphite flakes; G_r = graphite rosettes; G_n = graphite nodules; P = pearlite). From K. M. Ralls, T. H. Courtney, and J. Wulff, *Introduction to Materials Science and Engineering.* Copyright © 1976 by John Wiley & Sons, Inc. This material is used by permission of John Wiley & Sons, Inc.

As you can see, the process by which the iron–carbon alloy is processed and solidified is just as important as the overall stoichiometry. Although a discussion regarding phase transformations is more the realm of kinetic processes, it is nonetheless pertinent to summarize here the types of important ferrous alloys, particularly those in the *cast iron* categories. This is done in Figure 2.13.

As indicated in Table 2.4, cast irons contain more than 2% carbon by weight, compared to steels which usually have less than 1.5% carbon. Cast irons also differ microstructurally from steels in that a separate graphite phase typically exists. Graphite is seldom found in steels because the solid-state transformations that give rise to it are so slow that transformations yielding cementite always predominate. In cast irons, however, graphite can form directly from eutectic solidification and by solid-state transformations. Whether carbon is present as graphite or is tied up in cementite in cast irons depends on whether the solidification and cooling processes are carried out under close to equilibrium conditions, which favor graphite formation, or under highly nonequilibrium conditions, which favor Fe_3C formation. There is an analogy here with austenite decomposition that we have already described: Pearlite forms under near-equilibrium cooling conditions in preference to the less stable bainite, which forms under rapid cooling conditions. As shown in Figure 2.13, slow cooling rates favor graphite formation from the eutectic reaction. The graphite phase continues to grow with decreasing temperature as the carbon content of the iron-rich phase (γ or α) decreases. Under these conditions, the phases present in zones I, II, and III are austenite and liquid, austenite and graphite, and ferrite and graphite, respectively. Under normal conditions, the graphite is present as flakes, but if the melt is *desulfurized* (where sulfur is an impurity in the ore), the graphite will be present as nodules. The normal structure is called *ferritic gray cast iron*, and that for the desulfurized material is called *ferritic ductile cast iron*.

Under moderate cooling rates, graphite still forms in the eutectic reaction, but Fe_3C is one product of the eutectoid reaction. The final structure, consisting of locally interconnected graphite flakes dispersed in a pearlitic matrix, is called *pearlitic gray cast iron* or, if desulfurized, *pearlitic ductile cast iron*. White cast irons are formed under rapid cooling, and have similar structures to steel insofar as there is no graphite present. Thus, austenite and Fe_3C are present in zone II, and pearlite and Fe_3C are present in zone III. White casts can be converted into more usable structures by heating into zone II for fairly long time periods. This treatment causes graphite, in the shape of *rosettes*, to precipitate and the cementite is eliminated. Keep in mind that the cooling classifications of "slow," "medium," and "rapid," are arbitrary and that commercial processes can produce materials with a mixture of the various microstructures shown in Figure 2.13. We will discuss transformation processes and the effect of heating and cooling rates on them in Chapter 3.

2.2 THERMODYNAMICS OF CERAMICS AND GLASSES

2.2.1 Phase Equilibria in Ternary Component Systems

Three-component systems, or *ternary systems*, are fundamentally no different from two-component systems in terms of their thermodynamics. Phases in equilibrium must still meet the equilibrium criteria [Eqs. (2.14)–(2.16)], except that there may now be as many as five coexisting phases in equilibrium with each other. The phase rule still applies to ternary systems, and we will see that we can use an adaptation of the lever rule to three-component phase diagrams.

As with unary and binary component systems, the three components of a ternary system can be anything, even elements, as in our preceding descriptions. In fact, there are many important alloy systems that consist of three primary components, such as iron–carbon–silicon and aluminum–zinc–magnesium. However, the addition of one important element, oxygen, to any metallic binary component system suddenly transforms the discussion from that which is based primarily in metallurgy, to one which is more the realm of ceramics. For example, the aluminum–silicon binary metal alloy system becomes a ternary system of *aluminosilicates*, and an important class of ceramics, when oxygen is introduced as the third component, and compounds such as Al_2O_3 and SiO_2 can form. Thus, the discussion here will be limited to ceramics, but keep in mind that the principles of ternary phase diagrams can be applied to any three-component system, including three compounds. Again, let us consider only condensed phases.

2.2.1.1 Ternary Phase Diagrams.

In a ternary system, it is necessary to specify temperature, pressure, and two composition parameters to completely describe the system. Typically, pressure is fixed, so that there are three independent variables that are needed to fix the system: temperature and two compositions. The third composition is, of course, fixed by the first two. We could create a three-dimensional plot with three mutually perpendicular axes, as is usually the case in mathematics; however, it is more convenient, and graphically more appealing, to establish two compositional axes 60° apart from each other, with a third, redundant compositional axis, as in the form of an equilateral triangle (see Figure 2.14). The temperature axis is then constructed perpendicular to the plane of the triangle, if desired.

Figure 2.14 requires a bit of instruction and practice before it can be used correctly. First of all, the pure components—in this case A, B, and C—are located at the vertices

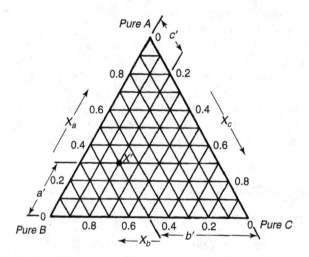

Figure 2.14 Illustration of how to express compositions on a three-component diagram. From O. F. Devereux, *Topics in Metallurgical Thermodynamics*. Copyright © 1983 by John Wiley & Sons, Inc. This material is used by permission of John Wiley & Sons, Inc.

of the equilateral triangle. The base of the triangle opposite the corresponding vertex refers to 0% of that component, so that the edge of the triangle connecting vertices B and C is 0 wt% A. Lines parallel to each edge constitute lines of constant composition for one of the components. For example, lines parallel to BC, which are spaced a uniform distance apart and become shorter in length as they approach the vertex A, are lines of constant wt% A. Sometimes these lines are every 10 wt%, sometimes every 20 wt%, and sometimes they are not there at all. Similarly, lines parallel to AC and AB represent lines of constant composition for components B and C, respectively. The triangular grid generated by these lines of constant composition helps us determine the overall composition of a point in the diagram. For example, point X' in Figure 2.14 is located at 30% A, 50% B, and 20% C. Obviously, the overall weight percentages of the components must add up to 100%. Not all compositions fall neatly on the intersection of three grid lines, of course, but when the grid is present, it is relatively easy to interpolate between gridlines.

Gridlines can clutter the phase diagram, however, especially when phases and boundaries start appearing in real systems, so that it is convenient to have a method for determining the composition of any point on the diagram without the assistance of gridlines. This is accomplished using the *center of gravity rule*, which states that the sum of the perpendicular distances from any point within an equilateral triangle to the three sides is constant and equal to the altitude of the triangle. With reference to Figure 2.15, this means that the quantity $(a + b + c)$ is a constant, regardless of where the point is located inside the triangle, and it represents the length from any edge to its opposite vertex.* Moreover, the relative amounts of the components are then given

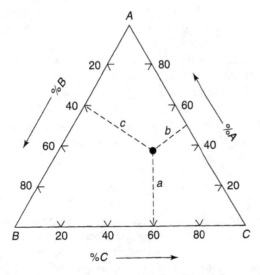

Figure 2.15 Illustration of center-of-gravity rule for determining compositions in ternary system.

*Don't let the different ways the axes are labeled between Figures 2.14 and 2.15 confuse you. For example, the labels for the gridlines of component. A are on the left in Figure 2.14, but on the right in Figure 2.15. Similar adjustments are made to the labels for components B and C. Both methods and both figures are entirely correct. You should convince yourself of this fact.

by the ratio of their respective line segments to the altitude:

$$\%A = \frac{100a}{(a+b+c)} \tag{2.43a}$$

$$\%B = \frac{100b}{(a+b+c)} \tag{2.43b}$$

$$\%C = \frac{100c}{(a+b+c)} \tag{2.43c}$$

Just as with the lever rule in binary phase diagrams, segments can be measured in any consistent units of length.

Cooperative Learning Exercise 2.3

Using the diagram in Figure 2.15:

Person 1: Lightly sketch in grid lines on the diagram and use them to determine the composition at the indicated point.

Person 2: Use the center-of-gravity equations (2.43) to calculate the composition at the indicated point.

Compare your answers.

Answer: 37% A; 21% B; 42% C

Now that we can determine overall compositions on a ternary diagram, we can return to the issue of representing temperature in the diagram. As mentioned earlier, the temperature axis is perpendicular to the compositional, triangular plane, as shown in Figure 2.16 for the case of three components that form a solid solution, α. Recall that in order to form solid solutions, the components must generally have the same crystal structure. Such substances are said to be *isomorphous*.

There are several important things to note in Figure 2.16. First, we are assuming a fixed pressure, as is the case in most condensed-phase diagrams. Second, the phase rule still applies. The single-phase regions, α or liquid, have three degrees of freedom, as given by the phase rule

$$F = C - \phi + N = 3 - 1 + 1 = 3 \tag{2.44}$$

so that composition of all three components can be varied continuously in these regions. In Figure 2.16, there is only one two-phase region, $L + \alpha$. Here, there are only $F = 3 - 2 + 1 = 2$ degrees of freedom. The upper surface of the two-phase region is the liquidus, and the lower surface of the region is the solidus. These phase boundary surfaces are indicated by selected solid and dashed lines, respectively. Finally, you should be able to visualize that each vertical face of the diagram represents a complete binary component phase diagram. For example, the C–B face in Figure 2.16 occurs only at $A = 0$ for all temperatures, and the resulting diagram is one similar to the binary component phase diagram for isomorphous substances shown in Figure 2.4. Similarly, the faces represented by A–B $(C = 0)$ and A–C $(B = 0)$ are independent

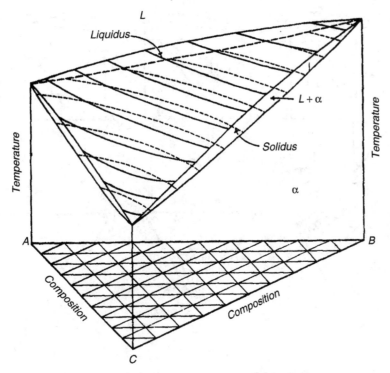

Figure 2.16 Temperature–composition space diagram of a ternary isomorphous system. Reprinted, by permission, from F. N. Rhines, *Phase Diagrams in Metallurgy*. Copyright © 1956 by McGraw-Hill Book Co.

binary phase diagrams. Thus, a ternary phase diagram contains not only all of the phase information we will be describing in this section, but three independent binary component phase diagrams as well.

Before moving on to more complicated ternary phase diagrams, it should be pointed out that the complete temperature–composition ternary component phase diagram can be "sectioned" in two important ways in order to simplify the representation. First, we can take horizontal "slices" of the diagram. Such sections are parallel to the base of the diagram and are, as such, necessarily at a fixed temperature, or *isotherm*. Examples of isothermal sections from a temperature–composition ternary phase diagram are shown in Figure 2.17. Note that at temperatures above the melting point of the highest-melting compound, the resulting diagram consists completely of a single-phase liquid region. As we decrease temperature and take a section below the liquidus, as in Figure 2.17a, part of the diagram contains liquid, part contains solid, and part contains the liquid + solid two-phase region. The liquidus and solidus lines can now be identified, and what now a two-phase region (L + α) with *two* degrees of freedom in Figure 2.16 is now a two-phase region with *one degree* of freedom, just as in a binary-component diagram, since the temperature has now been fixed and one degree of freedom has been lost. The *one degree* of freedom in this region *is a* composition, with the *other* compositional *variables* being fixed when the *first is* determined. Unlike true binary-component phase diagrams, the tie lines connecting the two *conjugate phases* (in this case liquid and α) that coexist at temperature T_1 cannot be drawn as completely

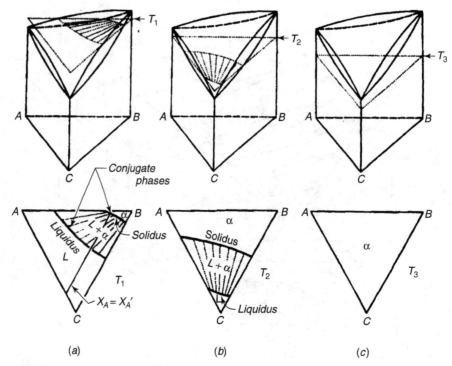

Figure 2.17 Isotherms through the ternary isomorphous phase diagram. Reprinted, by permission, from F.N. Rhines, *Phase Diagrams in Metallurgy*. Copyright © 1956 by McGraw-Hill Book Co.

horizontal or vertical lines, but must be determined experimentally. These tie lines are represented as dashed lines in the two-phase region. A similar type of diagram results from an isothermal section taken at T_2, as in Figure 2.17b, which is slightly lower than T_1. Finally, at T_3, the isotherm is completely within the single-phase solid region.

The second type of "slice" through the temperature-composition is a vertical section, known as an *isopleth*. Isopleths are sections of constant relative composition. For example, in Figure 2.18a, the line X–B is drawn on the base of the diagram and extended upwards along the temperature axis to generate the diagram shown below, known as a *pseudobinary phase diagram*, since it looks similar to a binary-component phase diagram, but is obviously different. Note, for example, that the liquidus and solidus lines do not converge at X on the pseudobinary phase diagram. This section has been selected such that A and C are always in the ratio of 1:1 at any point on the diagram. Obviously, the absolute amount of B can change, from 0% at X to 100% at B, thus changing the absolute amounts of A and C, but A and C always stay in the same proportion relative to one another. Vertical sections need not bisect a vertex. They are, however, normally taken in a logical manner—for example, a constant composition of one component or constant relative composition of two components. Figure 2.18b shows a section taken at a constant composition for A, as indicated by the line Y–Z, extended upward along the temperature axis. The resulting pseudobinary diagram is for a fixed composition of A, where the relative amounts of C and B are allowed to vary between pure C at Y and pure B at Z.

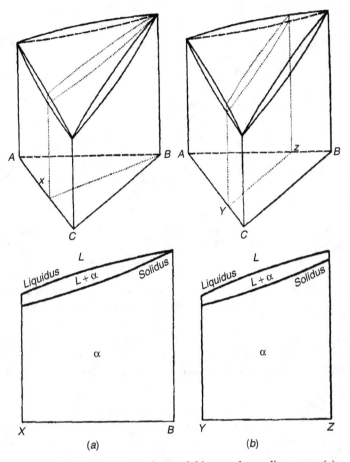

Figure 2.18 Examples of isopleths and pseudobinary phase diagrams. (a) constant A/C; (b) constant A. Reprinted, by permission, from F. N. Rhines, *Phase Diagrams in Metallurgy.* Copyright © 1956 by McGraw-Hill Book Co.

Just as in the case of binary systems, ternary systems can contain components that have only partial solubility on one another. This leads to eutectic diagrams, such as the one shown in Figure 2.19. This is the three-component analogue to Figure 2.5. Again, each face is a binary phase diagram, with binary eutectic melting points located at e_1, e_2, and e_3. The ternary eutectic melting point is located at point E, which is a "well" in the middle of the diagram. The lines connecting the binary eutectics with the ternary eutectics, e_1–E, e_2–E, and e_3–E, are part of the liquidus surface, separating the single-phase liquid region from the solid solution regions.

There are, of course, numerous combinations of solid solutions, phases, and transformations possible in these diagrams, and we will not attempt to address more than a few of the basic ones here. But keep in mind that these phase diagrams are generated in the same way that the unary and binary phase diagrams were generated: from free energy versus composition curves at various temperatures. The calculation of free energies of mixing for three components is directly analogous to that for two components, and the analogues for the minimization of free energy criteria as given by Eqs. (2.37) are appropriate.

Cooperative Learning Exercise 2.4

Work in groups or three. The melting points of components A, B, and C are 1000°C, 900°C, and 750°C, respectively. The constituent binary systems of the ternary system ABC each show complete solubility in both the liquid and the solid states. The following data refer to three binary alloys:

| Composition (wt%) | | | Temperature, °C | |
A	B	C	Liquidus	Solidus
50	50	—	975	950
50	—	50	920	850
—	50	50	840	800

Person 1: Sketch and label the AB binary temperature–composition phase diagram.
Person 2: Sketch and label the BC binary temperature–composition phase diagram.
Person 3: Sketch and label the AC binary temperature–composition phase diagram.
   Combine your information to sketch the ABC temperature–composition ternary phase diagram, assuming that the liquidus and solidus surfaces do not exhibit a maximum or minimum. Now take a vertical section joining 100% A to the mid-point of the BC binary face. Sketch the resulting pseudo-binary phase diagram.

Answers: Each binary phase diagram will look like Figure 2.4, with the melting point of each component end of the appropriate diagram. The ternary phase diagram is shown above, along with the vertical section joining the vertex at A to the 50% BC composition.

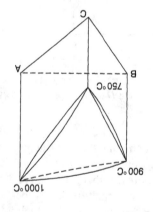

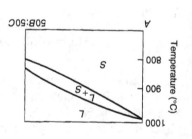

2.2.1.2 Ceramic Phase Diagrams.

Let us now examine a real three-component phase diagram. This could be three elements—for example, copper—with any of a number of other metallic elements that lead to an important class of alloys called *bronzes*. Copper–zinc–manganese alloys are *manganese bronzes*, copper–tin–nickel alloys are *nickel–tin bronzes*, copper–tin–lead alloys are *leaded tin bronzes*, and so on. But we are not limited only to metals in ternary phase diagrams. When we introduce nonmetallic elements, such as carbon, nitrogen, and oxygen, we enter a new realm of materials engineering and science—that of ceramics. The compounds formed in

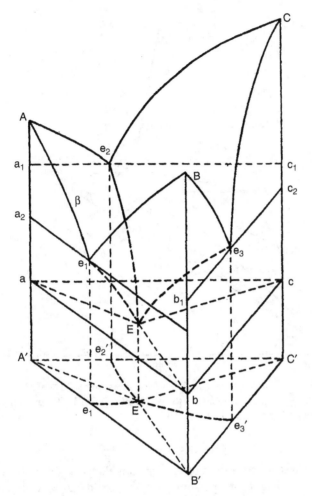

Figure 2.19 Perspective drawing of a ternary system with a simple eutectic and no ternary compound. Reprinted, by permission, from *Phase Diagrams for Ceramists*, Vol. 1, p. 15. Copyright © 1964, The American Ceramic Society.

these systems are carbides, nitrides, and oxides, respectively. For example, consider the copper–iron–oxygen (Cu–Fe–O) phase diagram shown in Figure 2.20. We know that this must be an isothermal section of a more extended diagram, such as that in Figure 2.19, and in this case it turns out to be a range of temperatures at which all the phases indicated are stable, namely, above 675°C.

Along the Fe–O binary, we find common ferrous oxides such as *hematite* (Fe_2O_3), *magnetite* (Fe_3O_4), and *wüstite* (FeO). The lines that connect these intermediate binary compounds with the third component are called *alkemade lines*, or alkemades. The alkemades divide ternary systems into composition triangles. For example, the alkemade connecting magnetite with Cu in Figure 2.20 could serve as the edge of a Fe–Cu–Fe_3O_4 phase diagram. In practice, any compound can be used as the vertex of the triangle. Such diagrams that illustrate the phase relationships between important compounds are called *subsystem* diagrams.

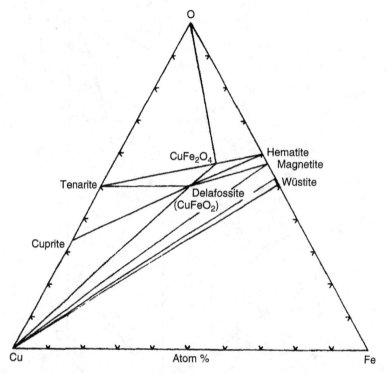

Figure 2.20 Stable assemblages above 675°C for the system Cu–Fe–O. Reprinted, by permission, from *Phase Diagrams for Ceramists*, p. 30, Fig. 2136, 1969 Supplement. Copyright © 1969, The American Ceramic Society.

If we were to isolate the portion of the diagram along the Fe–O binary and expand the phase diagram along the temperature axis, we would end up with a binary component phase diagram like that shown in Figure 2.21, which begins at the compound FeO (the phases are mostly the same from pure Fe to the compound FeO), then continue up to Fe_2O_3, which corresponds to about 30 wt% O. This is, in fact, how many oxide ceramic phase diagrams are presented—that is, not with oxygen as a continuously varying independent element, but as an integral part of metal oxide compounds that form when it reacts with the metallic components–in this case, ferrous oxides. After all, pure oxygen is in the gaseous form at the temperatures under consideration here. As a result, phase diagrams such as that in Figure 2.21 are often presented at a total pressure of 1 atm, with oxygen partial pressures superimposed on it in the form of *isobars*, or lines of constant partial pressure. The isobars help us follow the variation in equilibrium compositions with temperature under a fixed oxygen partial pressure in the system. This is important from an industrial standpoint, where the partial pressure can be controlled in manufacturing operations. For example, consider a small quantity of liquid iron at 1635°C, saturated with oxygen in a gas reservoir of oxygen partial pressure, $p_{o_2} = 10^{-8}$ atm. The volume of the gas reservoir is sufficiently large that any oxygen consumed by oxidation or produced by reduction of the iron or oxides has no effect on the total pressure of the system. The oxygen-saturated liquid iron, located at point *a* in Figure 2.21, is cooled slowly, such that equilibrium is maintained

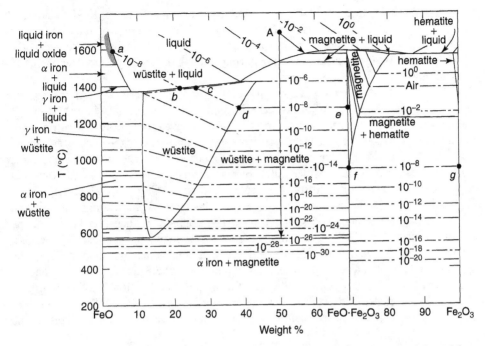

Figure 2.21 The $FeO-Fe_2O_3$ phase diagram showing oxygen isobars (in atm). Adapted from W. D. Kingery, H. K. Bowen, and D. R. Uhlmann, *Introduction to Ceramics.* Copyright © 1976 by John Wiley & Sons, Inc. This material is used by permission of John Wiley & Sons, Inc.

with the gas reservoir. Since we are cooling at a constant oxygen partial pressure, we must follow along the isobar, represented by the dotted line $a-b$. We are still in the single-phase liquid region, but the dissolved oxygen is now beginning to react with the iron to form FeO:

$$Fe(l) + \tfrac{1}{2}O_2(g) \longrightarrow FeO(l) \tag{2.45}$$

At point b, the liquid has mostly turned to wüstite, FeO, which is in the solid phase. From point b to point c, the remaining liquid becomes solid wüstite. A further decrease in temperature leads through the wüstite solid-phase field along the 10^{-8} atm isobar, until d is reached, at which point magnetite, Fe_3O_4, begins to form from the following oxidation reaction:

$$3FeO(s) + \tfrac{1}{2}O_2(g) \longrightarrow Fe_3O_4(s) \tag{2.46}$$

This reaction proceeds at 1275°C along the isobar to point e on the phase diagram. At this point, only magnetite remains. Magnetite remains a stable compound upon further cooling, to point f. Here, the formation of hematite, Fe_2O_3, becomes thermodynamically favorable at this partial pressure:

$$2Fe_3O_4(s) + \tfrac{1}{2}O_2(g) \longrightarrow 3Fe_2O_3(s) \tag{2.47}$$

This reaction proceeds at 875°C until hematite is the only remaining component at point g.

Equations (2.45) through (2.47) are typical oxidation reactions for the formation of metal oxides. The reverse reactions (steps *g−a*) are the reduction of metal oxides to form metal, such as is found in *smelting operations*. It should be apparent that the control of the oxygen partial pressure during heating is an important parameter in determining which phases will form.

As before, the free energy of mixing of the various components in the system are calculated, and the equilibrium criteria applied to determine which phases are stable at the indicated temperature. An analysis similar to the one just performed for the iron–oxygen system can be performed for the copper–oxygen binary system in Figure 2.20, for which there are two stable copper oxides, *cuprite* (cuprous oxide, Cu_2O) and *tenarite* (cupric oxide, CuO). Further inspection of Figure 2.20 shows that copper oxide and iron oxide form two stable compounds: $CuFe_2O_4$ and $CuFeO_2$.

Recall that we can take vertical "slices" of the ternary phase diagrams, as well as isothermal (horizontal) slices. If we take, for example, a slice that begins at the tenarite composition (CuO) and extends across to the hematite composition (Fe_2O_3), we would end up with a pseudobinary phase diagram, which, when plotted on the appropriate temperature–composition axes, would look like Figure 2.22. Note that the compound $CuFe_2O_4$ is present, here labeled as *spinel* (see Section 1.2.2.3), but there is much more phase and temperature information available to us. This is, in fact, how many metal oxide phase diagrams are presented. The most stable forms of the

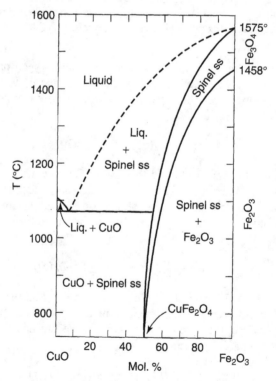

Figure 2.22 Binary component phase diagram for the system $CuOFe_2O_3$. Reprinted, by permission, from *Phase Diagrams for Ceramists*, p. 31, Fig. 2140, 1969 Supplement. Copyright © 1969 by The American Ceramic Society.

oxides, or at least the oxides of interest, are selected for the end compositions of the phase diagram, rather than plotting the elemental metals, along with oxygen, at the vertices of an equilateral triangle. In this way, a ternary phase diagram composed of three elements (Figure 2.20) can be represented as a binary phase diagram of two compounds (Figures 2.21 and 2.22).

This type of representation allows us to do something even more powerful. We can now take what would normally be a *quaternary* component system, composed of four elements, and represent it on a ternary phase diagram system by representing three of the elements as important compounds of the fourth. Again, this is done primarily with metal oxide systems, in which three of the components are metals and the fourth is oxygen. One of the most important ternary component phase diagram systems represented in this way is for the system $MgO-Al_2O_3-SiO_2$, shown in Figure 2.23.

This system is of great industrial importance because of the wide variety of naturally occurring compounds it contains. *Silica*, SiO_2, is found in nature in many forms, including sand and quartz. *Alumina*, Al_2O_3, is found as *bauxite*. Notice that we are using a new naming convention. Compounds that contain metallic elements as their root and end in "-a" are common industrial terms for oxides of that metal. *Magnesia* (MgO), *yttria* (Y_2O_3), and *zirconia* (ZrO_2) are a few more examples. This terminology is not

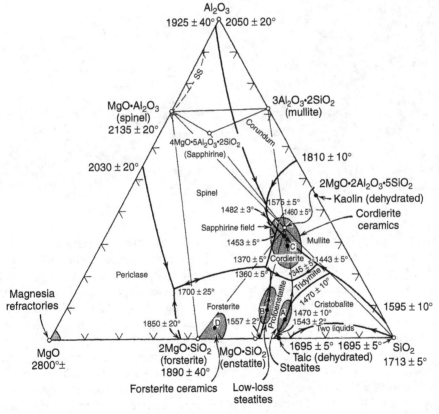

Figure 2.23 The $MgO-Al_2O_3-SiO_2$ phase diagram. Temperatures are in °C. From *Introduction to Ceramics*. W. D. Kingery, H. K. Bowen, and D. R. Uhlmann, Copyright © 1976 by John Wiley & Sons, Inc. This material is used by permission of John Wiley & Sons, Inc.

Example Problem 2.2

Ytrrium–barium–copper oxide (YBCO) superconductors were one of the first class of metal oxides to be found to superconduct (see Section 6.1.2.4) at liquid nitrogen temperatures (above about 77 K). Not all of the compounds of YBCO will superconduct, however, owing to the complex dependence of this phenomenon on the copper oxide lattice. For example, the "2–1–1" phase, named after the relative molar quantities of the three metals, Y_2BaCuO_5, does not superconduct, but the "1–2–3" phase, $YBa_2Cu_3O_7$, does. A portion of the Y_2O_3–BaO–CuO ternary phase diagram is shown below, with the corresponding pseudobinary phase diagram connecting the 1–2–3 compound with point F in the ternary phase diagram next to it. Determine the amount of 1–2–3 formed when a liquid of composition shown in the hatched region of the binary diagram (about 88 mol% $BaCu_3O_4$) is cooled to the vicinity of 960°C.

Answer: Application of the lever rule in the 1–2–3 + Liquid region shows that even in the best case (at the left edge of the hatched region) only about 3% of the 1–2–3 phase is formed, the rest being liquid of different composition. As the liquid is cooled further, more yttria-poor compounds are formed, such as barium cuprate (BC), that are difficult to

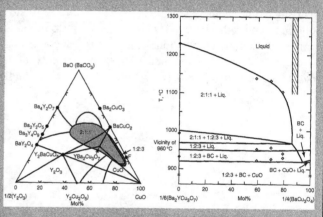

remove. Thus, melt processing of pure 1–2–3 superconductors is virtually impossible via this route. One solution has been to use nonequilibrium processing, which renders the equilibrium phase diagrams shown above useless.

Reprinted, by permission from T. P. Sheahen, *Introduction to High Temperature Superconductivity*, pp. 179, 180. Copyright © 1994 by Plenum Publishing.

universal, especially when the metal can have more than one common oxidation state. Many of the compounds along the Al_2O_3–SiO_2 binary, termed *aluminosilicates*, are the building blocks of most clays. *Kaolin*, for example, is a hydrated aluminosilicate of nominal composition $Al_2O_3 \cdot 2SiO_2 \cdot 2H_2O$, sometimes expressed as the compound $Al_2Si_2O_5(OH)_4$. It is used for making porcelain for valves, tubes, and fittings, as a refractory for bricks and furnace linings, and as a pigment filler in paints. *Talc* is a hydrated form of magnesium silicate, $3MgO \cdot 4SiO_2 \cdot H_2O$, and can be dehydrated to form *protoenstatite* crystals, $MgSiO_3$, in a silica matrix. *Cordierite* ceramics have variable compositions as shown in the shaded region of Figure 2.23, but have a nominal composition of $2MgO \cdot 2Al_2O_3 \cdot 5SiO_2$. Cordierite ceramics are particularly useful due to their low coefficient of thermal expansion (see Section 5.1.3.1) and thermal shock resistance. Many of the ceramics in this system are termed *refractories*, since they have very high melting points, generally above 1580°C, and are able to bear loads and resist corrosion at elevated temperatures.

2.2.1.3 *Ellingham Diagrams**. Equation (2.45) is an example of a formation reaction involving solid and gas reactants, leading to the formation of a solid metal oxide. The explicit free energy relations for this reaction were eliminated from the previous discussion for two principal reasons: (1) because the connection between thermodynamics and reaction kinetics is dealt with in Chapter 3 and (2) because the reaction involves a gaseous species, and we have purposely limited our discussion to condensed phases to this point. However, the formation of metal oxides by oxidation of their metallic precursors and its associated thermodynamics are important topics in both ceramics and physical metallurgy, such that a brief elaboration is in order. To do so effectively, we will need to discuss the thermodynamics of gases and introduce a quantity called the *thermodynamic activity*, or just activity. You may wish to look ahead to Chapter 3 for further discussion on reaction kinetics. The goal of this section is to develop a straightforward graphical representation of oxidation reactions that will allow us to determine which oxides form most favorably in mixtures of metals.

Consider the reaction equilibrium between a pure solid metal, M, its pure oxide, MO, and oxygen gas at temperature T and pressure P:

$$2M(s) + O_2(g) \longrightarrow 2MO(s) \tag{2.48}$$

It can be shown (see Chapter 3) that the standard state Gibbs free energy change for this reaction is given by the difference between the standard state free energies of the products minus the reactants and can be related to the partial pressure of the gaseous species:

$$\Delta G° = 2G°_{MO(s)} - G°_{O_2(g)} - 2G°_{M(s)} = -RT \ln P_{O_2}^{-1} \tag{2.49}$$

where P_{O_2} is the partial pressure of oxygen and R is the gas constant. Thus, in the case of reaction equilibrium involving only pure condensed phases and a gas phase, the Gibbs free energy change can be written solely in terms of the species that appear only in the gas phase. (You may recall from thermodynamics that this is due to that fact that the thermodynamic activity of condensed phases is usually taken as unity.) At any fixed temperature, the establishment of reaction equilibrium occurs at a unique value of oxygen partial pressure, which we will designate $P_{O_2,eqT}$. The equilibrium thus has one degree of freedom, as given by the phase rule ($\phi = 3$, two pure solids and a gas; $C = 2$, metal + oxygen; $N = 2$, temperature and pressure), $F = C - \phi + N = 2 - 3 + 2 = 1$.

At any temperature T, when the actual partial pressure of oxygen in a closed metal–metal oxide–oxygen system is greater than $P_{O_2,eqT}$, spontaneous oxidation of the metal will occur, thereby consuming oxygen and decreasing the oxygen partial pressure in the gas phase. When the actual oxygen pressure has been decreased to $P_{O_2,eqT}$, the oxidation ceases and reaction equilibrium is reestablished. Similarly, if the oxygen pressure in the closed vessel was originally less than $P_{O_2,eqT}$, spontaneous reduction of the oxide would occur [the reverse of Eq. (2.48)] until $P_{O_2,eqT}$ was reached. Metallurgical processes involving the reduction of oxide ores rely on the establishment and maintenance of an oxygen pressure less than $P_{O_2,eqT}$ in the reaction vessel.

The variation of the standard state Gibbs free energy change for the oxidation reaction at any temperature from experimentally measured variations in $P_{O_2,eqT}$ can be fitted to an equation of the form:

$$\Delta G° = A + BT \ln T + CT \tag{2.50}$$

For example, let us consider the oxidation of Cu to Cu_2O:

$$4Cu(s) + O_2(g) \longrightarrow 2Cu_2O(s) \tag{2.51}$$

The empirically fitted free-energy–temperature relationship (in joules) is

$$\Delta G°(J) = -333{,}900 + 14.2 \ln T + 247T \tag{2.52}$$

which can be approximated in linear form as

$$\Delta G°(J) = -333{,}000 + 141.3T \tag{2.53}$$

in the temperature range 298 to 1200 K. The error in using Eq. (2.53) instead of Eq. (2.52) is approximately 0.3% at 300 K and 1.4% at 1200 K. The linear approximation allows us to make a series of simple constructions that compare the free energy of formations for a variety of oxidation reactions.

H. J. T. Ellingham plotted the experimentally determined variations of $\Delta G°$ with temperature for the oxidation of a series of metals, and he found that the linear approximation was suitable when no change of state occurred. Thus, all free energy expressions for the oxidation of metals could be expressed by means of a simple equation of the form

$$\Delta G° = A + BT \tag{2.54}$$

where the constant A is identified with the temperature-independent standard enthalpy change for the reaction, $\Delta H°$, and the constant B is identified with the negative of the temperature-independent standard entropy change for the reaction, $\Delta S°$.

Consider the variation of $\Delta G°$ with T for the oxidation of silver:

$$4Ag(s) + O_2(g) \longrightarrow 2Ag_2O(s) \tag{2.55}$$

The plot of free energy with temperature for this reaction, using the correlation provided by Eq. (2.54), is called an Ellingham diagram and is shown in Figure 2.24. At 462 K, $\Delta G° = 0$, and pure solid silver and oxygen gas at 1 atm pressure are in equilibrium with silver oxide at this temperature. If the temperature is decreased to T_1, $\Delta G° < 0$, the metal phase becomes unstable relative to silver oxide and oxidizes spontaneously. The value of $P_{O_2,eqT}$ at this temperature is calculated using Eq. (2.49), and since $\Delta G°$ is a negative quantity at this temperature, $P_{O_2,eqT} < 1$ atm. Similarly, if the temperature of the system is increased to T_2, then $\Delta G° > 0$, and the oxide phase becomes unstable relative to silver metal and oxygen gas at 1 atm. In this instance, $P_{O_2,eqT} > 1$ atm. The value of $\Delta G°$ is thus a measure of the chemical affinity of the metal for oxygen, and the more negative its value at a given temperature, the more stable the oxide.

Ellingham diagrams can be made for any metal–metal oxide combination. When plotted together, the graphical representation of free energy versus temperature provides a useful method for evaluating the relative affinity of oxygen for metals. By way of example, let us first consider two general oxidation reactions for which the Ellingham lines intersect each other, as shown in Figure 2.25:

$$2A(s) + O_2(g) \longrightarrow 2AO(s)$$
$$B(s) + O_2(g) \longrightarrow BO_2(s) \tag{2.56}$$

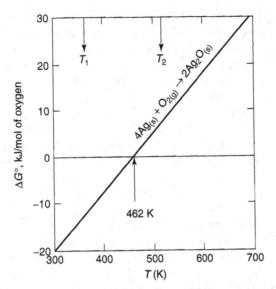

Figure 2.24 Ellingham diagram for the oxidation of silver. Reprinted, by permission, from D. R. Gaskell, *Introduction to Metallurgical Thermodynamics*, 2nd ed., p. 272. Copyright © 1981 by McGraw-Hill Book Co.

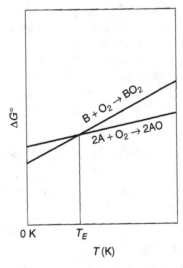

Figure 2.25 Ellingham diagram for two hypothetical oxidation reactions. Reprinted, by permission, from D. R. Gaskell, *Introduction to the Thermodynamics of Materials*, 3rd ed., p. 360. Copyright © 1973 by Taylor & Francis.

The two Ellingham plots intersect at temperature T_E. At temperatures less than T_E, A and BO_2 are stable with respect to B and AO. At temperatures above T_E, B and AO are stable. At T_E, A, B, AO, and BO_2 are in equilibrium with each other. Thus, if pure A were to be used as a reducing agent to reduce pure BO_2 to form B and pure AO, then the reduction would have to be conducted at temperature higher than T_E. This is the power of the Ellingham diagram: Not only can it be used to determine the stability of a single oxide at a given temperature, but it can provide relative stabilities of various oxides and indicate which metals can be used as reducing agents for a given metal oxide. You should note that in order to compare the stabilities of different oxides, the Ellingham diagrams must be drawn for oxidation reactions involving the consumption of the same number of moles of oxygen. The units of $\Delta G°$ for the oxidation reaction must then be given in energy per mole of O_2 consumed.

In order to avoid calculating the value of $P_{O_2,eqT}$ for all oxidation reactions, F. D. Richardson added a nomographic scale along the right-hand edge of the Ellingham diagram. The value of $P_{O_2,eqT}$ for any metal–metal oxide equilibrium is read off the graph as that value on the scale which is collinear with the points $\Delta G° = 0$, $T = 0$, and $\Delta G_T°$, $T = T$. Ellingham diagrams with this additional nomograph are often called Ellingham–Richardson diagrams. An Ellingham–Richardson diagram for the formation of common metal oxides is shown in Figure 2.26.

Finally, it should be noted that phase changes can be accommodated in the Ellingham diagram. When the temperature moves above the melting point of the metal or metal oxide, their corresponding standard states must change. For example, above 1100°C, copper metal is no longer solid, and the oxidation reaction of interest is:

$$4Cu(l) + O_2(g) \longrightarrow 2Cu_2O(s) \tag{2.57}$$

The Ellingham diagram reflects this change in state with a change in slope, albeit a small one for this example, above the melting point of copper, indicated by M. Melting points of all the metals are so indicated, and the melting points of the respective oxides with a boxed M. Boiling points of the metals, where appropriate, are indicated with a B, and the free energy line once again may undergo a change in slope. No metal oxides reach their boiling points at the temperatures of interest in the diagram.

The free energy line for the formation of carbon monoxide and carbon dioxide are included in this plot, as the reduction of metal oxides with carbon monoxide is common in smelting operations:

$$MO_2 + 2CO \longrightarrow M + 2CO_2 \tag{2.58}$$

Similar Ellingham–Richardson diagrams for the formation of nitrides and sulfides can be constructed and are available in the literature.

2.2.2 Interfacial Thermodynamics

Thermodynamics plays a fundamental role in practically all of the processes that are described in this book. Phase equilibria, which has been the focus of this chapter so far, is but one of these areas. Let us diverge from phase equilibria for a bit, and discuss a thermodynamic topic that will be of use in many of the subsequent chapters: *interfacial energy*. Interfacial energies will be developed using thermodynamic arguments primarily for liquids, but we will see that the results are general to solids as well. This point will be illustrated by application of interfacial energies to an important process in the densification of commercial materials called *sintering*. Later on, we will see that it has enormous utility in the description of many surface-related phenomena in materials science, such as fracture, adhesion, lubrication, and reaction kinetics.

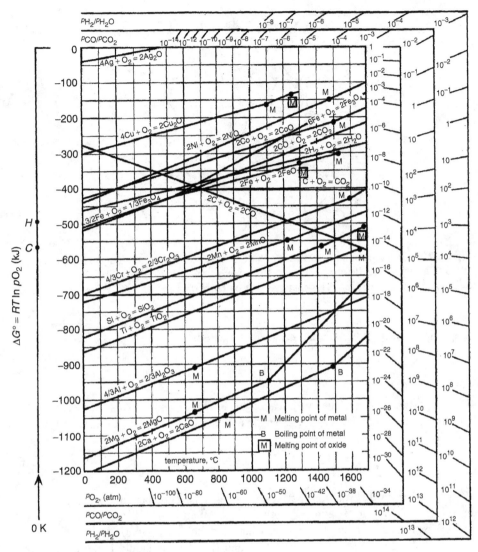

Figure 2.26 Ellingham–Richardson diagram for some common metal oxides. Reprinted, by permission, from D. R. Gaskell, *Introduction to the Thermodynamics of Materials*, 3rd ed., p. 370. Copyright © 1973 by Taylor & Francis.

2.2.2.1 Surface Energy. A *surface* is an inhomogeneous boundary region between two adjacent phases. As shown in Figure 2.27, atoms on the surface of a phase are necessarily different than those in the bulk. In particular, they have fewer nearest neighbors than the bulk, and they may be exposed to constituents from an adjacent phase. This generally means that less energy is required to remove an atom from a surface than to remove it from the bulk. Therefore, the potential energy of surface atoms is higher than bulk atoms. In turn, work is required to move atoms from the bulk to the surface. When this is done, new surface is created, and the surface area of the phase increases. The reversible work required to form the new surface, dW_s, is

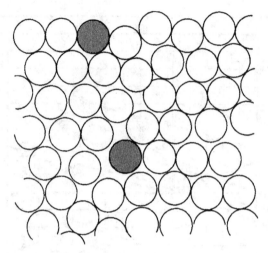

Figure 2.27 Schematic representation of surface and bulk atoms in a condensed phase. From W. D. Kingery, H. K. Bowen, and D. R. Uhlmann, *Introduction to Ceramics*. Copyright © 1976 by John Wiley & Sons, Inc. This material is used by permission of John Wiley & Sons, Inc.

proportional to the surface area, dA, that is created:

$$dW_S = \gamma dA \tag{2.59}$$

The proportionality constant, γ, is called the *surface energy*. When the bulk phase in question is a liquid, the surface energy is often called *surface tension*. Rearranging Eq. (2.59) gives:

$$\gamma = \frac{dW_S}{dA} \tag{2.60}$$

We can see from this relation that surface energy is work per unit area, so it should have units of J/m^2 in SI and ergs/cm^2 in cgs. Often you will see values of surface energy expressed in units of dyne/cm, where 1 dyne = 1 erg/cm.

With the aid of Eqs. (2.1), (2.7), (2.9), and (2.11), you should be able to prove to yourself that the reversible, non-pressure–volume work, dW_s, is equivalent to the free energy change, dG, so that Eq. (2.60) becomes, with proper use of partial differentials,

$$\gamma = \left(\frac{\partial G}{\partial A}\right)_{T,P,N_i} \tag{2.61}$$

This relationship identifies the surface energy as the increment of the Gibbs free energy per unit change in area at constant temperature, pressure, and number of moles. The path-dependent variable dW_s in Eq. (2.60) has been replaced by a state variable, namely, the Gibbs free energy. The energy interpretation of γ has been carried to the point where it has been identified with a specific thermodynamic function. As a result, many of the relationships that apply to G also apply to γ:

$$\gamma = H_s - T S_s \tag{2.62}$$

where the subscript s on the enthalpy and entropy indicates that these are surface properties. As mentioned earlier, the surface atoms are fundamentally different than bulk atoms, so that they have different enthalpies and entropies associated with them. Differentiation of Eq. (2.62) with respect to temperature at constant pressure gives

$$\left(\frac{\partial \gamma}{\partial T}\right)_P = -S_s \tag{2.63}$$

Substitution of Eq. (2.63) in (2.62) gives

$$\gamma = H_s + T\left(\frac{\partial \gamma}{\partial T}\right)_P \tag{2.64}$$

Equation (2.64) is useful from an experimental standpoint because the measurements of surface energies at various temperatures can, in principal, provide a measurement of the surface enthalpy. The surface enthalpy, H_s, can also be determined directly, because it is equivalent to the heat of sublimation or vaporization.

Since the surface energy is a direct result of intermolecular forces, its value will depend on the type of bond and the structural arrangement of the atoms. Normally, densely packed planes would have lower surface energies. Liquid hydrocarbons having only weak van der Waals forces have values of surface tension in the neighborhood of 15–30 dyn/cm, while liquids with polar forces and hydrogen bonds, like water, have surface tensions in the range of 3–72 dyn/cm. Fused salts and glasses with additional ionic bonds have surface energies from 100 to 600 dyn/cm, and molten metals have surface tensions from 100 to 3000 dyn/cm. Even higher surface energies can be found in covalent solids (see Example Problem 2.3). See Appendix 4 for values of solid and liquid surface energies for a variety of materials.

2.2.2.2 The LaPlace Equation.

The concept of surface energy allows us to describe a number of naturally occurring phenomena involving liquids and solids. One such situation that plays an important role in the processing and application of both liquids and solids is the pressure difference that arises due to a curved surface, such as a bubble or spherical particle. For the most part, we have ignored pressure effects, but for the isolated surfaces under consideration here, we must take pressure into account.

Consider a generic curved surface such as that found in a sphere or cylinder (see Figure 2.28). The curved surface has two principal radii of curvature, R_1 and R_2. The front of this surface is indicated by the line xx_1, and the back is represented by the line yy_1. Let us now move the surface out by a differential element, dz. The front of the new surface is now given by $\acute{x}\acute{x}_1$, and the back is indicated by $\acute{y}\acute{y}_1$. The work required to displace the surface this amount is supplied by a pressure difference, ΔP. The pressure acts on an area given by $(xx_1)(x_1y_1)$ moving through the differential element dz, such that the total pressure–volume work associated with extending the surface through dz is

$$W = \Delta P(xx_1)(x_1y_1)\, dz \tag{2.65}$$

The pressure–volume work must be counterbalanced by surface tension forces. The work required to move against surface tension forces is best calculated by breaking it

---

Example Problem 2.3

An alternative method for estimating surface energies is to calculate the work required to separate two surfaces of a crystal along a certain crystallographic plane. At 0 K, this work can be approximated as the energy required to break the number of bonds per unit area, or the energy of cohesion, E_{coh}. Since two surfaces are being formed in this cleavage process, the surface energy of a single surface is then

$$\gamma \approx 1/2 E_{coh}$$

Let us consider, for example, separation along the (111) plane in diamond. The lattice constant for diamond is $a = 3.56$ Å (see Table 1.11), so the number of atoms per square centimeter on the (111) surface is

$$\frac{2 \text{ atoms per plane(111)}}{\sqrt{3}a^2/2} = \frac{4}{\sqrt{3}(3.56 \times 10^{-8})^2}$$

$$= 1.82 \times 10^{15} \text{ atoms/cm}^2$$

From Appendix 1, the bond energy for C–C bonds is 348 kJ/mol (83.1 kcal/mol), so that the cohesive energy is

$$E_{coh} = \frac{(1.82 \times 10^{15} \text{ bonds/cm}^2)(348,000 \text{ J/mol})(10^7 \text{ erg/J})}{6.02 \times 10^{23} \text{ atoms/mol}}$$

$$= 10,500 \text{ erg/cm}^2$$

so that the surface energy is

$$\gamma = 1/2(10,500 \text{ erg/cm}^2) = 5250 \text{ erg/cm}^2$$

---

into two parts, W_1 and W_2. W_1 is the work required to move side xx_1 away from yy_1 a distance $(x_1 y_1/R_1)\, dz$ during the expansion:

$$W_1 = \frac{\gamma(xx_1)(x_1 y_1)\, dz}{R_1} \tag{2.66}$$

Similarly, the work required to move side xy away from $x_1 y_1$ through a distance $(xx_1/R_2)\, dz$ is W_2:

$$W_2 = \frac{\gamma(x_1 y_1)(xx_1)\, dz}{R_2} \tag{2.67}$$

Adding Eqs. (2.66) and (2.67) together to arrive at the work against surface tension, equating them with the pressure–volume work in Eq. (2.65) and simplifying leads to the *Laplace equation*:

$$\Delta P = \gamma \left(\frac{1}{R_1} + \frac{1}{R_2} \right) \tag{2.68}$$

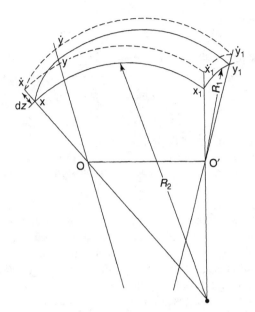

Figure 2.28 An element of a curved surface with principal radii R_1 and R_2. Reprinted, by permission, from J. F. Padday, in *Surface and Colloid Science*, E. Matijevic, ed., Vol. 1, p. 79. Copyright © 1969 by John Wiley & Sons, Inc.

The Laplace equation in this form is general and applies equally well to geometrical bodies whose radii of curvature are constant over the entire surface to more intricate shapes for which the Rs, are a function of surface position. In the instance of constant radii of curvature across the surface, Eq. (2.68) reduces for several common cases. For spherical surfaces, $R_1 = R_2 = R$, where R is the radius of the sphere, and Eq. (2.68) becomes:

$$\Delta P = \frac{2\gamma}{R} \tag{2.69}$$

For a cylindrical surface, R_1 (or R_2) is infinity, so the remaining radius, R, is the radius of the cylinder and:

$$\Delta P = \frac{\gamma}{R} \tag{2.70}$$

Finally, for a planar surface, both radii of curvature become infinity, and:

$$\Delta P = 0 \tag{2.71}$$

The pressure difference may also be numerically zero in the instance where the two principal radii of curvature lie on opposite sides of the surface, such as in the case of a saddle.

2.2.2.3 *The Young Equation.* The principle of balancing forces used in the derivation of the Laplace equation can also be used to derive another important equation in surface thermodynamics, the *Young equation.* Consider a liquid droplet in equilibrium

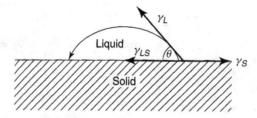

Figure 2.29   Schematic diagram of liquid droplet on solid surface. From Z. Jastrzebski, *The Nature and Properties of Engineering Materials*, 2nd ed., Copyright © 1976 by John Wiley & Sons, Inc. This material is used by permission of John Wiley & Sons, Inc.

with its own vapor and a flat, solid surface at constant temperature, as shown in Figure 2.29. The liquid–solid, liquid–vapor, and solid–vapor *interfacial surface energies* are defined as γ_{LS}, γ_{LV}, and γ_{SV}, respectively. Technically, the liquid–vapor and solid–vapor interfacial energies should be for the liquid and solid in equilibrium with their respective vapors, which are probably not the same. In practice, however, the vapor is usually a nonreactive gas, so γ_{LV} and γ_{SV} become the surface tension and solid surface energy in the gas of interest and are simply labeled γ_L and γ_S, respectively, as shown in Figure 2.29. These are both approximations. The surface tensions and surface energies of some common liquids and solids are listed in Appendix 4.

At equilibrium an angle θ, called the *contact angle*, is formed at the three-phase solid–liquid–gas junction. The contact angle can have values from zero to 180°. When $\theta = 0$, the liquid completely spreads of the solid surface, forming a thin monolayer (see Figure 2.30). Such a condition is termed *wetting*. The other extreme is called *nonwetting*, and it occurs when the entire liquid droplet sits as a sphere on the solid. All values of contact angle in between these two extremes are theoretically possible. Obviously, the chemical nature of the liquid, solid, and vapor determines the extent to which the liquid will wet the solid, and temperature has an important influence. Keep in mind that the contact angle under consideration here is an *equilibrium contact angle*. There are also *dynamic, receding*, and *advancing contact angles* associated with droplets as they spread and move on substrates. For now, the term contact angle will refer to the equilibrium condition.

The three interfacial surface energies, as shown at the three-phase junction in Figure 2.29, can be used to perform a simple force balance. The liquid–solid interfacial energy plus the component of the liquid–vapor interfacial energy that lies in the same direction must exactly balance the solid–vapor interfacial energy at equilibrium:

$$\gamma_L \cos \theta + \gamma_{SL} = \gamma_S \qquad (2.72)$$

This equation is called *Young's equation*; it is named after Thomas Young, who first proposed it in 1805. The derivation presented here in terms of force balances is simplistic, but there are more rigorous thermodynamic arguments to support its development.

In practice, the contact angle can be experimentally determined in a rather routine manner, as can the liquid surface tension and even the solid surface energy. The interfacial energy for the liquid–solid system of interest, γ_{SL}, can then be calculated using Young's equation. Alternatively, if γ_{SL}, γ_L, and γ_S are known as a function of temperature, the contact angle can be predicted at a specified temperature.

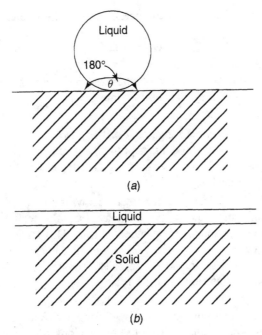

Figure 2.30 Schematic illustration of (a) nonwetting and (b) wetting of a liquid on a solid. From Z. Jastrzebski, *The Nature and Properties of Engineering Materials*, 2nd ed., Copyright © 1976 by John Wiley & Sons, Inc. This material is used by permission of John Wiley & Sons, Inc.

2.2.2.4 *Sintering and Densification.*

Curved surfaces and surface energies play a large role in the description of an important phenomenon in ceramic processing called *sintering*. Sintering is a process in which a particulate material is consolidated during heat treatment. Consolidation implies that within the material, particles have joined together into an aggregate that has strength. *Densification*, or the increase in density through the accompanying reduction in interparticle voids, often accompanies sintering, but there are some highly porous insulation products that can actually be less dense after they have been sintered. Nonetheless, we will examine sintering and densification simultaneously. We will also limit the discussion to solid state sintering.

The driving force for sintering is the reduction in the total free energy of the particulate system, ΔG, which is composed of free energy changes of volume, ΔG_V, boundaries, ΔG_b, and surfaces, ΔG_s:

$$\Delta G = \Delta G_V + \Delta G_b + \Delta G_s \tag{2.73}$$

In conventional sintering, the dominant term in the free energy reduction is that due to surface area reduction. Recall from Eq. (2.61) that the surface energy is the increment of the Gibbs free energy per unit change in area at constant temperature, pressure, and total number of moles

$$\gamma = \left(\frac{\partial G}{\partial A}\right) \tag{2.61}$$

such that if A_s is the surface area of the particles, then the free energy change due to surface area is

$$\Delta G_S = \gamma \Delta A_s \tag{2.74}$$

Three stages of solid-state sintering are recognized, although there is not a clear distinction between them, and they overlap to some extent.

- *Initial sintering* involves rearrangement of the powder particles and formation of a strong bond or neck at the contact points between particles (see Figure 2.31). The relative density of the powder may increase from 0.5 to 0.6 due mostly to an increase in particle packing.
- *Intermediate sintering* occurs as the necks grow, the number of pores decreases substantially, and shrinkage of the particle assembly occurs. Grain boundaries form, and some grains grow while others shrink. This stage continues while the pore channels are connected, called *open porosity*, but is considered over when the pores are isolated or in closed porosity. The majority of shrinkage occurs during intermediate sintering, and the relative density at the end of this stage is about 0.9.
- *Final-stage sintering* occurs as the pores become closed and are slowly eliminated by diffusion of vacancies from the pores along the grain boundaries. Little densification occurs in this stage, although grain sizes continue to grow.

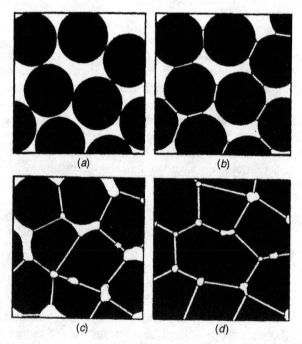

Figure 2.31 Development of ceramic microstructure during sintering: (a) Loose powder particles; (b) initial stage; (c) intermediate stage; and (d) final stage. From W. E. Lee and W. M. Rainforth, *Ceramic Microstructures*, p. 37. Copyright © 1994 by William E. Lee and W. Mark Rainforth, with kind permission of Kluwer Academic Publishers.

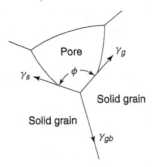

Figure 2.32 Schematic illustration of dihedral angle for solid–pore interaction.

During the initial stage, pore rounding causes a decrease in ΔG_s in proportion to the reduction of the surface area, as dictated by Eq. (2.74), but the formation of a grain boundary causes an increase in ΔG_b. The angle of intersection of the pore at the pore–grain boundary juncture (see Figure 2.32) is called the *dihedral angle*, ϕ, and is related to the relative interfacial energies of the grain boundary, γ_{gb} and the solid surface, γ_s:

$$\cos\left(\frac{\phi}{2}\right) = \frac{\gamma_{gb}}{2\gamma_s} \tag{2.75}$$

Values of ϕ range from about $105°$ to $113°$ in ceramics, implying that γ_{gb}/γ_s is 1.1 to 1.2, whereas in metals this ratio ranges from 0.25 to 0.5.

There is much more to be said concerning sintering, especially with regard to the role that diffusion plays. Many of the models that describe the sintering processes are based on the diffusivity of solid species, which are addressed in Section 4.3. The practical aspects of the sintering of metals and ceramics are described in Sections 7.1 and 7.2, respectively.

2.3 THERMODYNAMICS OF POLYMERS

Polymers don't behave like the atoms or compounds that have been described in the previous sections. We saw in Chapter 1 that their crystalline structure is different from that of metals and ceramics, and we know that they can, in many cases, form amorphous structures just as easily as they crystallize. In addition, unlike metals and ceramics, whose thermodynamics can be adequately described in most cases with theories of mixing and compound formation, the thermodynamics of polymers involves solution thermodynamics—that is, the behavior of the polymer molecules in a liquid solvent. These factors contribute to a thermodynamic approach to describing polymer systems that is necessarily different from that for simple mixtures of metals and compounds. Rest assured that free energy will play an important role in these discussions, just as it has in previous sections, but we are now dealing with highly inhomogeneous systems that will require some new parameters.

2.3.1 Solution Thermodynamics and Phase Separation

Most polymers are synthesized in solution. The solvent could be water, carbon tetra-chloride, benzene, or any number of aqueous or organic solvents. Many polymers are also processed in solution. They can be cast as solutions, and the solvent can be allowed to evaporate to leave only the polymer. So, it is critically important to understand how a polymer molecule interacts with the solvent. Again, intermolecular forces will play an important role here, but, as always, free energy will be used to determine the stability of any system we can create at any temperature.

Recall that the free energy of mixing for a binary component system is

$$\Delta G_{mix} = \Delta H_{mix} - T \Delta S_{mix} \tag{2.29}$$

and that for regular solutions, this relation becomes

$$\Delta G_{mix} = \alpha X_A X_B + RT(X_A \ln X_A + X_B \ln X_B) \tag{2.33}$$

The first term in Eq. (2.33) is the enthalpy of mixing, and the second term is the entropy of mixing. Let us examine each of these terms independently, beginning with the entropy of mixing.

For solutions of a polymer in a solvent, it can be shown, again through statistical thermodynamic arguments, that the entropy of mixing is given by

$$\Delta S_{mix} = -R(X_A \ln v_A + X_B \ln v_B) = -k_B(N_A \ln v_A + N_B \ln v_B) \tag{2.76}$$

Here k_B is Boltzmann's constant, or the gas constant per molecule, R/N_0, where N_0 is Avogadro's number (or $N_A + N_B$ for one mole of solution) and v_A and v_B are the volume fractions of solvent and polymer. Comparison of Eq. (2.76) with the entropy of mixing presented earlier in the chapter by Eq. (2.30) shows that they are similar in form, except that now the volume fractions of the components, v_A and v_B, are found to be the most convenient way of expressing the entropy change for polymers, rather than the mole fraction used for most small molecules. This change arises from the differences in size between the large polymer molecules and the small solvent molecules which would normally mean mole fractions close to unity for the solvent, especially when dilute solutions are being studied.

It can also be shown with related statistical arguments that the enthalpy of mixing for a regular polymer–solvent solution is given by

$$\Delta H_{mix} = k_B T \chi N_A v_B = RT \chi v_A v_B \tag{2.77}$$

where N_A is the number of solvent molecules, and v_A and v_B are the volume fraction of solvent and polymer molecules, respectively. Compare the second form of Eq. (2.77) with Eq. (2.31) and you will see that they have the same general form, except that the mole fractions have been replaced with volume fractions, just as for the entropy of mixing, and the interaction energy, α, is composed of the gas constant, R, temperature, T, and a new parameter, χ, which is called the *Flory–Huggins interaction parameter*, or just interaction parameter (do not confuse the interaction parameter with electronegativity, which goes by the same symbol!). Note that the interaction parameter is the interaction energy per solvent molecule per $k_B T$, and is inversely proportional to temperature. It is zero for ideal mixtures (zero enthalpy of mixing), positive for endothermic mixing, and negative for exothermic mixing. The interaction parameter is an important feature of polymer solution theory. It is an indication of how well the polymer interacts with the solvent. "Poor" solvents have values of χ close to 0.5,

Cooperative Learning Exercise 2.5

It can be shown that the volume fraction of polymer in solution, v_B, can be related to the weight fraction of polymer in solution, X_B, as follows:

$$v_B = \frac{X_B}{(1/x_n) + X_B[1 - (1/x_n)]}$$

where n is the number of repeat units in a polymer chain, or the degree of polymerization (cf. Section 1.3.5.1). (This is a good thing to sit down and prove to yourself sometime. Just recall that $X_B = N_B/(N_A + N_B)$ and $v_B/v_A = x_n N_B/N_A$.) For simplicity, let us assume a monodispersed polymer of $x_n = 500$; that is, all the polymer chains have exactly 500 repeat units.

First, work together to create a table of weight fractions in increments of 0.1 from 0.1 to 0.9; for example, $X_B = 0.1$, 0.2, and so on, and their corresponding volume fractions using the equation above. Then perform the following calculations separately.

Person 1: Use Eq. (2.30) to calculate values of $\Delta S_{mix}/R$ for the nine values of X_B and X_A. These values are the entropy of mixing for an ideal solution.

Person 2: Use Equation (2.76) to calculate values of $\Delta S_{mix}/R$ for the nine values of v_B and v_A. [*Hint*: Use the first form of Equation (2.76).] These values are the free energy of mixing for a real "athermal" polymer solution.

Combine your information to make a single, rough plot of $\Delta S_{mix}/R$ versus X_B. At what weight fraction is the difference in the entropy of mixing between an ideal and athermal polymer solution the greatest for this system?

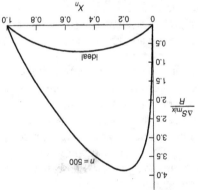

Adapted from P. Hiemenz, *Polymer Chemistry*, p. 520. Copyright © 1984 by Marcel Dekker, Inc.

Answer:
$X_B = 0.1$; $X_A = 0.9$; $v_B = 0.982$; $v_A = 0.018$; max difference occurs about $X_B = 0.2$. Polymer solutions are highly nonideal.

while "good" solvents have lower, or negative, values. The typical range for most synthetic polymer solutions is $0.3 < \chi < 0.6$ (see Table 2.5). A linear temperature dependence for χ is also predicted of the form $\chi = a + b/T$, which suggests that as the temperature increases, the solvating power of the liquid should increase.

Table 2.5   Some Polymer–Solvent Interaction Parameters at 25°C

Polymer	Solvent	Interaction Parameter χ
cis-Polyisoprene	Toluene $(V_1 = 106)^a$	0.391
	Benzene $(V_1 = 89.0)$	0.437
Polyisobutylene	Toluene	0.557
	Cyclohexane $(V_1 = 108)$	0.436
Butadiene-styrene	Benzene	0.442
71.5:28.5	Cyclohexane	0.489
Butadiene-acrylonitrile		
82:18	Benzene	0.390
70:30	Benzene	0.486
61:39	Benzene	0.564

^{a}Note that V_1 is in cubic centimeters per mole.
Source: F. Rodriguez, *Principles of Polymer Systems*, 2nd ed. Copyright © 1982 by McGraw-Hill Book Company.

Combining now the enthalpy of mixing and the entropy of mixing, along with substituting them into Eq. (2.29), gives us the free energy of mixing for a polymer solution:

$$\Delta G_{mix} = k_B T [\chi N_A v_B + N_A \ln v_A + N_B \ln v_B] \qquad (2.78)$$

From this expression, one can derive many useful relations involving experimentally obtainable quantities, such as *osmotic pressure*. We will use it to develop a phase diagram for a polymer solution.

Consider a binary component system consisting of a liquid, A, which is a poor solvent for a polymer, B. We already know that complete miscibility occurs when the Gibbs free energy of mixing is less than the Gibbs free energies of the components, and the solution maintains its homogeneity only as long as ΔG_{mix} remains less than the Gibbs free energy of any two possible coexisting phases. This situation is shown in Figure 2.33. As with our other phase diagrams, the free energy is plotted as a function of composition in the top part of the curve, and the resulting temperature–composition plot is shown in the bottom of the diagram. At high temperatures, T_5, the polymer and solvent are miscible, and a single phase is formed (Region I). As T decreases, the solution begins to separate into two phases. Phase separation, or *immiscibility*, begins to occur at the critical temperature, T_c. Here, the free energy curve begins to have maxima, minima, and inflection points, and the condition of phase stability as given by Eq. (2.36), namely, that the chemical potentials of the two phases (here labeled L1 and L2) be equal, is no longer met:

$$\mu_{L1} = \left(\frac{\partial \Delta G_{mix}}{\partial v_B} \right)_{T,P,n_{L2}} = \mu_{L2} = \left(\frac{\partial \Delta G_{mix}}{\partial v_B} \right)_{T,P,n_{L1}} \qquad (2.36)$$

(You should prove to yourself that it makes no difference if we differentiate with respect to component A or B, nor whether we differentiate with respect to mole or volume fraction, since the derivatives are set equal to zero.) The locus of points swept out by the stability condition shown in Eq. (2.36) at various temperatures is called the *binodal,* or *cloud point curve.* For example, just as we drew tangents to the

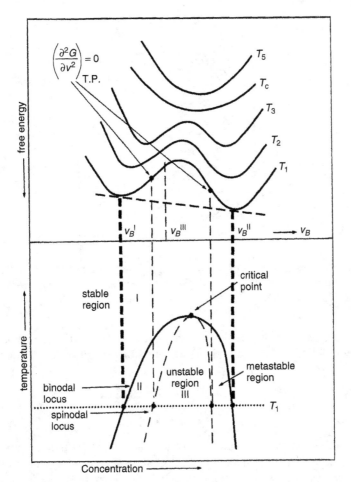

Figure 2.33 Polymer–solvent phase diagram showing binodal, spinodal, and miscibility gap. Reprinted, by permission, from J. M. G. Cowie, *Polymers: Chemistry & Physics of Modern Materials*, 2nd ed., P. 167. Copyright © 1991 by Chapman & Hall.

minima of the free energy curves in Figures 2.3 and 2.6 to map out the phase boundaries in solid–liquid equilibria, we can use tangents to the minima (line A–B) in a single free energy–composition plot for our polymer solution to establish the phase boundary—in this case, the binodal. A similar mapping of the inflection points, mathematically described by the second derivative of the free energy versus composition curve given in

$$\left(\frac{\partial^2 \Delta G_{mix}}{\partial v_B^2}\right) = 0 \tag{2.37}$$

generates the *spinodal*. The region between the binodal and spinodal curves (Region II) represents the region of two-phase metastability. Below the spinodal (Region III), the solution is unstable, phase separation occurs, and two liquid solutions are formed whose relative amounts are given by the lever rule. Note that as temperature increases, the solvating power of the solvent increases, and a single polymer–solvent phase becomes stable.

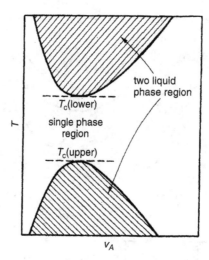

Figure 2.34   Schematic diagram of two phase regions resulting in UCST (*bottom*) and LCST (*top*). Reprinted, by permission, from J. M. G. Cowie, *Polymers: Chemistry & Physics of Modern Materials*, 2nd ed., p. 175: Copyright © 1991 by Chapman & Hall.

We can substitute our expression for the free energy of mixing in Eq. (2.78) into the stability criteria of (2.36) and (2.37) to solve for the critical interaction parameter at the onset of phase separation, χ_c

$$\chi_c = \frac{1}{2v_{A,c}^2} \qquad (2.79)$$

where $v_{A,c}$ is the volume fraction of the solvent at the critical point.

The instance we have considered here, that of a polymer in a poor solvent, results in an *upper critical solution temperature* (UCST) as shown in Figure 2.33. This occurs due to (a) decreased attractive forces between like molecules at higher temperatures and (b) increased solubility. For some systems, however, a decrease in solubility can occur, and the corresponding critical temperature is located at the minimum of the miscibility curve, resulting in a *lower critical solution temperature* (LCST). This situation is illustrated in Figure 2.34.

2.3.2   Cohesive Energy Density*

Solvent–polymer compatibility problems are often encountered in industry, such as in the selection of gaskets or hoses for the transportation of solvents. A rough guide exists to aid the selection of solvents for a polymer, or to assess the extent of polymer–liquid interactions. A semi empirical approach has been developed by Hildebrand based on the principle of "like dissolves like." The treatment involves relating the enthalpy of mixing to a *solubility parameter*, δ, and its related quantity, δ^2, called the *cohesive energy density*:

$$\Delta H_{mix} = (\delta_A - \delta_B)^2 v_A v_B \qquad (2.80)$$

For small molecules, such as solvents, the cohesive energy density is the energy of vaporization per unit volume. The value of the solubility-parameter approach is that

Table 2.6 Typical Values of the Solubility Parameter for Some Common Polymers and Solvents

Solvent	$\delta_A[(J/cm^3)^{1/2}]$	Polymer	$\delta_B[(J/cm^3)^{1/2}]$
n-Hexane	14.8	Polytetrafluoroethylene	12.7
Carbon tetrachloride	17.6	Poly(dimethyl siloxane)	14.9
Toluene	18.3	Polyethylene	16.2
2-Butanone	18.5	Polypropylene	16.6
Benzene	18.7	Polybutadiene	17.6
Cyclohexanone	19.0	Polystyrene	17.6
Styrene	19.0	Poly(methyl methacrylate)	18.6
Chlorobenzene	19.4	Poly(vinyl chloride)	19.4
Acetone	19.9	Poly(vinyl acetate)	21.7
Tetrahydrofuran	20.3	Poly(ethylene terephthalate)	21.9
Methanol	29.7	66-Nylon	27.8
Water	47.9	Polyacrylonitrile	31.5

Source: F. W. Billmeyer, *Textbook of Polymer Science*, 3rd ed. Copyright © 1984 by John Wiley & Sons, Inc.

δ can be estimated for both polymer and solvent. As a first approximation and in the absence of strong interactions such as hydrogen bonding, solubility can be expected if $\delta_A - \delta_B$ is less than 3.5 to 4.0, but not if appreciably larger. A few typical values of δ are given in Table 2.6. Values are readily available in the literature. Values of δ for polymers of known structure can be estimated using the *molar-attraction constants*, E of Table 2.7:

$$\delta_B = \frac{\rho \sum E}{M_0} \qquad (2.81)$$

where ρ is the polymer density, and M_0 is the repeat unit molecular weight.

2.3.3 Polymer Blends*

A *polymer blend* is created when two miscible polymers are mixed together. As in the case of composites, the impetus for creating polymer blends is to combine attributes of the two polymers to create a new material with improved performance. In practice, this is difficult with polymer–polymer solutions, since most common polymers do not mix well with one another to form homogeneous, one-phase solutions.

2.3.3.1 Flory–Huggins Approach. One explanation of blend behavior lies in the thermodynamics of the preceding section, where instead of a polymer–solvent mixture, we now have a polymer–polymer mixture. In these instances, the heat of mixing for polymer pairs (labeled 1 and 2) tends to be endothermic and can be approximated using the solubility parameter. The interaction parameter for a polymer–polymer mixture, χ_{12}, can be approximated by

$$\chi_{12} = \frac{V_0(\delta_1 - \delta_2)^2}{RT} \qquad (2.82)$$

Table 2.7 Molar Attraction Constants

Group	$E[(\text{J-cm}^3)^{1/2}/\text{mol}]$	Group	$E[(\text{J-cm}^3)^{1/2}/\text{mol}]$
$-CH_3$	303	NH_2	463
$-CH_2-$	269	$-NH-$	368
$>CH-$	176	$-N-$	125
$>C<$	65	$C\equiv N$	725
$CH_2=$	259	NCO	733
$-CH=$	249	$-S-$	429
$>C=$	173	Cl_2	701
$-CH=$ aromatic	239	Cl primary	419
$>C=$ aromatic	200	Cl secondary	425
$-O-$ ether, acetal	235	Cl aromatic	329
$-O-$ epoxide	360	F	84
$-COO-$	668	Conjugation	47
$>C=O$	538	cis	-14
$-CHO$	599	trans	-28
$(CO)_2O$	1159	Six-membered ring	-48
$-OH \rightarrow$	462	ortho	-19
OH aromatic	350	meta	-13
$-H$ acidic dimer	-103	para	-82

Source: F. W. Billmeyer, *Textbook of Polymer Science*, 3rd ed. Copyright © 1984 by John Wiley & Sons, Inc.

where the reference volume, V_0, typically assumes a value of 100 cm³/ mol. The critical value of χ_{12} at the onset of phase separation, $\chi_{12,c}$, can be estimated from

$$\chi_{12,c} = \frac{1}{2}\left(\frac{1}{x_1^{1/2}} + \frac{1}{x_2^{1/2}}\right)^2 \tag{2.83}$$

where x_i is the degree of polymerization for each polymer, related to its actual degree of polymerization, x_n, and the reference volume by

$$x_i = x_n\left(\frac{V_R}{V_0}\right) \tag{2.84}$$

with V_R the molar volume of the repeat unit.

Table 2.8 Complementary Groups Found in Miscible Polymer Blends

Group 1	Group 2
1. $-(CH_2-CH)-$ phenyl	$-(CH_2-CH)-$ $O-CH_3$
2. $-(CH_2-CH)-$ phenyl	$-(\overset{R}{\underset{R}{\bigcirc}}-O)-$
3. $-(CH_2-CR)-$ $O=C-OCH_3$	$-(CH_2-CF_2)-$
4. $-(CH_2-CR)-$ $O=C-OCH_3$	$-(CH_2-CH)-$ Cl
5. $-(R_1-O-\underset{O}{\overset{\parallel}{C}}-R_2-\underset{O}{\overset{\parallel}{C}}-O)-$	$-(CH_2-CH)-$ Cl
6. $-(CH_2-CH)-$ $O-\underset{O}{\overset{\parallel}{C}}-CH_3$	$\overset{\backslash}{\underset{ONO_2}{|}}$
7. $-(CH_2-CH)-$ $O=C-O-CH_3$	$\overset{\backslash}{\underset{ONO_2}{|}}$
8. $-\bigcirc-O-\bigcirc-\underset{O}{\overset{O}{\overset{\parallel}{\underset{\parallel}{S}}}}-$	$-(CH_2-CH_2-O)-$

Source: J. M. G. Cowie, *Polymers: Chemistry & Physics of Modern Materials*, 2nd ed. Copyright © 1991 by Chapman & Hall.

The situation will change if the enthalpy of mixing is negative, as this will encourage mixing. The search for miscible polymer blends has focused on combinations in which specific intermolecular interactions, such as hydrogen bonds, dipole–dipole interactions, ion–dipole interactions, or charge transfer complex formation, exist. It is now possible to identify certain groups or repeat units, which when incorporated in polymer chains tend to enter into these intermolecular attractions. A short list of some of these complementary groups is given in Table 2.8, where a polymer containing groups from column 1 will tend to form miscible blends with polymers containing groups from column 2.

2.4 THERMODYNAMICS OF COMPOSITES

Much of what we need to know about the thermodynamics of composites has been described in the previous sections. For example, if the composite matrix is composed of a metal, ceramic, or polymer, its phase stability behavior will be dictated by the free energy considerations of the preceding sections. Unary, binary, ternary, and even higher-order phase diagrams can be employed as appropriate to describe the phase behavior of both the reinforcement or matrix component of the composite system. At this level of discussion on composites, there is really only one topic that needs some further elaboration: a thermodynamic description of the interphase. As we did back in Chapter 1, we will reserve the term "interphase" for a phase consisting of three-dimensional structure (e.g., with a characteristic thickness) and will use the term "interface" for a two-dimensional surface. Once this topic has been addressed, we will briefly describe how composite phase diagrams differ from those of the metal, ceramic, and polymer constituents that we have studied so far.

2.4.1 Interphase Formation via Adhesion, Cohesion, and Spreading

We will see in subsequent chapters, particularly those concerning mechanical properties, that the physical properties of a composite are dictated to a large degree by the interphase between the matrix and reinforcement phases. The interphase plays an increasingly larger role as its proportion relative to the other constituents increases; for example, the surface-to-volume ratio of the reinforcement increases as its characteristic dimension decreases. As a result, it is important to understand the energetics of interphase formation. This is typically done by dividing the formation process into a combination of three possible processes: *adhesion, cohesion*, and *spreading*.

Let us consider two hypothetical phases in our composite, A and B, without specifying their physical state. They could be a polymer melt and a glass fiber reinforcement during melt infiltration processing, a metal powder and ceramic powder that are being subjected to consolidation at elevated temperature and pressure, or two immiscible polymer melts that will be co-extruded and solidified into a two-phase, three-dimensional object. In any case, the surface that forms between the two phases is designated AB, and their individual surfaces that are exposed to their own vapor, air, or inert gas (we make no distinction here) are labeled either A or B. The following three processes are defined as these surfaces interact and form:

Cohesion	No surface $\rightarrow$ 2A (or 2B) surfaces
Adhesion	1 AB surface $\rightarrow$ 1 A + 1 B surface
Spreading (B on A)	1 A surface $\rightarrow$ 1 AB + 1 B surface

These processes are schematically illustrated in Figure 2.35. We are interested in the free energy change that accompanies each process.

The separation process shown in Figure 2.35a consists of the formation of two new interfaces, each of unit cross-sectional area, at a location where no interface previously existed. The free energy change associated with the separation process comes directly from the definition of surface energy [Eq. (2.61)] where two surfaces of unit surface area are formed. With appropriate assumptions regarding constant temperature, pressure, and incompressible fluids, we can equate this free energy change with the

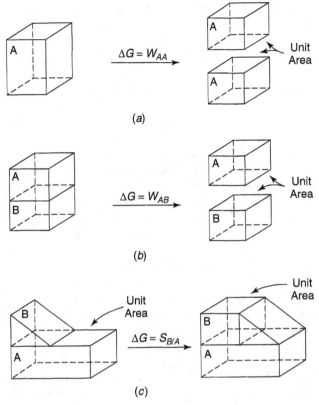

Figure 2.35 Schematic illustrations of the processes for which ΔG equals (a) the work of cohesion, (b) the work of adhesion, and (c) the work of spreading. Reprinted, by permission, from P. Heimenz, *Principles of Colloid and Surface Chemistry*, 2nd ed., p. 316. Copyright © 1986 by Marcel Dekker, Inc.

work required to form the two surface, which is called the *work of cohesion* and is labeled W_{AA}:

$$\Delta G = 2\gamma_A = W_{AA} \tag{2.85}$$

Here γ_A is the surface energy of component A.

For the case where A is separated from B, as shown in Figure 2.35b, we once again have the formation of two free surfaces, A and B, but must subtract from the energy associated with their formation the energy that was associated with the interface AB, γ_{AB}, in order to obtain the free energy change of this process, which is called adhesion. Again, equating work with the free energy change leads to the *work of adhesion*, W_{AB}.

$$\Delta G = \gamma_A + \gamma_B - \gamma_{AB} = W_{AB} \tag{2.86}$$

The work of adhesion measures the degree of attraction between the two phases. For the specific case where A is a solid (S) and B is a liquid (L), γ_A becomes the solid surface energy, γ_S, γ_B become the liquid surface tension, γ_L, and Young's equation, Eq. (2.72), can be inserted into Eq. (2.86) to obtain a relationship between the liquid–solid contact angle, θ, liquid surface tension and the work of adhesion:

$$W_{SL} = \gamma_L (1 + \cos\theta) \tag{2.87}$$

Finally, for the case of spreading of B on A shown in Figure 2.35c, the free energy change is

$$\Delta G = \gamma_{AB} + \gamma_B - \gamma_A \tag{2.88}$$

As usual with free energies, a negative value means that the process occurs spontaneously. We define a quantity called the *spreading coefficient*, $S_{B/A}$, which is the negative of the free energy of spreading, such that a thermodynamically favorable spreading process (negative free energy) leads to a positive value of the spreading coefficient. By combining Eqs. (2.85), (2.86), and (2.88), we note that the spreading coefficient can be related to the work of adhesion and work of cohesion:

$$S_{B/A} = W_{AB} - W_{BB} \tag{2.89}$$

If $W_{AB} > W_{BB}$, the A–B interaction is sufficiently strong to promote the wetting of A by B. Conversely, no wetting occurs if $W_{BB} > W_{AB}$, since the work required to overcome the attraction between two B molecules is not compensated for by the attraction between A and B. Hence, a negative spreading coefficient means that B will not spread over A.

Cooperative Learning Exercise 2.6

Consider the case of an adhesive joint created by bonding two poly(vinyl chloride) (PVC, component A) sheets together with an epoxy adhesive (component B). The surface energies for the PVC solid surface, liquid epoxy, and PVC/epoxy interface are 47 erg/cm^2, 41.7 erg/cm^2, and 4.0 erg/cm^2, respectively.

Person 1: Calculate the work of cohesion within the epoxy adhesive. This is the work required to separate epoxy bonds from one another.

Person 2: Calculate the work of adhesion between one surface of the PVC sheet and the liquid epoxy. This is the work required to separate the epoxy from the PVC.

Combine your results to calculate the spreading coefficient, $S_{B/A}$. Will the epoxy wet the PVC? Assuming that complete wetting can occur (and that a sufficiently thin layer of adhesive is applied), how will the adhesive joint fail when stressed; *i.e.*, will the epoxy or the PVC/epoxy interface fail first?

Answer: $W_{BB} = 2(41.7) = 83.4$ ergs/cm^2; $W_{AB} = 47 + 41.7 - 4 = 84.7$ erg/cm^2; $S_{B/A} = 84.7 - 83.4 = 1.3$ erg/cm^2; The epoxy will wet the PVC. Since $W_{AB} > W_{BB}$, cohesive failure will occur prior to adhesive failure, and the epoxy will fail before the interface.

2.4.2 Composite Phase Diagrams*

Unlike the unary, binary, and ternary phase diagrams of the previous sections, there are no standardized guidelines for presenting phase information in composite systems. This is due, in part, to the fact that even when two of the composite components are well mixed, the components are usually easily identified on the macroscopic scale. These macroscopic heterogeneities, while potentially beneficial from a functional standpoint, make it difficult to apply classical thermodynamics, which are based on interactions at the molecular level. Nevertheless, the free energy concept can be applied to composite systems, and phase diagrams can be generated from free energy expressions. For example, polymer–clay composites have been modeled in order to study their thermodynamic stability at various concentrations [1–4]. Clay particles are modeled as rigid disks dispersed in the polymer matrix. The composition of the mixture is characterized by the volume fraction of disks. Such factors as the disk length L, disk diameter D, polymer–disk interaction parameter χ, and disk orientation play important roles in the free energy expression. The free energy of mixing for the polymer–disk mixture has the following terms:

$$\Delta G_{mix} = \Delta G_{conf} + \Delta G_{ster} + \Delta G_{int} + \Delta G_{transl} \qquad (2.90)$$

where the various free energy terms are associated with conformational losses due to alignment of the disks, steric interactions between disks, nonspecific attractive forces between disks, and translational entropy of the disks, respectively. With appropriate assumptions and simplifications, the following free energy expression can be obtained:

$$\Delta G_{mix} = n_d[C + \ln \phi_d + \sigma - (b/v_d)(\rho - 1)\ln(1 - \phi_d) + \chi v_d \phi_d] + n_p \ln \phi_p \quad (2.91)$$

where n_d is the number of disks in a unit volume, n_p is the number of polymer molecules in a unit volume, v_d is the volume of an individual disk $(\pi D^2 L/4)$, ϕ_d is the volume fraction of disks in the composite such that the volume fraction of polymer ϕ_p is $(1 - \phi_d)$, σ is the conformational free energy (ΔG_{conf}), ρ is a function of the angle between two disks, and b is $\pi_2^2 D^3/16$. As with all the phase diagrams we have generated, the equilibrium conditions are applied; that is, the chemical potentials of any

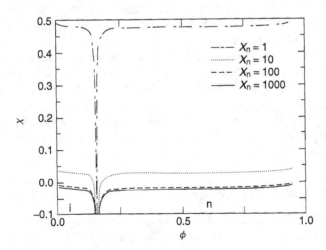

Figure 2.36 Phase diagram for clay disks and polymers of different degrees of polymerization, x_n. Reprinted with permission from Y. Lyatskaya and A. C. Balazs, *Macromolecules*, Vol. 31, p. 6676. Copyright © 1998 by the American Chemical Society.

two phases are equivalent at equilibrium. The chemical potentials are obtained directly from Eq. (2.91) by differentiation in accordance with Eq. (2.36). An example of the resulting phase diagram for the polymer–clay system is shown in Figure 2.36. Here the phase diagram is presented as a plot of the interaction parameter versus volume fraction of the disks. The diameter of the disks in this particular system is 30 arbitrary units, and their length is 1. The regions of disk–polymer miscibility lie below the boundaries. Within the regions of miscibility, i indicates an isotropic phase, and n indicates a nematic liquid crystalline phase (see Section 1.3.6.3). Other systems that have been studied include Al_2O_3 fibers in NiAl matrices [5], reinforced metal–matrix composites [6], Fe–TiC composites [7], and glass-fiber–polypropylene composites [8].

2.5 THERMODYNAMICS OF BIOLOGICS

Recall from Section 1.5 that tissue is composed not only of cells, but of an extracellular matrix, of which proteins are an important constituent. In this Section, we are concerned primarily with the thermodynamics that dictate protein behavior in biological systems. Let us begin by considering the insertion of a synthetic material into a tissue sample—for example, a metal catheter into the skin, or a metal–ceramic composite artificial hip into a human body. What happens? Does the foreign material stay isolated from the body? Obviously, it interacts with the tissue cells in some fashion. It can be argued that in the case of a hip implant, what we would like to see is cell adhesion—cells sticking to the material in an attempt to integrate it into the body. It turns out that one of the extracellular matrix proteins we described in Chapter 1, fibronectin, plays an important role in cell attachment and spreading. In this process, fibronectin spreads on the foreign surface, after which point the cell attaches to the fibronectin via a specific type of interaction called *receptor–ligand binding* (see Figure 1.94). What we are concerned with here is the process that allows a specific chemical constituent in a cell or a protein, such as fibronectin, called the *ligand*, to interact, or bind, to the corresponding specific chemical constituent in another cell membrane, called the *receptor*. In this process, the ligand, which is sort of a chemical "key," fits into the chemical "lock" of the receptor, which leads to a *receptor–ligand complex*. Receptor–ligand binding is a type of *specific binding*; that is, there is one receptor that is intended to interact with one ligand. There are other types of *nonspecific binding* that can occur and contribute to a small extent to the overall binding of the cell to the protein. Let us examine these two situations independently, by first describing the adhesion and spreading of a cell on a surface and then examining specific cell–cell interactions. The discussion at this point will be purely thermodynamic in nature. The kinetics of the process—that is, how fast or slow the adhesion and spreading occurs—is the subject of Chapter 3.

2.5.1 Cell Adhesion and Spreading on Surfaces

For cells the processes of adhesion and spreading on surfaces, are two distinctly separate processes. To date, only adhesion has been described in a thermodynamic way. Cell spreading, on the other hand, is difficult to describe in this manner, due to cellular activity, such as protein production and cytoskeleton transport which complicate the spreading process. As a result, the strength of adhesion does not correlate well with the amount of spreading the cell has undergone. Let us look first at the thermodynamics of cell adhesion on a substrate, then separately at the spreading process.

The free energy change for adhesion of cells from a liquid suspension onto a solid substrate is given by

$$G_{adh} = \gamma_{cs} - \gamma_{cl} - \gamma_{sl} \tag{2.92}$$

where the interfacial free energies are for cell–solid (γ_{cs}), cell–liquid (γ_{cl}), and solid–liquid (γ_{sl}) interfaces, respectively. If the free energy change of adhesion is negative, the adhesion process is spontaneous. If it is positive, the process is thermodynamically unfavorable. As a result, neither extremely hydrophilic nor extremely hydrophobic ($\gamma_{sl} < 40$ dyn/cm) substrates promote adhesion of cells such as fibroblasts. This trend, while still the subject of much experimental investigation and verification, is illustrated in Figure 2.37. Here, area A represents a nonadhesive zone between 20 and 30 dyn/cm critical surface energy. Area B represents biomaterials with good adhesive properties. Other factors such as surface charge, surface topography, porosity, and mechanical forces can greatly influence the ability of a cell to adhere to a surface.

Cooperative Learning Exercise 2.7

It has been observed that receptor–ligand binding processes are generally exothermic ($\Delta H < 0$). Protein adsorption on surfaces can be described with thermodynamic arguments similar to those used for cell adhesion—that is, with free energy arguments similar to those in Eqs. (2.12) and (2.92). For example, the enthalpies of adsorption for two proteins, HSA and RNase, on α-Fe$_2$O$_3$ and negatively charged polystyrene (PS-H), respectively, at various pH levels are given below.

Protein	Substrate	pH	Enthalpy (mJ/m^2)
HSA	α-Fe$_2$O$_3$	5	+1.9
		7	+7.0
RNase	PS-H	5	+4
		11	−2

Discuss with your neighbor how protein adsorption can still be spontaneous ($\Delta G < 0$), even though some of these processes are endothermic ($\Delta H > 0$). Look up values for the solid surface energies of the two substrates in Appendix 4 (you may have to select similar materials, but this is sufficient for these purposes). How do these enthalpies compare in magnitude to the respective surface energies?

Answer: According to Equation (2.12), the entropy must also be considered. In fact, many of these binding/adhesion processes are entropy-driven—even the exothermic ones. The surface energy of Fe$_2$O$_3$ is probably similar to FeO—about 500 erg/cm^2, and PS is around 33 erg/cm^2. Thus, the enthalpy of adsorption is much larger relative to the surface energy for the RNase/PS system than for the HSA/Fe$_2$O$_3$ system. Note that

$$1 \text{ mJ/m}^2 = 1 \text{ erg/cm}^2.$$

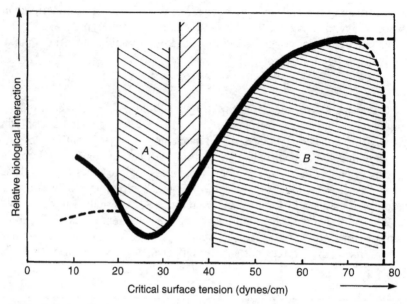

Figure 2.37   Relationship between the biological interactiveness and the critical surface tension of a biomaterial. Reprinted, by permission, from J. M. Schakenraad, Cells: Their Surfaces and Interactions with Materials, in *Biomaterials Science: An Introduction to Materials in Medicine*, B. D. Ratner, A. S. Hoffman, F. J. Schoen, and J. E. Lemons, p. 146. Copyright © 1996 by Academic Press.

Once the cell is adhered to the surface, it may undergo spreading. Cell spreading is the combined process of continued adhesion and cytoplasmic contractile meshwork activity. This activity involves the formation of *microfilaments* and *lamellar protrusions*. When a cell meets the membrane of another cell, further spreading is prohibited. Due to these complexities in the spreading process, it is difficult to give a purely thermodynamic explanation for cell spreading. Nonetheless, some correlation between the surface energy of the substrate and the ability of a cell to spread upon it has been noted (see Figure 2.38). Here, the presence of preadsorbed proteins (solid line in Figure 2.38) can affect the ability of the cell to spread relative to a surface free of preadsorbed proteins (dashed line in Figure 2.38).

2.5.2 Cell–Cell Adhesion

Consider two cells, capable of binding to one another (see Figure 2.39). The surface of a cell consists of a 100-Å-thick layer called the *glycocalyx*, which contains the surface receptors for binding. There are R_{1T} and R_{2T} total surface receptors, respectively, where R_1 is the ligand for R_2 and R_2 is the ligand for R_1. Let B represent the number of *receptor–ligand complexes* acting to bridge one cell to another, called *bridges*. The number of free receptors within and outside the cell–cell contact area are R_{ic} and R_{io}, respectively, where $i = 1, 2$. Assuming a constant number of total surface receptors on each cell, the conservation of receptors for cell i will be

$$R_{iT} = B + R_{ic} + R_{io} \tag{2.93}$$

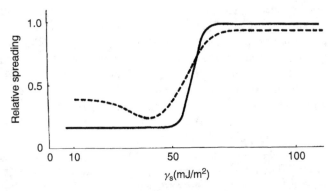

Figure 2.38 Influence of substrate surface energy on cell spreading with preadsorbed serum proteins (solid line) and without (dashed line). Reprinted, by permission, from J. M. Schakenraad Cells: Their Surfaces and Interactions with Materials, in *Biomaterials Science: An Introduction to Materials in Medicine*, B. D. Ratner, A. S. Hoffman, F. J. Schoen, and J. E. Lemons, p. 144. Copyright © 1996 by Academic Press.

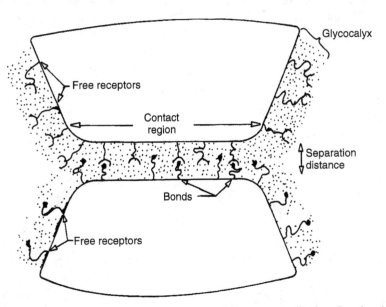

Figure 2.39 Schematic illustration of cell–cell and cell-substrate adhesion. Reprinted, by permission, from *Biophysical Journal*, **45**, 1051. Copyright © 1984 by the Biophysical Society.

The corresponding receptor densities are $n_{iT} = R_{iT}/A_i$, $n_b = B/A_c$, $n_{ic} = R_{ic}/A_c$, and $n_{io} = R_{io}/(A_i - A_c)$, where A_i is the total surface area of cell i and A_c is the contact area. We will assume that the distance between the cell membranes is constant and equal to s.

The Gibbs free energy for the system can expressed as

$$\Delta G = \sum_j (N_j n_j \mu_j) + A_c \Gamma(s) \tag{2.94}$$

where N_j is the number of receptor species (either within, R_{ic}, or outside, R_{io}, of the contact area or as a bridge, B), n_j is the density of each of the species (n_{ic}, n_{io}, or n_b), and μ_j is the chemical potential of each receptor state. The quantity $\Gamma(s)$ is called the *nonspecific interaction energy* and is the mechanical work that must be done against nonspecific repulsive forces to bring a unit area of membrane from an infinite separation to a cell–cell separation distance of s. The repulsive forces that must be overcome are steric and electrostatic in nature, whereas the attractive forces are of the van der Waals type (see Figure 2.40). At small separation distances, repulsive forces dominate the cell–cell interaction, and diminish at values of s greater than 200 Å. This is important because cell–cell and cell–surface adhesion occur with separation distances between 100 and 300 Å. As a result, repulsive forces dominate, and van der Waals forces are often neglected. So, the nonspecific interaction energy can be modeled empirically as a separation-distance-dependent net repulsion

$$\Gamma(s) = \frac{\sigma}{s} \exp\left(\frac{-s}{\delta_g}\right) \tag{2.95}$$

where σ is a measure of the glycocalyx stiffness and δ_g is its thickness. Substitution of (2.95) into (2.94) and expansion of the individual free energy terms of the

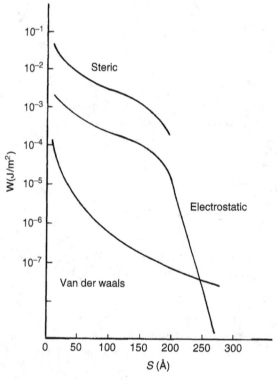

Figure 2.40   Nonspecific work of adhesion required to separate two like cells as a function of cell–cell separation distances, Reprinted, by permission, from P. Bongrand and G. I. Bell, *Cell Surface Dynamics: Concepts and Models*, A. S. Perelson, C. DeLisi, and F. W. Wiegel, ed., pp. 459–493. Copyright © 1984 by Marcel Dekker.

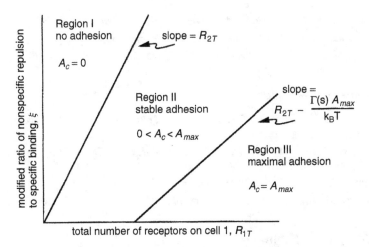

Figure 2.41   Predicted phase diagram for cell–cell adhesion. Reprinted, by permission, from D. A. Lauffenburger and J. J. Linderman, *Receptors*, p. 281. Copyright © 1993 by Oxford University Press.

receptor species as described above yields the final form of the free energy change for cell–cell adhesion:

$$\Delta G = R_{ic} n_{ic} \mu_{ic} + R_{io} n_{io} \mu_{io} + B n_b \mu_b + \frac{A_c \sigma}{s} \exp\left(\frac{-s}{\delta_g}\right) \qquad (2.96)$$

As with all free energy expressions in this chapter—minimization with respect to concentration, in this case—the concentration of receptor sites provides the information for the phase diagram, an example of which is shown in Figure 2.41. There are three regions for cell–cell adhesion in this diagram, according to the values R_{1T}, the total number of receptors on cell 1, and a parameter ξ, which is a dimensionless quantity that characterizes a modified ratio of nonspecific binding repulsion to binding. In region I, no stable adhesive interaction occurs between the cells. The attractive forces offered by the receptor–ligand bonds are not strong enough to overcome the repulsive forces resulting from cell–cell contact. In region II, stable cell–cell adhesion exists with a contact area less than A_{max}. Here, a balance exists between attractive and repulsive forces. In region III, stable cell–cell adhesion results and the contact area is A_{max}. Here, attractive forces outweigh repulsive forces and the area of contact is limited only by geometric and morphological constraints.

REFERENCES

Cited References

1. Lyatskaya, Y., and A. C. Balazs, *Macromolecules*, **31**, 6676 (1998).
2. Ginzburg, V. V., and A. C. Balazs, *Macromolecules*, **32**, 5681 (1999).
3. Ginzburg, V. V., C. Singh, and A. C. Balazs, *Macromolecules*, **33**, 1089 (2000).
4. Huh, J., V. V. Ginzburg, and A. C. Balazs, *Macromolecules*, **33**, 8085 (2000).
5. Hu, W., W. Wunderlich, and G. Gottstein, *Acta Mater.*, **44**(6), 2383 (1996).

6. Asthana, R., P. K. Rohatgi, and S. N. Tewari, *Proc. Adv. Mater.*, **2**, 1 (1992).

7. Popov, A., and M. M. Gasik, *Scripta Mater.*, **35**(5), 629 (1996).

8. Avalos, F., M. A. Lopez-Manchado, and M. Arroyo, *Polymer*, **39**(24), 6173 (1998).

Thermodynamics — General

Kyle, B. G. *Chemical and Process Thermodynamics*, 3rd ed., Prentice-Hall, Upper Saddle River, NJ, 1999.

Smith, J. M., van Ness, H. C., and M. Abbott, *Introduction to Chemical Engineering Thermodynamics*, 6th ed., McGraw-Hill, New York, 2001.

IUPAC Compendium of Chemical Terminology

www.iupac.org/publications/compendium/index.html

Thermodynamics of Materials

Gaskell, D. R., *Introduction to the Thermodynamics of Materials*, 3rd ed., Taylor & Francis, Washington, D.C., 1995.

http://www.tms.org/pubs/journals/JOM/9712/Kattner-9712.html

Thermodynamics of Metals and Alloys

Gaskell, D. R., *Introduction to Metallurgical Thermodynamics*, McGraw-Hill, New York, 1981.

Devereux, O. F., *Topics in Metallurgical Thermodynamics*, John Wiley & Sons, New York, 1983.

Rao, Y. K., *Stoichiometry and Thermodynamics of Metallurgical Processes*, Cambridge University Press, Cambridge, 1985.

http://cyberbuzz.gatech.edu/asm_tms/phase_diagrams/

Thermodynamics of Ceramics and Glasses

Kingery, W. D., H. K. Bowen and D. R. Uhlmann, *Introduction to Ceramics*, John Wiley & Sons, New York, 1976.

Scholze, H., *Glass*, Springer-Verlag, New York, 1991.

http://www.ceramics.nist.gov/webbook/glossary/ped/glossary.htm

Thermodynamics of Polymers

Cowie, J. M. G., *Polymers: Chemistry & Physics of Modern Materials*, 2nd ed., Chapman & Hall, New York, 1991.

Polymer Blends, D. R. Paul and C. B. Bucknall, eds., Vols. 1–2, Wiley-Interscience, New York, 2000.

Strobl, G., *The Physics of Polymers*, 2nd ed., Springer, Heidelberg, 1997.

Thermodynamics of Biologics

Matthews, F. L., and R. D. Rawlings, *Composite Materials: Engineering and Science*, CRC Press, Boca Raton, FL, 1999.

Jou, D., and J. E. Llebot, *Introduction to the Thermodynamics of Biological Processes*, Prentice-Hall, Englewood Cliffs, NJ, 1990.

Thermodynamic Data for Biochemistry and Biotechnology, H. J. Hintz, ed., Springer-Verlag, Berlin, 1986.

Molecular Dynamics: Applications in Molecular Biology, J. M. Goodfellow, ed., CRC Press, Boca Raton, FL, 1990.

Lauffenburger, D. A. and J. J. Linderman, *Receptors*, Oxford University Press, New York, 1993.

Computer Simulations

Rowley, R., *Statistical Mechanics for Thermophysical Property Calculations*, Prentice-Hall, Englewood Cliffs, NJ, 1994.

http://www.calphad.org

PROBLEMS

Level I

2.I.1 Draw schematic ΔG_{mix} versus X_{Ni} curves for the Cu-Ni system shown in Figure 2.4 at 1455, 1300, 1100, and 300°C.

2.I.2 Compare the relative efficiencies of H_2 and CO as reducing agents for metal oxides; for example, if cobalt oxide, CoO, were to be reduced to the pure metal, which would do a better job and at what temperatures?

2.I.3 At 200°C, a 50:50 Pb-Sn alloys exists as two phases, a lead-rich solid and a tin-rich liquid. Calculate the degrees of freedom for this alloy and comment on its practical significance.

2.I.4 For 1 kg of eutectoid steel at room temperature, calculate the amount of each phase (α and Fe_3C) present.

2.I.5 Calculate the amount of each phase present in 1 kg of a 50:50 Ni-Cu alloys at a) 1400°C, b) 1300°C, and c) 1200°C.

2.I.6 Calculate the amount of each phase present in 50 kg of free-cutting brass at 200°C. Neglect the presence of alloying elements less than 3% by weight.

2.I.7 Schematically sketch and label the resulting microstructure of an austenite phase containing 0.65 wt% C that is cooled to below 727°C. Callister, p. 292, problem 9.54 (d).

Level II

2.II.1 We will see in Chapter 5 that the strength of a material, σ, is the work required to separate surfaces divided by the separation distance. In this way, the cohesive strength, σ_c, is the work of cohesion, W_{BB}, divided by some characteristic intermolecular distance, x_{BB}. Similarly, the adhesive strength, σ_a, is the work of adhesion, W_{AB}, divided by some characteristic intermolecular distance, x_{AB}. (a) Assuming that $x_{BB} = x_{AB} \approx 5.0$ Å for both the epoxy adhesive and the PVC–epoxy interface, calculate the cohesive strength and adhesive strength of the PVC–epoxy composite joint in Cooperative Learning Exercise 2.6. Which has a higher strength, the epoxy with itself or the PVC–epoxy interface? Recall that 1 erg/cm^2 = 1 dyn/cm. (b) Now assume that the thickness of the interface,

x_{AB}, is increased. At what value of x_{AB} does mechanical failure switch from cohesive to adhesive? What does this tell you about the optimal thickness of an adhesive layer if it is to be stronger than the materials it binds together?

2.II.2 Consider 1 kg of a brass with composition 30 wt% Zn. (a) Upon cooling, at what temperature would the first solid appear? (b) What is the first solid phase to appear, and what is its composition? (c) At what temperature will the alloy completely solidify? (d) Over what temperature range will the microstructure be completely in the α-phase?

2.II.3 You have supplies of kaolin, silica, and mullite as raw materials. Using kaolin plus either silica or mullite, calculate the batch composition (in weight percent) necessary to produce a final microstructure that is equimolar in silica and mullite.

2.II.4 Hot isostatic pressing (HIP), in which an inert gas is applied at high pressure, can be used to consolidate and densify ceramic components. The driving force for pore shrinkage in such cases is the sum of the applied pressure, P_e, and the surface energy-driven component, as given by Eq. 2.69. Calculate the ratio of driving forces for HIPing at 35 MPa compared to conventional sintering. Assume a pore diameter of 5 μm and $\gamma_s = 700$ mN/m at the densification temperature.

2.II.5 The melting points of a series of poly(α-olefin) crystals were studied. All of the polymers were isotactic and had chains substituents of different bulkiness. The results are listed below. Use Eq. (2.12) to derive a relationship between the melting point, T_m, and the enthalpy and entropy of fusion, ΔH_f and ΔS_f, respectively. Use this relationship, plus what you know about polymer crystallinity and structure from Chapter 1, to rationalize the trend in melting point.

Substituent	$T_m(^\circ C)$
$-CH_3$	165
$-CH_2CH_3$	125
$-CH_2CH_2CH_3$	75
$-CH_2CH_2CH_2CH_3$	-55
$-CH_2CH-CH_2-CH_3$ $\quad\quad\vert$ $\quad\quad CH_3$	196
$\quad\quad CH_3$ $\quad\quad\vert$ $-CH_2-C-CH_2-CH_3$ $\quad\quad\vert$ $\quad\quad CH_3$	350

2.II.6 The diagram shown below qualitatively describes the phase relationships for Nylon-66, water, and phenol at $T > 70^\circ C$. Use this diagram to draw arrows that trace the following procedures: (1) Water is added to solution A until the system separates into solution B and C; (2) solution C is removed and water is added to solution B until the system separates into D and E; (3) phenol is added to solution C until a solution A′ results which is close to A in concentration. Would you expect solutions C and E, or B and D, to contain the higher-molecular-weight polymer? Briefly explain. Outline a strategy for Nylon fractionation based on steps 1–3. Some steps may be repeated as necessary.

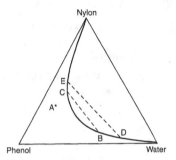

2.II.7 The following data were obtained for the work of adhesion between the listed liquids and carbon black. Use these data, together with the surface tensions of pure liquids from Appendix 4 (or from additional sources as necessary), to calculate the spreading coefficients for the various liquids on carbon black.

Liquid	W_{AB} (erg/cm^2)
Benzene	109.3
Toluene	110.2
CCl$_4$	112.4
CS$_2$	122.1
Ethyl ether	76.4
H$_2$O	126.8

Level III

2.III.1 The Maxwell relations in thermodynamics are obtained by treating a thermodynamic relation as an exact differential equation. Exact differential equations are of the form

$$dz = \left(\frac{dz}{dx}\right)_y dx + \left(\frac{dz}{dy}\right)_x dy = M\,dx + N\,dy$$

where $z = z(x, y)$ and M and N correspond to the partial derivatives of z with respect to x and y, respectively. A useful relation in exact differential equations is that

$$\left(\frac{\partial M}{\partial y}\right)_x = \left(\frac{\partial N}{\partial x}\right)_y$$

(Can you prove why this is so?) When x, y, and z are thermodynamic quantities, such as free energy, volume, temperature, or enthalpy, the relationship between the partial differentials of M and N as described above are called Maxwell relations. Use Maxwell relations to derive the Laplace equation for a sphere at constant temperature using the following thermodynamic expression for free energy, G:

$$dG = VdP - SdT + \gamma dA$$

where V is the volume of the sphere, P is the pressure, T is temperature, γ is the surface energy, and A is the sphere surface area.

2.III.2 Why does pore coalescence in a glassy material cause an increase in porosity and negative densification? (*Hint*: Consider two pores containing n moles of gas that become one pore during isothermal coalescence. The driving force for pore shrinkage is $2\gamma/r$. Use this information with the ideal gas law to calculate the change in pore radius.)

CHAPTER 3

Kinetic Processes in Materials

3.0 INTRODUCTION AND OBJECTIVES

Thermodynamics tells us whether or not a process is favorable; kinetics tells us how quickly that process will take place. The rate at which chemical and biological processes occur is of tantamount importance in many instances. For example, we all know that an exposed piece of iron will rust (oxidize) over time, but is this important if the piece will only be used once in the near future?

There are three categories of kinetic processes that are of concern to us: the rate at which materials are formed, the rate at which they are transformed, and the rate at which they decompose. In general, formation and decomposition are chemical processes, involving the reaction of two or more chemical species. Transformations, on the other hand, are usually physical processes, such as the melting of ice, and do not involve chemical reaction. We can also have situations in which both physical and chemical kinetic processes occur simultaneously. In all cases, our goal is to be able to describe the rate at which the process is occurring. There are two important concepts that help us accomplish our task: the thermodynamic concept of free energy, ΔG; and a related quantity called the activation energy, E_a. We will introduce these concepts and then see how one, or both, of them can be modified to help determine the rate of processes in the formation, transformation, and decomposition of different materials.

By the end of this chapter, you should be able to:

- Apply the law of mass action to a reaction to determine the equilibrium constant at a specified temperature.
- Calculate the activation energy and preexponential factor from kinetic data.
- Identify the dimensionality and calculate rates of a nucleation and growth process from kinetic data.
- Distinguish between the different types of corrosion in all material classes.
- Calculate the EMF of a galvanic cell from the half-cell potentials.
- Use the Nernst equation to calculate the potential of a half-cell not at unit activity.
- Distinguish between activation and concentration polarization.
- Distinguish between homogeneous and heterogeneous nucleation processes.
- Distinguish between stepwise and radical chain polymerizations.

An Introduction to Materials Engineering and Science: For Chemical and Materials Engineers,
by Brian S. Mitchell
ISBN 0-471-43623-2 Copyright © 2004 John Wiley & Sons, Inc.

- Calculate the degree of polymerization, number-average, and weight-average molecular weights given kinetic data for a polymerization reaction.
- Identify the different types of copolymers, and use the copolymer equation with appropriate kinetic data to determine which type of copolymer will form.
- Distinguish between the different types of polymerization processes, and identify the equipment and kinetics that are specific to each.
- Distinguish between the different types of polymer degradation.
- Distinguish between the different types of vapor deposition processes, and calculate the rate of deposition for each from kinetic data.
- Determine the rate of receptor–ligand binding from kinetic data.

3.0.1 The Law of Mass Action

We saw in Chapter 2 that an important thermodynamic quantity is the Gibbs free energy, ΔG. The specific functional relationship we use to describe the free energy will depend on whether we are studying a physical or a chemical transformation. For physical processes, such as phase transformations, the most useful form of the Gibbs free energy is its definition given in Chapter 2:

$$\Delta G = \Delta H - T\Delta S \tag{2.12}$$

In the case of reacting systems, however, the number of moles is not constant, and it is more convenient to consider the total free energy as a sum of individual free energy contributions. For this case, let us begin by considering the generalized reaction shown below, in which reactants A and B go to products C and D:

$$aA + bB \rightleftharpoons cC + dD \tag{3.1}$$

where a, b, c, and d are the *stoichiometric coefficients* of species A, B, C, and D, respectively, and can be represented generically by v_i. Recall that stoichiometric coefficients are positive for products and negative for reactants. There can, of course, be more reactants and/or products than those listed here, and the reaction may be reversible. The *Law of Mass Action* states that the velocity of the reaction at a given temperature is proportional to the product of the active masses of the reacting substances. In other words, the *forward reaction rate*, r_1, is a function of the concentrations of A and B raised to their corresponding exponents:

$$r_1 = k_1[A]^a[B]^b \tag{3.2}$$

Similarly, the *reverse reaction rate* is related to the amount of products, C and D, present

$$r_2 = k_2[C]^c[D]^d \tag{3.3}$$

In Eqs. (3.2) and (3.3), k_1 and k_2 are proportionality constants, or *rate constants*, for the forward and reverse reactions, respectively, and have units of inverse time (typically seconds) and an appropriate inverse concentration, depending on the manner in which the concentration of chemical species are represented and the magnitude of the stoichiometric coefficients.

At equilibrium, which is a thermodynamic state for which $\Delta G = 0$, the rate of the forward reaction must be equal to the rate of the reverse reaction, $r_1 = r_2$, and we can equate Eqs. (3.2) and (3.3) to give:

$$\frac{k_1}{k_2} = \frac{[A]^a[B]^b}{[C]^c[D]^d} = K \tag{3.4}$$

The constant K is the *equilibrium constant* for the reaction at constant temperature.

We can relate the equilibrium constant, K, to the free energy of the system. Recall that the free energy change for a reaction must be less than zero for it to proceed spontaneously, and any system whose overall free energy change is zero is in equilibrium. The free energy of a system is simply the sum of the free energy contributions of each of the components, which are measured using their chemical potentials, μ_i [see Eq. (2.13) for the definition of chemical potential], and the stoichiometric coefficients

$$\Delta G = \sum v_i \mu_i \tag{3.5}$$

When all the components are in their standard states (see Section 2.1.2), the standard Gibbs free energy, ΔG^0, is given by:

$$\Delta G^0 = \sum v_i \mu_i^0 \tag{3.6}$$

The chemical potentials in Eq. (3.5), μ_i, and the standard chemical potentials in Eq. (3.6), μ_i^0, are related to each other by the *activity* of each species, a_i:

$$\mu_i = \mu_i^0 + RT \ln a_i \tag{3.7}$$

where R is the gas constant and T is the absolute temperature. The activity of a species, then, is an indication of how far away from standard state that species is.

If we subtract Eq. (3.6) from Eq. (3.5) and substitute Eq. (3.7) into the result, we obtain a very useful relationship between the activity of the reactants and products and the free energy for a reacting system:

$$\Delta G - \Delta G^0 = RT \sum v_i \ln a_i = RT \ln \left(\prod_i a_i^{v_i} \right) \tag{3.8}$$

At equilibrium, then, $\Delta G = 0$, and the activities of the products and reactants, a_i, are related to their respective concentrations:

$$\Delta G^0 = -RT \ln \left\{ \frac{[A]^a[B]^b}{[C]^c[D]^d} \right\} \tag{3.9}$$

and we can identify the term in the logarithm from Eq. (3.4) as the equilibrium constant, K:

$$\Delta G^0 = -RT \ln K \tag{3.10}$$

Rearranging Eq. (3.10) to express K in terms of the standard free energy change gives

$$K = \exp \left(\frac{-\Delta G^0}{RT} \right) \tag{3.11}$$

Equation (3.11) gives the effect of temperature on the equilibrium constant. Note that a plot of $\ln K$ versus $1/T$ should give a straight line of slope $-\Delta G^0/R$.

3.0.2 The Activation Energy

Since the equilibrium constant, K, is a ratio of the forward to reverse reaction rate constants, k_1 and k_2, respectively, it is reasonable to assume that the forward and reverse rate constants also follow a relationship similar in form to Eq. (3.11). If we replace the standard free energy with some general energy term for activation, E_a, and include a proportionality constant called the *preexponential factor*, k_0, we get what is called an *Arrhenius expression* for the rate constant:

$$k = k_0 \exp(-E_a/RT) \tag{3.12}$$

We will not prove the Arrhenius relationship here, but it falls out nicely from statistical thermodynamics by considering that all molecules in a reaction must overcome an activation energy before they react and form products. The Boltzmann distribution tells us that the fraction of molecules with the required energy is given by $\exp(-E_a/RT)$, which leads to the functional dependence shown in Eq. (3.12).

Many of the processes we will study in this and other chapters have parameters that appear to follow the very general Arrhenius exponential relationship. The quantity E_a is called the *activation energy* and represents the energy barrier for the reaction under consideration. As shown in Figure 3.1, the reactants can be thought of as having to overcome an activation energy barrier of height E_a in order to achieve the activated complex state and proceed to products. The *heat of reaction*, ΔH, can be either negative (exothermic reaction) or positive (endothermic reaction), in which case the products C and D actually have a higher final energy than the reactants. Let us now see how a wide

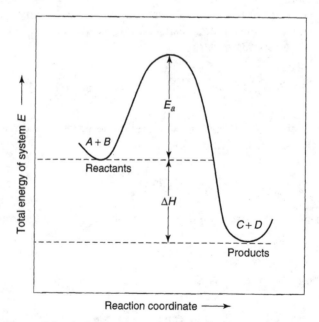

Figure 3.1 Activation energy barrier for a chemical reaction.

variety of kinetic processes in material formation, transformation, and degradation can be described using these relationships.

3.1 KINETIC PROCESSES IN METALS AND ALLOYS

We begin by examining the kinetics processes of formation reactions in metals and alloys. Although many of these substances are elements or binary mixtures of elements, they can be formed through complex reactions, such as the reduction of metal oxides. We will first examine the formation of an intermetallic compound.

3.1.1 Kinetics of Intermetallic Formation

We made a distinction between alloys and intermetallics in Chapter 1, indicating that intermetallics are more like compounds than alloys (which are more like mixtures), even though both contain two or more metallic components. Nonetheless, both are composed entirely of metallic elements, and a discussion of the formation of either would be appropriate here. We focus on the formation of an intermetallic, $TiAl_3$, to illustrate how kinetic parameters can be obtained from experimental observations.

Consider the dissolution of elemental titanium, Ti, in molten aluminum, Al, at various temperatures that lead to the formation of intermetallic $TiAl_3$ layers. Figure 3.2 shows the experimental data for the decrease in thickness of the titanium as a function of time, and Figure 3.3 shows the increase in thickness of the titanium aluminide layer with time. From these plots, the rate of dissolution and rate of formation can be determined, respectively, for each temperature. These values are presented in Cooperative Learning Exercise 3.1. Application of the Arrhenius equation (3.12) in both instances

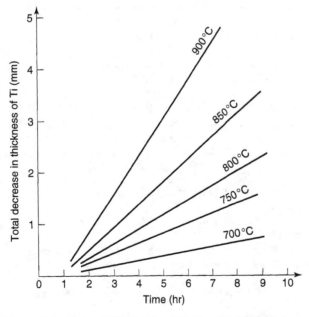

Figure 3.2 Rate of dissolution of Ti at various temperatures.

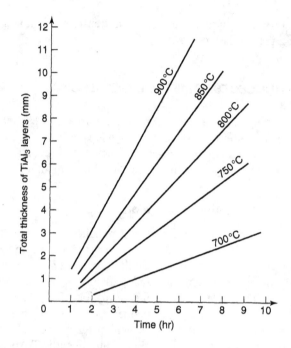

Figure 3.3 Rate of formation of TiAl$_3$ at various temperatures.

Cooperative Learning Exercise 3.1

Consider the following data for the dissolution of Ti in molten aluminum to form the intermetallic TiAl$_3$.

Temperature (°C)	Rate of Ti Dissolution, k_{Ti} (cm/s)	Rate of TiAl$_3$ Formation, k_{TiAl_3} (cm/s)
700	1.51×10^{-6}	1.042×10^{-5}
750	2.605×10^{-6}	1.98×10^{-5}
800	4.17×10^{-6}	2.777×10^{-5}
850	6.13×10^{-6}	3.798×10^{-5}
900	9.36×10^{-6}	6.805×10^{-5}

Person 1: Determine the activation energy, $E_{a,Ti}$, and preexponential factor, $k_{0,Ti}$, for the rate of dissolution of Ti.

Person 2: Determine the activation energy, $E_{a,TiAl_3}$, and preexponential factor, $k_{0,TiAl_3}$, for the rate of growth of TiAl$_3$.

Group: Compare your results and suggest a reason for your findings. What other processes could affect the rate of either of these reactions?

Answers: $E_{a,Ti} = 86$ kJ/mol; $k_{0,Ti} = 0.06$ $E_{a,TiAl_3} = 84$ kJ/mol; $k_{0,TiAl_3} = 5.3$

allows us to calculate the activation energy for both processes by plotting $\ln k_i$ versus $1/T$ (don't forget that T is in degrees Kelvin!), with the slope of each line equal to $-E_a/R$. The fact that the two activation energies calculated in Cooperative Learning Exercise 3.1 are essentially equivalent tells us that the rate of the chemical reaction between these two metals, and not the diffusion of either aluminum or titanium through the intermetallic layer formed on the surface of the solid titanium determines the overall rate of TiAl$_3$ formation. As we will see, diffusion is also an activated process and can ·be described using an Arrhenius-type relationship.

3.1.2 Kinetics of Phase Transformations in Metals and Alloys

3.1.2.1 Nucleation and Growth (Round 1). Phase transformations, such as the solidification of a solid from a liquid phase, or the transformation of one solid crystal form to another (remember allotropy?), are important for many industrial processes. We have investigated the thermodynamics that lead to phase stability and the establishment of equilibrium between phases in Chapter 2, but we now turn our attention toward determining what factors influence the rate at which transformations occur. In this section, we will simply look at the phase transformation kinetics from an overall rate standpoint. In Section 3.2.1, we will look at the fundamental principles involved in creating ordered, solid particles from a disordered, solid phase, termed *crystallization* or *devitrification*.

Consider the isothermal transformation of a disordered (amorphous) solid to an ordered (crystalline) solid. Obviously, there must be enough thermal energy to allow individual atoms to move around and reorient themselves, but assuming this is possible, it is generally found that *at constant temperature*, the amount of amorphous material transformed to crystalline material, dx (on a volume basis) per unit time, dt, is given by

$$\frac{dx}{dt} = nk(1-x)t^{n-1} \tag{3.13}$$

where the quantity $(1-x)$ is the fraction of amorphous material remaining, n is the reaction order, t is time, and k is the reaction rate constant as expressed by Eq. (3.12). The solution to this equation for the amount of amorphous materials transformed, x, at time t, is known as the *Johnson–Mehl–Avrami (JMA) equation* [1–3]:

$$x = 1 - \exp[-kt^n] \tag{3.14}$$

Taking the natural log of this equation twice yields a useful form of the JMA equation:

$$\ln[-\ln(1-x)] = \ln k + n\ln t \tag{3.15}$$

Thus, a plot of $\ln[-\ln(1-x)]$ versus $\ln t$ yields a straight line of slope n. The value of n is important not only because it indicates the reaction order, but also because it has some physical interpretation, depending on whether the crystals are growing at the surface of a material or in the bulk (sometimes called volume crystallization). See Table 3.1 for a description of the dimension of the crystals for various values of n. The intercept of this plot gives the rate constant, k. Obtaining k at different

Table 3.1 Values of the Growth Dimension, n (a.k.a Reaction Order), for Different Crystallization Processes

Mechanism	n
Bulk nucleation, 3-dimensional growth	4
Bulk nucleation, 2-dimensional growth	3
Bulk nucleation, 1-dimensional growth	2
Surface nucleation	1

isothermal temperatures and applying Eq. (3.12) gives the activation energy for the amorphous–crystalline transformation.

Most industrially relevant transformation processes are not isothermal; and even in a controlled laboratory environment, it is difficult to perform experiments that are completely isothermal. The kinetics of nonisothermal phase transformations are more complex, of course, but there are some useful relationships that have been developed that allow for the evaluation of kinetic parameters under nonisothermal conditions. One such equation takes into account the heating rate, ϕ usually in K/min, used in the experiment [4]:

$$\ln[-\ln(1-x)] = n \ln \phi + \ln k + \text{constant} \tag{3.16}$$

An equation developed by Kissinger and modified by others can be used to determine the activation energy using the temperature at which the transformation rate is a maximum, T_p, for various heating rates [5]:

$$\ln\left(\frac{\phi}{T_p^2}\right) = \text{constant} - \frac{E_a}{RT_p} \tag{3.17}$$

The temperature of maximum transformation rate is easily determined using either of two similar techniques called *differential scanning calorimetry* (DSC) or *differential thermal analysis* (DTA). These techniques are extremely useful in the kinetic study of both isothermal and nonisothermal phase transformations.

3.1.2.2 Martensitic Transformations.

Some phase transformations in the solid phase are so rapid that they approach the speed of sound. These *congruent phase transformations* (no change in composition) that involve the displacement of atoms over short distances, rather than diffusion (such as in the previous section), are termed *displasive* or *martensitic transformations*. The term "martensitic" comes from this type of transformation that is observed in the cooling of austenitic steel (see Section 2.1.3). Martensite is a supersaturated solution of carbon in ferrite with interstitial carbon atoms in the body-centered tetragonal structure that forms when face-centered cubic iron with carbon, called austenite, is rapidly cooled to about 260°C. The amount of austenite transformed to martensite is a function of the temperature only, varying from 0 to 100% between the temperatures M_s and M_f, as indicated in the accompanying time–temperature–transformation (TTT) curve shown in Figure 3.4.

Martensitic transformations occur in a variety of nonferrous alloy systems, particularly for allotropes that transform at low temperatures. The reasons for this are

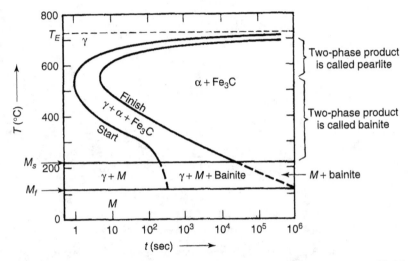

Figure 3.4 TTT diagram for eutectoid steel composition (0.80 wt% C). From K. M. Ralls, T. H. Courtney, and J. Wulff, *Introduction to Materials Science and Engineering.* Copyright © 1976 by John Wiley & Sons, Inc. This material is used by permission John Wiley & Sons, Inc.

that (a) the free energy difference between the high-temperature and low-temperature phases becomes more negative with decreasing temperature, and (b) the crystal structures of allotropes are relatively simple and similar to one another. Examples of some common martensitic transformations are given in Table 3.2.

The martensitic transformation plays an important role in a special class of alloys called *shape memory alloys*. These materials have the unique ability to "remember" their dimensions before deformation, and they return to their original shape upon undergoing the martensitic transition. This process is shown schematically in Figure 3.5 for the *one-way shape memory effect* (SME). There is also a *two-way shape memory effect* that we will not describe here. The shape memory effect is the result of a partial ordering of the structure as the BCC to CsCl structure transition takes place (see Figure 3.6a and3.6b). The martensitic phase (CsCl structure in this case) is easily deformed, and a distorted multidomain martensite phase results (Figure 3.6c). When the object is reheated, the austenitic phase (BCC in this case) reforms and returns to its original structure. The shape memory effect has been studied for applications ranging

Table 3.2 **Some Common Martensitic Transformations**

Alloy	Transition Temperature (°C)
FCC Co $\longrightarrow$ HCP Co	427
Cu–Zn–Al	−200 to +120
βTi $\longrightarrow$ αTi	883
Cu–Al–Ni	−200 to +170
50% Ni–50% Ti	−200 to +100

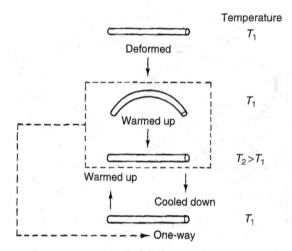

Figure 3.5 Illustration of the one-way shape memory effect.

from drive shafts that stiffen in response to increased shaft vibration [6], to orthodontic wires and other potential biomedical uses [7, 8].

3.1.3 Kinetics of Corrosion in Metals and Alloys

Corrosion is the deterioration of a material by reaction with its environment. Although the term is used primarily in conjunction with the deterioration of metals, the broader definition allows it to be used in conjunction with all types of materials. We will limit the description to corrosion of metals and alloys for the moment and will save the degradation of other types of materials, such as polymers, for a later section. In this section, we will see how corrosion is perhaps the clearest example of the battle between thermodynamics and kinetics for determining the likelihood of a given reaction occurring within a specified time period. We will also see how important this process is from an industrial standpoint. For example, a 1995 study showed that metallic corrosion costs the U.S. economy about $300 billion each year and that 30% of this cost could be prevented by using modern corrosion control techniques [9]. It is important to understand the mechanisms of corrosion before we can attempt to control it.

3.1.3.1 Types of Corrosion. There are many ways in which to categorize corrosion. One system classifies corrosion into two general categories: macroscopic and microscopic (see Figure 3.7). *Macroscopic corrosion* is basically "outside-in" corrosion and includes such phenomena as *pitting* and *galvanic* (two-metal) *corrosion*. *Microscopic corrosion* is more of an "inside-out" process, where such environmental influences as mechanical stress can cause intergranular corrosion and eventually cracking deep within the sample. Another way to classify corrosion is according to the type of environment—that is, wet versus dry corrosion. This is a useful classification scheme for describing why ordinary steel is relatively resistant to dry chlorine gas, but moist chlorine gas, and especially hydrochloric acid solutions, can cause severe damage to steel.

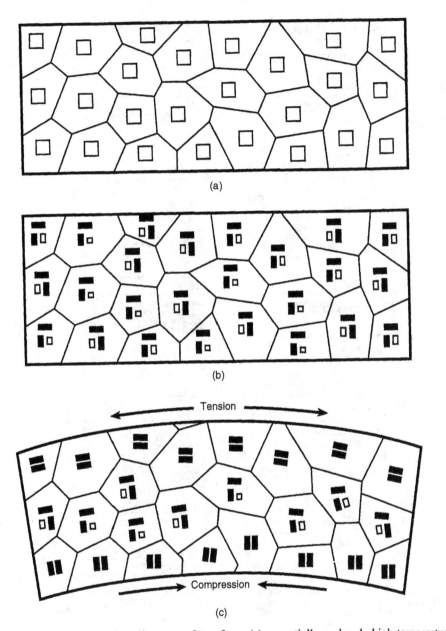

Figure 3.6 Shape-memory alloys transform from (a) a partially ordered, high-temperature austenitic phase to (b) a mixed austenite–martensite low-temperature state to (c) an ordered mixed-phase state under deformation.

For our purposes, the classification system is not as important as (a) recognizing that many different forms of corrosion exist and (b) understanding the fundamental kinetic processes behind these different types. To this end, we will first look at the important principles of corrosion and then see how they can be applied to some important types of corrosion.

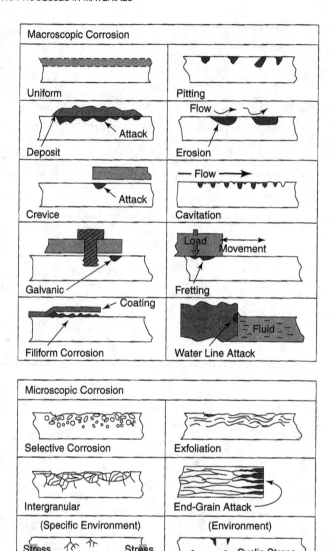

Figure 3.7   Types of corrosion. Reprinted, by permission, from P. Elliott, *Chem. Eng. Prog.*, **94**(5), 33. Copyright © 1998 by the American Institute of Chemical Engineers.

3.1.3.2   Electrochemical Reactions. Consider a simple *galvanic cell*, composed of two metal *electrodes*, zinc and copper, immersed in two different aqueous solutions of unit activity—in this case, 1.0 M $ZnSO_4$ and 1.0 M $CuSO_4$, respectively, connected by an electrical circuit, and separated by a *semipermeable membrane* (see Figure 3.8). The membrane allows passage of ions, but not bulk flow of the aqueous solutions from one side of the cell to the other. Electrons are liberated at the *anode* by the *oxidation* (increase in the oxidation number) of the zinc electrode:

$$Zn \longrightarrow Zn^{2+} + 2e^-$$ (3.18)

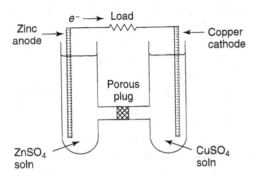

Figure 3.8   Schematic diagram of an electrochemical cell.

Electrons flow through the circuit connecting the two electrodes, and they are consumed at the *cathode* by the *reduction* (decrease in oxidation number) of copper ions in solution:

$$Cu^{2+} + 2e^- \longrightarrow Cu \tag{3.19}$$

The overall reaction, then, is the sum of these two *half-reactions*:

$$Zn + Cu^{2+} \longrightarrow Zn^{2+} + Cu \tag{3.20}$$

The sulfate ions that remain as a result of reaction (3.19) pass from right to left through the semipermeable membrane in order to balance the charges arising from the zinc ions forming at the anode. Current continues to pass through the circuit, and the zinc electrode slowly dissolves while the copper electrode slowly grows through the precipitation of copper. If we place a high-resistance voltmeter between the copper and zinc electrodes, we measure a potential difference of approximately 1.1 V.

The voltage we measure is characteristic of the metals we use. As an additional example, unit activity solutions of $CuCl_2$ and $AgCl$ with copper and silver electrodes, respectively, give a potential difference of about 0.45 V. We could continue with this type of measurement for all the different anode–cathode combinations, but the number of galvanic cells needed would be very large. Fortunately, the half-reactions for most metals have been calculated relative to a *standard reference electrode*, which is arbitrarily selected as the reduction of hydrogen:

$$2H^+ + 2e^- \longrightarrow H_2(g) \tag{3.21}$$

The hydrogen *electromotive force* (EMF) is defined as exactly zero, and all other half-cell EMFs are listed relative to the hydrogen electrode at 25°C (see Table 3.3). By convention, the half-cell EMFs are listed as reduction reactions with electrons on the left side of the reaction. All reactions must be reversible, of course, and the corresponding oxidation potential is found by reversing the reaction and reversing the sign on the voltage. For our Zn/Cu galvanic cell example, the EMF we predict for the cell is the sum of the two half-cell reactions. From Table 3.3, this should be the EMF for the reduction of copper (0.337 V) plus the EMF for the oxidation of zinc, which we obtain by reversing the reduction reaction and changing the sign (0.763). The sum of these two EMFs is 1.1 V, as measured earlier. It is important to note that the EMF potentials *do not* need to be multiplied by a factor when balancing half-reactions with different oxidation numbers (why?)—for example, a 2+ cathodic reaction with a 1+ anodic reaction. Test this with the Cu/Ag cell example above, where the measured EMF was about 0.45 V, using the data from Table 3.3.

Most environmental conditions do not allow for calculation of these potentials at 25°C; in this case, temperature effects cannot be easily neglected. We all know that a car battery has less "cranking power" on a cold winter morning than it does on a warm summer day. Temperature effects can be taken into account by returning to the free energy, which already accounts for temperature fluctuations. Equation (3.8) developed at the beginning of this chapter is where we start.

$$\Delta G - \Delta G° = RT \ln \left(\prod_i a_i^{v_i} \right) \tag{3.8}$$

Table 3.3 Standard EMF Potentials for Selected Metals

Electrode Reaction	Standard Potential, E (volts), 25°C
$Au^{3+} + 3e^- \longrightarrow Au$	1.50
$Pt^{2+} + 2e^- \longrightarrow Pt$	1.2
$Pd^{2+} + 2e^- \longrightarrow Pd$	0.987
$Hg^{2+} + 2e^- \longrightarrow Hg$	0.854
$Ag^+ + e^- \longrightarrow Ag$	0.800
$Cu^+ + e^- \longrightarrow Cu$	0.521
$Cu^{2+} + 2e^- \longrightarrow Cu$	0.337
$2H^+ + 2e^- \longrightarrow H_2$	0.000
$Pb^{2+} + 2e^- \longrightarrow Pb$	−0.126
$Sn^{2+} + 2e^- \longrightarrow Sn$	−0.136
$Ni^{2+} + 2e^- \longrightarrow Ni$	−0.250
$Co^{2+} + 2e^- \longrightarrow Co$	−0.277
$Cd^{2+} + 2e^- \longrightarrow Cd$	−0.403
$Fe^{2+} + 2e^- \longrightarrow Fe$	−0.440
$Ga^{3+} + 3e^- \longrightarrow Ga$	−0.53
$Cr^{3+} + 3e^- \longrightarrow Cr$	−0.74
$Zn^{2+} + 2e^- \longrightarrow Zn$	−0.763
$Mn^{2+} + 2e^- \longrightarrow Mn$	−1.18
$Zr^{4+} + 4e^- \longrightarrow Zr$	−1.53
$Ti^{2+} + 2e^- \longrightarrow Ti$	−1.63
$Al^{3+} + 3e^- \longrightarrow Al$	−1.66
$Mg^{2+} + 2e^- \longrightarrow Mg$	−2.37
$Na^+ + e^- \longrightarrow Na$	−2.71
$Ca^{2+} + 2e^- \longrightarrow Ca$	−2.87
$K^+ + e^- \longrightarrow K$	−2.93
$Li^+ + e^- \longrightarrow Li$	−3.05

(Left margin, top to bottom: Corrosion Less Likely ↑ ; Corrosion More Likely ↓)

Source: B. D., Craig, *Fundamental Aspects of Corrosion Films in Corrosion Science*, Copyright © 1991 by Plenum Publishing.

For electrochemical systems, we do not have pressure–vapor work that is normally associated with gaseous systems, nor do we have significant thermal effects like heats of fusion or formation. The result is that our usual description of free energy using enthalpies and entropies is not the best representation of free energy here. Instead, the work being generated is electrochemical work that can be best represented by the following relationship:

$$\Delta G = -nEF \tag{3.22}$$

where n is the number of electrons taking part in the reaction, F is a constant called the Faraday constant (96,500 coulombs/mol), and E is the potential of the half-reaction in volts. The corresponding standard state free energy is

$$\Delta G^\circ = -nE^\circ F \tag{3.23}$$

Substitution of Eqs. (3.22) and (3.23) into Eq. (3.8) results in the well-known *Nernst equation*:

$$E - E^\circ = \frac{-RT}{nF} \ln \left(\prod_i a_i^{v_i} \right) \tag{3.24}$$

The Nernst equation tells us the electrical potential of a half-cell when the reactants are not at unit activity.

3.1.3.3 Exchange Current Density.

Let us now return to our electrochemical cell shown in Figure 3.8. This cell is a combination of two half-cells, with the oxidation reaction occurring at the anode and the reduction reaction occurring at the cathode resulting in a net flow of electrons from the anode to the cathode. Equilibrium conditions dictate that the rate of oxidation and reduction, r_{oxid} and r_{red}, be equal, where both rates can be obtained from *Faraday's Law*:

$$r_{oxid} = r_{red} = \frac{i_0}{nF} \tag{3.25}$$

As in Eq. (3.22), F is the Faraday constant, n is the number of electrons taking part in the reaction, but i_0 is a new quantity called the *exchange current density*. These rates have units of mol/cm$^2 \cdot$s, so the exchange current density has units of A/cm^2. Typical values of i_0 for some common oxidation and reduction reactions of various metals are shown in Table 3.4. Like reversible potentials, exchange current densities are influenced by temperature, surface roughness, and such factors as the ratio of oxidized and reduced species present in the system. Therefore, they must be determined experimentally.

3.1.3.4 Polarization.

The net current flow produced in a cell results in a deviation of each half-cell potential from the equilibrium value listed in Table 3.3. This deviation from equilibrium is termed *polarization*, the magnitude of which is given the lowercase greek symbol eta, η and is called the *overpotential*, $E - E^\circ$. There are two primary types of polarization: *activation polarization* and *concentration polarization*.

Activation polarization is an activated process and possesses an activation barrier, just as described earlier in Figure 3.1. There are two parts to the activation process, either one of which can determine the rate of the reaction at the electrode.

Table 3.4 Some Exchange Current Densities

Reaction	Electrode	Solution	i_0, A/cm^2
$2H^+ + 2e^- \rightleftharpoons H_2$	Al	$2N$ H_2SO_4	10^{-10}
$2H^+ + 2e^- \rightleftharpoons H_2$	Au	$1N$ HCl	10^{-6}
$2H^+ + 2e^- \rightleftharpoons H_2$	Cu	$0.1N$ HCl	2×10^{-7}
$2H^+ + 2e^- \rightleftharpoons H_2$	Fe	$2N$ H_2SO_4	10^{-6}
$2H^+ + 2e^- \rightleftharpoons H_2$	Hg	$1N$ HCl	2×10^{-12}
$2H^+ + 2e^- \rightleftharpoons H_2$	Hg	$5N$ HCl	4×10^{-11}
$2H^+ + 2e^- \rightleftharpoons H_2$	Ni	$1N$ HCl	4×10^{-6}
$2H^+ + 2e^- \rightleftharpoons H_2$	Pb	$1N$ HCl	2×10^{-13}
$2H^+ + 2e^- \rightleftharpoons H_2$	Pt	$1N$ HCl	10^{-3}
$2H^+ + 2e^- \rightleftharpoons H_2$	Pd	$0.6N$ HCl	2×10^{-4}
$2H^+ + 2e^- \rightleftharpoons H_2$	Sn	$1N$ HCl	10^{-8}
$O_2 + 4H^+ + 4e^- \rightleftharpoons 2H_2O$	Au	$0.1N$ NaOH	5×10^{-13}
$O_2 + 4H^+ + 4e^- \rightleftharpoons 2H_2O$	Pt	$0.1N$ NaOH	4×10^{-13}
$Fe^{3+} + e^- \rightleftharpoons Fe^{2+}$	Pt		2×10^{-3}
$Ni \rightleftharpoons Ni^{2+} + 2e^-$	Ni	$0.5N$ $NiSO_4$	10^{-6}

Source: M. G. Fontana, *Corrosion Engineering,* p. 457, 3rd ed. Copyright © 1986 by McGraw-Hill Book Co.

The slowest step, or *rate-determining step*, can be either (a) electron transfer at the electrode–solution interface or (b) formation of atoms at the electrode surface. The activation polarization component of the overpotential, η_a, is related to the actual rate of oxidation or reduction, i, and the exchange current density:

$$\eta_a = \pm\beta \log(i/i_0) \tag{3.26}$$

This equation is often called the *Tafel equation.* Here, β is a constant, representing the expression $2.3RT/\alpha nF$, where R, T, n, and F have been defined previously, and α is the *symmetry coefficient,* which describes the shape of the rate-controlling energy barrier. Equation (3.26) is represented graphically in Figure 3.9, the resulting plot of which is called an *Evans diagram.* The linear relationship between overpotential and current density is normally valid over a range of about ±50 mV, but linearity is assumed for most Evans diagrams. The reaction rate, as represented by the current density, changes by one order of magnitude for each 100-mV change in overpotential. Note also that at $\eta_a = 0$, the rates of oxidation and reduction are equal, and the current density is given by the exchange current density, $i = i_0$.

The second type of polarization, concentration polarization, results from the depletion of ions at the electrode surface as the reaction proceeds. A concentration gradient builds up between the electrode surface and the bulk solution, and the reaction rate is controlled by the rate of diffusion of ions from the bulk to the electrode surface. Hence, the *limiting current* under concentration polarization, i_L, is proportional to the *diffusion coefficient* for the reacting ion, D (see Section 4.0 and 4.3 for more information on the diffusion coefficient):

$$i_L = \frac{DnFC_0}{x} \tag{3.27}$$

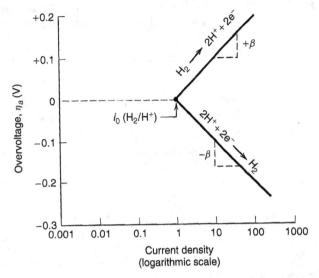

Figure 3.9 Evans diagram showing effect of activation polarization on overpotential for a hydrogen electrode. Reprinted, by permission, from W. Callister, *Materials Science and Engineering: An Introduction*, p. 574, 5th ed. Copyright © 2000 by John Wiley & Sons, Inc.

Cooperative Learning Exercise 3.2

Iron metal corrodes in acidic solution according to the following reaction:

$$Fe + 2H^+ \longrightarrow Fe^{2+} + H_2$$

Given that the rates of oxidation and reduction of the half-reactions are controlled by activation polarization only, that $\beta_{Fe} = +0.07$ and $\beta_{H_2} = -0.08$, and that the exchange current densities for both the oxidation of Fe and reduction of hydrogen in acidic solution are identical, use the data in Tables 3.3 and 3.4 to determine the following quantities. Recall that the potential for each half-cell is the sum of the equilibrium potential and the corresponding overpotential, in this case, η_a. The exchange current density for iron in acidified solution is 10^{-8} A/cm^2.

Person 1: Derive an expression for the electrical potential, E_H for the reduction of hydrogen as a function of the current density, i.

Person 2: Derive an expression for the electrical potential for the oxidation of iron, E_{Fe} as a function of the current density, i.

Group: Recognize the fact that at equilibrium, $E_H = E_{Fe}$ and combine your results to determine the current density, i, for the cell, and use this value to calculate the corrosion rate of iron, r_{Fe}.

---

where n and F have been defined previously, C_0 is the concentration of reacting ions in the bulk solution, and x is the thickness of the *depleted region* near the electrode, or the distance through which the ions must travel from the bulk to the electrode, termed the *Nernst diffusion layer*. The overpotential for concentration polarization, η_c, is:

$$\eta_c = \frac{2.3RT}{nF} \log\left(1 - \frac{i}{i_L}\right) \tag{3.28}$$

Both activation and concentration polarization typically occur at the same electrode, although activation polarization is predominant at low reaction rates (small current densities) and concentration polarization controls at higher reaction rates (see Figure 3.10). The combined effect of activation and concentration polarization on the current density can be obtained by adding the contributions from each [Eqs. (3.26) and (3.28)], with appropriate signs for a reduction process only to obtain the *Butler–Volmer equation*:

$$\eta_{reduction} = -\beta \log\left(\frac{i}{i_0}\right) + \frac{2.3RT}{nF} \log\left(1 - \frac{i}{i_L}\right) \tag{3.29}$$

Equation (3.29) allows us to determine the kinetics of most corrosion reactions from only three parameters: β, i_0, and i_L.

3.1.3.5 Corrosion Rate.

The current densities give us an estimate of the rate of the electrochemical reactions involved in corrosion, but they do not give us a direct measurement of the corrosion rate. A simple, empirical expression can be used to quantify the rate of corrosion in terms of a quantity called the *corrosion penetration rate* (CPR):

$$CPR = \frac{KW}{\rho At} \tag{3.30}$$

where W is the weight loss of the sample per unit time, t. The variables ρ and A represent the density of the material and the exposed surface area of the sample, respectively. The constant K varies with the system of units being used; $K = 534$ to

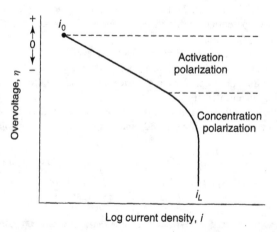

Figure 3.10   Effect of activation and concentration polarization on overpotential. Reprinted, by permission, from W. Callister, *Materials Science and Engineering: An Introduction*, 5th ed., p. 576. Copyright © 2000 by John Wiley & Sons, Inc.

give *CPR* in mils per year (*mpy*, where 1 mil = 0.001 inch), and $K = 87.6$ to give *CPR* in mm/y.

3.2   KINETIC PROCESSES IN CERAMICS AND GLASSES*

3.2.1   Nucleation and Growth (Round 2)

We described martensitic-type transformations in metals in Section 3.1.2.2, described spinodal decomposition of polymers in Chapter 2, and initiated a description of crys-tallization processes in Section 3.1.2.1. We wish to elaborate on crystallization by identifying two primary phenomena called *nucleation* and *growth*. Nucleation and growth kinetics are composed of two separate processes. Nucleation is the initial forma-tion of small particles of the product phase from the parent phase. The resulting nuclei are often composed of just a few molecules. The growth step involves the increase in size of the nucleated particles. Many natural phenomena follow nucleation and growth kinetics, such as ice formation. Unlike martensitic transformations, both spin-odal decomposition and nucleation and growth processes involve diffusion, but there are some subtle differences. Figure 3.11 shows a comparison based on composition and spatial extent between a spinodal transformation and a nucleation and growth trans-formation. Spinodal transformations are large in spatial extent but involve relatively small concentration differences throughout the sample, whereas the nucleation process involves the formation of small domains of composition very different from the parent phase, which then grow in spatial extent via concentration-gradient-driven diffusion.

Let us first consider the liquid–solid phase transformation. At the melting point (or more appropriately, fusion point for a solidification process), liquid and solid are in equilibrium with each other. At equilibrium, we know that the free energy change for the liquid–solid transition must be zero. We can modify Eq. (2.11) for this situation

$$\Delta G_v = \Delta H_f - T_m \Delta S_f = 0 \tag{3.31}$$

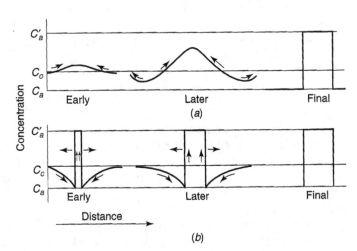

Figure 3.11   Schematic comparison of dimensional changes that occur in (a) spinodal and (b) nucleation and growth transformation processes. From W. D. Kingery, H. K. Bowen, and D. R. Uhlmann. *Introduction to Ceramics*, Copyright © 1976 by John Wiley & Sons, Inc. This material is used by permission of John Wiley & Sons, Inc.

where ΔG_v is the free energy change per unit volume for fusion, ΔH_f is the enthalpy of fusion and is related to the strength of interatomic or intermolecular forces of the solid, ΔS_f is the entropy of fusion, representing the degree of disorder in the solid, and T_m is the melting point. We can solve Eq. (3.31) for the entropy change of fusion:

$$\Delta S_f = \frac{\Delta H_f}{T_m} \tag{3.32}$$

If we assume that the entropy and enthalpy are relatively independent of tempera-ture, we can drop the subscript "m" on temperature in Eq. (3.32), thereby obtaining a more general relationship between the entropy and enthalpy of fusion as a function of temperature. Substitution of this modified form of Eq. (3.32) into Eq. (3.31) gives the following relation for the free energy:

$$\Delta G_v = \Delta H_f \left[\frac{(T_m - T)}{T_m} \right] \tag{3.33}$$

This expression will be very useful in the discussion of nucleation and growth kinetics.

In the nucleation step, there must be sites upon which the crystals can form. This is similar to "seeding" the clouds to cause water to precipitate (rain). There are two sources for these nucleating particles: homogeneous and heterogeneous agents.

3.2.1.1 Homogeneous Nucleation.

In *homogeneous nucleation*, critical nuclei (nuclei with enough molecules to initiate growth) form without the aid of a foreign agent. This happens very rarely in practice, but is useful for theoretical discussions. Consider the formation of a small, solid, spherical particle from the parent (liquid) phase. There is a free energy change, ΔG, associated with this process. Much like the potential energy function between two atoms, there are two competing forces that con-tribute to the overall free energy change: (1) the free energy change due to a volume change (a negative contribution) and (2) the free energy change due to formation of an interphase (a positive contribution). These forces are quantified in Eq. (3.34):

$$\Delta G = \left(\frac{4\pi r^3}{3} \right) \Delta G_v + 4\pi r^2 \gamma \tag{3.34}$$

where ΔG_v is the free energy change per unit volume for the phase change (which must be negative); r is the nucleated particle radius; γ is the *interfacial energy* (often called "surface tension" for a liquid–gas interface), which has units of energy/area (erg/cm^2) or force/length (dyne/cm). The first term in Eq. (3.34) is the free energy change associated with the volume change, and the second term is the free energy change associated with the interphase formation.

As with any activated process, there is a free energy barrier, ΔG^*, or "critical free energy" that must be surmounted in order to nucleate a particle. This is illustrated in Figure 3.12 as a plot of free energy versus the particle size, r. Once the overall free energy change decreases past ΔG^*, the nucleation process becomes spontaneous. This occurs as enough molecules gather to form a nucleus with critical radius, r^*. We can solve for r^* by taking the derivative of ΔG in Eq. (3.34) with respect to r, setting

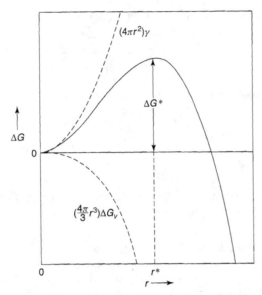

Figure 3.12   Representation of activation energy barrier for nucleation. From K. M. Ralls, T. H. Courtney, and J. Wulff, *Introduction to Materials Science and Engineering.* Copyright © 1976 by John Wiley & Sons, Inc. This material is used by permission John Wiley & Sons, Inc.

Cooperative Learning Exercise 3.3

Work with a neighbor. Consider Eq. (3.34), which describes the free energy change associated with the homogeneous nucleation of a solid, spherical particle of radius, r, from a parent liquid phase:

$$\Delta G = \left(\frac{4\pi r^3}{3} \right) \Delta G_v + 4\pi r^2 \gamma$$

Recall that this equation could be minimized with respect to particle radius to determine the critical particle size, r^*, as given by Eq. (3.35). This critical radius could then be used to determine the height of the free energy activation energy barrier, ΔG^*, as given by Eq. (3.36). A similar derivation can be performed for a cubic particle with edge length, a.

Person 1: What is the equivalent expression for the first term in Eq. (3.34) for the cube? Recall that ΔG_v is the free energy change per unit volume.

Person 2: What is the equivalent expression for the second term in Eq. (3.34) for the cube? Recall that γ is the interfacial energy for the liquid–solid interface.

Combine your results to arrive at an equivalent expression for Eq. (3.34) for the cube. Minimize this equation to determine the critical cube length, a^*. Substitute this value back into your expression to determine the critical free energy barrier for the cube, ΔG_v^*. Why is ΔG_v^* higher for a cube than for a sphere? What does this tell you about the likelihood of cubes nucleating rather than spheres?

Answer: $a^3 \Delta G_v$; $6a^2\gamma$; $\Delta G/\partial a = -4\gamma/\Delta G_v$; $\Delta G^* = 32\gamma^3/\Delta G_v^2$. The barrier is lower for spherical particles so their formation is more likely.

it equal to zero (the slope of ΔG vs. r in Figure 3.12 is zero at $\Delta G = \Delta G^*$), and replacing r by r^* to get

$$r^* = \frac{-2\gamma}{\Delta G_v} \tag{3.35}$$

The height of the activation energy barrier, ΔG^*, can now be found by evaluating Eq. (3.34) at r by r^* with the aid of Eq. (3.35)

$$\Delta G^* = \frac{16\pi\gamma^3}{3\Delta G_v^2} \tag{3.36}$$

To summarize the nucleation stage, then, particles that achieve a critical radius $r > r^*$ may enter the growth stage of this phase transformation process.

3.2.1.2 Heterogeneous Nucleation. In *heterogeneous nucleation*, the critical nuclei form with the assistance of nucleating sites. These sites can be impurities, container walls, cracks, or pores and may either be naturally present or intentionally added to promote nucleation. The effect of nucleating agents is to decrease ΔG^*.

Unlike homogeneous nucleation, heterogeneous nucleation involves an interface between two compositionally different materials, so we must account for the interaction of the parent phase with the nucleating particle. This is accomplished through the introduction of the contact angle, θ, at the three-phase interphase of the parent liquid, its solid phase, and the nucleating agent (see Figure 3.13 and Section 2.2.2.3). Without derivation, we present the free energy barrier for heterogeneous nucleation, ΔG_s^*, as

$$\Delta G_s^* = \Delta G^* \frac{(2 + \cos\theta)(1 - \cos\theta)^2}{4} \tag{3.37}$$

where ΔG^* is the free energy barrier for homogeneous nucleation and the subscript "s" on ΔG_s^* indicates heterogeneous or "surface" nucleation.

3.2.1.3 Nucleation Rate. We can now define an overall nucleation rate:

$$\mathcal{N} = \nu n_s n^* \tag{3.38}$$

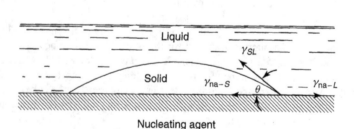

Figure 3.13 Schematic diagram of contact angle formed at three phase intersection during heterogeneous nucleation. From K. M. Ralls, T. H. Courtney, and J. Wulff, *Introduction to Materials Science and Engineering*. Copyright © 1976 by John Wiley & Sons, Inc. This material is used by permission John Wiley & Sons, Inc.

where N is the number of critical nuclei that form per unit volume per unit time, n_s is the number of molecules in contact with critical nucleus, n^* is the number of critical size clusters per unit volume, and v is the collision frequency of single molecules with the nuclei (see Figure 3.14). The number of critical nuclei, n^* can be related to the free energy barrier, ΔG^*, through an Arrhenius-type expression:

$$n^* = n_0 \exp(-\Delta G^*/k_B T) \tag{3.39}$$

where n_0 is the number of single molecules per unit volume, k_B is Boltzmann's constant, and T is absolute temperature. Similarly, the collision frequency, v, can be expressed by an Arrhenius relation:

$$v = v_0 \exp(-\Delta G_m/k_B T) \tag{3.40}$$

where v_0 is the molecular jump frequency, and ΔG_m is the activation energy for transport across the nucleus–matrix interface, which is related to short-range diffusion. Putting these expressions back into Eq. (3.38) gives

$$N = v n_s n_0 \exp(-\Delta G^*/k_B T) \exp(-\Delta G_m/k_B T) \tag{3.41}$$

For heterogeneous nucleation, we simply use ΔG_s^* instead of ΔG^*:

$$N = v_0 n_s n_0 \exp(-\Delta G_s^*/k_B T) \exp(-\Delta G_m/k_B T) \tag{3.42}$$

Though Eqs. (3.41) and (3.42) are somewhat cumbersome, they describe the rate at which nuclei form as a function of temperature. Note that in both equations, n_s and n_0 have dimensions of number per unit volume, and v_0 has units of time^{-1}.

What would be more useful are forms of Eqs. (3.41) and (3.42) that contain some measurable, physical properties of the system. Free energies hardly fit this description. The danger in doing this is that we have to make some simplifications and assumptions. Let us begin by recalling that the free energy for homogeneous nucleation is given by Eq. (3.36):

$$\Delta G^* = \frac{16\pi \gamma^3}{3(\Delta G_v)^2} \tag{3.36}$$

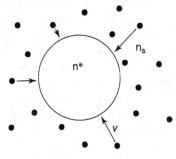

Figure 3.14 Schematic illustration of a nucleation site formation.

Recall also that for liquid–solid transitions at temperatures close to T_m, ΔG_v is given by Equation (3.33):

$$\Delta G_v = \frac{\Delta H_v (T_m - T)}{T_m} \tag{3.33}$$

where ΔH_v is the heat of transformation per unit volume. Substitution of Eqs. (3.33) and (3.36) into the first exponential in Eq. (3.42) gives us the temperature dependence of the nucleation rate for homogeneous nucleation:

$$\mathbb{N} = v_0 n_s n_0 \exp\left[\frac{-16\pi \gamma^3 T^2}{3 k_B T \Delta H_v^2 (T_m - T)^2} \right] \exp\left(\frac{-\Delta G_m}{k_B T} \right) \tag{3.43}$$

While this may appear even more cumbersome than Eq. (3.41), it contains some parameters that are directly measurable such as the interfacial surface energy, γ, and the heat of fusion, ΔH_v, but more importantly, it contains the temperature difference $(T_m - T)$, which is the degree of undercooling—that is, how far the temperature is below the melting (solidification) point!

The two exponentials in Eq. (3.43) compete against each other. As T decreases below the melting point:

- $(T_m - T)$ becomes larger (remember that this term is squared, so it is always a positive value), the term inside the first exponential becomes smaller, but because it is negative, the entire exponential gets bigger and the *nucleation rate increases* due to this exponential—*the driving force for nucleation becomes greater.*
- The term in the second exponential becomes more negative, the exponential gets smaller and the *nucleation rate decreases* due to this exponential term—*diffusion becomes more difficult as the temperature decreases and particles cannot migrate to the nucleation surface.*

These generalizations are true for both homogeneous and heterogeneous nucleation. As a result, we would expect that the two competing exponentials give rise to a maximum nucleation rate at some temperature below the melting point, and this is indeed the case, as illustrated in Figure 3.15. It is also logical that heterogeneous nucleation has a higher absolute nucleation rate than homogeneous nucleation and that heterogeneous occurs at a higher temperature (lower degree of undercooling) than homogeneous nucleation.

3.2.1.4 Growth.

The second portion of nucleation and growth is the growth process. Here the tiny nuclei grow into large crystals through the addition of molecules to the solid phase. There are two primary types of crystal growth: thermally activated (diffusion controlled) and diffusionless (martensitic). We have already discussed martensitic transformations, so only *thermally activated crystal growth* will be considered here. The development of the proper growth rate expression is highly dependent upon the type of phase transformation—that is, crystallization from the melt, vapor phase, or dilute solution. We will simply use a general form of the growth rate expression which is based upon an Arrhenius-type expression:

$$\mathbb{R} = A[1 - \exp(\Delta G / k_B T)] \tag{3.44}$$

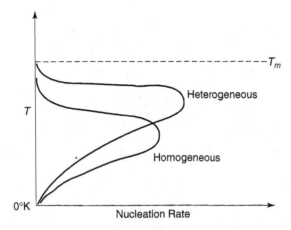

Figure 3.15 Effects of temperature and undercooling on homogeneous and heterogeneous nucleation rates.

where $\mathbb{R}$ is the growth rate, ΔG is the molar free energy difference between product and parent phase, and A is a preexponential factor. The form of the preexponential factor, A, depends on the type of theory one wishes to employ. It can be directly related to the liquid phase viscosity, as in $A = \nu a_0$, where ν is the *frequency factor* for transport across the interphase

$$\nu = \frac{k_B T}{3\pi a^3 \eta} \tag{3.45}$$

in which η is the viscosity of the liquid phase, and a_0 is the molecular diameter; or A can be related to the diffusivity of atoms across the interphase, as in $A = KD$, where K is a constant and the diffusivity of atoms jumping across interphase, D, is given by

$$D = D \exp\left(\frac{-\Delta H^0}{k_B T}\right) \tag{3.46}$$

These two expression are actually quite similar since D and η are interrelated (see Section 4.3). Other expressions exist as well for such phenomena as surface growth and screw dislocation growth.

3.2.1.5 *Overall Phase Transformation Rate.* Finally, the overall transformation rate, $\mathbb{P}$ is given by the product of the nucleation and growth expressions [Eqs. (3.41) or (3.42), and (3.44), respectively):

$$\mathbb{P} = \mathcal{N} \times \mathbb{R} \tag{3.47}$$

The overall transformation rate is shown qualitatively in Figure 3.16. Notice that the maximum nucleation rate occurs at a lower temperature than the maximum growth rate, and that the maximum transformation rate may not be at either of these two rate maxima. Note also that there is some finite transformation rate, even at very low

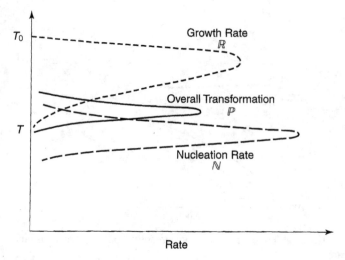

Figure 3.16 Schematic representation of transformation rates involved in crystallization by nucleation and growth kinetics.

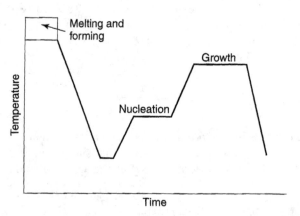

Figure 3.17 Schematic time-temperature cycle for the controlled crystallization of a glass-ceramic body.

temperatures. This is why some glasses can crystallize over very long periods of time. As long as there is some molecular motion, there is a probability of crystallization taking place.

Finally, it should be pointed out that the breaking down of the liquid to solid phase transformation into two separate steps, nucleation and growth, is not entirely artificial. An interesting application of the distinct nature of the nucleation and growth processes is found in the formation of glass ceramics. As shown in Figure 3.17, certain glass-forming inorganic materials can be heat-treated in very controlled ways in order to affect the structure of the crystals that form. For instance, a glass can be rapidly cooled below its melting point, then heated to the maximum nucleation rate, which (recall) is below both the melting point and the maximum growth rate. At this temperature and with sufficient time, many small nuclei will be formed. The glass, now replete with many tiny nuclei, is heated to the maximum growth rate temperature and held there while the crystals grow. The crystals can only grow so large, though, since they soon run into another growing crystal from a neighboring nucleation site. In this way, the size of the crystals, or *crystallites*, in the glass can be controlled, and a *glass ceramic* is formed. Glass ceramics are unique in that they are crystalline materials, yet they are transparent in many cases (see Section 1.2.5). They also possess unique physical properties, such as low thermal expansion, due to the large amount of interphase relative to the same material made of larger crystals. In the extreme, crystallites on the order of 10^{-9} meters in diameter can be formed, resulting in so-called *nanostructured ceramics*. These materials, like glass ceramics, have unique physical properties, which may one day lead to improved ductility and thermal properties in ceramics.

3.2.2 Kinetics of Corrosion in Ceramics and Glasses

Although the term corrosion is normally associated with metals, the definition presented in Section 3.1.3 is equally applicable to ceramic materials. The cost to industry due to ceramic corrosion is similarly large, over $1 billion annually. For example, the cost to replace the refractory insulation in a glass melting furnace might be about $10 million, nearly 60% of which is related to replacing the insulation itself, with the remainder due to labor costs. As with metallic corrosion, types of ceramic corrosion can be grouped a number of different ways, but classification according to the type of attacking medium (i.e., liquid vs. gaseous corrosion) is a useful distinction. We will investigate these two methods of attack separately.

3.2.2.1 Kinetics of Ceramic and Glass Corrosion by Liquids. The corrosion of a solid, crystalline ceramic material by a liquid occurs through the formation of a reaction product between the solid and the solvent. This reaction product is generally less soluble than the bulk ceramic solid, and it may form an attached surface layer on the solid. This type of corrosion in a liquid is called *indirect dissolution, incongruent dissolution*, or *heterogeneous dissolution*. The rate-limiting step in this type of ceramic corrosion can be the interfacial layer formation, diffusion through the interfacial layer, or diffusion through the liquid. *Direct dissolution, congruent dissolution*, or *homogeneous dissolution* occurs when the solid, crystalline ceramic material dissolves directly into the liquid through dissociation or reaction. The concentrations of the ceramic species in solution, along with the corresponding diffusion coefficients of the species, determine the rate of direct dissolution.

The type of attack that occurs in liquid media is highly dependent on the chemical nature of the liquid—that is, molten metal, molten ceramic, or aqueous solution. We will consider two industrially important cases: attack by molten metals and attack by aqueous media. The attack of most metal oxide ceramics by molten metals involves a simple exchange of one metal ion for another. For example, silicon dioxide in contact with molten aluminum is susceptible to the following corrosion reaction:

$$3SiO_2(s) + 2Al(l) \longrightarrow 2Al_2O_3(s) + 3Si(l) \tag{3.48}$$

The ease with which this exchange can occur is given by the free energy of reaction, ΔG, which is the summation of the free energy of formation for the individual oxides, presented here at 300 K in kJ/mol:

$$3SiO_2(s) \longrightarrow 3O_2(g) + 3Si(l) \qquad \Delta G_f = 2568$$
$$\frac{3O_2(g) + 4Al(l) \longrightarrow 2Al_2O_3(s)}{3SiO_2(s) + 4Al(l) \longrightarrow 2Al_2O_3(s) + 3Si(l)} \qquad \begin{array}{l} \Delta G_f = -3164 \\[6pt] \Delta G_f = -596 \end{array} \qquad (3.49)$$

Values for formation reactions can be found in the JANAF tables [10]. Another way in which a molten metal can attack an oxide is through compound formation, such as the formation of spinel:

$$4Al_2O_3(s) + 3Mg(l) \longrightarrow 3MgAl_2O_4(s) + 2Al(l) \qquad (3.50)$$

This reaction has a lower free energy of formation (-218 kJ/mol at $1000°C$) compared to the simple oxidation–reduction reaction (-118 kJ/mol) analogous to Eq. (3.48) above. It is also possible that the molten metal can completely reduced the metal oxide ceramic, forming an alloy and gaseous oxygen:

$$M_xO_y(s) \xrightarrow{\text{liquid metal}} xM(solution) + \tfrac{y}{2}O_2(g) \qquad (3.51)$$

Ceramic materials in contact with aqueous liquids are also susceptible to attack. This is of particular importance in the realm of civil engineering, where materials of construction that contain minerals routinely come in contact with groundwater and soil. Minerals dissolve into aqueous solutions through the diffusion of leachable species into a stationary thin film of water, about 110 μm thick

$$\text{Mineral } A + nH^+ + mH_2O \rightleftharpoons \text{Mineral } B + qM^+ \qquad (3.52)$$

where M^+ is the soluble species. The equilibrium constant for this equation is given by Eq. (3.4):

$$K = \frac{[M^+]^q}{[H^+]^n[H_2O]^m} \qquad (3.53)$$

where the solids species (minerals) have unit activity. In dilute solutions and at one atmosphere pressure, the activity of water is also approximately unity. At higher pressures, the activity of water is approximately proportional to the pressure. It is evident from this relationship that the dissolution of ceramics is highly dependent upon the pH ($[H^+]$) of the aqueous medium. Generally, at low pH values (acidic environments) the equilibrium shifts to the right, and corrosion is promoted, whereas high pH values (basic environments) retard corrosion by shifting the equilibrium back to the left.

Not all ceramic materials behave the same at a given pH, however. As the material begins to dissolve, ions form at the surface, water molecules orient themselves accordingly, and an *electrical double layer* is established, as shown in Figure 3.18. The first layer of charged ions and oriented water molecules is called the *inner Helmholtz plane* (IHP), and the second layer of oppositely charged particles is called the *outer Helmholtz*

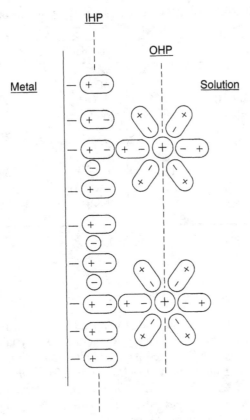

Figure 3.18 Formation of the electrical double layer of a surface in solution, showing the inner Helmholtz plane (IHP) and outer Helmholtz plane (OHP). Reprinted, by permission, from B. D. Craig, *Fundamental Aspects of Corrosion Films in Corrosion Science*, p. 4. Copyright © 1991 by Plenum Press.

plane (OHP). This arrangement should remind you of a capacitor; and indeed, this is how this system is often modeled. The potential established between these two layers decreases as we move away from the surface into the solution. The extent to which this double layer is established and the magnitude of the charge on the surface are directly related to the pH of the surrounding solution. The pH value for which there is no net surface change is called the *isoelectric point* (IEP). An empirical formula for determining the IEP of oxides has been established [11]:

$$IEP = 18.6 - \frac{11.5z}{d} \tag{3.54}$$

where z is the oxidation state of the cation in the film, and d is the distance (in angstroms) of an adsorbed proton from the cation via the oxygen ion. Additional values of IEPs for some typical oxides and hydroxides are given in Table 3.5.

Related to the attack of polycrystalline ceramic materials by aqueous media is the hydrolysis of silicate glasses. The following relationship has been developed to describe the effect of time and temperature on the acid corrosion (10% HCl) of silicate

Table 3.5 Some Isoelectric Points (IEP) for Selected Oxides and Hydroxides

Material	IEP	Material	IEP
α-Al_2O_3	5–9.2	ZnO	8.8
γ-Al_2O_3	8.0	TiO_2	4.7–6.2
γ-$Al(OH)_3$	9.25	$Pb(OH)_2$	11.0
α-$Al(OH)_3$	5.0	$Mg(OH)_2$	12
CuO	9.5	SiO_2	2.2
Fe_2O_3	8.6	ZrO_2	6.7
Fe_3O_4	6.5	SnO_2	4.3
CdO	12	WO_3	0.4
Ta_2O_5	2.9	M_2O_5, MO_3	<0.5
MO_2	0–7.5	M_2O_3	6.5–10.4
MO	8.5–12.5	M_2O	>11.5

M = metal cation.

glasses [12]:

$$W = at^{b_1} \exp(-b_2/T) \tag{3.55}$$

where W is the weight loss, t is time, T is temperature, and a, b_1, and b_2 are experimentally determined coefficients. Like aqueous attack of crystalline ceramics, glass dissolution rates are a strong function of pH. Unlike crystalline ceramics, however, there are two competing mechanisms in glass dissolution: *ion exchange* and *matrix dissolution* (see Figure 3.19). In ion exchange the metal–oxygen bond is coordinated to hydrogen ions, whereas in matrix dissolution the metal–oxygen bond is coordinated to hydroxyl ions. Ultimately, the corrosion of glasses affects their mechanical properties and determines their useful service lifetime on an industrial scale.

3.2.2.2 *Kinetics of Ceramic and Glass Corrosion by Gases.* The attack of ceramics and glasses by gaseous reactants is much more prevalent than corrosion due to liquids. In addition to the chemical composition of the attacking medium, the geometry of the ceramic solid is of tantamount importance to its corrosion resistance. That is, a

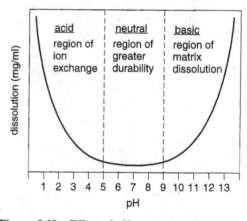

Figure 3.19 Effect of pH upon glass dissolution.

porous or highly permeable substrate will be much more susceptible to attack by gases than a theoretically dense one. Pore size, pore size distribution, surface-to-volume ratio, and diffusivity all play important roles in this process.

Gas or, more appropriately, vapor attack can result in the formation of either a solid, liquid, or gas, depending on the reaction conditions and the type of product formed. For example, Na_2O vapors can attack SiO_2 to form liquid sodium silicate. In some cases, the vapor may enter the ceramic due to a thermal gradient, then condense and dissolve the ceramic by liquid attack.

The corrosion rate can be limited by any number of steps in the dissolution process:

- Diffusion of the gas to the solid.
- Adsorption of the gas molecule onto the solid surface.
- Surface diffusion of the adsorbed gas.
- Decomposition of reactants at surface sites.
- Reaction at the surface.
- Removal of products from reaction sites.
- Surface diffusion of products.
- Desorption of gas molecules from the surface.
- Diffusion of products away from the solid.

Notice that most of the possible rate-limiting steps involve movement of atoms—diffusion, adsorption, and desorption—and only one direct chemical reaction is involved. In general, the total pressure of the system will determine which step is rate-limiting. At low total pressures ($<10^{-4}$ atm), gas transport is "line-of-sight," and the surface reaction may be rate-controlling. At intermediate pressures (10^{-4}–10^{-1} atm), bulk diffusion is usually rate-limiting, while diffusion through the reaction boundary layer, formed as products accumulate at the surface, controls the overall rate at high pressures ($>10^{-1}$ atm). We will reserve the majority of our description of diffusion in solids to Chapter 4; but for now, recall that the diffusion coefficient is given by D and has units of cm^2/sec. For gases, the concentration is often represented by the partial pressure of component A, P_A, which gives the functional dependence of corrosion rate on pressure, as described previously.

When the rate of the chemical reaction occurring at the surface is the rate-limiting step, the principles we have described to this point apply. The reaction rate can have any order, and the gas reacts with the ceramic substrate to produce products. Although our discussion to this point has focused on oxide ceramics, there are a number of non-oxide ceramics, such as carbides, nitrides, or borides, that are of importance and that undergo common decomposition reactions in the presence of oxygen. These ceramics are particularly susceptible to corrosion since they are often used at elevated temperatures in oxidizing and/or corrosive environments. For example, metal nitrides can be oxidized to form oxides:

$$2M_xN_y(s) + O_2(g) \longrightarrow 2M_xO(s) + yN_2(g) \tag{3.56}$$

In this case, the reaction rate, r, is a function of the partial pressure of the gaseous species, P_i, since solids are taken to have unit activity:

$$r = k\frac{P_{N_2}^y}{P_{O_2}} \tag{3.57}$$

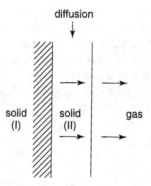

Figure 3.20 Diffusion of reaction products through boundary layer.

Note from reaction (3.56) that the chemical composition of the surface is changing. As this surface layer builds over the course of the reaction, diffusion of species back through this *surface reaction layer*, or *boundary layer*, can become the rate-limiting step (see Figure 3.20). Again, the intricacies of boundary layer diffusion will be addressed in Chapter 4, but for now, we can look at the kinetics associated with the growth of this boundary layer. For thin reaction films (<5 nm in thickness), the thickness of the film, x, changes with time, t, according to one of three general forms: logarithmic, inverse logarithmic, or asymptotic

$$x = k_1 \log t \tag{3.58a}$$

$$\frac{1}{x} = k_2 - k_3 \log t \tag{3.58b}$$

$$x = k_4[1 - \exp(-k_5 t)] \tag{3.58c}$$

For thick films (>5 nm), the film thickness follows either a parabolic or rectilinear relationship with time

$$x^2 = k_6 t \tag{3.59a}$$

$$x = k_7 \tag{3.59b}$$

In Eqs. (3.58) and (3.59), the k_i are the reaction rate constants. We will see in Chapter 4 that many solid-state ceramic processes involve simultaneous mass transport (diffusion), thermal transport, and reaction.

3.3 KINETIC PROCESSES IN POLYMERS

The previous sections dealt primarily with phase transformations and corrosion in materials. Polymers also undergo phase transformations. For example, there are many polymers that utilize nucleation and growth kinetics to transform from amorphous to crystalline polymers. The kinetics of these transformations are very similar, in principle, to the preceding descriptions for glasses, so it is not necessary to duplicate that material here. Polymers also are susceptible to corrosion, but the term *degradation* is more widely utilized for deterioration in polymers. The kinetics of polymer degradation will be discussed in Section 3.3.3. Before we get to degradation, we should first discuss formation of polymers. Here we find kinetic processes that are unique to polymers. These are primarily formation reactions whose rates can vary widely depending on such variables as temperature and concentration. This is not the case with most metals. Thermodynamics tells us whether an alloy will form between two metals; and if that process is favorable, it generally proceeds spontaneously at a very rapid rate. The same is true of most ceramics and glasses: When the temperature is high enough, the reactants go to products—not only rapidly, but sometimes explosively. Polymer systems are more complex, in that the reacting molecules and products are much larger than those in metallic and ceramic systems, leading to mass- and heat-transport limitations that can affect the kinetics. It is now not enough to know that a reaction is thermodynamically favorable—we must look closely at the actual rate at which polymers are formed. This is the subject of Section 3.3.1. The effect of different types of processing condition on these reactions will then be discussed in Section 3.3.2.

3.3.1 Kinetics of Polymerization

Polymerization reactions are generally grouped into *stepwise* (also known as *condensation*) reactions and *radical chain* (sometimes called *addition*) reactions, depending upon the nature in which the monomers react with each other or with growing polymer chains to form higher-molecular-weight structures. Condensation reactions are characterized by the reaction of two functional groups that result in the loss of a small molecule. Thus, the repeat unit has a different molecular formula than the monomer from which it was formed. The small molecule formed during condensation is often water, though the specific formation of water is not a necessity for a condensation reaction. In contrast, addition polymerizations are those in which no loss of a small molecule occurs. This type of polymerization reaction usually requires an active center, such as a free radical or an ion, to proceed. Analysis of the kinetics for these two classes of reactions reveals some interesting differences not only in how the polymer is formed, but in how the reactions can be controlled.

3.3.1.1 *Kinetics of Stepwise Polymerization.* Consider the reaction of hexamethylene diamine and adipic acid:

$$
\mathrm{H_2N-(CH_2)_6-NH_2} + \mathrm{HO-\overset{\overset{O}{\|}}{C}-(CH_2)_4-\overset{\overset{O}{\|}}{C}-OH}
$$

Hexamethylene
diamine   Adipic acid

$$
k_r \Big\updownarrow k_f \tag{3.60}
$$

$$
\mathrm{-[HN-(CH_2)_6-\overset{H}{N}-\overset{\overset{O}{\|}}{C}-(CH_2)_4-\overset{\overset{O}{\|}}{C}]_n-} + \mathrm{H_2O}
$$

Nylon 66

Note first of all that there is the characteristic loss of a small molecule for stepwise reactions, which in this case is water, and that the loss of water results in an amide (HN-CO) linkage; hence the resulting polymer is called a *polyamide*. This particular polyamide is very common and is better known by its trade name *nylon*. There are many types of nylons, depending on the exact formula of the diamine and diacid used to form the polyamide, but in this case, there are six carbon units between the nitrogens in the diamine and there are six carbon units (including the carbons which contain the double-bonded oxygens) in the diacid, so this polyamide is called *Nylon 66*.

The kinetics of this reaction are analyzed by monitoring the disappearance of one of the reactants, or the formation of the product. If we choose to follow the disappearance of the adipic acid, the rate expression for a catalyzed reaction* is

$$\frac{-d[COOH]}{dt} = k[COOH][NH] \tag{3.61}$$

If we start with equal amounts of diamine and diacid, then for the duration of the reaction we have $[COOH] = [NH] = c$, and we have a second-order differential equation:

$$\frac{-dc}{dt} = kc^2 \tag{3.62}$$

that is easily separated and integrated:

$$\frac{1}{c} = kt + \text{constant} \tag{3.63}$$

A convenient parameter for following the progress of these reactions is the fraction of functional groups reacted at any time, called the *extent of reaction*, $p = (c_0 - c)/c_0$. Here, c_0 is the initial concentration. In terms of the extent of reaction, the rate expression becomes

$$c_0 kt = \frac{1}{1 - p} + \text{constant} \tag{3.64}$$

The quantity $1/(1 - p)$ is called the *number-average degree of polymerization*, $\bar{x}_n$, which represents the initial number of structural units present in the monomer relative to the total number of molecules, both monomer and polymer chains, present at any time t:

$$\bar{x}_n = c_0 kt + \text{constant} \tag{3.65}$$

Equation (3.65) shows us that the degree of polymerization increases linearly with time, an observation that has been verified experimentally.

3.3.1.2 Kinetics of Addition Polymerization. As the name suggests, addition polymerizations proceed by the addition of many monomer units to a single active center on the growing polymer chain. Though there are many types of active centers, and thus many types of addition polymerizations, such as anionic, cationic, and coordination polymerizations, the most common active center is a radical, usually formed at

---

*In the absence of an added strong acid, the reaction is self-catalyzed by a second molecule of the adipic acid, so the reaction is second order in acid: $\dfrac{-d[COOH]}{dt} = k[COOH]^2[NH]$.

Table 3.6   Characteristics of Stepwise and Addition Polymerization Reactions

Stepwise Polymerization	Addition Polymerization
Any two molecular species present can react	Only growth reaction adds repeating units one at a time to the chain
Monomer disappears early in the reaction	Monomer concentration decreases steadily throughout the reaction
Polymer molecular weight rises steadily throughout the reaction	High polymer is formed immediately, polymer molecular weight changes little throughout the reaction
Long reaction times necessary to obtain high molecular weight	Long reaction times do not affect molecular weight much, but do increase yield

a double bond in the monomer, such as in vinyl chloride (see Table 3.6). As a result, addition polymerizations are sometimes called *free radical polymerizations*, or even *vinyl polymerizations*, although it should be noted that each of these terms is increasingly narrow in its definition. However, analysis of a vinyl polymerization provides a good example of the kinetics of addition polymerizations.

In the 1930s, Paul Flory showed that radical polymerizations generally consist of three distinct steps: initiation, propagation, and termination. The free radical must first be initiated, then must propagate through the addition of monomers to the chain, and must eventually be terminated, either through consumption of the monomer or through addition of an agent that kills the free radicals. There are many ways to *initiate* radicals, such as ultraviolet radiation, electrochemical initiation, or oxidation–reduction reactions, but the most common is through addition of an initiator that transfers its radicals to the monomer to begin the polymerization. Organic peroxides, such as benzoyl peroxide, are common initiators that can transfer their radical to a monomer unit:

$$\phi-\overset{\overset{\displaystyle O}{\|}}{C}-\overset{\overset{\displaystyle O}{\|}}{C}-O-O-\overset{\overset{\displaystyle O}{\|}}{C}-C-\phi \xrightarrow{k_d} 2\ \phi-\overset{\overset{\displaystyle O}{\|}}{C}-C-O\cdot$$

$$k_a\Big| H_2C{=}CHCl \qquad\qquad (3.66)$$

$$2\ \phi-\overset{\overset{\displaystyle O}{\|}}{C}-C-O-CH_2-\overset{\overset{\displaystyle H}{|}}{\underset{\underset{\displaystyle Cl}{|}}{C}}\cdot$$

where ϕ is a phenol group. Most of these initiators have efficiencies of between 60% and nearly 100%, with recombination of the radical pairs the most common cause of low efficiency. If the fraction of radicals formed in the dissociation step that lead to successful radical chains is represented by f, then the rate of initiation is given by the change in initiated monomer concentration with time:

$$\frac{d[\text{M}\bullet]}{dt} = 2fk_d[\text{I}] \qquad\qquad (3.67)$$

where [I] is the concentration of the initiator, in this case benzoyl peroxide, and [M•] is the radical concentration, in this case the initiated vinyl chloride monomer.

Once the monomer radical has been initiated, high polymer is formed through addition of monomer units to the radical in the *propagation* phase of the polymerization. Each time a monomer unit is added, the radical transfers to the end of the chain to allow the polymerization to continue:

$$
\begin{array}{c}
\text{H}\\
|\\
\text{R}-(\text{H}_2\text{C}-\text{CHCl})_x-\text{CH}_2-\text{C}\cdot \quad + \quad \text{H}_2\text{C}=\text{CHCl}\\
|\\
\text{Cl}
\end{array}
$$

$$\downarrow k_p$$

$$
\begin{array}{c}
\text{H}\\
|\\
\text{R}-(\text{H}_2\text{C}-\text{CHCl})_{x+1}-\text{CH}_2-\text{C}\cdot\\
|\\
\text{Cl}
\end{array}
$$

(3.68)

The rate expression for propagation is given by the rate of monomer disappearance:

$$\frac{-d[\text{M}]}{dt} = k_p[\text{M}][\text{M}\bullet] \tag{3.69}$$

where [M] is the instantaneous monomer concentration. It is assumed that the rate constant for propagation, k_p, is the same for all propagation reactions in which a monomer is added to a growing radical chain, regardless of the molecular weight of the chain.

The propagation reactions proceed at multiple radical sites until the monomer is exhausted, or until one of two common types of *termination* reactions occur: either *combination* or *disproportionation*.

$$
\begin{array}{c}
\text{H} \quad\quad \text{H}\\
| \quad\quad\quad |\\
\cdots\text{CH}_2-\text{C}\cdot \ + \ \cdot\text{C}-\text{CH}_2\cdots \ \xrightarrow{\ k_{tc}\ } \ \cdots\text{CH}_2-\text{C}-\text{C}-\text{CH}_2\cdots\\
| \quad\quad\quad |\\
\text{Cl} \quad\quad \text{Cl}
\end{array}
$$

(3.70)

termination by combination

$$
\begin{array}{c}
\text{H} \quad\quad \text{H}\\
| \quad\quad\quad |\\
\cdots\text{CH}_2-\text{C}\cdot \ + \ \cdot\text{C}-\text{CH}_2\cdots \ \xrightarrow{\ k_{td}\ } \ \cdots\text{CH}_2-\text{C}-\text{H} \ + \ \text{C}=\text{CH}\cdots\\
| \quad\quad\quad |\\
\text{Cl} \quad\quad \text{Cl}
\end{array}
$$

(3.71)

termination by disproportionation

Each of these termination reactions has a similar rate expression that is second order in radical concentration, so the overall termination rate can be expressed as the rate of disappearance of the radicals due to both types of termination:

$$\frac{-d[\text{M}\bullet]}{dt} = 2k_t[\text{M}\bullet]^2 \tag{3.72}$$

Other reactions can also take place, such as chain transfer reactions, in which the radical is transferred from a growing polymer chain to a small molecule such as the initiator or a solvent molecule. These so-called *chain transfer agents* are often deliberately added to the reaction mix to either control the rate of the reaction or create branched polymers. The rate of the reaction can also be controlled with *inhibitors* or *retarders*. These additives also serve to help control the molecular weight of the polymer chain. In this instance, the rate of disappearance of monomer is given by an equation analogous to (3.69), except that the rate constant is now $k_{tr,i}$; and instead of adding monomer to continue chain propagation, M• now transfers its radical to some agent, A, which may be an initiator, a solvent, a monomer, or a chain-transfer agent:

$$\frac{-d[M]}{dt} = \sum k_{tr,i}[A][M\bullet] \qquad (3.73)$$

In the steady-state condition, the rate of initiation is equivalent to the rate of radical termination, and the radical concentration, [M•], remains essentially constant with time. Reactions (3.67) and (3.72) can be set equal at steady state to solve for [M•],

$$[M\bullet] = \left(\frac{fk_d[I]}{k_t}\right)^{1/2} \qquad (3.74)$$

which is then substituted back into Eq. (3.69) to obtain the overall rate expression for addition polymerization:

$$\frac{-[M]}{dt} = k_p \left(\frac{fk_d[I]}{k_t}\right)^{1/2}[M] \qquad (3.75)$$

As was the case for condensation-type reactions, there is a number-average degree of polymerization, $\bar{x}_n$, where

$$\bar{x}_n = \frac{\text{rate of growth}}{\sum \text{rates of reactions leading to dead polymer}} \qquad (3.76)$$

The rate of growth is simply given by Eq. (3.75), and the sum of the reaction rates leading to dead polymer come from termination [Eq. (3.72)] and chain transfer [Eq. (3.73)] reactions

$$\bar{x}_n = \frac{k_p[M\bullet][M]}{\Upsilon k_t[M\bullet]^2 + \sum_i k_{tr,i}[A][M\bullet]} \qquad (3.77)$$

where $\Upsilon = 1$ for termination by coupling and 2 if termination is by disproportionation, since two dead polymer molecules are formed in the latter reaction but only one dead polymer molecule is formed in coupling. Equation (3.75) is more easily simplified by inverting it to obtain an expression for $1/\bar{x}_n$:

$$\frac{1}{\bar{x}_n} = \frac{\Upsilon k_t[M\bullet]^2 + \sum_i k_{tr,i}[A][M\bullet]}{k_p[M\bullet][M]} = \frac{\Upsilon k_t[M\bullet]}{k_p[M]} + \sum_i \left(\frac{k_{tr,i}[A]}{k_p[M]}\right) \qquad (3.78)$$

Cooperative Learning Exercise 3.4

Work in groups of three. Equation 3.75 can be re-written to reflect the fact experimental determination of polymerization rates typically yield a single rate constant, called the *apparent rate constant*, k_{app}, which is a composite of the three different rate constants, k_d, k_t and k_p. If we call the rate of disappearance of monomer concentration, $-d[M]/dt$ the rate of polymerization, R_p, then we can rewrite Equation 3.75 as

$$R_p = k_{app}[M][I]^{1/2} \text{ or } \ln R_p = \ln k_{app} + \ln[M] + 0.5 \ln[I]$$

This relationship will allow us to estimate the variation in polymerization rate with temperature. Taking the derivative and treating the monomer and initiator concentrations as constants with respect to temperature gives

$$d \ln R_p = d \ln k_{app} \text{ or } (dR_p)/R_p = d \ln k_{app}$$

From here, we can introduce temperature dependence by assuming that k_{app} is given by an Arrhenius-type expression (Equation 3.12).

1. Work together to derive the following expression for the temperature-dependence of the polymerization rate from the previous expression:

$$\frac{dR_p}{R_p} = \frac{E_{a,p} + \dfrac{E_{a,d}}{2} - \dfrac{E_{a,t}}{2}}{RT^2} dT$$

where $E_{a,p}$, $E_{a,d}$, and $E_{a,t}$ are the activation energies for propagation, initiator decomposition, and termination, respectively.

2. Use the data below for the activation energies of some typical addition polymerization reactions at 60°C to calculate an average activation energy for propagation, termination, and initiator decomposition. Each person should calculate one of the three average activation energies.

Monomer	$E_{a,p}$ (kJ/mol)	$E_{a,t}$ (kJ/mol)	Initiator	T(°C)	$E_{a,d}$ (kJ/mol)
Acrylonitrile	16.2	15.5	2,2' Azobisisobutyroni-	70	123.4
Methyl acrylate	29.7	22.2	trile in		
Methyl methacrylate	26.4	11.9	benzene		
Styrene	26.0	8.0	t-Butyl peroxide in	100	146.9
Vinyl acetate	18.0	21.9	benzene		
2-Vinyl pyridine	33.0	21.0	Benzoyl peroxide in	70	123.8
			benzene		
			t-Butyl hydroperoxide in benzene	169	170.7

3. Combined your answers to estimate the percent rate of change in the polymerization rate ($dR_p/R_p \times 100$) for each 1°C change in temperature ($dT = 1$) at 50°C.

Answer: $E_{a,p} = 24.9$ kJ/mol; $E_{a,t} = 16.8$ kJ/mol; $E_{a,d} = 141$ kJ/mol; $dR_p/R_p = \{[24.9+141/2-16.8/2](10_3)(1)/(8.314)(323)_2\} \times 100 = 10\%$°C^{-1}

Once again, elimination of [M•] using Eq. (3.74) gives a general expression for the instantaneous degree of polymerization as a function of rate constants and concentrations:

$$\frac{1}{\bar{x}_n} = \frac{\Upsilon f k_d k_t^{1/2}[\text{I}]}{k_p[\text{M}]} + \sum_i \left(\frac{k_{tr,i}[\text{A}]}{k_p[\text{M}]} \right) \tag{3.79}$$

The first term in Eq. (3.79) is the inverse of the degree of polymerization that would occur in the absence of chain transfer, $(\bar{x}_n)_0$. The second term represents the effect of different chain transfer agents on the molecular weight.

3.3.1.3 Kinetics in the Polymerization of Copolymers.
Recall from Chapter 1 that a copolymer is a polymer composed of two or more different repeat units—the result of the simultaneous polymerization of two or more monomers. Though copolymers can be formed by any of the reaction mechanisms described above, the most interesting copolymerizations are of the free radical type. This is because the reactivity of the radicals can vary depending on the structure of the monomer to which they are attached. Such changes in reactivity between monomers are not generally found in condensation-type reactions; that is, a carboxylic acid group is a carboxylic acid group, independent of the chain or monomer to which it is attached. In chain polymerizations, however, the monomers can add to the growing chain in different ways and with different rates. Let us examine the simple system of two monomers M_1 and M_2 that are initiated to form radicals $M_1\bullet$ and $M_2\bullet$. There are four ways in which a monomer can add to a growing chain and four corresponding rate expressions:

$$M_1\bullet + M_1 \xrightarrow{k_{11}} M_1. \tag{3.80a}$$

$$M_1\bullet + M_2 \xrightarrow{k_{12}} M_2. \tag{3.80b}$$

$$M_2\bullet + M_1 \xrightarrow{k_{21}} M_1. \tag{3.80c}$$

$$M_2\bullet + M_2 \xrightarrow{k_{22}} M_2. \tag{3.80d}$$

The rate of disappearance of each monomer is then given by

$$\frac{-d[\text{M}_1]}{dt} = k_{11}[\text{M}_1\bullet][\text{M}_1] + k_{21}[\text{M}_2\bullet][\text{M}_1] \tag{3.81}$$

$$\frac{-d[\text{M}_2]}{dt} = k_{12}[\text{M}_1\bullet][\text{M}_2] + k_{22}[\text{M}_2\bullet][\text{M}_2] \tag{3.82}$$

If we define the monomer *reactivity ratio* for monomer 1 and 2, r_1 and r_2, respectively, as the ratio of rate constants for a given radical adding to its own monomer to the rate constant for it adding to the other monomer ($r_1 = k_{11}/k_{12}$ and $r_2 = k_{22}/k_{21}$ see Table 3.7 for typical values), then we arrive at the following relationship known as the *copolymer equation*:

$$\frac{d[\text{M}_1]}{d[\text{M}_2]} = \frac{[\text{M}_1](r_1[\text{M}_1] + [\text{M}_2])}{[\text{M}_2]([\text{M}_1] + r_2[\text{M}_2])} \tag{3.83}$$

Table 3.7 Typical Monomer Reactivity Ratios in the Addition Polymerization of Copolymers

Monomer 1	Monomer 2	r_1	r_2	T(°C)
Acrylonitrile	1,3-Butadiene	0.02	0.3	40
	Methyl methacrylate	0.15	1.22	80
	Styrene	0.04	0.4	60
	Vinyl acetate	4.2	0.05	50
	Vinyl chloride	2.7	0.04	60
1,3-Butadiene	Methyl methacrylate	0.75	0.25	90
	Styrene	1.35	0.58	50
	Vinyl chloride	8.8	0.035	50
Methyl methacrylate	Styrene	0.46	0.52	60
	Vinyl acetate	20	0.015	60
	Vinyl chloride	10	0.1	68
Styrene	Vinyl acetate	55	0.01	60
	Vinyl chloride	17	0.02	60
Vinyl acetate	Vinyl chloride	0.23	1.68	60

[a]Data from Lewis J. Young, Copolymerization Reactivity Ratios, pp. II-105–II-386 in J. Brandrup and E. H. Immergut, eds., with the collaboration of W. McDowell, *Polymer Handbook*, 2nd ed., Wiley-Interscience, New York, 1975.

Though this is a useful relationship, there are some special cases that are more easily described by introducing two types of ratios. The first ratio is F_1, the rate of change of monomer M_1 mole fraction:

$$F_1 = \frac{d[M_1]/dt}{d[M_1]/dt + d[M_2]/dt} \tag{3.84}$$

Similarly,

$$F_2 = \frac{d[M_2]/dt}{d[M_1]/dt + d[M_2]/dt} \tag{3.85}$$

The second useful ratio is the instantaneous concentration of a given monomer, f_1 or f_2:

$$f_1 = \frac{[M_1]}{[M_1] + [M_2]}, \qquad f_2 = \frac{[M_2]}{[M_1] + [M_2]} \tag{3.86}$$

It is important to distinguish between the concentration ratio, f_i, and the rate of change of concentration, F_i, since if monomers M_1 and M_2 are consumed at different rates, then $F_i \neq f_i$. Substitution of these ratios F_1, F_2, f_1, and f_2 into the copolymer equation gives

$$\frac{F_1}{F_2} = \frac{(r_1 f_1/f_2) + 1}{(r_2 f_2/f_1) + 1} \tag{3.87}$$

or

$$\frac{F_1}{1 - F_1} = \frac{[r_1 f_1/(1 - f_1)] + 1}{[r_2(1 - f_1)/f_1] + 1} \tag{3.88}$$

This form of the copolymer equation allows us to identify several simplifying cases:

- k_{12} and k_{21} are negligible; that is, a radical prefers to react with its own monomer, and no copolymer is formed—only homopolymer.

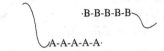

- k_{11} and k_{22} are negligible; that is, a radical prefers to react with the other monomer, $F_1 = F_2 = 0.5$,

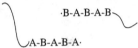

  and a perfectly alternating copolymer is formed.
- $k_{11} \approx k_{12}$ and $k_{22} \approx k_{21}$; $r_1 = r_2 = 1$; $F_1 = f_1$, and a perfectly random copolymer is formed.

  ·B-A-B-A-A⌇

  ⌇A-B-B-B-A·

- $r_1 r_2 = 1$, so that Eq. (3.88) becomes

$$\frac{F_1}{1 - F_1} = \frac{r_1 f_1}{1 - f_1} \qquad (3.89)$$

3.3.2 Polymerization Processes*

For the engineer, a knowledge of the reaction kinetics is merely a piece of what is usually a much larger problem involving the design of reactor components and the production and processing of large quantities of product—in this case, polymers. As a result, it is useful to discuss not only how the kinetic principles developed so far are applied to different methods of polymerization, but how problems associated with real processes, such as heat generation and removal, can in turn affect the polymerization kinetics (see Table 3.8). We will look at four common polymerization processes: *bulk polymerization, solution polymerization, suspension polymerization,* and *emulsion polymerization.* Though most of these processes are used primarily for addition-type polymerizations, applications involving condensation-type polymerizations will be noted where appropriate.

3.3.2.1 Bulk Polymerization. Monomer and polymer (with traces of initiator) are the only constituents in bulk polymerizations. Obviously, the monomer must be soluble in the polymer for this type of process to effectively proceed. Bulk polymerization, also called *mass* or *block polymerization,* can occur in stirred-tank reactors, or can be unstirred, in which instance it is called *quiescent bulk polymerization.* The primary difficulty with bulk polymerizations is that as the polymerization proceeds and more polymer is formed, the viscosity increases, thermal conductivity decreases, and heat removal becomes difficult.

Table 3.8 Comparison of Polymerization Processes

Type	Advantages	Disadvantages
Batch bulk	Minimum contamination	Poor heat control
	Simple equipment	Broad MW distribution
Continuous bulk	Better heat control	Requires stirring, separation, and
	Narrower MW distribution	recycle
Solution	Good heat control	Not useful for dry polymer
	Direct use of solution possible	
		Solvent removal difficult
Suspension	Excellent heat control	Requires stirring
	Direct use of suspension possible	Contamination by stabilizer possible
		Additional processing (washing, drying) required
Emulsion	Excellent heat control	Contamination by emulsifier likely
	Narrow MW distribution	Additional processing (washing, drying) required
	High MW attainable	
	Direct use of emulsion possible	

Source: Adapted from F. W. Billmeyer, *Textbook of Polymer Science*, Wiley, New York.

There is an increase in temperature as heat removal slows, along with a corresponding increase in the reaction rate. This phenomenon is known as the *autoacceleration* or *Trommsdorff effect* and can lead to catastrophic results if not properly controlled. Even with successful control of the reaction, it is difficult to remove the traces of remaining monomer from the polymer due to decreased diffusion. Similarly, it is difficult to get the reactions to proceed to completion due to limited monomer mobility.

A bulk polymerization reactor can be as simple as a tube into which the reactants are fed and from which the polymer mixture emerges at the end; it can be more of a traditional, continuous stirred-tank reactor (CSTR), or even a high-pressure autoclave-type reactor (see Figure 3.21). A bulk polymerization process need not be continuous, but it should not be confused with a batch reaction. There can be batch bulk polymerizations just as there are continuous bulk polymerizations processes.

3.3.2.2 *Solution Polymerization.* By adding a solvent to the monomer–polymer mixture, heat removal can be improved dramatically over bulk reactions. The solvent must be removed after the polymerization is completed, however, which leads to a primary disadvantage of *solution polymerization*. Another problem associated with radical chain polymerizations carried out in solution is associated with chain transfer to the solvent. As we saw in Section 3.3.1.2, chain transfer can significantly affect the molecular weight of the final polymer. This is particularly true in solution polymerization, where there are many solvent molecules present. In fact, chain transfer to solvent often dominates over chain transfer to other types of molecules, so that Eq. (3.79) reduces to

$$\frac{1}{\bar{x}_n} = \left(\frac{1}{\bar{x}_n}\right)_0 + \frac{C_S[S]}{[M]} \tag{3.90}$$

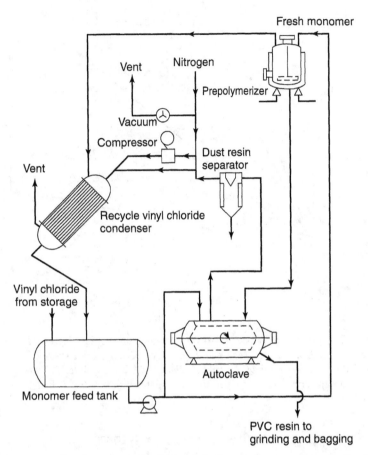

Figure 3.21 Schematic diagram of vinyl chloride bulk, continuous polymerization. Reprinted, by permission, from A. Krause, *Chem. Eng.*, **72**, p. 72. Copyright © 1965 by McGraw-Hill.

where the first term once again describes the contributions of propagation to the molecular weight, [S] and [M] are concentrations of the solvent and monomer, respectively, and C_S is the ratio of rate constants $k_{tr,s}/k_p$. This equation shows that a plot of $1/\bar{x}_n$ versus [S]/[M] should yield a straight line, the intercept for which gives the degree of polymerization in the absence of chain transfer to solvent. This relationship has been borne out experimentally for polystyrene and several solvent systems (see Figure 3.22).

The reactor equipment used for solution polymerizations is typically glass-lined stainless steel. An example of solution polymerization is the reaction of ethylene in isooctane with a chromia silica alumina catalyst initiator (see Figure 3.23) to form polyethylene. Typical reaction conditions for this polymerization are 150–180°C and 2.1–4.8 MPa (300–700 psi).

3.3.2.3 Suspension Polymerization.
When water is the solvent and the monomer is insoluble in water, *suspension polymerization* is often employed. The monomer and initiator form small droplets within the aqueous phase, and polymerization proceeds via bulk polymerization. Droplet sizes are typically between 10 and 1000 μm

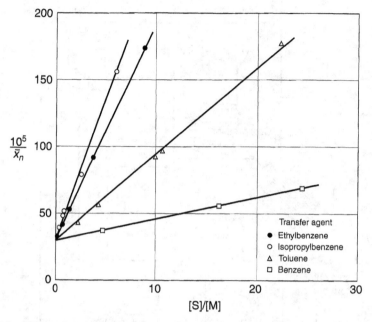

Figure 3.22   Effect of chain transfer to solvent on number average degree of polymerization for polystyrene [data from R. A. Gregg and F. R. Mayo, *Discuss Faraday Soc.*, **2**, 328 (1947).] From F. W. Billmeyer, *Textbook of Polymer Science*, 3rd ed. Copyright © 1984 by John Wiley & Sons, Inc. This material is used by permission of John Wiley & Sons, Inc.

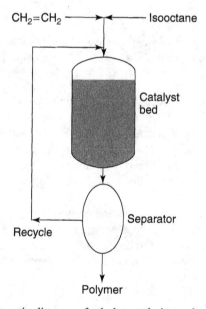

Figure 3.23   Schematic diagram of ethylene solution polymerization process.

in diameter, so that heat transfer to the suspending medium and subsequently to the reactor walls is very efficient throughout the course of the reaction. The result of the polymerization within each of the droplets is a polymer "bead." Though some agglomeration of particles takes place, the addition of a suspending agent such as poly(vinyl alcohol) minimizes coalescence, and the final polymer bead diameters are typically 100–1000 µm. The polymer beads are recovered by filtration, which is followed by a washing step. Suspension polymerization can be carried to nearly 90% completion at moderate pressures of 600 kPa (88 psi) and 80°C with stirring at about 150 to 300 rpm. A schematic diagram of the suspension polymerization of polystyrene is shown in Figure 3.24, and one for methyl methacrylate is shown in Figure 3.25.

3.3.2.4 Emulsion Polymerization. *Emulsion polymerization* is similar to suspension polymerization in that the monomer is insoluble in the solvent, water. The similarity ends there. In emulsion polymerization, the initiator is water-soluble, the particles in which the propagation step occurs are smaller, typically 0.05–5 µm in diameter, and the final beads are not usually filtered from the aqueous solution. The *emulsifier*, or *surfactant*, plays a very important role in emulsion polymerization. It is usually a long-chained hydrocarbon with a hydrophylic "head" and hydrophobic "tail." This charge polarization causes the tails to surround the monomer molecules, with the heads pointing outward into the water phase, forming a spherical droplet called a *micelle* (see Figure 3.26). As radicals initiate in the aqueous phase, they react with what few monomer molecules are dissolved in the water. As the radicals slowly grow and become more hydrophobic, they find it more stable to be inside of the micelles. Upon entering a micelle, the radicals cause rapid propagation and the micelle expands.

It can be shown that the degree of polymerization in emulsion polymerization can be derived from general polymerization kinetics [Eq. (3.79)] by neglecting chain transfer:

$$\frac{1}{\overline{x}_n} = \frac{2fk_d[I]}{k_p[M^*][N^*]} \tag{3.91}$$

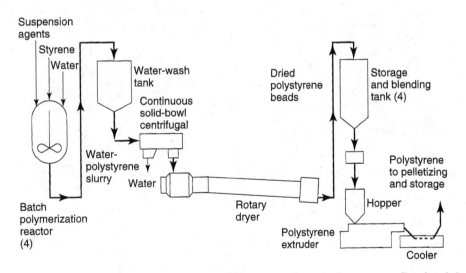

Figure 3.24 Schematic diagram of styrene suspension polymerization process. Reprinted, by permission, from *Chem. Eng.*, **65**, 98. Copyright © 1958 by McGraw-Hill.

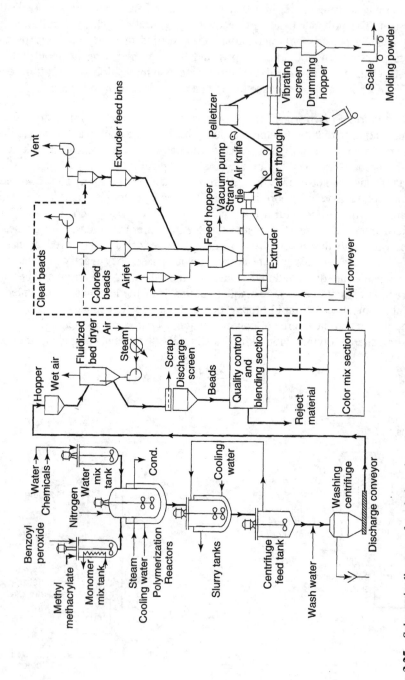

Figure 3.25 Schematic diagram of methyl methacrylate suspension polymerization process. Reprinted, by permission, from E. Guccione, *Chem. Eng.*, **73**, 138. Copyright © 1966 by McGraw-Hill.

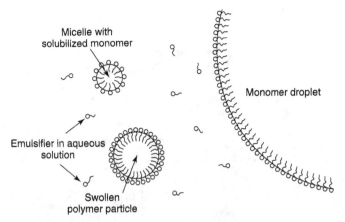

Figure 3.26 Schematic representation of micelle formation in emulsion polymerization.

where the superscript "*" on the monomer concentration indicates that this is the concentration of monomer in the micelle, not the solution. Since the initiator is soluble in water, its concentration remains based upon the total volume of water present. In Eq. (3.91), $[N^*]$ is the concentration of micelle-swollen polymer particles.

The final product of emulsion polymerization is an *emulsion*—a stable, heterogeneous mixture of fine polymer beads in an aqueous solution, sometime called a *latex emulsion*. Water-based paints, for example, can be formed from the emulsion polymerization of vinyl acetate. In this process, 1 m^3 of water containing 3% poly(vinyl alcohol) and 1% surfactant are heated to 60°C in a reaction vessel (see Figure 3.27) The temperature rises to around 80°C over a 4 to 5 hour period as monomer and an aqueous persulfate solution are added. The rate at which heat can be removed limits the rate at which monomer can be added.

Copolymerizations can also be carried out in an emulsion reaction. For example, *nitrile rubber* (NBR) can be made from the emulsion polymerization of acrylonitrile and butadiene:

$$H_2C\!\!=\!\!C\!-\!CN \ + \ H_2C\!\!=\!\!CH\!-\!HC\!\!=\!\!CH_2$$
$$\quad \textit{Acrylonitrile} \qquad\qquad \textit{Butadiene}$$

$$\downarrow \qquad\qquad\qquad\qquad\qquad\qquad (3.92)$$

$$[-H_2C\!-\!CH\!-\!CH_2\!-\!CH\!\!=\!\!CH\!-\!CH_2\!-\!]_n$$
$$\qquad\qquad |$$
$$\qquad\quad CN$$

As the acrylonitrile content in nitrile rubber increases, so does the resistance to non-polar solvents.

3.3.2.5 Gas-Phase Polymerization.
Not all polymerization reactions are carried out in the liquid phase. Polyethylene, for example, can be polymerized not only in solution as discussed in Section 3.3.2.2, but also through gas-phase polymerization (see Figure 3.28). Significant economic advantage is gained by eliminating costly solvent and catalyst recovery equipment. A fluidized bed reactor is used, into which purified

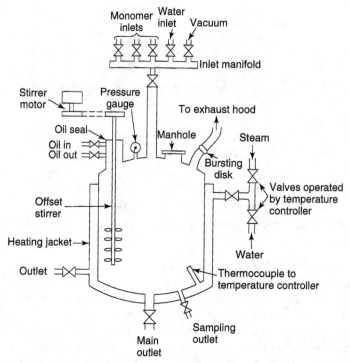

Figure 3.27 Schematic diagram of a typical emulsion polymerization reactor. Reprinted, by permission, from J. A. Brydson, *Plastic Materials*, p. 160. Copyright © 1966 by Van Nostrand.

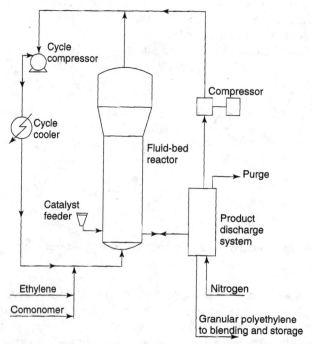

Figure 3.28 Schematic diagram of ethylene gas phase polymerization process, Reprinted, by permission, from *Chem. Eng.*, **80**, 71. Copyright © 1979 by McGraw-Hill.

ethylene and a modified chromia on silica powdered catalyst are fed. The reactor pressure is about 2 MPa (290 psi) and the temperature is 85–100°C. Heat is removed as gas is circulated using external coolers. The growing particles remain in the reactor 3–5 hours; emerging with an average diameter of about 500 μm. Small amounts of catalyst are used, so catalyst removal is often not necessary. Hydrogen is used as a chain transfer agent, and branching can be introduced by the addition of 1-butene.

3.3.3 Kinetics of Polymer Degradation

Degradation of polymers may be considered as any type of undesired modification of a polymer chain backbone, side groups, or both. These modifications are usually chemical in nature; that is, they involve the breaking of primary bonds, leading to reduced molecular weight and altered physical properties. The causes of degradation may be environmental (such as heat, light, or atmospheric pollutants), induced, (as in rubber mastication), or a result of processing due to heat and/or air. We will be concerned primarily with *environmental degradation*. Even within environmental degradation, there are six primary classes of polymer degradation: thermal, oxidative, radiative, chemical, mechanochemical, and biological. We will look briefly at the mechanisms of each of these types of degradative processes and present without derivation their related kinetic expressions.

3.3.3.1 Thermal Degradation. In *thermal degradation*, a polymer chain—in this case, a vinyl polymer chain—*depolymerizes*, or *unzips*. Just as in the formation reactions, there is initiation, chain transfer, and termination. Once a free radical has been initiated through some sort of *random* or *chain-end scission* process, the *depropagation* step proceeds, and monomer fragments split off from the polymer chain in a stepwise manner (see Eq. 3.93). The monomer fragments are usually assumed to be volatile.

$$R-(H_2C-CHCl)_{x+1}-CH_2-\underset{\underset{Cl}{|}}{\overset{\overset{H}{|}}{C\cdot}} \xrightarrow{k_{dp}} R-(H_2C-CHCl)_x-CH_2-\underset{\underset{Cl}{|}}{\overset{\overset{H}{|}}{C\cdot}} \tag{3.93}$$

$$+$$

$$H_2C=CHCl$$

It can be shown that the change in polymer weight, W, with time, t, for thermal degradation is given by

$$\frac{-dW}{dt} = \frac{2k_{ir}b(\bar{x}_n)_0\ W^{(2b-1)/b}}{W_0^{(b-1)/b}} \tag{3.94}$$

where k_{ir} is the rate constant for random initiation of radicals, W_0 is the initial weight of the polymer, $(\bar{x}_n)_0$ is the initial degree of polymerization, and b is a parameter that depends on the degree of polydispersity ($b = 1$ for a monodispersed system). The value of b is a measure of the breadth of the molecular weight distribution curve. For example, when the molecular weight distribution follows the exponential distribution, $b = 2$. According to Eq. (3.94), this corresponds to a reaction order of 3/2. As b increases, the reaction order approaches a limiting value of 2.

3.3.3.2 *Oxidative Degradation.* There is a wealth of literature available on the *oxidatative degradation* of polymers. Unfortunately, there is no single mathematical treatment that is applicable to all polymer types. As a result, the development of an appropriate kinetic expression is highly dependent upon the type of polymer being considered and the presence of additives, inhibitors, and antioxidants. For the sake of simplicity, we will consider a general case of autoxidation in a polyolefin in the absence of additives. It will be assumed that the diffusion of oxygen is rapid and is therefore not the rate-limiting kinetic step.

As in thermal degradation, the process begins with an initiation step, in this case the formation of radicals where X denotes a source of free radicals, such as polymer chains, which then react with atmospheric oxygen:

$$X \xrightarrow{k_1} R\bullet \tag{3.95a}$$

$$R\bullet + O_2 \xrightarrow{k_2} RO_2\bullet \tag{3.95b}$$

The oxidized radicals can then react with another polymer chain, RH, to form peroxides:

$$RO_2\bullet + RH \xrightarrow{k_3} RO_2H + R\bullet \tag{3.96}$$

Reactions (3.95) and (3.96) are the propagation steps, although the former is very rapid, with a negligible activation energy, whereas the latter is slower, with an activation energy of about 1.7 kJ/mol (7 kcal/mol). Three types of termination reactions ensue, all of which are very rapid, with activation energies of less than 1 kJ/mol (3 kcal/mol):

$$2R\bullet \xrightarrow{k_4} \text{Products} \tag{3.97a}$$

$$RO_2\bullet + R\bullet \xrightarrow{k_5} \text{Products} \tag{3.97b}$$

$$2RO_2\bullet \xrightarrow{k_6} \text{Products} + O_2 \tag{3.97c}$$

In the presence of oxygen, the first two termination reactions can be neglected. The oxidation rate can then be calculated from the rate-limiting propagation step, Eq. (3.96):

$$\frac{d[RO_2H]}{dt} = k_3[RO_2\bullet][RH] \tag{3.98}$$

A steady-state approximation allows us to solve for the intermediate free-radical concentration, $[RO_2\bullet]$, to obtain

$$\frac{d[RO_2H]}{dt} = \frac{k_3 k_1^{1/2}[X]^{1/2}[RH]}{k_6^{1/2}} \tag{3.99}$$

3.3.3.3 *Radiative Degradation. Radiative degradation* in polymers can give rise to both low-molecular-weight species, as in thermal degradation, or crosslinked structures, which are insoluble and infusible. Though crosslinking actually leads to an increase in the molecular weight of the sample, it is still considered a degradative process since the resulting product has unfavorably altered physical properties. There are generally two types of radiative degradation processes in polymers, grouped according to the amount of radiant energy involved and which result in different types of degradation. *Photolysis* occurs when ultraviolet light (wavelength $\lambda = 10^2 - 10^4$ Å) imparts energy on the sample of the order $10^2 - 10^3$ kJ/mol. *Radiolysis* occurs when higher radiant waves ($\lambda = 10^{-3} - 10^2$ Å) impart energy of the order $10^5 - 10^{10}$ kJ/mol to the sample. Each of these types attacks the polymer chain in a different fashion.

3.3.3.4 Chemical Degradation.

For most applications in the chemical process industry, *chemical degradation* is the most important type of polymer degradation. It is also the largest class overall, encompassing degradation due to both gaseous and liquid species on all polymer classes. Even within a polymer class, such as polyolefins, slight changes in the chain chemistry can lead to enormous changes in chemical compatibility. As a result, the presentation here must be severely limited. The intents will be to provide some insight into how the chemical degradation occurs, and to present practical information in guiding the selection of an appropriate polymer for a specific chemical environment.

In addition to oxygen, the most important gaseous agents that lead to polymer chemical degradation are nitrogen dioxide (NO_2), sulfur dioxide (SO_2), and ozone (O_3). Nitrogen dioxide attacks hydrocarbons by removing a tertiary hydrogen atom:

$$NO_2(g) + RH \longrightarrow HNO_2 + R\bullet \tag{3.100}$$

The resulting radical then reacts further with nitrogen dioxide,

$$NO_2(g) + R\bullet \longrightarrow RNO_2 \tag{3.101}$$

which then leads to products via chain scission and/or crosslinking. Polystyrene undergoes such a degradation process, the degree of polymerization for which changes according to the following empirical relationship

$$\frac{1}{\bar{x}_n} - \frac{1}{(\bar{x}_n)_0} = kt \tag{3.102}$$

where the rate constant, k, follows an Arrhenius-type relationship of the form

$$k = P_{NO_2}[41.4 \exp(-3870/RT)] \text{hr}^{-1} \tag{3.103}$$

Here P_{NO_2} is the partial pressure of nitrogen dioxide in pascals (1 Pa = 0.000145 psi), and the activation energy is 3870 kJ/mol. Other saturated polymers are less susceptible to attack by NO_2 than most unsaturated polymers such as synthetic rubbers (polyisoprene, polybutadiene, and butyl rubber). The presence of oxygen also tends to accelerate degradation by NO_2. The reaction of sulfur dioxide with saturated polymers is complex, but appears to be activated by ultraviolet radiation.

3.3.3.5 Mechanochemical Degradation.

Mechanochemical degradation occurs in polymers as the result of an applied mechanical force. This type of degradation is quite common in machining processes such as *grinding, ball milling*, and *mastication*.

In general, the degradation processes are similar to the other types of degradation, in that radicals are formed and cause subsequent reactions, such as oxidation and chain transfer, that can lead to either crosslinking or a reduction in molecular weight, though molecular weight reduction is the most common result. There are some unique aspects of mechanochemical degradation, however. In contrast to thermal degradation, radical formation is nonrandom, and the location of chain scission depends on the manner in which the forces are applied to the polymer. Also, the molecular weight decreases rapidly from an initial degree of polymerization, $(\bar{x}_n)_0$, for polymers under mechanical shear, then level off to a limiting degree of polymerization, $(\bar{x}_n)_\infty$ (Figure 3.29).

This exponential decrease in molecular weight and degree of polymerization with time can be expressed by an empirical relationship of the form

$$\frac{(\bar{x}_n)_0 - (\bar{x}_n)_t}{(\bar{x}_n)_0 - (\bar{x}_n)_\infty} = 1 - \exp(-kt) \tag{3.104}$$

where $(\bar{x}_n)_t$ is the degree of polymerization at time t. Typical values of k for common polymers of various molecular weights are given in Table 3.9.

In addition to a reduction in molecular weight, mechanical forces generally lead to an increase in solubility, a narrowing of the molecular weight distribution, decreased tensile strength, decreased crystallinity, and increased plasticity. These effects are highly dependent upon the machining conditions. In particular, higher machining temperatures result in non-Arrhenius behavior since there is a corresponding decrease in viscosity. The presence of radical scavengers and the type of machining equipment employed

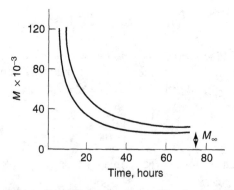

Figure 3.29   Effect of machining time on molecular weight of polymer.

Table 3.9   Rate Constants for Mechanochemical Chemical Degradation of Polymers

Polymer	Molecular Weight	k, hr^{-1}
Poly(vinyl alcohol)	4,000	0.0237
Polystyrene	7,000	0.0945
Poly(methyl methacrylate)	9,000	0.1200
Poly(vinyl acetate)	11,000	0.0468

also play significant roles in determining the extent of mechanochemical degradation during machining.

3.3.3.6 *Biological Degradation.*

In contrast to the degradation processes discussed to this point, *biological degradation* in polymers is oftentimes advantageous. In fact, there is a class of polymers that have been developed to enhance their susceptibility to attack by biological agents in order to improve their biodegradability and reduce their long-term impact on the environment. Other important applications of biodegradable polymers are as biomaterials, such as *resorbable sutures*, and for *controlled drug delivery* devices.

There are generally two types of polymeric materials that are susceptible to biological agents such as water, enzymes, and microbes: (a) natural materials like collagen, cellulose, and starch and (b) synthetic polymers. Though natural polymers represent a large fraction of all biodegradable polymers, particularly for biomedical uses like catgut sutures, we will concentrate on synthetic polymers, since they represent the class of materials where most of the advancement is still possible. It is the synthetic polymers that can be optimized in terms of mechanical properties and rate of degradation for biological and biomedical applications.

Substitutes for petroleum based polymers used in packaging (e.g., polystyrene, polyethylene, and polypropylene) represent one of the largest potential markets for biodegradable polymers. Some natural polymers have been used as substitutes, but their processability and mechanical properties have limited their use. In addition, factors such as cost, recycling, and alternative strategies like incineration have limited the need for the development of synthetic biodegradable polymers. Work continues in these areas, but it is the realm of synthetic polymers for biomedical applications that represents one of the more important areas of new material development. It is also the area where the kinetics of the degradation processes have been most widely studied. Though several classes of biodegradable synthetic polymers have been developed, including polyamides based on 4, 4'-spirobibutyrolactone [13] and poly(hydroxylalkanoates), it is the poly(α-esters) that have been most widely utilized for biomedical applications. Let us look at a specific example of synthetic polymers for resorbable suture material.

The most common synthetic biodegradable polymers for suture material and their corresponding weight loss in aqueous solution are listed in Table 3.10. Of these, poly(glycolic acid), PGA, poly(lactic acid), PLA, and copolymers of these two polyesters are the most widely used for resorbable suture material. PGA is a tough,

Table 3.10 Comparative Weight Loss for Some Common Biodegradable Polymers, Given as Time for 10% Weight Loss at 37°C and pH 7.4

Polymer	$t_{10}(\text{hr})$
Poly(L-lactic acid) (PLLA 8%)-co-Poly(glycolic acid) (PGA 92%)	450
Poly(glycolic acid) (PGA)	550
Poly(p-dioxanone) (PDS)	1200
Poly(3-hydroxybutyrate-co-3-valerate) PHBV (20% hydroxyvalerate, HV)	4.7% @ 5500
PHBV (12% HV)	5.6% @ 5500
Poly(3-hydroxybuturate) HB (0% HV)	1.8% @ 2500

Source: W. Amass, A. Amass, and B. Tighe, *Polym. Int.*, **47**, 89–144 (1998).

crystalline, inelastic material with a T_m of 230°C and T_g of 36°C. There are two types of poly(lactic acid): poly(L-lactic acid), PLLA, which is a tough, crystalline, inelastic material with $T_m = 180$°C, $T_g = 67$°C; and poly(D-lactic acid), PDLA, which is a tough, amorphous, inelastic material with $T_g = 57$°C. The *in vitro* and molecular weight loss data for PLLA show that over a 16-week period, only 10–15% of the mass is lost, and the molecular weight is reduced to ~50% of its original value [14]. Similar studies for PGA show that mass loss begins in the fourth week and is completed in 10–12 weeks (see Figure 3.30 for two different types of commercial PGA sutures), and the molecular weight loss data (Figure 3.31) shows a steady decrease in the molecular weight of the primary peak.

The degradation mechanisms of these polymers are quite complex. Factors influencing the biological stability include the molecular weight (high MW gives stability), the presence of residual monomer and low-molecular-weight compounds, temperature, % crystallinity (high crystallinity decreases the rate of biodegradation), and extent of sterilization by β- and γ-radiation (both lead to degradation). Polyesters generally undergo a

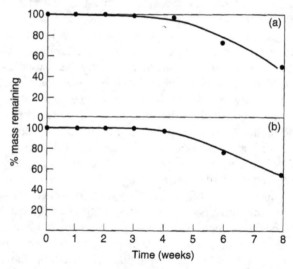

Figure 3.30   Percent weight loss for two types of PGA sutures, (a) Dexon and (b) Dexon-s, as function of exposure time to a buffered solution (pH 7, 37°C). Reprinted, by permission, from J. Feijin, in *Polymeric Biomaterials*, p. 68. Copyright © 1986 by Martinus Nijhoff Publishers.

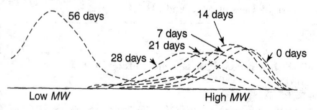

Figure 3.31   Shift in molecular weight distribution for a commercial PGA suture material (Dexon) as a function of exposure time to a buffered solution (pH 7, 37°C). Reprinted, by permission, from J. Feijin, in *Polymeric Biomaterials*, p. 68. Copyright © 1986 by Martinus Nijhoff Publishers.

two-stage hydrolysis process. In the first stage, water diffuses into the amorphous regions, producing random hydrolytic scission at the susceptible ester linkage. This leads to an apparent increase in the percent crystallinity of the polymer during this stage. After most of the amorphous regions have been eroded, hydrolysis of the crystalline regions commences. A method for determining the mode of hydrolysis has been developed by Shih [15], in which the mole fraction of monomer, m_1, and the degree of polymerization, α, are determined by a technique like nuclear magnetic resonance (NMR) spectroscopy. If completely random scission of the backbone bonds occurs, $m_1 = \alpha^2$, and if exclusive chain-end unzipping occurs, $m_1 = \alpha$. In addition to hydrolysis, these polymers can undergo degradation due to the effect of enzymes, chemicals, and radiation.

3.4   KINETIC PROCESSES IN COMPOSITES*

Due to the fact that industrial composites are made up of combinations of metals, polymers, and ceramics, the kinetic processes involved in the formation, transformation, and degradation of composites are often the same as those of the individual components. Most of the processes we have described to this point have involved condensed phases—liquids or solids—but there are two gas-phase processes, widely utilized for composite formation, that require some individualized attention. *Chemical vapor deposition* (CVD) and *chemical vapor infiltration* (CVI) involve the reaction of gas phase species with a solid substrate to form a heterogeneous, solid-phase composite. Because this discussion must necessarily involve some of the concepts of transport phenomena, namely diffusion, you may wish to refresh your memory from your transport course, or refer to the specific topics in Chapter 4 as they come up in the course of this description.

3.4.1   Kinetics of Chemical Vapor Deposition*

Chemical vapor deposition (CVD) involves the reaction of gaseous reactants to form solid products. There are two general types of CVD processes: (a) the thermal decomposition of a homogeneous gas to form a solid and (b) the chemical reaction of two or more gaseous species to form a solid. Both types of CVD reactions are used industrially to form a variety of important elements and compounds for semiconductor, superconductor, and ceramic coating applications.

An example of the first kind can be found in the decomposition of trichloromethylsilane to give silicon carbide and hydrogen chloride:

$$(CH_3)SiCl_3(g) \longrightarrow SiC(s) + 3HCl(g) \tag{3.105}$$

The reaction of titanium tetrachloride with boron trichloride and hydrogen to give solid titanium diboride is an example of the second type of CVD reaction:

$$TiCl_4(g) + 2BCl_3(g) + 5H_2(g) \longrightarrow TiB_2(s) + 10HCl(g) \tag{3.106}$$

Reaction (3.106) illustrates the primary advantage of a CVD process. Solid titanium diboride, TiB_2, melts at 3325°C, yet it can be produced via reaction (3.106) at 1027°C or lower on a suitable substrate by the CVD process. Additionally, the diffusion of gas phase reactants is generally rapid, allowing for the efficient coating of high-surface-area (i.e., porous), substrates. The mechanisms of these CVD reactions can be highly complex, often involving interactions with solid surface sites, and the combination of heat and momentum transport coupled with reaction kinetics can lead to complex models (see Figure 3.32). Let us look at some simple, yet important, systems in order to gain a better understanding of the kinetics of CVD processes.

Consider the growth of a germanium (Ge) film in the production of a semiconducting thin film [16]. The germanium can be deposited onto a substrate through the reaction of $GeCl_4$ with hydrogen gas:

$$GeCl_4(g) + H_2(g) \longrightarrow Ge(s) + 2HCl(g) + Cl_2(g) \tag{3.107}$$

The proposed mechanism for this overall reaction involves a gas-phase dissociation step

$$GeCl_4(g) \rightleftharpoons GeCl_2(g) + Cl_2(g) \tag{3.108}$$

two adsorption processes, where S is a surface site on the solid

$$GeCl_2(g) + S \; \overset{K_A}{\rightleftharpoons} \; GeCl_2 \bullet S \tag{3.109}$$

$$H_2(g) + 2S \; \overset{K_H}{\rightleftharpoons} \; 2H \bullet S \tag{3.110}$$

and a surface reaction

$$GeCl_2 \bullet S + 2H \bullet S \; \overset{k_s}{\longrightarrow} \; Ge(s) + 2HCl(g) + 2S \tag{3.111}$$

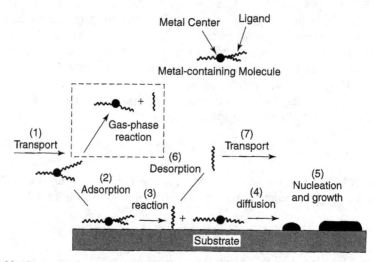

Figure 3.32 Processes involved in chemical vapor deposition. From *Chemistry of Advanced Materials: An Overview*, L. V. Interrante and M. J. Hampden-Smith, eds. Copyright © 1998 by John Wiley & Sons, Inc. This material is used by permission of John Wiley & Sons, Inc.

The rate-limiting step is believed to be the surface reaction between adsorbed molecular hydrogen and germanium dichloride (3.111):

$$\frac{d[\text{Ge}]}{dt} = k_s [\text{GeCl}_2 \bullet \text{S}][\text{H} \bullet \text{S}]^2 \tag{3.112}$$

where the concentration of adsorbed surface species is often represented by their *fractional coverage*, f_i, or the fraction of the surface occupied by species i:

$$\frac{d[\text{Ge}]}{dt} = k_s f_{GeCl_2 \bullet S} f_{H \bullet S}^2 \tag{3.113}$$

The *deposition rate*, or *film growth rate*, is usually expressed in nm/s, as is the reaction rate constant, k_s. Deposition rates can be expressed in mol/m$^2\bullet$s by multiplying by the molar density of solid germanium.

It can be shown that through appropriate substitution of Eqs. (3.108) through (3.110) and use of the fractional area balance ($f_v + f_{GeCl_2} + f_H = 1$, where f_v is the fraction of vacant surface sites), the deposition rate can be expressed in terms of the partial pressures (concentrations) of the reactants as

$$\frac{d[\text{Ge}]}{dt} = \frac{k' P_{GeCl_2} P_{H_2}}{(1 + K_A P_{GeCl_2} + \sqrt{K_H P_{H_2}})^3} \tag{3.114}$$

where k' is the product of the two adsorption equilibrium constants, K_A and K_H, and the reaction rate constant k_s.

Another example of an industrially relevant CVD process is the production of tungsten (W) films on various substrates for (a) wear and corrosion protection and (b) diffusion barriers in electronic devices. One CVD method for depositing tungsten is through the reduction of WF$_6$ with hydrogen, which follows the overall reaction:

$$\text{WF}_6(g) + 3\text{H}_2(g) \longrightarrow \text{W}(s) + 6\text{HF}(g) \tag{3.115}$$

This reaction is actually slightly endothermic ($\Delta H = 88$ kJ/mol), but the large net increase in entropy and the nonequilibrium nature of most CVD processes lead to significant tungsten deposition. As with the Ge example, the deposition mechanism involves adsorption steps and surface reactions. At low pressures and under conditions of excess hydrogen gas, the deposition rate follows the general form:

$$\frac{d[\text{W}(s)]}{dt} \propto P_{H_2}^{1/2} P_{WF_6}^0 \tag{3.116}$$

Ideally, the functional dependence of the growth rate on the pressure (concentration) of the gaseous species would be determined by analysis of all the mechanistic steps, as was the case for our Ge example and as could be done for the W example. This is not always possible, however, due to insufficient information on such factors as reaction rate constants, adsorption energies, and preexponential factors. As a result, deposition rate data are often determined experimentally and are used to either support or refute

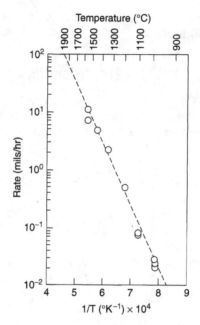

Figure 3.33 Arrhenius plot for the deposition of Si_3N_4. Reprinted, by permission, from F. Galasso, *Chemical Vapor Deposited Materials*, p. 23. Copyright © by CRC Press.

suspected rate-limiting steps. Arrhenius-type plots are often made by plotting the logarithm of the measured deposition rate versus inverse temperature (see Figure 3.33 for the formation of Si_3N_4 from SiF_4 and NH_3), sometimes for various total gas pressures (see Figure 3.34 for the formation of B_4C from BCl_3, CH_4, and H_2).

There are numerous materials, both metallic and ceramic, that are produced via CVD processes, including some exciting new applications such as CVD diamond, but they all involve deposition on some substrate, making them fundamentally composite materials. There are equally numerous modifications to the basic CVD processes, leading to such exotic-sounding processes as *vapor-phase epitaxy* (VPE), *atomic-layer epitaxy* (ALE), *chemical-beam epitaxy* (CBE), *plasma-enhanced CVD* (PECVD), *laser-assisted CVD* (LACVD), and *metal–organic compound CVD* (MOCVD). We will discuss the specifics of CVD processing equipment and more CVD materials in Chapter 7.

3.4.2 Kinetics of Chemical Vapor Infiltration*

Chemical vapor infiltration (CVI) is similar to CVD in that gaseous reactants are used to form solid products on a substrate, but it is more specialized in that the substrate is generally porous, instead of a more uniform, nominally flat surface, as in CVD. The porous substrate introduces an additional complexity with regard to transport of the reactants to the surface, which can play an important role in the reaction as illustrated earlier with CVD reactions. The reactants can be introduced into the porous substrate by either a diffusive or convective process prior to the deposition step. As infiltration proceeds, the deposit (matrix) becomes thicker, eventually (in the ideal situation) filling the pores and producing a dense composite.

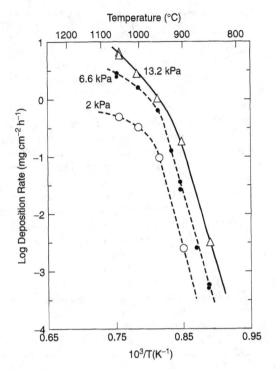

Figure 3.34 Arrhenius plot for the deposition of B_4C at various total pressures. Reprinted, by permission, from R. Naslain, CVI composites, in *Ceramic-Matrix Composites*, R. Warren, ed., pp. 199–244. Copyright © 1992 by Chapman & Hall, London.

There are five general categories of CVI processes, depending on which type of transport process they utilize (diffusion or convection) and whether there is an imposed thermal gradient or not (see Figure 3.35). Type I, known as *isothermal CVI* (ICVI) has no thermal gradients imposed, and reagents enter the substrate via diffusion. The disadvantage of this process is that the entire substrate is at the reaction temperature, so that pores on the surface can close off before the inner pores are completely filled. This problem is circumvented with a *thermal gradient*, as in Type II CVI. Here, the reactants contact the substrate in a cold region, and diffuse to the heated reaction zone. Type III and Type IV CVI are the forced convection analogues of Type I and Type II, respectively, where the reactants are now forced into the substrate by convection instead of being allowed to diffuse. Type V CVI is specialized, in that not only are the reactants forced into the substrate, but gaseous products are also forcefully removed in a cyclical manner.

The isothermal reaction/diffusion system (Type I) is the easiest to model and provides us with some insight into the competition between kinetics and diffusion. Consider the case of CVI in a filament bundle, as shown schematically in Figure 3.36. The regions between the fibers can be approximated as cylindrical pores with diameter r_0 and depth L. Ideally, reactants will diffuse into the pore and create a uniform concentration throughout the pore length before they react to form product, thereby filling the pore uniformly from bottom to top. This is often not the case, and there can be a concentration gradient, as shown in Figure 3.36, leading to nonuniform deposition.

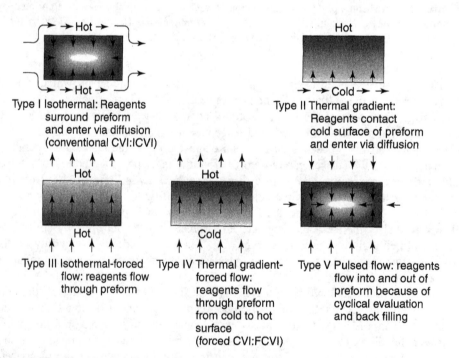

Figure 3.35 Classification of chemical vapor infiltration processes. From *Carbide, Nitride, and Boride Materials Synthesis and Processing*, A. W. Weimer, ed. p. 563. Copyright © 1997 by Chapman & Hall, London, UK, with kind permission of Kluwer Academic Publishers.

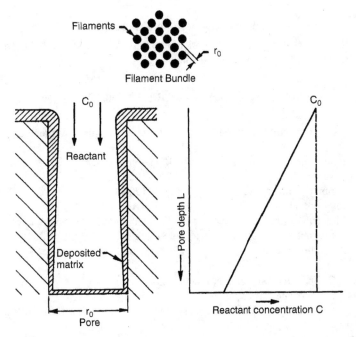

Figure 3.36 Schematic diagram of pore-filling process in the chemical vapor infiltration of a fiber bundle. From *Carbide, Nitride, and Boride Materials Synthesis and Processing*, A. W. Weimer, ed., p. 566, Copyright © 1997 by Chapman & Hall, London, with kind permission of Kluwer Academic Publishers.

Eventually, the top of the pore will close off, and reactants will not be able to find their way to the bottom of the pore to complete the densification process.

The competition between kinetics and transport processes can be characterized using one of two dimensionless groups. The first is called the *Dahmköhler number, Da,* $k_1 r_0 / D$, where k_1 is the reaction rate constant for the rate-limiting step, r_0 is the pore radius, and D is the diffusivity of the gaseous reactants. A related dimensionless group is the *Thiele modulus,* $\Phi^2 = 2k_1 L^2 / r_0 D$, where the additional parameter, L, is the pore length. Both groups represent a ratio of the kinetic deposition rate to the gas phase mass transport rate, or, similarly, the diffusion time to the reaction time. For very small values of either dimensionless group, gas transport is very rapid and the reaction rate is very slow. This is the ideal situation, with uniform deposition along the pore length the result. In contrast, deposition is rapid for large values of Da and Φ^2, but diffusion is slow, so there is a decreasing concentration of reactants down the pore length and a nonuniform density results. Mixed control occurs when the reaction and transport rates are comparable ($Da \approx \Phi^2 \approx 1$).

The CVI procedure can lead to some very interesting composite materials. Lackey et al. [17] have produced composites of carbon fibers in a laminated SiC/carbon matrix (see Figure 3.37). The alternating SiC/carbon layers were produced by CVI of two different reactions: the decomposition of methyltrichlorosilane (MTS) in the presence of hydrogen; and the decomposition of propylene in hydrogen, respectively. There is also an excellent opportunity to model these processes with the intent at gaining better control of the CVI process [18].

Cooperative Learning Exercise 3.5

Consider the thermal decomposition of trichloromethylsilane (TMS) in the presence of hydrogen gas to give SiC in a porous substrate:

$$(CH_3)SiCl_3(g) + H_2(g) \xrightarrow{k_1} SiC(s) + 3HCl(g) + H_2(g)$$

The following operating conditions are typically used for this CVI process:

Pore diameter	1–10 μm
Pore length	100–1000 μm
Ambient temperature	800–1200°C
Inlet pressure	5–200 torr
Molecular diffusivity	726 cm^2/s@1000°C and 5 torr
Preexponential reaction constant	262 cm/s
Activation energy	120 kJ/mol

Person 1: Estimate the Dahmköhler number, *Da*.
Person 2: Estimate the Thiele modulus, Φ^2.
Compare your results. Is this CVI process reaction-controlled, diffusion-controlled, or a combination of both?

Answer: For T = 1273 K, r_0 = 0.0005 cm, and L = 0.1 cm, *Da* is about 10^{-9}, and Φ^2 is about 10^{-4}, both of which indicate a reaction-controlled CVI process.

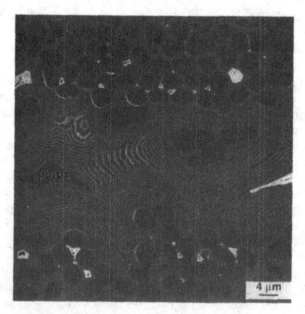

Figure 3.37 CVI-laminated SiC (light) and carbon (dark) matrix layers surrounding carbon fibers. Reprinted, by permission, from W. J. Lackey, S. Vaidyaraman, and K. L. More, *J. Am. Ceram. Soc.*, **80**, 113. Copyright © 1997 by The American Ceramic Society.

3.5   KINETIC PROCESSES IN BIOLOGICS*

To attempt to describe the kinetics of formation, adaptation, and decomposition of tissue is to try and detail the processes inherent to life itself. This presents a rather daunting challenge, even if we limit ourselves to human tissue. Luckily, other disciplines such as biology have performed this task for us. The factors involved in the reproduction and growth of living tissue, as well as the kinetics of energy production, respiration, and excretion in cells, are not of concern to us here. What we are hoping to understand are the kinetic processes that will eventually allow us to seamlessly integrate biological and synthetic systems. We aim to describe not only the building blocks of biological systems (as we did in Chapter 1), and whether these building blocks are thermodynamically compatible with each other (as we did in Chapter 2), but also how rapidly the incorporation of synthetic materials into living systems can proceed. We will describe the kinetics of one such process here, namely, the kinetics of receptor–ligand binding in cells.

3.5.1   Kinetics of Cell Surface Receptor–Ligand Binding*

Recall from Sections 1.5 and 2.5 that there are types of specific binding in and between cells called receptor–ligand binding (see Figure 3.38). The receptor–ligand interaction phenomenon is an extremely important one—one that goes far beyond simple cell adhesion. Such other cellular functions as growth, secretion, contraction, and motion are known to be receptor-mediated cell behaviors. Moreover, an understanding of the receptor–ligand interaction affords us the potential to control these cellular functions. For example, cell biologists can chemically alter the structure of the receptor and/or the ligand in order to modify the association and dissociation rate constants.

Figure 3.38 suggests that we can write a reversible reaction for the formation of the receptor–ligand complex

$$R + L \underset{k_r}{\overset{k_f}{\rightleftharpoons}} C \tag{3.117}$$

and that the time rate of change of the complex density is given by

$$\frac{d[C]}{dt} = k_f[R][L] - k_r[C] \tag{3.118}$$

Let us continue by considering the simplest case, where the ligand concentration remains essentially constant at its initial value of $[L]_0$, and the concentration of unbound

R   L   C

Figure 3.38   Schematic representation of receptor (R)–ligand (L) binding to form a receptor/ligand complex (C).

receptors is given by the difference between the total concentration of receptors, $[R]_T$ (assumed to be constant), and the concentration of receptors in the complex, $[R_T - C]$:

$$\frac{d[C]}{dt} = k_f[R_T - C][L]_0 - k_r[C] \tag{3.119}$$

The use of dimensionless groups will simplify the solution to Eq. (3.119). Let the concentration of the complex, $[C]$, be scaled to the total receptor concentration, such that $u = [C]/[R]_T$ and the dimensionless time, τ, is given by $\tau = k_r/t$. The ratio of the reverse to forward rate constants is simply the dissociation equilibrium constant (the inverse of our "normal" equilibrium constant), $K_D = k_r/k_f$, so that Eq. (3.119) now becomes

$$\frac{du}{d\tau} = \frac{(1-u)[L]_0}{K_D} - u \tag{3.120}$$

which we can solve using the boundary condition that $u = u_0$ at $\tau = 0$ to give

$$u(t) = u_0 \exp\left[-\left(1 + \frac{L_0}{K_D}\right)\tau\right] + \frac{L_0 K_D}{1 + \frac{L_0}{K_D}}\left\{1 - \exp\left[-\left(1 + \frac{L_0}{K_D}\right)\tau\right]\right\} \tag{3.121}$$

Though this equation seems a bit unwieldy, it is a rather straightforward description of the variation of the scaled complex concentration as a function of time for various L_0/K_D ratios. For example, if $u_0 = 0$, as is the case for the beginning of the binding procedure where there is no complex yet formed, a series of plots results as shown in Figure 3.39 for various L_0/K_D ratios. Such a plot is general for different types of receptor–ligand pairs. Some typical values of the rate constants and receptor concentrations for investigated receptor–ligand pairs are shown in Table 3.11.

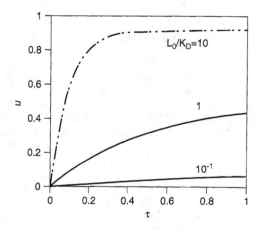

Figure 3.39   The change in dimensionless receptor/ligand binding complex concentration, u, with dimensionless time, τ, for various values of L_0/K_D. Reprinted, by permission, from D. A. Lauffenburger and J. J. Linderman, *Receptors*, p. 21. Copyright © 1993 by Oxford University Press, Inc.

Table 3.11 Some Receptor–Ligand Binding Parameters

Receptor	Ligand	Cell Type	R_T (#/cell)	k_f ($M^{-1}\,min^{-1}$)	k_r (min^{-1})	K_D (M)
Transferrin	Transferrin	HepG2	5×10^4	3×10^6	0.1	3.3×10^{-8}
Fc_γ	2.4G2 Fab	Mouse macrophage	7.1×10^5	3×10^6	0.0023	7.7×10^{-10}
Chemotactic peptide	FNLLP	Rabbit neutrophil	5×10^4	2×10^7	0.4	2×10^{-8}
Interferon	Human interferon-α_{2a}	A549	900	2.2×10^8	0.072	3.3×10^{-10}
TNF	TNF	A549	6.6×10^3	9.6×10^8	0.14	1.5×10^{-10}
β-adrenergic	Hydroxybenzylpindolol	Turkey erythrocyte	—	8×10^8	0.08	1×10^{-10}
α_1-adrenergic	Prazosin	BC3H1	1.4×10^4	2.4×10^8	0.018	7.5×10^{-11}
Insulin	Insulin	Rat fat cells	1×10^5	9.6×10^6	0.2	2.1×10^{-8}
EGF	EGF	Fetal rat lung	2.5×10^4	1.8×10^8	0.12	6.7×10^{-10}
Fibronectin	Fibronectin	Fibroblasts	5×10^5	7×10^5	0.6	8.6×10^{-7}
Fc_g	IgE	Human basophils	—	3.1×10^6	0.0015	4.8×10^{-10}
IL-2 (heavy chain)	IL-2	T lymphocytes	2×10^3	2.3×10^7	0.015	6.5×10^{-10}
IL-2 (light chain)			1.1×10^4	8.4×10^8	24	2.9×10^{-8}
IL-2 (heterodimer)			2×10^3	1.9×10^9	0.014	7.4×10^{-12}

HepG2 = human hepatoma cell line; 2.4G2 Fab = Fab portion of 2.4G2 antibody against receptor; FNLLP = N-formylnorleucylleucylphenylalanine; A549 = human lung alveolar carcinoma; TNF = tumor necrosis factor; hydroxybenzylpindolol is an antagonist to the receptor; EGF = epidermal growth factor; IgE = immunoglobulin E; IL-2 = interleukin 2; prazosin is an antagonist to the receptor; BC3H1 = smooth muscle-like cell line; RBL = rat basophilic leukemia cell line.

Source: D. A. Lauffenburger and J. J. Linderman, *Receptors*, p. 30. Copyright © 1993 by Oxford University Press, Inc.

REFERENCES

Cited References

1. Avrami, M., Kinetics of phase change I: General theory, *J. Chem. Phys.*, **7**, 1103 (1939).
2. Avrami, M., Kinetics of phase change II: Transformation–time relations for random distribution of nuclei, *J. Chem. Phys.*, **8**, 212 (1940).
3. Avrami, M., Kinetics of phase change III: Granulation, phase change, and microstructure, *J. Chem. Phys.*, **9**, 177 (1941).
4. Matusita, K., and S. Sakka, Kinetics study of the crystallization of glass by differential scanning calorimetry, *Phys. Chem. Glasses*, **20**, 81 (1979).
5. Kissinger, H. E., Reactive kinetics in differential thermal analysis, *Anal. Chem.*, **29**, 1702 (1957).
6. Baz, A., and T. Chen, Torsional stiffness of nitinol-reinforced composite drive shafts, *Composites Eng.*, **3**, 1119 (1993).
7. http://www.seas.upenn.edu/~zshinar/researchb.html
8. http://aml.seas.ucla.edu/
9. Lewis, T. H., Take a deeper look at cathodic protection, *Chem. Eng. Prog.*, **95**(6), 55 (1999).
10. *JANAF Thermochemical Tables*, D. R. Stull, and H. Prophet, project directors, U.S. National Bureau of Standards, Washington, D.C., 1971.
11. Parks, G. A., *Chem. Rev.*, **65**, 177 (1965).
12. Hogenson, D. K., and J. H. Healy, Mathematical treatment of glass corrosion data, *J. Am. Cer. Soc.*, **45**(4), 178 (1962).
13. Vanderbilt, D. P., J. P. English, G. L. Fleming, F. W. McNeely, D. R. Cowsar, and R. L. Dunn, Biodegradable polyamides based on 4, 4′-spirobibutyrolactone, in *Progress in Biomedical Polymers*, C. G. Gebelein and R. L. Dunn, eds., Plenum Press, New York, 1990, p. 249.
14. Feijen, J., Biodegradable polymers for medical purposes, in *Polymeric Biomaterials*, E. Piskin and A. S. Hoffman, eds., NATO ASI Series, Martinus Nijhoff Publishers, Dordrecht, 1986, p. 62.
15. Shih. C., *Pharmaceut. Res.*, **12**, 2036 (1995).
16. Ishii, H., and Y. Takahashi, *J. Electrochem. Soc.*, **135**, 1539 (1988).
17. Lackey, W. J., S. Vaidyaraman, and K. L. More, Laminated C–SiC matrix composites produced by CVI, *J. Am. Ceram. Soc.*, **80**, 113 (1997).
18. Gupte, S. M., and J. A. Tsamopoulos, Densification of porous materials by chemical vapor infiltration, *J. Electrochem. Soc.*, **136**, 2–555 (1989).

General

Gaskell, David R., *Introduction to the Thermodynamics of Materials*, 3rd ed., Taylor and Francis, Washington, D.C., 1995.

Hill, C. G., *An Introduction to Chemical Engineering Kinetics and Reactor Design*, John Wiley & Sons, New York, 1977.

Fogler, H. S., *Elements of Chemical Reaction Engineering*, 3rd ed., Prentice-Hall, Upper Saddle River, NJ, 1999.

Schmidt, L. D., *The Engineering of Chemical Reactions*, Oxford University Press, New York, 1998.

Metal Formation

Engineering Metallurgy, F. T. Sisco, ed., Pitman Publishing, New York, 1957.

Metal Phase Transformations

Haasen, P., *Physical Metallurgy*, Cambridge University Press, London, 1978.

Machlin, E. S., *Thermodynamics and Kinetics Relevant to Materials Science*, Giro Press, New York, 1991.

Metal Degradation

Craig, B. D., *Fundamental Aspects of Corrosion Films in Corrosion Science*, Plenum Press, New York, 1991.

Fontana, M. G., *Corrosion Engineering*, 3rd edition, McGraw-Hill, New York, 1986.

Corrosion Mechanisms in Theory and Practice, P. Marcus, edi. Marcel Dekker, New York, 2002.

Ceramic Formation

Weimer, A. W., *Carbide, Nitride and Boride Materials Synthesis and Processing*, Chapman Hall, London, 1997.

Ceramic Phase Transformations

Kingery, W. D., H. K. Bowen, and D. R. Uhlmann, *Introduction to Ceramics*, John Wiley & Sons, New York, 1976.

Ceramic Degradation

McCauley, R. A., *Corrosion of Ceramics*, Marcel Dekker, New York, 1995.

Polymer Formation

Billmeyer, F. W., Jr., *Textbook of Polymer Science*, 3rd edition, John Wiley & Sons, New York, 1984.

Rodriguez, F., *Principles of Polymer Systems*, 2nd ed., McGraw-Hill, New York, 1982.

Polymer Degradation

Reich, L., and S. S. Stivala, *Elements of Polymer Degradation*, McGraw-Hill, New York, 1971.

Polymers of Biological and Biomedical Significance, S. Shalaby, Y. Ikada, R. Langer, and J. Williams, eds., ACS Symposium Series 540, ACS, Washington, D.C., 1994.

Progress in Biomedical Polymers, C. G. Gebelein and R. L. Dunn, eds., Plenum Press, New York, 1988.

CVD/CVI

Interrante, L. V., and M. J., Hampden-Smith, *Chemistry of Advanced Materials: An Overview*, Wiley-CVH, New York, 1998.

Galasso, F. S., *Chemical Vapor Deposited Materials*, CRC Press, Boca Raton, FL, 1991.

Cell-Surface Receptor–Ligand Binding

Lauffenberger, D. A., and J. J. Linderman, *Receptors: Models for Binding, Trafficking and Signaling*, Oxford University Press, New York, 1993.

PROBLEMS

Level I

3.I.1 One-half of an electrochemical cell consists of a pure nickel electrode in a solution of Ni^{2+} ions; the other is a cadmium electrode immersed in a Cd^{2+} solution. If the cell is a standard one, write the spontaneous overall reaction and calculate the voltage that is generated. Adapted from Callister, Example Problem 18.1, page 569 (part of problem).

3.I.2 A piece of corroded steel plate was found in a submerged ocean vessel. It was estimated that the original area of the plate was 64.5 cm^2 and that approximately 2.6 kg had corroded away during the submersion. Assuming a corrosion penetration rate of 200 mpy for this alloy in seawater, estimate the time of submersion in years. Adapted from Callister, Problem 18.13, p. 602 (some numbers converted to cgs).

3.I.3 For the following pairs of alloys that are coupled in seawater, predict the possibility of corrosion; and if corrosion is possible, note which alloy will corrode: (a) Al/Mg; (b) Zn/low carbon steel; (c) brass/Monel; (d) titanium/304 stainless steel; and (e) cast iron/316 stainless steel. Clearly state any assumptions you make about compositions of alloys. Callister, Problem 18.10, p. 602 (editorial changes only).

3.I.4 It is known that the kinetics of recrystallization for some alloy obeys the JMA equation and that the value of n is 2.5. If, at some temperature, the fraction recrystallized is 0.40 after 200 minutes, determine the rate of recrystallization at this temperature. Callister, problem 10.4, p. 323.

3.I.5 Consider the condensation polyesterification reaction between ethylene glycol, $HO-(CH_2)_2-OH$, and terephthalic acid, $HOOC-Ph-COOH$, each of which has an initial concentration of 1.0 mol/liter. Calculate the number average degree of polymerization at 1, 5, and 20 hours. The forward reaction rate constant for the polymerization reaction is 10.0 liter/mol•hr, and second-order, catalyzed kinetics can be assumed.

Level II

3.II.1 Compute the cell potential at 25°C for the electrochemical cell in Problem 3.I.1 if the Cd^{2+} and Ni^{2+} concentrations are 0.5 and 10^{-3} M; respectively. Is the spontaneous reaction direction still the same as for the standard cell? Callister, Example Problem 18.1, page 569 (second part of problem).

3.II.2 The corrosion potential of iron immersed in deaerated acidic solution of pH = 3 is −0.720 V as measured at 25°C with respect to the normal calomel electrode. Calculate the corrosion rate of iron in millimeters per year, assuming the Tafel slope of the cathodic polarization curve, β_c, equals 0.1 V/decade and the hydrogen ion exchange current, i_0, is $0.1 \times 10^{-6} A/cm^2$.

3.II.3 In an emulsion polymerization, all the ingredients are charged at time $t = 0$. The time to convert various amounts of monomer to polymer is indicated in the following table. Predict the time for 30% conversion. Polymerization is "normal"—that is, no water solubility or polymer precipitation.

Fraction of Original Monomer Charge Converted to Polymer	Time (hours)
0.05	2.2
0.12	4.0
0.155	4.9

3.II.4 Construct a plot analogous to Figure 3.39 for the dissociation of the receptor–ligand complex; that is, for $u_0 = 1$. Make plots for $L_0/K_D = 10$, 1, and 0.1.

Level III

3.III.1 As a pure liquid is supercooled below the equilibrium freezing temperature (T_m), the free energy difference between the solid and liquid becomes increasingly negative. The variation of ΔG_v with temperature is given approximately by $\Delta G_v = (\Delta H_m/T_m)(T_m - T)$, where $\Delta H_m(< 0)$ is the latent heat of solidification. Using this expression, show that $\exp(-N_o\Delta G^*/RT)$ is a maximum when $T = T_m/3$. Why does the maximum nucleation rate occur at a temperature greater than $T_m/3$?

3.III.2 A review article on the CVD processes used to form SiC and Si_3N_4 by one of the pioneers in this area, Erich Fitzer [Fitzer, E., and D. Hegen, Chemical vapor deposition of silicon carbide and silicon nitride—Chemistry's contribution to modern silicon ceramics, *Angew. Chem. Int. Ed. Engl.*, **18**, 295 (1979)], describes the reaction kinetics of the gas-phase formation of these two technical ceramics in various reactor arrangements (hot wall, cold

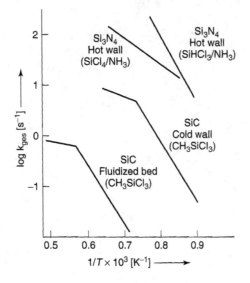

wall, and fluidized bed) for several precursor combinations, as indicated in the accompanying diagram. (a) Determine the activation energy(ies) for each ceramic/reactor/precursor type. (b) Comment on what causes the break in two of the curves at high temperatures.

3.III.3   Consider three 10-m^3 reactors operating at 70°C. Each contains 1000 kg of styrene monomer. Additional ingredients initially are as follows:

	Reactor I	Reactor II	Reactor III
Inert liquid	8.00 m^3 benzene	8.00 m^3 water	8.00 m^3 water
Initiator	1.00 kg benzoyl peroxide	1.00 kg benzoyl peroxide	1.00 kg potassium persulfate
Polystyrene	10.0 kg polymer in solution	10.0 kg polymer dissolved in the monomer	10.0 kg polymer in stable latex form
Additive	—	1.00 kg poly(vinyl alcohol)	—

In each of the three reactors, does $-d[M]/dt$ (initial value) increase, decrease, or remain essentially the same when each of the following changes is made? (a) Amount of initiator is doubled; (b) amount of monomer is halved; (c) temperature is increased to 90°C; (d) amount of initial polymer is doubled. Also, predict the effect of each change on the $\bar{x}_n$ of the polymer being formed initially.

CHAPTER 4

Transport Properties of Materials

4.0 INTRODUCTION AND OBJECTIVES

Perhaps the single largest subject area that differentiates engineering from the sciences is that of transport phenomena. For those of you who have had little or no background in transport processes, you may find some supplemental reading to be of great help in this chapter. There are many excellent textbooks and references on transport phenomena, including the landmark text by Bird, Stewart, and Lightfoot, which first brought the analogies between momentum, heat, and mass transfer to engineering students in a rigorous fashion. We will rely heavily on these analogies in this chapter; and without going into any great depth of coverage on any one transport topic, we will see how transport properties of liquid and solid materials go a long way towards describing many important phenomena in materials selection, design, and processing.

The analogy to which we refer is fundamentally quite simple. It states that the rate of flow per unit area, called a *flux*, of a fundamental quantity such as momentum, heat, or mass is proportional to a corresponding *driving force* which causes the flow to occur.

$$\text{Flux} = \text{``constant''} \times \text{Driving force} \tag{4.1}$$

In this chapter, we will focus mostly on the "constant" in Eq. (4.1), and not the flux or driving force, since these are specific to a process. The proportionality constant is a material property, however, which we can measure, correlate with process parameters, and hopefully predict. The term "constant" is in quotations, because we will see that in some instances, it is really a function of the driving force. It is important to understand (a) the driving forces under which this proportionality constant operates and (b) the effect the constant has on transport phenomena.

The transport process about which most of us have an intuitive understanding is heat transfer so we will begin there. In order for heat to flow (from hot to cold), there must be a driving force, namely, a *temperature gradient*. The heat flow per unit area (Q/A) in one direction, say the y direction, is the heat flux, q_y. The temperature difference per unit length for an infinitesimally small unit is the temperature gradient, dT/dy. According to Eq. (4.1), there is then a proportionality constant that relates these two quantities, which we call the *thermal conductivity*, k. Do not confuse this quantity with

An Introduction to Materials Engineering and Science: For Chemical and Materials Engineers,
by Brian S. Mitchell
ISBN 0-471-43623-2 Copyright © 2004 John Wiley & Sons, Inc.

rate constants of the previous chapter, or with the Boltzmann constant, k_B. The form of Eq. (4.1) for heat transfer, then, is called *Fourier's Law*:

$$q_y = -k\frac{dT}{dy} \tag{4.2}$$

The minus sign in Eq. (4.2) arises due to the fact that in order for there to be heat flow in the $+y$ direction, the temperature gradient in that direction must be negative—that is, lower temperature in the direction of heat flow. If the temperature gradient is expressed in units of K/m, and the heat flux is in J/m^2· s, then the thermal conductivity has units of J/K·m·s, or W/m·K. A related quantity is the *thermal diffusivity*, which is often represented by the lowercase Greek letter alpha, α. Thermal diffusivity is defined as $k/\rho C_p$, where k is the thermal conductivity, ρ is the density, and C_p is the *heat capacity* at constant pressure per unit mass. We will see in a moment why the term "diffusivity" is used to describe this parameter. We will generally confine our descriptions in this chapter to thermal conductivity.

The equations for one-dimensional momentum and mass flow are directly analogous to Fourier's Law. A *velocity gradient*, dv_x/dy, is the driving force for the bulk flow of momentum, or *momentum flux*, which we call the *shear stress* (shear force per unit area), τ_{yx}. This leads to *Newton's Law of Viscosity*:

$$\tau_{yx} = -\mu\frac{dv_x}{dy} \tag{4.3}$$

The proportionality constant, μ, is the *viscosity*. When the velocity gradient is cm/s·cm, or simply s^{-1}, we call it the *shear rate* because of the units of inverse time. Often, the lowercase Greek letter gamma with an overdot, $\dot{\gamma}$, is used to represent the shear rate, dv_x/dy. The shear stress of x-directed moment in the y direction, τ_{yx}, typically has units of g/cm·s^2. Hence, viscosity has cgs units of g/cm·s, or *poise*. In the SI system, the unit of Pa·s is often used. Note that 1 Pa·s $= 10$ poise. In this chapter we will have a lot to say about viscosity and how it varies for different materials in their molten or liquid states.

Finally, if we have a *concentration gradient* of one component (A) in a binary mixture (A + B), dc_A/dy, this leads to the bulk motion of molecules, known as the *molar flux* of component A in the y direction, J_{Ay}^*, which can be related to the concentration gradient using *Fick's Law*:

$$J_{Ay}^* = -D_{AB}\frac{dc_A}{dy} \tag{4.4}$$

The proportionality constant for mass transport is called the *molecular diffusivity* of A in B, and it has cgs units of cm^2/s when concentration is mol/cm^3 and distance is in centimeters. As a result, the molar flux has units of mol/cm^2·s. You should now recognize why the term "thermal diffusivity" is used in certain versions of Eq. (4.2).

You should be able to see the similarities in Eqs. (4.2) through (4.4). We will use these relationships a great deal in this chapter, so you should understand them fully and do some background reading if they are unfamiliar to you, but we will again concentrate on the proportionality constants: thermal conductivity, viscosity, and molecular diffusivity.

By the end of this chapter, you should be able to:

- Locate or calculate thermal conductivities, heat capacities, and specific heats for any material.
- Locate or calculate viscosities for any liquid or molten material.
- Locate or calculate the binary-component diffusivity or self-diffusion coefficient for a gas, dilute solute, or melt.
- Estimate the diffusivity of a gas in a porous solid.
- Differentiate between Newtonian and non-Newtonian fluids, and know which behavior is appropriate for a certain type of material in the fluid state.
- Convert between thermal conductivity, thermal conductance, thermal resistivity, and thermal resistance for a material of specific geometry.
- Calculate permeabilities for substances, given diffusivity and solubility data.
- Calculate partition coefficients across membranes.

Finally, you will notice that there are a few things that are different in this chapter. It is primarily organized by the type of transport processes, then secondarily organized by material class—unlike the previous chapters, which were primarily organized by material type alone. This is done to add continuity to the transport subtopics while maintaining the analogies between them within this chapter. Also, much of the material here is indicated as "optional"; that is, the information on momentum and mass transport properties are not taught to any great extent in a "traditional" materials science-oriented course. This doesn't mean they aren't important, even to an introductory course in materials. In this case, it means that they are perhaps best read in conjunction with the chapter on materials processing (Chapter 7), since the driving forces arise due to the type of processing equipment being used—for example, e.g., casting, extrusion, or CVD processing. In any case, as engineers, we need to keep the similarities between the types of transport processes in mind as we study them, regardless of the relative extent to which each one is covered.

4.1 MOMENTUM TRANSPORT PROPERTIES OF MATERIALS*

In this section, we will examine the microscopic origins of viscosity for each material type, and we will see how viscosity is much more complex than simply serving as a proportionality constant in Eq. (4.3). Ultimately, we will find that viscosity is not a constant at all, but a complex function of temperature, shear rate, and composition, among other things.

4.1.1 Momentum Transport Properties of Metals and Alloys: Inviscid Systems

Most metallic systems above their melting point have very low viscosities, typically less than about 100 poise. Liquids with viscosities below this level are generally termed *inviscid*, that is, they are water-like in their fluidity. This is a relative statement, of course, since their viscosities are distinctly nonzero, but for all practical purposes the pressure required to cause these fluids to flow or overcome surface tension forces, for example, is vanishingly small. Nevertheless, the relative simplicity of metallic melts allows us to explore, at least briefly, the molecular origins of viscosity. We will then explore the temperature and composition dependences of metal and alloy viscosities.

4.1.1.1 *The Molecular Origins of Viscosity.*

Viscosity, like all of the transport properties in this chapter, has its roots in the kinetic theory of gases. Though we concentrate on condensed phases in this book, it is important to look at the viscosity of gases so that we may progress to the more complex topic of liquid viscosities, and so that we might see the analogies with the molecular origins of thermal conductivity and molecular diffusion in subsequent sections.

We already have much of the information we need to derive an expression for the viscosity of a gas. Recall from Chapter 1 that the atoms or molecules of a gas interact with one another. This interaction can be described in terms of repulsive and attractive energies, as given by Eq. (1.11). Recall also that the force, F, can be related to the energy, U, by Eq. (1.13). A potential energy function that has been extensively studied is the Lennard-Jones potential of Eq. (1.12). We can modify the form of Eq. (1.12) slightly to give the following potential energy function

$$U(r) = 4\varepsilon \left[\left(\frac{\sigma}{r} \right)^{12} - \left(\frac{\sigma}{r} \right)^{6} \right] \tag{4.5}$$

where σ is the *collision diameter* of the molecule and ε is the maximum energy of attraction between a pair of molecules. These two parameters are known for many substances and can be approximated for others using fluid critical properties, boiling point, or melting point. From this potential energy function, a relationship for viscosity (in poise) can be derived for a gas at low density:

$$\mu = 2.6693 \times 10^{-5} \frac{\sqrt{MT}}{\sigma^2 \Omega} \tag{4.6}$$

where T is the temperature of the gas in degrees Kelvin, M is the molecular weight of the gas, σ is in Å, and Ω is a dimensionless quantity that correlates with the dimensionless temperature, $k_B T/\varepsilon$. We will see that the dimensionless temperature plays a role in thermal conductivity and mass diffusivity as well. For rigid spheres (no attraction or repulsion), $\Omega = 1$, so that it represents a deviation of the real molecules from a rigid-sphere model. Note that the viscosity of gases at low density increases with temperature and that there is no dependence on pressure in this regime. So, we see that, in theory, there is indeed a link between the potential energy function of a gas and its viscosity.

For liquids, even simple monatomic liquids such as molten metals, this link between molecular interactions and a bulk property such as viscosity is still there, in principle, but it is difficult to derive a relationship from fundamental interaction energies that is useful for a wide variety of systems. Nonetheless, theoretical expressions for liquid viscosities do exist. Some are based on statistical mechanical arguments, but the model that is most consistent with the discussion so far is that of nonattracting hard spheres in a dense fluid. The viscosity (in Pa·s) of such a liquid is

$$\mu = 3.8 \times 10^{-8} \frac{\sqrt{MT}}{V^{2/3}} \left[\frac{PF^{4/3}\left(1 - \dfrac{PF}{2}\right)}{(1 - PF)^3} \right] \tag{4.7}$$

where M, T are the molecular weight and temperature as before, V is the molar volume of the liquid, and PF is the *packing fraction* or *packing density* of spheres in the liquid, which can be related to the collision diameter, σ, as

$$PF = \frac{\pi n_0 \sigma^3}{6} \tag{4.8}$$

where n_0 is the average number density (N/V). Some viscosities for elemental liquid metals as calculated from Eq. (4.7) are compared with experimental values in Table 4.1.

Notice the similarity between the relationship for liquid viscosity [Eq. (4.7)] and that for gaseous viscosity [Eq. (4.6)]. They both have a square root dependence on temperature and molecular weight and depend on the inverse square of the collision diameter [can you prove this for Eq. 4.7)?]. So, at least in principle, there is a fundamental relationship between the structure of a liquid and its viscosity.

Table 4.1   Comparison of Calculated [using Eq. (4.7) with PF = 0.45] and Experimental Values of Liquid Metal Viscosities Near Their Melting Points

Metal	μ_{cal} (mPa·s)	μ_{exp} (mPa·s)
Na	0.62	0.70
Mg	1.69	1.25
Al	1.79	1.2–4.2
K	0.50	0.54
Fe	4.55	6.92
Co	4.76	4.1–5.3
Ni	4.76	4.5–6.4
Cu	4.20	4.34
Zn	2.63	3.50
Ga	1.63	1.94
Ag	4.07	4.28
In	1.97	1.80
Sn	2.11	1.81
Sb	2.68	1.43
Au	5.80	5.38
Hg	2.06	2.04
Tl	2.85	2.64
Pb	2.78	2.61
Bi	2.54	1.63

Source: T. Lida and R. I. L. Guthrie, *The Physical Properties of Liquid Metals*. Copyright © 1988 Oxford University Press.

4.1.1.2 Temperature Dependence of Pure Metal Viscosity.

Practically speaking, empirical and semiempirical relationships do a much better job of correlating viscosity with useful parameters such as temperature than do equations like (4.7). There are numerous models and their resulting equations that can be used for this purpose, and the interested student is referred to the many excellent references listed at the end of this chapter. A useful empirical relationship that we have already studied, and that is applicable to viscosity, is an Arrhenius-type relationship. For viscosity, this is

$$\mu = A \exp\left(\frac{H_\mu}{RT}\right) \tag{4.9}$$

where A and H_μ are constants. There are several methods for evaluating these two constants, but the simplest expressions relate them to the melting point of the metal, T_m. (We know that this makes some fundamental sense, because the melting point is related to the intermolecular attraction between atoms.) The preexponential factor is given by

$$A(\text{cpoise}) = \frac{5.7 \times 10^{-2}\sqrt{MT_m}}{V_m^{2/3} \exp(H_\mu/RT_m)} \tag{4.10}$$

The activation energy for viscous flow, H_μ, for "normal" metals is

$$H_\mu = 1.21 T_m^{1.2} \tag{4.11}$$

and the activation energy for semi-metals is

$$H_\mu = 0.75 T_m^{1.2} \tag{4.12}$$

The two relationships for activation energy as a function of melting point are shown graphically in Figure 4.1.

4.1.1.3 Composition Dependence of Alloy Viscosity.

Attempts have been made to calculate the viscosity of a dilute liquid alloy from a theoretical standpoint, but with little success. This is primarily due to the fact that little is known about the interaction of dissimilar atoms in the liquid state. Empirical relationships for the viscosity of dilute liquid alloys have been developed, but these are generally limited to specific alloy systems—for example, mercury alloys with less than 1% impurities. The viscosities of binary liquid alloys have been empirically described using a quantity called the *excess viscosity*, μ^E (not to be confused with the excess chemical potential), which is defined as the difference between the viscosity of the binary mixture (alloy), μ_A, and the weighted contributions of each component, $x_1\mu_1$ and $x_2\mu_2$:

$$\mu^E = \mu_A - (x_1\mu_1 + x_2\mu_2) = -2(x_1\mu_1 + x_2\mu_2)\frac{H^E}{RT} \tag{4.13}$$

where x_1 and x_2 are the mole (atomic) fractions of each component, and H^E is the enthalpy of mixing. This expression indicates that the sign of the excess viscosity (either $-$ or $+$) depends only on the sign of the enthalpy of mixing, which can also

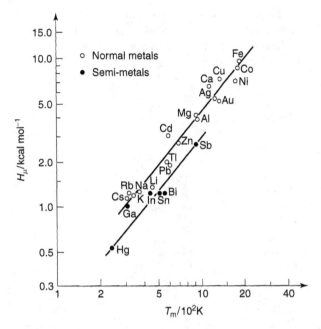

Figure 4.1 Dependence of activation energy for viscous flow on melting point for "normal" metals and semi-metals. Reprinted, by permission, from T. Iida and R. I. L. Guthrie, *The Physical Properties of Liquid Metals*, p. 187. Copyright © 1988 by Oxford University Press.

be either negative or positive. This relationship is only true some of the time, as borne out by experimental data (see Table 4.2).

4.1.2 Momentum Transport Properties of Ceramics and Glasses

4.1.2.1 *Newtonian Viscosity in Glasses.* As we saw in Chapter 1, the structure of glasses is fundamentally different from metals. Unlike metals and alloys, which can be modeled as hard spheres, the structural unit in most oxide glasses is a polyhedron, often a tetrahedron. As a result, the response of a structural unit to a shear force is necessarily different in molten glasses than in molten metals. The response is also generally more complicated, such that theoretical descriptions of viscosity must give way completely to empirical expressions. Let us briefly explore how this is so.

Consider the melting of a common glass such as vitreous silica, SiO_2. Recall that the structural unit in the silicates is the $(SiO_4)^{4-}$ tetrahedron (see Figure 1.43). As heat is provided to the glass, Si–O bonds begin to rupture, and some structural units become mobile. Upon reaching the melting point, a certain portion of broken bonds is sufficient to cause flow in response to shear, but not all the bonds have been broken. For this reason, the SiO_2 melt has a high viscosity. As temperature is increased further above the melting point, more bonds are broken and the viscosity decreases. For silicate glasses, this change in viscosity can cover many order of magnitudes. Thus, viscosity–temperature plots for glasses are often given as semilogarithmic plots, as in Figure 4.2.

The large range of viscosities for glasses is related to its structure. For example, compared to a highly mobile liquid like water, the *relaxation time* is much longer in glasses. Relaxation time is the time needed for the structural components to adjust to

Table 4.2 Relationship Between Excess Viscosity and Difference in Ionic Radii $|d_1 - d_2|$ and Enthalpy of Mixing for Various Binary Liquid Alloys

Alloy	$\|d_1 - d_2\|$ (Å)	μ^E	H^E
Au–Sn	0.66	−	−
Ag–Sb	0.64	−	−
Ag–Sn	0.55	−	−, +[a]
Al–Cu	0.46	−	−
Au–Cu	0.46	0, −	−
K–Na	0.38	−	+
Cd–Sb	0.35	−	+, −[a]
Cu–Sb	0.34	−	−
Ag–Cu	0.30	−	+
Hg–In	0.29	−, +[a]	−
Cu–Sn	0.25	−	−
Al–Zn	0.24	−	+
Cd–Bi	0.23	−	+
K–Hg	0.23	+	−
Pb–Sb	0.22	−	+, −[a]
Mg–Pb	0.19	+, −[a]	−
Na–Hg	0.15	+	−
Al–Mg	0.15	+	−
Cd–Pb	0.13	−	+
Pb–Sn	0.13	0	+
Sb–Bi	0.12	−	+
Ag–Au	0.11	+	−
Fe–Ni	0.10	−, +[a]	−
In–Bi	0.07	+	−
Mg–Sn	0.06	+	−
In–Pb	0.03	−	+
Sn–Zn	0.03	−	+
Sn–Bi	0.03	+	+

[a] In some instances the signs are composition-dependent.
Source: T. Iida and R. I. L. Guthrie, *The Physical Properties of Liquid Metals*. Copyright © 1988 by Oxford University Press.

the deformation forces resulting from viscous flow. In general, this is the same order of magnitude as the time required for a structural unit to move approximately one of its own diameters. The smaller the structural units and the less strongly they interact with each other, the shorter the relaxation time. For water, the relaxation time is of the order 10^{-12} s. For glasses near their softening points, this relaxation time could be around 10^2 s. It is interesting to note that almost all glasses have a inflection point in the log μ versus temperature curve at about 10^{15} Pa·s (10^{16} poise, just beyond where the data end in Figure 4.2), above which the viscosity begins to level off. This is the point at which brittleness begins to set in.

There are a number of named temperature regimes that have been assigned to glasses, within which certain characteristic melt properties, primarily related to

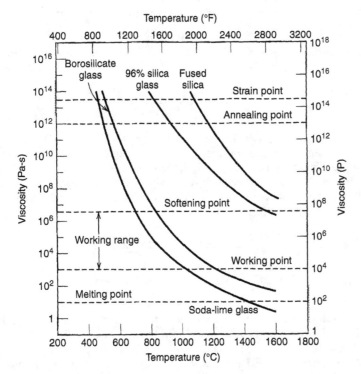

Figure 4.2 Dependence of viscosity on temperature for several silica-based glasses. Reprinted, by permission, from W. Callister, *Materials Science and Engineering: An Introduction*, 5th ed., p. 425. Copyright © 2000 by John Wiley & Sons, Inc.

formability, have been identified. From highest to lowest temperature (as is found in a cooling operation), these temperatures include the *melting point* (~10 Pa·s), where the glass is fluid enough to be considered a liquid; the *working point* (~10^3 Pa·s), where the glass is easily deformed; the *softening point* (~10^6 Pa·s), where a glass piece can be handled without significant dimensional change; the *annealing point* (~10^{12} Pa·s), where the relaxation time is on the order 10^3 s; and the *strain point* (~10^{13} Pa·s), where fracture occurs before plastic deformation in response to an applied force (see Chapter 5 for a comparison of fracture and plastic deformation). Other terminology used in the glass industry includes the *fictive temperature*, where the properties of the glass are the same as those of the melt in metastable equilibrium; the *devitrification temperature*, where the glass transforms from a noncrystalline state to one that is crystalline; and the *Littleton temperature*, where a glass rod heated at 5–10 K/m will deform at a rate of 1 mm/min under its own weight.

Most viscosity–temperature relationships for glasses take the form of an Arrhenius expression, as was the case for binary metal alloys. The Vogel–Fulcher–Tammann (VFT) equation is one such relationship.

$$\mu(T) = A \exp\left[\frac{E_\mu}{R(T - T_0)}\right] \tag{4.14}$$

where E_μ is the activation energy for viscosity, A is the preexponential factor, R is the gas constant, and T_0 is an additional adjustable parameter specific to a glass.

Cooperative Learning Exercise 4.1

Three-parameter models such as the VFT equation (4.14) can be fit to data by first placing the equation in the following form:

$$\ln \mu = A + \frac{E_\mu}{R(T - T_0)}$$

then using three pairs of dependent/independent variable values: $(\ln \mu_1, T_1)$; $(\ln \mu_2, T_2)$, and $(\ln \mu_3, T_3)$. With three such pairs of data, you can create three independent equations and solve them simultaneously using simple algebraic substitution or using a spreadsheet (see Problem 4.III.2 at the end of the chapter). Alternatively, the parameters can be determined sequentially using the following relations, which are represented for the VFT equation, where L_i is shorthand notation for $\ln \mu_i$:

$$A = \frac{(L_1 - L_2)(L_3 T_3 - L_2 T_2) - (L_2 - L_3)(L_2 T_2 - L_1 T_1)}{(L_1 - L_2)(T_3 - T_2) - (L_2 - L_3)(T_2 - T_1)}$$

$$T_0 = \frac{L_1 T_1 - L_2 T_2 - A(T_1 - T_2)}{L_1 - L_2} \qquad E_\mu / R = (T_i - T_0)(L_i - A)$$

Use the following viscosity–temperature data to complete the following exercise: $\ln \mu_1 = 13.3$, $T_g = 550°C$; $\ln \mu_2 = 7.6$, Littleton temperature $= 715°C$; $\ln \mu_3 = 4.0$, working point $= 1000°C$.

Person 1: Determine the VFT parameters A, E_μ/R, and T_0 from the data provided by solving the three independent equations created when substituting each data pair into the linearized form of the VFT equation.

Person 2: Determine the VFT parameters A, E_μ, and T_0 from the data provided by using the three relationships provided above.

Compare your answers!

Answer: $\ln \mu = -1.386 + 38030.3/(T - 288.8)$

Viscosity–temperature plots can be generated for glasses with three adjustable parameters using the VFT equation.

4.1.2.2 Non-Newtonian Viscosities in Ceramic Slurries and Suspensions.

As the melting point of inorganic materials such as oxides, borides, and carbides increases, melt processing becomes less practical due, for example, to increased energy requirements, difficulty in containment, or reduced formability. As a result, high-melting-point ceramics are often processed in the form of slurries, suspensions, or pastes. The distinction between a paste, slurry, or suspension depends upon a number of factors, such as (a) whether the ceramic particles, liquid phase, and trapped gases are continuous or discontinuous and (b) the value of particle parameters such as packing fraction (PF), degree of pore saturation (DPS), and mechanical compressibility (X), as summarized in Table 4.3. In a powder composed of spherical particles, the packing fraction (PF) is the fraction of the bulk volume occupied by spheres. As described in Section 1.1.1, the packing fraction for simple cubic is 0.52, and 0.74 for close-packed structures.

Table 4.3 Characteristics of Ceramic/Liquid Mixture Consistency States[a]

Consistency	Particle Structure	Gas	Liquid	PF	DPS	X
Powder	C	C	D	L	<1	H
Granules	C	C	D	L-M	<1	M
Plastic	C	D	C	M-H	<1	L
Paste	C	D	C	M-L	=1	0
Slurry	D	D	C	M-L	>1	0
Suspension	D	D	C	L	>1	0

[a]C = continuous; D = discontinuous; H = high; M = moderate; L = low.
Source: J. S. Reed, *Principles of Ceramics Processing*, 2nd ed. Copyright © 1995 by John Wiley & Sons, Inc.

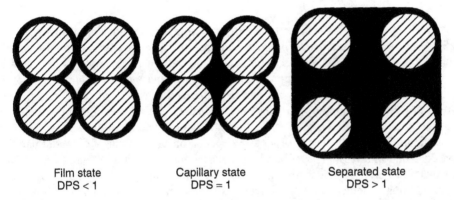

Film state	Capillary state	Separated state
DPS < 1	DPS = 1	DPS > 1

Figure 4.3 Liquid distribution in a powder leading to varying degree of pore saturation (DPS).

The DPS is a measure of the pore saturation and is defined as the ratio of volume of the suspending liquid to pore volume (see Figure 4.3). The mechanical compressibility of a collection of particulates, X, is defined as $(\Delta V/V_0)/\Delta P$, where ΔV is the volume change resulting from an incremental pressure change ΔP, and V_0 is the initial volume. As more liquid is added to the particles, the mixture becomes essentially incompressible, and X approaches zero.

These distinctions are important from a processing standpoint, since the transfer of shear forces within the mixture changes significantly. These differences are shown schematically in Figure 4.4. A dry, bulk powder consists of discrete particles and random agglomerates produced by attractive van der Waals, electrostatic, or magnetic interparticle forces. Here, the flow resistance depends upon random particle attraction and agglomeration, and it tends to decrease with decreasing particle size. This occurs because the surface force relative to the gravitational force increases with decreasing particle size. When a wetting liquid of low viscosity is added to a dry powder, *agglomerates* form. Agglomerates that are purposefully made in this way are called *granules*, and the free-flowing material is referred to as being *granular*. The resistance to shearing flow is still low in these mixtures. On further increase of the liquid content,

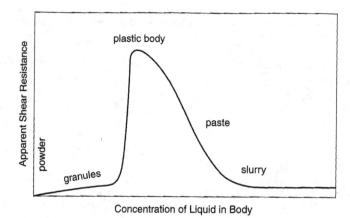

Figure 4.4   Processing consistency produced on mixing a powder with a wetting liquid. From J. S. Reed, *Principles of Ceramics Processing*, 2nd ed. Copyright © 1995 by John Wiley & Sons, Inc. This material is used by permission of John Wiley & Sons, Inc.

nearly all of the powder will be transformed into granules with dry surfaces. At this point, only a small amount of liquid is required to cause nearly total agglomeration into a cohesive mass. The cohesive mass is very resistant to shearing, as indicated by the peak in Figure 4.4. The response of this cohesive mass to shear is plastic-like, so it is often called a *plastic body*. In the agglomerate, plastic, and paste states, shear stress may be transmitted by particles in contact with one another. From this point on, resistance to shear decreases as more liquid is added, and the liquid phase becomes continuous. The flow behavior of a paste depends on the content of colloidal particles and coagulation forces. A *paste* becomes a *slurry* upon further addition of liquid. Now the material may flow under its own weight. A dilute slurry is termed a *suspension*.

In addition to temperature, the viscosity of these mixtures can change dramatically over time, or even with applied shear. Liquids or solutions whose viscosity changes with time or shear rate are said to be *non-Newtonian*; that is, viscosity can no longer be considered a proportionality constant between the shear stress and the shear rate. In solutions containing large molecules and suspensions contain nonattracting anisometric particles, flow can orient the molecules or particles. This orientation reduces the resistance to shear, and the stress required to increase the shear rate diminishes with increasing shear rate. This behavior is often described by an empirical power law equation that is simply a variation of Eq. (4.3), and the fluid is said to be a *power law fluid*:

$$\tau = K\dot{\gamma}^n \tag{4.15}$$

Here, K is sometimes referred to as the *consistency index* and has units that depend on the value of the *power law index*, n—for example, $N \cdot s^n/m^2$. The power law index is itself dimensionless. Typical values of K and n are listed in Table 4.4. In general, the power law index is independent of both temperature and concentration, although fluids tend to become more Newtonian (n approaches 1.0) as temperature increases and concentration decreases. The consistency factor, however, is more sensitive to temperature and concentration. To correct for temperature, the following relationship is often used:

$$K_\theta = K_0 m_\theta / m_0 \tag{4.16}$$

Table 4.4 Power-Law Parameters for Selected Fluids

Fluid	K $(\text{N·s}^n/\text{m}^2)$	n
23.3% Illinois yellow clay in water[1]	5.55	0.229
0.67% carboxymethylcellulose (CMC) in water[1]	0.304	0.716
1.5% CMC in water[1]	3.13	0.554
3.0% CMC in water[1]	9.29	0.566
33% lime in water[1]	7.18	0.171
10% napalm in kerosene[1]	4.28	0.520
4% paper pulp in water[1]	20.0	0.575
54.3% cement rock in water[1]	2.51	0.153
0.5% hydroxymethylcellulose (HMC) in water (293 K)[2]	0.84	0.509
2.0% HMC in water (293 K)[2]	93.5	0.189
1.0% polyethylene oxide in water (293 K)[2]	0.994	0.532
Human blood (293 K)[3]	0.00384	0.890
Ketchup[3]	33.2	0.242
Soybean oil[3]	0.04	1.0
Sewage sludge[3] (1.7–6.2 wt% sol.) $w = $ wt% solids	$0.00113w^{3.4}$	0.817–0.843
Orange juice[3]	1.89	0.680
Corn syrup (48.4 total solids)[3]	0.0053	1.0

Source: (1) Bird, R. B., W. E. Stewart, and E. N. Lightfoot, *Transport Phenomena*, John Wiley & Sons, New York, 1960., (2) *Dynamics of Polymeric Liquids*, Bird, R. B., R. C. Armstrong, and O. Hassager, Vol. 1, John Wiley & Sons, New York, 1977., (3) Johnson, A. T., *Biological Process Engineering*, Wiley-Interscience, New York, 1999.

where K_θ is the consistency index at temperature θ, K_0 is the consistency factor at the reference temperature, and μ_θ and μ_0 are the corresponding viscosities of the suspending fluid, which is usually, but not always, water.

For a power-law fluid, the viscosity is now a function of shear rate. Often times, a non- Newtonian viscosity is designated with the lowercase Greek letter eta, η, to differentiate it from a Newtonian viscosity, μ:

$$\eta = K\dot{\gamma}^{n-1} \qquad (4.17)$$

In general, we will use μ to represent Newtonian viscosity, and η to represent non-Newtonian viscosity. Recognize, of course, that in certain limits and under certain conditions, non-Newtonian fluids can behave in a Newtonian manner. We will not attempt to switch between the two representations for viscosity under these circumstances.

The value of n in Eqs. (4.15) and (4.17) can be greater than or less than 1.0. For $n > 1.0$, the viscosity increases with shear rate, in what is a called a *shear-thickening* fluid. Moderately concentrated suspensions containing large agglomerates and concentrated, deflocculated slurries of ceramic powders behave in this manner. When $n < 1.0$, a *shear-thinning* fluid results, and less stress is required to incrementally increase the shear rate. Suspensions containing nonattracting, anisometric particles, along with liquids or solutions containing large molecules, are typically shear thinning, because laminar flow tends to orient the particles and reduces the resistance to shear. We will see in the next section that polymeric materials can be both shear-thinning and shear-thickening. These two types of power-law behavior are summarized graphically in the lower half of Figure 4.5. Notice that at low shear rates and over

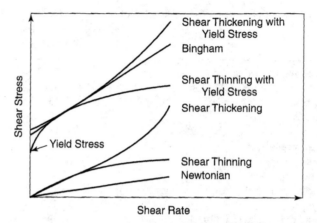

Figure 4.5 Variation of shear stress with shear rate for different models of flow behavior. From J. S. Reed, *Principles of Ceramics Processing*, 2nd ed. Copyright © 1995 by John Wiley & Sons, Inc. This material is used by permission of John Wiley & Sons, Inc.

limited shear rate regimes, shear-thinning and shear-thickening fluids can behave in a psuedo-Newtonian manner.

The top three curves in Figure 4.5 represent the behavior of yet another class of non-Newtonian fluids that exhibit a *yield stress* (note how their curves begin further up on the shear stress axis than the previous group of curves). In ceramic systems, slurries that contain bonded molecules or strongly attracting particles require a finite stress called the yield stress, τ_y, to initiate flow. Once the yield stress has been overcome, the fluid can behave in either a Newtonian or non-Newtonian manner. Fluids that possess a yield stress, yet have a constant proportionality between stress and strain (shear rate) above the yield stress, are termed *Bingham fluids* (or Bingham plastics for organic macromolecules) and obey the general relationship:

$$\tau = \tau_y + \eta_p \dot{\gamma} \tag{4.18}$$

Bingham fluids that are either shear-thinning or shear-thickening above their yield stresses have corresponding power-law expressions incorporated into their viscosity models.

Further non-Newtonian types of behavior appear when time is introduced as a variable. Fluids whose viscosity decreases with time when sheared at a constant rate are called *thixotropic*, whereas fluids whose viscosity increases with time when sheared at a constant rate are termed *rheopectic*. Thixotropic behavior is represented schematically in Figure 4.6. Note that the two viscosities (slope of shear stress vs. shear rate) η_{P1} and η_{P2} are different, depending upon the time they were sheared at a given rate.

With this background of non-Newtonian behavior in hand, let us examine the viscous behavior of suspensions and slurries in ceramic systems. For dilute suspensions on noninteracting spheres in a Newtonian liquid, the viscosity of the suspension, η_S, is greater than the viscosity of the pure liquid medium, η_L. In such cases, a *relative viscosity*, η_r, is utilized, which is defined as η_S/η_L. For laminar flow, η_r is given by the *Einstein equation*

$$\eta_r = 1 + 2.5 f_p^v \tag{4.19}$$

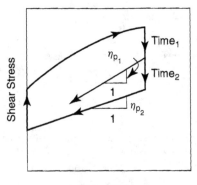

Shear Rate

Figure 4.6   Thixotropic behavior, in which time at constant shear rate affects viscosity. From J. S. Reed, *Principles of Ceramics Processing*, 2nd ed. Copyright © 1995 by John Wiley & Sons, Inc. This material is used by permission of John Wiley & Sons, Inc.

Cooperative Learning Exercise 4.2

Agglomeration in a slurry causes a change in the packing factor at which flow is blocked, f_{cr}^v, from 0.6 to 0.45 and causes a change in the hydrodynamic shape factor from 3.5 to 2.5. In both cases, the volume fraction of dispersed particles is 0.40.

Person 1: Calculate the relative viscosity for $f_{cr}^v = 0.6$ and $K_H = 3.5$.
Person 2: Calculate the relative viscosity for $f_{cr}^v = 0.45$ and $K_H = 2.5$.

Compare your answers. How would you expect these values to change if the shear rate were increased?

Answers: $\eta_r(0.6, 3.5) = 4.7; \eta_r(0.45, 2.5) = 9.0$

Increased shear rate would help prevent particle agglomeration, effectively increasing f_{cr}^v and reducing the relative viscosity.

---

where f_p^v is the volume fraction of dispersed spheres. A more general relationship for the reduced viscosity allows for nonspherical particles, through the use of an apparent *hydrodynamic shape* factor, K_H:

$$\eta_r = 1 + K_H f_p^v \tag{4.20}$$

In suspensions of particles with an *aspect ratio* (length to diameter) greater than 1, particle rotation during flow results in a large effective hydrodynamic volume, and $K_H > 2.5$ (see Figure 4.7). At particle volume fractions above about 5–10%, interaction between particles during flow causes the viscosity relationship to deviate from the Einstein equation. In such instances, the reduced viscosity is better described by the following relationship:

$$\eta_r = [1 - f_p^v / f_{cr}^v]^{-K_H f_p^v} \tag{4.21}$$

Here, f_p^v is once again the volume fraction of dispersed spheres, and K_H is the apparent hydrodynamic shape factor, but a new parameter, f_{cr}^v, is introduced which accounts

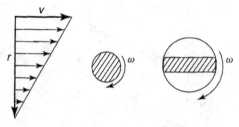

Figure 4.7   Effect of hydrodynamic shape factor on effective particle size in a flow field. From J. S. Reed, *Principles of Ceramics Processing*, 2nd ed. Copyright © 1995 by John Wiley & Sons, Inc. This material is used by permission of John Wiley & Sons, Inc.

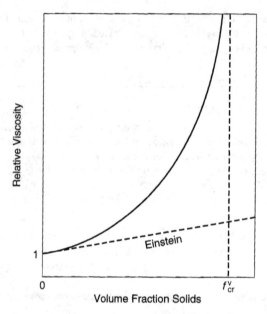

Figure 4.8   Effect of increasing the powder concentration in a slurry on relative viscosity. From J. S. Reed, *Principles of Ceramics Processing*, 2nd ed. Copyright © 1995 by John Wiley & Sons, Inc. This material is used by permission of John Wiley & Sons, Inc.

for the packing factor at which flow is blocked. This parameter can be determined graphically, as in Figure 4.8. The effect of volume fraction of dispersed particles on the relative viscosity for several different apparent hydrodynamic shape factors, as determined by Eq. (4.21), is shown in Figure 4.9. There are numerous other equations and correlations that attempt to predict the viscosity of suspensions and slurries. Some take into account adsorbed binder on the particles, and some attempt to predict such things as yield stress and the effects of time on rheological behavior. In most cases, these fluids behave in a highly non-Newtonian manner, and rheological properties can be predicted only for very narrow temperature and concentration ranges.

This description of non-Newtonian behavior in ceramic slurries and suspensions leads us directly into a description of viscosity in polymeric materials, which are oftentimes highly non-Newtonian in their behavior.

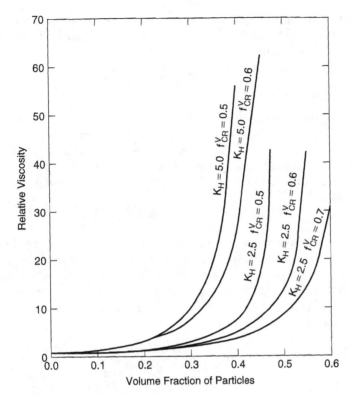

Figure 4.9 Dependence of relative viscosity on solids concentrations for several values of apparent hydrodynamic shape factors. From J. S. Reed, *Principles of Ceramics Processing*, 2nd ed. Copyright © 1995 by John Wiley & Sons, Inc. This material is used by permission of John Wiley & Sons, Inc.

4.1.3 Momentum Transport Properties of Polymers

Unlike metals, in which momentum transport properties are generally limited to the molten state, and ceramics, in which momentum transport properties are primarily (but not always) described by the solution state, polymers can be found in either the solution or molten state. As a result, many of the principles that have been previously discussed apply to polymers, especially with regard to non-Newtonian behavior. There are, however, a few viscosity-related concepts, exclusive to polymers, that we will describe here.

4.1.3.1 Viscosity of Polymer Solutions. Polymer solutions are very common. Glues, pastes, and paints are just a few examples of commercially available aqueous suspensions of organic macromolecules. Also, recall from Chapter 3 that certain types of polymerization reactions are carried out in solution to assist in heat removal. The resulting polymers are also in solution, and their behavior must be fully understand in order to properly transport them and effect solvent removal, if necessary.

Unless they are formed directly in solution, polymers dissolve in two general steps. First, the polymer absorbs solvent to form a swollen *gel*. The second step involves the dissolution of the gel into the solution, sometimes called a *sol*. Networked or thermoset polymers may only undergo gel formation, if they dissolve at all. Once the gel is in solution, the solution viscosity is generally described as it is for ceramic suspensions and slurries, with the relative viscosity

$$\eta_r = \frac{\eta}{\eta_s} \tag{4.22}$$

where η is the true solution viscosity, and η_s is the pure solvent viscosity. Keep in mind that not all polymer solutions are aqueous; that is, many polymers dissolve in nonpolar and/or organic solvents, depending on their chemical structure. Equation 4.22 shows us that the relative viscosity for pure solvents is exactly 1.0.

Oftentimes, a Taylor series expansion in polymer concentration, c, is used for the relative viscosity

$$\eta_r = 1 + [\eta]c + k'[\eta]^2 c^2 + \cdots \tag{4.23}$$

where $[\eta]$ is called the *intrinsic viscosity* of the solution, and k' is called the *Huggins coefficient*. The expansion is often truncated at the second-order term. The intrinsic viscosity has units of reciprocal concentration. Both intrinsic viscosity and the Huggins coefficient are independent of concentration., although they can be functions of shear rate. You should be able to see from Eqs. (4.22) and (4.23) that the following relationship holds for the intrinsic viscosity:

$$[\eta] = \lim_{c \to 0} \left(\frac{\eta - \eta_s}{c\eta_s} \right) \tag{4.24}$$

Cooperative Learning Exercise 4.3

We are given the following measurements of relative viscosities of solutions of polyisobutylene ($\overline{M}_w = 2.5 \times 10^6$) in decalin and in a 70% decalin and 30% cyclohexanol mixture at 298 K.

Person 1: Calculate the intrinsic viscosity, $[\eta]$, for the polyisobutylene sample in decalin.
Person 2: Calculate the intrinsic viscosity, $[\eta]$, for the polyisobutylene sample in decalin-cyclohexanol.

Compare your answers. Which is the "better" solvent?

Decalin		Decalin-Cyclohexanol	
$c(g/cm^3)$	η_r	$c(g/cm^3)$	η_r
0.000408	1.227	0.000409	1.155
0.000543	1.311	0.000546	1.210
0.000815	1.485	0.000818	1.326
0.001304	1.857	0.001169	1.488
0.001630	2.125	0.001637	1.740

Source: Adapted from *Dynamics of Polymeric Liquids*

Answers: Decalin 5.1×10^2 cm^3/g; decalin-cyclohexanol 3.1×10^2 cm^3/g.

It may seem confusing to introduce yet another type of viscosity. In fact, there are many other types of viscosity (see Table 4.5), but we will concentrate on the intrinsic viscosity for polymer solutions. The intrinsic viscosity is important because it has been found to correlate well with a number of important parameters, including polymer molecular weight and type of solvent. Note that if Eq. (4.23) is truncated to the first two terms, the intrinsic viscosity is the same as the reduced specific viscosity, η_{sp}, of Table 4.5.

Polymer solvents are generally characterized as being either a *good solvent* or a *poor solvent*. A good solvent is defined as a solvent in which polymer–solvent interactions are more favorable than polymer–polymer interactions. Conversely, in a poor solvent, polymer–polymer interactions are more favorable than polymer–solvent interactions. So, in a good solvent, the polymer chains will tend to uncoil in order to interact more with the solvent, thereby dissolving more effectively. The Huggins coefficient has values in the range 0.3–0.4 for good solvents, with higher values for poor solvents.

Intrinsic viscosity has also been found to correlate with the molecular weight of the polymer (see Figure 4.10) according to the Mark–Houwink relationship:

$$[\eta] = K'\overline{M}_v^a \tag{4.25}$$

Table 4.5   Summary of Viscometric Terms

Symbol	Name	Common Units
η	Solution viscosity	Poise or Pa·s
η_s	Solvent viscosity	Poise or Pa·s
$\eta_r = \eta/\eta_s$	Relative viscosity	Dimensionless
$\eta_{sp} = \eta_r - 1$	Specific viscosity	Dimensionless
$(\ln \eta_r)/c$	Inherent viscosity	dl/g
$\eta_{sp}/c = \eta_{sr}$	Reduced viscosity	dl/g

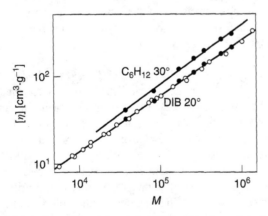

Figure 4.10   Intrinsic viscosity-molecular weight relationship for polyisobutylene in diisobutylene (DIB) and cyclohexane. Reprinted, by permission, from G. Strobl, *The Physics of Polymers*, 2nd ed., p. 295. Copyright © 1997 by Springer-Verlag.

Table 4.6   Selected Values of Parameters for Mark–Houwink Equation

Polymer	Solvent	Temperature (°C)	$K' \times 10^5$	a
Cellulose triacetate	Acetone	25	8.97	0.90
SBR rubber	Benzene	25	54	0.66
Natural rubber	Benzene	30	18.5	0.74
	n-Propyl ketone	14.5	119	0.50
Polyacrylamide	Water	30	68	0.66
Polyacrylonitrile	Dimethyl formamide	25	23.3	0.75
Poly(dimethylsiloxane)	Toluene	20	20.0	0.66
Polyethylene	Decalin	135	62	0.70
Polyisobutylene	Benzene	24	107	0.50
	Benzene	40	43	0.60
	Cyclohexane	30	27.6	0.69
Poly(methyl methacrylate)	Toluene	25	7.1	0.73
Polystyrene				
Atactic	Toluene	30	11.0	0.725
Isotactic	Toluene	30	10.6	0.725
Poly(vinyl acetate)	Benzene	30	22	0.65
	Ethyl n-butyl ketone	29	92.9	0.50
Poly(vinyl chloride)	Tetrahydrofuran	20	3.63	0.92

Source: F. Rodriguez, *Principles of Polymer Systems*, 2nd ed. Copyright © 1982 by McGraw-Hill Book Company.

where K' and a are constants that depend on the particular polymer-solvent system, and $\overline{M}_v$ is the viscosity-average molecular weight (cf. Table 1.23). The constant a is often called the *Mark–Houwink exponent*, and lies in the range of 0.5–0.8 (see Table 4.6), which leads to the result that the viscosity-average molecular weight lies between the number-average and weight-average molecular weights. When $a = 1.0$, $\overline{M}_v = \overline{M}_w$. Many other important characteristics of the polymer can be predicted using intrinsic viscosity, including chain stiffness and hydrodynamic coil radius, further emphasizing the importance of intrinsic viscosity.

4.1.3.2 Viscosity of Polymer Melts and Blends.

Much regarding the shear-dependent viscosity of polymer melts has already been described in the context of cermamic slurries (cf. Section 4.1.2.2). There is an immeasurable amount of additional information on this topic, and much research has been conducted on predicting the rheological properties of polymer melts from their structure using what are termed *constitutive equations*. A review of these equations, even a perfunctory one, is beyond the scope of this text. However, we know enough about polymeric structure and fluid flow to describe two topics on polymer melts that have direct utility with regard to the processing of polymeric materials. The first deals with the effect of chain entanglements on polymer melt viscosity, and the second has to do with the viscosity of polymer melt blends. A third relevant topic, viscoelasticity, will be reserved until Chapter 5 when mechanical properties and elastic response are described.

The high viscosities of polymeric melts can be attributed to the high level of intermolecular bonding between polymer chains. In fact, the *interchain forces* can be stronger than primary bonds in the chain—for example, C–C backbone bonds, to the point that thermal degradation often occurs before vaporization in polymers. This phenomenon is chain-length-dependent; that is, the longer the chain, the more inter-atomic bonding that can occur. Polymer molecules with strong intermolecular forces such as hydrogen bonding or dipoles have stronger intermolecular bonds than primary bonds when the degree of polymerization, $\bar{x}_n$, exceeds about 200. Polymer molecules with weak intermolecular forces, such as dispersion forces, require longer chains, on the order of $\bar{x}_n > 500$, in order for intermolecular forces to dominate.

Other than intermolecular forces, *chain entanglements* play an important role in determining polymer melt viscosity. As shown in Figure 4.11, there is a *critical chain length*, here represented by the weight-average degree of polymerization, $\bar{x}_w$, above which chain entanglements are significant enough to retard flow and increase the melt viscosity. Below this point, chain segments are too short to allow for significant entan-glement effects. Typical values of the critical chain length are 610 for polyisobutylene, 730 for polystyrene, and 208 for poly(methyl methacrylate) [1]. Generally, the critical chain length is lower for polar polymers than for nonpolar polymers, again due to intermolecular forces.

Intermolecular forces also play an important role in determining the compatibility of two or more polymers in a *polymer blend* or *polymer alloy*. Although the distinction between a polymer blend and a polymer alloy is still the subject of some debate, we will use the convention that a polymer alloy is a single-phase, homogeneous mate-rial (much as for a metal), whereas a blend has two or more distinct phases as a result of polymer–polymer immiscibility (cf. Section 2.3.3). In general, polymers are

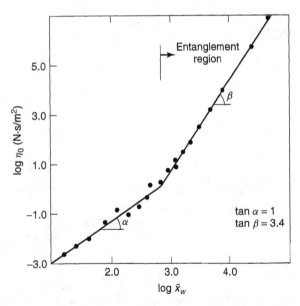

Figure 4.11   Effect of polymer chain entanglements on melt viscosity. From Z. Tadmor and C. G. Gogos, *Principles of Polymer Processing*, Copyright © 1979 by John Wiley & Sons, Inc. This material is used by permission of John Wiley & Sons, Inc.

incompatible with each other, so that most polymer–polymer mixtures are blends, in which the primary phase is continuous, and the secondary phase exists as droplets dispersed throughout the primary phase. As the droplet sizes become smaller, the two phases become indistinguishable, and an alloy begins to form. There are some immiscible polymer blends that form *bi-continuous structures* as well.

Needless to say, the rheological properties of polymer mixtures are complex and nearly impossible to predict. Figure 4.12 shows the viscosity of a natural rubber (NR)/poly(methyl methacrylate) (PMMA) blend (top curve) as a function of percentage NR [2]. For comparison, the predictions of four common equations are shown. The equations are as follows:

$$\eta_{mix} = \eta_1 \phi_1 + \eta_2 \phi_2 \tag{4.26}$$

$$\eta_{mix} = \eta_2 + \frac{\phi_1}{1/(\eta_1 - \eta_2) + \phi_2/2\eta_2} \tag{4.27}$$

$$\eta_{mix} = \eta_1 + \frac{\phi_2}{1/(\eta_2 - \eta_1) + \phi_1/2\eta_1} \tag{4.28}$$

$$\ln \eta_{mix} = \frac{\phi_1(\alpha - 1 - \gamma\phi_2)\ln \eta_1 + \alpha\phi_2(\alpha - 1 + \gamma\phi_1)\ln \eta_2}{\phi_1(\alpha - 1 - \gamma\phi_2) + \alpha\phi_2(\alpha - 1 + \gamma\phi_1)} \tag{4.29}$$

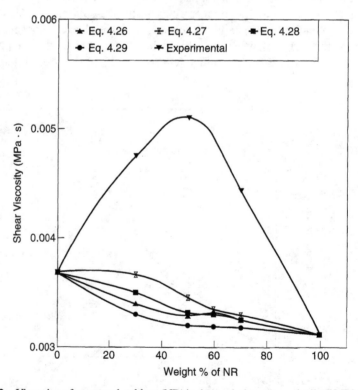

Figure 4.12 Viscosity of a natural rubber (NR)/poly(methyl methacrylate) (PMMA) polymer blend and predictions of Eq. (4.26) through (4.29) at a shear rate of 333 s^{-1}. Adapted from Z. Oommen, S. Thomas, C. K. Premalatha, and B. Kuriakose, *Polymer*, **38**(22), 5611–5621. Copyright © 1997 by Elsevier.

where η_1 and η_2 are the viscosities of components 1 and 2, and ϕ_1 and ϕ_2 are their corresponding volume fractions. In Eq. (4.29), α and γ are adjustable parameters [3].

Other models are based upon the immiscibility of polymer blends described above, and they model the system as Newtonian drops of the dispersed polymer with concentration ϕ_1 in a Newtonian medium of the second polymer with concentration $\phi_2 = 1 - \phi_1$. There exists some concentration, $\phi_{1I} = \phi_{2I} - 1$, at which phase inversion takes place; that is, at sufficiently high concentration, the droplet phase suddenly becomes continuous, and the second phase forms droplets. The phase inversion concentration has been shown to correlate with the viscosity ratio, $\lambda = \eta_1/\eta_2$, and the intrinsic viscosity for at least a dozen polymer alloys and blends:

$$\lambda = \left[\frac{\phi_m - \phi_{2I}}{\phi_m - \phi_{1I}} \right]^{[\eta]\phi_m} \tag{4.30}$$

where ϕ_m is the maximum packing volume fraction, which is about 0.84 for most polymer alloys and blends. Much work remains to be done in this important area of blend viscosity that is a vital component of polymer processing and characterization.

4.1.4 Momentum Transport Properties of Composites

Given the vast number of possible matrix–reinforcement combinations in composites and the relative inability of current theories to describe the viscosity of even the most compositionally simple suspensions and solutions, it is fruitless to attempt to describe the momentum transport properties of composite precursors in a general manner. There are, however, two topics that can be addressed here in an introductory fashion: flow properties *of* matrix/reinforcement mixtures; and flow of matrix precursor materials *through* the reinforcement. In both cases, we will concentrate on the flow of molten polymeric materials or precursors, since the vast majority of high-performance composites are polymer-based. Furthermore, the principles here are general, and they apply to the fluid-based processing of most metal–, ceramic–, and polymer–matrix composites.

4.1.4.1 *Viscosity of Polymer-Fiber Melts.* For most liquids with suspended particles that are processed into reinforced matrix composites, the viscosity can be described by the modified Einstein equation given in Eq. (4.20), which accounts for the variation in particle shape using the hydrodynamic shape factor, K_H. However, fiber-reinforced polymers are so ubiquitous and commercially important that some empirical studies have been performed that are worth describing here. Recent studies have included carbon [4], glass [5–7], and polyester [8] fibers in polyethylene [7], polypropylene [4, 5], polycarbonate [4, 5], polyamide [5, 6], and polyurethane [8] matrices. In addition to composition, reinforcing fibers can vary by length and diameter, but are normally classified as either long fibers (length more than about 5 mm) or short fibers (length less than about 5 mm, with a corresponding length/diameter ratio of 500–800). The effect of fiber fillers on polymer viscosity is generally the same, regardless of fiber length. As shown in Figure 4.13, the effect of the fibers is to cause more pronounced shear-thinning in the polymer (see 5 and 10 wt% curves) than is present without the fibers (0% curve), with plateaus at high and low shear rates. This is due to the tendency of fibers to align along the direction of flow, especially at high shear

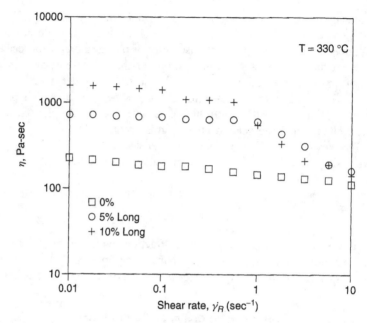

Figure 4.13 Effect of fiber content on shear viscosity for long glass fibers in polycarbonate. Reprinted, by permission, from J. P. Green and J. O. Wilkes, *Polym. Eng. Sci.*, **32**(21), 1670. Copyright © 1995 Society of Plastics Engineers.

rates, thus counteracting any viscosity-building effect due to suspended particles that exists at low shear rates. The extent of shear thinning and the existence of a low-shear-rate plateau depend on the aspect ratio, fiber modulus, fiber concentration, and the viscosity of the suspending fluid. The low shear rate plateau is usually not observed for suspensions of fibers with small aspect ratios at high concentrations in low-viscosity fluids.

Power-law expressions are still used to describe such polymer–fiber melts. Typical power-law parameters for selected fiber–polymer systems are shown in Table 4.7. Semiempirical expressions based on Eq. (4.23) have also been developed, as well as models based on energy dissipation. A complete review of these correlations is beyond the scope of this text, and the interested reader is referred to reference 9 for a more complete review of viscosity in fiber-reinforced polymer melts.

4.1.4.2 Viscosity of Polymers Through Fiber Preforms.

The flow of molten polymers, sometimes called resins if they are network polymers, through a fibrous preform is integral to a number of important composite processes, such as melt infiltration. In addition to nonreactive processes, there are a number of important reactive processing techniques such as *reaction injection molding* (RIM), *resin transfer molding* (RTM), and *reactive liquid composite molding* (RLCM) in which the polymer precursors flow and polymerize *in situ* as they flow through the fibrous preform. These techniques are described in more detail in Chapter 7, but it is worth describing the viscosity effects here first.

In the general case of a resin filling a fiber-packed mold, such as is found in melt infiltration (see Figure 4.14), fluid flow is equivalent to flow through porous

Table 4.7   Power-Law Parameters for a Fiber-Filled Polyamide

Sample	$K\,(\mathrm{N\cdot s^n/m^2})$	n
Polyamide (PA) 6.6	213.1	0.81
PA 6.6 with 0.5 wt% blowing agent	109.2	0.80
30 vol% long glass-fiber-filled PA	1709.5	0.74
40 vol% long glass-fiber-filled PA	2400.6	0.73
30 vol% long glass-fiber-filled PA with 0.5 wt% blowing agent	1173.3	0.75

Source: S. F. Shuler, D. M. Binding, and R. B. Pipes, *Polymer Composites*, **15**(6), (1994), 427.

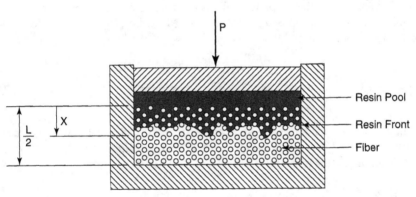

Figure 4.14   Schematic diagram of the melt infiltration process for flow past a unidirectional fiber assembly.

media or packed columns, both of which have been studied for a long time. Typically, *Darcy's Law* is used for flow through porous media. This model describes the flow of Newtonian fluids in porous media, and it states that the volumetric flow rate, Q, through a constant-area packed channel is proportional to the cross-sectional area, A, and the pressure drop across the channel, ΔP, and is inversely proportional to the length of the channel, L, and the viscosity of the fluid, μ:

$$Q = K_D \frac{A}{\mu}\left(\frac{\Delta P}{L}\right)$$

(4.31)

where K_D is called the *permeability*, and has units of length squared. Not surprisingly, Eq. (4.31) has the same form as the Hagen–Poiseuille equation for volumetric flow in a cylindrical tube, with a correction factor for the permeability of the media (K_D), which in this case is fibers.

The permeability is given by the *Kozeny–Carman relationship*

$$K_D = \frac{\varepsilon^3}{(1-\varepsilon)^2}\left(\frac{1}{S^2 k}\right)$$ (4.32)

where ε is the *porosity*, S the specific surface of the medium particle ($6/D$ for spheres, $4/D$ for cylinders), D is the diameter, and k is the so-called *Kozeny constant*. For the specific case of flow normal to the axial direction of a parallel array of fibers with diameter, D, as is found in the melt infiltration example, the permeability becomes $K_D = D^2\varepsilon^3/[80(1-\varepsilon)^2]$.

For the case of simultaneous reaction and flow, as is present in reactive molding, an expression is needed that explicitly relates the viscosity to both the temperature and the extent of reaction. The viscosity increases with time as the reaction proceeds (see Figure 4.15, top diagram). The temperature, however, has opposing effects; it will decrease the viscosity as the temperature is raised, but will also increase the reaction rate, leading to increased viscosity. To separate these effects, the polymerization reaction kinetics must be monitored independently. A viscosity–concentration correlation can be constructed by taking *isochrones* (t_1 and t_2 in Figure 4.15) of viscosity and extent of reaction and then replotting them on a viscosity–conversion diagram. The following relationship, proposed by Castro and Macosko [10], has been found useful in modeling the reactive molding process:

$$\mu = A_\mu \exp(E_\mu/RT)\left(\frac{C_g^*}{C_g^* - C^*}\right)^{A+BC^*}$$ (4.33)

Here A_μ and E_μ are the Arrhenius-type preexponential and activation energy terms, R is the gas constant, T is absolute temperature, C_g^* is the solidification (gel) composition, and C^* is the extent of reaction, $(C_0 - C)/C_0$.

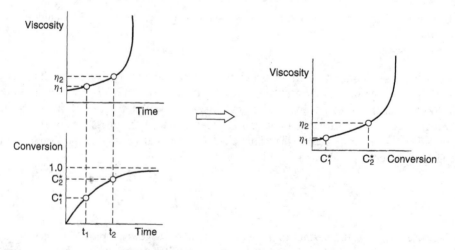

Figure 4.15 Mapping of viscosity rise versus time plot into viscosity–conversion plot. Reprinted, by permission, from P. K. Mallick, *Composites Engineering Handbook*, p. 507. Copyright © 1997 by Marcel Dekker, Inc.

Cooperative Learning Exercise 4.4

Darcy's Law can be utilized for non-Newtonian as well as Newtonian fluids, by using an effective viscosity in place of the Newtonian viscosity:

$$v_0 = \frac{Q}{A} = \frac{K_D \Delta P}{\eta_{eff} L}$$

where v_0 is the superficial velocity of the fluid flow front. In addition to the shear-rate dependent non-Newtonian viscosities described earlier in this chapter, there are other models that are stress-dependent. One such model is the *Ellis Model*, in which the ratio of the zero-shear-rate viscosity, η_0, to the effective viscosity, η_{eff}, for the case of flow through porous media is given by

$$\frac{\eta_0}{\eta_{eff}} = 1 + \frac{4}{\alpha + 3} \left(\frac{\tau_{Rh}}{\tau_{1/2}} \right)^{\alpha - 1}$$

where $\tau_{1/2}$ is the shear stress at which $\eta_{eff} = \eta_0/2$, τ_{Rh} is the wall shear stress based on the hydrodynamic radius, and α is a constant that is the inverse of the power-law exponent, n.

Person 1: Calculate the permeability, K_D, for the case of a fiber preform made from unidirectional fibers of diameter 10 μm, and a porosity of 60%.

Person 2: Derive an expression for the effective viscosity for a 1.0% solution of poly(ethylene oxide) at 293 K, which has a $\eta_0 = 2.37$ N·s/m^2, $\tau_{1/2} = 0.412$ N/m^2. Consult Table 4.4 for the appropriate value of α.

Combine your answers to come up with an expression for the superficial velocity of a 1.0% solution of poly(ethylene oxide) within a mold consisting of 10-μm-diameter unidirectional fibers with 60% porosity.

$$v_0 = (7.12 \times 10^{-13} + 1.27 \times 10^{-12} \tau_{Rh}^{0.88}) \frac{\Delta P}{L}$$

$$\text{Answers: } K_D = 1.7 \times 10^{-12}; \quad \eta_{eff} = \frac{2.37}{1 + 1.82 \tau_{Rh}^{0.88}};$$

4.1.5 Momentum Transport Properties in Biologics

Many biological fluids are aqueous-based suspensions of particles. Such is the case with perhaps the most important of all biological fluids: blood. Blood is essentially a suspension of red blood cells in plasma. The concentration of red blood cells in blood is described as the *hematocrit*, or fraction of the blood volume represented by blood cells. The viscosity of human blood as a function of hematocrit is shown in Figure 4.16. Note that the viscosity of blood is substantially lower in small tubes than in large tubes, especially at high hematocrit values. This is due to the tendency of red blood cells to be carried in midstream (think "velocity profile" and "fully developed flow"!) such that in small blood vessels, the fluid at the wall is nearly completely plasma, which has a lower viscosity than that of whole blood (compare 0% to "normal value" hematocrit for large tubes in Figure 4.16). This has important implications to human anatomy. If the heart had to be large and powerful enough to propel blood throughout the body with "normal viscosities" such as are found in large tubes (e.g., arteries), then there would hardly be room in the chest cavity for lungs and other important organs.

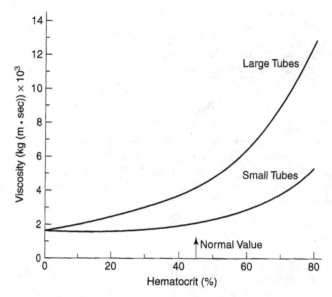

Figure 4.16   Viscosity of blood as a function of percentage blood volume represented by red blood cells (hematocrit). From *Biological Process Engineering*, by A. T. Johnson; Copyright © 1999 by John Wiley & Sons, Inc. This material is used by permission of John Wiley & Sons, Inc.

Other biological fluids can be modeled as suspensions of particles in a solvent, such as was used for the description of suspensions and slurries in Section 4.1.2.2, namely, that the relative viscosity of the suspension is related to the hydrodynamic shape factor, K_H, and the volume fraction of particles, f_p^v, as in Eq. (4.20):

$$\eta_r = 1 + K_H f_p^v \tag{4.20}$$

The biggest difference between biological particles and ceramic particles in the application of Eq. (4.20) is that while most ceramic particles are spherical ($K_H = 2.5$), most biological particles can be modeled as either *prolate ellipsoids* or *oblate spheroids* (or ellipsoids). Ellipsoids are characterized according to their shape factor, a/b, for which a and b are the dimensions of the semimajor and semiminor axes, respectively (see Figure 4.17). In a prolate ellipsoid, $a > b$, whereas in an oblate ellipsoid, $b > a$. In the extremes, $a \gg b$ approximates a cylinder, and $b \gg a$ approximates a disk, or platelet.

The hydrodynamic shape factor and axial ratio are related (see Figure 4.18), but are not generally used interchangeably in the literature. The axial ratio is used almost exclusively to characterize the shape of biological particles, so this is what we will utilize here. As the ellipsoidal particle becomes less and less spherical, the viscosity deviates further and further from the Einstein equation (see Figure 4.19). Note that in the limit of $a = b$, both the prolate and oblate ellipsoid give an intrinsic viscosity of 2.5, as predicted for spheres by the Einstein equation.

In practice, it is the viscosity that is experimentally determined, and the correlations are used to determine axial ratios and shape factors. The viscosity can be determined by any number of techniques, the most common of which is light scattering. In addition to ellipticity, *solvation* (particle swelling due to water absorption) can have an effect on

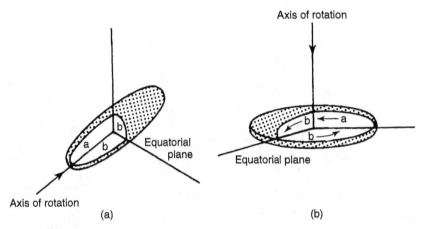

Figure 4.17 (a) Prolate and (b) oblate ellipsoids of revolution, showing the relationship between the semiaxes and the axis of revolution. Reprinted, by permission, from P. Hiemenz, *Principles of Colloid and Surface Chemistry*, 2nd ed., p. 26. Copyright © 1986 by Marcel Dekker, Inc.

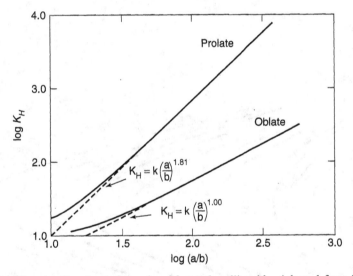

Figure 4.18 Theoretical values of the shape factor for ellipsoids. Adapted from F. H. Silver and D. L. Christiansen, *Biomaterials Science and Biocompatibility*, p. 150. Copyright © 1999 by Springer-Verlag.

the viscosity of biological suspensions. An example of how both solvation and particle ellipticity can affect solution viscosity is shown in Figure 4.20.

Finally, for biological molecules that are macromolecules, such as most proteins, Eq. (4.23) can also be used to relate the relative viscosity to the intrinsic viscosity of the solution and the macromolecule concentration:

$$\eta_r = 1 + [\eta]c + k'[\eta]^2 c^2 + \cdots \tag{4.23}$$

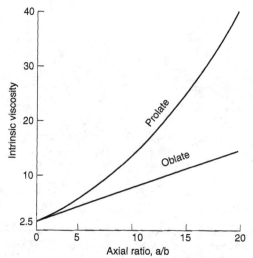

Figure 4.19 Intrinsic viscosity as a function of axial ratio for oblate and prolate ellipsoids. Intrinsic viscosity in mL/g. Reprinted, by permission, from P. Hiemenz, *Polymer Chemistry*, p. 596. Copyright © 1984 by Marcel Dekker, Inc.

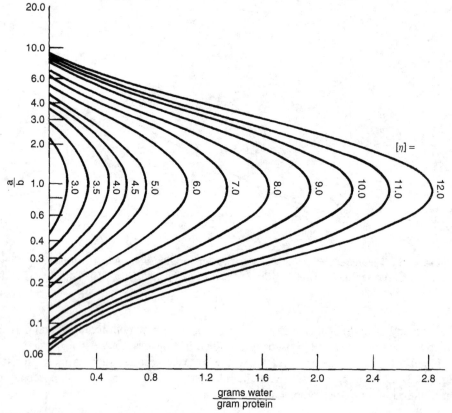

Figure 4.20 Variation of intrinsic viscosity of aqueous protein solutions with axial ratio and extent of solvation. Reprinted, by permission, from P. Hiemenz, *Polymer Chemistry*, p. 598. Copyright © 1984 by Marcel Dekker, Inc.

Cooperative Learning Exercise 4.5

Some plots of reduced specific viscosity (see Table 4.5) as a function of concentration for various samples of vitreous hyaluronan bovine protein are shown at right. Hyaluronan may be modeled as a prolate ellipsoid, but it is unknown whether it is a flexible or rigid chain.

Person 1: Pick one of the curves in the plot, and determine the intrinsic viscosity by applying Eq. (4.23) and the definition of reduced specific viscosity. Use this value of intrinsic viscosity to estimate the axial ratio of the particle using Figure 4.19. Note that you may have to assume a linear approximation to prolate ellipsoid curve and extrapolate to obtain an estimate for *a/b*.

Person 2: Use the *a/b* ratio estimated by Person 1 to estimate the diffusivity of this protein. The diffusivity of prolate ellipsoids, D_{AB}, in units of cm^2/s, can be estimated using the following relationship

$$D_{AB} = \frac{(2.15 \times 10^{-13} \ cm^3/s)[\ln(2a/b)]}{a}$$

For ease of calculation, assume that the hyaluronan protein has a chain length, $a = 215$ nm.

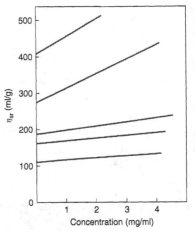

Reprinted, by permission, from F. H. Silver and D. L. Christiansen, *Biomaterials Science and Biocompatibility*, p. 151. Copyrigyt © 1999 by Springer-Veralg.

Compare the intrinsic viscosity and diffusivity results to come up with some conclusion regarding the shape and rigidity of this protein; *i.e.*, rigid chain or flexible chain.

Answers: The intrinsic viscosity of all the plots are determined by extrapolating back to zero concentration, and are found to range from about 110 to 410 mL/g. According to Figure 4.19, this corresponds to *a/b* values on the order of 100 or so. When this value is plugged into the equation for diffusivity, one obtains a value of about 10^{-12} m²/s. As described in the text, the intrinsic viscosity is relatively high, and the diffusivity is relatively low, leading us to believe that the hyaluronan chain is somewhat rigid.

An example of how both particle concentration and axial ratio can be used to describe their effect on solution viscosity—in this case the reduced viscosity—is shown in Figure 4.21. Here, the deviations from the Einstein model ($a/b = 1$ at low concentrations) are high, since it is based upon a dilute solution of nonsolvated, spherical particles.

Later in the chapter, we will see that there is a relationship between particle shape and size, intrinsic viscosity, and the ability of the particle to diffuse through the solution, as characterized by the *diffusion coefficient*, or *diffusivity*. In general, molecules with high values of intrinsic viscosity (>100 ml/g) and relatively low diffusivities ($<10^{-9}$ m²/s) are rigid or semiflexible rods. Molecules with both low diffusivities and intrinsic viscosities (~1 ml/g) are flexible rods, and molecules with high diffusivities (>10^{-9} m²/s) and low intrinsic viscosities are roughly spherical. Consult Section 4.3.5 for more information on diffusivities of biologics.

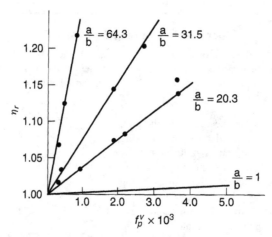

Figure 4.21 Effect of axial ratio and particle concentration on relative viscosity. Data are for tobacco mosaic virus particles. Adapted from M. A. Lauffer, *J. Am. Chem. Soc.*, **66**, 1188. Copyright © 1944 by The American Chemical Society, Inc.

4.2 HEAT TRANSPORT PROPERTIES OF MATERIALS

Of the three general categories of transport processes, heat transport gets the most attention for several reasons. First, unlike momentum transfer, it occurs in both the liquid and solid states of a material. Second, it is important not only in the processing and production of materials, but in their application and use. Ultimately, the thermal properties of a material may be the most influential design parameters in selecting a material for a specific application. In the description of heat transport properties, let us limit ourselves to conduction as the primary means of transfer, while recognizing that for some processes, convection or radiation may play a more important role. Finally, we will limit the discussion here to theoretical and empirical correlations and trends in heat transport properties. Tabulated values of thermal conductivities for a variety of materials can be found in Appendix 5.

4.2.1 Heat Transport Properties of Metals and Alloys

4.2.1.1 Conduction, Convection, and Radiation. We begin this section with a brief review of the three primary modes of heat transfer: conduction, convection, and radiation.

  In *conduction*, heat is conducted by the transfer of energy of motion between adjacent molecules in a liquid, gas, or solid. In a gas, atoms transfer energy to one another through molecular collisions. In metallic solids, the process of energy transfer via "free electrons" is also important. In *convection*, heat is transferred by bulk transport and mixing of macroscopic fluid elements. Recall that there can be *forced convection*, where the fluid is forced to flow via mechanical means, or *natural (free) convection*, where density differences cause fluid elements to flow. Since convection is found only in fluids, we will deal with it on only a limited basis. *Radiation* differs from conduction and convection in that no medium is needed for its propagation. As a result, the form of Eq. (4.1) is inappropriate for describing radiative heat transfer. Radiation is the transfer of energy through space by means of electromagnetic waves. The laws that govern the movement and absorption of electromagnetic waves such as light govern the transfer of heat by radiation. We will cover some of these topics later in Chapter 6.

4.2.1.2 The Molecular Origins of Thermal Conductivity.

In an analogous fashion to the derivation from the potential energy function for the viscosity of a dilute gas at low density [see Eq. (4.6), the thermal conductivity of a gas, k, based on the rigid sphere model, can be derived for a gas (in W/m·K):

$$k = 8.32 \times 10^{-2} \frac{\sqrt{T/M}}{\sigma^2 \Omega} \tag{4.34}$$

where T is the absolute temperature in degrees kelvin, M is the atomic weight of the monatomic gas, and σ and Ω are the same Lennard-Jones parameters used in Eq. (4.6). Notice the dependence on the same properties as viscosity—namely, atomic weight, molecular diameter, and temperature—although the thermal conductivity has an inverse dependence on the square root of molecular weight instead of the proportional dependence for viscosity. There is no variation of thermal conductivity with pressure for a rigid sphere, ideal, monatomic gas at low pressure, but the model does predict that thermal conductivity will be zero at absolute zero.

It can be shown that thermal conductivity and viscosity are related in this case as follows:

$$k = \frac{15R\mu}{4M} = \frac{5\hat{C}_v\mu}{2} \tag{4.35}$$

where M is again the atomic weight of the gas, R is the gas constant in appropriate units [1.9872 cal/mol·K for thermal conductivity (in cal/cm·K·s) and 831.4 J·cm/m·mol·K for thermal conductivity (in W/m·K)], and μ is the viscosity of the gas in poise (g/cm·s). The quantity, $\hat{C}_v$, is called the *heat capacity* per unit mass at constant volume and has the same units as R/M. In the final section of this chapter, we will see how both the viscosity and thermal conductivity (thermal diffusivity) are related to the molar diffusivity.

As with viscosity, the theoretical predictions become more complex as the atoms themselves become more complex and more dense. For a polyatomic gas, the thermal conductivity is given by extension of Eq. (4.35):

$$k = \frac{15R\mu}{4M} \left(\frac{4C_v}{15R} + \frac{3}{5} \right) \tag{4.36}$$

where R, μ, and M are as in Eq. (4.35) (with M the molecular weight of the polyatomic gas rather than the atomic weight). Here, C_v is the molar heat capacity, obtained by multiplying the per mass heat capacity by the molecular weight, $M\hat{C}_v$. Note that for an ideal monatomic gas, $C_v = 3R/2$, and Eq. (4.36) reduces to (4.35).

Equation (4.36) provides a simple method for estimating an important heat transfer dimensionless group called the *Prandtl number*. Recall from general chemistry and thermodynamics that there are two types of molar heat capacities, C_v, and the constant pressure heat capacity, C_p. For an ideal gas, $C_v = 3C_p/5$. The Prandtl number is defined as $Pr = \hat{C}_p \mu / k$. Substitution of Eq. (4.36) into this relation, along with use of the relationship between C_v and C_p, gives the following for polyatomic gases:

$$Pr = \frac{C_p}{C_p + 1.25R} \tag{4.37}$$

Cooperative Learning Exercise 4.6

The molecular weight of O_2 is 32.000; its molar heat capacity C_p at 300 K and low pressure is 29.37 J/mol·K. The Lennard-Jones parameters for molecular oxygen are $\sigma = 3.433$ Å and $\Omega = 1.074$.

Person 1: Estimate the viscosity of molecular oxygen at low pressure and 300 K. Then estimate the Prandtl number, Pr, using Eq. (4.37).

Person 2: Use Person 1's estimate of viscosity to estimate the thermal conductivity of molecular oxygen at low pressure and 300 K using Eq. (4.36). Note that you will need to modify Eq. (4.36) to include C_p instead of C_v. Beware of units and molar heat capacities!

Combine your answers to give an estimate for the Prandtl number, Pr, based on the definition $Pr = \hat{C}_p \mu / k$. How does this compare with your estimation given by Eq. (4.37)?

Answers: Person 1: $\mu = 2.066 \times 10^{-4}$ g/cm·s; Pr = 0.74. Person 2: $k = 0.023$ W/m·K.
Combined: Pr = 0.66.

The introduction of heat capacity into the relationships for thermal conductivity and the Prandtl number gives us an opportunity to make a clarification regarding these two quantities. Thermal conductivity is a true heat transport property; it describes the ability of a material to *transport* heat via conduction. Heat capacity, on the other hand, is a thermodynamic quantity and describes the ability of a material to *store* heat as energy. The latter, while not technically a transport property, will nonetheless be described in this chapter for the various materials types, due in part to its theoretical relationship to thermal conductivity, as given by Eq. (4.35) and (4.36), and, more practically, because it is often used in combination with thermal conductivity as a design parameter in materials selection.

4.2.1.3 Thermal Conductivities of Liquids.
As was the case with viscosity, it is difficult to derive useful relationships that allow us to estimate thermal conductivities for liquids from molecular parameters. There is a theoretical development by Bridgman, the details of which are presented elsewhere [11], which assumes that the liquid molecules are arranged in a cubic lattice, in which energy is transferred from one lattice plane to the next at sonic velocity, v_s. This development is a reinterpretation of the kinetic theory model used in the last section, and with some minor modifications to improve the fit with experimental data, the following equation results:

$$k = 2.8(V/N)^{-2/3} k_B v_s \tag{4.38}$$

where V/N is the volume per molecule and k_B is Boltzmann's constant. This equation is limited to liquids well above their critical densities. The sonic velocity in liquids can be estimated by

$$v_s = \sqrt{\frac{C_p}{C_v}\left(\frac{\partial p}{\partial \rho}\right)_T}$$

(4.39)

where the quantity $(\partial p/\partial)_T$ may be obtained from isothermal compressibility measurements or from an equation of state, and (C_p/C_v) is approximately unity for liquids away from their critical point.

More frequently, empirical relationships are used to estimate liquid thermal conductivities. A linear relationship with temperature is adequate, since liquid thermal conductivities do not vary considerably with temperature, and hardly at all with pressure.

Some experimental data for thermal conductivities of pure liquid metals are shown in Figure 4.22, and data for some binary metallic alloys are shown in Figure 4.23. These values will, in general, be lower than the thermal conductivities of the corresponding solid forms of the metals and alloys.

4.2.1.4 *Thermal Properties of Metallic Solids.*

In the preceding sections, we saw that thermal conductivities of gases, and to some extent liquids, could be related to viscosity and heat capacity. For a solid material such as an elemental metal, the link between thermal conductivity and viscosity loses its validity, since we do not normally think in terms of "solid viscosities." The connection with heat capacity is still there, however. In fact, a theoretical description of thermal conductivity in solids is derived directly from the kinetic gas theory used to develop expressions in Section 4.2.1.2.

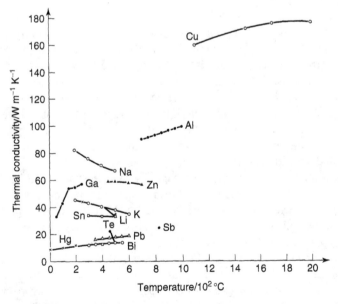

Figure 4.22 Thermal conductivity of some pure liquid metals as a function of temperature. Reprinted, by permission, from T. Iida and R. I. L. Guthrie, *The Physical Properties of Liquid Metals*, p. 241. Copyright © 1988 by Oxford University Press.

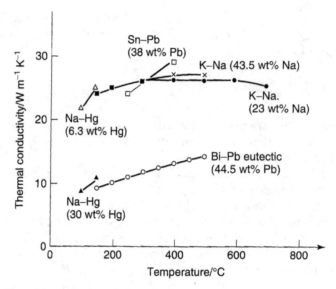

Figure 4.23 Thermal conductivity of some liquid alloys as a function of temperature. Reprinted, by permission, from T. Iida, and R. I. L. Guthrie, *The Physical Properties of Liquid Metals*, p. 241. Copyright © 1988 by Oxford University Press.

Equation (4.34), for example, can be derived from a simple expression for the thermal conductivity of a monatomic gas:

$$k = \frac{1}{3}\rho\hat{C}_v\bar{v}l \tag{4.40}$$

where ρ is the gas density, $\hat{C}_v$ is the constant volume heat capacity per unit mass (sometimes called the *specific heat*), $\bar{v}$ is the *mean molecular speed* (sometimes called the average particle velocity), and l is a quantity called the *mean free path*, which is a measure of the average distance traveled by the molecules between collisions.

We saw in Section 4.2.1.2 that the molar heat capacity, C_v, equals $3R/2$ for an ideal monatomic gas. This is because almost all of the energy of the gas is kinetic in nature; that is, an increase in temperature results in a direct increase in the translational kinetic energy of the atoms. For polyatomic gases, additional energy states are available, (e.g., vibrational and rotational, such that the gas can store energy in forms other than translational motion), and a higher heat capacity than that predicted by $3R/2$ results. In a solid, the atoms are similar to a vibrating diatomic molecule, where half of the thermal energy is the kinetic energy ($3R/2$) and half is the average potential energy ($3R/2$). In this simplistic approach, the heat capacity should then be roughly $3R$ for solids. This approximation is known as the *Law of Dulong–Petit*, and it is applicable to many nonmetallic solids at and above room temperature. It does not apply at low temperatures because the energy of atomic vibrations is quantized and the number of vibrational modes that are excited decreases as the temperature decreases below some value for each material. This temperature is known as the *Debye temperature*, θ_D, above which the heat capacity approaches the value of Dulong and Petit ($3R$), and below which the heat capacity decreases monotonically with decreasing temperature

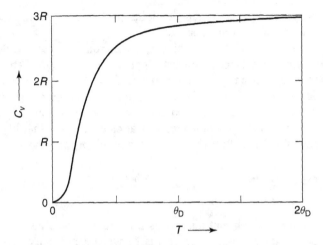

Figure 4.24 Molar heat capacity as a function of temperature, based on the Debye model. From K. M. Ralls, T. H. Courtney, and J. Wulff, *Introduction to Materials Science and Engineering*. Copyright © 1976 by John Wiley & Sons, Inc. This material is used by permission John Wiley & Sons, Inc.

Table 4.8   Representative Values of Debye Temperature for Selected Elements

Element	θ_D(K)	Element	θ_D(K)	Element	θ_D(K)
Ag	225	Fe	467	Pt	229
Al	418	Ge	366	Si	658
Au	165	Hg	60–90	Sn (gray)	212
Be	1160	In	109	Sn (white)	189
Bi	117	La	132	Ta	231
C (diamond)	2000	Mg	406	Ti	278
Ca	219	Mo	425	Tl	89
Cd	300	Nb	252	V	273
Co	445	Ni	456	W	379
Cr	402	Pb	94.5	Zn	308
Cu	339	Pd	275	Zr	270

Source: C. Kittel, *Introduction to Solid State Physics*, p. 132, 2nd ed. Copyright © 1957 by John Wiley & Sons, Inc.

(see Figure 4.24 and Table 4.8 for values of θ_D). In this region ($T \ll \theta_D$), the heat capacity is due to lattice vibrations (hence, a subscript l is added), and is given by:

$$C_{vl} = \frac{12\pi^4 R}{5} \left(\frac{T}{\theta_D} \right)^3 \tag{4.41}$$

Most nonmetals follow Eq. (4.41) at low temperatures, as illustrated in Figure 4.24. Most metals, however, show a linear decrease in heat capacity with temperature as absolute zero is approached. We require a slightly different model to describe the low-temperature behavior of C_v in metals.

Recall from Figure 1.15 that metals have free electrons in what is called the valence band and have empty orbitals forming what is called the conduction band. In Chapter 6, we will see how this electronic structure contributes to the electrical conductivity of a metallic material. It turns out that these same electronic configurations can be responsible for thermal as well as electrical conduction. When electrons act as the thermal energy carriers, they contribute an electronic heat capacity, C_{ve}, that is proportional to both the number of valence electrons per unit volume, n, and the absolute temperature, T:

$$C_{ve} = \frac{\pi^2 n k_B^2 T}{2 E_F} \tag{4.42}$$

where k_B is Boltzmann's constant and E_F is the energy of the Fermi level, which is characteristic of an element, but in general is between 3 and 5 eV. The contribution of the electronic heat capacity to the overall heat capacity is usually quite small, but below about 4 K the electronic contribution becomes larger than the lattice contribution [Eq. (4.41)]. In this low-temperature regime, the total heat capacity is a sum of both the electronic and lattice contributions:

$$C_v = C_{vl} + C_{ve} = \alpha T^3 + \gamma T \tag{4.43}$$

where α is given by the coefficients in Eq. (4.41), and γ is given by the coefficients in Eq. (4.42).

Returning now to thermal conductivity, Eq. (4.40) tells us that any functional dependence of heat capacity on temperature should be implicit in the thermal conductivity, since thermal conductivity is proportional to heat capacity. For example, at low temperatures, we would expect thermal conductivity to follow Eq. (4.43). This is indeed the case, as illustrated in Figure 4.25. In copper, a pure metal, electrons are the primary heat carriers, and we would expect the electronic contribution to heat capacity to dominate the thermal conductivity. This is the case, with the thermal conductivity varying proportionally with temperature, as given by Eq. (4.42). For a semiconductor such as germanium, there are less free electrons to conduct heat, and lattice conduction dominates—hence the T^3 dependence on thermal conductivity as suggested by Eq. (4.41).

This competition between electrons and the heat carriers in the lattice (phonons) is the key factor in determining not only whether a material is a good heat conductor or not, but also the temperature dependence of thermal conductivity. In fact, Eq. (4.40) can be written for either thermal conduction via electrons, k_e, or thermal conduction via phonons, k_p, where the mean free path corresponds to either electrons or phonons, respectively. For pure metals, $k_e/k_p \approx 30$, so that electronic conduction dominates. This is because the mean free path for electrons is 10 to 100 times higher than that of phonons, which more than compensates for the fact that C_{ve} is only 10% of the total heat capacity at normal temperatures. In disordered metallic mixtures, such as alloys, the disorder limits the mean free path of both the electrons and the phonons, such that the two modes of thermal conductivity are more similar, and $k_e/k_p \approx 3$. Similarly, in semiconductors, the density of free electrons is so low that heat transport by phonon conduction dominates.

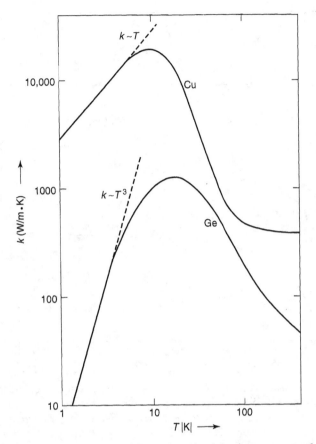

Figure 4.25 Temperature dependence of thermal conductivity for a pure metal (Cu) and a non-metal (Ge), illustrating different temperature dependences at low temperature. From K. M. Ralls, T. H. Courtney, and J. Wulff, *Introduction to Materials Science and Engineering*. Copyright © 1976 by John Wiley & Sons, Inc. This material is used by permission John Wiley & Sons, Inc.

So, in general, highly ordered materials such as pure metals, most pure ceramics, and crystalline polymers exhibit increasing thermal conductivity with decreasing temperature due to increases in both the electronic and phonon mean free paths at low temperatures. At sufficiently low temperatures, the mean free paths become constant, and the variation in heat capacity, as given by Eqs. (4.41) and (4.42), dominate the thermal conductivity. Disordered materials, on the other hand, such as solid–solution alloys, inorganic glasses, and glassy polymers, exhibit decreasing thermal conductivity with decreasing temperature. This is because the electronic and phonon mean free paths are relatively small at all temperatures, and temperature variation in heat capacity dominates the temperature dependence of thermal conductivity (see Figure 4.24). These trends are summarized in Figure 4.26. Note that highly ordered materials such as copper, α-iron, polyethylene, and BeO have significant regions in which thermal conductivity decreases with temperature. Disordered materials such as PMMA, glass, and stainless steel tend to increase thermal conductivity with temperature. The lowering of thermal conductivity with alloying is an important design consideration, as illustrated in Figure 4.27. Note that as more alloying elements are added to iron, the thermal

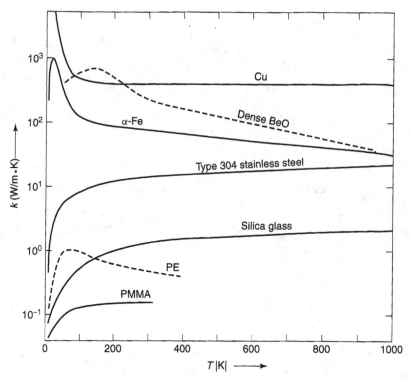

Figure 4.26 Variation of thermal conductivity with temperature for ordered (Cu, BeO, Fe, and PE) and disordered (SS, silica glass, and PMMA) with temperature. From K. M. Ralls, T. H. Courtney, and J. Wulff, *Introduction to Materials Science and Engineering.* Copyright © 1976 by John Wiley & Sons, Inc. This material is used by permission John Wiley & Sons, Inc.

conductivity drops at a given temperature. In the subsequent sections, we expand upon these generalizations for the remaining material classes.

4.2.2 Heat Transport Properties of Ceramics and Glasses

4.2.2.1 Heat Transport Properties of Ceramics. As a means of further illustrating the importance of the theoretical developments for heat capacity and thermal conductivity in the previous section, Figure 4.28 shows that the general trend of increasing heat capacity with temperature and the ultimate achievement of $C_v = 3R$ shown in Figure 4.24 holds true for crystalline ceramic materials as well. The primary result of heat capacity theory as applied to ceramics is that the $3R$ value is achieved in the neighborhood of 1000°C for most oxides and carbides. Further increases in temperature do not strongly affect this value, and it does not depend strongly on the crystal structure. However, heat capacity does depend on porosity. Consequently, the heat energy required to raise the temperature of porous insulating fire brick is much lower than that required to raise the temperature of the dense fire brick material, even though its thermal conductivity is lower. This is one of the useful properties of insulating materials for furnaces, which can be rapidly heated and cooled due to their porosity. Other factors affecting heat capacity in ceramics are phase transformations, such as the $\alpha-\beta$ quartz transition, which result in step changes to the heat capacity, and the

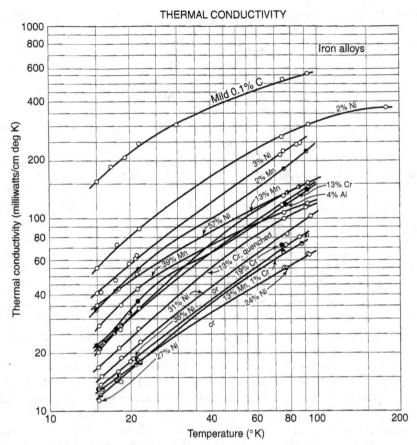

Figure 4.27 Thermal conductivity of some iron alloys. Reprinted, by permission, from C. Kittel, *Introduction to Solid State Physics*, 2nd ed., p. 151. Copyright © 1957 by John Wiley & Sons, Inc.

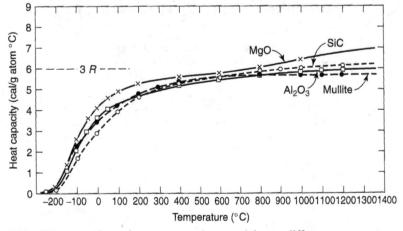

Figure 4.28 Heat capacity of some ceramic materials at different temperatures. From W. D. Kingery, H. K. Bowen, and D. R. Uhlmann, *Introduction to Ceramics*. Copyright © 1976 by John Wiley & Sons, Inc. This material is used by permission of John Wiley & Sons, Inc.

presence of point defects (cf. Section 1.1.3), which tend to increase the value of the heat capacity.

The discussion of the previous section would also lead us to believe that since most ceramics are poor electrical conductors (with a few notable exceptions) due to a lack of free electrons, electronic conduction would be negligible compared to lattice, or phonon, conduction. This is indeed the case, and we will see that structural effects such as complexity, defects, and impurity atoms have a profound effect on thermal conductivity due to phonon mean free path, even if heat capacity is relatively unchanged.

Materials with complex structures possess more thermal phonon scattering, which reduces the mean free path and lowers the thermal conductivity, as given by Eq. (4.40). For example, the magnesium aluminate spinel shown in Figure 1.41 has a lower thermal conductivity than either Al_2O_3 or MgO, although each has a similar heat capacity and structure. Similarly, the complex structure of mullite has a much lower thermal conductivity than magnesium aluminate spinel. Structure can also affect the temperature dependence of thermal conductivity, since in complex structures the phonon mean free path tends to approach lattice dimensions at high temperatures. Similarly, anisotropic crystal structures can have thermal conductivities that vary with direction, as illustrated for quartz (cf. Figure 1.44) in Table 4.9. This variation tends to decrease with increasing temperature, since anisotropic crystals become more symmetrical as the temperature is raised. Crystallite size also has an effect on thermal conductivity, though its effect is evident only at higher temperatures due to electronic conductivity, as shown in Figure 4.29 for three common ceramic materials.

Finally, in the same way that structural complexity lowers conductivity, the addition of impurities and variations in composition can affect thermal conductivity. As shown in Figure 4.30, the thermal conductivity is a maximum for simple elementary oxide and carbide structures, such as BeO. As the cation atomic number increases and the difference in atomic weight between the components becomes larger, the thermal conductivity decreases. In solid solutions, the additional phonon scattering for low concentrations of impurities is directly proportional to the volume fraction of impurity, and the impurity effect is relatively independent of temperature. The effect of solid–solution impurities over an extended range of compositions depends on mass difference, size difference, and binding energy difference of the constituents. In addition, temperature effects can be magnified in solid solutions, as illustrated in Figure 4.31. For the

Table 4.9 Thermal Conductivity of Quartz

Temperature (°C)	Thermal Conductivity (W/m·K)		
	Normal to c-Axis	Parallel to c-Axis	Ratio
0	0.67	1.13	1.69
100	0.50	0.79	1.58
200	0.42	0.63	1.50
300	0.35	0.50	1.43
400	0.31	0.42	1.35

Source: W. D. Kingery, H. K. Bowen, and D. R. Uhlmann, *Introduction to Ceramics*. Copyright © 1976 by John Wiley & Sons, Inc.

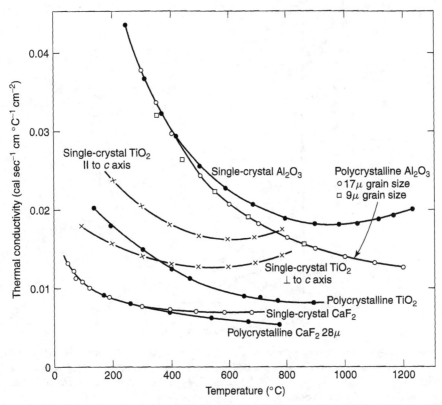

Figure 4.29 Thermal conductivity of single-crystal and polycrystalline Al_2O_3, TiO_2, and CaF. Multiply by 418.7 to obtain k in units of W/m·K. From W. D. Kingery, H. K. Bowen, and D. R. Uhlmann, *Introduction to Ceramics*. Copyright © 1976 by John Wiley & Sons, Inc. This material is used by permission of John Wiley & Sons, Inc.

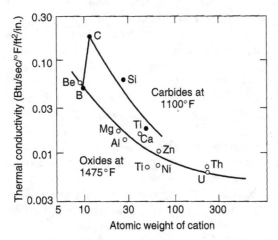

Figure 4.30 Effect of cation atomic number on thermal conductivity of some oxides and carbides. Multiply by 519.2 to obtain k in units of W/m·K. From W. D. Kingery, H. K. Bowen, and D. R. Uhlmann, *Introduction to Ceramics*. Copyright © 1976 by John Wiley & Sons, Inc. This material is used by permission of John Wiley & Sons, Inc.

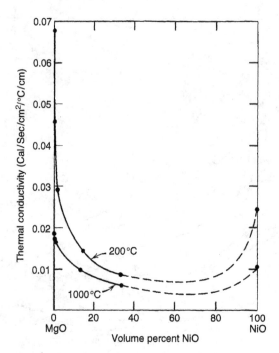

Figure 4.31   Thermal conductivity in the solid solution system MgO–NiO. Multiply by 418.7 to obtain k in units of W/m·K. From W. D. Kingery, H. K. Bowen, and D. R. Uhlmann, *Introduction to Ceramics.* Copyright © 1976 by John Wiley & Sons, Inc. This material is used by permission of John Wiley & Sons, Inc.

MgO–NiO solid solution shown, the effect of thermal conductivity lowering is even more pronounced at 200°C than at 1000°C.

One final note is appropriate for this section. Due to the fact that many oxide ceramics are used as insulating materials, the term *thermal resistivity* is often used instead of thermal conductivity. As will be the case with electrical properties in Chapter 6, resistivity and conductivity are merely inverses of one another, and the appropriateness of one or the other is determined by the context in which it is used. Similarly, *thermal conductance* is often used to describe the thermal conductivity of materials with standard thicknesses (e.g., building materials). Thermal conductance is the thermal conductivity divided by the thickness $(C = k/L)$, and *thermal resistance* is the inverse of the product of thermal conductance and area $(R = 1/C \cdot A)$.

4.2.2.2 Heat Transport Properties of Glasses. As was described in Chapter 1, glasses are structurally different from crystalline ceramics, and we would expect this to affect the heat capacity. This is true, and in general the heat capacities of most oxide glasses approach 0.7 to 0.95 of the theoretical $3R$ value at the low-temperature end of the glass transition. On passing through the glass transition, the heat capacity generally increases by a factor of 1.3 to 3.0, as a result of increased configurational entropy. Eventually, the heat capacity of glasses reaches a value of $n3R$, where n is the number of atoms in the compound; thus $n = 3$ for SiO_2 and $C_p \approx C_v \approx 78$ J/mol·K above 1000°C. A number of empirical correlations exist for estimating the heat capacity of glasses and related thermodynamic quantities such as enthalpy. One such correlation is presented here, and others can be found in reference 14. The molar heat capacity of a multicomponent glass up to about 900°C can be estimated from the molar composition, x_i (entered in mol%), according to

$$C_p = \frac{1}{100}\left(\sum a_i x_i + T \sum b_i x_i + \frac{\sum c_i x_i}{T^2} + \frac{\sum d_i x_i}{T^{1/2}}\right) \qquad (4.44)$$

where T is the temperature in K, and the factors a_i to d_i are presented in Table 4.10.

Based on the discussion of the previous section, we would also expect that structural complexity and disorder present in glasses would tend to decrease the thermal conductivity. This is true, especially when alkalis are added to sodium silicate glasses. Figure 4.32 shows the effect of replacing SiO_2 in Na_2O–SiO_2 glass with other oxides.

Table 4.10   Factors Used to Determine Molar Heat Capacities of Glasses Using Eq. (4.44)

Oxide	a_i	$10^3 b_i$	$10^{-5} c_i$	d_i	Validity Region (K)
Na_2O	70.884	26.110	−3.5820	0	270–1170
K_2O	84.323	0.731	−8.2980	0	270–1190
MgO	46.704	11.220	−13.280	0	270–1080
CaO	39.159	18.650	−1.5230	0	270–1130
Al_2O_3	175.491	−5.839	−13.470	−1370.0	270–1190
SiO_2	127.200	−10.777	4.3127	−1463.9	270–1600
TiO_2	64.111	22.590	−23.020	0	300–800
FeO	31.770	38.515	−0.012	0	300–800
Fe_2O_3	135.250	12.311	−39.098	0	300–800

Source: H. Scholze, *Glass*. Copyright © 1991 by Springer-Verlag.

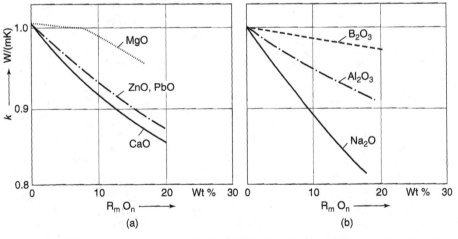

Figure 4.32   Change in thermal conductivity at 0°C of an Na_2O–SiO_2 glass (18–82 wt%) as SiO_2 is replaced proportionally by weight with other oxides. Reprinted, by permission, from H. Scholze, *Glass*, p. 363. Copyright © 1991 by Springer-Verlag.

These results reflect the fact that the stronger the bonds in the glass are, the higher the thermal conductivity. The dependence of thermal conductivity on bond strength allows an empirical expression to be developed in which the thermal conductivities of glass components are additive:

$$k = 10^{-1} \sum k_i x_i \tag{4.45}$$

where x_i is the weight fraction of component i, and k_i is its corresponding thermal conductivity factor, representative values for which can be found in Table 4.11. An alternative set of factors at higher temperatures has been developed by Primenko [13] and can be found elsewhere [14]. As a rough approximation, thermal conductivity of a glass can be calculated from the density, ρ, by using

$$k = \frac{a}{\rho} + b \tag{4.46}$$

where the pair values for constants a and b of 2.09 and 0.17, or 2.30 and 0.21, are valid for temperatures between 0°C and 100°C.

4.2.3   Heat Transport Properties of Polymers

As with other disordered materials, the thermal conductivities of polymers are low due to phonon scattering. As a result, even though polymers tend to have heat capacities of the same order of magnitude as metals (1.5 to 3.5 J/g·K), their thermal conductivities (0.1 to 1.0 W/m·K) are 1000 times lower than metals. Polymers, therefore, are generally good insulators, as long as their use temperature is below their thermal stability temperature. Few correlations for heat capacity and thermal conductivity of polymers

Table 4.11   Factors Used to Determine Thermal Conductivity of Glasses Based on Composition Using Eq. (4.45)

Oxide	k_i (W/m·K)[a]
Li_2O	−9.29
Na_2O	−4.79
K_2O	2.17
MgO	21.73
CaO	13.06
SrO	8.63
BaO	2.89
B_2O_3	8.216
Al_2O_3	13.61
SiO_2	13.33
TiO_2	−31.38
ZnO	8.00
PbO	2.68

[a] Values from Ammar et al. [12].
Source: H. Scholze, *Glass*. Copyright © 1991 by Springer-Verlag.

Cooperative Learning Exercise 4.7

Consider a glass with the following composition (in wt%): 72 SiO_2, 15 Na_2O, 11 CaO, and 2 Al_2O_3, and a density of 2.458 g/cm^3.

Person 1: Estimate the thermal conductivity at 30°C using Eq. (4.45).

Person 2: Estimate the thermal conductivity at 30°C using Eq. (4.46).

Compare your answers. What are some ways you could experimentally verify these estimates?

Thermal conductivities can be determined using any setup that would allow Fourier's Law of conduction to be applied. For example, a plate of the glass could be placed on top of a heat sink (such as a large copper block), heated on one side, and the temperature difference at each face measured to determine the temperature gradient. Once the heat flow is measured, the proportionality constant, k, can be calculated.

Answers: $k_1 = 1.06$; $k_2 = 1.02$

---

exist. As for all the materials in this chapter, representative values of thermal conductivity can be found in Appendix 5.

The heat capacity of polymers at constant pressure increases with temperature up to the melting point. In practice, there is a sharp increase in heat capacity at the onset of melting, followed by a sharp drop at the end of the melting range (see Figure 4.33). There is then a slight increase in heat capacity with temperature in the melt.

The thermal conductivity of polymers is molecular-weight-dependent, and it increases with increasing molecular weight. Orientation also has an effect, and it can have a strong influence on thermal conductivity in highly crystalline polymers such as HDPE. As is evident from Figure 4.34, the thermal conductivity of most polymeric materials is, with a few notable exceptions for highly crystalline polymers such as polyethylene, less dependent upon temperature and, as a result, can only be influenced significantly through the incorporation of fillers. These effects will be discussed in the next section. Foaming with air or some other gas can also decrease thermal conductivity (see Figure 4.35). This

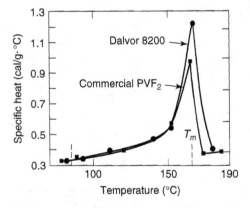

Figure 4.33 Specific heat of two commercial polyvinylidene fluoride samples with temperature. From Z. Tadmor and C. G. Gogos, *Principles of Polymer Processing*, Copyright © 1979 by John Wiley & Sons, Inc. This material is used by permission of John Wiley & Sons, Inc.

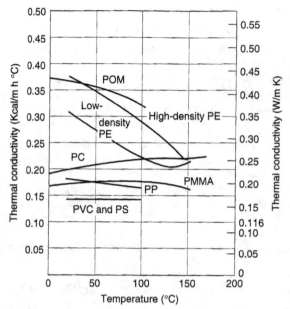

Figure 4.34 Temperature dependence of thermal conductivity for selected polymers. Reprinted, by permission, from P. C. Powell and A. J. I. Housz, *Engineering with Polymers*, p. 288. Copyright © 1998 by P. C. Powell and A. J. I. Housz.

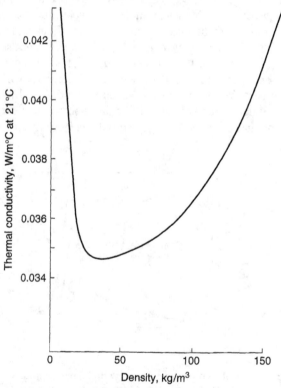

Figure 4.35 Thermal conductivity of expanded polystyrene beads as a function of density. Reprinted, by permission, from The Dow Chemical Company. Copyright © 1966.

has the same effect as the introduction of porosity in ceramic materials such as fire brick, due to the introduction of a lower thermal conductivity material (in this case, a gas) into the structure.

Cooperative Learning Exercise 4.8

Recall from the beginning of the chapter that a related quantity to the thermal conductivity is the *thermal diffusivity*, α, which is defined as $k/\rho C_p$, where k is the thermal conductivity, ρ is the density and C_p is the *heat capacity* at constant pressure per unit mass, or specific heat. Below are these thermal properties for polycarbonate.

Person 1: Determine the density and thermal diffusivity for polycarbonate at 350 K from the plots.

Person 2: Determine the thermal conductivity and heat capacity for polycarbonate at 350 K from the plots.

Combine your information to calculate the thermal diffusivity at 350 K. How does this compare to the value obtained from the plot? Can you estimate the melting point of this polycarbonate from these plots?

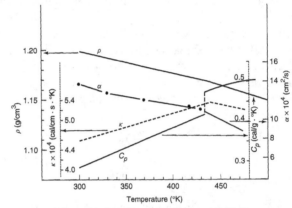

Reprinted, by permission, from Z. Tadmor and C. G. Gogos. *Principles of Polymer Processing*, p. 132. Copyright © 1979 by John Wiley & Sons, Inc.

Answers: $\rho = 1.19$ g/cm^3; $\alpha = 12 \times 10^{-4}$ cm^2/s; $k = 4.8 \times 10^{-4}$ cal/m·s·K; $C_p = 0.33$ cal/g·K $\alpha_{calc} = k/\rho C_p = 4.8 \times 10^{-4}$ cal/m·s·K/(1.19 g/cm^3 × 0.33 cal/g·K) = 12 × 10^{-4} cm^2/s, same as on graph. $T_m \approx 434$ K = 162.8°C, as determined from graphs where discontinuity exists; thermal properties change!

4.2.4 Heat Transport Properties of Composites

As was the case with momentum transport, the heat transport properties of heterogeneous systems are difficult to correlate and virtually impossible to predict. There are two topics worthy of note, however, namely, the heat transport properties of filled composites, and the thermal conductivity of laminar composites.

4.2.4.1 Thermal Conductivity of Filled Composites.

As we saw at the end of the previous section, the presence of gases in a material's structure can dramatically affect the heat transport properties. We expand upon this principle here, again concentrating on polymeric materials that have been reinforced with various fibers and fillers and that may also contain porous gas structures, as is found in foams.

To a first approximation, the heat capacity in these heterogeneous, multiphase systems, $\overline{C}_p$ can be approximated by a weighted average of the component heat capacities:

$$\overline{C}_p = \sum w_i C_{pi} \tag{4.47}$$

where w_i is the weight fraction of each component. Since the relative heat capacities (per gram) of gas, polymer, and metal are 2:4:1, the heat capacity of polymer tends to dominate in these types of multicomponent systems, acting as a heat sink during processing.

Due to the dependence on mean free path as described in Eq. (4.40), the thermal conductivity of heterogeneous systems is impossible to predict on heat capacity alone. As in previous sections, we do know that disorder tends to decrease thermal conductivity due to mean free path considerations, and this is indeed the case for fillers with high thermal conductivities, such as copper and aluminum in epoxy matrices (see Table 4.12). The thermal conductivity of the epoxy matrix increases only modestly due to the addition of even high percentages of thermally conductive fillers.

One way to deal with heterogeneous materials is to assume that the composite has an effective thermal conductivity, k_{eff}, which can then be used in heat transport expressions as previously performed for homogeneous materials. There are several expressions for k_{eff} that have been developed, depending upon the type and amount of reinforcing material. For spherical particles of small volume fractions, ϕ, we have

$$\frac{k_{eff}}{k_0} = 1 + \frac{3\phi}{\left(\dfrac{k_1 + 2k_0}{k_1 - k_0}\right) - \phi} \tag{4.48}$$

where k_1 and k_0 are the thermal conductivities of the sphere and matrix phase materials, respectively. For large volume fractions, the thermal conductivity of the composite can be estimated with:

$$\frac{k_{eff}}{k_0} = 1 + \frac{3\phi}{\left(\dfrac{k_1 + 2k_0}{k_1 - k_0}\right) - \phi + 1.569\left(\dfrac{k_1 - k_0}{3k_1 - 4k_0}\right)\phi^{10/3} + \cdots} \tag{4.49}$$

Table 4.12   Thermal Conductivity of Epoxy Resin Filled with Various Compounds

| | | Thermal Conductivity (W/m·K) | | |
Filler	Vol% Filler	Filler	Epoxy Resin	Composite
Aluminum, 30 mesh	63	208	0.20	2.5
Sand, coarse grain	64	1.17	0.20	1.0
Mica, 325 mesh	24	0.67	0.20	0.5
Alumina, tabular	53	30.3	0.20	1.0
Alumina, 325 mesh	53	30.3	0.23	1.4
Silica, 325 mesh	39	1.17	0.23	0.77
Copper powder	60	385	0.23	1.6
Alumina (tabular/325 mesh)	45/20	30.3	0.20	2.5

Source: F. Rodriguez, *Principles of Polymer Systems*, 2nd ed. Copyright © 1982 by McGraw-Hill Book Co.

Generally, the thermal interaction of spheres that can occur at high volume fractions is negligible, such that the simpler Eq. (4.48) is often used. For nonspherical reinforcement, such as aligned, continuous fibers, the thermal conductivity is anisotropic—that is, not the same in all directions. Rayleigh showed that the effective thermal conductivity along the direction of the fiber axis is

$$\frac{k_{\text{eff},z}}{k_0} = 1 + \left(\frac{k_1 - k_0}{k_0}\right)\phi \tag{4.50}$$

whereas the effective thermal conductivity in the perpendicular directions are identical:

$$\frac{k_{\text{eff},x}}{k_0} = \frac{k_{\text{eff},y}}{k_0} = 1 + \frac{2\phi}{\left(\dfrac{k_1 + k_0}{k_1 - k_0}\right) - \phi + \left(\dfrac{k_1 - k_0}{k_1 - k_0}\right)(0.30584\phi^4 + 0.013363\phi^8 \ldots)} \tag{4.51}$$

Additional expressions exist for nonspherical reinforcements and for solids containing gas pockets. The interested reader is referred to reference 15 for more information.

4.2.4.2 Thermal Conductivity of Laminar Composites.

In the case of laminar composites or layered materials (cf. Figure 1.74), the thermal conductance can be modeled as heat flow through plane walls in a series, as shown in Figure 4.36. At steady state, the heat flux through each wall in the x direction must be the same, q_x, resulting in a different temperature gradient across each wall. Equation (4.2) then becomes

$$q_x = -k_A \frac{dT}{dx_A} = -k_B \frac{dT}{dx_B} = -k_C \frac{dT}{dx_C} \tag{4.52}$$

where the minus sign simply indicates the direction of heat flow, depending upon how the temperature gradient is defined. For the case of constant thermal conductivity with temperature, which is true for small temperature gradients, and a constant cross-sectional heat flow area, A, the heat flow in the x direction, $Q_x = q_x \cdot A$ becomes

$$Q_x = k_A A \frac{(T_1 - T_2)}{\Delta x_A} = k_B A \frac{(T_2 - T_3)}{\Delta x_B} = k_C A \frac{(T_3 - T_4)}{\Delta x_C} \tag{4.53}$$

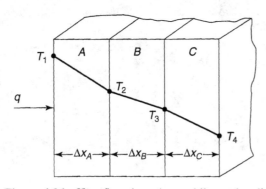

Figure 4.36 Heat flow through a multilayered wall.

Example Problem 4.1

Question: Derive an expression analogous to Eq. (4.54) for steady-state heat flow through concentric cylinders, as shown at right. All cylinders are of the same length, L. Assume that thermal conductivity is constant with temperature, but that the material in each concentric ring is different.

Answer: Fourier's Law in cylindrical coordinates can be written as

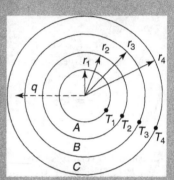

$$q = \frac{Q}{A} = -k \frac{dT}{dr}$$

The cross-sectional area normal to heat flow, A, is not constant, and varies with radius as $A = 2\pi r L$, where L is the length of the cylinders. For the inner cylinder, substitution of the area expression into Fourier's Law and separation of variables gives

$$\frac{Q_A}{2\pi L} \int_{r_1}^{r_2} \frac{dr}{r} = -k_A \int_{T_1}^{T_2} dT$$

and integration results in

$$Q_A = \frac{k_A 2\pi L}{\ln(r_1/r_2)}(T_1 - T_2)$$

Multiply the numerator and denominator by $(r_2 - r_1)$, and defining a log-mean area, A_{lm} as follows

$$A_{A,lm} = \frac{(2\pi L r_2) - (2\pi L r_1)}{\ln(2\pi L r_2/2\pi L r_1)} = \frac{A_2 - A_1}{\ln(A_2/A_1)}$$

the heat flow through the inner cylinder becomes

$$Q = \frac{k A_{A,lm}(T_1 - T_2)}{(r_2 - r_1)} = \frac{(T_1 - T_2)}{(r_2 - r_1/k A_{A,lm})} = \frac{(T_1 - T_2)}{R_A}$$

where R_A is the resistance, which for the cylindrical geometry is defined as

$$R_A = \frac{(r_2 - r_1)}{k_A A_{A,lm}}$$

Finally, we recognize that at steady state, the heat flow through all the cylinders must be the same, $Q = Q_A = Q_B = Q_C$, so that

$$Q = \frac{(T_1 - T_2)}{R_A} = \frac{(T_2 - T_3)}{R_B} = \frac{(T_3 - T_4)}{R_C}$$

As was the case for the multilayer wall, the internal wall temperatures T_2 and T_3 can be eliminated through combination of the three equations to yield one equation

$$Q = \frac{(T_1 - T_4)}{R_A + R_B + R_C} = \frac{(T_1 - T_4)}{\sum R_i}$$

Note that this is the same result as Eq. (4.54), except that the resistance is defined differently.

Each of the three relations in Eq. 4.53 can be solved for the temperature difference, and the equations can be added to eliminate the internal surface temperatures, T_2 and T_3, such that only the external surface temperatures, T_1 and T_4 remain. The heat flow is then expressed as follows:

$$Q_x = \frac{(T_1 - T_4)}{\dfrac{\Delta x_A}{k_A A} + \dfrac{\Delta x_B}{k_B A} + \dfrac{\Delta x_C}{k_C A}} = \frac{(T_1 - T_4)}{R_A + R_B + R_C} \tag{4.54}$$

Cooperative Learning Exercise 4.9

Work in groups of three. Your plant is replacing asbestos used to insulate low-pressure steam pipes with a new multilayer composite material composed of a 2.5-cm-thick, polystyrene foam ($k = 0.029$ W/m·K) and a 2-mm-thick, outer, protective layer of polyethylene ($k = 0.33$ W/m·K). Assume that the pipe carries saturated steam at 130°C and that the outer surface of the insulation is 30°C. The piping is stainless steel ($k = 14$ W/m·K), with ID = 0.0254 m and OD = 0.0508 m. Use the equations developed in Example Problem 4.1 to solve the following problems, assuming a pipe length of 1 m.

Person 1: Calculate the log-mean heat transfer area for the stainless steel pipe, then use this to calculate the thermal resistance through the pipe.

Person 2: Calculate the log-mean heat transfer area for the polystyrene foam only, then use this to calculate the thermal resistance through the foam.

Person 3: Calculate the log-mean heat transfer area for the outer polyethylene layer, then use this to calculate the thermal resistance through the polyethylene.

Combine your resistance to calculate the heat transfer rate through the 1-m section of pipe. Which of the resistances dominates the calculation?

Answers: $A_{ss,lm} = 0.115$ m^2; $R_{ss} = 0.008$ K/W
$A_{ps,lm} = 0.229$ m^2; $R_{ps} = 3.76$ K/W
$A_{pe,lm} = 0.323$ m^2; $R_{pe} = 0.019$ K/W
$Q = 26.4$ W; polystyrene foam dominates

---

where $R_i = \Delta x_i / (k_i \cdot A)$ is called the *thermal resistance*, as introduced at the end of Section 4.2.2.1. The total thermal resistance is simply the sum of the individual resistances, $R_{tot} = \sum R_i$. These "R values" are often reported in lieu of thermal conductivity values for common insulating materials, such as fiberglass and rock wool, for which thickness and sizes have been standardized, typically 1-ft^2 area and 1-inch thickness. Selected thermal conductivities for these and other insulating materials can be found in Appendix 5.

4.2.5 Heat Transport Properties of Biologics

Heat storage in biological systems is much the same as for the previous material classes. For example, the heat capacity of biological fluids, which are heterogeneous suspensions and mixtures, can be approximated by Eq. (4.47). In instances where the biological fluid or food substance is an aqueous-based suspension or solution of fat and solids, the specific heat (heat capacity per unit weight in J/g·K) can be approximated as

$$C_p = 4.2(x_{H_2O} + 0.5x_{fat} + 0.3x_{solids}) \tag{4.55}$$

where x_i is the mass fraction of water, fat, and solids, respectively.

Heat transfer in biological systems, on the other hand, is often a complex combination of conduction, convection, and radiation. For example, the insulating value of fur and feathers in animals is certainly measurable, but is based on more than just the thermal conductivity of the biological materials constituting the fur or feathers, since gases become entrapped in these often times porous materials, and heat is being radiated to and from the surroundings. As a result, the description of heat transfer in biological systems is much more empirical, and it involves a separate set of terminologies from the traditional, engineering-oriented set of terms. Nonetheless, the principles are the same, and we relax our restriction on heat transfer via conduction only so as to better describe heat transfer in "biological units"—animals and plants.

4.2.5.1 Thermal Conductance in Biological Systems. As introduced at the end of Section 4.2.2.1, thermal conductance is the thermal conductivity divided by the thickness $(C = k/L)$, and *thermal resistance* is the inverse of the product of thermal conductance and area $(R = 1/C \cdot A)$. These terms are widely used to describe the insulating value of skin coverings such as fur and clothing in biological systems. As mentioned in the introduction to this section, the insulating value of fur and clothing is derived from more than thermal conductivity considerations. It involves the reduction of permeability to water vapor transmission, the disruption of convective heat transport from the surface, and an increase in surface area. Nonetheless, the insulating value of clothing has been standardized in terms of thermal conductance, and is expressed in units of *clo*. A clo is the amount of insulation that would allow 6.45 N·m/s of heat to transfer to the surroundings from a 1.0 m^2 area of skin across a 1°C temperature difference between the skin and the environment. The higher the clo value of clothing, the less heat will be transferred from the skin. We will continue to use common engineering units for our analysis.

For clothing, Fourier's Law [Equation (4.2)], can be expressed as

$$q_y = \frac{Q}{A} = -k_{cl}\frac{dT}{dy} \tag{4.2}$$

which at steady state and for constant thermal properties, can be expressed in terms of conductance of clothing, C_{cl}, as

$$Q = \frac{-k_{cl} A \Delta T}{\Delta y} = C_{cl} A(T_{sk} - T_{\infty}) \tag{4.56}$$

where T_{sk} is the temperature of the skin and T_{∞} is the temperature of the surrounding environment. Some representative values for conductance of common items of clothing are given in Table 4.13. Clothing is often worn in layers, and the principles represented by Eq. (4.54)—that is, the additivity of resistances—apply here as well. Due to the added complexities of layered clothing in terms of simultaneous series and parallel heat conduction, the following empirical relationships have been developed for the thermal resistance of clothing, R_{cl} (in K/W), in men and women:

$$R_{cl,men} = \frac{0.494 \sum(1/C_{cl}) + 0.0119}{A} \tag{4.57}$$

$$R_{cl,women} = \frac{0.524 \sum(1/C_{cl}) + 0.0053}{A} \tag{4.58}$$

where A is the surface area in m². Clothing surface area is based on a 15–25% increase over the surface area of a nude body, A (in m²), as determined by the formula of DuBois:

$$A = 0.07673 W^{0.425} H_t^{0.725} = 0.2025 m^{0.425} H_t^{0.725} \tag{4.59}$$

where W is the body weight in newtons, H_t is the height in meters, and m is the body mass in kilograms. Mean skin temperature, T_{sk}, is also affected by the presence of clothing. The following empirical relationship can be employed and is essentially independent of metabolic rate up to four or five times the rate at rest. For clothed humans,

$$T_{sk} = 25.8 + 0.267 T_{\infty} \tag{4.60}$$

where T_{∞} is again the surrounding environmental temperature, in degrees Celsius.

Table 4.13 Conductance Values for Selected Pieces of Clothing

Clothing	$C_{cl}(W/m^2 \cdot K)$
Undergarments	34–130
Thermal undergarments	18
Shirts, vests, and blouses	22–46
Pants and Slacks	15–25
Sweaters and jackets	17–38
Sock	65–161
Stockings, panty hose	640
Sandals	320
Boots	81

Source: A. T. Johnson, *Biological Process Engineering*. Copyright © 1999 by John Wiley & Sons, Inc.

Fur on animals is treated much the same way as clothing on humans. Some typical values of conductance for furs and feathers are presented in Table 4.14. The total surface area of an animal, A (in m^2), can be estimated by

$$A = 0.09m^{2/3} \tag{4.61}$$

where m is the animal's body mass, in kilograms. The values in Table 4.14 include the effects of convection and conduction, although radiation was not a significant contributor in these instances. We will see in the next section how the effects of convection can be combined with thermal conductivity to describe heat transfer in these systems.

4.2.5.2 Convective Heat Transfer in Biological Systems*.

Unlike the materials of the previous sections, for which processing temperatures, thermal conductivities, and processing environments vary greatly, even within a material subclass (e.g., the heat transport processes involved in creating SiC via chemical vapor deposition are much different from those involved in forming SiC by a solid-state reaction), biological materials generally function within a relatively narrow temperature range, have thermal conductivities within an order of magnitude of each other, and function primarily in only one gaseous environment—air. As a result, we are able to briefly discuss here the role of convection in heat transfer in biological systems. To be sure, convection plays an important role in the processing and application of all the other material classes as well, and as you probably are already aware, the dimensionless groups and principles discussed here are general to heat transport in all systems. We will revisit some of these topics in Chapter 7, when the processing of materials is described.

Table 4.14 Selected Values of Thermal Conductance for Fur and Feathers[a]

Animal	Condition	m (kg)	Thickness (mm)	C (W/m²·K)
Rabbit	Back	2.5	9	2.4
Horse	Flank	650	8	4.1
Pig	Flank	100	5	2.8
Chicken	Well-feathered	2	25	1.6
Cow	—	600	10	3.2
Dog	Winter	50	44	1.2
Polar bear	In air	400	63	1.2
Polar bear	In water	400	57	27
Beaver	In air	26	44	1.2
Beaver	In water	26	40	14
Squirrel	Summer	—	3	9.0
Monkey	Summer	—	8	3.2

[a] Includes skin and still air layer.
Source: A. T. Johnson, *Biological Process Engineering*. Copyright © 1999 by John Wiley & Sons, Inc.

Convection involves the transfer of heat by means of a fluid, including gases and liquids. Typically, convection describes heat transfer from a solid surface to an adjacent fluid, but it can also describe the bulk movement of fluid and the associate transport of heat energy, as in the case of a hot, rising gas. Recall that there are two general types of convection: *forced convection* and *natural (free) convection*. In the former, fluid is forced past an object by mechanical means, such as a pump or a fan, whereas the latter describes the free motion of fluid elements due primarily to density differences. It is common for both types of convection to occur simultaneously in what is termed *mixed convection*. In such instance, a modified form of Fourier's Law is applied, called *Newton's Law of Cooling*, where the thermal conductivity is replaced with what is called the *heat transfer coefficient, h_c*:

$$q_y = \frac{Q_y}{A} = h_c dT \qquad (4.62)$$

Unlike Fourier's Law, Eq. (4.62) is purely empirical—it is simply the definition for the heat transfer coefficient. Note that the units of h_c (W/m^2·K) are different from those for thermal conductivity. Under steady-state conditions and assuming that the heat transfer area is constant and h_c is not a function of temperature, the following form of Eq. (4.62) is often employed:

$$Q_y = h_c A \Delta T \qquad (4.63)$$

where ΔT is generally taken as the difference between the solid surface temperature and the fluid temperature far from the surface (bulk fluid temperature).

Heat transfer coefficients cannot, in general, be calculated from first principles, and are normally derived from experiments with specific materials, fluids, geometries, and conditions. The use of dimensionless groups helps simplify this analysis. A complete review of dimensionless groups and their use in heat transfer analysis is not appropriate here. We will limit our discussion to three appropriate dimensionless groups: the Reynolds number, Re; Prandtl number, Pr; and Nusselt number, Nu. The *Reynolds number* is a ratio of inertial forces to viscous forces and is given by Re $= Dv\rho/\mu$, where D is the characteristic diameter in which a fluid is flowing, v is the fluid velocity, μ is the fluid viscosity, and ρ is its density. The Prandtl number was introduced in Section 4.2.1.2 as a ratio of momentum diffusivity to thermal diffusivity and is defined as Pr $= \hat{C}_p \mu/k$, where μ is again the fluid viscosity, k is its thermal conductivity, and $\hat{C}_p$ is the heat capacity. The Reynolds and Prandtl numbers provide a means for correlating measurable fluid properties (μ, ρ, k, and $\hat{C}_p$) and geometric considerations (D, v) with the heat transfer coefficient, h_c, in terms of the *Nusselt number*, Nu, which is defined as

$$\text{Nu} = \frac{h_c D}{k} \qquad (4.64)$$

where D is again a characteristic diameter and k is the thermal conductivity. Most correlations involve Pr and Re raised to an appropriate power:

$$\text{Nu} = \text{constant Re}^n \text{Pr}^m \qquad (4.65)$$

Example Problem 4.2 (Adapted from Johnson)

The metabolic rate of animals, M (in W), can be approximated by $M = 3.39m^{0.75}$, where m is the body mass of the animal in kilograms. About 15% of this heat is lost from the respiratory system, and another 50% is lost from the legs, tail, and ears. The remainder must be transported away from the body through the skin. Estimate the thermal conductivity of a 5-mm-thick hair coat of a 600-kg cow if the skin surface temperature is at $32°C$ and the ambient temperature is $15°C$. Assume that the cow is standing still in a field, with a breeze of 0.5 m/s. For flow across cylinders where $1000 < \text{Re} < 50,000$, the following correlation for Nusselt number may be used:

$$\text{Nu} = 0.26(\text{Re})^{0.6}(\text{Pr})^{0.3}$$

Answer: The heat generated by the cow is

$$M = 3.39(600 \text{ kg})^{0.75}, \text{ so } Q = 0.35(411) = 144 \text{ N} \cdot \text{m/s}$$

The total surface area of the cow, as given by Eq. (4.61) is

$$A = 0.09(600 \text{ kg})^{2/3} = 6.40 \text{ m}^2$$

We might estimate that about 20% of this surface area is in the legs and tail, which have already been accounted for in terms of heat loss, so that the remainder, about 5.12 m^2, is the surface area of the body. We can further model the cow's body as roughly cylindrical in shape and can estimate that its radius is 40 cm. This allows us to calculate the length of the cylinder (including the surfaces at both ends) as

$$A = 2\pi r L + 2\pi r^2 = 2\pi(0.4 \text{ m})L + 2\pi(0.4 \text{ m})^2 = 5.12 \text{ m}^2 \Rightarrow L = 1.64 \text{ m}$$

We will assume forced convection in order to calculate dimensionless groups. Physical properties of air must be estimated, and we will use the ambient air temperature to estimate them. The following physical properties can be found in tables or calculated:

$$\rho(15°C) = 1.226 \text{ kg/m}^3, \qquad \mu(15°C) = 1.8 \times 10^{-5} \text{ kg/m·s},$$

$$C_p = 1.0048 \text{ kJ/kg·K}, \qquad k = 0.0253 \text{ W/m·K}$$

$$\text{Pr} = (1004.8 \text{ J/kg·K})(1.8 \times 10^{-5} \text{ kg/m·s})/(0.0253 \text{ J/m·K·s}) = 0.7$$

$$\text{Re} = (0.8 \text{ m})(0.5 \text{ m/s})(1.226 \text{ kg/m}^3)/(1.8 \times 10^{-5} \text{ kg/m·s}) = 27244$$

$$\text{Nu} = 0.26(27,244)^{0.6}(0.7)^{0.3} = 107 = h_c D/k$$

Solving for h_c from definition of Nu (see Eq. 4.64), we obtain

$$h_c = 107(0.0253 \text{ J/m·K·s})/(0.80 \text{ m}) = 2.17 \text{ W/K}$$

The total thermal resistance comes from a combination of forced convection from the body and thermal conduction through the skin:

$$R_{tot} = R_{\text{conduction}} + R_{\text{convection}} = L_{\text{hair}}/k_{\text{hair}}A + 1/h_c A = \Delta T/Q$$

Substituting values from above and solving for k_{hair} gives

$$(0.005 \text{ m})/k_{hair}(5.12 \text{ m}^2) + 1/(2.17 \text{ W/K})(5.12 \text{ m}^2) = (32 - 15 \text{ K})/(144 \text{ N·m/s}),$$

$$k_{hair} = 0.035 \text{ W/m·K}$$

The values of m and n vary greatly depending on whether the flow is turbulent or laminar and depending on the specific geometry involved in heat transfer—for example, flow from smooth-walled tubes, or non-Newtonian fluid flow. We will not review all of these correlations here. We simply note that with proper descriptions of the surrounding fluid and the geometry of the biological system, values for heat transfer coefficients can be estimated and used in the analysis of heat transfer in biological materials. In particular, these correlations are useful in estimating physical properties such as thermal conductivity in biological systems where direct measurement may be difficult, as illustrated in Example Problem 4.2.

4.3 MASS TRANSPORT PROPERTIES OF MATERIALS*

We bring the analogies of momentum, heat, and mass transfer in materials to a close with a description of mass transport, often called *mass diffusion*, or simply *diffusion*. Many of the phenomena we have already studied depend upon diffusion. For example, most phase transformations (excluding "diffusionless" transformations such as martensitic) rely on the movement of atoms relative to one another. Corrosion cannot occur unless there is a movement of atoms or ions to a surface that allows the appropriate chemical reactions to take place. And, as we will see in Chapter 7, some processing methods such as chemical vapor deposition are wholly dependent upon the transport of chemical reactants to the location where reaction conditions allow products to form. This is why the term "diffusion-controlled" is often used to describe the kinetics of a process—the reaction is rapid relative to the time it takes to transport reactants to and/or products from the reaction site. For these reasons, it is important to study diffusion in materials, even though the driving forces are complex and the mass transport properties of a material are highly dependent upon the diffusing species of interest.

Unlike momentum and heat transport, the entity being transported is not the same in all diffusion situations. We must describe not only the material in which the transport is taking place, but also *what* is being transported (see Appendix 6, for example). The movement of hydrogen gas through metals, for example, is much different from the movement of liquid water through metals. As a result, the systems we will address will be highly specific and not at all general. Nonetheless, we can begin as in the previous sections of this chapter, with a description of mass transport properties from a fundamental viewpoint, and extend this to liquids and solids, with the recognition that as systems become more complex, theoretical predictions must give way to empiricisms and correlations.

4.3.1   Mass Transport Properties of Metals and Alloys

Recall from the beginning of this chapter that the transport of mass at steady state is governed by Fick's Law, given in one-dimension by Equation (4.4):

$$J_{Ay}^* = -D_{AB}\frac{dc_A}{dy} \tag{4.4}$$

where J_{Ay}^* is the molar flux of component A in the y direction, dc_A/dy is the concentration gradient of A in the y direction, and D_{AB} is the molecular diffusivity, or *diffusion coefficient*, of A in component B. As in the previous sections, we will concentrate on the material-related properties of transport, which in this section is the molecular diffusivity. We begin with a description of the molecular origins of mass diffusivity, and we then proceed to correlations and predictive relationships for the specific material classes.

4.3.1.1   The Molecular Origins of Mass Diffusivity.   In a manner directly analogous to the derivations of Eq. (4.6) for viscosity and Eq. (4.34) for thermal conductivity, the diffusion coefficient, or mass diffusivity, D, in units of m²/s, can be derived from the kinetic theory of gases for rigid-sphere molecules. By means of summary, we present all three expressions for transport coefficients here to further illustrate their similarities.

$$\mu = 2.6693 \times 10^{-5}\frac{\sqrt{MT}}{\sigma^2\Omega} \tag{4.6}$$

$$k = 8.32 \times 10^{-2}\frac{\sqrt{T/M}}{\sigma^2\Omega} \tag{4.34}$$

$$D = 2.6280 \times 10^{-7}\frac{\sqrt{T^3/M}}{P\sigma^2\Omega} \tag{4.65}$$

As before, M is the molecular weight of the rigid sphere, T is the absolute temperature, and σ and Ω are the same Lennard-Jones parameters used in Eq. (4.6) (note that technically, $\Omega = 1$ for the rigid-sphere model). The pressure, P, in Eq. (4.65) is in units of atmospheres. Equation (4.65) gives the so-called *self-diffusion coefficient*, or diffusivity for a sphere diffusing through a gas of the same spheres (i.e., D_{AA}). Fick's Law [cf. Eq. (4.4)] related the molar flux to the concentration gradient in a mixture of gases, A and B. In this case, the diffusion coefficient for the rigid-sphere model is

$$D_{AB} = 2.6280 \times 10^{-7}\frac{\sqrt{T^3(M_A + M_B)/2M_AM_B}}{P\sigma_{AB}^2\Omega_{AB}} \tag{4.66}$$

where M_A and M_B are the molecular weights of species A and B, respectively; $\sigma_{AB} = \frac{1}{2}(\sigma_A + \sigma_B)$, and Ω_{AB} is again approximately unity for rigid spheres. For rigid sphere gases, we note that the diffusivity of species B through A should be the same as the diffusivity for species A through B, that is, $D_{AB} = D_{BA}$.

4.3.1.2 *Mass Diffusivity in Liquid Metals and Alloys.* The hard-sphere model of gases works relatively well for self-diffusion in monatomic liquid metals. Several models based on hard-sphere theory exist for predicting the self-diffusivity in liquid metals. One such model utilizes the hard-sphere packing fraction, *PF*, to determine D (in cm^2/s):

$$D = \sigma C_{AW} \left(\frac{\pi RT}{M} \right)^{1/2} \frac{(1-PF)^3}{8PF(2-PF)} \tag{4.67}$$

where T and M are the absolute temperature and molecular weight, as before; R is the gas constant, σ is the collision diameter of the previous equations, and C_{AW}, the *Adler–Wainwright correction factor*, is a function of PF that can be determined from Figure 4.37. The packing fraction can be calculated using Eq. (4.8) as before. Recall from Table 4.1 that $PF = 0.45$ near the melting point. The values predicted by Eq. (4.67) for several liquid metals are compared with experimental values in Table 4.15.

Several empirical expressions also exist, including one that correlates self-diffusivity (in cm^2/s) with melting point, T_m:

$$D = 3.5 \times 10^{-6} \left(\frac{T_m}{M} \right)^{1/2} V_m^{1/3} \tag{4.68}$$

where M is the molecular weight and V_m is the atomic volume at the melting point. The self-diffusivity values predicted by Eq. (4.68) for several liquid metals at their melting point are compared in Table 4.15 with values from the semiempirical hard sphere model of Eq. (4.67) and experimental values.

Notice the similarity in form of Eq. (4.65) with Eq. (4.7), which used the packing fraction to estimate the viscosity. In this case, however, the dependences are the inverse of packing fraction and molecular weight to those in Eq. (4.7). In other words, whatever factors that change viscosity will have the opposite effect on diffusion. As viscosity goes up, diffusivity goes down; the inverse is true as well. This relationship is stated

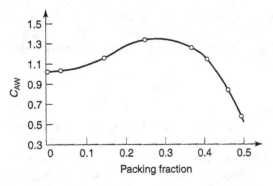

Figure 4.37   The Adler–Wainwright correction, C_{AW}, as a function of packing fraction, PF. Reprinted, by permission, from T. Iida and R. I. L. Guthrie, *The Physical Properties of Liquid Metals*, p. 209. Copyright © 1988 by Oxford University Press.

Table 4.15 Comparison of Experimental Melting Point Self-Diffusivities (in Units of 10^{-9} m^2/s) with Correlations for Selected Liquid Metals

Metal	Experimental	Eq. (4.67)	Eq. (4.68)
Li	5.76–6.80	7.01	6.72
Na	3.85–4.23	4.24	4.10
K	3.59–3.76	3.85	3.71
Cu	3.97	3.40	3.18
Zn	2.03–2.06	2.55	2.45
Ga	1.60–1.71	1.73	1.64
Rb	2.22–2.68	2.68	2.58
Ag	2.55–2.56	2.77	2.68
Cd	1.78	2.00	1.94
In	1.68–1.69	1.77	1.72
Sn	2.05–2.31	1.96	1.86
Hg	0.93–1.07	1.07	0.93
Pb	1.68–2.19	1.67	1.60
Al	—	—	4.87
Ni	—	—	3.90
Mg	—	—	5.63
Sb	—	—	2.66
Cs	—	—	2.31
Fe	—	—	4.16

Source: T. Iida, and R. I. L. Guthrie, *The Physical Properties of Liquid Metals*. Copyright © 1988 by Oxford University Press.

explicitly in the *Stokes–Einstein equation*

$$D = \frac{k_B T}{4\pi \mu r} \tag{4.69}$$

where k_B is Boltzmann's constant, T is absolute temperature, μ is the viscosity of the "solvent" (in this case, the molten metal), and r is the radius of the diffusing hard sphere (in this case, a molten metal atom). Equation (4.69) is the Stokes–Einstein equation for self-diffusion. Another form of the Stokes–Einstein equation can be developed for dilute solutions of large spheres in small, spherical molecules, which leads to a binary diffusion coefficient:

$$D_{AB} = \frac{k_B T}{6\pi \mu r} \tag{4.70}$$

Equation (4.70) is a starting point in the determination of diffusivities in liquid metal alloys, but in most real systems, experimental values are difficult to obtain to confirm theoretical expressions, and pair potentials and molecular interactions that exist in liquid alloys are not sufficiently quantified. Even semiempirical approaches do not fare well when applied to liquid alloy systems. There have been some attempts to correlate diffusivities with thermodynamic quantities such as partial molar enthalpy and free energy of solution, but their application has been limited to only a few systems.

4.3.1.3 Mass Diffusivity in Solids. Fick's Law [Eq. (4.4)] implies that a concentration gradient, dc/dy, is necessary for diffusion and molar flow to occur. In solid metals such as pure nickel, no concentration gradient exists, yet radioactive tracer experiments tell us that diffusion still takes place. As illustrated in Figure 4.38, a thin layer of radioactive nickel placed at one end of a pure nickel bar will slowly diffuse through the pure metal, a process that can be monitored easily by radiation detection. This tells us that even though no concentration gradient exists (the chemical potentials of radioisotopes are identical, hence there is no difference in concentration across the interface), there is still atomic movement. We know that above absolute zero, atomic vibrations occur in solids. It is these small amplitude vibrations that allow some atoms to overcome an activation energy barrier and move from one position to another in the lattice. This type of random walk process allows atomic interchange to take place—neighboring atoms simply switch places due to thermal vibrations.

Experiments such as the one illustrated in Figure 4.38 not only give us self-diffusion coefficients for certain substances, but as the temperature of the experiment is varied, they give us the temperature dependence of the process and a measurement of the activation energy barrier to diffusion. Diffusion in solid systems, then, can be modeled as an activated process; that is, an Arrhenius-type relationship can be written in which an activation energy, E_a, and temperature dependence are incorporated, along with a preexponential factor, D_0, sometimes called the *frequency factor*:

$$D = D_0 \exp(-E_a/RT) \tag{4.71}$$

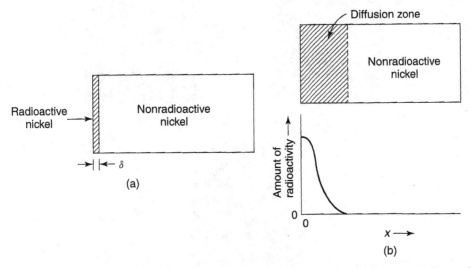

Figure 4.38 Schematic illustration of a self-diffusion experiment in which (a) a thin layer of radioactive nickel is deposited on one surface of a nonradioactive nickel specimen. After heating and time (b), the radioactive nickel has diffused into the sample, as monitored with a radioactive detector. From K. M. Ralls, T. H. Courtney, and J. Wulff, *Introduction to Materials Science and Engineering*. Copyright © 1976 by John Wiley & Sons, Inc. This material is used by permission John Wiley & Sons, Inc.

Since this is an empirical expression, it can be applied to almost any diffusion process, but it does not tell us much about the diffusion mechanism. In general, three types of processes have been identified for diffusion in solids: an *interchange mechanism* (sometimes called an *exchange mechanism*), a *vacancy mechanism*, and an *interstitial mechanism*. These mechanisms are illustrated in Figure 4.39. A fourth mechanism, *ring interchange*, is possible, but rarely observed, and will not be considered here. In most metals, the vacancy mechanism (Figure 4.39a) dominates, because it requires a relatively small amount of energy, and there are a sufficient number of vacancies in most metals. The interstitial mechanism (Figure 4.39b) is possible in metals since a self-interstitial is a relatively mobile species requiring shorter distances of atomic movement, but the equilibrium number of self-interstitials is relatively small at almost all temperatures. Finally, the direct interchange mechanism (Figure 4.39c) is possible in metals, but requires an extremely high activation energy.

Activation energies for self-diffusion are expected to be in the range of 232–261 kJ/mol for the vacancy mechanism and 492–615 kJ/mol for the interstitial mechanism. Observed values are typically in the range of 164–203 kJ/mol. Some representative values for self-diffusion activation energies, as well as frequency factor in Eq. (4.71), D_0, are given in Table 4.16. Bear in mind that D_0 contains the temperature-independent portions of diffusion, such as entropic effects, whereas E_a reflects all activation energy processes, including both atomic mobility and vacancy formation in the case of diffusion via the vacancy mechanism. Notice in Table 4.16 that there is a general trend of increasing activation energy for diffusion with increasing

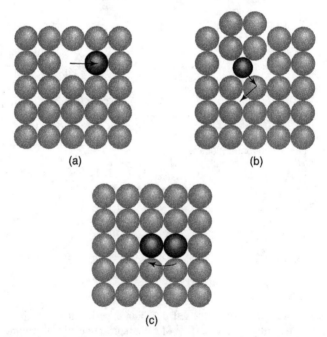

(a)   (b)

(c)

Figure 4.39 Illustration of (a) vacancy, (b) interstitial, and (c) interchange (exchange) mechanisms in atomic diffusion. From K. M. Ralls, T. H. Courtney, and J. Wulff, *Introduction to Materials Science and Engineering*. Copyright © 1976 by John Wiley & Sons, Inc. This material is used by permission John Wiley & Sons, Inc.

Table 4.16   Frequency Factors, Activation Energies and Melting Points for Self-Diffusion in Selected Metals

Species	D_0 (cm^2/s)	E_a (KJ/mol)	T_m (°C)
Pb	6.6	117	327
Zn ($\parallel$ to c axis)	0.046	85	419
Zn ($\perp$ to c axis)	91	130	419
Ag	0.89	192	962
Au	0.16	222	1064
Cu	11	239	1085
Ni	—	293	1455
Co	0.31	280	1494
α-Fe	2300	306	1538
γ-Fe	5.8	310	—
Pt	0.048	233	1772

Source: B. Chalmers, *Physical Metallurgy*. Copyright © 1959 by John Wiley & Sons, Inc.

melting point. Recall from Chapter 1 that intermolecular forces have a profound influence on a number of physical properties, such as melting point. The more strongly atoms are attracted to one another, the more energy it requires to pull them apart. This is reflected in both the melting point and the activation energy for diffusion, in which atoms must overcome attractive forces in order to move away from one another.

In alloys, attractive forces still play an important role, but the situation is slightly different insofar as solute atoms are sufficiently small relative to the host atom to allow the interstitial mechanism to dominate (see Figure 4.40). As a result, hydrogen, carbon, nitrogen, and oxygen diffuse interstitially in most metals with relatively low activation energy barriers. The diffusion of carbon in iron has been particularly well-studied in conjunction with the phase transformations in the C–Fe system (cf. Figure 2.8). An example of the excellent fit provided by Equation (4.71) to experimental data in the C–Fe system is shown in Figure 4.41. Notice how the activation energy for this interstitial diffusion process is about 85 kJ/mol, which is much smaller than for vacancy-driven diffusion, due to the small size of the carbon atoms relative to iron. Not only do larger atoms, such as other metals, have difficulty diffusing interstitially, but also atomic interactions between dissimilar atoms greatly affect the activation energy for diffusion.

In addition to concentration and random walk diffusion, which are termed *ordinary diffusion*, there are other types of diffusion that can be important under certain circumstances. Atomic mobility can be brought about by thermal and pressure gradients. There are also such structurally driven diffusion processes as *dislocation diffusion*, *surface diffusion*, and *grain-boundary diffusion*. The latter is particularly important in polycrystalline materials where grain boundaries predominate. The grain boundaries provide a path of lower activation energy for movement than through the bulk solid. Moreover, grain boundary diffusion is less temperature-variant than bulk, or volume, diffusion in these materials, as illustrated for single crystalline and polycrystalline silver in Figure 4.42.

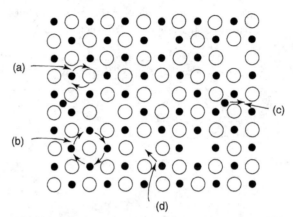

Figure 4.40 Illustration of diffusion mechanisms in alloys and ionic solids: (a) interchange (exchange); (b) ring rotation (rare); (c) interstitial migration; and (d) vacancy migration. From W. D. Kingery, H. K. Bowen, and D. R. Uhlmann, *Introduction to Ceramics.* Copyright © 1976 by John Wiley & Sons, Inc. This material is used by permission of John Wiley & Sons, Inc.

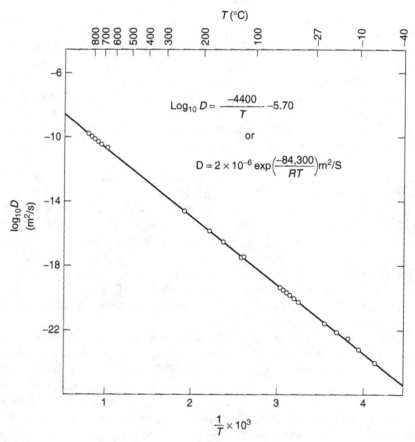

$$\text{Log}_{10}\, D = \frac{-4400}{T} - 5.70$$

or

$$D = 2 \times 10^{-6}\, \exp\!\left(\frac{-84{,}300}{RT}\right) \text{m}^2/\text{S}$$

Figure 4.41 Variation with temperature of the diffusivity for carbon in BCC-Fe ($\alpha - $Fe). Activation energy is in units of J/mol. Reprinted, by permission, from D. R. Gaskell, *An Introduction to Transport Phenomena in Materials Engineering*, p. 519. Copyright © 1992 by Macmillan Publishing Co.

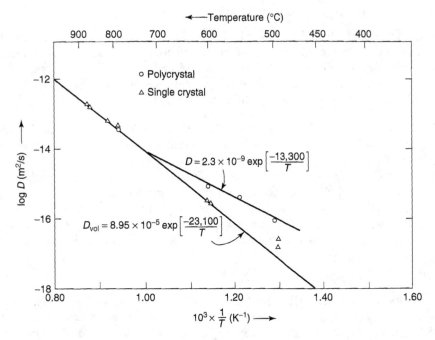

Figure 4.42   Self-diffusion coefficient in single and polycrystalline silver, illustrating the effect of grain boundary diffusion, especially at lower temperatures. From K. M. Ralls, T. H. Courtney, and J. Wulff, *Introduction to Materials Science and Engineering.* Copyright © 1976 by John Wiley & Sons, Inc. This material is used by permission John Wiley & Sons, Inc.

Cooperative Learning Exercise 4.10

Consider the diffusivity data presented in Figure 4.42 for the self-diffusion of silver in polycrystalline and single-crystalline form. Use the correlations provided to calculate the following quantities.

Person 1: Calculate the self-diffusivity in polycrystalline silver (grain boundary diffusion), D_{GB}, at 500°C in m^2/s. What is the activation energy for this process in kJ/mol?

Person 2: Calculate the self-diffusivity in single-crystalline silver (volume diffusion), D_{vol}, at 500°C in m^2/s. What is the activation energy for this process in kJ/mol?

Combine your answer to get the ratio of grain boundary diffusion to volume diffusion for silver at 500°C. How would this change at higher temperatures? How would this change as the grain size decreased further?

Answers: $D_{GB}(500°C) = 2.3 \times 10^{-9} \exp(-13,300/773) = 7.8 \times 10^{-17} \ m^2/s$
$E_a = (13,300 \ K)(8.314 \ J/mol \cdot K) = 111 \ kJ/mol$
$D_{vol}(500°C) = 8.95 \times 10^{-5} \exp(-23,100/773) = 9.4 \times 10^{-18} \ m^2/s$ $E_a = (23,100 \ K)(8.314 \ J/mol \cdot K) = 192 \ kJ/mol$
$D_{GB}/D_{vol} = 7.8/0.94 = 8.3$. We would expect this ratio to decrease at higher temperature, eventually approaching 1.0 as volume diffusivity increases due to its stronger temperature dependence. This ratio should get larger with decreasing grain size as there are more grain boundaries per unit volume.

4.3.2 Mass Transport Properties of Ceramics and Glasses

As with thermal conductivity, we see in this section that disorder can greatly affect the mechanism of diffusion and the magnitude of diffusivities, so that crystalline ceramics and oxide glasses will be treated separately. Finally, we will briefly describe an important topic relevant to all material classes, but especially appropriate for ceramics such as catalyst supports—namely, diffusion in porous solids.

4.3.2.1 *Diffusion in Crystalline Oxides.* We wish to investigate the individual terms in Eq. (4.71) (i.e., D_0 and E_a), in more detail as they apply to crystalline oxide ceramics. To do so, we look first at a model ionic system, potassium chloride, KCl, realizing that the results will be general to all ceramics that have close-packed anion lattices and in which diffusion occurs by a vacancy mechanism.

Diffusion of K^+ ions in KCl occurs by interchange of the potassium ions with cation vacancies (see Figure 4.40d). It makes sense, then, that the diffusivity of potassium ions in KCl, $D_{K,Cl}$ is both a function of the potassium ion mobility, D_i, and the cation vacancy concentration, $[V'_k]$:

$$D_{K,Cl} = D_i[V'_K] \tag{4.72}$$

Let us first concentrate on the potassium ion mobility, D_i, in Eq. (4.72). As with any activated process, the potassium ions must overcome an activation energy barrier, which we will notate as a free energy $\Delta G^\dagger$, to move to the cation vacancy site, as illustrated in Figure 4.43. Recall from Section 3.0.2 that the fraction of ions with the required energy to overcome the barrier is given by the Boltzmann distribution and depends on $\exp(-\Delta G^\dagger/k_B T)$. It can be shown, then, that

$$D_i = \gamma \lambda^2 v \exp(-\Delta G^\dagger/k_B T) \tag{4.73}$$

where γ is a nearest-neighbor factor on the order of unity, λ is the jump distance, and v is the frequency factor, which for solids has a value of about 10^{13} s^{-1}.

Turning now to the vacancy concentration in Eq. (4.72), recall from Chapter 1, and Eq. (1.48) in particular, that the concentration of vacancies is related to the free energy of formation of the vacancy, so that for a cation–anion vacancy pair, or Schottky defect, the concentration of vacancies, $[V'_k]$ is given by

$$n_V/N = [V'_k] = \exp(-\Delta G_s/2k_B T) \tag{4.74}$$

where n_V is the number of defects (vacancies), N is the total number of lattice sights, T is the absolute temperature, k_B is Boltzmann's constant, and ΔG_s is the free energy change necessary to form a Schottky vacancy pair. We know that the free energy can be broken down into entropic and enthalpic components [Eq. (2.12)], such that

$$[V'_k] = \exp(-\Delta G_s/2k_B T) = \exp\left(\frac{-\Delta S_s T + \Delta H_s}{2k_B T}\right) \tag{4.75}$$

If we perform a similar breakdown of $\Delta G^\dagger$ in Eq. (4.73) and substitute Eqs. (4.73) and (4.75) into (4.72), we obtain the following relationship for the diffusivity of potassium ions in KCl:

$$D_{K,Cl} = \gamma \lambda^2 v \exp\left(\frac{\Delta S^\dagger + \Delta S_s/2}{k_B}\right) \exp\left(\frac{-\Delta H^\dagger - \Delta H_s/2}{k_B T}\right) \tag{4.76}$$

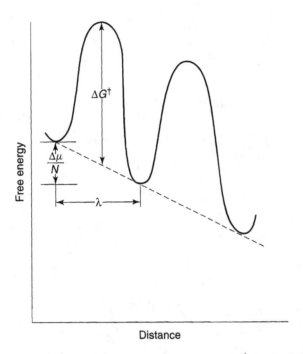

Figure 4.43 Diffusion in a potential gradient $\Delta\mu$, where $\Delta G^\dagger$ is the height of the activation energy barrier and λ is the jump distance. From W. D. Kingery, H. K. Bowen, and D. R. Uhlmann, *Introduction to Ceramics.* Copyright © 1976 by John Wiley & Sons, Inc. This material is used by permission of John Wiley & Sons, Inc.

The important parts of Eq. (4.76) are the exponential terms. The first exponential, which contains the entropies associated with potassium ion movement and vacancy formation, respectively, form the temperature-independent contributions to D_0, as discussed in the previous section. The second exponential, which contains the enthalpies of the two processes and the temperature dependence, form the activation energy, E_a, and temperature dependence of Eq. (4.71).

To complete our discussion on diffusion in KCl, then, Eq. (4.76) tells us that a plot of $D_{K,Cl}$ versus $1/T$ should result in a straight line with slope of $(\Delta H^\dagger/k_B + \Delta H_s/2k_B)$. As shown in Figure 4.44, this is the case only at high temperatures, where the *intrinsic properties* of the material dominate. In the lower-temperature region, impurities within the crystal fix the vacancy concentration. This is called the *extrinsic region*, where the diffusion coefficient is a function of divalent cation impurity concentration, such as calcium. The insert in Figure 4.44 shows that the cation vacancy concentration becomes fixed at low temperatures, and this is reflected in the fact that the slope of the diffusivity curve becomes $\Delta H^\dagger/k_B$ only. The change in slope occurs at the point where the intrinsic defect concentration (attributable only to temperature) is comparable with the extrinsic defect concentration (attributable to impurities). When the Schottky formation enthalpy is in the range of 628 kJ/mol, as is typical for BeO, MgO, CaO, and Al_2O_3, the crystal must have an impurity concentration smaller than 10^{-5} before intrinsic diffusion can be observed at 2000°C. It is thus unlikely that intrinsic diffusion will be observed in these oxides because impurity levels of only parts per million are sufficient to control the vacancy concentrations.

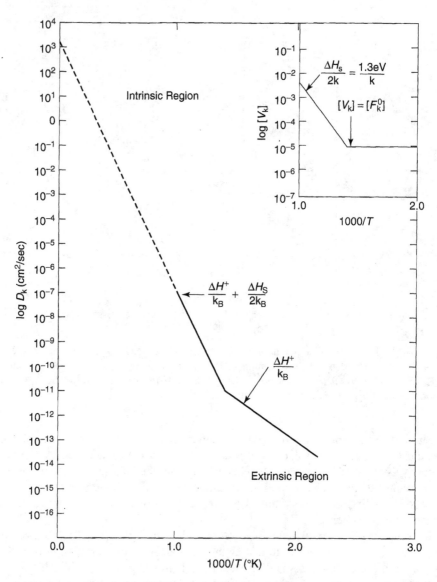

Figure 4.44 Diffusivity as a function of temperature for KCl with 10^{-5} atom fraction divalent cation impurities. Insert plot shows the variation of vacancies with temperature. Adapted from W. D. Kingery, H. K. Bowen, and D. R. Uhlmann, *Introduction to Ceramics*. Copyright © 1976 by John Wiley & Sons, Inc. This material is used by permission of John Wiley & Sons, Inc.

The experimental diffusion coefficients for most oxide ceramics do not generally show this intrinsic/extrinsic behavior illustrated in KCl. This is probably due to the fact that diffusion data are normally collected over a relatively small temperature range, as illustrated for a number of oxides in Figure 4.45. However, some diffusion in oxides is clearly extrinsic, such as for the diffusivity of oxygen in oxides containing the fluorite structure, such as UO_2, ThO_2, and ZrO_2. In these compounds, the addition of divalent or trivalent cation oxides, such as CaO, has been shown to fix the concentration of oxygen

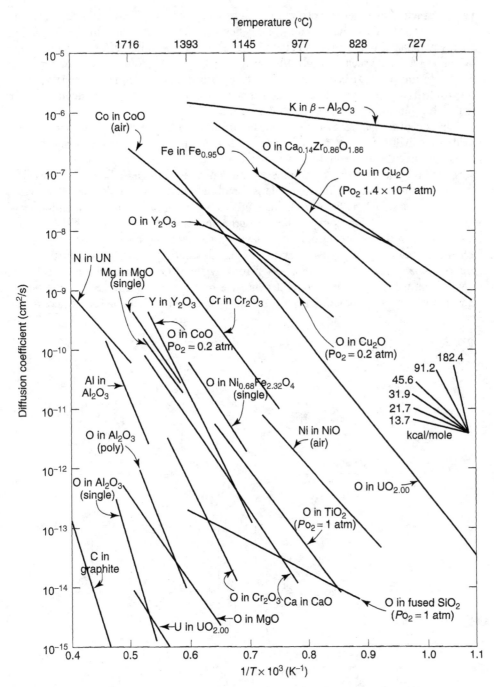

Figure 4.45 Diffusivities in some common ceramics as a function of temperature. The activation energy can be estimated from the slope of the insert. From W. D. Kingery, H. K. Bowen, and D. R. Uhlmann, *Introduction to Ceramics*. Copyright © 1976 by John Wiley & Sons, Inc. This material is used by permission of John Wiley & Sons, Inc.

vacancies and make their concentration independent of temperature. Consequently, the oxygen ion diffusivity has a temperature dependence related only to the activation energy associated with oxygen ion mobility (121 kJ/mol). Cation diffusion coefficients, on the other hand, such as Mg diffusion in MgO, are more difficult to categorize as wholly intrinsic or extrinsic. The relatively high activation energies in these systems suggest that high- temperature measurements may be intrinsic and independent of minor impurities, but it is difficult to generalize.

Intrinsic diffusion behavior is dominant in nonstoichiometric oxides such as zinc oxide or cobaltous oxide in equilibrium with oxidizing or reducing atmospheres. In these systems, the interstitial and vacancy concentrations are a function of the partial pressure in the surrounding environment. The dependence on partial pressure is related to whether the nonstoichiometric oxide is metal-deficient or oxide-deficient. In metal-deficient oxides such as FeO, NiO, and MnO, the concentration of vacant cation sites can be very large. For example, $Fe_{1-x}O$ contains 5–15% vacant iron sites. In this case, the solubility of oxygen in the metal oxide becomes an issue, and the metal ion diffusivity, D_M, can be a function of oxygen partial pressure, as illustrated in Figure 4.46a. The free energy of solution for oxygen, ΔG_O, must then be considered, which in turn leads to a functional dependence of D_M and vacancy concentration on the enthalpy of solution for oxygen, ΔH_O, as shown in Figure 4.46b, whereas the

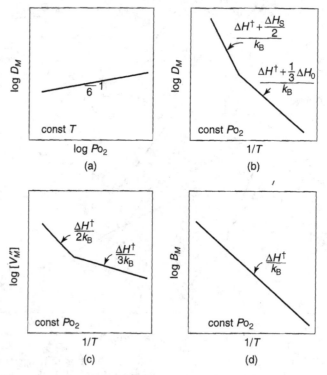

Figure 4.46 Schematic representation of the variation of metal ion diffusivity as a function of (a) oxygen partial pressure and (b) temperature, and (c) the metal ion vacancy concentration and (d) mobility with temperature. From W. D. Kingery, H. K. Bowen, and D. R. Uhlmann, *Introduction to Ceramics.* Copyright © 1976 by John Wiley & Sons, Inc. This material is used by permission of John Wiley & Sons, Inc.

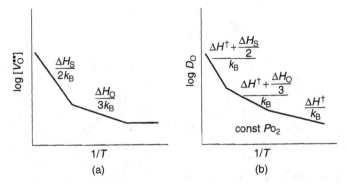

Figure 4.47 Schematic representation of temperature dependence of (a) oxygen vacancy and (b) oxygen diffusivity in oxygen-deficient oxide ceramics. From W. D. Kingery, H. K. Bowen, and D. R. Uhlmann, *Introduction to Ceramics*. Copyright © 1976 by John Wiley & Sons, Inc. This material is used by permission of John Wiley & Sons, Inc.

metal ion mobility, B_M, remains a function of the activation energy barrier only, as illustrated in Figure 4.46d.

A similar situation arises in oxygen-deficient oxides, such as CdO or Nb_2O_5, except that now three temperature regimes are possible, as shown in Figure 4.47. In the low-temperature regime, oxygen vacancy concentration is controlled by impurities, and extrinsic behavior is observed. At intermediate temperatures, the oxygen vacancy concentration changes due to a change in oxygen solubility with temperature. This is nonstoichiometric behavior. At high temperatures, thermal vacancies become dominant, and intrinsic behavior dominates.

Chemical effects also occur in crystalline oxides; that is, impurity atoms diffuse at varying rates through oxides. In all cases, cation diffusion is much faster than oxygen diffusion, but similar cations, such as Ca and Mg, can behave very differently in different oxides. This is due to differences in chemical potential and activity, and mobility differences.

As in metals, dislocation, surface, and grain boundary diffusion can be important. Grain boundary diffusion, in particular, can lead to diffusion coefficients that are 10^3 to 10^6 times greater than the bulk diffusion coefficient. However, the effect is not the same for both cations and anions. In some systems, only the ion that is expected to be present in excess at the grain boundary will exhibit enhanced grain boundary diffusion.

4.3.2.2 Diffusion in Glasses — Solubility and Permeability. The principles developed in previous sections apply to noncrystalline solids as well, but the terminology used is often different, owing to the applications for which diffusion in oxide glasses are important. There are two such situations that will be described here: (a) diffusion of gases through glass and (b) diffusion of cations through glass. The former is important in many applications, such as high-vacuum use, where glass is a common material of construction. The latter is important in glass containers, where cations in the glass can leach out over time depending on the solution it contains. Yet another application, the diffusion of gases through molten glasses, is important from an industrial standpoint in the fining of glasses, but is beyond the scope of this text.

In both gas and cation diffusion, the movement of atoms through solid glasses is often described in terms of the *permeability*, P_M, which is directly proportional to the diffusivity, D:

$$P_M = SD \qquad (4.77)$$

where S is the *solubility*. In this case, the permeability is defined as the volume of gas at standard temperature and pressure (STP) passing per second through a unit area of glass of thickness 1 cm, with a 1 atm pressure difference across the glass. The solubility is defined as the volume of gas at STP dissolved in a unit volume of glass per atmosphere of external gas pressure. The diffusivity is defined as before. It should be clear from Eq. (4.77) that determination of two of the parameters fixes the third. Solubility increases with temperature according to an Arrhenius-type relationship, where the heat of solution serves as the activation energy. Since diffusivity also follows an Arrhenius expression, as described in the previous sections, one would expect that the permeability would also show a similar temperature dependence, and this is indeed the case. An example of the logarithmic permeability dependence on inverse temperature is shown in Figure 4.48. The differences in permeabilities among the various glasses can be explained by considering the network-modifying cations as "hole blockers" in the glass network. Thus, the permeability would be expected to increase with an increase in the concentration of network-forming constituents such as SiO_2, B_2O_3, and P_2O_5 (cf. Section 1.2.4). Such an increase has been shown experimentally. As for the diffusivity of gases in glasses, the activation energy increases with increasing size of the gas molecule, as illustrated in Figure 4.49. For these gases, the frequency factor, D_0, is in the range of 10^{-4} to 10^{-3} cm^2/s.

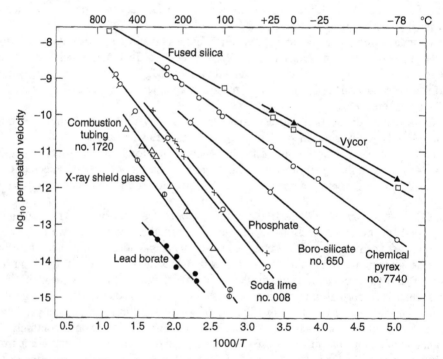

Figure 4.48 Permeability of helium through various glasses. From W. D. Kingery, H. K. Bowen, and D. R. Uhlmann, *Introduction to Ceramics*. Copyright © 1976 by John Wiley & Sons, Inc. This material is used by permission of John Wiley & Sons, Inc.

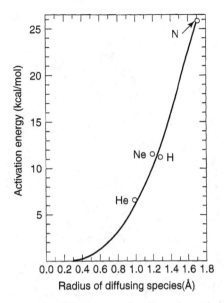

Figure 4.49   Activation energy for diffusion as a function of atomic radius of diffusing species in silica glass. From W. D. Kingery, H. K. Bowen, and D. R. Uhlmann, *Introduction to Ceramics*. Copyright © 1976 by John Wiley & Sons, Inc. This material is used by permission of John Wiley & Sons, Inc.

In the case of oxygen transport, diffusion may take place either as molecular or network diffusion, although molecular diffusion apparently takes place some seven orders of magnitude faster than network diffusion. The movement of water through glasses can also be important, particularly in the dehydroxylation of hydrated glasses such as vitreous silica. The hydroxyl content has been shown to greatly affect the properties of glasses, including reductions in viscosity and glass transition temperatures, and increases in phase separation and crystallization rates with increasing hydroxyl content. Some data from a typical dehydroxylation experiment for vitreous silica are shown in Figure 4.50. Here, t is time and L is the sample thickness. The value of $t^{1/2}L$ at 50% hydroxyl removal is shown in the plot. This value is significant, because the diffusivity can be estimated according to the following relationship:

$$D = \frac{0.049}{(t/L^2)_{50\%}}$$ (4.78)

where $(t/L^2)_{50\%}$ is the value of t/L^2 at removal of exactly 50% of the hydroxyl groups from the glass.

For the movement of cations in glasses, as the amount of modifying cations in silicate glasses is increased, the activation energy for their diffusive transport decreases, resulting in an increase in the diffusivity. The increase in cation concentration presumably leads to a break up of the network and a decrease in the average interionic separation. Divalent cations diffuse much more slowly at a given temperature than monovalent cations, and their activation energies are much larger. Diffusivities of modifying cations in rapidly cooled glasses are generally higher than those in well-annealed glasses. These differences, often as large as an order of magnitude, reflect differences in

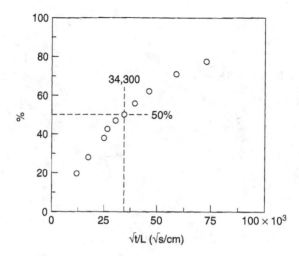

Figure 4.50 Experimental dehydroxylation curve for vitreous silica at constant temperature. Reprinted, by permission, from J. E. Shelby, in *Experimental Techniques of Glass Science*, C. J. Simmons and O. H. El-Bayoumi ed. p. 374. Copyright © 1993, The American Ceramic Society.

specific volumes of the glasses, with the larger-volume, more open structure possessing the larger diffusivity.

The process of cation movement in glasses is important for a number of processes, such as ion exchange and dissolution. Many univalent cations in glasses can be substituted for one another, such as the complete substitution of Na ion by Ag ions, even at low temperatures. Ions such as Li^+ and K^+ can also be substituted for Na^+, although stresses can build up in the glass due to cation radius differences. This procedure can actually be utilized to strengthen glasses, as will be discussed in Chapter 5. In the dissolution of glasses, ions in solution attack the glass network through leaching of network modifiers, such as Na^+. This is particularly prevalent in acidic aqueous solutions, where the countermigration of H^+ ions into the network accelerates the leaching process, although an exact 1:1 exchange of Na^+:H^+ does not necessarily take place. Water is involved in the process as well. A theoretical description of this process, or even one in terms of diffusion theory, is difficult, since the composition of the diffusing matrix is changing with time. A simple formula for determining the amount of leached-out alkalis, $[Q]$, is to assume a power-law dependence with time, $[Q] = kt^\alpha$, where α typically has values between 0.45 and 0.9. A square-root dependence on leaching time (i.e., $\alpha = 0.5$) is predicted by theory, but is not necessarily experimentally observed. The properties and development of alkali- and hydroxyl-resistant glasses are important areas of investigation, and the interested reader is referred to the many fine texts on these subjects listed at the end of this chapter.

4.3.2.3 *Diffusion in Porous Solids.*

To this point, we have primarily concentrated on the atomic-level movement of atoms involved in diffusion. There are macroscopic effects to consider as well. For example, liquids and gases moving through porous solids can often travel faster through the pores, which are also filled with either gas or liquid, than through the solid particles. If the pores are small relative to the mean free path of the diffusing species (Figure 4.51a), then the diffusion rate can be influenced strongly

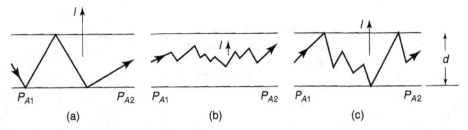

Figure 4.51 Diffusion in pores: (a) Knudsen diffusion, (b) Fickian diffusion, and (c) transitional diffusion.

by collisions with the pore walls in a process known as *Knudsen diffusion*. Knudsen diffusion is rare for diffusing liquids since their mean free paths are so small. It can be significant, however, for gases, especially at low pressures or high temperatures. Knudsen diffusion is characterized by the Knudsen number, $Kn = l/d$, where l is the mean free path of the diffusing species, and d is the pore diameter in the same units as l. For $Kn < 0.01$, Fickian diffusion occurs (Figure 4.51b). For $Kn > 100$, Knudsen diffusion dominates. In this case, a Knudsen mass diffusivity, D_{KA} (in m^2/s), is defined as follows:

$$D_{KA} = 48.5d\sqrt{T/M_A} \tag{4.79}$$

where d is again the pore diameter (in meters), T is the absolute temperature in degrees kelvin, and M_A is the molecular weight of the diffusing species, A, in g/mol. It is interesting to note that for a multicomponent gas, D_{KA} is completely independent of the diffusivity in the other gaseous species, since A collides with the walls of the pore and not with the other gases. Each gas will have its own Knudsen diffusion coefficient in this case. When $Kn \sim 1.0$, neither Fickian nor Knudsen diffusion dominates (Figure 4.51c), and *transition diffusion* occurs. The diffusivity in this case can be estimated in a number of ways, including a vectorial combination of the two diffusion coefficients.

In the specific instance of A diffusing through a catalytic pore and reacting at the end of the pore to form gaseous component B, equimolar counterdiffusion can be assumed, and an *effective transition region diffusivity*, D'_{NA}, is independent of concentration and can be calculated from the Knudsen and binary diffusion coefficients:

$$D'_{NA} = \frac{1}{1/D_{AB} + 1/D_{KA}} \tag{4.80}$$

This simplified diffusivity is sometimes used for diffusion in porous catalysts even when equimolar counterdiffusion is not occurring. This greatly simplifies the equations. When no reactions are occurring, the diffusivity is a function of concentration (in terms of mole fraction, x_A):

$$D_{NA} = \frac{1}{(1 - \alpha x_A)/D_{AB} + 1/D_{KA}} \tag{4.81}$$

Cooperative Learning Exercise 4.11

A gas mixture at a total pressure of 0.10 atm absolute and 298 K is composed of N_2 (component A) and He (component B). The mixture is diffusing through an open capillary having a diameter of 5×10^{-7} m. The mole fraction of N_2 at one end is $x_{A1} = 0.8$ and at the other end is $x_{A2} = 0.2$. The mean free paths for nitrogen and helium at 298 K are 654 and 1936 Å, respectively (1 Å $= 10^{-10}$ m); and their collision diameters, σ, are 3.75 and 2.18 Å, respectively.

Person 1: Calculate the Knudsen number for N_2. Is the diffusion molecular, Knudsen, or transitional? Estimate the Knudsen diffusivity for nitrogen, D_{KA}, under these conditions.

Person 2: Estimate the hard-sphere binary diffusivity for nitrogen in helium, D_{AB}, under these conditions. You may assume that $\Omega_{AB} = 1.0$ in Eq. (4.66).

Combine your information to estimate the transitional diffusivity for nitrogen, D'_{NA}. You can use an average concentration for nitrogen in your calculation. How does it compare to the two diffusivities you calculated individually? How does it compare to the transitional diffusivity if there were equimolar counterdiffusion—that is reaction at the end of the pore?

Answers: Kn $= 0.13$ for N_2 $D'_{NA} = 6.3 \times 10^{-5}$ m²/s $D'_{NA} = 6.95 \times 10^{-5}$ m²/s: a difference of about 9%
$D_{KA} = 7.9 \times 10^{-5}$ m²/s $D_{AB} = 5.8 \times 10^{-4}$ m²/s

---

where α is the flux ratio factor, which for flow in an open system is related to the molecular weights of the two components:

$$\alpha = 1 - \sqrt{\frac{M_A}{M_B}} \tag{4.82}$$

When chemical reaction occurs, α must be defined in terms of molar fluxes. For the case of equimolar counterdiffusion, $\alpha = 1$, and Eq. (4.81) reduces to (4.80).

In real porous solids, the pores are not straight, and the pore radius can vary. Two parameters are used to describe the diffusion path through real porous solids: the *void fraction*, ε, defined as the ratio of pore area to total cross-sectional area, and the *tortuosity*, τ, which corrects for the fact that pores are not straight. The resulting effective diffusivity is then

$$D_{A,eff} = D_{AB}(\varepsilon/\tau) \tag{4.83}$$

Tortuosity values range from about 1.5 to over 10. A reasonable range of values for many commercial porous solids is about 2–6.

4.3.3 Mass Transport Properties of Polymers

4.3.3.1 Diffusion in Polymer Solutions. A dilute polymer in a low-molecular weight solvent can be modeled as bead-spring chain (see Figure 4.52a). Each chain is a linear arrangement of N beads and $N - 1$ Hookean springs (see Chapter 5). The beads are characterized by a *friction coefficient*, represented by the lowercase Greek letter zeta, ζ, which describes the resistance to bead motion through the solvent. The model also allows for hydrodynamic interactions between the beads and the solvent

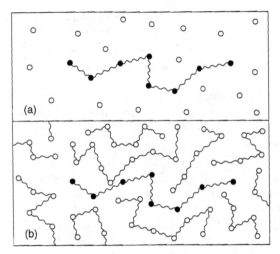

Figure 4.52 Single-molecule bead spring models for (a) dilute polymer solution and (b) polymer melt. From R. B. Bird, W. E. Stewart, and E. N. Lightfoot, *Transport Phenomena*, 2nd ed. Copyright © 2002 by John Wiley & Sons, Inc. This material is used by permission of John Wiley & Sons, Inc.

molecules. The general result, which holds true for most solutions, relates the diffusivity to the friction coefficient:

$$D_{AB} = \frac{k_B T}{\zeta} \tag{4.84}$$

The form of Eq. (4.84) should look familiar. For large, spherical particles in a low-molecular-weight solvent, $\zeta = 6\pi \mu r$, where μ is the viscosity of the pure solvent and r is the large particle radius, and Eq. (4.84) becomes Equation (4.70) a form of the Stokes-Einstein equation, which gives the binary diffusion coefficient, D_{AB},

$$D_{AB} = \frac{k_B T}{6\pi \mu r} \tag{4.70}$$

Equation (4.69) for self-diffusion can be arrived at in a similar way through proper derivation of the friction coefficient. Equation (4.70) does not take into account hydrodynamic interactions. It is also necessary to come up with an equivalent radius, r, for a polymer chain, which can be difficult, especially when the conformation is such that the chain is extended, and does not form a sphere at all. Nonetheless, a *radius of gyration*, r_g, is often used to characterize polymer chains in solution, and the resulting friction coefficient is $\zeta = 4\pi \mu r_g$.

More rigorous treatments include hydrodynamic interaction effects and relate the diffusivity directly to the molecular weight of the polymer chain. The general result is that the diffusivity of polymer chains in solution is related to the molecular weight of the polymer, M:

$$D_{AB} \propto M^{-\nu} \tag{4.85}$$

The value of ν can vary. Theory predicts that the diffusivity of polymer chains in solution should be proportional to the inverse square root of the number of beads,

$N^{-1/2}$, and since N is proportional to M, we have $D_{AB} \propto M^{-1/2}$. In "good solvents," or solvents in which interactions between the polymer chain and solvent are favorable, and the chain takes on an extended structure, we have $v = 3/5$. If hydrodynamic interactions are ignored, we expect $D_{AB} \propto 1/M$, and $v = 1.0$.

4.3.3.2 *Diffusion in Polymer Melts.* We expect the diffusion situation to be more complex in polymer melts, since chains become entangled and it is difficult for them to move past one another. We further expect that chain entanglements should increase with the molecular weight of the polymer chain. A model has been developed that predicts these molecular weight effects in polymer melts. The *reptation model* assumes that an isolated polymer chain is contained in a hypothetical tube, as shown schematically in Figure 4.53. Under the constraints imposed by the tube, two types of chain motion are possible. First, there is a conformational change of the chain taking place within the tube, along with a "reptation" motion that causes the chain to translate through the tube, much like a snake wriggling through the grass. Eventually, this motion will carry the chain completely out of the hypothetical tube along what is termed the "*primitive path.*" The time-dependent evolution of the primitive path leads to chain disentanglement, and it is the basis for determination of the diffusion coefficient. The diffusivity in polymer melts is predicted by this model to vary inversely with the square of molecular weight—that is, $v = 2$ in Eq. (4.85). This dependence has been observed experimentally, as shown in Figure 4.54 for a series of deuterated polyethylene (PE) chains in a PE melt. A non-entangled polymer melt is expected to follow the model of non-interacting spheres, and $v = 1.0$. There can be a transition from non-entangled to entangled melts, depending on time and conditions.

4.3.3.3 *Permeability and Partition Coefficients.* In the solid form, many polymers are used in packaging. The diffusion of gases through the packaging material can be important in a number of applications, especially where food and beverages are concerned. Polymeric films can also be used as *membranes*, such as in reverse osmosis applications, in fuel cells, or as replacements for biological materials in artificial kidneys, intestines, or other organs. Membranes are thin layers of materials used to control access of gases, liquids, or solids in solutions. We briefly describe here the

Figure 4.53   Schematic illustration of the reptation process in polymer melts, showing chain entanglements (light arrows), the "wriggling motion" of the polymer chain (darker arrows), and the primitive path of the polymer chain (dark line). Reprinted, by permission, from G. Strobl, *The Physics of Polymers*, 2nd ed., p. 283. Copyright © 1997 by Springer-Verlag.

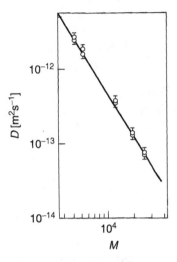

Figure 4.54   Diffusivity at 176°C for polyethylene of different molecular weights, M. The continuous line corresponds to $D_{AB} \propto M^{-2}$. Reprinted, by permission, from G. Strobl, *The Physics of Polymers*, 2nd ed., p. 286. Copyright © 1997 by Springer-Verlag.

transport of molecules through nonporous polymeric films, primarily for membrane applications. There are also porous membranes, which allow particles smaller than the pore diameter to pass through while retaining the larger particles, but the principles of diffusion in porous solids of the previous section are applicable there.

In nonporous membranes, diffusion occurs as it would in any other nonporous solid. However, the molecular species must first dissolve into the membrane material. This step can oftentimes be slower than the diffusion, such that it is the rate-limiting step in the process. As a result, membranes are not characterized solely in terms of diffusion coefficients, but in terms of how effective they are in promoting or limiting both solubilization and diffusion of certain molecular species or solutes. When the solute dissolves in the membrane material, there is usually a concentration discontinuity at the interface between the membrane and the surrounding medium (see Figure 4.55). The equilibrium ratio of the solute concentration in one medium, c_1, to the solute concentration in the surrounding medium, c_2, is called the *partition coefficient*, K_{12}, and can be expressed in terms of either side of the membrane. For the water–membrane–water example illustrated in Figure 4.55,

$$K_{12} = \frac{c_{m1}}{c_{w1}} = \frac{c_{m2}}{c_{w2}} \tag{4.86}$$

The partition coefficient is thus dimensionless. Values for partition coefficients vary greatly, depending on the solute and the membrane material.

The concept of permeability, P_M, described first in Section 4.3.2.2 also applies to membranes. Equation (4.77) relates the permeability to the diffusion coefficient and solubility. Some representative values of permeabilities for common gases in common polymer films are given in Table 4.17. The units of permeability in Table 4.17 are obtained when diffusivity is in units of m²/s, and gas solubility is in units of m³ gas·m²/(m³ solid·N). Note that carbon dioxide permeabilities are generally 3–4 times

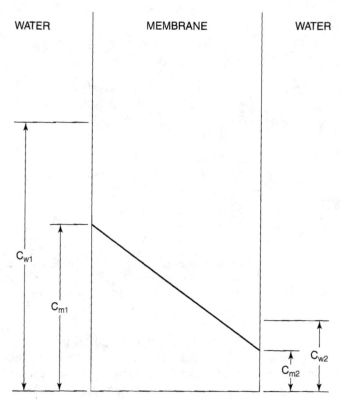

Figure 4.55   Schematic illustration of concentration drop at a membrane interface, and concentration gradient across the membrane. From A. T. Johnson, *Biological Process Engineering*. Copyright © 1999 by John Wiley & Sons, Inc. This material is used by permission of John Wiley & Sons, Inc.

Table 4.17   Permeabilities of Selected Polymers

Film Type	Permeability [10^{16} m^4/(s N)]		
	Oxygen	Carbon Dioxide	Water Vapor
Low-density polyethylene	0.45–1.5	0.88–8.8	0.97–3.8
Linear low-density polyethylene	0.80–1.1	—	2.6–5.0
Medium-density polyethylene	0.30–0.95	0.88–4.4	1.3–2.4
High-density polyethylene	0.059–0.46	0.45–1.1	0.65–1.6
Polypropylene	0.15–0.73	0.88–2.4	0.65–1.7
Polyvinyl chloride	0.071–0.26	0.49–0.93	—
Polyvinyl chloride, plasticized	0.0088–0.086	0.088–6.3	>1.3
Polystyrene	0.023–0.088	1.1–3.0	18–25
Ethylene vinyl acetate copolymer (12%)	0.091–1.5	4.0–6.1	9.7
Ionomer	0.40–0.86	1.1–2.0	3.6–4.9
Ruber hydrochloride	0.015–0.15	0.059–0.59	>1.3
Polyvinylidene chloride	0.00091–0.0030	0.0067	0.24–0.81

Source: A. T. Johnson, *Biological Process Engineering*. Copyright © 1999 By John Wiley & Sons, Inc.

higher than for oxygen, due primarily to the higher solubility of carbon dioxide in most hydrocarbon materials.

To obtain the volume flow rate of gases through these films, a modified form of Fick's Law may be used

$$\dot{V} = \frac{P_M A \Delta P}{L} \tag{4.87}$$

Cooperative Learning Exercise 4.12

A 0.15-mm-thick film is required for use in packaging a pharmaceutical product at 30°C. The partial pressure of O_2 is 0.21 atm outside the package and is 0.01 atm inside the package. Use permeability data from Table 4.17, and assume that the resistance to diffusion outside and inside the membrane are negligible compared to the resistance of the film.

Person 1: Calculate the volume flow rate per unit area of oxygen through a linear low-density polyethylene film.

Person 2: Calculate the volume flow rate per unit area of oxygen through a low-density polystyrene film.

Compare your answers. Which is better for this type of application? Which do you think would be preferable under high relative humidity, assuming that water vapor could potentially damage the product?

Answers: (Average values of P_M used)
$\dot{V}/A = 1.3 \times 10^{-8}$ m^3/s·m^2 for polyethylene
$\dot{V}/A = 7.5 \times 10^{-10}$ m^3/s·m^2 for polystyrene
PS is better for this application (lower flow rate of oxygen across barrier), but its permeability for water is higher than PE.

---

where $\dot{V}$ is the volumetric flow rate of gas through the film in m^3/s, P_M is the permeability, A is the membrane cross-sectional area in m^2, L is the membrane thickness in meters, and ΔP is the partial pressure difference of the diffusing gas across the membrane in Pa.

The permeability is affected not only by temperature, but also by relative humidity. For example, cellophane is an excellent oxygen blocker when dry, but becomes poor when moist. The values for permeability in Table 4.17 are strictly for flat materials. Real permeabilities can vary greatly due to flexural cracks, overlapping, and delamination that are common in real materials. There have been attempts, however, to model the diffusion in strained materials [16].

4.3.4 Mass Transport Properties of Composites

Little is known about the mass transport properties of reinforced-composite materials. Certainly, there are no new relations or concepts that govern estimations of diffusivities that have not already been discussed. In most polymer–matrix composites, the transport properties of the polymer play an important role in diffusion through the composite. For example, hydrophilic polymers such as epoxy readily absorb water from the atmosphere. Thermoplastic polymers absorb relatively little moisture since they are more hydrophobic, but are more susceptible to uptake of organic solvents.

Structural effects can also be very important in diffusion through composites. For example, when pores are created either during formation or through de-bonding between phases, diffusion in porous solids as described in Section 4.3.2.3 can occur. More importantly, atomic mobility can occur along the matrix–reinforcement interphase and can lead to composite degradation. Not only is diffusion important in this case, but such transport processes as *wicking* and *capillary action* can occur, especially when liquids are involved. For this reason, *environmental barrier coatings* (EBCs) are often employed, which provide only an outer layer of protection, in this case a diffusion barrier, but contribute nothing from a structural standpoint. Recall from Chapter 1 that this is an example a complementary composite effect.

Environmental barrier coatings are a type of laminar composite. As with heat transfer, diffusion in laminar composites can be modeled as steady state diffusion through a composite wall, as illustrated in Figure 4.56. Here, hydrogen gas is in contact with solid material A at pressure P_1 and in contact with solid B at pressure P_2. At steady state, the molar flux of hydrogen through both walls must be the same (i.e., $J_{H,Ax}^* = J_{H,Bx}^*$), and Fick's Law [Eq. (4.4)] in the x direction becomes

$$D_{HA}\frac{c_1 - c_1^*}{\Delta x_1} = D_{HB}\frac{c_2^* - c_2}{\Delta x_2} \qquad (4.88)$$

The concentrations in the solid phases, c_1^* and c_2^*, are determined by the solubilities and diffusivities of hydrogen in A and B, and so they are not equal. The thermodynamic activity of hydrogen has a single value at the interface, however. (Refer to Section 3.0.1 for a description of thermodynamic activity.) Hence, the treatment of diffusion flux in a composite wall is simplified by considering activity gradients rather than concentration gradients. If the dissolution of hydrogen gas in the solid follows the reaction

$$\tfrac{1}{2}H_2(g) \Longleftrightarrow H \text{ (dissolved)} \qquad (4.89)$$

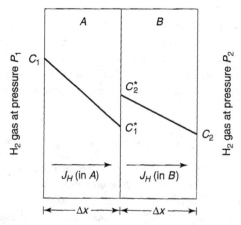

Figure 4.56 Schematic representation of concentration profiles of hydrogen diffusing through a composite wall. Reprinted, by permission, from D. R. Gaskell, *An Introduction to Transport Phenomena in Materials Engineering*, p. 498. Copyright © 1992 by Macmillan Publishing.

then the Law of Mass Action [Eq. (3.4)] tells us that the equilibrium constant for the reaction, K, is given by

$$K = \frac{[\text{H } (dissolved)]}{[H_2(g)]^{1/2}} \tag{4.90}$$

Recall that the concentration (and activity) of molecular hydrogen in the gas phase can be represented by its partial pressure, P_{H_2}, and the concentration of dissolved hydrogen is represented as c_H (either c_1 or c_2) in Figure 4.56, so that the equilibrium constant becomes

$$K = \frac{c_H}{P_{H_2}^{1/2}} \tag{4.91}$$

There is an equilibrium constant for the dissolution of hydrogen in each solid phase, K_A and K_B, respectively. Similarly, the activity of hydrogen at the interface, a_H^* is related to the concentration at the interface and equilibrium constants, $c_1^* = K_A a_H^*$ and $c_2^* = K_B a_H^*$, so that Eq. (4.88) becomes

$$D_{HA}K_A \frac{P_1^{1/2} - a_H^*}{\Delta x_1} = D_{HB}K_B \frac{a_H^* - P_2^{1/2}}{\Delta x_2} \tag{4.92}$$

and we can solve for the activity of the interface to obtain

$$a_H^* = \frac{D_{HA}K_A \Delta x_2 P_1^{1/2} + D_{HB}K_B \Delta x_1 P_2^{1/2}}{D_{HA}K_A \Delta x_2 + D_{HB}K_B \Delta x_1} \tag{4.93}$$

4.3.5 Mass Transport Properties of Biologics

As was the case for composites, there is little new in the way of fundamental concepts for mass transport in biologics that has not already been presented. However, it is possible to briefly describe extensions of some previously introduced topics that are of particular importance to biological materials—namely, diffusion of nonspherical molecules in solution, diffusion through biological membranes, and convective mass transfer in biological systems.

4.3.5.1 Diffusion in Biological Fluids.
Recall from Section 4.3.3.1 that polymer molecules in dilute solutions can be characterized by a friction coefficient, ζ, which describes the resistance to motion through the solvent, and that the diffusivity of these molecules can be related to the friction coefficient by

$$D_{AB} = \frac{k_B T}{\zeta} \tag{4.84}$$

Recall also from Section 4.1.5 that many biological molecules in solution are not spherical, and can be modeled as either prolate or oblate ellipsoids (see Figure 4.17). A further factor affecting diffusion of these ellipsoids in dilute aqueous solutions is that the molecules can become solvated, resulting in an effective radius that is different than the unsolvated molecules. These concepts of asymmetry and solvation are illustrated in Figure 4.57. Accordingly, there are friction coefficients for the unsolvated Stoke's

---

Cooperative Learning Exercise 4.13

Consider the diffusion of hydrogen through a composite wall comprising Ni and Pd of identical thicknesses at 400°C. The hydrogen gas pressure in contact with the Ni is 1.0 atm, and that in contact with the Pd is 0.1 atm. The following data on each metal are provided:

	Ni	Pd
Atomic weight, kg/kg-mol	58.71	106.4
Density at 400°C, kg/m^3	8770	11,850
Diffusivity of H at 400°C, m^2/s	4.9×10^{-10}	6.4×10^{-9}
Solubility of H at 400°C and 1 atm hydrogen pressure, atomic ppm	250	10,000

Person 1: Calculate the equilibrium constant for the dissolution of hydrogen in nickel, K_{Ni}.

Person 2: Calculate the equilibrium constant for the dissolution of hydrogen in palladium, K_{Pd}.

(*Hint*: Determine concentration in each metal by converting solubility to weight and volume.)

Combine your answers to calculate the activity of hydrogen at the interface.

Person 1: Use the activity of hydrogen at the interface to calculate the concentration of hydrogen in nickel at the interface.

Person 2: Use the activity of hydrogen at the interface to calculate the concentration of hydrogen in palladium at the interface.

Combine your information to calculate the flux of hydrogen (in kg/m^2·s) through the composite wall if both metals have a thickness of 1 mm.

$$\textit{Answers:} \ K_{Ni} = 0.0374/(1.0)^{0.5} = 1.11; \ K_{Pd} = 0.0374/(1.0)^{0.5} = 1.11; \ a_{H} = 0.318$$
$$c_{*}^{1} = 0.0374(0.318) = 0.0119 \ \text{kg/m}^3; \ c_{*}^{2} = 1.11(0.318) = 0.353 \ \text{kg/m}^3$$
$$J_{H-Ni} = 4.9 \times 10^{-10}(0.0374 - 0.0119)/0.001 = 1.25 \times 10^{-8} \ \text{kg/m}^2 \cdot \text{s}.$$

(Adapted from Gaskell)

---

sphere, ζ_0, the actual particle, ζ, and the equivalent solvated spheres, ζ^*. The ratio of the friction coefficients, ζ/ζ^*, then is a measure in the increase in ζ due to particle asymmetry, and the ratio ζ^*/ζ_0 is a measure of the increase in friction coefficient due to solvation. The first ratio can be related to the axial ratios (b/a) of the ellipsoids. For prolate ellipsoids $(b/a < 1)$

$$\frac{\zeta}{\zeta^*} = \frac{[1 - (b/a)^2]^{1/2}}{(b/a)^{2/3} \ln\left\{\dfrac{1 + [1 - (b/a)^2]^{1/2}}{b/a}\right\}} \tag{4.94}$$

and for oblate ellipsoids $(b/a > 1)$

$$\frac{\zeta}{\zeta^*} = \frac{[(b/a)^2 - 1]^{1/2}}{(b/a)^{2/3} \tan^{-1}[(b/a)^2 - 1]^{1/2}} \tag{4.95}$$

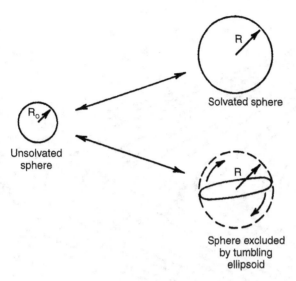

Figure 4.57 Schematic illustration of effective radius of solvated ellipsoids. Reprinted, by permission, from P. Hiemenz, *Polymer Chemistry: The Basic Concepts*, p. 626. Copyright © 1984 by Marcel Dekker, Inc.

The second ratio, ζ^*/ζ_0, is related to the density and mass of the solvated and unsolvated molecules:

$$\frac{\zeta^*}{\zeta_0} = \left[1 + \frac{m_{1b}\rho_2}{m_2\rho_1}\right]^{1/3} \tag{4.96}$$

where m_{1b} is the mass of bound solvent, m_2 is the mass of the solute molecule, and ρ_1 and ρ_2 are the density of the solvent and solute, respectively. Note that multiplication of the two ratios gives the third ratio, ζ/ζ_0. A graphical summary of the variation in this ratio with axial ratio and concentration for aqueous protein dispersions is given in Figure 4.58. With Eqs. (4.94)–(4.96) and Eq. (4.84), it is possible to estimate the diffusivity for solvated ellipsoids.

4.3.5.2 *Diffusion through Biological Membranes.*

Diffusion through polymeric membranes has already been described, and while there are many examples of membrane diffusion in biological systems, such as dialysis, reverse osmosis, and cell metabolism, the principles described previously are still applicable to biological materials. We take a moment here to elaborate on only one example topic, namely, the role of skin as a biological membrane. In a way, skin is a type of environmental barrier coating (cf. Section 4.3.4). Its purpose is to separate an organism from its environment and to provide a barrier to certain substances. Fortunately (and unfortunately), skin is not totally impervious to all substances, particularly air and water. Some animals can obtain a reasonably large fraction of their oxygen requirements through the skin. For example, humans obtain about 2% of the resting oxygen demand through the skin. Passive diffusion of water vapor through the skin is about 6% of the maximum sweat capacity in humans. Water can also be absorbed through the skin, where the mass diffusivity is about 2.5 to 25×10^{-12} m^2/s.

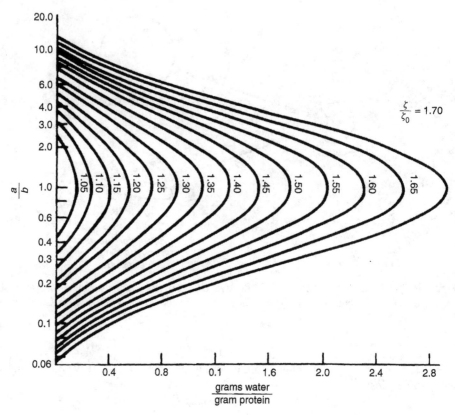

Figure 4.58 Effect of axial ratio, a/b, and concentration on friction coefficient ratio for aqueous protein dispersions. Reprinted, by permission, from P. Hiemenz, *Principles of Colloid and Surface Chemistry*, 2nd ed., p. 88. Copyright © 1986 by Marcel Dekker, Inc.

Cooperative Learning Exercise 4.14

The diffusivity of human hemoglobin at 20°C is 6.9×10^{-11} m^2/s. The viscosity of water at 20°C is approximately 0.01 poise, and the radius of a hemoglobin molecule in solution is approximately 2.64×10^{-9} m.

Person 1: Estimate the experimental friction coefficient, ζ, from the experimental diffusivity for human hemoglobin using Equation 4.84.

Person 2: Estimate the unsolvated Stoke's sphere friction coefficient, ζ_0, using the Stokes–Einstein relationship for spheres, $\zeta_0 = 6\pi\mu r$.

Combine your answers to obtain the ratio of friction coefficients, ζ/ζ_0. If the axial ratio of hemoglobin is $a/b = 1.2$, use Figure 4.58 to estimate the extent of hydration for hemoglobin in solution.

water/gram hemoglobin.

$\zeta/\zeta_0 = 5.86/4.98 = 1.18$; from Figure 4.58, fraction solvation is about 0.5 grams

$6\pi(10^{-3})(2.64 \times 10^{-9}) = 4.98 \times 10^{-11}$ kg/s

Answers: $\zeta = (1.38 \times 10^{-23})(293)/(6.9 \times 10^{-11}) = 5.86 \times 10^{-11}$ kg/s; $\zeta_0 =$

In a manner similar to a laminar composite material, skin is composed of two major layers: the nonvascular epidermis and the highly vascularized (blood-vessel-containing) dermis. The outer layer of the epidermis is the stratum corneum, which is about 10 μm thick and is composed of compacted dead skin cells. This is the greatest barrier to permeability of penetrating substances. The stratum corneum provides the greatest barrier against hydrophilic compounds, while the viable epidermis and dermis are most resistant to lipophilic compounds. Attempts have been made to increase the permeability of the skin, as in the case of transdermal drug delivery. For example, the diffusivity of the stratum corneum to certain compounds can increase as much as tenfold when the layer is hydrated. Hydrating agents such as dimethyl sulfoxide (DMSO) and surfactants may be added to patches to enhance drug movement through the stratum corneum. Some values of permeability to various commercial solvents of female breast epidermis (300–600 μm thick) are given in Table 4.18. Note that these units of permeability are different than those in Table 4.17 for polymers. Units of m/s can be obtained by multiplying values in Table 4.18 by the appropriate solvent liquid density.

4.3.5.3 Convective Mass Transfer in Biological Systems*.

In a manner analogous to convective heat transfer (cf. Section 4.2.5.2), mass transfer can occur not only by static diffusion, but by convection as well. Convective mass transfer occurs when mass is moved in bulk, either as a component in a moving medium or by itself due to pressure or surface tension gradients. As in the case with convective heat transfer, we briefly describe here the dimensionless groups that are involved in convective mass transfer, not as a thorough analysis of the topic, but rather to provide a means for estimating important physical parameters for the materials of interest—specifically, diffusivity. Although the focus here is on convective mass transfer in biologics, primarily due to some important processes such as drying and membrane separations that

Table 4.18   Skin Permeability to Various Commercial Solvents

Solvent	Permeability $[10^{-7}$ kg$/($m^2 s$)]$
Dimethyl sulfoxide (DMSO)	489
N-Methyl-2-pyrrolidone	475
Dimethyl acetamide	297
Dimethyl formamide	272
Methyl ethyl ketone	147
Methylene chloride	66.7
Water	41.1
Ethanol	31.4
Butyl acetate	4.4
Gammabutyrolactone	3.1
Toluene	2.2
Propylene carbonate	1.9
Sulfolane	0.6

Source: A. T. Johnson, *Biological Process Engineering*. Copyright © 1999 by John Wiley & Sons, Inc.

are important in biological processes, the discussion here is general to all materials classes.

The mass transfer analog to the Prandtl number, Pr, is the *Schmidt number*, Sc:

$$\text{Sc} = \frac{\mu}{\rho D_{AB}} \tag{4.97}$$

where μ is the fluid viscosity, ρ the fluid density, and D_{AB} the mass diffusivity of dilute solute through the fluid. The Schmidt number is the ratio of momentum diffusivity to mass diffusivity, just as the Prandtl number is the ratio of momentum diffusivity to thermal diffusivity. The Nusselt number [Eq. (4.64)] analog in convective mass transport is the *Sherwood number*, Sh:

$$\text{Sh} = \frac{k_x l}{c D_{AB}} \tag{4.98}$$

where k_x (in molar units) is the mass transfer analog to the heat transfer coefficient, h_c, [see Eq. (4.94)], l is a characteristic length, c is the concentration (in molar units), and D_{AB} is again the diffusivity. The characteristic length, l, may be a distance along a straight path of mass transfer, or it may be a diameter or radius for mass transfer to and from circular or spherical geometries.

Just as there are correlations for the Nusselt number based on the Reynolds and Prandtl numbers [Eq. (4.96)], there are empirical correlations for the Sherwood number as a function of Reynolds and Schmidt numbers, which are of the general form

$$\text{Sh} = \text{constant } \text{Re}^n \text{Sc}^m \tag{4.98}$$

The values of n and m vary depending on the system and the geometry, but typical values are $n = m = 0.5$ for creeping flow around a sphere in a gas–liquid system, and $n = m = 0.33$ for creeping flow around a sphere in liquid–solid systems.

REFERENCES

Cited References

1. Cowie, J. M. G., *Polymers: Chemistry and Physics of Modern Materials*, 2nd ed., p. 252, Chapman and Hall, New York, 1991.
2. Oommen, Z., S. Thomas, C. K. Premalatha, and B. Kuriakose, Melt rheological behavior of natural rubber/poly(methyl methacrylate)/natural rubber-g-poly(methy methacrylate) blends, *Polymer*, **38**(22), 5611–5621 (1997).
3. Sood, R., M. Kulkarni, A. Dutta, and R. A. Mashelkar, *Polym. Eng. Sci.*, **28**, 20 (1988).
4. Caneiro, O. S., and J. M. Maia, Rheological behavior of (short) carbon fiber/thermoplastic composites. Part II: The influence of matrix type, *Polym. Composites*, **21**(6), 970 (2000).
5. Green, J. P., and J. O. Wilkes Steady State and Dynamic Properties of Concentrated Fiber-Filled Thermoplastics, *Pol. Eng. Sci.*, **32**(21), 1670 (1995).
6. Shuler, S. F., D. M. Binding, and R. B. Pipes, Rheological behavior of two- and three-phase fiber suspensions, *Polym. Composites*, **15**(6), 427 (1994).
7. Jamil, F. A., M. S. Hameed, and F. A. Stephan, Rheological, mechanical, and thermal properties of glass-reinforced polyethylenes, *Polym.-Plast. Technol. Eng.*, **33**(6), 659 (1994).

8. Suhara, F., and S. K. N. Kutty, Rheological properties of short polyester fiber-polyurethane elastomer composite with different interfacial bonding agents, *Polym. Plast. Technol. Eng.*, **37**(1), 57 (1998).

9. *Two Phase Polymer Systems*, L. A. Utracki, ed., Hanser, Munich, 1991, p. 305.

10. Castro, J. M., and C. W. Macosko, *AIChE J.*, **28**, 250 (1982).

11. Bird, R. B., W. E. Stewart, and E. N. Lightfoot, *Transport Phenomena*, 2nd ed., John Wiley & Sons, New York, 2001, p. 279.

12. Ammar, M. M., et al., Thermal conductivity of silicate and borate glasses, *J. Am. Ceram. Soc.*, **66**, C76–C77 (1983).

13. Primenko, V. I., Theoretical method of determining the temperature dependence of the thermal conductivity of glasses, *Glass Ceram.*, **37**, 240–242 (1980).

14. Scholze, H., *Glass: Nature, Structure and Properties*, Springer-Verlag, New York, 1991, p. 364.

15. Bird, R. B., W. E. Stewart, and E. N. Lightfoot, *Transport Phenomena*, 2nd ed., John Wiley & Sons, 2001, p. 282.

16. Hinestroza, J., D. De Kee, and P. Pintauro, Apparatus for studying the effect of mechanical deformation on the permeation of organics through polymer films, *Ind. Eng. Chem. Res.*, **40**, 2183 (2001).

Transport Phenomena

Bird, R. B., W. E. Stewart, and E. N. Lightfoot, *Transport Phenomena*, 2nd ed., John Wiley & Sons, New York, 2001.

Hirschfelder, J. O., C. F. Curtiss, and R. B. Bird, *Molecular Theory of Gases and Liquids*, John Wiley & Sons, New York, 1954.

Gaskell, D. R., *An Introduction to Transport Phenomena in Materials Engineering*, Macmillan, New York, 1992.

Transport Properties of Metals and Alloys

Iida, T., and R. I. L. Guthrie, *The Physical Properties of Liquid Metals*, Clarendon Press, Oxford, 1988.

Transport Properties of Ceramics and Glasses

Scholze, H., *Glass: Nature, Structure and Properties*, Springer-Verlag, New York, 1991.

Experimental Techniques of Glass Science, C. J. Simmons and O. H. El-Bayoumi, ed., American Ceramic Society, Westerville, OH, 1993.

Transport Properties of Polymers

Polymer Blends and Alloys, M. J. Folkes and P. S. Hope, eds., Blackie Academic & Professional, New York, 1993.

Two-Phase Polymer Systems, L. A. Utracki, ed., Hanser, New York, 1991.

Polymer Blends and Composites, J. A. Manson and L. H. Sperling, Plenum, New York, 1976.

Dynamics of Polymeric Liquids, Bird, R. B., R. C. Armstrong, and O. Hassager, Vol. 1, John Wiley & Sons, New York, 1977.

Transport Properties of Composites

Composites Engineering Handbook, P. K. Mallick, editor, Marcel Dekker, New York, 1997.

Transport Properties of Biologics

Johnson, A. T., *Biological Process Engineering*, Wiley-Interscience, New York, 1999.

Lightfoot, E. N., *Transport Phenomena and Living Systems*, John Wiley & Sons, New York, 1974.

PROBLEMS

Level I

4.I.1 Estimate the viscosity of liquid chromium at its melting temperature of 2171 K.

4.I.2 Estimate the viscosity of liquid iron at 1600°C. The density of liquid iron at 1600°C is 7160 kg/m^3.

4.I.3 Various molecular weight fractions of cellulose nitrate were dissolved in acetone and the intrinsic viscosity was measured at 25°C:

$$M \times 10^{-3}$$

(g/mol) 77 89 273 360 400 640 846 1550 2510 2640

$[\eta]$ (dl/g) 1.23 1.45 3.54 5.50 6.50 10.6 14.9 30.3 31.0 36.3

Use these data to evaluate the Mark–Houwink constants for this system.

4.I.4 Agglomeration in a slurry causes a change in f_{cr}^{ν} from 0.6 to 0.45 and a change in K_H from 3.5 to 2.5. Calculate and compare the viscosity for each slurry when $f_p^{\nu} = 0.40$.

4.I.5 Calculate the thermal resistance of a 2-mm-thick, 1 cm^2 section of human skin. The thermal conductivity of human skin is 0.627 N/K·s, the specific heat is 3470 N·m/kg·K, and the density is 1100 kg/m^3.

4.I.6 A plate of iron is exposed to a carburizing (carbon-rich) atmosphere on one side and a decarburizing (carbon-deficient) atmosphere on the other side at 700°C. If a condition of steady state is achieved, calculate the diffusion flux of carbon through the plate if the concentration of carbon at positions of 5 and 10 mm beneath the carburizing surface are 1.2 and 0.8 kg/m^3, respectively. Assume a diffusion coefficient of 3×10^{-11} m^2/s at this temperature. Callister, p. 97, Example Problem 5.1.

Level II

4.II.1 The thermal conductivity of a material varies with temperature as

$$\ln k = 0.01T + 0.5$$

where k has units of W/m·K, and T has units of °C. Heat flows by conduction through a plane slab of this material of thickness 0.1 m, the left face of which is at 100°C and the right face of which is at 0°C. Calculate (a) the mean thermal conductivity of this material in the range 0–100°C, (b) the heat flux through the slab, and (c) the temperature at $x = 0.05$ m.

4.II.2 Hydrogen dissolved in iron obeys *Sievert's Law*; that is, at equilibrium we have

$$\tfrac{1}{2}H_2(g) \;\Longleftrightarrow\; [H]_{dissolved\ in\ iron}$$

where [H] is the concentration of atomic hydrogen in the iron, in ppm by weight, in equilibrium with hydrogen gas at the pressure P_{H_2}. At 400°C and a hydrogen pressure of 1.013×10^5 Pa (1 atm), the solubility of hydrogen in iron is 3 ppm by weight. Calculate the flux of hydrogen through an iron tank of wall thickness 0.001 m at 400°C. The density of iron 400°C is 7730 kg/m³, and the diffusivity of hydrogen is 10^{-8} m²/s. Clearly state any assumptions you make.

4.II.3 A hemodialysis membrane with an effective area of 0.06 m², thickness of 50 μm, and permeability of 2.96×10^{-14} m² is used to filter urea and other impurities from the blood. The viscosity of blood plasma is 1.2×10^{-3} N·s/m². What is the expected filtration rate if the transmembrane pressure is 2.25 Pa?

4.II.4 Fick's Second Law of Diffusion relates the change in concentration of a diffusing species with time to the diffusion coefficient and the concentration gradient for non-steady-state diffusion:

$$\frac{\partial c_A}{\partial t} = D\frac{\partial^2 c_A}{\partial x^2}$$

The solution to this partial differential equation depends upon geometry, which imposes certain boundary conditions. Look up the solution to this equation for a semi-infinite solid in which the surface concentration is held constant, and the diffusion coefficient is assumed to be constant. The solution should contain the error function. Report the following: the boundary conditions, the resulting equation, and a table of the error function.

Level III

4.III.1 A relationship for the viscosity of a simple liquid metal near its melting point, T_m, was proposed by Andrade:

$$\mu = 1.68 \times 10^{-4}\frac{\sqrt{MT_m}}{V_m^{2/3}}$$

where μ is in mPa·s and V_m is the atomic volume at the melting point. Show that this relationship gives approximately the same value for viscosity (within a factor of about 0.3) as the hard-sphere model when a packing fraction of 0.45 is used. Use this formula to estimate the viscosity of molten silver near its melting point. How does this compare to an experimental value? Cite your references.

4.III.2 Use a spreadsheet to solve for the three constants in the Vogel–Fulcher–Tammann equation using the data in Cooperative Learning Exercise 4.1. Show your work.

4.III.3   Determine the VFT equation for a pure SiO_2 melt based on the following three viscosity–temperature points:

Glass transformation temperature $= 1203°C$ $\ln \mu = 13.3$
Littleton temperature $= 1740°C$ $\ln \mu = 7.6$
Working point $= 2565°C$ $\ln \mu = 4.0$

4.III.4   In Cooperative Learning Exercise 4.4, you derived an expression for the superficial velocity of a 1% solution of poly(ethylene oxide) through a bed of 10-μm-diameter fibers with 60% porosity. You were left with a parameter, τ_{Rh}, in the expression, which is the shear stress at the wall, based upon the hydrodynamic radius, R_h. The mean hydrodynamic radius is defined as the ratio of the particle cross section to the "wetted perimeter," which is conveniently related to the void fraction, or porosity, ε, and the wetted surface, a, as $R_h = \varepsilon/a$. The wetted surface, in turn, is related to the porosity and the "specific surface," S, as $a = S(1 - \varepsilon)$. Finally, the specific surface is directly related to geometrical dimensions of the particle and is defined as the ratio of particle surface area to particle volume. Complete the derivation of an expression for the superficial velocity by calculating the specific surface for cylinders (this result is general for all cylindrical geometries) and calculating a value of τ_{Rh} for the case of 60% porosity. Your final expression for v_0 should have only the pressure drop across the fiber bed, ΔP, and the bed length, L.

4.III.5   Derive an expression for steady-state heat flow through three parallel planes as shown below. In this instance, the total heat flow, Q_{tot}, is a sum of the individual heat flows, Q_i. The cross-sectional areas of each plane, A_i, are not necessarily equivalent.

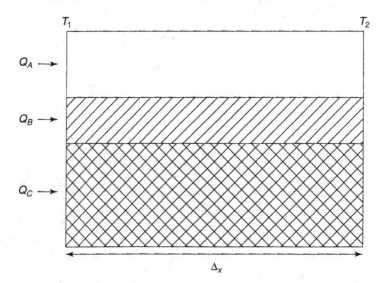

4.III.6   Composite membranes are used where the membrane of choice for a particular separation must be strengthened by addition of a second material. Assume a hollow fiber membrane of inside diameter 1.2 mm and outside diameter 1.3 mm. Mass diffusivity of the target species through this membrane is 3.0×10^{-10} m^2/ s. A second material is layered on the outside of the first material to add strength. Mass diffusivity of the same target species through this material is 18×10^{-10} m^2/ s. What thickness of the outside material can be added if the total diffusion does not exceed the resistance of the inner material by more than 25%?

CHAPTER 5

Mechanics of Materials

5.0 INTRODUCTION AND OBJECTIVES

The engineering principles of thermodynamics, kinetics, and transport phenomena, as well as the chemistry and physics of molecular structure, serve as the basis for the remaining topics in this book. In this chapter, we look at what for many applications is the primary materials selection criterion: mechanical properties. As in the previous chapter, we focus primarily on the properties of materials, but discuss briefly the mechanics, both in the fluid and solid states, that give rise to the properties. There is a great deal of new terminology in this chapter, and it is cumulative—take time to understand all the definitions before proceeding to the next section.

We seek to understand the response of a material to an applied stress. In Chapter 4, we saw how a fluid responds to a shearing stress through the application of Newton's Law of Viscosity [Eq. (4.3)]. In this chapter, we examine other types of stresses, such as tensile and compressive, and describe the response of solids (primarily) to these stresses. That response usually takes on one of several forms: elastic, inelastic, viscoelastic, plastic (ductile), fracture, or time-dependent creep. We will see that Newton's Law will be useful in describing some of these responses and that the concepts of stress (applied force per unit area) and strain (change in dimensions) are universal to these topics.

By the end of this chapter, you should be able to:

- Determine strength and modulus from a stress–strain diagram.
- Qualitatively differentiate between stress–strain diagrams for different material classes.
- Describe the differences between the Kelvin and Maxwell models for viscoelasticity and identify their corresponding equations.
- Calculate Poisson's ratio.
- Relate flexural, bulk, and tensile moduli.
- Distinguish between ultimate strength, yield strength, and tensile strength.
- Distinguish between ductile and brittle materials.
- Understand the role of strain rate on mechanical properties.

An Introduction to Materials Engineering and Science: For Chemical and Materials Engineers,
by Brian S. Mitchell
ISBN 0-471-43623-2 Copyright © 2004 John Wiley & Sons, Inc.

- Identify the different types of mechanical creep and calculate creep rate.
- Apply time–temperature superposition principles to polymer moduli and calculate shift factors.
- Describe temperature, molecular weight, and strain-rate effects on the mechanical properties of polymers.
- Calculate axial and transverse moduli and strengths for unidirectional, continuous and discontinuous, fiber-reinforced composites and laminate composites.
- Describe the structure of soft and hard biologics that give rise to their unique mechanical properties.
- Use mechanical properties to help select appropriate candidate materials for a specific application.

5.1 MECHANICS OF METALS AND ALLOYS

In this section, we will describe the response of a material to an applied stress in terms of its atomic structure. Initially, we will focus on a single crystal of a metallic element, then work our way into topics related to polycrystalline materials.

5.1.1 Elasticity

5.1.1.1 The Molecular Origins of Elasticity. Recall from Section 1.0.4 that atoms are held together by interatomic bonds and that there are equations such as Eq. (1.13) that relate the interatomic force, F, to the potential energy function between the atoms, U, and the separation distance, r:

$$F = -\frac{dU}{dr} \tag{1.13}$$

It should make sense then, that as we attempt to pull atoms apart or force them further together through an applied stress, we can, at least in principle, relate how relatively difficult or easy this is to the potential energy function. We can develop a quantitative description of this process of pulling atoms apart, provided that we do so over small deformations, in which the deformation is wholly recoverable; that is, the atoms can return back to their original, undeformed positions with no permanent displacement relative to one another. This is called an *elastic* response. The term "elastic" here does not imply anything specific to polymers in the same way that the more everyday use of the term does. It is used in the same sense that it is in physics and chemistry—a completely recoverable deformation.

A convenient way to visualize the elastic bond between two atoms is to place an imaginary spring between the atoms, as illustrated in Figure 5.1. For small atomic deformations, either away from each other with what is called a *tensile force* (shown in Figure 5.1) or toward each other with what is called a *compressive force* (not shown), the force stored in the spring causes the atoms to return to the undeformed, equilibrium separation distance, r_0, where the forces are zero (cf. Figure 1.3). Attractive forces between the atoms counteract the tensile force, and repulsive forces between the atoms counteract the compressive forces.

Example Problem 5.1

Derive an expression for the elastic modulus of an ionic solid in terms of the potential energy function parameters a, b, and n.

Solution: According to Eq. (5.5), the elastic modulus, E, is proportional to the stiffness of the theoretical springs that model the bonds between atoms in the solid, S_0. According to Equation (5.1), S_0 is in turn proportional to the second derivative of the potential energy function, U, with respect to interatomic separation distance, r.

We begin with the potential energy function of Eq. (1.11), and recognize that for an ionic solid, $m = 1$, such that the potential energy function becomes

$U = \dfrac{-a}{r} + \dfrac{b}{r^n}$. The strategy, then, is to differentiate U twice with respect to r, and evaluate at $r = r_0$.

The first derivative is $F = \dfrac{dU}{dr} = \dfrac{a}{r^2} - \dfrac{nb}{r^{n+1}}$, and the second derivative is $\dfrac{dF}{dr} = \dfrac{-2a}{r^3} +$

$\dfrac{n(n+1)b}{r^{n+2}}$. All that is left now is to evaluate dF/dr at $r = r_0$. Equation (1.16) provides us with the equilibrium separation distance as a function of the potential energy parameters, which for the case of $m = 1$ becomes $r_0 = \left(\dfrac{nb}{a}\right)^{\frac{1}{n-1}}$. Substitution of r_0 into dF/dr with appropriate simplification gives the desired result:

$$E \propto S_0 = \left(\frac{dU}{dr}\right)_{r=r_0} = \frac{-2a}{\left(\dfrac{nb}{a}\right)^{\frac{3}{n-1}}} + \frac{n(n+1)b}{\left(\dfrac{nb}{a}\right)^{\frac{n+2}{n-1}}}$$

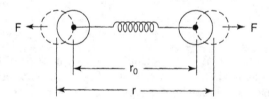

Figure 5.1 Schematic illustration of "bead-and-spring" model of atomic force between atoms. Reprinted, by permission, from M. F. Ashby and D. R. H. Jones, *Engineering Materials 1*, 2nd ed., p. 44. Copyright © 1996 by Michael F. Ashby and David R. H. Jones.

The stiffness of the spring, S_0, is a measure of how much force must be overcome to pull the atoms apart, and it is related to the potential energy function as

$$S_0 = \left(\frac{d^2 U}{dr^2}\right)_{r=r_0} \tag{5.1}$$

Notice that the stiffness is the slope (first derivative) of the force versus interatomic separation distance plot in the vicinity of r_0, as shown in Figure 1.3. So, for small deformations (i.e., $r \approx r_0$) we can relate the stiffness to the applied force, F:

$$S_0 = \frac{dF}{dr} \approx \frac{\Delta F}{\Delta r} = \frac{F - F_0}{r - r_0} = \frac{F}{r - r_0} \qquad (5.2)$$

since $F = F_0 = 0$ at $r = r_0$ (see Figure 1.3).

Now consider a square array of N springs (bonds), as would be found in a single crystal, and illustrated in Figure 5.2. The total tensile force exerted across the unit area, F/A, is represented by the Greek lowercase sigma, σ_t, where the subscript "t" indicates a tensile force. It is the product of the stiffness of each spring, S_0, the number of springs per unit area, N/A, and the distance the springs are extended, $r - r_0$

$$\sigma_t = (N/A)S_0(r - r_0) \qquad (5.3)$$

The number of springs per unit area is simply $1/r_0^2$, (see Figure 5.2), so that Eq. (5.3) becomes

$$\sigma_t = \left(\frac{S_0}{r_0}\right)\left(\frac{r - r_0}{r_0}\right) \qquad (5.4)$$

We have written Eq. (5.4) with variables grouped as they are in order to define two very important quantities. The first quantity in parentheses is called the *modulus*—or in this case, the *tensile modulus*, E, since a tensile force is being applied. The tensile modulus is sometimes called *Young's modulus, elastic modulus,* or *modulus of elasticity,* since it describes the elastic, or recoverable, response to the applied force, as represented by the springs. The second set of parentheses in Eq. (5.4) represents the tensile *strain*, which is indicated by the Greek lowercase epsilon, ε_t. The strain is defined as the displacement, $r - r_0$, relative to the initial position, r_0, so that it is an indication of relative displacement and not absolute displacement. This allows comparisons to be made between tensile test performed at a variety of length scales. Equation (5.4) thus becomes

$$\sigma_t = E\varepsilon_t \qquad (5.5)$$

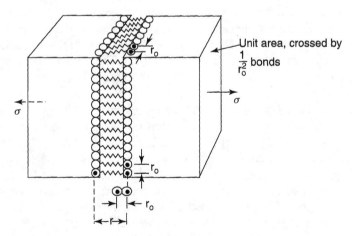

Figure 5.2 Illustration of multiple bead-and-springs in tensile separation of atomic planes. Reprinted, by permission, from M. R. Ashby and D. R. H. Jones, *Engineering Materials 1*, 2nd ed., p. 59, Copyright © 1996 by Michael F. Ashby and David R. H. Jones.

Note that the strain is dimensionless, but is often given in units of length/length (e.g., in./in. or cm/cm) to indicate the length scale of the displacement. Modulus then has the same units of stress, which are typically Pa (N/m^2) or psi ($lb_f/in.^2$). The stresses required to separate bonds are surprisingly high, and the resulting extensions are vanishingly small, such that the values of modulus are often reported in MPa (10^6 Pa) and GPa (10^9 Pa), or kpsi (10^3 psi, sometimes called "ksi") and Mpsi (10^6 psi).

Cooperative Learning Exercise 5.1

Work in groups of three. Look up the modulus of elasticity for each of the following substances in Appendix 7, and compute the tensile stress, σ_t, that arises when a cylindrical sample of each is strained to 0.1% of their original length. Assume that all deformations are perfectly elastic.

Person 1: Tungsten (W)   *Person 2*: Tungsten carbide (WC)   *Person 3*: Polycarbonate

Compare your answers. Assuming that all samples have the same geometry (cross-sectional area), which of the three will require the greatest load to achieve this amount of strain?

Answers: $\sigma_t(W) = (400 \text{ GPa})(0.001) = 400 \text{ MPa}; \sigma_t(WC) = (689)(0.001) = 689 \text{ MPa}; \sigma_t(\text{Polycarbonate}) = (2.4 \text{ GPa})(0.001) = 2.4 \text{ MPa}. WC will require the greatest load (force).$

Equation (5.5) is known as *Hooke's Law* and simply states that in the elastic region, the stress and strain are related through a proportionality constant, E. Note the similarity in form to Newton's Law of Viscosity [Eq. (4.3)], where the shear stress, τ, is proportional to the strain rate, $\dot{\gamma}$. The primary differences are that we are now describing a solid, not a fluid, the response is to a tensile force, not a shear force, and we do not (yet) consider time dependency in our tensile stress or strain.

The relationship between the potential energy and the modulus is an important one. If we know the potential energy function, even qualitatively or relatively to another material, we can make inferences as to the relative magnitude of the elastic modulus. Figure 5.3 shows the interrelationship between the potential energy (top curves), the force (represented per unit area as stress in the bottom curves), and tensile modulus (slope of force curves for two hypothetical materials). The deeper the potential energy well, the larger the modulus (compare Figure 5.3a to 5.3b).

5.1.1.2 Generalized Hooke's Law*.

The discussion in the previous section was a simplified one insofar as the relationship between stress and strain was considered in only one direction: along the applied stress. In reality, a stress applied to a volume will have not only the normal forces, or forces perpendicular to the surface to which the force is applied, but also shear stresses in the plane of the surface. Thus there are a total of nine components to the applied stress, one normal and two shear along each of three directions (see Figure 5.4). Recall from the beginning of Chapter 4 that for shear stresses, the first subscript indicates the direction of the applied force (outward normal to the surface), and the second subscript indicates the direction of the resulting stress. Thus, τ_{xy} is the shear stress of x-directed force in the y direction. Since this notation for normal forces is somewhat redundant—that is, the x component of an

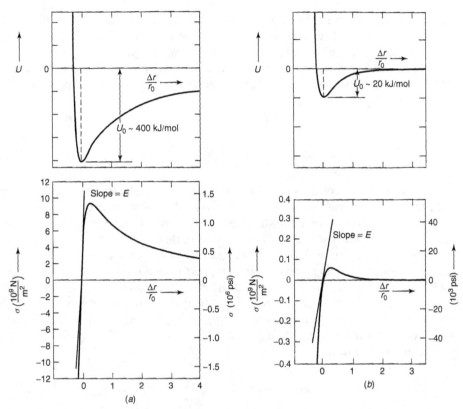

Figure 5.3   Schematic diagrams showing potential energy (*top*) and applied stress (*bottom*) versus linear strain for crystalline solids with (a) strong bonds and (b) weak bonds. From K. M. Ralls, T. H. Courtney, and J. Wulff, *Introduction to Materials Science and Engineering.* Copyright © 1976 by John Wiley & Sons, Inc. This material is used by permission John Wiley & Sons, Inc.

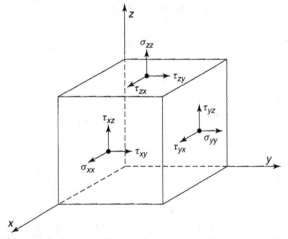

Figure 5.4   Distribution of shear and normal stresses acting on a volume element. From Z. Jastrzebski, *The Nature and Properties of Engineering Materials*, 2nd ed. Copyright © 1976 by John Wiley & Sons, Inc. This material is used by permission of John Wiley & Sons, Inc.

x-directed force is by definition a normal force and not a shear force—the subscripts are sometimes simplified, and $\sigma_{xx} = \sigma_x$, $\sigma_{yy} = \sigma_y$, and $\sigma_{zz} = \sigma_z$. We will maintain the double subscript notation for completeness for the time being.

It is often convenient to consider these nine stress components in the form of a *stress tensor*, σ, which can be represented by an array of the stress components:

$$
\sigma \equiv \begin{bmatrix} \sigma_{xx} & \tau_{xy} & \tau_{xz} \\ \tau_{yx} & \sigma_{yy} & \tau_{yz} \\ \tau_{zx} & \tau_{zy} & \sigma_{zz} \end{bmatrix} \tag{5.6}
$$

In this array, the stress components in the first row act on a plane perpendicular to the x axis, the stress components in the second row act on a plane perpendicular to the y axis, and the stress components in the third row act on a plane perpendicular to the z axis. The stress matrix of Eq. (5.6) is symmetric, that is, the complementary shear stress components are equivalent—for example, $\tau_{xy} = \tau_{yx}$, and so on.

There is a corresponding *strain tensor* (which is not presented here due to its complexity), and for each stress component there is a corresponding strain component. Hence, there are three normal strains, ε_{xx}, ε_{yy}, and ε_{zz}, and six shear strains, which reduce to three due to matrix symmetry: γ_{xy}, γ_{yz}, and γ_{xz}. There is a linear relationship between stress and strain for each component, giving rise to the following six equations, known as the generalized form of Hooke's Law:

$$
\begin{aligned}
\sigma_{xx} &= Q_{11}\varepsilon_{xx} + Q_{12}\varepsilon_{yy} + Q_{13}\varepsilon_{zz} + Q_{14}\gamma_{xy} + Q_{15}\gamma_{xz} + Q_{16}\varepsilon_{yz} \\
\sigma_{yy} &= Q_{21}\varepsilon_{xx} + Q_{22}\varepsilon_{yy} + Q_{23}\varepsilon_{zz} + Q_{24}\gamma_{xy} + Q_{25}\gamma_{xz} + Q_{26}\varepsilon_{yz} \\
\sigma_{zz} &= Q_{31}\varepsilon_{xx} + Q_{32}\varepsilon_{yy} + Q_{33}\varepsilon_{zz} + Q_{34}\gamma_{xy} + Q_{35}\gamma_{xz} + Q_{36}\varepsilon_{yz} \\
\tau_{yz} &= Q_{41}\varepsilon_{xx} + Q_{42}\varepsilon_{yy} + Q_{43}\varepsilon_{zz} + Q_{44}\gamma_{xy} + Q_{45}\gamma_{xz} + Q_{46}\varepsilon_{yz} \\
\tau_{xz} &= Q_{51}\varepsilon_{xx} + Q_{52}\varepsilon_{yy} + Q_{53}\varepsilon_{zz} + Q_{54}\gamma_{xy} + Q_{55}\gamma_{xz} + Q_{56}\varepsilon_{yz} \\
\tau_{xy} &= Q_{61}\varepsilon_{xx} + Q_{62}\varepsilon_{yy} + Q_{63}\varepsilon_{zz} + Q_{64}\gamma_{xy} + Q_{65}\gamma_{xz} + Q_{66}\varepsilon_{yz}
\end{aligned} \tag{5.7}
$$

The *elastic coefficients*, Q_{ij}, can also be summarized in a 6×6 array:

$$
\begin{bmatrix} Q_{11} & Q_{12} & \cdots & Q_{16} \\ Q_{21} & Q_{22} & \cdots & Q_{26} \\ \vdots & \vdots & \vdots & \vdots \\ Q_{16} & Q_{26} & \cdots & Q_{66} \end{bmatrix} \tag{5.8}
$$

It can be shown that for the cross-terms $Q_{21} = Q_{12}$, $Q_{31} = Q_{13}$, and so on, so that of the initial 36 values, there are only 21 independent elastic constants necessary to completely define an anisotropic volume without any geometrical symmetry (not to be confused with matrix symmetry). The number of independent elastic constants decreases with increasing geometrical symmetry. For example, orthorhombic symmetry has 9 elastic constants, tetragonal 6, hexagonal 5, and cubic only 3. If the body is isotropic, the number of independent moduli can decrease even further, to a limiting

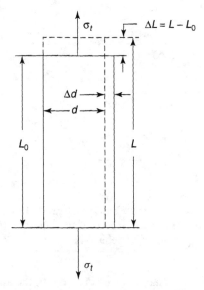

Figure 5.5 Schematic illustration of tensile strain and corresponding lateral strain. From Z. Jastrzebski, *The Nature and Properties of Engineering Materials*, 2nd ed. Copyright © 1976 by John Wiley & Sons, Inc. This material is used by permission of John Wiley & Sons, Inc.

number of *two* independent moduli, corresponding to the *two* forms of stress: *normal* and shear.

5.1.1.3 Moduli and Poisson's Ratio.

Let us examine the elastic deformation of a material on a more macroscopic level. Figure 5.5 shows a typical test material under tension on a scale somewhat larger than that represented in Figure 5.4. A stress, σ_t is once again applied, resulting in a strain ε_t, which we define as $\Delta L/L_0$. For small deformations, the stress and strain are related by Eq. (5.5). Notice, however, that in order for the material to (ideally) maintain a constant volume, the diameter, d, must decrease by an increment $\Delta d = d - d_0$. There is no requirement, of course, that volume be maintained, but in most real systems, there is a decrease in the cross section with elongation. This response is characteristic of a material and is given by *Poisson's ratio*, ν, which is defined as the ratio of the induced lateral strain to the axial strain:

$$\nu = -\frac{\Delta d/d_0}{\Delta L/L_0} \tag{5.9}$$

The minus sign in Eq. (5.9) is to account for the fact that Δd as defined above is usually negative. Thus, Poisson's ratio is normally a positive quantity, though there is nothing that prevents it from having a negative value. For constant volume deformations (such as in polymeric elastomers), $\nu = 0.5$, but for most metals, Poisson's ratio varies between 0.25 and 0.35. Values of Poisson's ratio for selected materials are presented in Appendix 7.

Poisson's ratio provides a basis for relating the lateral response of a material to the axial response. By definition, the forces that cause lateral deformation in a tensile experiment must be shear forces, since they are normal to the direction of the applied

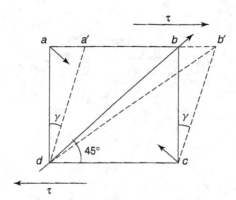

Figure 5.6 Schematic illustration of shear strain and stress. From Z. Jastrzebski, *The Nature and Properties of Engineering Materials*, 2nd ed. Copyright © 1976 by John Wiley & Sons, Inc. This material is used by permission of John Wiley & Sons, Inc.

stress. In a generalized shear situation, as in Figure 5.6, the shear stress, τ, results in a shear strain, γ, which we must define slightly differently than in the tensile experiment. Here, the shear strain is the amount of shear, aa', divided by the distance of which the shear has occurred, ad, so $\gamma = aa'/ad$. There is still a proportionality between stress and strain (for small deformations), but now we call the proportionality constant the *shear modulus*, G:

$$\tau = G\gamma \tag{5.10}$$

The shear modulus and elastic modulus are related through Poisson's ratio:

$$G = \frac{E}{2(1 + v)} \tag{5.11}$$

In addition to the tensile and shear moduli, a *compressive modulus*, or *modulus of compressibility*, K, exists to describe the elastic response to compressive stresses (see Figure 5.7). The compressive modulus is also sometimes called the *bulk modulus*. It is the proportionality constant between the compressive stress, σ_c, and the bulk strain, represented by the relative change in bulk volume, $\Delta V/V_0$.

The compressive modulus and elastic modulus are also related through Poisson's ratio:

$$K = \frac{E}{3(1 - 2v)} \tag{5.12}$$

From Eqs. (5.11) and (5.12), we can derive a relationship that relates all three moduli to one another:

$$E = \frac{9GK}{3K + G} \tag{5.13}$$

Equations (5.11), (5.12), and (5.13) allow us to apply some convenient "rules of thumb." For most hard materials, $v \approx 1/3$, such that $K \approx E$ and $G \approx (3/8)E$. For elastomers, putties, gels, and colloidal systems, the compressive modulus is much higher than the other moduli, and the system is considered "incompressible." In this case, $E \approx 3G$.

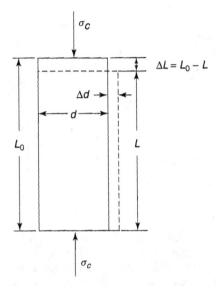

Figure 5.7 Schematic illustration of compressive strain and lateral strain. From Z. Jastrzebski, *The Nature and Properties of Engineering Materials*, 2nd ed. Copyright © 1976 by John Wiley & Sons, Inc. This material is used by permission of John Wiley & Sons, Inc.

Cooperative Learning Exercise 5.2

A tensile load of 5600 N is applied to along the long axis (z direction) of cylindrical brass rod that has an initial diameter of 10 mm. The resulting deformation is entirely elastic, and it results in a diameter reduction (x direction) of 2.5×10^{-3} mm.

Person 1: Calculate the lateral strain, ε_x.

Person 2: Calculate the axial strain, ε_z. Assume that the modulus of elasticity for brass is 97 GPa.

Combine your information to calculate Poisson's ratio for this material.

Person 2: Use this value of Poisson's ratio to calculate the compressive modulus, K.

Person 1: Use the value of Poisson's ratio to calculate the shear modulus, G.

Use Eq. (5.13) to recalculate E from G and K to confirm your results.

Answers: $\varepsilon_x = -(2.5 \times 10^{-3} \text{ mm})/(10 \text{ mm}) = -2.5 \times 10^{-4}$; in order to calculate ε_z, we must first calculate the applied stress, $\sigma = (5600 \text{ N})/[\pi(5 \times 10^{-3} \text{ m})^2] = 71.3 \times 10^6$ Pa, then use Hooke's Law and the elastic modulus to calculate the axial strain. $\varepsilon_z = \sigma/E = (71.3 \text{ MPa})/(97000 \text{ MPa}) = 7.35 \times 10^{-4}$; $\nu = -\varepsilon_x/\varepsilon_z = -(-2.5)/(7.35)$ = 0.34, $K = 97 \text{ GPa}/\{3[1 - (2)(0.34)]\} = 101 \text{ GPa}$; $G = 97 \text{ GPa}/\{2[1 + (0.34)]\}$ = 36.2 GPa; $E = 9(36.2)(101)/[3(101) + 36.2] = 97.0 \text{ GPa}$. It checks out!

5.1.2 Ductility

To this point, we have limited the discussion to small strains—that is, small deviations from the equilibrium bond distance, such that all imposed deformations are completely recoverable. This is the elastic response region, one that virtually all materials possess (see Figure 5.8). What happens at larger deformations, however, is dependent to some

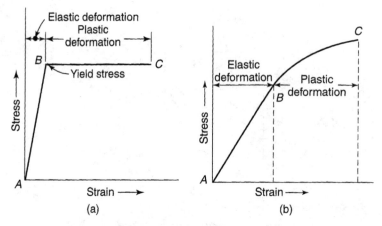

Figure 5.8   Illustration of (a) ideal elastic deformation followed by ideal plastic deformation and (b) typical elastic and plastic deformation in rigid bodies. From Z. Jastrzebski, *The Nature and Properties of Engineering Materials*, 2nd ed., Copyright © 1976 by John Wiley & Sons, Inc. This material is used by permission of John Wiley & Sons, Inc.

extent on the type of material under consideration. Beyond the elastic region, we enter a realm of nonrecoverable mechanical response termed *permanent deformation*. There are two primary forms of permanent deformation: *viscous flow* and *plastic flow*. We will reserve the discussion of viscous flow to the appropriate section on mechanical properties of polymers, since this is where viscous flow primarily occurs. Other types of nonelastic responses to applied forces, such as *fracture* and *creep*, will also be delayed until later sections. We concentrate now on a type of nonrecoverable deformation called plastic flow, or, more commonly in the terminology of metals and alloys, *ductility*. As illustrated in Figure 5.8, the plastic deformation region ideally occurs immediately following the elastic response region, and it results in elongation (strain) without additional stress requirement, hence the flat region represented by segment BC in Figure 5.8a. More practically, there are microstructural phenomena that lead to the behavior shown in Figure 5.8b, namely, an increase in stress with % strain in the plastic deformation region. In the following section, we explore the molecular basis for these observations.

5.1.2.1   The Molecular Origins of Ductility — Slip in Single Crystals.   Plastic deformation in single crystals occurs by one of two possible mechanisms: slip or twinning. *Slip* is the process by which bulk lattice displacement occurs by the motion of crystallographic planes relative to one another (see Figure 5.9a). The surface on which slip takes place is often a plane, known as the *slip plane*. The direction of motion is known as the *slip direction*. The visible intersection of a slip plane with the outer surface of the crystal is known as a *slip band*. *Twinning* is the process in which atoms subjected to stress rearrange themselves so that one part of the crystal becomes a mirror image of the other part. Every plane of atoms shifts in the same direction by an amount proportional to its distance from the twinning plane (see Figure 5.9b). Twinning usually requires a higher shear stress than does slip, and it occurs primarily in BCC and HCP crystal structures. We will focus primarily on slip to describe the plastic deformation process.

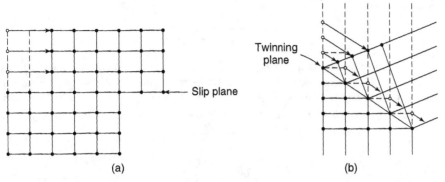

Figure 5.9 Schematic illustration of plastic deformation in single crystals by (a) slip and (b) twinning. From Z. Jastrzebski, *The Nature and Properties of Engineering Materials*, 2nd ed. Copyright © 1976 by John Wiley & Sons, Inc. This material is used by permission of John Wiley & Sons, Inc.

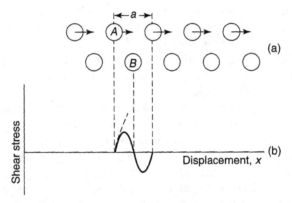

Figure 5.10 Schematic illustration of (a) relative shear of two planes of atoms in a strained material and (b) shear stress as a function of relative displacement of the planes from their equilibrium positions. Reprinted, by permission, from C. Kittel, *Introduction to Solid State Physics*, 2nd ed., p. 517. Copyright © 1957 by John Wiley & Sons, Inc.

Based on the concepts of intermolecular forces and shear modulus introduced in the previous section, it is relatively easy to estimate the theoretical stress required to cause slip in a single crystal. We call this the *critical shear stress*, σ_{cr}. Refer to Figure 5.10a, and consider the force required to shear two planes of atoms past each other. In the region of small elastic strains, the stress, τ, is related to the displacement, x, relative to the initial interplanar spacing, d, according to a modified form of Eq. (5.10) for the shear modulus, G:

$$\tau = G\frac{x}{d} \tag{5.14}$$

As a first approximation, we may represent the stress-displacement relation by a sine function (Figure 5.10b), since when atom A in Figure 5.10a is displaced to the point where it is directly over atom B in the plane below, an unstable equilibrium exists and the stress is zero, so

$$\tau = \frac{Ga}{2\pi d} \sin(2\pi x/a) \tag{5.15}$$

where a is the interatomic spacing in the direction of shear. Note that the slope (represented by the dashed line) of the sine wave in Figure 5.10b is simply the shear modulus, G. For small displacements, x, Eq. (5.15) reduces to Eq. (5.14). The critical shear stress, τ_{cr}, at which the lattice becomes unstable and begins to plasticly deform is given by the maximum value of τ in Eq. (5.15), or

$$\tau_{cr} = \frac{Ga}{2\pi d} \tag{5.16}$$

If we make the approximation that $a \approx d$, then

$$\tau_{cr} \approx \frac{G}{2\pi} \approx \frac{G}{6} \tag{5.17}$$

that is, the critical shear stress to plastic flow should be of the order of 1/6 the shear modulus. In reality, the measured critical shear stress is much smaller than this for most single-crystal metals. Further theoretical considerations reduced this value to $G/30$, but even this is much larger than the critical shear stress observed in most metals. Clearly, we cannot simply extend the theory of elasticity to large strains to determine shear strength—there are other considerations.

The primary consideration we are missing is that of crystal imperfections. Recall from Section 1.1.4 that virtually all crystals contain some concentration of defects. In particular, the presence of dislocations causes the actual critical shear stress to be much smaller than that predicted by Eq. (5.17). Recall also that there are three primary types of dislocations: edge, screw, and mixed. Although all three types of dislocations can propagate through a crystal and result in plastic deformation, we concentrate here on the most common and conceptually most simple of the dislocations, the edge dislocation.

The movement of an edge dislocation in a single crystal in response to an applied shear stress is illustrated in Figure 5.11. An edge dislocation can move by slip only in its slip plane, which is the plane containing the line of the dislocation and the Burger's vector (cf. Figure 1.31). Dislocation densities range from 10^2 to 10^3 dislocations/cm^2 in the best silicon single crystals to as high as 10^{11} or 10^{12} dislocations/cm^2 in heavily deformed metal crystals. As we will see later on, heat treatment can reduce the dislocation density in metals by about five orders of magnitude.

Beside dislocation density, dislocation orientation is the primary factor in determining the critical shear stress required for plastic deformation. Dislocations do not move with the same degree of ease in all crystallographic directions or in all crystallographic planes. There is usually a preferred direction for slip dislocation movement. The combination of slip direction and slip plane is called the *slip system*, and it depends on the crystal structure of the metal. The slip plane is usually that plane having the most dense atomic packing (cf. Section 1.1.1.2). In face-centered cubic structures, this plane is the (111) plane, and the slip direction is the [110] direction. Each slip plane may contain more than one possible slip direction, so several slip systems may exist for a particular crystal structure. For FCC, there are a total of 12 possible slip systems: four different (111) planes and three independent [110] directions for each plane. The

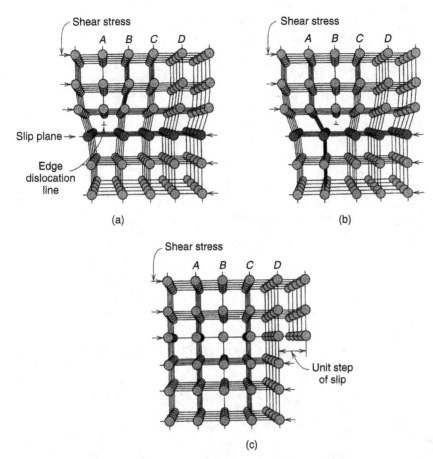

Figure 5.11   Schematic illustration edge dislocation motion in response to an applied shear stress, where (a) the extra half-plane is labeled as *A* (cf. Figure 1.28), (b) the dislocation moves one atomic distance to the right, and (c) a step forms on the crystal surface as the extra half-plane exits the crystal. Reprinted, by permission, from W. Callister, *Materials Science and Engineering: An Introduction*, 5th ed., p. 155. Copyright © 2000 by John Wiley & Sons, Inc.

possible slip systems for BCC and HCP metals are listed in Table 5.1. Body-centered cubic structures have a high number of slip systems and, as a result, undergo extensive plastic deformation; that is, they are highly ductile. The opposite is true of HCP structures, which have relatively few slip systems. For reference, the corresponding twinning systems are listed in Table 5.2.

The final factor in determining the critical shear stress for plastic deformation is how the preferred slip planes are oriented relative to the applied stress. Figure 5.12 illustrates how a test sample may be oriented such that the preferred slip plane is not necessarily along the direction of the applied force, F. Shear forces will still exist, and their magnitude will depend not only on the applied stress, but on orientation. The angle ϕ in Figure 5.12 represents the angle between the normal to the slip plane and the applied stress direction. The angle λ represents the angle between the slip direction and the applied stress direction. Remember that the slip plane and the slip direction are not the same. It is also important to note that the three vectors in Figure 5.12 are

Table 5.1 Slip Systems in FCC, BCC, and HCP Metals

Metals	Slip Plane	Slip Direction	Number of Slip Systems
Face-Centered Cubic			
Cu, Al, Ni, Ag, Au	(111)	[1$\bar{1}$0]	12
Body-Centered Cubic			
α-Fe, W, Mo	(110)	[$\bar{1}$11]	12
α-Fe, W	(211)	[$\bar{1}$11]	12
α-Fe, K	(321)	[$\bar{1}$11]	24
Hexagonal Close-Packed			
Cd, Zn, Mg, Ti, Be	(0001)	[11$\bar{2}$0]	3
Ti, Mg, Zr	(10$\bar{1}$0)	[11$\bar{2}$0]	3
Ti, Mg	(10$\bar{1}$1)	[11$\bar{2}$0]	6

Table 5.2 Twinning Systems in FCC, BCC, and HCP Metals

Crystal Structure	Twinning Planes	Twinning Directions
fcc	(111)	[11$\bar{2}$]
bcc	(112)	[11$\bar{1}$]
hcp	(10$\bar{1}$2)	[10$\bar{1}$1]

not coplanar in general, that is, $\phi + \lambda \neq 90°$. The *resolved shear stress*, τ_r, then is the component of the applied stress, σ_t, acting along the particular slip plane of interest and can be determined as

$$\tau_r = \sigma_t \cos \phi \cos \lambda \qquad (5.18)$$

As mentioned previously, a number of slip systems can operate simultaneously, but there will be one system that has the orientation which affords it the largest resolved shear stress of all the slip systems in operation. This system will be the one that has the maximum geometric factor, $(\cos \phi \cos \lambda)_{max}$, since the applied stress is the same for all slip systems. So, the *maximum resolved shear stress*, $\tau_{r,max}$, is given by:

$$\tau_{r,max} = \sigma_t (\cos \phi \cos \lambda)_{max} \qquad (5.19)$$

As the applied stress, σ_t, increases, the maximum resolved shear stress increases according to Eq. (5.19), finally reaching a critical value, called the *critical resolved shear stress*, τ_{cr}, at which slip along the preferred plane begins and plastic deformation commences. We refer to the applied stress at which plastic deformation commences as

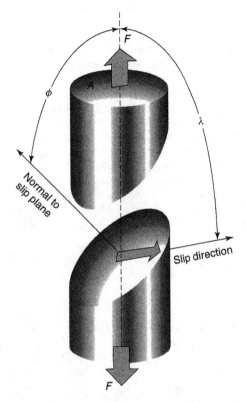

Figure 5.12 Schematic illustration of the relationship between tensile axis, slip plane, and slip direction used in calculating the resolved shear stress for a single crystal. Reprinted, by permission, from W. Callister, *Materials Science and Engineering: An Introduction*, 5th ed., p. 160. Copyright © 2000 by John Wiley & Sons, Inc.

the *yield stress*, σ_y. The yield stress can be determined from Eq. (5.19) in a form of *Schmid's Law*, which states that slip takes place along a given slip plane and direction when the corresponding component of shear stress reaches a critical value:

$$\sigma_y = \frac{\tau_{cr}}{(\cos\phi\cos\lambda)_{max}} \tag{5.20}$$

The minimum stress necessary to introduce yielding occurs when a single crystal is oriented such that $\phi = \lambda = 45°$, for which

$$\sigma_y = 2\tau_{cr} \tag{5.21}$$

Equations (5.20) and (5.21) are valid for an applied tensile or compressive stress and can be used in the case of twinning as well. However, τ_{cr} for twinning is usually greater than τ_{cr} for shear. Some values of the critical resolved shear stress for slip in some common metals, and their temperature dependence, are shown in Figure 5.13. Note that the critical resolved shear stress for HCP and FCC (close-packed) structures rises only modestly at low temperatures, whereas that for the BCC and rock salt structures increases significantly as temperature decreases.

Cooperative Learning Exercise 5.3

Consider a single crystal of BCC Mo oriented such that a tensile stress of 52 MPa is applied along a [010] direction. From Table 5.1, we see that the slip plane for BCC Mo is (110) and the slip direction is [$\bar{1}$11].

Person 1: Determine the angle, ϕ, between the normal to the slip plane and the applied stress direction. You will find it helpful to draw a (110) plane in a BCC unit cell, and draw the normal to this plane. Once you have determined this direction, you should be able to easily determine the angle it makes with the [010] direction by inspection.

Person 2: Determine the angle, λ, between the slip direction and the applied stress direction. You will find it helpful to draw a cubic cell with the [010] and [$\bar{1}$11] directions as two sides of a triangle, the third side being a face diagonal. At this point, Table 1.8 may be helpful in establishing a relationship between the length of these sides in terms of the lattice parameter, a. You can then use some simple geometry to find λ.

Combine your information to calculate the resolved shear stress, τ_r, using Eq. (5.18).

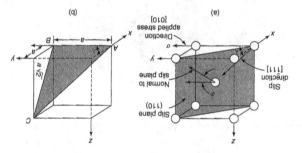

Reprinted, by permission, from W. Callister, Materials Science and Engineering: An Introduction, p. 162, 5th ed. Copyright © 2000 by John Wiley & Sons, Inc.

Answers: (see figures) $\phi = 45°$; $\lambda = \tan^{-1}(2)^{0.5} = 54.7°$;
$\tau_r = 52$ MPa$(\cos 45°)(\cos 54.7°) = 21.3$ MPa

5.1.2.2 Slip in Polycrystals — Strength and Hardening.

The yield stress, σ_y, sometimes called the *yield strength*, as introduced by Eqs. (5.20) and (5.21), is an important material property, particularly in ductile materials. It not only represents the change from a recoverable, elastic response to a plastic response and the accompanying permanent deformation (see Figure 5.8), but also represents an important design parameter in materials selection. Obviously, when a material undergoes permanent deformation, it may not be useful for its specified purpose and may need to be replaced. As a result, the yield strength may be an even more important mechanical property than the modulus for certain applications.

There are no theoretical expressions for the yield strength beyond what has already been described in terms of the critical resolved shear stress in single crystals. This problem is further complicated by the fact that most real materials are polycrystalline—that is, composed of many randomly oriented crystallites, or grains, with corresponding grain boundaries where the grains meet. As a result, only certain grains may be oriented favorably to allow slip to begin, but neighboring grains may not be so oriented, and the stress required to initiate plastic flow increases substantially in polycrystalline materials. It has been shown that at least five independent slip systems

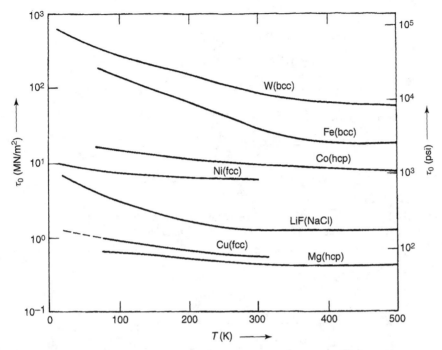

Figure 5.13   Temperature variation of critical resolved shear stress for single-crystal metals of different crystal structures. From K. M. Ralls, T. H. Courtney, and J. Wulff, *Introduction to Materials Science and Engineering*. Copyright © 1976 by John Wiley & Sons, Inc. This material is used by permission John Wiley & Sons, Inc.

must be mutually operative for a polycrystalline solid to exhibit ductility. This means that some of the metals in Table 5.1, particularly the HCP metals such as zinc, do not have sufficient slip systems to undergo significant plastic deformation when in the polycrystalline structure.

Since the grain boundaries essentially act as barriers to slip in adjacent crystals (see Figure 5.14), it makes sense that the yield strength should depend on grain size. This is indeed the case, and the *Hall–Petch relationship* shows an inverse square-root dependence of yield strength on grain size, d:

$$\sigma_y = \sigma_0 + kd^{-1/2} \tag{5.22}$$

where σ_0 and k are empirically determined constants. Equation (5.22) has proved very useful for describing the variation in yield strength with grain size over small ranges of grain sizes, as illustrated for a copper–zinc–brass alloy in Figure 5.15. It is not valid, however, over large size ranges, nor for all grain sizes. The Hall–Petch relationship, and others like it, have found recently renewed interest in the area of nanostructured materials, since they predict enormously large yield strengths as grains approach the *nanocrystalline* range ($d \approx 10^{-9}$ m, or 1 nm).

The presence of grain boundaries also affects slip in the material even after plastic deformation has commenced. A phenomenon known as *strain hardening*, or *work hardening* or *cold working*, is the result of constrained dislocation mobility and increased

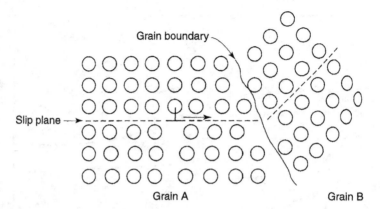

Figure 5.14 Illustration of grain boundaries acting as barriers to slip in polycrystalline materials. Reprinted, by permission, from W. Callister, *Materials Science and Engineering: An Introduction*, p. 166, 5th ed. Copyright © 2000 by John Wiley & Sons, Inc.

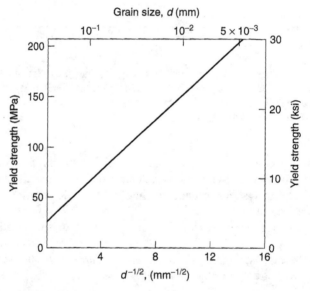

Figure 5.15 Influence of grain size on yield strength for a 70–30 Cu–Zn brass alloy. Reprinted, by permission, from W. Callister, *Materials Science and Engineering: An Introduction*, 5th ed., p. 167. Copyright © 2000 by John Wiley & Sons, Inc.

dislocation density as a material is strained. It is easy to observe this phenomenon; in fact, you have probably used it many times in your life without even knowing it. Simply take a small piece of metal such as a copper tube or an aluminum paper clip, and bend it back and forth several times. In addition to becoming harder to bend and generating heat, the metal will eventually break. Why would a seemingly ductile metal suddenly break? This is strain hardening. As the metal is "worked" back and forth, the dislocations that allow slip and ductility to occur eventually increase in density and decrease in mobility such that the critical resolved shear stress necessary to continue plastic flow increases beyond the applied stress. At that point, either you cannot work the piece anymore or brittle fracture (see Section 5.2.2) occurs. The critical resolved shear stress can be related to the *dislocation density*, ρ (not to be confused with the mass density):

$$\tau_{cr} = \tau_{cr,0} + A\rho^{1/2} \tag{5.23}$$

where A is an empirically determined constant that is approximately the product of the shear modulus, G, and the Burger's vector magnitude, b, and $\tau_{cr,0}$ is the critical resolved shear stress in the absence of interfering dislocations. This relationship has been borne out experimentally, as shown in Figure 5.16 for both single- and poly-crystalline copper. The dislocation density in a metal increases with deformation or cold work, due to dislocation multiplication or the formation of new dislocations. Consequently, the average distance between dislocations decreases, and the motion of dislocations is limited by the other dislocations around it. Thus, as shown in Eq. (5.23), as the dislocation density increases, the stress necessary to deform the metal increases.

In addition to grain size control and strain hardening, a single-phase metal can be strengthened (yield strength increased) by a number of thermal and chemical methods, including solid solution formation and *dispersion strengthening*. Solute atoms intro-duced in the formation of solid solutions tend to segregate to dislocations because the total elastic strain energy associated with a dislocation and the solute atom is reduced. As a result, alloys are almost always stronger than pure metals because impurity atoms that go into solid solution ordinarily impose lattice strains (either compressive or ten-sile, depending on the relative size of the impurity atom) on the surrounding atoms. In this way, lattice strain field interactions between dislocations and these impurity atoms restrict dislocation movement. These same lattice interactions exist during plastic

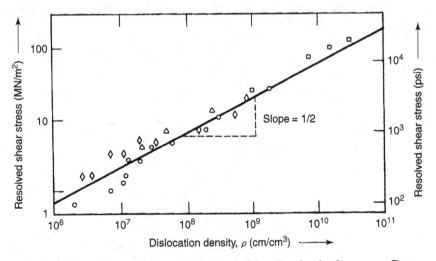

Figure 5.16 Resolved shear stress as a function of dislocation density for copper. Data are for ☐ polycrystalline copper; ○ single-crystal copper with one slip system operative; ◇ single-crystal copper with two slip systems operative; and △ single-crystal copper with six slip systems operative. From K. M. Ralls, T. H. Courtney, and J. Wulff, *Introduction to Materials Science and Engineering*. Copyright © 1976 by John Wiley & Sons, Inc. This material is used by permission John Wiley & Sons, Inc.

deformation, too, so a greater stress is necessary to first initiate and then continue plastic deformation in solid-solution-strengthened alloys. Besides relative size of the solute atoms, the effective elastic modulus of a solute atom also determines its effectiveness as a solid-solution strengthener.

The same effect on dislocation movement during slip can be accomplished by two other techniques known as *precipitation hardening* and *dispersion strengthening*. In both cases, the principle is the same: Fine dispersions of heterogeneities impede dislocation motion. In precipitation hardening, an alloy is subjected to a solution heat treatment (see Figure 5.17). First, an alloy of suitable composition (line x–x in Figure 5.17) is heated to a temperature above the solvus line but below the solidus line (point 1), and it is held at this temperature to allow complete homogenization of the solid solution. The alloy is quenched to form a supersaturated solution (point 2). The quenched alloy is then reheated and held at this temperature, called the *aging temperature* (point 3), for a specified period of time to allow the nonequilibrium second phase to precipitate out. Precipitation hardening is a very important method of strengthening many solid-solution alloys such as aluminum and magnesium alloys, copper–beryllium alloys, high nickel-based alloys, and some stainless steels. In dispersion-strengthened alloys, the atmosphere can be controlled during melt processing to form fine compound materials that do not dissolve readily in the base metal. These dispersed particles then act to restrict dislocation movement. A common example is *oxide-dispersion-strengthened* (ODS) *alloys* in which oxygen reacts with the parent metal to form higher-melting-point metal oxide particles. These materials are further described in Section 5.4.1.

5.1.2.3   *Recovery in Polycrystals — Annealing and Recrystallization.*   Deformation in polycrystalline materials can lead to not only strain hardening and an increase in dislocation density, as described above, but also a change in grain shape. In some instances, all of these processes can be reversed through heat treatment in what is generically called *annealing*. Annealing refers to a heat treatment in which a material

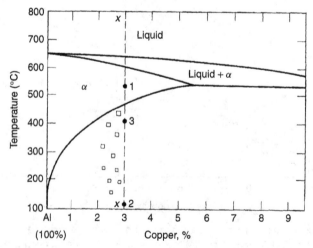

Figure 5.17   Partial Al–Cu equilibrium phase diagram illustrating precipitation hardening. From Z. Jastrzebski, *The Nature and Properties of Engineering Materials*, 2nd ed., Copyright © 1976 by John Wiley & Sons, Inc. This material is used by permission of John Wiley & Sons, Inc.

is exposed to an elevated temperature for an extended period of time and then slowly cooled. The annealing process usually consists of three stages: (1) heating to the desired temperature; (2) holding or "soaking" at that temperature; and (3) cooling at a slow rate, usually to room temperature. Time is an important parameter in each step. During annealing, several processes can occur that lead to what is termed "recovery," which is a restoration of properties and structures that are more similar to those found in the pre-cold-worked states. We concentrate here on those recovery processes during annealing that are specific to metals and alloys, and we return in later sections to describe annealing in other materials, particularly inorganic glasses.

Annealing in metals can first lead to *stress relaxation* in which stored internal strain energy due to plastic deformation is relieved by thermally activated dislocation motion (see Figure 5.18). Because there is enhanced atomic mobility at elevated temperatures, dislocation density can decrease during the recovery process. At still higher temperatures, a process known as *recrystallization* is possible, in which a new set of

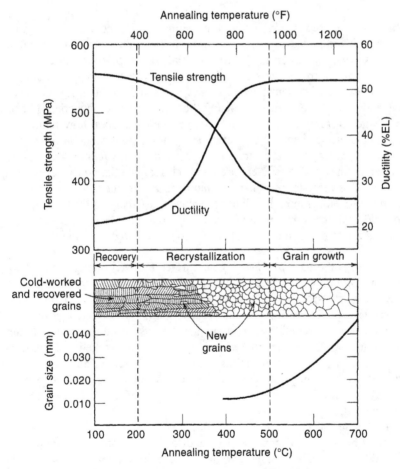

Figure 5.18 Influence of annealing temperature on tensile strength and ductility of a brass alloy. Grain structures during recovery, recrystallization, and grain growth are shown schematically. Reprinted, by permission, from W. Callister, *Materials Science and Engineering: An Introduction*, 5th ed., p. 175. Copyright © 2000 by John Wiley & Sons, Inc.

Table 5.3 Recrystallization and Melting Points of Some Common Metals and Alloys

Metal	Recrystallization Temperature		Melting Temperature	
	°C	°F	°C	°F
Lead	−4	25	327	620
Tin	−4	25	232	450
Zinc	10	50	420	788
Aluminum (99.999 wt%)	80	176	660	1220
Copper (99.999 wt%)	120	250	1085	1985
Brass (60 Cu–40 Zn)	475	887	900	1652
Nickel (99.99 wt%)	370	700	1455	2651
Iron	450	840	1538	2800
Tungsten	1200	2200	3410	6170

strain-free and equiaxed (having equal dimensions in all directions) grains are formed. These recrystallized grains are similar to those in the pre-cold-worked metal and have relatively low dislocation densities. The recrystallization process is a diffusion-driven phenomenon similar to the crystallization processes described in Section 3.1.2. As shown in Figure 5.18, the tensile strength decreases and the ductility increases during recrystallization. It is worth reemphasizing that this is a time-dependent phenomenon, too. The degree of recrystallization and recovery depends upon how long the strained material is held at the specified temperature. Typically, the times necessary to obtain complete recrystallization are on the order of hours, and a *recrystallization temperature* is often defined as the temperature at which recrystallization reaches completion in one hour. The recrystallization temperature is about one-third to one-half of the absolute melting temperature, T_m. For pure metals, the recrystallization temperature is close to $0.3T_m$, but for commercial alloys it may be as high as $0.7T_m$. Recrystallization temperatures for some common metals and alloys are listed in Table 5.3.

5.1.2.4 Strength and Hardness. One further property that is related to plastic deformation is *hardness*. Hardness is a measure of a material's resistance to localized plastic deformation—for example, from a scratch or small indentation. A description of hardness is included in this section because hardness is related to binding forces of molecules, just as strength and modulus are. We also used the term "hardening" to describe the increase in tensile strength as a metal is cold-worked. Hence, hardness is a recoverable quantity during annealing, just as yield strength and ductility are. The principles of strengthening through solid solutions and dispersants also apply here, such that alloys and dispersion-strengthened metals tend to be harder than the pure metals as well.

An unfortunate aspect of hardness is that it is difficult to quantify in an absolute manner. Most hardness scales are relative and fairly qualitative, and there is a proliferation of different hardness scales. Some of the more common scales are shown in Figure 5.19. In a typical hardness test, a hard indenter of a standard shape is pressed into the surface of a material under a specified load, causing first elastic and then plastic deformation. The resulting area of the indentation or depth of indentation is measured and assigned a numerical value. The value depends upon the apparatus and scale used.

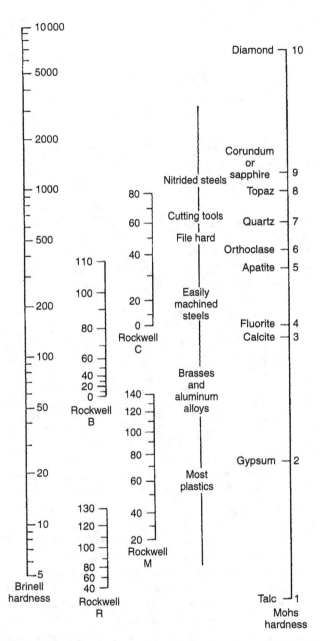

Figure 5.19 Comparison of some common hardness scales. From Z. Jastrzebski, *The Nature and Properties of Engineering Materials*, 2nd ed., Copyright © 1976 by John Wiley & Sons, Inc. This material is used by permission of John Wiley & Sons, Inc.

One of the simplest scales is the Mohs scale, in which a material is assigned a place on a scale depending only upon what it will scratch and what will scratch it, relative to the hardest (diamond) and softest (talc) materials known.

Since both hardness and strength are relative measures of resistance to plastic deformation, we would expect them to be proportional to each other. This is indeed the

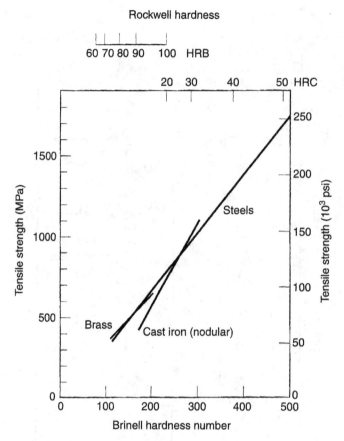

Figure 5.20 Relationship between hardness and tensile strength for steel, brass and cast iron. Reprinted, by permission, from W. Callister, *Materials Science and Engineering: An Introduction*, 5th ed., p. 140. Copyright © 2000 by John Wiley & Sons, Inc.

case, but since absolute values of hardness are hard to determine, it is difficult to assign a universal equation to this relationship. The relationship between tensile strength and Brinell hardness is shown in Figure 5.20 for steel, brass, and cast iron.

In the past two decades, more standardized measurements of hardness have been made using a technique called *nanoindentation*. This technique is essentially the same as that described above for traditional hardness testers, except that the indenters are much smaller (about 100 nm), as is the indentation area. Developments in instrumentation have made possible the measurements of loads in the nanonewtons range, and displacements of about 0.1 nm can be accurately measured [1]. The result of a nanoindentation experiment, shown schematically in Figure 5.21b, is a load versus displacement curve, an example of which is shown in Figure 5.21a. In Figure 5.21a, h_{max} represents the displacement at the peak load, P_{max}; and h_c, the contact depth, is defined as the depth of the indenter in contact with the sample under load. The final displacement, h_f, is the displacement after complete unloading. The quantity, S, is called the initial unloading contact stiffness and is determined from the slope of the initial portion of the unloading curve, dP/dh, as indicated in Figure 5.21a.

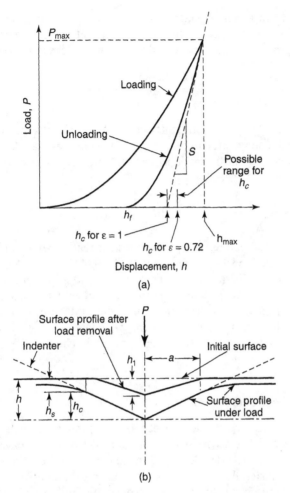

Figure 5.21 (a) A typical load-displacement curve and (b) the deformation pattern of an elastic–plastic sample during and after nanoindentation. Reprinted, by permission, from X. Li and B. Bhushan, *Materials Characterization*, Vol. 48, p. 13. Copyright © 2002 by Elsevier Science, Inc.

Nanoindentation hardness, H, is defined as the indentation load divided by the projected contact area of the indentation. From the load–displacement curve, hardness at the peak load can be determined as

$$H = \frac{P_{max}}{A} \tag{5.24}$$

where A is the projected contact area. The hardness thus has units of N/m^2, or GPa. For an indenter with a known geometry, A is a function of the contact depth, h_c. For a perfect triangular-shaped, diamond-tipped indenter known as a *Berkovich indenter*, the projected contact area is

$$A = 24.56h_c^2 \tag{5.25}$$

Often times corrections must be made to A for imperfect tips and blunting effects.

The modulus from a nanoindentation test is often reported in terms of a *reduced elastic modulus*, E_r, to take into account the fact that at this size scale, the elastic response of the probe tip, E_i, as well as the modulus of the test material, E, must be considered:

$$E_r = \frac{1 - v^2}{E} + \frac{1 - v_i^2}{E_i} \tag{5.26}$$

where v and v_i are Poisson's ratio for the test material and the indenter material, respectively. For diamond, which is a common indenter material, $E_i = 1141$ GPa and $v_i = 0.07$. The reduced elastic modulus can be obtained from a load–displacement curve via the stiffness, S, according to

$$S = 2\beta E_r \sqrt{A/\pi} \tag{5.27}$$

where A and E_r are as before, and β is a constant that depends on the geometry of the indenter. For a Berkovich-type indenter, $\beta = 1.034$.

There are many experimental details to consider in nanoindentation, but the end result is that both hardness and elastic modulus can be determined relatively routinely on very small samples, including thin films and nanocomposites. What is more, the hardness measurement is quantitative and reproducible. A typical hardness–modulus–indentation depth curve is shown in Figure 5.22 for an Si(100) plane.

5.1.3 Thermomechanical Effects

In addition to the temperature dependence of the properties such as strength and modulus, which we will discuss individually for each material class, there are two fundamental topics that are often described in the context of heat transfer properties or thermodynamics of materials—for example, thermal conductivity or specific heat—but are related more to mechanical properties because they involve dimensional changes. These two properties, *thermoelasticity* and *thermal expansion*, are closely related, but will be described separately.

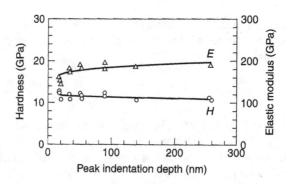

Figure 5.22 Hardness and elastic modulus as a function of indentation depth at various peak loads for Si(100). Reprinted, by permission, from X. Li and B. Bhushan, *Materials Characterization*, Vol. 48, p. 14. Copyright © 2002 by Elsevier Science, Inc.

5.1.3.1 *Thermal Expansion.* Thermal expansion is a measure of the expansion or contraction of a solid when it is heated or cooled. In the most general case, the *coefficient of thermal expansion*, α, sometimes called the volume thermal expansion coefficient, is defined as the change in volume per unit volume, V, with temperature, T, at constant pressure, P:

$$\alpha_V = \frac{1}{V_0}\left(\frac{dV}{dT}\right)_P \tag{5.28}$$

where we use the subscript V to indicate volume thermal expansion. In many materials, the change of dimensions in any one direction may be different from that in the other two directions, depending on the crystallographic directions of the measurement. The change in linear dimensions with temperature along any one direction define the *linear thermal expansion coefficient*, α_L:

$$\alpha_L = \frac{1}{L_0}\left(\frac{dL}{dT}\right)_P \tag{5.29}$$

For materials that are *isotropic*, that is, have the same properties in all directions, it can be shown that $\alpha_V = 3\alpha_L$. A material that has different properties in different directions is said to be *anisotropic*. Thus, a linear expansion coefficient, if no direction of measurement is explicitly stated, implies an isotropic material. Conversely, a volume thermal expansion coefficient implies an anisotropic material, and one should exercise caution when deriving linear thermal expansion coefficients from volume-based measurements.

Like the melting point, specific heat, and elastic modulus, thermal expansion can be directly related to the binding force between molecules, and thus the potential energy function, as illustrated in Figure 5.23. Just as it affected the energy required to separate bonds during melting or separate bonds under an applied force, the bond strength between atoms affects the ability of atoms to deviate from their equilibrium positions due to increased thermal energy. Due to its origins in lattice vibration, the thermal expansion coefficient shows the same temperature dependence as heat capacity, and at lower temperatures it approaches zero as absolute zero is reached. It is, however, possible for a material to have a negative value of thermal expansion coefficient.

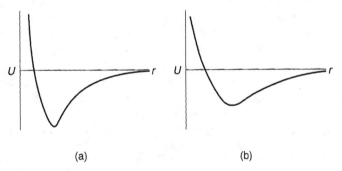

(a)   (b)

Figure 5.23 Potential energy diagrams for materials with (a) high melting point, high elastic modulus, and low thermal expansion coefficient and (b) low melting point, low elastic modulus, and high thermal expansion coefficient.

5.1.3.2 *Thermoelastic Effect.* A mechanical phenomenon that involves the thermal expansion coefficient is the thermoelastic effect, in which a material is heated or cooled due to mechanical deformation. The thermoelastic effect is represented by the following relation:

$$\left(\frac{dT}{d\varepsilon}\right)_s = \frac{-V_m \alpha E T}{C_V} \tag{5.30}$$

where $(dT/d\varepsilon)_S$ is the change in temperature with strain at constant entropy (adiabatic stretching or cooling), V_m is the molar volume, α is the thermal expansion coefficient, E is the tensile modulus, and C_v is the molar heat capacity at constant volume. Most materials have a positive value of α; that is, they expand upon heating, so that $(dT/d\varepsilon)_S$ is negative and the material cools upon straining. When α is negative, the material heats upon straining. This is the case for many polymer elastomers, such as a rubber band, which heats up upon being stretched.

5.1.4 Stress–Strain Diagrams

We turn our attention now to some practical aspects of mechanical property determinations. The important quantities such as modulus, strength, and ductility are typically summarized in graphical form on a *stress–strain diagram*. The details of how the experiment is performed and how the stress–strain diagram is generated are described for some common types of applied forces below.

5.1.4.1 *The Tensile Test.* A typical tensile test is performed in what is known as a universal testing apparatus, which is shown schematically in Figure 5.24. The specimen to be tested is typically in the shape of a "dog bone" such that the specimen is wider at the ends than in the center to facilitate mounting in the specimen grips. The sample cross section can be any geometry, as long as it is uniform along the distance between the grips, known as the *gauge length*. Circular and rectangular cross sections are the most common geometries. The gauge length can be any length, and is not standardized, though gauge lengths of 1 inch (25 mm) and 2 inches (50 mm) are common. One *grip*, typically the top grip, is attached to a nonmovable *load cell* that is mounted to the test frame. The load cell measures the force being imparted to the sample. The other grip is attached to a moveable portion of the frame called the *crosshead*. The crosshead moves downward at a specified rate, called the *crosshead speed*, which imparts a known *extension rate* to the sample. These rates are normally very slow, on the order of fractions of an inch per minute. The absolute extension in the sample could be estimated from the extension rate and time, but it is typically measured much more accurately with what is called an *extensometer* (eks′ ten **säm′** ə ter).

The tensile test is typically destructive; that is, the sample is extended until it plasticly deforms or breaks, though this need not be the case if only elastic modulus determinations are desired. As described in the previous section, ductile materials past their yield point undergo plastic deformation and, in doing so, exhibit a reduction in the cross-sectional area in a phenomenon known as *necking*.

The "raw data" from a tensile test are the load versus elongation measurements made by the load cell and the extensometer, respectively. To eliminate sample geometry effects, the extension is divided by the initial length to obtain the dimensionless strain (which is occasionally multiplied by 100 and reported as % elongation for samples

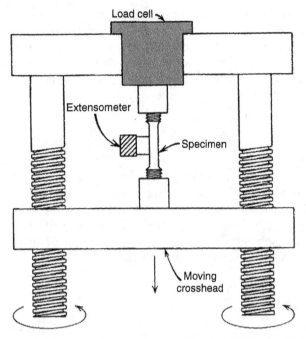

Figure 5.24 Schematic diagram of a tensile test. From K. M. Ralls, T. H. Courtney, and J. Wulff, *Introduction to Materials Science and Engineering.* Copyright © 1976 by John Wiley & Sons, Inc. This material is used by permission John Wiley & Sons, Inc.

Cooperative Learning Exercise 5.4

Work with a neighbor. The tensile testing machine shown in Figure 5.24 can be used to measure the tensile properties of materials ranging from the most flexible and weak rubber to the strongest and stiffest steel. In order to measure the tensile properties of these vastly different materials, some things on the machine or the way a test is performed may need to be changed. What would you have to change in order to run these two very different types of materials? Name at least three things you can think of.

Answers: Load cell; elongation rate; type of grip; type of extensometer.

(Courtesy of Brian P. Grady)

---

that undergo large deformations), and the load is divided by the cross-sectional area to obtain the stress. Stress is plotted versus strain, and the result is a *stress–strain diagram.*

The presence of necking presents a dilemma. The cross-sectional area is integral to the calculation of stress, yet it has a decreasing value as the experiment proceeds, especially where necking is prevalent. Two types of stress are usually encountered in this case. If the actual cross-sectional area corresponding to the load measurement is used, the *true stress* is obtained. This requires that the cross-sectional area be continuously monitored and matched to the corresponding load measurements. A more convenient method is to divide the measured load by the initial cross sectional area, which is a constant, to obtain what is termed the *engineering stress*, $\sigma_{eng} = F/A_0$. There is a corresponding true strain, ε_{true}, which is the sum of all the instantaneous length changes divided by the instantaneous length (see Figure 5.5), which for small deformations is related to the instantaneous cross-sectional area, A:

$$\varepsilon_{true} = \ln\left(\frac{L_0 + \Delta L}{L_0}\right) \approx \ln\left(\frac{A_0}{A}\right) \tag{5.31}$$

Obviously, the engineering stress will be less than the true stress (since A_0 is always greater than or equal to A), and the engineering strain will be greater than the true strain at a given load for similar reasons. The result is that the true stress–true strain and engineering stress–engineering strain plots look different, as illustrated in Figure 5.25. It is customary to use engineering stress and strain for small deformations, and this is the convention we will use throughout the rest of the book, except where noted.

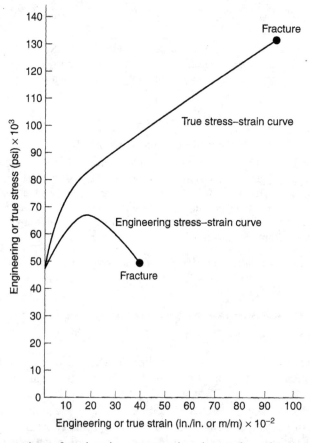

Figure 5.25 Comparison of engineering stress-engineering strain and true stress–true strain plots. Reprinted, by permission, from J. F. Shackelford, *Introduction to Materials Science for Engineers*, 5th ed., p. 192. Copyright © 2000 by Prentice Hall, Inc.

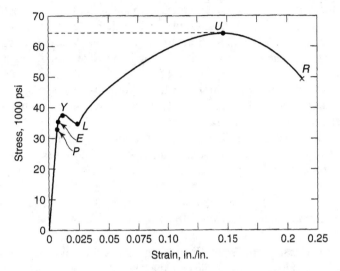

Figure 5.26   Stress–strain diagram for mild steel, illustrating different types of stress. From Z. Jastrzebski, *The Nature and Properties of Engineering Materials*, 2nd ed. Copyright © 1976 by John Wiley & Sons, Inc. This material is used by permission of John Wiley & Sons, Inc.

Subscripts will be dropped from stress and strain, and the use of engineering values will be assumed.

5.1.4.2 *Modulus and Strength Determinations.*

Once the stress–strain diagram has been generated, a number of important mechanical properties for a material can be obtained. Consider the stress–strain diagram for mild steel in Figure 5.26. Several points have been labeled for ease of identification. The elastic region of the diagram is easily identifiable. It begins at the origin and is the constant slope region of the graph. Here, the tensile modulus, E, is given by Eq. (5.5), and it can be determined as the slope of the stress–strain curve in the elastic region. Where the elastic region ends is difficult to determine exactly. In Figure 5.26, two points have been labeled that can correspond to the end of the elastic region. The first, point P, is called the *proportional limit*, and it signals the end of the mathematical relationship between stress and strain that leads to a constant slope—that is, the proportional region. Within the region OP, Hooke's Law applies. The second point, point E, indicates the *elastic limit*. Up to point E, the strain is fully recoverable. The implication, however, is that while the response in the region OE may be elastic, there may not be a perfect Hookean response from point P to point E. In reality, it is difficult to differentiate between the proportional and elastic limits for most materials, and either one will produce an adequate determination of the tensile modulus.

Beyond point E, the material begins to plasticly deform, and at point Y the yield point is achieved. The stress at the yield point corresponds to the yield strength, σ_y [see Eq. (5.20)]. Technically, point Y is called the *upper yield point*, and it corresponds to the stress necessary to free dislocations. The point at which the dislocations actually begin to move is point L, which is called the *lower yield point*. After point L, the material enters the ductile region, and in polycrystalline materials such as that of Figure 5.26, strain hardening occurs. There is a corresponding increase in the stress required to maintain plastic flow, and a maximum stress value is often achieved in this region, here at point U. The stress at point U is called the ultimate strength, σ_u, or the *ultimate tensile strength* (UTS). Sometimes the ultimate tensile strength is a more important design parameter than the yield strength, because it indicates the maximum load that a material will bear, regardless of whether it deforms or not. It should be pointed out that not all materials strain harden, so the ultimate strength can be located at any point along the stress–strain curve; it is simply the maximum strength. We will see examples for which the maximum stress occurs at the yield point, in which case $\sigma_u = \sigma_y$. The final point in Figure 5.26 is point R, which represents the *breaking point*, or *rupture point* or *fracture point*, of the material. The stress at point R is called the *fracture strength*, σ_f. Again, we will see an example of stress–strain diagrams for which the fracture strength can vary position; it is simply the point at which the material physically separates into two or more pieces. If the material is not extended to the breaking point, then no fracture strength can be determined.

5.1.4.3 *Yield Stress and Proof Stress*.
In Figure 5.26, we noted that there was a distinct point, point P, which indicated the end of the proportional region in the stress–strain diagram, up to which point Hooke's Law applied. Unfortunately, most ductile metals do not have a true proportional limit, nor a true yield point. A common example of the stress–strain diagram in a tensile experiment is shown in Figure 5.27. Instead of having an upper and lower yield point, the modulus changes slowly and

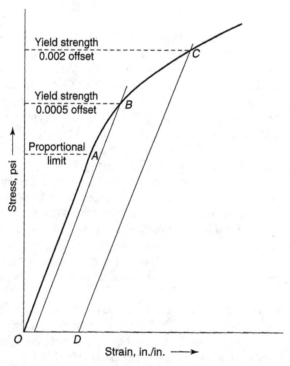

Figure 5.27 Stress–strain diagram indicating method for determining proof stress. From Z. Jastrzebski, *The Nature and Properties of Engineering Materials*, 2nd ed. Copyright © 1976 by John Wiley & Sons, Inc. This material is used by permission of John Wiley & Sons, Inc.

continuously until the ductile region is attained. This makes it difficult to determine the proportional limit and modulus, as well as the yield point and yield strength.

We must create an artificial convention for determining where the elastic region ends and the ductile region begins. This is achieved with what is known as the *proof stress*. The proof stress is defined as the stress required to produce a certain definite permanent strain. This is determined by the offset method. A line (DC in Figure 5.27) is drawn parallel to the straight-line portion of the stress–strain diagram (OA in Figure 5.27), starting at a predetermined strain for point D. Typically, this is a value of 0.2% strain (0.002), and the distance OD is known as the "0.2% offset." The yield point is then taken as the point where the line DC intersects the stress–strain diagram, or point C. The stress at point C is the proof stress, which is then taken as the yield stress and is referred to as the "yield stress at 0.2% offset." The modulus is still calculated in the straight-line portion of the curve, as indicated by line OA. It should be noted that other types of offsets can be used, such as the 0.05% offset indicated by the line ending at point B. The type of offset used should be clearly indicated in the yield strength determination for these materials.

5.1.4.4 Ductility, Toughness, and Resilience.
Several additional useful quantities can be determined from a stress–strain diagram. These are illustrated in Figure 5.28. In addition to the tensile modulus, yield strength, and ultimate strength as described previously, the ductility, which we have previously used in a qualitative sense to describe a metal that undergoes plastic deformation, can be given a quantitative value by reporting the percent elongation at failure. This point is indicated by point 4 in Figure 5.28. Note that in most cases, there is an elastic recovery after fracture (the slope is equal to that given by point 1) that must be accounted for in determining the final elongation. Recall that elongation is

$$\%\text{Elongation} = \left(\frac{L_f - L_0}{L_0} \right) \times 100 \tag{5.32}$$

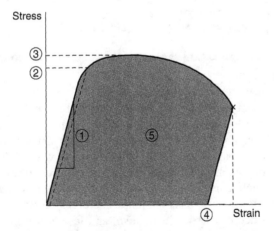

Figure 5.28 Stress–strain diagram showing (1) modulus, (2) yield strength, (3) ultimate tensile strength, (4) ductility, and (5) toughness. Note the use of proof stress in determination of yield stress. Reprinted, by permission, from J. F. Shackelford, *Introduction to Materials Science for Engineers*, 5th ed., p. 190. Copyright © 2000 by Prentice-Hall, Inc.

The % elongation at fracture is thus dependent upon the initial gauge length, which should be cited when reporting values of ductility. Typically, a gauge length of 50 mm (2 in.) is used. Another quantity that is sometimes used to express ductility is the % *reduction in area* (%RA):

$$\%RA = \left(\frac{A_0 - A_f}{A_0}\right) \times 100 \tag{5.33}$$

Cooperative Learning Exercise 5.5

Work in groups of four. A tensile specimen of 0.505-in. diameter and 2-in. gauge length is subjected to a load of 10,000 lb_f that causes it to elasticly deform at constant volume to a gauge length of 2.519 in.

Person 1: Calculate the engineering strain.

Person 2: Calculate the true strain

Person 3: Calculate the % elongation.

Person 4: Calculate the percentage reduction in area, %RA. (You may want to consult Person 2 on this.)

Compare your answers. How are they similar? How are they different?

Persons 1, 3: Determine the engineering stress. Use this value and the engineering strain to estimate the elastic modulus.

Person 2, 4: Determine the true stress. Use this value and the true strain to estimate the elastic modulus.

Compare your answers. Why are the moduli determined with these two methods the same or different?

Answers: $\varepsilon = (2.519 - 2)/2 = 0.2595$; $\varepsilon_{true} = \ln(2.519/2) = 0.2307$; %elongation = (0.2595) × 100 = 25.95%; %RA = $(1 - A_f/A_0) \times 100 = (1 - 0.79397) \times 100 = 20.6$.

$\sigma = 10{,}000\ lb_f/[\pi(0.505in/2)^2] = 49.9$ kpsi; $E = 49.9$ kpsi/0.2595 = 192.4 Mpsi

$\sigma_{true} = 10{,}000(1.259)/[\pi(0.505in/2)^2] = 62.9$ kpsi; $E = 62.9/0.2307 = 272.5$ Mpsi.

As illustrated in Figure 5.25, the true stress–strain and engineering stress–strain curves can be quite different, especially at high elongations such as this. In fact, relations such as (5.31) are probably not valid at these large elongations. Furthermore, the elastic limit has probably been exceeded at this point, despite our assumptions.

Two remaining quantities, resilience and toughness, are related to the ability of a material to absorb energy. *Toughness*, as indicated in Figure 5.28, is the area under the stress–strain curve up to the point of fracture. Thus, it is both a measure of a material's strength and ability to deform plastically. A material with low strength and high ductility may possess more toughness than one with high strength and low ductility. This is an important design consideration in many applications. The continued use of metals in automobile manufacturing (for frames and other structural components) is the result, in part, of safety considerations. There are certainly stronger materials than aluminum, less dense materials than steel, and less costly materials than either, but the combination of properties, including the ability to absorb energy during impact, is what keeps metals attractive for these applications.

A quantity related to toughness is the *resilience*, which is the ability of a material to absorb energy when it is deformed elastically, and then, upon unloading, to recover

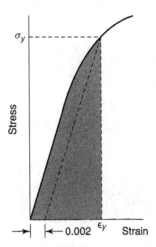

Figure 5.29 Schematic illustration of how modulus of resilience (shaded area) is determined. Reprinted, by permission, from W. Callister, *Materials Science and Engineering: An Introduction*, 5th ed., p. 130. Copyright © 2000 by John Wiley & Sons, Inc.

the energy. Resilience is quantified with the *modulus of resilience, U_r*, which is the strain energy per unit volume required to stress a material from an unloaded state up to the point of yielding. As shown in Figure 5.29, this is simply the area under the curve to the yield point. Again, note how the yield point is determined from the 0.2% offset in metals. Assuming a linear elastic region up to the yield point (which we know is an approximation in most situations), the modulus of resilience can be found from the yield strength, σ_y, and modulus, E:

$$U_r = \frac{\sigma_y^2}{2E} \tag{5.34}$$

5.1.4.5 *Other Types of Stress–Strain Experiments.* In addition to the tensile test described above, a number of other mechanical tests can be performed to obtain specific mechanical properties. The compressive modulus, K [see Eq. (5.12)], is obtained from a stress–strain diagram generated in a compression experiment. Although the testing apparatus shown in Figure 5.24 is shown for a tensile test, it can be used for any number of tests, including compression, with the use of proper fixtures. A schematic of a typical compression fixture is shown in Figure 5.30. The sample sits between two platens, which are designed to simply reverse the direction of the applied load, such that the top platen moves downward and the bottom platen moves upward, thus compressing the sample in between them. A stress–strain diagram is again generated, and the compressive modulus, and compressive strength, σ_c, which is typically the strength at fracture for compression, can be determined. Compression tests are common for load-bearing materials such as cement and concrete, and the devices used to test them can be enormous. Compression tests can, in principle, be conducted with a tensile-type setup in which the sample is mounted between two grips as in Figure 5.24, and the crosshead is moved upward instead of downward. In fact, this is used for some cyclical loading experiments in which tension and compression are alternated at a specific frequency, usually over small deformations. This configuration

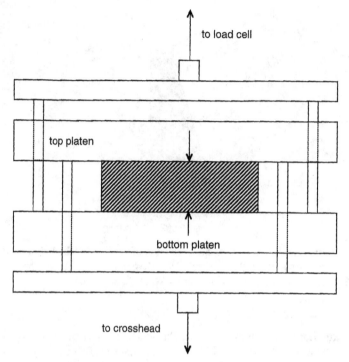

to load cell

top platen

bottom platen

to crosshead

Figure 5.30 Schematic diagram of compression fixture for testing apparatus shown in Figure 5.24.

can present some practical problems, however, particularly with regard to nonrigid samples and with sample slippage in the grips.

Other types of fixtures can be used for other types of tests. Figure 5.31 illustrates a type of test known as a *three-point bending test,* in which the sample of rectangular or circular cross section is mounted between three knife points as shown. This geometry produces tensile stresses at the bottom of the sample and compressive stresses at the top of the sample. Since most samples subjected to this test (like concrete) are weaker in tension than in compression, the specimen fails—literally breaks into two—following the formation of a nearly vertical crack, called the flexural crack, near the center. The load at failure, F_f, can then be used to calculate the *modulus of rupture,* also known as the *bend strength,* fracture strength, or flexural strength, σ_{fs}:

$$\sigma_{fs} = \frac{MC}{I} \tag{5.35}$$

where M is the maximum moment ($F_f L/4$), I is the *moment of inertia,* and c is the distance from the neutral axis to the extreme fiber in tension. The relationships for M, I, and c for both rectangular and circular cross sections, along with the corresponding equations for σ_{fs}, are given in Figure 5.31.

The flexural strength will depend on sample size, since with increasing size there is an increase in the probability of the existence of a crack-producing flaw, along with a corresponding decrease in the flexural strength. This test, and others like it such as the four-point bend, are common for brittle materials such as ceramics, which we

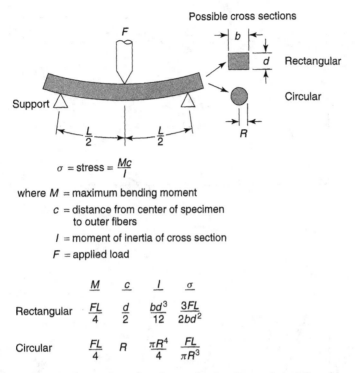

Figure 5.31 Schematic illustration of a three-point bend experiment for either rectangular or circular cross sections. Reprinted, by permission, from W. Callister, *Materials Science and Engineering: An Introduction*, 5th ed., p. 409, Copyright © 2000 by John Wiley & Sons, Inc.

will describe in the next section. A number of other specialized tests exist to measure specific properties such as *impact strength* and *fracture strength*. The interested reader is referred to the texts (listed at the end of this chapter) on mechanics of specific materials for further information.

5.1.5 Mechanical Properties of Metals and Alloys

After a somewhat lengthy introduction into the theoretical and experimental aspects of mechanical properties, which we will see apply to all material classes, we come finally to the specifics of mechanical properties of metals and alloys.

5.1.5.1 Temperature Dependence of Strength.
The modulus and strength of selected metals and alloys are listed in Appendix 7. However, there are some generalizations that can be made and trends to be observed in the mechanical properties of metals and alloys.

At elevated temperatures, the thermal recovery processes described in Section 5.1.2.3 can occur concurrently with deformation, and both strength and strain hardening are consequently reduced. The latter effect results in decreasing the difference between yield and tensile strengths until at sufficiently high temperatures, they are essentially equal. At lower temperatures, temperature has a marked influence on deformation in crystalline materials. Temperature can affect the number of active slip systems in some

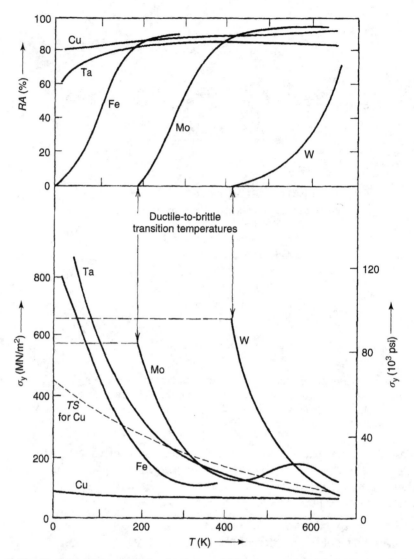

Figure 5.32 Variation of percent reduction in area (%RA, top graph) and yield strength (bottom graph) with temperature for selected metals. From K. M. Ralls, T. H. Courtney, and J. Wulff, *Introduction to Materials Science and Engineering.* Copyright © 1976 by John Wiley & Sons, Inc. This material is used by permission John Wiley & Sons, Inc.

metals, particularly HCP metals (see Section 5.1.2.1), and this determines the ductility of polycrystalline metals such as Zn and Mg. Furthermore, dislocations move less readily as temperature is decreased. As a result, yield and tensile strengths of polycrystalline materials increase and their ductility, measured in percent reduction in area, %RA [see Eq. (5.33)], decreases with decreasing temperature, as illustrated in Figure 5.32. Below a certain temperature known as the *brittle-to-ductile transition temperature*, metals can behave in a brittle manner; that is, they fracture before they undergo extensive plastic deformation. This is particularly true for BCC metals below the brittle-to-ductile transition temperature. However, FCC metals seldom show such a transition and their ductility decreases only slightly with decreasing temperature. This phenomenon is related to the temperature dependence of the critical resolved shear stress previously shown in Figure 5.13.

5.1.5.2 Strain-Rate Dependence of Strength.

The plastic behavior of crystalline materials also depends on tensile strain rate, $\dot{\varepsilon}$. To this point, we have basically assumed that tests are performed at a constant strain rate that does not vary much from experiment to experiment. Strain rate is an important parameter, however, as illustrated in Figure 5.33. Here, the tensile strength (*TS*, or σ_u) for polycrystalline copper is shown both as a function of temperature and strain rate. The temperature dependence is as discussed in the previous section; that is, *TS* increases with decreasing temperature. The strain rate dependence on tensile strength is more pronounced at elevated temperatures and, as illustrated by the logarithmic coordinates in Figure 5.33, can be described by the relationship

$$\sigma_u = K\dot{\varepsilon}^m \tag{5.36}$$

where K is a proportionality constant and m is a temperature-dependent measure of the strain-rate sensitivity of a material. A similar relationship holds for the yield stress, σ_y. The value of m is usually small at low temperatures but can be appreciable at high temperatures. For FCC metals at low temperatures, m is of the order 0.01. At low temperatures, BCC metals are more strain rate sensitive, but m is still only about 0.1. In the next section, we describe materials that have large strain-rate exponents on the order of 0.5 and that can undergo great extensions.

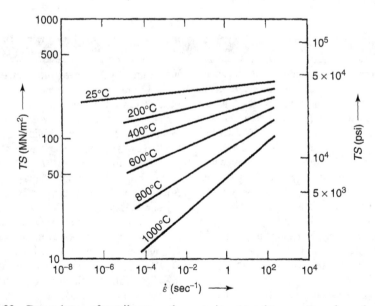

Figure 5.33 Dependence of tensile strength on strain rate and temperature for polycrystalline copper. From K. M. Ralls, T. H. Courtney, and J. Wulff, *Introduction to Materials Science and Engineering.* Copyright © 1976 by John Wiley & Sons, Inc. This material is used by permission John Wiley & Sons, Inc.

Cooperative Learning Exercise 5.6

Using the data in Figure 5.33, select two strain rates for which data are provided.

Person 1: Determine the value of m in Eq. (5.36) for copper at 1000°C.

Person 2: Determine the value of m in Eq. (5.36) for copper at 25°C.

Compare your results. Do they support the claim that strain-rate dependence is more pronounced at elevated temperatures? What would the value of m be if there were no strain-rate dependence?

Answer: In both cases, Eq. (5.36) can be used to obtain $\ln(\sigma_2/\sigma_1) = m \ln(\varepsilon_2/\varepsilon_1)$, from which $m(1000°) \approx 0.1$; $m(25°) \approx 0.02$. There is indeed more dependence at higher temperatures. The value of m would be zero for no dependence.

5.1.5.3 Superelasticity and Superplasticity.

Superplastic materials are polycrystalline solids that have the ability to undergo large uniform strains prior to failure. For deformation in uniaxial tension, elongations to failure in excess of 200% are usually indicative of superplasticity, although several materials can attain extensions greater than 1000%. There are two main types of superplastic behavior: (a) *micrograin* or *microstructural superplasticity* and (b) *transformation* or *environmental superplasticity*. Micrograin superplasticity is exhibited by materials with a fine grain size, usually less than 10 μm, when they are deformed within the strain rate range 10^{-5} to 10^{-1}/s at temperatures greater than $0.5T_m$, where T_m is the melting point in degrees Kelvin. Superplastic deformation is characterized by low flow stresses; this, combined with the high uniformity of plastic flow, has led to considerable commercial interest in the superplastic forming of components.

Observations of what appeared to be superplastic behavior were initially made in the late 1920s with a maximum of 361% for the Cd–Zn eutectic at 20°C and strain rates of ~10^{-8}/s and 405% at 120°C and a strain rate of ~10^{-6}/s. Jenkins [2] also reported a maximum of 410% for the Pb-Sn eutectic at room temperature but at strain rates of ~10^{-6}/s. However, the most spectacular of the earlier observations was that by Pearson in 1934 [3]. While working on eutectics, he reported a tensile elongation of 1950% without failure for a Bi-Sn alloy.

Following these observations there was little further interest in the Western world in what was clearly regarded as a laboratory curiosity. Later, and unrelated, studies carried out in the USSR discovered the same phenomenon and the term *superplasticity*

was coined by Bochavar and Sviderskaya in 1945 [4] to describe the extended ductility observed in Zn-Al alloys.

The word superplasticity is derived from the Latin prefix *super* meaning excess and the Greek word *plastikos* which means to give form to; *sverhplastichnost* was used in the original Russian language paper of Bochavar and was subsequently translated as superplasticity in Chemical Abstracts (1947). In 1962 Underwood [5] reviewed previous work on superplasticity and it was this paper, along with the subsequent work of Backofen and his colleagues reported in 1966 [6], which was the forerunner of the present expanding scientific and technological interest in superplasticity.

Source: http://callisto.my.mtu.edu:591/notes/intro.html

The most important mechanical characteristic of a superplastic material is its high strain-rate sensitivity of flow stress, as given previously in Eq. (5.36). For superplastic behavior, m would be greater than or equal to 0.5; for the majority of superplastic materials, m lies in the range 0.4 to 0.8. The presence of a neck in a material subject to tensile straining leads to a locally high strain rate and, for a high value of m, to a sharp increase in the flow stress within the necked region. Hence the neck undergoes strain rate hardening that inhibits its further development. Thus, a high strain-rate sensitivity results in a high resistance to neck development and the high tensile elongations characteristic of superplastic materials.

There are two main types of superplastic alloys: *pseudo-single phase* and *microduplex*. In the former class of material, a combination of hot and cold working and heat treatment is employed to develop a fine-scale distribution of dispersoids so that on recrystallization the alloy will have a grain size less than 5 μm. Ideally, the dispersion of particles will also prevent any further grain growth during superplastic deformation; however, most of the second phase particles have usually redissolved into solution at the superplastic forming temperatures. The precipitation-strengthened aluminum alloys based on the Al–Cu, Al–Mg, Al–Zn–Mg and the Al–Li and Al–Cu–Li alloys can be included in this group. Some typical properties of the Al-based superplastic alloys are presented in Table 5.4. Other materials include dispersion-strengthened copper alloys

Table 5.4   Properties of Selected Alloys Exhibiting Superplasticity

Alloy Composition	Temperature of Applicability (°C)	Superplasticity (% Elongation)	m
Al-Based Alloys (Balance Al)			
6%Cu–0.4%Zr–0.3%Mg	400–480	1800	0.45–0.7
5.5%Zn–2.0%Mg–1.5% Cu–0.2%Cr	510–530	1400	0.5–0.8
2.7%Cu–2.2%Li–0.7% Mg–0.12%Zr	510–530	800	0.4–0.6
4.8%Cu–1.3%Li–0.4% Mg–0.4%Ag–0.14%Zr	470–530	1000	0.45
5%Ca–5%Zn	450–550	600	0.4–0.5
4.7%Mg–0.7%Mn–0.15%Cr	480–550	670	0.4–0.65
2.5%Li–1.2%Cu–0.6% Mg–0.1%Zr	500–540	1000	0.4–0.6
Ti-Based Alloys (Balance Ti)			
6%Al–4%V	790–940	1400	0.6–0.8
5.8%Al–4%Sn–3.5%Zr–0.5% Mo–0.3%Si–0.05%C	950–990	400	0.35–0.65
4%Al–4%Mo–2%Sn–0.5%Si	810–930	1600	0.48–0.65
4.5%Al–3%V–2%Fe–2%Mo	750–830	700	0.5–0.55
14%Al–20%Nb–3%V–2%Mo	940–980	1350	0.4–0.6
6%Al–2%Sn–4%Zr–2%Mo	880–970	900	0.5–0.7

Source: J. Pilling, Superplasticity on the Web, http://callisto.my.mtu.edu:591/FMPro?-db = sp&-format = sp.html&-view

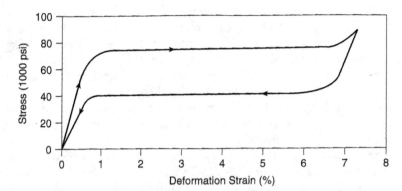

Figure 5.34 Example of superelasticity in an Ni–Ti alloy. A deformation of 7% is fully recovered. (ShapeMemory Applications, Inc.).

where silicide or aluminide particles are used to limit grain growth. The aluminum alloys, which from a commercial viewpoint are the most important of the pseudo-single-phase materials, can be further subdivided into those which are recrystallized prior to superplastic forming and those which acquire their fine grain structure only after a limited amount of deformation at the forming temperature. The microduplex materials are thermomechanically processed to give a fine grain or phase size. Grain growth is limited by having a microstructure that consists of roughly equal proportions of two or more chemically and structurally different phases. This latter group of materials includes α/β titanium alloys, α/γ stainless steels, α/β and $\alpha/\beta/\kappa$ copper alloys, and eutectics. Some properties of selected titanium-based superplastic alloys are presented in Table 5.4.

A property related to superplasticity is *superelasticity*, or *psuedoelasticity*. Many of the same types of alloys that possess superplasticity also possess superelasticity, particularly the titanium-based alloys, and these are the same alloys that were described in Section 3.1.2.2 as shape memory alloys. Recall that these materials have a transformation temperature (cf. Figure 3.5) below which they are in an austenitic–martensitic state and above which they are in the austenitic state. If the material is tested just above its transformation temperature to austenite, the applied stress transforms the austenite to martensite and the material exhibits increasing strain at constant applied stress; that is, considerable deformation occurs for a relatively small applied stress. When the stress is removed, the martensite reverts to austenite and the material recovers its original shape. This effect makes the alloy appear extremely elastic; in some cases, strains of up to about 10% can be fully recovered (see Figure 5.34). Superelastic shape memory alloys are used in the nearly indestructible frames of eyeglasses, and they makes excellent springs. Many biomedical applications use superelastic wires and tubes, including catheters and guide wires for steering catheters, and orthodontics.

5.2 MECHANICS OF CERAMICS AND GLASSES

In the previous sections, we saw how materials can behave in elastic and ductile manners and how plastic flow past the yield point could be considered a form of destructive failure due to permanent deformation. In this section, we will see that there

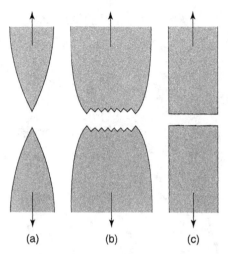

Figure 5.35   Schematic illustration of (a) rupture, (b) ductile fracture, and (c) brittle fracture. Reprinted, by permission, from W. Callister, *Materials Science and Engineering: An Introduction*, 5th ed., p. 186. Copyright © 2000 by John Wiley & Sons, Inc.

are additional types of destructive failure. In particular, we will concentrate on two types of failure known as brittle fracture and creep.

These types of failure are common in, but not limited to, brittle materials such as ceramics and glasses. Brittle materials are characterized by their lack of plastic deformation, as illustrated in Figure 5.35. The ductile materials such as metals have a distinct yield point, after which plastic flow begins, and the process of cross-sectional area reduction known as necking occurs (cf. Section 5.1.4.1). The necking phenomenon is illustrated in Figures 5.35a and 5.35b, after which point failure can occur by either highly ductile fracture (known as rupture) or moderately ductile fracture. In brittle materials, little or no plastic deformation precedes the fracture, as in Figure 5.35c.

In terms of the stress–strain diagram, brittle materials exhibit primarily an elastic response, with little or no strain beyond the elastic limit prior to fracture (Figure 5.36). The modulus may be higher or lower than a ductile material. As illustrated earlier by point R in Figure 5.26, the stress at fracture is known as the fracture strength, σ_f. We will devote the next section to a development of theoretical expressions for fracture strength. A description of creep will then be given, followed by an overview of the mechanical properties of glasses and ceramics.

5.2.1   Fracture Mechanics

5.2.1.1   The Molecular Origins of Fracture. Let us first calculate the theoretical fracture strength of a material—that is, the energy required to separate the atomic planes of a solid. Consider pulling a bar of unit cross section, much as we did schematically in Figure 5.10, except that instead of a shear stress, τ, as was used there, we now are applying a tensile stress, σ, in an attempt to pull the atomic planes apart. Cohesive stresses between atomic planes, σ, vary with the change in interplanar separation, x, where $x = a - a_0$. Here, a_0 is the original (equilibrium) interplanar spacing, for which the corresponding stress is zero, and a is the interplanar spacing at any applied stress.

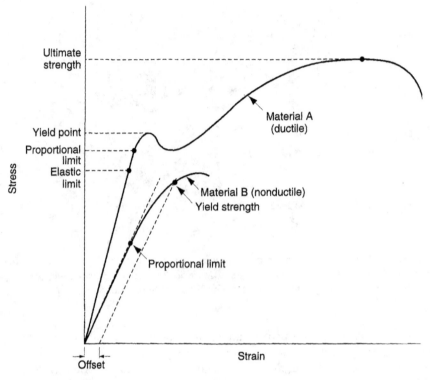

Figure 5.36 Comparison of typical stress–strain diagrams for ductile (top curve) and brittle (bottom curve) materials. Reprinted, by permission, from S. Somayaji, *Civil Engineering Materials*, 2nd ed., p. 24. Copyright © 2001 by Prentice-Hall, Inc.

The variation of the tensile stress with interplanar separation can be approximated by a sine curve of wavelength λ (see Figure 5.37), much like we did in Figure 5.10 for an applied shear stress [cf. Eq. (5.15)]:

$$\sigma = \sigma_c \sin\left(\frac{2\pi x}{\lambda}\right) \tag{5.37}$$

The quantity σ_c is called the *theoretical cleavage strength* (we will use the terms fracture and cleavage interchangeably here) and it is the maximum stress required to separate, or cleave, the planes. Since cleavage occurs at an interplanar separation of $x = \lambda/2$, the work per unit area required to separate planes, also known as the *strain energy*, is the area under the curve in Figure 5.37 up to that point,

$$W_\varepsilon = \int_0^{\lambda/2} \sigma \, dx \tag{5.38}$$

or, if we substitute the expression for stress from Eq. (5.37) into Eq. (5.38),

$$W_\varepsilon = \int_0^{\lambda/2} \sigma_c \sin\left(\frac{2\pi x}{\lambda}\right) dx = \frac{\lambda \sigma_c}{\pi} \tag{5.39}$$

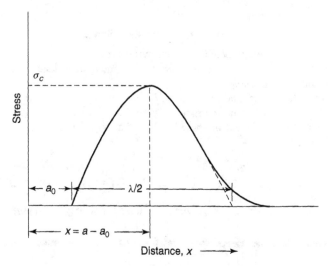

Figure 5.37 Sine-wave representation of stress variation with interatomic separation distance for two atomic planes. From Z. Jastrzebski, *The Nature and Properties of Engineering Materials*, 2nd ed. Copyright © 1976 by John Wiley & Sons, Inc. This material is used by permission of John Wiley & Sons, Inc.

Recall from Section 2.4.1 that there is work associated with creating any new surface. The fracture of the solid at $\lambda/2$ creates two new surfaces where none existed before (see Figure 5.35c), so that the work of cohesion between the two surfaces, W_ε, is given by Eq. (2.85), $W_\varepsilon = 2\gamma$, where γ is the surface energy of the solid. We can equate this work with the work calculated in Eq. (5.39) to give

$$\frac{\sigma_c \lambda}{\pi} = 2\gamma \qquad (5.40)$$

We also know that to a first approximation, Hooke's Law applies to the elastic region, which for brittle materials is effectively up to the point of cleavage, so that the stress at cleavage, σ_c, is given by Hooke's Law as

$$\sigma_c = E\varepsilon_c \qquad (5.41)$$

where E is the elastic modulus, and ε_c is the strain at cleavage, $(a - a_0)/a_0$. The interplanar separation $x = a - a_0$ can be estimated as $\lambda/2$ at cleavage, so that Eq. (5.41) becomes

$$\sigma_c = \frac{E\lambda}{2a_0} \qquad (5.42)$$

Solving Eq. (5.40) for λ and substituting into Eq. (5.42) gives

$$\sigma_c^2 = \frac{E\gamma\pi}{a_0} \qquad (5.43)$$

Cooperative Learning Exercise 5.7

Use values of modulus from Appendix 7, along with surface energy from Appendix 4, to estimate the following quantities. In both cases, take an interplanar distance of 10^{-8} cm as an order-of-magnitude estimate.

Person 1: Calculate the σ_c and the ratio σ_c/E for MgO.

Person 2: Calculate the σ_c and the ratio σ_c/E for Au.

Compare your answers. Based on these results, what would you expect the theoretical ratio of σ_c/E to be?

Answers: $\sigma_c(\text{MgO}) = [(210 \times 10^9 \text{ N/m}^2)(1.0 \text{ J/m}^2)/(10^{-10} \text{ m})]^{1/2} = 56.5 \text{ GPa} \approx E/4$;

$\sigma_c \approx E/6$

$\sigma_c(\text{Au}) = [(77 \times 10^9 \text{ N/m}^2)(1.4 \text{ J/m}^2)/(10^{-10} \text{ m})]^{1/2} = 32.8 \text{ GPa}; \sigma_c \approx E/2.5$;

generally $E/10 < \sigma_c < E/10$.

---

or, since $\sqrt{\pi}$ is a numerical factor close to 1 (we are looking for only order of magnitude effects here), we have

$$\sigma_c \approx \sqrt{\frac{E\gamma}{a_0}}$$

(5.44)

The functional dependences of Eq. (5.44) are important: The cleavage strength is proportional to the square root of both modulus and surface energy and is inversely proportional to the square root of the equilibrium interplanar distance. These relationships return in subsequent sections. For now, work through the Cooperative Learning Exercise 5.7 to see some important generalizations about the theoretical cleavage strength and modulus.

5.2.1.2 Fracture in Real Brittle Materials — Griffith Theory. In the Cooperative Learning Exercise 5.7, we estimated that the theoretical cleavage strength might lie somewhere between the value for the modulus and $E/10$. In reality, the fracture strength of most ceramic materials is about $E/1000$. What is wrong with Eq. (5.44)? We took experimental values for modulus and surface energy, which, granted, may not be very accurate, but are not off by a combined four orders of magnitude necessary to cause these errors. The problem is with the denominator. Just as dislocations caused the theoretical shear strength to differ from the yield strength, there are structural imperfections that cause the experimental fracture strength to differ from the theoretical cleavage strength.

It was proposed by Griffith in 1920 that microscopic cracks exist both within and on the surface of all real materials, which are deleterious to the strength of any material that does not possess ductility. The presence of cracks whose longest dimension is perpendicular to the direction of the applied tensile stress gives rise to especially large stress concentrations at the crack tip, where the real localized stress can approach the theoretical strength of the material due to a small area over which it is applied (see Figure 5.38). It can be shown that the maximum stress at the tip of a crack, σ_m, is

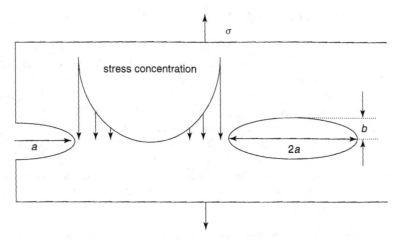

Figure 5.38   The Griffith model for micro-crack induced brittle failure. Cracks at the surface have a length of a, whereas internal cracks have a length of $2a$.

related to the applied tensile stress, σ_0, and the geometry of the crack

$$\sigma_m = 2\sigma_0 \left(\frac{a}{\rho} \right)^{1/2}$$ (5.45)

where ρ is the radius of curvature of the crack tip, b^2/a, where b is the height of the crack and a is its length as defined in Figure 5.38, and a is the length of a surface crack or *half* the length of an interior crack (see Figure 5.38). Do not confuse the length of the crack with the interplanar spacing; the same terminology is used for a reason, as will be shown momentarily, but the values are certainly not the same.

For a crack to propagate, $\sigma_m > \sigma_c$; that is, the maximum stress at the crack tip must exceed the theoretical cleavage strength. Since the radius of curvature cannot be less than the order of an atomic diameter, σ_m must always be finite, but can attain σ_c even though the bulk of the material is under fairly low applied stress. The stress required to cause crack propagation is then the stress that leads to fracture. Using surface energy and work arguments similar to those in the development of the theoretical cleavage strength from last section, it can be shown that the fracture strength according to this crack propagation model of Griffith is

$$\sigma_f = \left(\frac{2\gamma E}{\pi a} \right)^{1/2}$$ (5.46)

where γ is again the surface energy of the material, E is its elastic modulus, and a is now one-half the length of an internal crack, or the length of a surface crack. Notice the similarity in form between Eqs. (5.46) and (5.44). As we surmised, it was not our values of modulus or surface energy that were erroneous in the application of Eq. (5.44), but rather the size parameter. Instead of interplanar separations on the order of angstroms (Å, 10^{-10} m) as in Eq. (5.44), it is actually cracks, on the order of microns (10^{-6} m), that lead to fracture as in Eq. (5.46) and account for the nearly hundred fold (square root of 10^4) difference in the observed and calculated values of the fracture strength estimated from the elastic modulus referred to at the beginning of this section.

Equation (5.46) is an important result. It shows clearly that fracture strength decreases as the size of the crack increases, and it also shows that fracture strength can be related directly to surface energy and elastic modulus. The Griffith equation applies only to completely brittle materials. For materials that exhibit some plastic deformation prior to fracture, the surface energy term may be replaced by a term $\gamma + \gamma_p$, where γ_p is the plastic deformation energy associated with crack extension. Values of γ_p can be up to 1000 times greater than γ, such that the latter can often be disregarded for materials that undergo some plastic deformation prior to fracture. Finally, the Griffith equation tells us that if we can reduced crack sizes, or eliminate them altogether, the fracture strength of a material can be manipulated. We will see that this has important applications in glass science.

Cooperative Learning Exercise 5.8

The crack detection limit for a device that inspects steel structural beams is 3 mm. A structural steel with a fracture toughness of 60 MPa $\cdot$ m$^{0.5}$ has no detectable surface cracks. Assume a geometric factor of unity to determine:

Person 1: The load at which this structural steel may undergo fast fracture if a surface crack just at the detection limit goes undetected.

Person 2: The % increase in crack size required for this structural steel to undergo fast fracture at an applied load of 500 MPa.

Answer: $\sigma_c = 618$ MPa; $a = 4.58$ mm (52.7% increase).

5.2.1.3 Failure Modes and Fracture Toughness.
Once the critical fracture stress has been achieved and the crack begins to propagate, it can do so in one of three primary modes (see Figure 5.39). Mode I is a tensile, or opening, mode. Modes II and III are sliding and tearing modes, respectively. We will limit our present description to the opening mode, though you should be aware that corresponding values for the quantities developed in this section apply to the other two modes as well.

The parameter K_1, known as the mode I *stress intensity factor*, is used to characterize the distribution of stresses at the crack tip and is a function of the applied stress, the crack size (length), and the sample geometry. Stress intensity factors are conveniently defined in terms of the stress components on the crack plane. In most cases, fracture is dominated by tensile loading, so that the important stress intensity factor is K_1. The value of K_1 at which fracture occurs is the critical value of the stress intensity factor at which propagation initiates, K_{1c}, also known as the *fracture toughness*. Do not confuse the fracture toughness with the toughness defined in Section 5.1.4.4, which was the area under the stress–strain curve. The fracture toughness is specific to fracture in brittle materials, and it indicates the conditions of flaw size and stress necessary for brittle fracture. It is defined as

$$K_{1c} = Y\sigma_c\sqrt{\pi a} \tag{5.47}$$

where σ_c is the theoretical cleavage strength, a is crack length as defined in Figure 5.38 (length of a surface crack, or half the length of an internal crack), and Y is a geometric

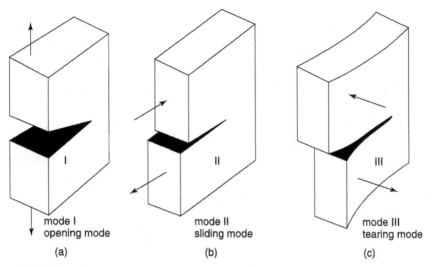

mode I
opening mode

mode II
sliding mode

mode III
tearing mode

(a)

(b)

(c)

Figure 5.39 The three modes of crack surface displacement: (a) opening mode; (b) sliding mode; and (c) tearing mode.

factor that depends on the sample geometry and the manner of load application. The units of fracture toughness are thus MPa·m$^{1/2}$, or psi·in$^{1/2}$. The magnitude of K_{1c} diminishes with increasing strain rate, decreasing temperature, and increasing grain size.

5.2.1.4 *Fatigue*. A failure phenomenon that is closely related to brittle fracture is caused by repeated exposure to cyclic stresses. This type of failure, called *fatigue*, results in fracture below the ultimate strength of ductile materials, σ_u, or below the fracture strength, σ_f, of brittle materials. Even in ductile materials, the result of fatigue is brittle-like failure, insofar as there is little or no plastic deformation prior to failure. This process, like fracture, is caused by the initiation and propagation of cracks. Fatigue is estimated to comprise approximately 90% of all metal failure, though polymers and ceramics are also susceptible to it. Fatigue is particularly important because is can cause catastrophic failure in even the largest of structures due to small, virtually unnoticeable stress fluctuations over long periods of time.

Though stresses, especially environmental stresses, are generally not uniform, there are four categories of cyclical stresses to consider. The first is called a *reversed stress cycle* (Figure 5.40a), in which the stress alternates equally between compressive and tensile loads. A *repeated stress cycle* occurs when the maximum and minimum stress of a reversed stress cycle are asymmetric relative to the zero stress level, as in Figure 5.40b. Other types of cyclical stresses include a *fluctuating stress cycle*, which has a maximum and minimum tensile *or* compressive load (but not both), and a *random stress cycle*, in which the stresses are neither periodic nor uniform in the magnitudes. In most cases, the stress cycle is characterized by the *mean stress*, $\bar{\sigma} = (\sigma_{\max} + \sigma_{\min})/2$, the *range of stress*, $\sigma_r = \sigma_{\max} - \sigma_{\min}$, and the *stress amplitude*, $S = (\sigma_{\max} - \sigma_{\min})/2$. Note that for a reversed stress cycle, the mean stress is zero, so that the stress amplitude is usually a more meaningful quantity.

Fatigue tests are normally performed by applying one of the stress cycles described above until the test specimen fractures. The *number of cycles to failure*, N_f, at a

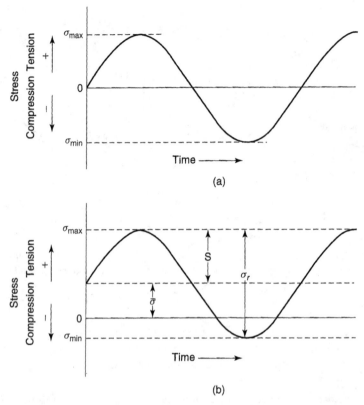

Figure 5.40 Variation of stress with time that accounts for fatigue failure by (a) a reversed stress cycle and (b) a repeated stress cycle. Reprinted, by permission, from W. Callister, *Materials Science and Engineering: An Introduction*, 5th ed., p. 210. Copyright © 2000 by John Wiley & Sons, Inc.

specified stress amplitude, S, sometimes called the *fatigue life*, then characterizes the material. For some materials, a *fatigue limit* is reached (see Figure 5.41), which is defined as the value of S below which no failure will occur, regardless of the number of cycles, N. The *fatigue strength*, sometimes called the *endurance limit*, is defined as the stress level at which failure will occur for some specified number value of N_f—for example, 10^7 cycles.

As mentioned earlier, fatigue is the result of crack initiation and propagation. Crack growth under fatigue is estimated by the *Paris equation*:

$$\frac{da}{dN} = A(\Delta K)^m \tag{5.48}$$

where a is once again the crack length as defined earlier (length of a surface crack or half the length of an internal crack), N is the number of cycles (not necessarily to failure), A and m are material-dependent constants, and ΔK is the range of stress intensity factor at the crack tip, $K_{max} - K_{min}$. Utilization of Eq. (5.47) shows that

$$\Delta K = Y(\sigma_{max} - \sigma_{min})\sqrt{\pi a} \tag{5.49}$$

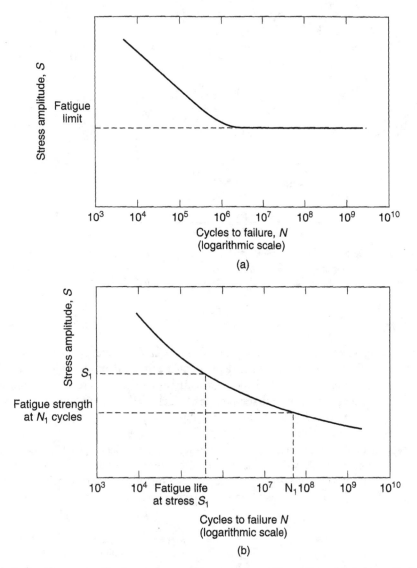

Figure 5.41 Stress amplitude versus cycles to failure illustrating (a) fatigue limit and (b) fatigue life. Reprinted, by permission, from W. Callister, *Materials Science and Engineering: An Introduction*, 5th ed., p. 212. Copyright © 2000 by John Wiley & Sons, Inc.

The values of A and m are determined using the Paris equation by plotting da/dN at various values of ΔK on logarithmic coordinates. The values of da/dN are determined from plots of crack length versus number of cycles, as shown in Figure 5.42. The corresponding crack lengths and stress ranges can then be used to determine ΔK using Eq. (5.49). Typical values of m range between 1 and 6. It should be noted that the Paris equation applies only to a specific region of crack formation. As shown in Figure 5.42, the crack growth rate is initially small, and fatigue cracks can be nonpropagating. At high crack growth rates, crack growth is unstable. Only at intermediate values of crack growth rates does the Paris equation apply.

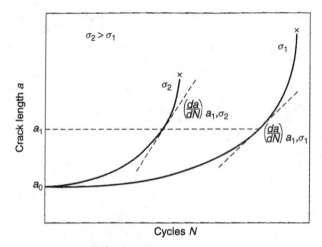

Figure 5.42   Determination of crack growth rate, *da/dN* from crack length versus number of cycles data. Reprinted, by permission, from W. Callister, *Materials Science and Engineering: An Introduction*, 5th ed., p. 217. Copyright © 2000 by John Wiley & Sons, Inc.

5.2.2   Creep

Fatigue is an example of the influence of time on the mechanical properties of a material. Another example of a time-dependent mechanical property is *creep*. Creep, sometimes called *viscoplasticity*, is defined as time-dependent deformation under constant stress, usually at elevated temperatures. Elevated temperatures are necessary because creep is typically important only above $T_{mp}/2$, where T_{mp} is the absolute melting point of the material.

A typical creep experiment involves measuring the extent of deformation, called the *creep strain*, ε, over extended periods of time, on the order of thousands of hours, under constant tensile loads and temperature. The resulting plot of creep strain versus time (Figure 5.43) shows the resulting *creep rate*, $\dot{\varepsilon} = d\varepsilon/dt$, which is the slope of the

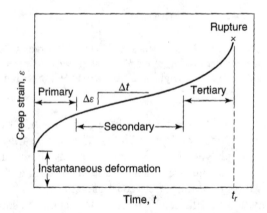

Figure 5.43   Representative creep curve illustrating primary, secondary, and tertiary creep. Reprinted, by permission, from W. Callister, *Materials Science and Engineering: An Introduction*, 5th ed., p. 226. Copyright © 2000 by John Wiley & Sons, Inc.

curve at a specified time. A typical creep curve exhibits three distinct regions, with different functional dependences of strain on time for each. The first region is called *primary creep*, or *transient creep*, in which the primary creep rate, $\dot{\varepsilon}_p$, continuously decreases with time. A power-law relationship can be applied in this region:

$$\dot{\varepsilon}_p \propto t^n \tag{5.50}$$

where t is time and n is a negative number. This region represents the strain-hardening region of creep. A similar power law expression can be used in the *tertiary creep* regime to describe the increase in tertiary creep rate, $\dot{\varepsilon}_t$, with time, except that n is now (usually) a positive number. Values of n range from -2 to 20 for tertiary creep, with values of $n = 4$ being typical. The creep rate increases during tertiary creep until failure, termed *rupture*, occurs. The stage of creep that usually occupies the longest time is *secondary creep*, or *steady-state creep*, which is characterized by a constant creep rate, $\dot{\varepsilon}_s$. In this area, strain hardening is balanced by thermal recovery.

The variation of creep with time as a function of both load and temperature is illustrated in Figure 5.44. Arrhenius-type relationships have been developed for steady-state creep as a function of both variables such as

$$\dot{\varepsilon}_s = K\sigma^m \exp\left(\frac{-E_c}{RT}\right) \tag{5.51}$$

where σ is the applied stress, R is the gas constant, T is absolute temperature, K and m are material constants, and E_c is called the activation energy for creep.

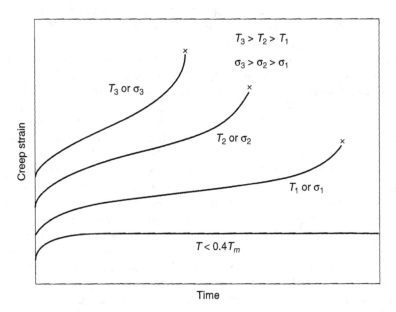

Figure 5.44 Influence of applied stress and temperature on creep behavior. Reprinted, by permission, from W. Callister, *Materials Science and Engineering: An Introduction*, 5th ed., p. 227. Copyright © 2000 by John Wiley & Sons, Inc.

Cooperative Learning Exercise 5.9

A proposed jet engine turbine blade material exhibits a steady-state creep rate of 0.0001 in/in-sec at 500°C under a stress of 20,000 psi. The activation energy for creep is 40 kcal/mol.

    Person 1: At what temperature will this creep rate be doubled at the same stress level?

    Person 2: At what temperature will this creep rate be tripled at the same stress level?

    Compare your answers. Agree upon a temperature at which you both estimate these blades will operate, and calculate the corresponding creep rate.

Answer: In both instances, use a ratio of Eq. (5.51) at the two different temperatures to solve for $T(0.0002) = 521°C$; $T(0.0003) = 534°C$. In military jet engines, the inlet gas temperatures can be in excess of 1400°C. At the stated loads, the creep rate is over 10^6 times that at 500°C!

5.2.3 Mechanical Properties of Ceramics and Glasses

In this section, we describe the elastic, ductile, fracture, fatigue, and creep properties that are specific to certain classes of ceramic and glass materials, with primary emphasis on those mechanical properties that make the material favorable for certain applications from a design point of view.

5.2.3.1 Crystalline Ceramics. The moduli of ceramic materials are generally higher than those for metals, being in the range of 70 to 500 GPa. For example, the elastic modulus as a function of temperature for a common engineering ceramic, polycrystalline Al_2O_3, is shown in comparison to moduli for three common metals in Figure 5.45. Note that the modulus is relatively independent of temperature, which is common in polycrystalline ceramics.

The moduli for single crystals can be even higher than for the polycrystalline analogues, depending upon the direction of orientation of the crystal relative to the applied stress. For example, the maximum modulus for MgO lies along the [111] direction and has a value of $E_{[111]} = 336$ GPa, whereas the minimum value is along the [100] direction, for which $E_{[100]} = 245$ GPa. As is the case for most polycrystalline materials, the modulus of a polycrystalline sample of MgO lies somewhere in between these two values, around 310 GPa.

One may not normally think of ceramics as having significant ductile responses, since they are mostly brittle materials that normally fracture around their yield point, but slip is a definite possibility, especially in single crystal ceramics. As was the case for metals, plastic deformation in crystalline ceramics occurs by a slip dislocation mechanism (cf. Section 5.1.2.1). The slip systems in some common ceramic crystals are shown in Table 5.5. Note that the number of independent slip systems is low compared to the metallic structures of Table 5.3, especially the BCC metals.

The yield strength and ductility of single crystals are highly dependent upon temperature and strain rate. These effects are illustrated in Figure 5.46 for single crystal Al_2O_3. Note the presence of both an upper yield stress and lower yield stress, much as there was for metals, and that they both decrease exponentially with increasing temperature and decreasing strain rate. This drop in yield point values can be explained

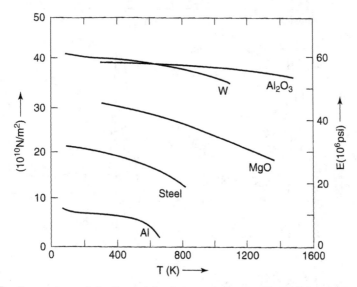

Figure 5.45   Comparison of elastic modulus between polycrystalline Al$_2$O$_3$ and some common metals as a function of temperature. From K. M. Ralls, T. H. Courtney, and J. Wulff, *Introduction to Materials Science and Engineering*. Copyright © 1976 by John Wiley & Sons, Inc. This material is used by permission John Wiley & Sons, Inc.

Table 5.5   Slip Systems in Some Ceramic Crystals

Crystal	Slip System	Number of Independent Systems	Comments
C (diamond), Si, Ge	(111)[1$\bar{1}$0]	5	At $T > 0.5T_m$
NaCl, LiF, MgO, NaF	(110)[1$\bar{1}$0]	2	At low temperatures
NaCl, LiF, MgO, NaF	(110)[1$\bar{1}$0]		At high temperatures
	(001)[1$\bar{1}$0]	5	
	(111)[1$\bar{1}$0]		
TiC, UC	(111)[1$\bar{1}$0]	5	At high temperatures
PbS, PbTe	(001)[1$\bar{1}$0]	3	
	(110)[001]		
CaF$_2$, UO$_2$	(001)[1$\bar{1}$0]	3	
CaF$_2$, UO$_2$	(001)[1$\bar{1}$0]		At high temperatures
	(110)	5	
	(111)		
C (graphite), Al$_2$O$_3$, BeO	(0001)[11$\bar{2}$0]	2	
TiO$_2$	(101)[10$\bar{1}$]	4	
	(110)[001]		
MgAl$_2$O$_4$	(111)[1$\bar{1}$0]	5	
	(110)		

Source: W. D. Kingery, H. K. Bowen and D. R. Uhlmann, *Introduction to Ceramics*. Copyright © 1976 by John Wiley & Sons, Inc.

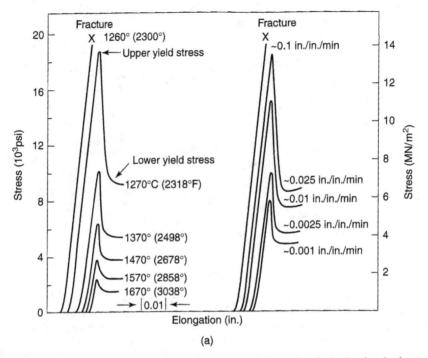

Figure 5.46 Effect of temperature and strain rate on deformation behavior in single-crystal Al$_2$O$_3$. From W. D. Kingery, H. K. Bowen, and D. R. Uhlmann, *Introduction to Ceramics*. Copyright © 1976 by John Wiley & Sons, Inc. This material is used by permission of John Wiley & Sons, Inc.

in terms of dislocation multiplication, rather than dislocation movement. Notice again how the modulus is relatively unaffected by either temperature or strain rate. At low temperatures and high strain rates, brittle behavior is observed. Finally, the alloying of single-crystal Al$_2$O$_3$ with different cations such as Fe, Ni, Cr, Ti, and Mg leads to a strengthening effect, much as precipitation hardening in metals.

As was the case for metals, polycrystalline ceramics show a dependence of mechanical properties on grain size, as given earlier by the Hall–Petch relation in Eq. (5.22), which in this case applies to fracture strength, σ_f:

$$\sigma_f = \sigma_0 + kd^{-1/2} \tag{5.22}$$

Remember that in brittle materials, fracture occurs as a result of the most severe flaw (crack) in the sample. Hence, the correlations of strength with average grain size must reflect processing techniques that produce samples in which the size of the most severe flaw scales with the average grain size. In cases where the sizes of the initial flaws are limited by the grains and scale with the grain size, the strengths vary with a modified form of Eq. (5.22):

$$\sigma_f = kd^{-1/2} \tag{5.52}$$

This behavior is reflected at larger grain sizes in Figure 5.47, whereas Eq. (5.22) is followed at smaller grain sizes.

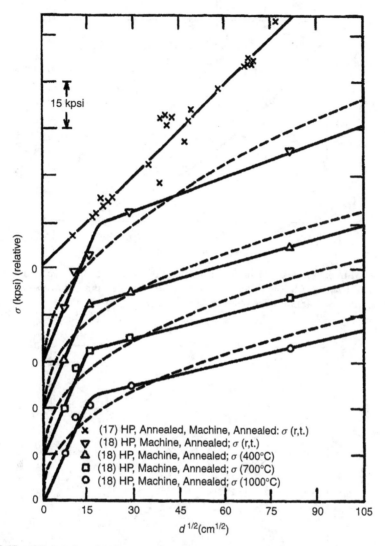

Figure 5.47 Effect of grain size on strength in polycrystalline Al_2O_3 for various process-ing methods (HP = hot pressed, rt = room temperature). From *Introduction to Ceramics*, by W. D. Kingery, H. K. Bowen and D. R. Uhlmann, Copyright © 1976 by John Wiley & Sons, Inc. This material is used by permission of John Wiley & Sons, Inc.

Typically, the most severe flaws are associated with porosity, surface damage, abnor-mal grains, or foreign inclusions. The effect of porosity on both modulus and strength is illustrated in Figures 5.48 and 5.49, respectively. The solid curve in Figure 5.48 represents the following equation, which holds for some ceramics:

$$E = E_0(1 - 1.9P - 0.9P^2) \tag{5.53}$$

where E_0 is the modulus of the nonporous material, and P is the volume fraction porosity. A similar relationship can be used to estimate the flexural strength of a

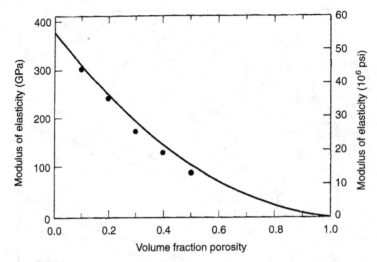

Figure 5.48 Effect of porosity on the elastic modulus of Al_2O_3 at room temperature. Reprinted, by permission, from W. Callister, *Materials Science and Engineering: An Introduction*, 5th ed., p. 413. Copyright © 2000 by John Wiley & Sons, Inc.

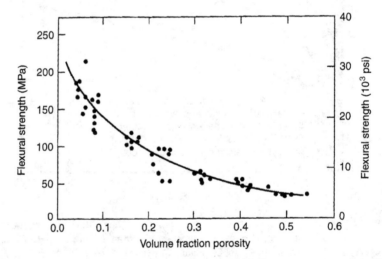

Figure 5.49 Effect of porosity on the flexural strength of Al_2O_3 at room temperature. Reprinted, by permission, from W. Callister, *Materials Science and Engineering: An Introduction*, 5th ed., p. 413. Copyright © 2000 by John Wiley & Sons, Inc.

porous ceramic, σ_{fs}, as a function of porosity:

$$\sigma_{fs} = \sigma_0 \exp(-nP) \qquad (5.54)$$

where σ_0 and n are experimentally determined constants. This relationship is represented by the solid line in Figure 5.49.

The creep properties of ceramics are of particular importance, especially at high temperatures. In general, the principles of Section 5.2.2 still apply, but the ceramic

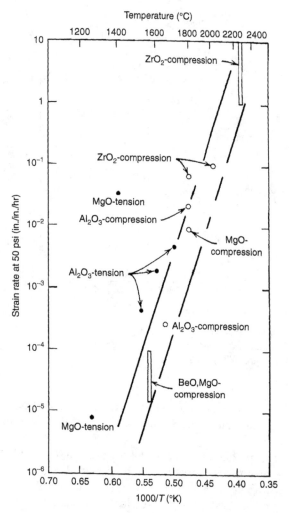

Figure 5.50 Low-stress creep rates of several polycrystalline oxides. From W. D. Kingery, H. K. Bowen, and D. R. Uhlmann, *Introduction to Ceramics*. Copyright © 1976 by John Wiley & Sons, Inc. This material is used by permission of John Wiley & Sons, Inc.

creep mechanisms can be very complex, and factors such as diffusional creep, grain boundary creep, and substructure formation must be considered in a complete description of creep in ceramics. Porosity and grain size are also factors in creep, as they were in modulus and strength determinations. A summary of these effects may be seen graphically in Figure 5.50, in which the creep rates of several polycrystalline oxide ceramics are presented as a function of temperature. All of the values fall within a band, represented by the dashed lines, and variations in creep rate values for a given material (e.g., Al_2O_3 in tension or compression) can be attributed to differences in microstructure (pores, grain size, etc.).

5.2.3.2 *Glasses.* Due to the disordered structure, the modulus and strength of most inorganic glasses are less than those in most crystalline ceramics (see Figure 5.51).

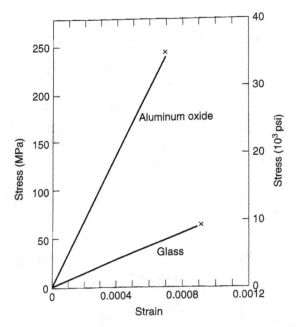

Figure 5.51   Typical stress–strain diagrams for aluminum oxide and glass. Reprinted, by permission, from W. Callister, *Materials Science and Engineering: An Introduction*, 5th ed., p. 410. Copyright © 2000 by John Wiley & Sons, Inc.

However, due to other favorable properties, such as optical transparency and ease of formability, glasses find wide industrial use and, as a result, have been the subject of much investigation.

The mechanical properties of silicate- based glasses have been extensively studied due to their widespread use as containers, reinforcing fibers, and optical fibers. Consequently, there is a wealth of information on the influence of many variables on the mechanical properties of these glasses. We shall concentrate on (a) composition and temperature effects on modulus and strength and (b) an important method of strength enhancement in glass fibers.

A number of empirical expressions exist for estimating the modulus of multicomponent glasses based upon composition. One particularly useful one that provides an absolute value for the modulus, E, is based upon the density of the glass, ρ, and the packing density, V_t, dissociation energy, U_i, mole %, p_i, and molecular weight, M_i, of the components:

$$E = \frac{0.2\rho \sum V_i p_i \sum U_i p_i}{\sum M_i p_i} \tag{5.55}$$

Values for V_i and U_i for a number of metal oxide glass components are listed in Table 5.6. In a similar manner, the compressive modulus, K, and Poisson's ratio, ν, can be determined using these parameters:

$$K = V_t^2 \sum U_i p_i \tag{5.56}$$

Table 5.6   Factors Used to Determine Elastic Modulus in Multicomponent Glasses Using Eq. (5.55)

Oxide	V_i (mole %)	U_i (kJ/cm³)
Li_2O	8.0	80.4
Na_2O	11.2	37.3
K_2O	18.8	23.4
BeO	7.0	125.6
MgO	7.6	83.7
CaO	9.4	64.9
SrO	10.5	48.6
BaO	13.1	40.6
B_2O_3	20.8	77.8
Al_2O_3	21.4	134.0
SiO_2	14.0	64.5
TiO_2	14.6	86.7
ZrO_2	15.1	97.1
P_2O_5	34.8	62.8
As_2O_5	36.2	54.8
ZnO	7.9	41.5
CdO	9.2	31.8
PbO	11.7	17.6

Source: H. Scholze, *Glass.* Copyright © 1991 by Springer-Verlag.

and

$$v = 0.5 - \frac{0.139}{V_t} \tag{5.57}$$

respectively, where

$$V_t = \frac{\rho \sum V_i p_i}{\sum M_i p_i} \tag{5.58}$$

Strength in glasses is also composition-dependent, as shown in Figure 5.52. Correlations similar to Eqs. (5.55)–(5.58) for the variation of strength with composition do exist; however, it is the effect of surface cracks that more profoundly influences strength in real glasses. There are two important techniques that can be used to improve the strength in glasses: *thermal tempering* and *chemical tempering*. Tempering utilizes the fact that glasses, and crystalline ceramics for that matter, can be strengthened by developing a state of compression on their surfaces.

Tempered glasses are useful because failure normally occurs under an applied tensile stress, and failure in ceramics and glasses is almost always initiated at the surface. When a permanent compressive stress, called a *residual compressive stress*, is placed on a surface, either through thermal or chemical means, the applied stress must first overcome this residual compression before the surface is brought into tension under which failure can occur (see Figure 5.53). Notice that the residual stress is compressive in nature at the surface of the plate and is tensile in the center (shaded areas). When

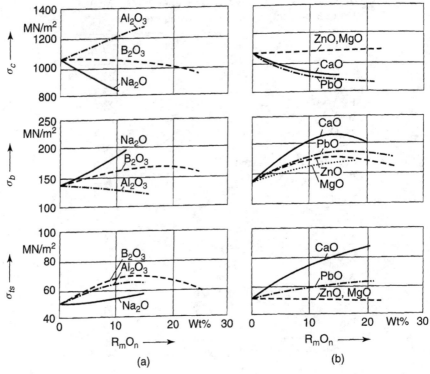

Figure 5.52 Change in compressive strength, σ_c, bending strength, σ_b, and tensile strength, σ_t, of a Na_2O–SiO_2 glass (20–80%) as SiO_2 is replaced proportionately by weight with other oxides. Reprinted, by permission, from H. Scholze, *Glass*, p. 273. Copyright © 1991 by Springer-Verlag.

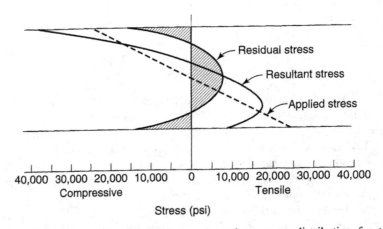

Figure 5.53 Residual stress, applied stress, and resultant stress distribution for transverse loading of a tempered glass plate. From W. D. Kingery, H. K. Bowen, and D. R. Uhlmann, *Introduction to Ceramics.* Copyright © 1976 by John Wiley & Sons, Inc. This material is used by permission of John Wiley & Sons, Inc.

a transverse load (dash line) is placed on the plate (recall that a bending-type load creates a tensile force at the bottom and a compressive force at the top), the resultant stress (solid line) creates a large compressive force at the top of the plate, which it can accommodate, but the tensile force at the bottom of the plate is counteracted by the residual compressive stress, such that the resultant stress is only mildly tensile in nature, and well below the fracture stress. A similar situation occurs for glass fibers (cylinders) in tension, which possess compressive residual stresses on their surfaces.

In thermal tempering, the residual compressive stresses are introduced into the surface of the glass through thermal treatment. A glass initially at a uniform temperature above the glass transition temperature, T_g, is rapidly cooled, or *quenched*, to a new temperature by the use of cold air or oil baths. The outside of the glass, which initially cools more rapidly than the interior, becomes rigid while the interior is still molten. The temperature difference between the surface and the midplane of the glass rapidly reaches a maximum upon quenching, and the interior contracts at a greater rate than the rigid outside due to thermal expansion differences. The stresses induced in the later stages of cooling remain. The resulting stress profile is nearly parabolic in shape (see Figure 5.53), with the magnitude of the compressive stresses at the surface being approximately twice the maximum tension in the interior of the glass. Obviously, heat transfer rate and the heat transfer coefficient at the surface play important roles in determining the residual stress distribution.

Another way to introduce compressive stresses on the surface of glasses is to change the chemical composition of the surface through the selective replacement of surface atoms in a process called chemical tempering. This process increases the molar volume of the surface relative to the interior. This is normally done through the substitution of large ions for small. Diffusivity is an important consideration in this process, since the ion being inserted into the surface of the glass must have sufficient mobility to displace the existing ion. The most common example of chemical tempering is the replacement of Na^+ ions in sodium-silicate glasses with K^+ ions.

The compressive stress on the surface associated with the fractional change in molar volume, $\Delta V/V$, can be approximated by

$$\sigma = \frac{E(\Delta V/V)}{3(1-2\nu)} \tag{5.59}$$

where E is the tensile modulus and ν is Poisson's ratio. The compression depth of chemically strengthened glass is controllable, but in practice is limited to a few hundred microns.

In a related glass processing technique called *etching*, the surface of the processed glass is treated with an acid such as hydrofluoric acid, HF, which removes a thin, outer layer of the glass. With this procedure, cracks can be eliminated, or at least reduced in length. As the crack length decreases, the strength of the glass increases in accordance with Eq. (5.46).

In some instances, residual stresses are not desirable, for example, when the glass is to be used as an optical component. Residual stresses can be removed through a processes called *annealing*, in which the glass is heated to a uniform temperature near the transformation temperature and held for a specific amount of time. The stresses are resolved due to increased atomic mobility at elevated temperatures. The glass is then slowly cooled through the critical viscosity range at a rate sufficiently slow to keep residual stresses from developing. The glass is then rapidly cooled.

5.2.3.3 Cement and Concrete*. A great irony of the materials engineering and science disciplines is that the single largest material of construction used in terms of volume—namely, *cement*—finds little or no technical coverage in an introductory text. To assist in remedying this situation, we present here some introductory information on cements and concretes, with an emphasis on their mechanical properties, so that proper consideration can be given them during materials selection and design. First, we must review some terminology.

HISTORICAL HIGHLIGHT

The Romans pioneered the use of hydraulic, or water-cured, cement. Its unique chemical and physical properties produced a material so lasting that it stands today in magnificent structures like the Pantheon. Yet the formula was forgotten in the first few centuries after the fall of the Roman Empire and wasn't rediscovered until 1824 as Portland cement. One Roman version was based on a burned mixture of two major components: volcanic ash—called *pozzolana*—from Mount Vesuvius, which destroyed Pompeii and nearby towns in A.D. 79; and calcium carbonate, the stuff of seashells, chalk, and limestone. Adding water to these sets off a complex set of chemical reactions that convert the gritty pasty stuff into what is essentially artificial stone. The nineteenth-century rediscovery of Roman cement, the aforementioned Portland cement, is made from a combination of burned limestone, clay, and water. It is the single most heavily used human-made material on earth.

Source: *Stuff*, I. Amato, p. 33

The term "cement" refers to a substance that can be used to hold together aggregates such as sand or stone. A cement can, in principle, be any kind of adhesive; it need not be inorganic, and it may be a single compound or a mixture. In this section, we will use the term to refer to a specific inorganic mixture of calcium, aluminum, and silicon oxides known as *hydraulic cements*. Hydraulic cements set through reaction with water, as opposed to *air-set cements* that are principally organic and set through a loss of water.

Hydraulic cements are so named because they *set* (initial stiffening) and *harden* (develop strength) through a series of *hydration reactions*. The starting materials of limestone or chalk ($CaCO_3$), gypsum ($CaSO_4$), kaolin (Al_2O_3), shale or sand (SiO_2), and slags containing various metal oxides such as MgO and FeO are mixed and burned to form a fused mass called a *clinker*, which is then ground to form cement powder. During the burning process, decomposition of the carbonates occurs:

$$CaCO_3 \longrightarrow CaO + CO_2 \qquad (5.60)$$

A shorthand notation is often used to designate the oxides (e.g., C for CaO and A for Al_2O_3; see Table 5.7), and it is also used to designate the compounds formed between the components during heating, such as calcium aluminate (CA) and tricalcium silicate (C_3S):

$$CaO + Al_2O_3 \longrightarrow CaAl_2O_4 \equiv CA \qquad (5.61)$$

$$3CaO + SiO_2 \longrightarrow Ca_3SiO_5 \equiv C_3S \qquad (5.62)$$

Table 5.7 Shorthand Notation of Important Oxides in Cement Compositions

C	CaO	F	Fe_2O_3	N	Na_2O
A	Al_2O_3	M	MgO	K	K_2O
L	Li_2O	T	TiO_2	P	P_2O_5
S	SiO_2	H	H_2O	f	FeO
$\overline{C}$	CO_2	$\overline{S}$	SO_3		

The primary components of Portland cement prior to the hydrolysis reactions are C_3S ($3CaO–SiO_2$, tricalcium silicate); C_2S ($2CaO–SiO_2$, dicalcium silicate); C_3A ($3CaO–Al_2O_3$, tricalcium aluminate); and C_4AF ($4CaO–Al_2O_3–Fe_2O_3$, tetracalcium aluminoferrite).

Upon addition of water, the hydration reactions initiate, and the hydraulic cement begins to gain strength. This process is very complex, but the strengthening effect is due basically to the formation of three types of hydration products: *colloidal products* such as $C_2S \cdot xH_2O$, which have a size of less than 0.1 μm; *submicrocrystalline products* such as $Ca(OH)_2$, Al^{3+}, Fe^{3+}, and SO_4^{2-} phases with sizes from 0.1 to 1 μm; and *microcrystalline products*, primarily of $Ca(OH)_2$, with particle sizes greater than 1 μm. The most common type of hydraulic cement, *Portland cement*, usually contains mostly colloidal products.

The setting and hardening processes can take many days, and the ultimate strength of the cement may not be realized for years in some instances. This is illustrated in Figure 5.54 for several types of Portland cement, where the compressive strength is plotted as a function of time. Recall that most applications of cement such as buildings and bridges call for compressive loads, such that compressive modulus and strength are the operative mechanical properties.

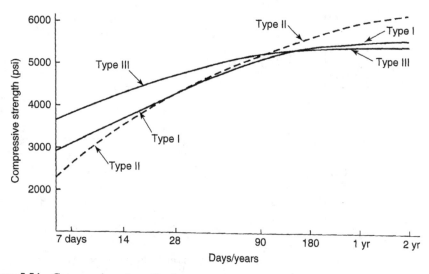

Figure 5.54 Compressive strength of concrete as it develops with time. Reprinted, by permission, from S. Somayaji, *Civil Engineering Materials*, 2nd ed., p. 127. Copyright © 2001 by Prentice-Hall, Inc.

There are different types of Portland cement, depending upon the intended application. Type I is General Purpose; Type II is Moderate Sulfate-Resistant; Type III is High Early Strength for applications where the cement must set rapidly; Type IV is Low Heat of Hydration, which is required for applications such as dams since the hydration reactions are highly exothermic; and Type V is Sulfate-Resistant. Sulfate resistance is important when using cement that may be exposed to some sulfur-containing groundwaters and seawaters since sulfur tends to expand and disintegrate cement.

The material of construction known as *concrete* is obtained by combining cement with *aggregates* and *admixtures* in addition to the water required for hydration. The aggregates take up about 60–75% of the volume of the concrete. Aggregates are classified according to size as either fine (about the size of sand) or coarse (about the size of gravel). The aggregate size and quantity is selected to control the consistency, workability, and ultimate mechanical properties of the concrete. The effect of aggregate size on the compressive strength of concrete is shown in Figure 5.55. The increase in strength that comes from coarser gradation is due solely to a reduction in water requirements for the reduced amount of cement. Other important effects on the ultimate mechanical properties of concrete are cure temperature, air content, water-to-cement ratio, and, of course, the use of reinforcements.

5.2.3.4 *Specific Properties.*

One aspect of ceramics that makes them particularly attractive for structural applications is their low density relative to most structural metals, especially for high-temperature applications. One way to illustrate this is to report mechanical properties such as modulus and strength on a per unit weight basis, resulting in quantities known as *specific properties*. Hence, the tensile modulus divided by the density is known as the *specific modulus*, and the ultimate strength divided by the density is known as the *specific strength*. Table 1.31 shows the specific moduli and strengths of some common ceramics in fiber form, as well as values for some common

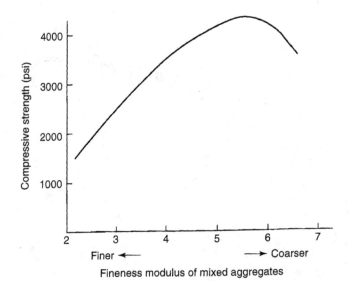

Figure 5.55 Effect of aggregate size on compressive strength of concrete. Reprinted, by permission, from S. Somayaji, *Civil Engineering Materials*, 2nd ed., p. 127. Copyright © 2001 by Prentice-Hall, Inc.

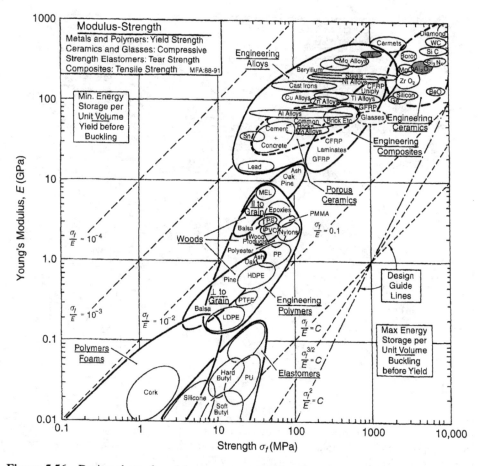

Figure 5.56 Design chart of modulus versus strength (see inset for type of strength determination). Reprinted, by permission, from M. F. Ashby, *Materials Selection in Mechanical Design*, 2nd ed., p. 42. Copyright © 1999 by Michael F. Ashby.

metals for comparison. This effect is most clearly seen by utilizing the well-known design charts of Ashby [7]. The use of Ashby's design charts are elaborated upon in Chapter 8, but we preview their use here with two very important charts. The first is a plot of elastic modulus (Young's modulus) versus strength (either yield strength, compressive strength, tensile strength, or tear strength, as indicated) in Figure 5.56. Materials generally fall into groups (e.g., engineering ceramics, engineering alloys, etc.) with individual materials listed where possible. Figure 5.57 is the corresponding plot in terms of specific properties; that is, the modulus and strength values of Figure 5.57 have been divided by density. In both charts, materials at the upper right-hand side of the diagram have the best combination of modulus and strength.

We select here an extreme example to illustrate the importance of specific properties. Suppose we are looking for a refractory material. Tungsten, W, is a well-known refractory metal with a melting point 3387°C. Its region of modulus and strength values are highlighted in Figure 5.56. A similar ceramic material in terms of modulus and strength would be Al_2O_3, also highlighted in Figure 5.55, with a melting point of

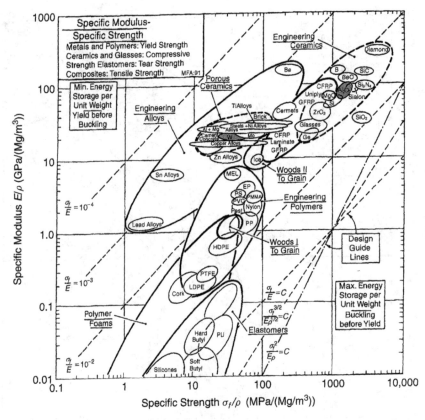

Figure 5.57 Design chart of specific modulus versus specific strength (see inset for type of strength determination). Reprinted, by permission, from M. F. Ashby, *Materials Selection in Mechanical Design*, 2nd ed., p. 44. Copyright © 1999 by Michael F. Ashby.

around 2050°C. Now locate these materials in Figure 5.57. Notice how W has moved downward and to the left relative to Al_2O_3, due solely to its high density (19.3 g/cm³) compared to Al_2O_3 (3.98 g/cm³). Similar arguments apply to the use of polymers, for which densities are even lower. These materials are described in the next section.

5.3 MECHANICS OF POLYMERS

In this section, we describe the mechanical properties of a class of materials that continues to grow in terms of use in structural applications. As issues related to energy consumption and global warming continue to increase demands for lightweight, recyclable materials, the development of new polymers and the characterization of recycled polymers will continue to dominate research and development efforts in this area.

In terms of the mechanical behavior that has already been described in Sections 5.1 and Section 5.2, stress–strain diagrams for polymers can exhibit many of the same characteristics as brittle materials (Figure 5.58, curve A) and ductile materials (Figure 5.58, curve B). In general, highly crystalline polymers (curve A) behave in a brittle manner, whereas amorphous polymers can exhibit plastic deformation, as in

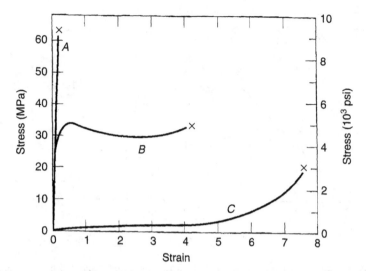

Figure 5.58 The stress–strain behavior of brittle polymer (curve A), ductile polymer (curve B), and highly elastic polymer (curve C). Reprinted, by permission, from W. Callister, *Materials Science and Engineering: An Introduction*, 5th ed., p. 475. Copyright © 2000 by John Wiley & Sons, Inc.

curve B. We will see that these phenomena are highly temperature-dependent, even more so with polymers than they are with metals or ceramics. We will also see that due to the unique structures of some crosslinked polymers, recoverable deformations at high strains up to the point of rupture can be observed in materials known as elastomers (Figure 5.58, curve C).

Despite the similarities in brittle and ductile behavior to ceramics and metals, respectively, the elastic and permanent deformation mechanisms in polymers are quite different, owing to the difference in structure and size scale of the entities undergoing movement. Whereas plastic deformation (or lack thereof) could be described in terms of dislocations and slip planes in metals and ceramics, the polymer chains that must be deformed are of a much larger size scale. Before discussing polymer mechanical properties in this context, however, we must first describe a phenomenon that is somewhat unique to polymers—one that imparts some astounding properties to these materials. That property is *viscoelasticity*, and it can be described in terms of fundamental processes that we have already introduced.

5.3.1 Viscoelasticity

As the term implies, viscoelasticity is the response of a material to an applied stress that has both a *viscous* and an *elastic* component. In addition to a recoverable elastic response to an applied force, polymers can undergo permanent deformation at high strains, just as was the case for metals and some glasses, as described previously. The mechanism of permanent deformation is different in polymers, however, and can resemble liquid-like, or viscous flow, just like we described in Chapter 4. Let us first develop two important theoretical models to describe viscoelasticity, then describe how certain polymers exhibit this important property.

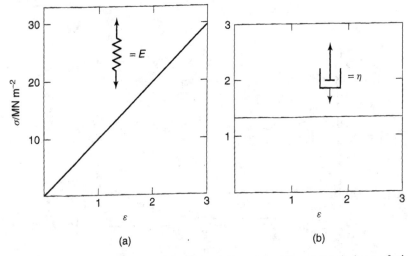

Figure 5.59 Stress–strain behavior of (a) spring of modulus E and (b) dashpot of viscosity η. Reprinted, by permission, from J. M. G. Cowie, *Polymers: Chemistry & Physics of Modern Materials*, p. 277, 2nd ed. Copyright © 1991 by J. M. G. Cowie.

In both models of viscoelasticity, the elastic response is represented by a Hookean *spring* of elastic modulus E (Figure 5.59a), and the viscous flow is represented by a piston-like device known as a *dashpot* of viscosity η (Figure 5.59b). Note that we use the non-Newtonian viscosity, η, rather than the Newtonian viscosity, μ, because we make no assumptions about the viscous behavior of the polymer. We can also generalize these models for both tension and shear by using either the elastic modulus, E, or the shear modulus, G. In fact, we will use these moduli more interchangeably than we have previously. Keep in mind that both the spring and dashpot are completely imaginary devices, much as the "frictionless pulley" of introductory Physics, and are utilized only to assist in the development of the two models of viscoelasticity.

5.3.1.1 *The Maxwell Model.* The first model of viscoelasticity was proposed by Maxwell in 1867, and it assumes that the viscous and elastic components occur in series, as in Figure 5.60a. We will develop the model for the case of shear, but the results are equally general for the case of tension. The mathematical development of the *Maxwell model* is fairly straightforward when we consider that the applied shear stress, τ, is the same on both the elastic, τ_e, and viscous, τ_v, elements,

$$\tau = \tau_e = \tau_v \tag{5.63}$$

since any applied stress is transmitted completely through the spring, once it is extended, to the dashpot below. The total shear strain, γ, is the sum of the elastic strain, γ_e, and the viscous strain, γ_v,

$$\gamma = \gamma_e + \gamma_v \tag{5.64}$$

since the total displacement is the sum of the displacements in the spring and the dashpot, which act independently. We can take the time derivative of Eq. (5.64) to

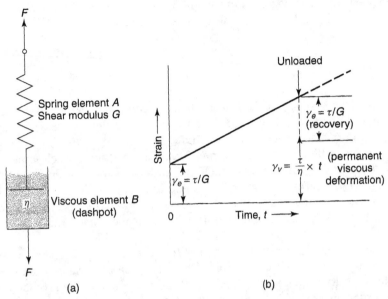

Figure 5.60 (a) Maxwell spring and dashpot in series model of viscoelasticity and (b) constant stress conditions resulting in time-dependent strain. From Z. Jastrzebski, *The Nature and Properties of Engineering Materials*, 2nd ed. Copyright © 1976 by John Wiley & Sons, Inc. This material is used by permission of John Wiley & Sons, Inc.

obtain the total strain rate, $\dot{\gamma} = d\gamma/dt$:

$$\frac{d\gamma}{dt} = \frac{\gamma_e}{dt} + \frac{\gamma_v}{dt} \qquad (5.65)$$

Recall from Eq. (5.10) that the shear strain can be related to the shear stress through the shear modulus, G, according to Hooke's Law, where we now add subscripts to differentiate the elastic quantities from the viscous quantities:

$$\tau_e = G\gamma_e \qquad (5.10)$$

This relation can be differentiated with respect to time and solved for the strain rate:

$$\frac{d\gamma_e}{dt} = \frac{d\tau_e/dt}{G} \qquad (5.66)$$

Recall also from Section 4.0 that the viscous shear rate, $\dot{\gamma}_v$, can be related to the viscous shear stress through the viscosity, η, according to Newton's Law of Viscosity, Eq. (4.3):

$$\tau_v = \eta \frac{d\gamma_v}{dt} \qquad (4.3)$$

where the subscript "v" has once again been added to differentiate the various shear stresses and strains. [In this version of Eq. (4.3), the minus sign has been incorporated into the direction of the shear stress through a change in coordinate system for ease of mathematical manipulation.] Equation (4.3) can be solved for the shear rate and substituted into Eq. (5.65), along with Eq. (5.66), to yield

$$\frac{d\gamma}{dt} = \frac{1}{G}\frac{d\tau_e}{dt} + \frac{1}{\eta}\tau_v \tag{5.67}$$

The utility of the Maxwell model is that it predicts permanent deformation due to flow at any applied stress. As shown in Figure 5.60b, at zero time a stress is applied and the spring deforms according to Hooke's Law. Immediately following zero time, the dashpot begins to flow under the restoring forces of the spring, which loses its stored energy to the dashpot for dissipation. If the stress is kept constant, then $d\tau/dt = 0$, and Eq. (5.67) becomes Newton's Law of Viscosity, and the Maxwell body is indistinguishable from a liquid. Once the stress is removed, the spring recovers, again according to Hooke's Law, and the permanent deformation that remains is due solely to the viscous flow.

Under conditions of constant shear, $d\gamma/dt = 0$, Eq. (5.67) becomes an ordinary differential equation, which can be solved by separation of variables and integration using the boundary conditions $\tau = \tau_0$ at $t = 0$ and $\tau = \tau$ at $t = t$ to give the following relation for the shear stress, τ, as a function of time

$$\tau = \tau_0 \exp\left(\frac{-Gt}{\eta}\right) \tag{5.68}$$

Under constant shear conditions, Eq. (5.68) predicts that the stress will exponentially disappear, so that after a time $t = \eta/G$, the stress will have decayed to $1/e$ of its original value. This time is referred to as the *relaxation time*, t_{rel}, for the material, so that Eq. (5.68) becomes

$$\tau = \tau_0 \exp\left(\frac{-t}{t_{rel}}\right) \tag{5.69}$$

This *stress relaxation* process is important, and we shall revisit it after developing the second model of viscoelasticity.

5.3.1.2 The Kelvin—Voigt Model.
A similar development can be followed for the case of a spring and dashpot in parallel, as shown schematically in Figure 5.61a. In this model, referred to as the *Kelvin–Voigt model* of viscoelasticity, the stresses are additive

$$\tau = \tau_e + \tau_v \tag{5.70}$$

and the strain is distributed

$$\gamma = \gamma_e = \gamma_v \tag{5.71}$$

Hooke's Law and Newton's Law can once again be substituted, this time into Eq. (5.70), to give

$$\tau = G\gamma + \eta\frac{d\gamma}{dt} \tag{5.72}$$

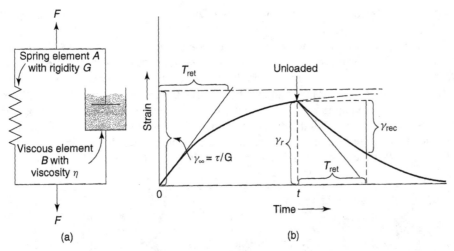

Figure 5.61 (a) Kelvin–Voigt spring and dashpot in parallel model of viscoelasticity and (b) resulting time-dependent strain. From Z. Jastrzebski, *The Nature and Properties of Engineering Materials*, 2nd ed. Copyright © 1976 by John Wiley & Sons, Inc. This material is used by permission of John Wiley & Sons, Inc.

Under conditions of constant stress, $d\tau/dt = 0$, Eq. (5.72) can be solved to give a relation for the total strain as a function of time:

$$\gamma = \frac{\tau}{G}\left[1 - \exp\left(\frac{-Gt}{\eta}\right)\right] \tag{5.73}$$

When the Kelvin–Voigt element is stressed under these constant stress conditions, part of the energy is stored in the spring, while the remainder of the energy is dissipated in the dashpot. Eq. (5.73) indicates that when the sample is subjected to the stress, τ, its elastic response is retarded by the viscous element and the strain does not appear instantaneously but increases gradually, asymptotically approaching its final value of γ_∞, the strain at infinite time, as shown in Figure 5.61b. Upon releasing the stress, recovery will occur because of the elasticity of the spring, but it is again retarded by the dashpot. The strain vanishes gradually and the sample slowly returns to its original shape, theoretically at infinite time. This time-dependent relaxation process is an example of *anelasticity*, which is time-dependent elastic behavior. Although all materials exhibit some anelasticity, it is generally neglected in everything but polymers, where it can be significant. In fact, the entire time-dependent elastic behavior described in this section as viscoelasticity is a form of anelasticity.

Practically, the time needed to achieve the equilibrium after removal of the stress is called the *retardation time*, t_{ret}, which is equal to the ratio η/G, as in the Maxwell model. Equation (5.73) can then be written

$$\gamma_r = \frac{\tau}{G}\left[1 - \exp\left(\frac{-t}{t_{ret}}\right)\right] \tag{5.74}$$

where γ_r is the strain at time t. If the retardation time is small, the recovery will take place at finite time. The Kelvin–Voigt model illustrates the behavior of materials exhibiting a retarded elastic deformation, as occurs in such phenomena as the *elastic aftereffect, creep recovery,* and *elastic memory.*

When a material is subjected to a tensile or compressive stress, Eqs. (5.63) through (5.74) should be developed with the shear modulus, G, replaced by the elastic modulus, E, the viscosity, η, replaced by a quantity known as *Trouton's coefficient* of viscous traction, λ, and shear stress, τ, replaced by the tensile or compressive stress, σ. It can be shown that for incompressible materials, $\lambda = 3\eta$, because the flow under tensile or compressive stress occurs in the direction of stress as well as in the two other directions perpendicular to the axis of stress. Recall from Section 5.1.1.3 that for incompressible solids, $E = 3G$; therefore the relaxation or retardation times are λ/E.

Example Problem 5.2

In Cooperative Learning Exercise 5.9, you studied the effect of temperature on the creep rate of an engine turbine blade. Calculate the viscosity in poise that can be assigned to this material at 500°C under a stress of 20,000 psi.

Solution

For the case of continuous deformation under a constant tensile stress, we can write, by analogy to Newton's Law of Viscosity [Eq. (4.3)], a relationship between tensile strain rate, $\dot{\varepsilon}$, tensile stress, σ, and Trouton's coefficient, λ, as

$$\sigma = \lambda\dot{\varepsilon}$$

For incompressible materials, Trouton's coefficient can be related to $\lambda = 3\eta$, so that

$$\eta = \frac{\sigma}{3\dot{\varepsilon}}$$

From CLE 5.9, $\dot{\varepsilon} = 0.0001$ in./in.-s at 500°C, and $\sigma = 20,000$ psi, such that

$$\eta = \frac{(20,000 \text{ psi})(6.895 \times 10^4 \text{ dyn/cm}^2 \cdot \text{psi})}{3(0.0001 \text{s}^{-1})}$$

$$\eta = 4.6 \times 10^{12} \text{ poise}$$

5.3.1.3   The Four-Element Model*. The behavior of viscoelastic materials is complex and can be better represented by a model consisting of four elements, as shown in Figure 5.62. We will not go through the mathematical development as we did for the Maxwell and Kelvin–Voigt models, but it is worthwhile studying this model from a qualitative standpoint.

The total deformation in the four-element model consists of an instantaneous elastic deformation, delayed or retarded elastic deformation, and viscous flow. The first two deformations are recoverable upon removal of the load, and the third results in a permanent deformation in the material. Instantaneous elastic deformation is little affected by temperature as compared to retarded elastic deformation and viscous deformation, which are highly temperature-dependent. In Figure 5.62b, the total viscoelastic deformation is given by the curve OABDC. Upon unloading (dashed curve DEFG),

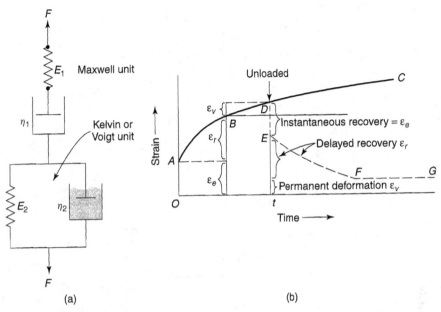

Figure 5.62 (a) Four-element spring and dashpot model of viscoelasticity and (b) resulting strain-dependent time diagram. From Z. Jastrzebski, *The Nature and Properties of Engineering Materials*, 2nd ed. Copyright © 1976 by John Wiley & Sons, Inc. This material is used by permission of John Wiley & Sons, Inc.

instantaneous elastic, ε_e, and retarded elastic, ε_r, strains are recovered, but viscous flow, ε_v, results in permanent deformation.

The four-element model is particularly useful for describing creep in polymers. The initial, almost instantaneous, elongation produced by the application of the stress is inversely proportional to the modulus of the material; that is, a polymer with a low modulus stretches considerably more than a material in the glassy state with a high modulus. This rapid deformation, shown as OA in Figure 5.62b, is followed by a region of creep, A to B, initially fast, but eventually slowing down to a constant rate represented by B to D (or B to C for no unloading). This is the steady-state, or secondary, creep rate of Figure 5.43. As a result of the inverse dependence on modulus and time-dependence of strain, creep data for polymers are usually reported in terms of the *creep compliance*, defined as the ratio of the relative elongation at a specified time to the stress. Though notations vary, the creep compliance is usually represented by $J(t)$, where for the constant-stress tensile experiment shown in Figure 5.62 we have:

$$J(t) = \frac{\varepsilon(t)}{\sigma_0} \qquad (5.75)$$

Notice that the compliance is inversely proportional to the modulus. Equations for the time-dependent strain can be developed for the four-element model shown above, or for any combination of elements that provide a useful model. The corresponding time-dependent compliance can then be determined using Eq. (5.75), or the time-dependent modulus using an analogous equation.

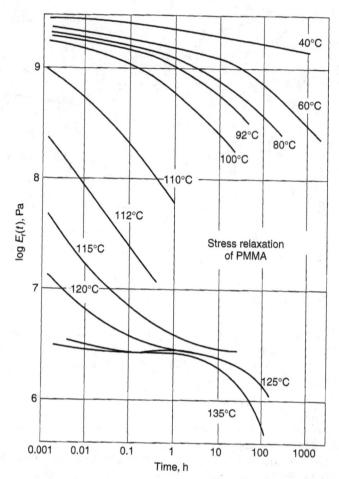

Figure 5.63 Time-dependent modulus for unfractionated poly (methyl methacrylate). Reprinted, by permission, from F. Rodriguez, *Principles of Polymer Systems*, 2nd ed., p. 216. Copyright © 1982 by Hemisphere Publishing Corporation.

5.3.1.4 *Time-Temperature Superposition — The WLF Equation.* Recall that the stress relaxation, t_{rel}, for the Maxwell model is simply the ratio of the viscosity to the modulus. These values are easily measured in an oscillating-load apparatus, much like that described for fatigue testing in Section 5.2.1.4. The time-dependent tensile modulus for a typical amorphous polymer, poly(methyl methacrylate), measured by stress relaxation at various temperatures is shown in Figure 5.63. We could take a cross section of the modulus-time curve at some fixed relaxation time, say 0.001 hours, 1 hour, and 100 hours, and replot the data as log E versus temperature, as in Figure 5.64. In doing so, we notice two reasonably well-defined plateaus at all relaxation times. The first, at $E = 10^{10}$ Pa, is the region of glassy polymer behavior. The second plateau, at $E = 10^6$ Pa, is the region of rubbery polymer behavior. Between the two plateaus is the glass transition, which occurs at $T = T_g$. Notice that T_g is dependent on relaxation rate. At still higher temperatures, rubbery flow and then liquid-like flow occur.

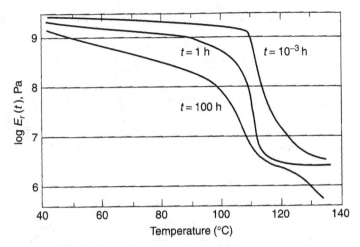

Figure 5.64 Modulus–temperature master curve based on cross sections of Figure 5.63. Reprinted, by permission, from F. Rodriguez, *Principles of Polymer Systems*, 2nd ed., p. 217. Copyright © 1982 by Hemisphere Publishing Corporation.

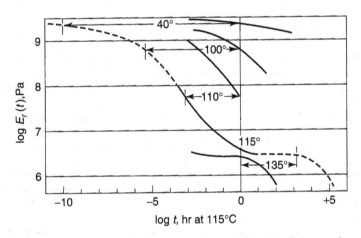

Figure 5.65 Modulus–time master curve based on WLF-shift factors using data from Figure 5.63 with a reference temperature of 114°C. Reprinted, by permission, from F. Rodriguez, *Principles of Polymer Systems*, 2nd ed., p. 217. Copyright © 1982 by Hemisphere Publishing Corporation.

Another method for obtaining a composite curve is to shift the curves in Figure 5.63 relative to some reference temperature, T_R, as shown in Figure 5.65. For example, the log E versus time curve at 115°C in Figure 5.63 could be selected as the reference curve; that is, $T_R = 115°C$. Curves at 110°C, 100°C, and 40°C could be shifted to the left, for example, and the curve at 135°C could be shifted to the right, as illustrated in Figure 5.63. The result is a modulus–time master curve, and the process of shifting the curves relative to a reference point is called the *time–temperature superposition principle*. This is not simply a graphical exercise. The process, which has been borne out experimentally for a number of amorphous polymers, is carried out by calculating a *shift factor*, a_T, according to the *Williams–Lendel–Ferry (WLF) equation*:

$$\log a_T = \frac{-17.44(T - T_g)}{51.6 + T - T_g} \tag{5.76}$$

The WLF equation holds over the temperature range from T_g to about $T_g + 100$ K. The constants in Eq. (5.76) are related to the free volume. This is a procedure analogous to the one we used to generate time–temperature–transformation (TTT) diagrams for metallic phase transformations in Section 3.1.2.2.

Cooperative Learning Exercise 5.10

Work in groups of three. The shift factor, a_T, in the WLF Equation [Eq. (5.76)], is actually a ratio of stress relaxation times, t_{rel}, in the polymer at an elevated temperature, T, relative to some reference temperature, T_0, and can be related via an Arrhenius-type expression to the activation energy for relaxation, E_{rel} as

$$a_T = \exp\left[\frac{E_{rel}}{R}\left(\frac{1}{T} - \frac{1}{T_0}\right)\right] \tag{CLE 1.1}$$

You have developed a new semicrystalline polymer, which has a typical activation energy for relaxation of $E_{rel} = 120$ kJ/mol. You wish to know the creep compliance for 10 years at 27°C. You know that, in principle, you can obtain the same information in a much shorter period of time by conducting your compliance tests at a temperature above 27°C.

Person 1: Calculate the shift factor and the required test duration if you performed the tests at 32°C.

Person 2: Calculate the shift factor and the required test duration if you performed the tests at 77°C.

Person 3: Calculate the shift factor and the required test duration if you performed the tests at 180°C.

Compare your answers. Which is the most reasonable temperature for running the compliance tests? What are some things that you should check before deciding on a final temperature?

Answers: $a_T(305) = 0.45$, 4.5 years; $a_T(350) = 1.03 \times 10^{-3}$, 91 hrs; $a_T(423) = 0.9 \times 10^{-7}$, 27 sec; 77°C gives the most reasonable test duration, 32°C is too long and 180°C may be too short to collect sufficient data. You should determine T_g and/or T_m, to make sure that the test is neither exactly on the glass transition temperature, nor above the crystalline melting point.

The time–temperature superposition principle has practical applications. Stress relaxation experiments are practical on a time scale of 10^1 to 10^6 seconds (10^{-2} to 10^3 hours), but stress relaxation data over much larger time periods, including fractions of a second for impacts and decades for creep, are necessary. Temperature is easily varied in stress relaxation experiments and, when used to shift experimental data over shorter time intervals, can provide a master curve over relatively large time intervals, as shown in Figure 5.65. The master curves for several crystalline and amorphous polymers are shown in Figure 5.66.

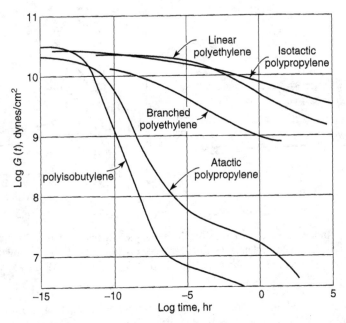

Figure 5.66 Stress–relaxation master curve for several polymers. From F. W. Billmeyer, *Textbook of Polymer Science*, 3rd ed. Copyright © 1984 by John Wiley & Sons, Inc. This material is used by permission of John Wiley & Sons, Inc.

5.3.2 Mechanical Properties of Polymers

To a much greater extent than either metals or ceramics, the mechanical properties of polymers show a marked dependence on a number of parameters, including temperature, strain rate, and morphology. In addition, factors such as molecular weight and temperature relative to the glass transition play important roles that are not present in other types of materials. Needless to say, it is impossible to cover, even briefly, all of these effects. We concentrate here on the most important effects that can affect selection of polymers from a mechanical design point of view.

5.3.2.1 *Temperature and Strain Rate Effects.* Polymers such as polystyrene and poly(methyl methacrylate), PMMA, with a high tensile modulus at ambient temperatures fall into the category of brittle materials, since they tend to fracture prior to yielding. "Tough" polymers, or those that undergo some deformation prior to failure, can be typified by cellulose acetate. These two categories of behavior are illustrated in Figure 5.67, along with their respective temperature dependences.

It can be seen from Figure 5.67 that the effect of temperature on the characteristic shape of the curves is significant. As the temperature increases, both the rigidity and the yield strength decrease, while the elongation generally increases. For cellulose acetate, there is a ductile–brittle transition at around 273 K, above which the polymer is softer and tougher than the hard, brittle polymer below 273 K. For PMMA, the hard, brittle characteristics are retained to a much higher temperature, but it eventually reaches a soft, tough state at about 320 K. Thus, if the requirements of high rigidity and toughness are to be met, the temperature is important.

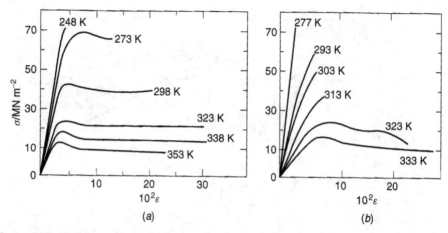

Figure 5.67 Influence of temperature on the stress–strain response of (a) cellulose acetate and (b) poly(methyl methacrylate). Reprinted, by permission, from J. M. G. Cowie, *Polymers: Chemistry & Physics of Modern Materials*, 2nd ed., p. 283. Copyright © 1991 by J. M. G. Cowie.

As was illustrated in Figure 5.46 for a ceramic material, temperature and strain rate have similar, if not opposite, effects on the stress–strain relationship in polymers. Figure 5.68 shows the effect of increasing strain rate on the stress–strain diagram for polyethylene. Compare this effect with decreasing temperature for the polymers in Figure 5.67, where similar effects are observed—that is, increasing yield strength and modulus. Despite the similarities in yield strength behavior with temperature and strain rate between polymers, metals, and ceramics, the mechanisms are quite different, as alluded to earlier. Specifically, the necking of polymers is affected by two physical factors that are not normally significant in metals. The first is the dissipation of mechanical energy as heat, which can raise the temperature in the neck, causing significant softening. The magnitude of this effect increases with strain rate. The second is deformation resistance of the neck, which has a higher strain rate than the surrounding polymer, and results in the strain-rate dependence of yield strength, as shown in Figure 5.68. The relative importance of these two opposing effects depends on the material, the length and thickness of the specimen, and the test conditions, especially the strain rate, which can vary greatly in different forming operations.

In semicrystalline polymers such as polyethylene, yielding involves significant disruption of the crystal structure. Slip occurs between the crystal lamellae, which slide by each other, and within the individual lamellae by a process comparable to glide in metallic crystals. The slip within the individual lamellae is the dominant process, and leads to molecular orientation, since the slip direction within the crystal is along the axis of the polymer molecule. As plastic flow continues, the slip direction rotates toward the tensile axis. Ultimately, the slip direction (molecular axis) coincides with the tensile axis, and the polymer is then oriented and resists further flow. The two slip processes continue to occur during plastic flow, but the lamellae and spherullites increasingly lose their identity and a new fibrillar structure is formed (see Figure 5.69).

A model has been developed by Eyring to describe the effects of temperature and strain rate on flow stress in polymers. The fundamental idea is that a segment of a polymer must pass over an energy barrier in moving from one position to another in the solid. This is an activated process, similar to the many activated processes that we

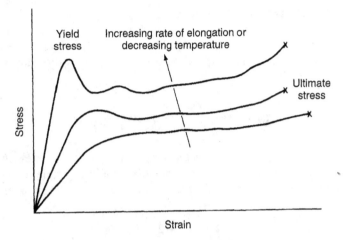

Figure 5.68   Effect of increasing strain rate and decreasing temperature on stress–strain curves for polyethylene. Reprinted, by permission, from F. Rodriguez, *Principles of Polymer Systems*, p. 249, 2nd ed. Copyright © 1982 by Hemisphere Publishing Corporation.

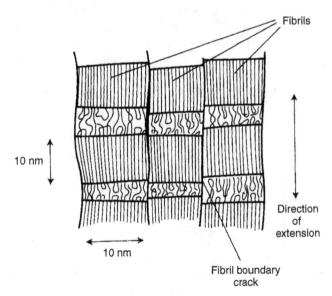

Figure 5.69   Fibrillar structure of an oriented polymer crystal as the result of an applied tensile force. Reprinted, by permission, from N. G. McCrum, C. P. Buckley, and C. B. Bucknall, *Principles of Polymer Engineering*, 2nd ed., p. 72. Copyright © 1997 by Oxford University Press.

have already described and that was first illustrated in Figure 3.1. There is a forward rate and a reverse rate across the barrier, as for all activated processes. The significant point introduced by Eyring is that the application of a shear stress modifies the barrier height, so that in the direction of stress, the rate of segment jumping, formerly slow, now becomes fast enough to give rise to measurable strain change. The final form of the Eyring equation rests on the reasonable assumptions that the imposed strain rate is proportional to the net rate at which segments jump preferentially in the direction of the shear stress and that the dominant shear stress in a tensile test is the maximum shear stress, which at yield is $\sigma_y/2$. It should be of no surprise that the final Eyring result contains an Arrhenius-type expression based on the activation energy barrier for segment jumping, ΔH:

$$\dot{\varepsilon}_y = \dot{\varepsilon}_0 \exp\left(\frac{-\Delta H}{RT}\right) \exp\left(\frac{\sigma_y V^*}{2RT}\right) \tag{5.77}$$

where V^* is the activation volume, R is the gas constant, T is absolute temperature, $\dot{\varepsilon}_y$ is the strain rate at yield, and $\dot{\varepsilon}_0$ is a constant. It is usually desirable to investigate the effect of changing strain rate on yield stress, so that Eq. (5.77) is rewritten in the form

$$\frac{\sigma_y}{T} = \left(\frac{2}{V^*}\right)\left[\frac{\Delta H}{T} + 2.303 R \log\left(\frac{\dot{\varepsilon}_y}{\dot{\varepsilon}_0}\right)\right] \tag{5.78}$$

Figure 5.70 shows plots of σ_y/T against $\log \dot{\varepsilon}_y$ for polycarbonate for a series of temperatures between 21.5°C and 140°C, where the T_g of polycarbonate is 160°C. Note that the predictions of the Eyring model are obeyed, namely, that σ_y/T increases linearly with $\log \dot{\varepsilon}_y$, and at constant $\log \dot{\varepsilon}_y$, σ_y/T increases with decreasing temperature. Thus, the yield strength is dependent on both strain rate and temperature.

The effect of temperature relative to the glass transition, T_g, is illustrated further with the shear modulus, G. The variation of shear modulus with temperature for three common engineering polymers is shown in Figure 5.71. In general, the temperature dependence is characterized by a shallow decline of $\log G$ with temperature with

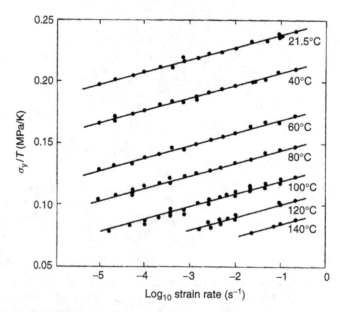

Figure 5.70 Eyring plot for polycarbonate. Reprinted, by permission, from N. G. McCrum, C. P. Buckley, and C. B. Bucknall, *Principles of Polymer Engineering*, 2nd ed., p. 191. Copyright © 1997 by Oxford University Press.

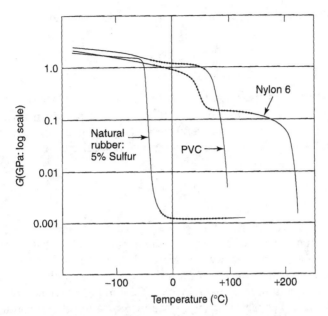

Figure 5.71 Dependence of shear modulus on temperature for three common engineering polymers: crosslinked natural rubber, amorphous polyvinyl chloride (PVC), and crystalline Nylon 6. The typical use temperatures are indicated by dotted lines. Reprinted, by permission, from N. G. McCrum, C. P. Buckley, and C. B. Bucknall, *Principles of Polymer Engineering*, 2nd ed., p. 154. Copyright © 1997 by Oxford University Press.

one or two abrupt drops. The shallow decline is due to thermal expansion effects. As the temperature increases, the molecules in the solid move further apart. This effect occurs in all solids. The abrupt drops in modulus are attributable to viscoelastic relaxation processes due to some type of molecular movement—for example, a side group rotation or backbone movement. Note that the modulus of all three polymers at $-200°C$ is in the range of 1 to 5 GPa. For natural rubber, which is slightly crosslinked, the modulus falls abruptly on heating from -50 to $-25°C$. This is the *glass-to-rubber transition*. The modulus then levels off at a value that depends upon the degree of crosslinking. We will describe the mechanical properties of elastomers in more detail later. For polyvinyl chloride (PVC), a small viscoelastic relaxation occurs at $-40°C$, followed by a major drop at $70°C$ attributable to the glass-to-rubber transition. The Nylon 6 (polycaprolactam) sample has a crystallinity of about 50% and shows a series of relaxations when heated. The major relaxations at $50°C$ and $220°C$ are attributable to the glass-to-rubber and crystalline melting transitions, respectively. This behavior is typical of crystalline polymers, which normally possess several relaxations, whereas amorphous polymers such as PVC typically have only one transition.

5.3.2.2 *Molecular Weight and Composition Effects.*

As we saw in the previous section, degree of crystallinity has an effect on the mechanical behavior and characteristics of polymers. A related factor is molecular weight. To the extent that molecular weight affects the glass transition (see Figure 1.72) and the use temperature relative to T_g, in turn, affects the mechanical properties such as shear modulus (see Figure 5.71), molecular weight has an effect on mechanical properties. The relationship between T_g

and the number-average molecular weight can be approximated by

$$T_g = T_g^\infty - \frac{k}{M_n} \tag{5.79}$$

where T_g^∞ is the glass transition temperature at infinite molecular weight, and k is about 2×10^5 for polystyrene and 3.5×10^5 for atactic poly(α-methyl styrene). The form of this empirical equation is useful for relating a number of polymer properties, including tensile strength, to the number-average molecular weight

$$\text{Property} = a - \frac{b}{M_n} \tag{5.80}$$

where a and b are constants, although the actual variation with molecular weight sometimes depends on an average between the number- and weight-average molecular weights. If a polymer exhibits a yield point and then undergoes extensive elongation before tensile failure, its ultimate tensile strength increases with increasing molecular weight. This effect is illustrated in Figure 5.72 for polyethylene, where molecular weight is represented as melt viscosity, since there is a direct correlation between the two (see Figure 4.11). A similar plot can be made for tensile modulus.

The effects of crystallinity on tensile strength are also included in Figure 5.72 in terms of the polymer density. As the crystalline content of the polymer rises, so does its density, and there is a trend toward higher tensile strength for more highly crystalline polymers. Taken together, crystallinity and molecular weight influence a number of mechanical properties, including hardness, flexural fatigue resistance, elongation at tensile break, and even impact strength. The chance of brittle failure is reduced by

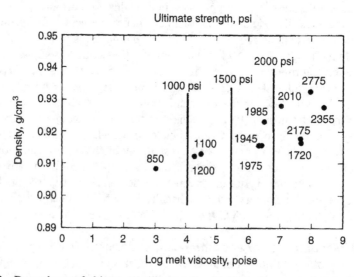

Figure 5.72 Dependence of ultimate tensile strength of polyethylene on molecular weight (melt viscosity) and crystallinity (density). From F. W. Billmeyer, *Textbook of Polymer Science*, 3rd ed. Copyright © 1984 by John Wiley & Sons, Inc. This material is used by permission of John Wiley & Sons, Inc.

raising molecular weight, which increases brittle strength, and by reducing crystallinity. For amorphous polymers, impact strength is found to depend on the weight-average molecular weight. As the degree of crystallinity decreases with temperature during the approach to the crystalline melting point, T_m, stiffness and yield strength decrease correspondingly. These factors often set limits on the temperature at which a polymer is useful for mechanical purposes.

Equation (5.80) implies that the tensile strength of a mixture of polymer components (blend or alloy) can be related to the weight average of the component tensile strengths, $\sigma_{t,i}$:

$$\bar{\sigma}_t = \sum w_i \sigma_{t,i} \tag{5.81}$$

where w_i is the weight fraction of component i. In most real polymer blends, the relationship between composition and mechanical properties is much more complicated than Eq. (5.81) would suggest. As described in Section 2.3.3, the ultimate structure of a blend—that is, one phase or two phase, and the resulting microstructure—depends on how the polymers interact. For example, some mechanical properties of a polycarbonate/poly(methyl methacrylate) blend are shown in Figure 5.73. These blends are invariably two-phase structures. At both 70% and 90% PC, the blends exhibit improved tensile modulus, tensile strength, and yield strength, relative to pure PC,

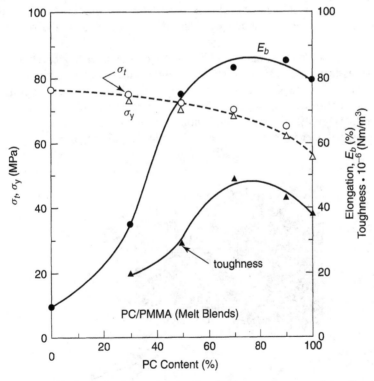

Figure 5.73 Toughness, elongation at break, E_b, yield strength, σ_y, and tensile strength σ_t as a function of composition for a PC/PMMA blend. Reprinted, by permission, from T. Kyu, J. M. Saldanha, and M. J. Kiesel, in *Two-Phase Polymer Systems*, L. A. Utracki, ed., p. 271. Copyright © 1991 by Carl Hanser Verlag.

without sacrificing elongation at break. Hence, the area under the stress–strain curve, or toughness, is greater for these two blends than for pure PC. The toughness enhancement can be attributed to the brittle–ductile transition of the brittle phase, but the diameter of the dispersed phase must be smaller than 1 μm to observe this effect. As the amount of PMMA is further increased (% PC decreases), a significant drop-off in elongation at break and toughness is observed, even though the yield and tensile strengths do not change significantly. This is a reflection of the brittle behavior of PMMA, as first illustrated in Figure 5.67b. This example should serve as an illustration only. The properties of specific polymer–polymer blends are highly variable, and experimentation should give way to generalizations in these cases.

As first introduced in Section 4.2.3, polymer foams are formed when the second phase is a gaseous one such as air or CO_2. The primary benefit of structural foams in relation to solid polymers is the increased specific modulus. This benefit can only be realized when the foam has the proper skin-core structure, as illustrated in Figure 5.74. As a result of manipulations during processing, an injected molded polypropylene foam, for example, will have a nonuniform density through its cross section, with larger, nearly spherical gas cells in the center and small, elongated cells nearer the surface. This form of cell morphology reflects the changing shear and thermal history experienced by the foaming polymer during processing. The outer layer, or skin, contains very few pores. Without the proper skin formation, this functionally graded material will not exhibit adequate mechanical response, since it is the outer skin that bears the initial load, particularly in bending tests. This is illustrated in Figure 5.75, where the effect of gas cell size on specific modulus is less of an effect than is the removal of the outer skin.

In terms of copolymerization, the addition of a comonomer to a crystalline polymer usually causes a marked loss in crystallinity, unless the second monomer crystallizes isomorphously with the first. Crystallinity typically decreases very rapidly, accompanied by reductions in stiffness, hardness, and softening point, as relatively small amounts (10–20 mol%) of the second monomer are added. In many cases, a rigid, fiber-forming polymer is converted to a highly elastic, rubbery product by such minor

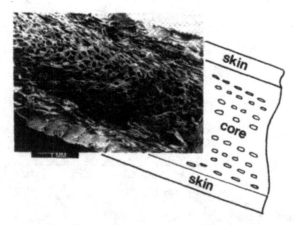

Figure 5.74 Skin-core morphology of an injection-molded polypropylene structural foam. Reprinted, by permission, from P. R. Hornsby, in *Two-Phase Polymer Systems*, L. A. Utracki, ed., p. 102. Copyright © 1991 by Carl Hanser Verlag.

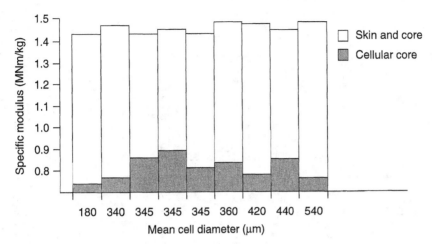

Figure 5.75 Effect of gas cell size on the specific flexural modulus of 6-mm-thick polypropylene structural foam. Samples were tested at $23°C$ with and without their outer skins. Reprinted, by permission, from P. R. Hornsby, in *Two-Phase Polymer Systems*, L. A. Utracki, ed., p. 115. Copyright © 1991 by Carl Hanser Verlag.

modifications. The dependence of mechanical properties on copolymer composition in systems that do not crystallize results primarily from changes in intermolecular forces as measure by cohesive energy (cf. Section 2.3.2). Higher cohesive energy results in higher stiffness and hardness and generally improves mechanical properties. The effect of temperature on the mechanical properties of amorphous copolymers depends on structure, as shown in Figure 5.76. Polymer A and polymer B are two amorphous homopolymers with mechanical properties as indicated. A random copolymer of the two has a comparable modulus–temperature spectrum to the two homopolymers, exhibiting a nearly three-decade drop in shear modulus, with differences only in the relaxation temperature. The relaxation temperature is high for polymer B, low for polymer A, and intermediate to the two for the copolymer. However, the block and graft versions of the copolymer exhibit two relaxation temperatures, corresponding to the two phases of A and B that form in these copolymers. One relaxation process is typical of the polystyrene phase, while the other is typical of the polybutadiene phase. By varying the proportions of comonomer, it is possible to obtain a low styrene content, rubbery solid with glassy inclusions, or a high styrene content, glassy solid with rubbery inclusions.

Finally, the use of low-molecular-weight species to improve flow properties called *plasticizers* normally reduces stiffness, hardness, and brittleness. Plasticization is usually restricted to amorphous polymers or polymers with a low degree of crystallinity because of the limited compatibility of plasticizers with highly crystalline polymers. Other additives, such as antioxidants, do not affect the mechanical properties significantly by themselves, but can substantially improve property retention over long periods of time, particularly where the polymer is subject to environmental degradation.

5.3.2.3 Elastomers. In addition to the numerous ways we have grouped polymers (cf. Section 1.3.2), polymers can also be grouped according to mechanical behavior into the categories plastics, fibers, and elastomers. Though we will not elaborate upon

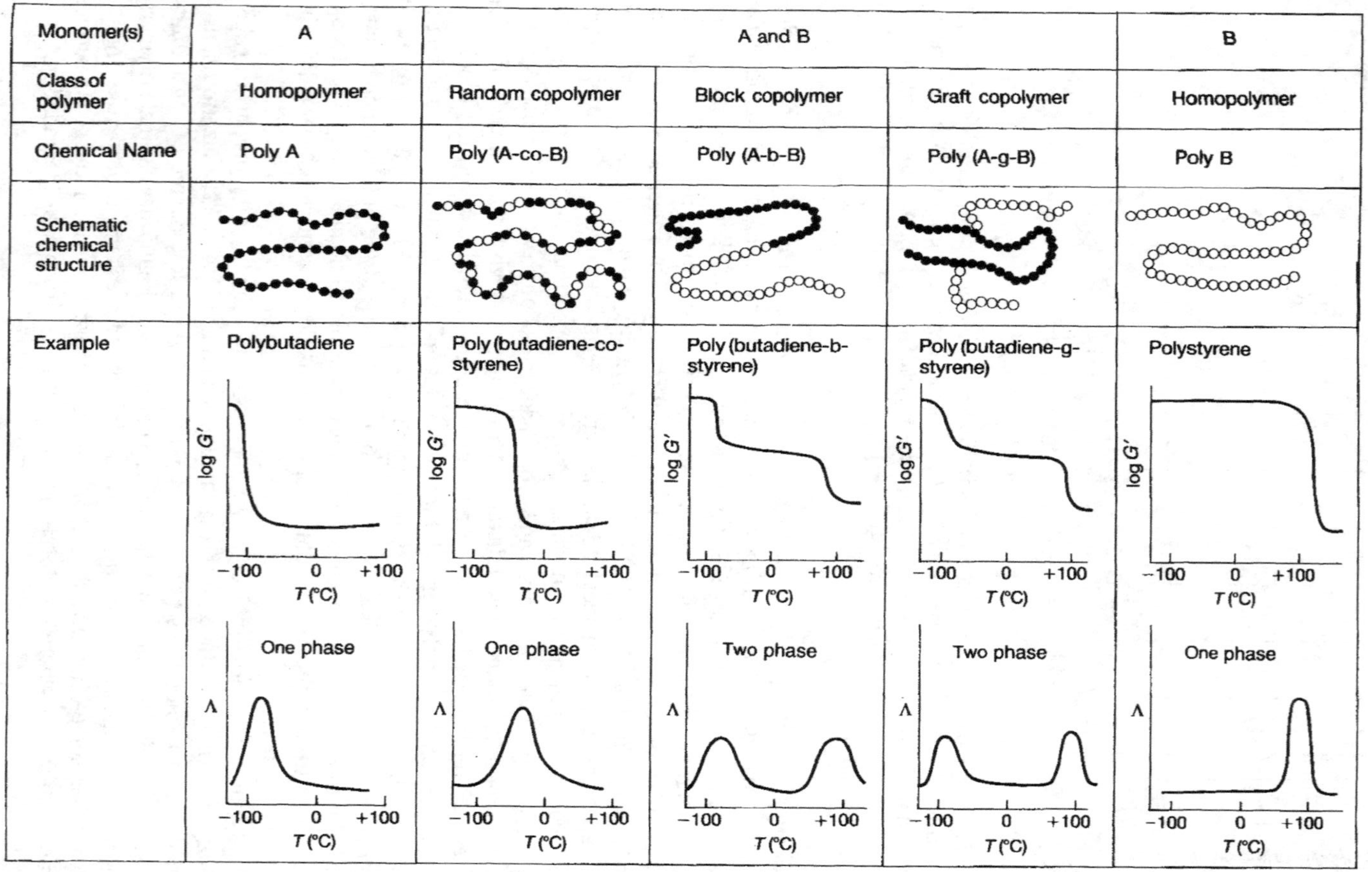

Figure 5.76 Effect of temperature on shear modulus for random, block, and graft copolymers. Bottom curves are the derivative of the log G curves. Reprinted, by permission, from N. G. McCrum, C. P. Buckley, and C. B. Bucknall, *Principles of Polymer Engineering*, 2nd ed., p. 173. Copyright © 1997 by Oxford University Press.

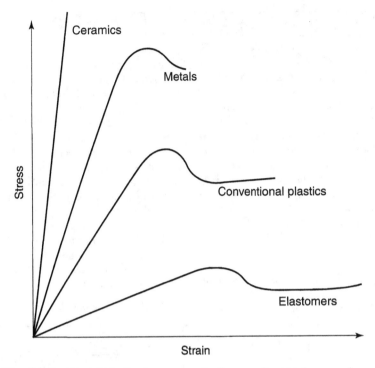

Figure 5.77 Comparison of idealized stress–strain diagrams for metals, amorphous polymers, and elastomers.

these rather non-specific distinctions here, we simply note that elastomers exhibit some unique mechanical behavior when compared to conventional plastics and metals, as in Figure 5.77. The most notable characteristics are the low modulus and high deformations. Elastomers readily undergo deformation and exhibit large, reversible elongations under small applied stresses.

Although a large number of synthetic elastomers are now available, natural rubber must still be regarded as the standard elastomer because of the excellently balanced combination of desirable qualities. The most important synthetic elastomer is styrene–butadiene rubber (SBR), which is used predominantly for tires when reinforced with carbon black. Nitrile rubber (NR) is a random copolymer of acrylonitrile and butadiene and is used when an elastomer is required that is resistant to swelling in organic solvents.

Elastomers exhibit this behavior due to their unique, crosslinked structure (cf. Section 1.3.2.2). It has been found that as the temperature of an elastomer increases, so does the elastic modulus. The elastic modulus is simply a measure of the resistance to the uncoiling of randomly oriented chains in an elastomer sample under stress. Application of a stress eventually tends to untangle the chains and align them in the direction of the stress, but an increase in temperature will increase the thermal motion of the chains and make it harder to induce orientation. This leads to a higher elastic modulus. Under a constant force, some chain orientation will take place, but an increase in temperature will stimulate a reversion to a randomly coiled conformation and the elastomer will contract.

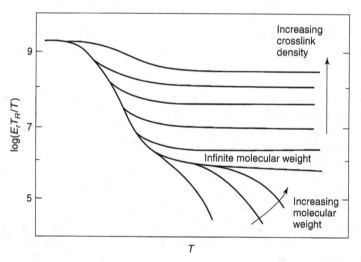

Figure 5.78 Qualitative effects of increasing molecular weight and crosslink density on modulus where T_R is any reference temperature. Reprinted, by permission, from F. Rodriguez, *Principles of Polymer Systems*, 2nd ed., p. 221. Copyright © 1982 by Hemisphere Publishing Corporation.

Unlike non-crosslinked polymers—for which mechanical properties vary little with molecular weight alone, as long as the molecular weight exceeds that of a polymer segment—elastomers, and network polymers in general, exhibit an increase in elastic modulus with crosslink density, as shown in Figure 5.78. Recall that a network polymer technically has an infinite molecular weight. For a finite molecular weight, the crosslinks are temporary ones and are due to entanglements.

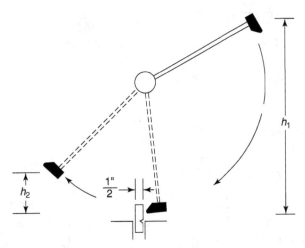

Figure 5.79 Izod (notched) impact test. Reprinted, by permission, from F. Rodriguez, *Principles of Polymer Systems*, 2nd ed., p. 235. Copyright © 1982 by Hemisphere Publishing Corporation.

5.3.2.4 *Other Polymer Mechanical Properties.* In addition to the tension, compression, shear, hardness, fatigue, and creep tests already described, there are a few tests that are common for polymers, although they can be applied to any material, in principle. When polymers are used as films, particularly in packaging applications, their resistance to tearing is important. In one test of *tear strength*, a specimen is torn apart at a cut made by a sharp blade. Energy is provided by a falling pendulum, and the work done is measured by the residual energy of the pendulum. Tear strength and tensile strength are closely related. An *impact test* is also performed with a swinging pendulum. The impact strength is measured when the pendulum is allowed to hit the specimen, often at a carefully placed cut called a *notch*. This type of test is often called a *notched-impact test*, or an *Izod impact test* for the specific type of apparatus (see Figure 5.79). In the impact test, the original and final heights of the hammer, h_1 and h_2, determine the strength of the sample of thickness d which is held in a vise.

For an adhesive bond, the *peel strength* (normal to the bonded surface) or the shear strength (in the plane of the bonded surface) may be measured. For other applications, the stress supportable in compression before yielding may be the most important parameter, as in a rigid foam, for example.

A *torsional pendulum* (Figure 5.80) is often used to determine dynamic properties. The lower end of the specimen is clamped rigidly and the upper clamp is attached to the inertia arm. By moving the masses of the inertia arm, the rotational momentum of inertia can be adjusted so as to obtain the required frequency of rotational oscillation. The *dynamic shear modulus, G'*, can be measured in this manner. A related device is the *dynamic mechanical analyzer* (DMA), which is commonly used to evaluate the dynamic mechanical properties of polymers at temperatures down to cryogenic temperatures.

Abrasion resistance usually takes the form of a *scratch test*, in which the material is subjected to many scratches, usually from contact with an abrasive wheel or a stream of falling abrasive material. The degree of abrasion can be determined by loss of weight for severe damage, but is usually measured by evidence of surface marring.

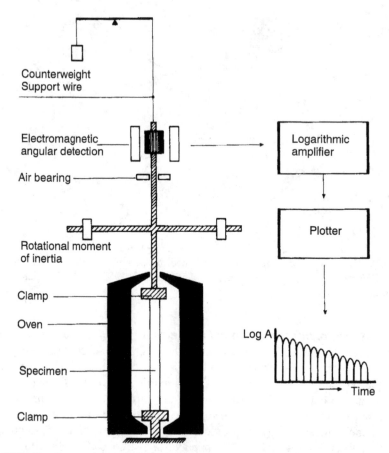

Figure 5.80   Torsion pendulum for the determination of shear modulus and damping as functions of temperature at frequencies around 1 Hz. Reprinted, by permission, from N. G. McCrum, C. P. Buckley, and C. B. Bucknall, *Principles of Polymer Engineering*, 2nd ed., p. 133. Copyright © 1997 by Oxford University Press.

Friction, hardness, and abrasion resistance are related closely to the viscoelastic properties of polymers.

5.4   MECHANICS OF COMPOSITES

The effect of dispersoids on the mechanical properties of metals has already been described in Section 5.1.2.2. In effect, these materials are composites, since the dispersoids are a second phase relative to the primary, metallic matrix. There are, however, many other types of composite materials, as outlined in Section 1.4, including laminates, random-fiber composites, and oriented fiber composites. Since the chemical nature of the matrix and reinforcement phases, as well as the way in which the two are brought together (e.g., random versus oriented), vary tremendously, we shall deal with specific types of composites separately. We will not attempt to deal with all possible matrix–reinforcement combinations, but rather focus on the most common and industrially important composites from a mechanical design point of view.

The starting point for these descriptions will be the law of mixtures, which was first introduced in Eq. (1.62) and which can be used to describe, to a first approximation, the composite property, P, that results from a combination of the reinforcement and matrix properties, P_r and P_m, respectively:

$$P_c = P_r V_r + P_m V_m \qquad (5.82)$$

where V_r and V_m are the volume fractions of the reinforcement and matrix components, respectively. This relationship will not apply in all instances, but will serve as a starting point for descriptions of specific composite systems.

5.4.1   Mechanical Properties of Particle-Reinforced Composites

5.4.1.1   Dispersion Strengthened Metals and Alloys. In this section, we elaborate upon the dispersion-strengthened metals first described in Section 5.1.2.2. The hard particle constituent is a small component in comparison to the softer matrix phase, seldom exceeding 3% by volume, and the particles are only a few micrometers in size. Dispersion-hardened alloys differ from precipitation-hardened alloys in that the particle is added to the matrix, usually by nonchemical means. As a result, the characteristics of the particles largely control the properties of the composite, particularly with regard to strength. The finer the interparticle spacing for a given alloy series, the better the ductility. Cold working is generally required to achieve high strength levels.

The production method for dispersion-strengthened metals has a considerable effect on the properties of the final product. The particular production method depends on the materials involved and includes such techniques as surface oxidation of ultrafine powders, internal oxidation of dilute solid–solution alloys, mechanical mixing of fine metal powders, direct production from liquid metals, and colloidal processing. The surface oxidation of ultrafine powders method is used to form sintered aluminum powder (SAP) alloys, in which each aluminum particle is surrounded by a thin oxide layer. The powder is then compacted, sintered, and cold-worked to form a strong, dispersion-strengthened alloy. The unique characteristics of SAP alloys are their high-temperature strength retention. Instead of softening at a temperature about half the melting point, recrystallization and grain growth are avoided and an appreciable fraction of the material's low-temperature strength persists up to 80% of its melting temperature (see Figure 5.81). The SAP technique has also been employed in the production of lead, tin and zinc oxide-dispersion strengthened alloys.

In the colloidal technique, the size and distribution of a dispersed thoria (ThO phase is controlled to produce dispersion-strengthened alloys, primarily with nicke' the metallic phase. The so-called TD (thoria-dispersed) nickel has modest streng' room temperature, but retains this strength nearly to its melting point. TD nicke to 4 times stronger than pure nickel in the 870–1315°C range, and oxidation res' of the alloy is better than that of nickel at 1100°C.

Other dispersion-strengthened systems include Al_2O_3-reinforced copper, ' iron and TiO_2 and MgO-reinforced copper. In general, the strength of (strengthened materials is not particularly impressive at low temperatures or the metal phase melting point. The primary benefit, as discussed prev' the high-temperature strength retention. This effect is summarized in F'

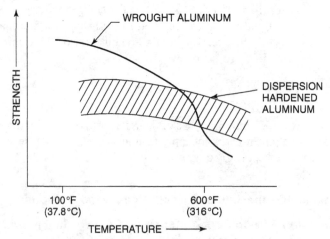

Figure 5.81   Strength of dispersion-strengthened aluminum compared with wrought aluminum. Reprinted, by permission, from M. Schwartz, *Composite Materials Handbook*, 2nd ed., p. 1.29. Copyright © 1992 McGraw-Hill Book Co.

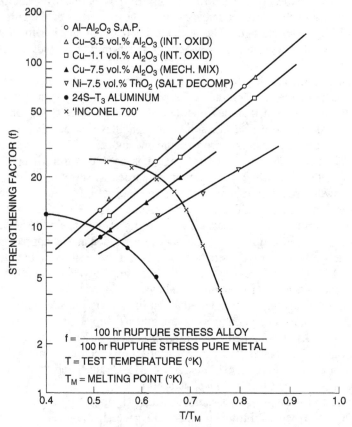

5.82   The strength of different dispersion-strengthened alloys relative to the pure metal as a function of relative temperature. Preparation techniques include sintered aluminum powder (SAP), internal oxidation, and salt decomposition. Reprinted, by permission, [Ke]lly, *Composite Materials*, p. 62. Copyright © 1966 by American Elsevier, Inc.

a number of dispersion-strengthened materials in comparison to pure aluminum and Inconel metals.

5.4.1.2   *Cermets.*

Recall from Section 1.4.5.2 that a cermet is a particular composite that is similar to a dispersion-strengthened metal, except that the grains are generally larger (cf. Figure 1.78). The metal matrix usually accounts for up to 30% of the total volume, though a wide a variety of compositions are possible. In fact, in oxide-based cermets, either the oxide or the metal may be the matrix phase. For example, the thermal shock resistance of a 28% Al_2O_3–72% Cr is good, but can be greatly enhanced by reversing the proportions of the constituents. In the same system, composition reversal reduces the elastic modulus at room temperature by a factor of 10, from 361 to 32.4 GPa. Other oxide-based cermets include Al_2O_3 in stainless steel and in a Cr–Mo alloy. The oxide-based cermets have been extensively applied as a tool material in high-speed cutting of difficult to machine materials.

Carbide-based cermets fall into three general categories: tungsten carbide-based, chromium carbide-based, and titanium carbide-based. Tungsten carbide is widely used as a cutting-tool material and is high in rigidity, compressive strength, hardness, and abrasion resistance. Cobalt is added up to 35% by volume to assist in bonding, and the properties of the cermet vary accordingly. As with the oxide-based cermets, increasing binder volume produces such property improvements as improved ductility and toughness. Compressive strength, hardness, and tensile modulus are reduced as binder content increases. Combined with superior abrasion resistance, the higher impact strength cermets result in die life improvements of up to 7000%. Chromium carbide-based cermets offer phenomenal resistance to oxidation, excellent corrosion resistance, relatively high thermal expansion, relatively low density, and the lowest melting point of the stable carbides. Also important is the similarity of thermal expansion to that of steel. Titanium carbide, used principally for high-temperature applications, has good oxidation and thermal-shock resistance, good retention of strength at elevated temperatures, and a high elastic modulus. Because of the relatively poor oxidation resistance of cobalt at elevated temperatures, nickel is more commonly used as the matrix phase. Finally, refractory cermets such as tungsten-thoria and nickel-barium carbonate are used in high-temperature applications.

5.4.1.3   *Rubber-Reinforced Polymer Composites.*

One of the major developments leading to plastics with outstanding toughness has been the production of rubber-modified glassy polymers such as the acrylonitrile butadiene–styrene (ABS) resins made by polymerizing a continuous glassy matrix in the presence of small rubber particles. Only a small amount (5–15%) of rubber is needed to effect this improvement, and the optimum size of the rubber particles is 1–10 μm. The mechanism by which toughness is developed in these materials can be explained in terms of the crack propagation mechanism first described in Section 5.2.1.2. Most glassy thermoplastics undergo a specific type of crack formation and propagation called *crazing*, in which a network of fibrils connect the two bulk surfaces in a polymeric crack, as illustrated in Figure 5.83. The incorporation of rubber particles into the glassy matrix serves to bridge the microvoids and arrest crazing and crack propagation, resulting in increased toughness. This effect is illustrated in Figure 5.84, in which the ductility is improved in polystyrene (PS) through the addition of small rubber particles to produce high-impact polystyrene (HIPS).

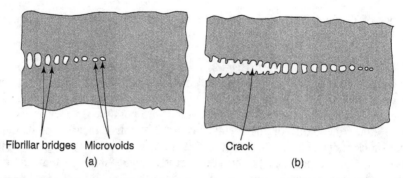

Fibrillar bridges   Microvoids   Crack
             (a)   (b)

Figure 5.83 Schematic illustrations of (a) a craze showing microvoids and fibrillar bridges and (b) a craze followed by a crack. Reprinted, by permission, from W. Callister, *Materials Science and Engineering: An Introduction*, 5th ed., p. 493. Copyright © 2000 by John Wiley & Sons, Inc.

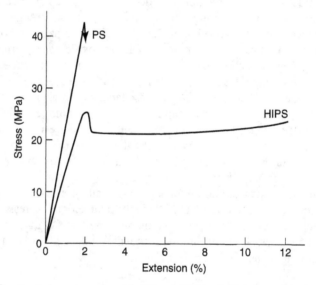

Figure 5.84 Stress–strain curves for polystyrene (PS) and high-impact polystyrene (HIPS). Reprinted, by permission, from N. G. McCrum, C. P. Buckley, and C. B. Bucknall, *Principles of Polymer Engineering*, 2nd ed., p. 200. Copyright © 1997 by Oxford University Press.

5.4.2 Mechanical Properties of Fiber-Reinforced Composites

As first described in Section 1.4.2, there are a number of ways of further classifying fiber–matrix composites, such as according to the fiber and matrix type—for example, glass-fiber-reinforced polymer composites (GFRP) or by fiber orientation. In this section, we utilize all of these combinations to describe the mechanical properties of some important fiber-reinforced composites. Again, not all possible combinations are covered, but the principles involved are applicable to most fiber-reinforced composites. We begin with some theoretical aspects of strength and modulus in composites.

Recall from Section 1.4, and from Figure 1.75 in particular, that the fiber reinforcement can be classified as either continuous or discontinuous. In the next two sections we describe the mechanical properties of these two classes of fiber reinforcement in composites, with the emphasis on reinforcement of polymer–matrix composites. Generally, the highest strength and stiffness are obtained with continuous reinforcement. Discontinuous fibers are used only when manufacturing economics dictate the use of a process where the fibers must be in this form—for example, injection molding. We begin our description of discontinuous fiber-reinforced composites with some concepts related to the distribution of stresses that are independent of fiber composition. These developments will be limited to unidirectionally aligned fibers (cf. Figure 1.76)—that is, fibers with their axes aligned. First, we will concentrate on continuous, unidirectional fibers, and then the strength and modulus of discontinuous, unidirectional fiber-reinforced composites will be analyzed.

5.4.2.1 *Fiber–Matrix Coupling.*

Consider the case of fibers that are so long that the effects of their ends can be ignored. Furthermore, consider the case where these continuous fibers are all aligned along a preferred direction, as illustrated in Figure 5.85. Three orthogonal axes, 1, 2, and 3, are defined within the composite. When loads are applied to the composite block, we can assume that the block deforms as if the stress and strain are uniform within each component. For example, if a stress σ_1 is applied parallel to axis 1, as in Figure 5.86a, the fibers and matrix are approximately coupled together in parallel; that is, the fibers and matrix elongate equally in the direction of σ_1. The fiber and matrix axial strains along axis 1—ε_{f1} and ε_{m1}, respectively—equal the total strain in the composite:

$$\varepsilon_{f1} = \varepsilon_{m1} = \varepsilon_1 \tag{5.83}$$

Fibers extend the entire length of the composite, so that at any section the area fractions occupied by fibers and matrix equal their respective volume fractions, V_f and $V_m = 1 - V_f$. The total stress, σ_1, must then equal the weighted sum of stresses in fibers and matrix, σ_{f1}, and σ_{m1}, respectively:

$$\sigma_1 = V_f \sigma_{f1} + (1 - V_f)\sigma_{m1} \tag{5.84}$$

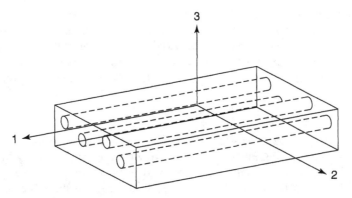

Figure 5.85 Definition of axes in a continuous, unidirectional fiber-reinforced composite. Reprinted, by permission, from N. G. McCrum, C. P. Buckley, and C. B. Bucknall, *Principles of Polymer Engineering*, 2nd ed., p. 258. Copyright © 1997 by Oxford University Press.

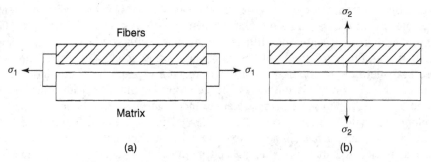

Figure 5.86   Fiber–matrix coupling in continuous, unidirectional fiber-reinforced composites: (a) Coupling for tensile stress parallel to fiber axis. (b) Coupling for stress perpendicular to fiber axis. Reprinted, by permission, from N. G. McCrum, C. P. Buckley, and C. B. Bucknall, *Principles of Polymer Engineering*, 2nd ed., p. 258. Copyright © 1997 by Oxford University Press.

This is the rule of mixtures for stress, as given in Eq. (5.82). Fibers and matrix are assumed to carry pure axial tension, with no stress in the 2–3 plane; that is, $\sigma_{f2} = \sigma_{f3} = \sigma_{m2} = \sigma_{m3} = 0$. We treat the fibers and matrix as acting in a purely elastic manner and neglect the viscoelastic effects of the polymer matrix. Hooke's Law then applies to both the fiber and the matrix

$$\sigma_{f1} = E_f \varepsilon_{f1} \tag{5.85}$$

$$\sigma_{m1} = E_m \varepsilon_{m1} \tag{5.86}$$

as well as to the composite, where E_1 is the composite modulus

$$\sigma_1 = E_1 \varepsilon_1 \tag{5.87}$$

Substitution of the stresses from Eqs. (5.86) to (5.87) into Eq. (5.84) and division by ε_1 yields a prediction for the axial composite tensile modulus:

$$E_1 = V_f E_f + (1 - V_f) E_m \tag{5.88}$$

This, too, is the simple rule of mixtures of Eq. (5.82) applied to modulus. Care must be exercised to use the correct fiber modulus, E_1 in the case of anisotropic fibers such as carbon and Kevlar®. In Eq. (5.88), it is the axial tensile modulus that must be used. The second term in Eq. (5.88) usually makes only a small contribution, since $E_m \ll E_f$. Except at very low fiber volume fractions, E_1 is then given to a good approximation by

$$E_1 \approx V_f E_f \tag{5.89}$$

In contrast, when a stress acts in the 2–3 plane, as in Figure 5.86b, the matrix plays a crucial load-bearing role. Fibers and matrix now couple approximately in series, and the whole tensile force is assumed to be carried fully by both the fibers and matrix. The tensile forces in the fiber and matrix, σ_{f2} and σ_{m2}, are therefore equal to each other and to the overall stress in the composite, σ_2:

$$\sigma_{f2} = \sigma_{m2} = \sigma_2 \tag{5.90}$$

Displacement in the fibers and matrix in the direction of σ_2 can be approximated as being additive. The total strain is therefore the weighted sum of strains in the fibers and matrix:

$$\varepsilon_2 = V_f \varepsilon_{f2} + (1 - V_f)\varepsilon_{m2} \tag{5.91}$$

Again, the stress in each component is approximated as being pure uniaxial tension; that is, there is no stress in the 1–3 plane, and $\sigma_{f1} = \sigma_{f3} = \sigma_{m1} = \sigma_{m3} = 0$. Once again, Hooke's Law applies to the fiber, matrix, and composite, and when the strains from these relationships are substituted into Eq. (5.91), an expression for the transverse tensile modulus of the composite, E_2 is obtained:

$$E_2 = \frac{E_f E_m}{(1 - V_f)E_f + V_f E_m} \tag{5.92}$$

The axial and transverse tensile moduli for a continuous, unidirectional glass-fiber-reinforced epoxy matrix composite as predicted by Eqs. (5.88) and (5.92) are given as a function of volume fraction fiber, V_f, in Figure 5.87. Since $E_f \gg E_m$, Eq. (5.92) reduces to the approximate expression:

$$E_2 \approx \frac{E_m}{1 - V_f} \tag{5.93}$$

which indicates that fibers now play the subordinate role for transverse tensile loads in continuous, unidirectional fiber-reinforced composites.

Similar arguments can be applied to Poisson's ratio, and the free contraction parallel to axis 2 when a tensile stress is applied parallel to axis 1, ν_{12}, is found to obey the rule of mixtures

$$\nu_{12} = V_f \nu_f + (1 - V_f)\nu_m \tag{5.94}$$

The other Poisson's ratio, ν_{21} applying to the 1–2 plane does not equal ν_{12}, but can be found from ν_{12}, E_1, and E_2.

When a shear stress acts parallel to the fibers, τ_{12}, the composite deforms as if the fibers and matrix were coupled in series. Hence, the shear strain, γ_{12}, can be found from a relation similar to Eq. (5.91), and the corresponding shear modulus for the composite, G_{12}, is

$$G_{12} = \frac{G_f G_m}{(1 - V_f)G_f + V_f G_m} \tag{5.95}$$

The same models of mechanical coupling can be used to predict the coefficient of linear thermal expansion in the composite, α_1, based on the moduli, and thermal expansion coefficients of the fiber and matrix, α_f and α_m, respectively:

$$\alpha_1 = \frac{V_f E_f \alpha_f + (1 - V_f)E_m \alpha_m}{V_f E_f + (1 - V_f)E_m} \tag{5.96}$$

Here, α_f is assumed to be independent of direction within the fiber, but for highly anisotropic fibers such as carbon and Kevlar®, allowances must be made for their variation in thermal expansion coefficient with direction. The thermal expansion coefficient perpendicular to the fiber axis, α_2, can also be derived:

$$\alpha_2 = V_f \alpha_f (1 + \nu_f) + (1 - V_f)\alpha_m (1 + \nu_m) - \alpha_1 \nu_{12} \tag{5.97}$$

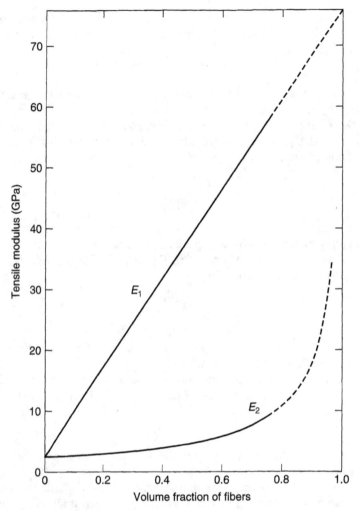

Figure 5.87 Predicted tensile moduli for continuous, unidirectional glass-fiber-reinforced epoxy matrix composite. Reprinted, by permission, from N. G. McCrum, C. P. Buckley, and C. B. Bucknall, *Principles of Polymer Engineering*, 2nd ed., p. 259. Copyright © 1997 by Oxford University Press.

It is usually the case that $\alpha_f \ll \alpha_m$ and $E_f \gg E_m$, such that Eqs. (5.96) and (5.97) predict widely differing coefficients of thermal expansion parallel and perpendicular to the fibers. This effect is illustrated in Figure 5.88, where the thermal expansion coefficients in the axial and transverse direction for a continuous, unidirectional glass-fiber-reinforced epoxy matrix composite are plotted as a function of fiber volume fraction. It is clear that the fibers are highly effective at reducing thermal expansion in the aligned direction, but the thermal expansion at right angles to the fibers is not only higher than in the axial direction, but may actually exceed either of the component expansion coefficients. A further difference between α_f and α_m is that microscopic thermal stresses are generated, which may significantly affect the strength of the composite.

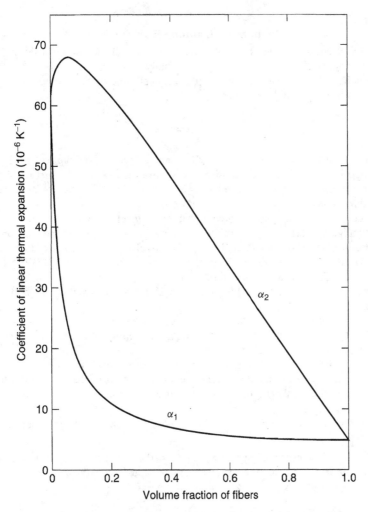

Figure 5.88 Predicted thermal expansion coefficients in the axial and transverse direction for continuous, unidirectional glass-fiber-reinforced epoxy matrix composite. Reprinted, by permission, from N. G. McCrum, C. P. Buckley, and C. B. Bucknall, *Principles of Polymer Engineering*, 2nd ed., p. 265. Copyright © 1997 by Oxford University Press.

Finally, we turn our attention to the prediction of strength in continuous, undirectional fiber composites. Consider again the case in which a tensile stress acts parallel to the fibers (cf. Figure 5.86a). The sequence of events varies, depending on which of the two components is more brittle—that is, which extension at fracture is smaller, ε_f^* or ε_m^*, where the asterisk indicates fracture (failure). Let us consider each case independently.

In the case where $\varepsilon_f^* < \varepsilon_m^*$, as in the case of carbon-reinforced epoxy, we employ the simple assumption of parallel coupling between the fibers and the matrix, for stress parallel to axis 1. We further assume that the fibers and the matrix fail independently, at the same stresses and strains as when the tensile tests are carried out on the pure materials. Our analysis begins by superimposing the fiber and matrix stress–strain

---

Cooperative Learning Exercise 5.11

Recall from Section 5.1.3.1 that thermal expansion is the change in dimensions of a solid due to heating or cooling. As a result, a thermally-induced strain can result when the material is heated, as dictated by Eq. (5.29). For small deformations at constant pressure, we can thus approximate the strain as

$$\varepsilon = \frac{\Delta L}{L} = \alpha_T \Delta T$$

where ΔT is the temperature change due to heating or cooling. For unidirectionally aligned fiber composites, then there will be a different strain in the axial and transverse directions for a given temperature change, since the coefficients of thermal expansion are different in the two directions.

Consider, then, a composite material that consists of 60% by volume continuous, uniaxially-aligned, glass fibers in an epoxy matrix. Take the tensile modulus and Poisson's ratio of glass to be 76 GPa and 0.22, respectively, and of the epoxy to be 2.4 GPa and 0.34, respectively. Take the coefficients of thermal expansion to be 5×10^{-6} K^{-1} and 60×10^{-6} K^{-1} for glass and epoxy, respectively.

Person 1: Calculate the thermal expansion coefficient in the axial direction, α_1 for this composite.

Person 2: Calculate Poisson's ratio for the composite, ν_{12}.

Combine your information to calculate the thermal expansion coefficient in the transverse direction, α_2 for this composite.

The composite is now subjected to a temperature rise of 100 K.

Person 1: Calculate the strain in axial direction as a result of the temperature rise.

Person 2: Calculate the strain in the transverse direction as a result of the temperature rise.

Compare your results. In which direction will the greatest percentage change in dimensions occur?

Answers: $\alpha_1 = 6.13 \times 10^{-6}$ K^{-1}; $\nu = 0.268$; $\alpha_2 = 34.2 \times 10^{-6}$ K^{-1}. $\varepsilon_1 = 6.13 \times 10^{-4}$; $\varepsilon_2 = 3.42 \times 10^{-3}$. The greatest percentage change occurs in the transverse direction, although the absolute change will depend upon the original dimensions.

---

curves, multiplied by their respective volume fractions, as in Figure 5.89. The total stress, σ_1, at strain, ε_1, is given by Eq. (5.84) as the sum of the two components (dashed lines), which in Figure 5.89 have already been multiplied by their respective volume fractions. The result is the solid line in Figure 5.89a and 5.89b. As strain is increased, a point is reached around ε_f^* where the fibers begin to break, and their contribution to the total stress falls to zero. What happens next depends on the volume fraction of fibers, V_f. At low V_f, (Figure 5.89a), there is sufficient matrix to carry the tensile load even after all the fibers have failed. The total stress at strain ε_1 is, from this point on, just the contribution from the matrix alone, $(1 - V_f)\sigma_m$. On further straining, the composite fails when the matrix, which now contains broken fibers, finally fails at $\varepsilon_1 = \varepsilon_m^*$. On the other hand, at high V_f, when the fibers break, there is insufficient matrix to maintain the stress. The point of failure of the fibers then marks the failure of the composite (Figure 5.89b). In this case, the stress just prior to failure is the sum of $V_f\sigma_f^*$ and $(1 - V_f)\sigma_m'$, where $\sigma_m' = \sigma_m\varepsilon_f^*$, the stress carried by the matrix at the failure strain of the fibers. So, there are two criteria for the strength at failure. At low

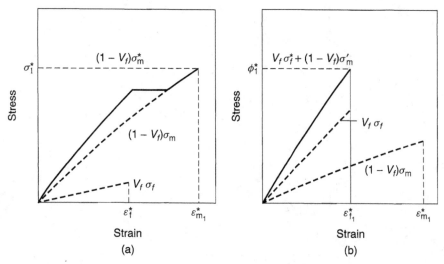

Figure 5.89 Schematic illustration of stress–strain curves for continuous, unidirectional fiber-reinforced composites containing brittle fibers in a ductile matrix. Contributions from fibers and matrix are shown as dashed lines at (a) low fiber volume fractions and (b) high fiber volume fractions. Adapted from N. G. McCrum, C. P. Buckley, and C. B. Bucknall, *Principles of Polymer Engineering*, 2nd ed., p. 267. Copyright © 1997 by Oxford University Press.

V_f

$$\sigma_1^* = (1 - V_f)\sigma_m^* \tag{5.98}$$

and at high V_f

$$\sigma_1^* = V_f\sigma_f^* + (1 - V_f)\sigma_m' \tag{5.99}$$

These two estimates for the fracture strength are shown schematically in Figure 5.90 as two intersecting straight lines. The greater of the two represents the actual strength of the composite and is indicated by a solid line. Figure 5.90 shows an important fact: A minimum fiber fraction, $V_{f,\text{min}}$, is required before the fibers exert any strengthening effect. The minimum fiber fraction can be found from Eq. (5.99):

$$V_{f,\text{min}} = \frac{\sigma_m^* - \sigma_m'}{\sigma_f^* - \sigma_m'} \tag{5.100}$$

For example, the minimum fiber fraction for a carbon-fiber-reinforced epoxy matrix composite is 0.03.

In the second case where $\varepsilon_f^* > \varepsilon_m^*$, two criteria for the failure of the composite are obtained. The analysis is similar to the previous case, but is omitted here. At low V_f

$$\sigma_1^* = V_f\sigma_f' + (1 - V_f)\sigma_m^* \tag{5.101}$$

and at high V_f

$$\sigma_1^* = V_f\sigma_f^* \tag{5.102}$$

The plots of tensile strength versus volume fraction fibers corresponding to those in Figure 5.90 for $\varepsilon_f^* < \varepsilon_m^*$ are shown in Figure 5.91 for $\varepsilon_f^* > \varepsilon_m^*$. Again, the solid line

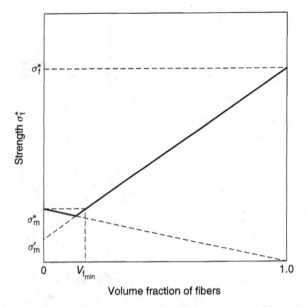

Figure 5.90   Schematic illustration of tensile strength as a function of fiber volume fraction for a continuous, unidirectional fiber-reinforced, ductile matrix composite. Adapted from N. G. McCrum, C. P. Buckley, and C. B. Bucknall, *Principles of Polymer Engineering*, 2nd ed., p. 268. Copyright © 1997 by Oxford University Press.

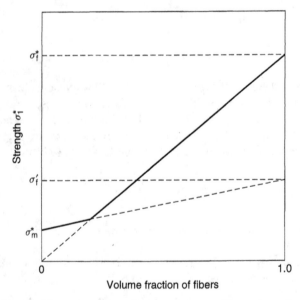

Figure 5.91   Schematic illustration of tensile strength versus fiber volume fraction for continuous, unidirectional ductile fibers in a brittle matrix. Adapted from N. G. McCrum, C. P. Buckley, and C. B. Bucknall, *Principles of Polymer Engineering*, 2nd ed., p. 269. Copyright © 1997 by Oxford University Press.

represents the behavior of the composite. Equations (5.98)–(5.99) and (5.101)–(5.102) have been found to predict well the experimental tensile strengths of brittle fiber/ductile matrix and ductile fiber/brittle matrix composites at a variety of fiber volume fractions in undirectional, continuous-fiber-reinforced composites.

5.4.2.2 Fiber End Effects*.
When considering discontinuous fibers, it is necessary to take into account the fiber ends, which are weak points in the composite. Here stresses tend to concentrate. The critical issue is how stresses get transferred from the low-strength, low-modulus matrix to the high-strength, high-modulus fiber reinforcement. Consider an isolated short fiber in a polymer matrix under tension, as illustrated in Figure 5.92. It must be assumed that negligible stress gets transferred to the fibers across their end faces. It is the shear stress at the fiber–matrix interface, τ_i, that performs the vital task of transmitting tension to the fiber. It can be shown that the relationship between the axial tension in the fiber, σ_f, and the interfacial shear stress is

$$\frac{d\sigma_f}{dx} = \frac{-4\tau_i}{d} \tag{5.103}$$

where d is the fiber diameter and $d\sigma_f/dx$ is the stress distribution along the fiber axis direction, x. Predicting σ_f and τ_i involves stress analysis of the fiber–matrix composite under load. A simplified model assumes perfect bonding between the fibers and the matrix, and it results in the following prediction of σ_f as a function of distance x, along the fiber, measured from its center:

$$\sigma_f = E_f \varepsilon_f \left\{ 1 - \left[\frac{\cosh\left(\frac{na2x}{l}\right)}{\cosh(na)} \right] \right\} \tag{5.104}$$

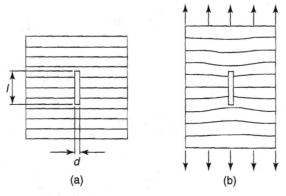

(a)   (b)

Figure 5.92 Schematic illustration of fiber in polymer matrix (a) in the undeformed state and (b) under a tensile load. Horizontal lines are shown to demonstrate strain distribution. Reprinted, by permission, from N. G. McCrum, C. P. Buckley, and C. B. Bucknall, *Principles of Polymer Engineering*, 2nd ed., p. 242. Copyright © 1997 by Oxford University Press.

where a is the fiber aspect ratio (length to diameter, l/d) and n represents a dimensionless group of constants:

$$n = \sqrt{\frac{2G_m}{E_f \ln(2R/d)}} \tag{5.105}$$

where $2R$ is the distance from the fiber to its nearest neighbor, d is again the fiber diameter, and G_m and E_f are the shear modulus and elastic modulus of the matrix and fiber, respectively. Equation (5.104) can be substituted into Eq. (5.103) and differentiated to obtain the interfacial shear stress:

$$\tau_i = \frac{nE_f\varepsilon_f}{2} \frac{\sinh\left(\dfrac{na2x}{l}\right)}{\cosh(na)} \tag{5.106}$$

Equation (5.106) shows that the critical parameter dictating how rapidly stress builds up along the fiber is the product na. This is illustrated in Figure 5.93, where graphs of σ_f and τ_i/n are plotted versus x/l for three values of na. These plots show that for the most efficient stress transfer to the fibers, the product na should be as high as possible. This confirms the desirability of a high aspect ratio, but also shows that n should be high. The ratio of moduli, G_m/E_f, should therefore be as high as possible. Typical values encountered in real composites are $a = 50$, $n = 0.24$, so $na = 12$ for 30 vol% glass fibers in a nylon matrix.

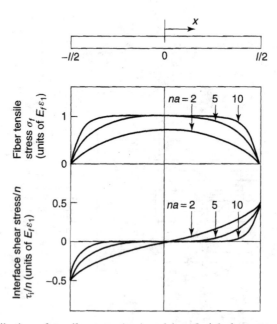

Figure 5.93 Distribution of tensile stress (*top*) and interfacial shear stress (*bottom*) along a short fiber. Reprinted, by permission, from N. G. McCrum, C. P. Buckley, and C. B. Bucknall, *Principles of Polymer Engineering*, 2nd ed., p. 276. Copyright © 1997 by Oxford University Press.

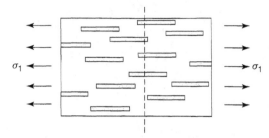

Figure 5.94 Schematic illustration of aligned discontinuous fibers in polymer matrix. Reprinted, by permission, from N. G. McCrum, C. P. Buckley, and C. B. Bucknall, *Principles of Polymer Engineering*, 2nd ed., p. 277. Copyright © 1997 by Oxford University Press.

The most important effects of the fiber tensile stress falling away to zero at the ends are reductions in the axial tensile modulus and strength of the composite. Consider the idealized composite shown in Figure 5.94, where the short fibers are in parallel, and an imaginary (dash) line is drawn across the composite at right angles to the fibers. It can be shown that the stress carried by the composite, σ_1, is given by

$$\sigma_1 = V_f \bar{\sigma}_f + (1 - V_f)\sigma_m \qquad (5.107)$$

where V_f is the fiber volume fraction, σ_m is the matrix stress, and $\bar{\sigma}_f$ is the mean fiber stress, given by

$$\bar{\sigma}_f = \frac{1}{l} \int_{-l/2}^{l/2} \sigma_f \, dx \qquad (5.108)$$

Substitution of Eq. (5.104) into (5.108) and integration yields

$$\bar{\sigma}_f = E_f \varepsilon_1 \left\{ 1 - \left[\frac{\tan h(na)}{na} \right] \right\} = E_f \varepsilon_1 \beta_1 \qquad (5.109)$$

The axial tensile modulus, E_1, is obtained by dividing the stress σ_1 by the strain ε_1:

$$E_1 = \beta_1 V_f E_f + (1 - V_f) E_m \qquad (5.110)$$

Note the similarity of this relationship to the equation of mixing given at the beginning of this section by Eq. (5.82). The factor β_1 is called the *fiber length correction factor*, and it corrects the fiber modulus for the shortness of the fibers. It is plotted in Figure 5.95 as a function of the product na. When na becomes very large, β_1 approaches one, as is expected since this limit is the case of continuous fibers, and Eq. (5.110) reduces to Eq. (5.82). When na falls below about 10, β_1 is significantly less than one.

As indicated in Figure 5.93, the interfacial shear stress, τ_i, is concentrated at the fiber ends. With increasing strain, these are the locations where the interface first fails and *debonding* of the matrix from the fiber begins. This occurs when τ_i reaches the interfacial shear strength, the magnitude of which is determined by three factors: the strength of the chemical bond between the fibers and the matrix; the friction between the fibers and the matrix resulting from matrix pressure introduced during cooling;

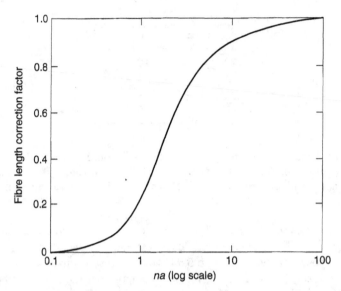

Figure 5.95 Fiber length correction factor for modulus, β_1, as a function of the geometric factor, na. Adapted from N. G. McCrum, C. P. Buckley, and C. B. Bucknall, *Principles of Polymer Engineering*, 2nd ed., p. 277. Copyright © 1997 by Oxford University Press.

and the shear strength of the matrix. With further straining, the debonding regions of the fiber slip within the matrix. Slippage is resisted by a constant friction stress, τ_i^*, and the stress distributions switch from those shown in Figure 5.93 to those shown in Figure 5.96. Since the fiber ends now carry a constant interfacial shear stress, the tensile stress has a constant gradient, given by Eq. (5.103), and stress builds up linearly from each end. To a good approximation, only the stresses shown in the shaded area of Figure 5.96 need to be considered. The length of the debonded region, δ, can be found by applying Eq. (5.103) to either end of the fiber and rearranging:

$$\delta = \frac{E_f \varepsilon_1 d}{4\tau_i^*} \tag{5.111}$$

As straining proceeds, the debonded region grows with increasing strain. At the point of fiber breakage, the strain is $\varepsilon_1 = \varepsilon_f^* = \sigma_f^*/E_f$, and the debonded regions have lengthened to:

$$\delta^* = \frac{\sigma_f^* d}{4\tau_i^*} \tag{5.112}$$

However, the debonded region can only reach this value if the fibers are longer than a critical length, l_c, which is the sum of the debonded regions at each end of the fiber:

$$l_c = 2\delta^* = \frac{\sigma_f^* d}{2\tau_i^*} \tag{5.113}$$

If the fibers are shorter than this critical length, failure of the composite occurs after the debonded regions have extended along the full length of the fiber. The entire

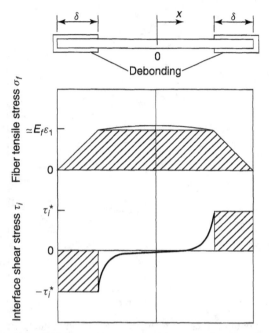

Figure 5.96 Distribution of fiber tensile stress and interfacial shear stress after fiber–matrix debonding has commenced. Reprinted, by permission, from N. G. McCrum, C. P. Buckley, and C. B. Bucknall, *Principles of Polymer Engineering*, 2nd ed., p. 280. Copyright © 1997 by Oxford University Press.

fiber is then merely slipping within the matrix and it cannot be loaded to its failure stress; the fiber is not properly performing its reinforcing function. As a result, l_c is an important parameter, and efforts are made to produce composites in which the fiber length exceeds l_c.

The axial strength of the composite can be calculated by applying Eq. (5.107). In the most common case, the matrix is more ductile than the fibers, such that the matrix critical strain exceeds the fiber critical strain, $\varepsilon_m^* > \varepsilon_f^*$. At low volume fraction of fibers, the fibers located with their ends within a distance δ^* of the failure plane do not fracture, but rather simply pull out of the matrix. This movement is resisted by the friction stress, τ_i^*. The proportion of fibers for which this occurs is l_c/l, and the mean stress resisting pull-out is $\sigma_f^*/2$. The fibers therefore contribute a mean stress of $\bar{\sigma}_f = \sigma_f^* l_c / 2l$ and from Eq. (5.107) the strength becomes

$$\sigma_1^* = \frac{l_c}{2l} V_f \sigma_f^* + (1 - V_f)\sigma_m^* \tag{5.114}$$

At high volume fractions of fibers, the mean fiber stress at failure is obtained from the trapezoidal-shaped distribution of tensile stress illustrated in Figure 5.97, with $\varepsilon_1 = \sigma_f^*/E_f$ and $\delta = \delta^* = l_c/2$, such that

$$\bar{\sigma}_f = \sigma_f^* \left(1 - \frac{l_c}{2l}\right) \tag{5.115}$$

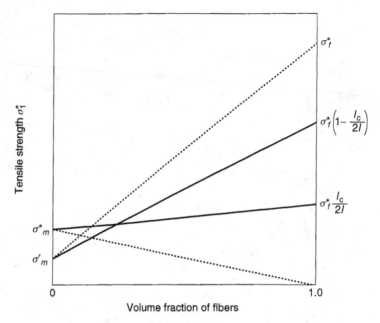

Figure 5.97   Plot of composite tensile strength versus fiber volume fraction for an aligned, short fiber-reinforced polymer matrix composite. The dotted lines show the corresponding values for continuous fibers for comparison purposes. Reprinted, by permission, from N. G. McCrum, C. P. Buckley, and C. B. Bucknall, *Principles of Polymer Engineering*, 2nd ed., p. 281. Copyright © 1997 by Oxford University Press.

and the strength is again given by Eq. (5.107):

$$\sigma_1^* = \left(1 - \frac{l_c}{2l}\right) V_f \sigma_f^* + (1 - V_f)\sigma_m' \tag{5.116}$$

where σ_m' indicates the stress carried by the matrix at the failure strain of the fibers, $\sigma_m \varepsilon_f^*$. The equations for the composite tensile strength in the two extremes of low volume fraction fibers and high volume fraction fibers, Eqs. (5.114) and (5.116), respectively, are plotted as a function of volume fraction of fibers in Figure 5.97. Here, the dashed lines represent the low V_f and high V_f limits, with the greater of the two values shown as a solid line. The solid line represents the predicted strength of the composite. For comparison purposes, the corresponding strength equations at low- and high-volume fractions of fibers are shown as dotted lines for continuous-fiber composites. Figure 5.97 clearly shows that reducing the fiber length has two major effects on the failure of composites—namely, that the strength is reduced (except at very low volume fractions) and there is a wider range of volume fractions over which failure is matrix-controlled.

5.4.2.3 *Composite Moduli: Halpin–Tsai Equations**. Derivations of estimates for the effective moduli (tensile E, bulk K, and shear G) of discontinuous-fiber-reinforced composite materials are extremely complex. The basic difficulty lies in the complex, and often undefined, internal geometry of the composite. The problem has been approached in a number of ways, but there are three widely recognized

---

Cooperative Learning Exercise 5.12

An injection molded bar contains 20% by volume of short carbon fibers in a matrix of nylon 6,6. Take the tensile strengths of the carbon fibers and the nylon to be 3200 MPa and 70 MPa, respectively, and assume the shear strength of the carbon fiber-nylon interface to be 32 MPa. The tensile moduli of nylon 6,6 and carbon fibers are 2.7 GPa and 230 GPa, respectively. Assume that the fibers are all of length 400 μm and diameter 6 μm and are perfectly aligned along the axis of the bar.

Person 1: Calculate the strain at failure in the matrix, ε_m^*, from the matrix properties.

Person 2: Calculate the strain at failure in the fiber, ε_f^*, from the fiber properties. Compare your answers. Will the failure occur in the matrix or the fiber first?

Person 1: Calculate the matrix stress at failure based on which component fails first.

Person 2: Calculate fiber critical length l_c.

Combine your information to calculate the composite tensile strength assuming that V_f is a high value.

Answers: $\varepsilon_m^* = \sigma_m^*/E_m = 70/2700 = 0.026$; $\varepsilon_f^* = \sigma_f^*/E_f = 3200/230000 = 0.0139$

$\varepsilon_f^* > \varepsilon_m^*$, fiber fails prior to matrix, so strain at failure is 0.0139.

$\sigma_m^* = E_m \varepsilon_c^* = 2700(0.0139) = 37.5 \text{ MPa}; \quad l_c = \dfrac{\sigma_f^* d}{2\tau^*} = \dfrac{(3200)(6)}{2(32)} = 300 \ \mu\text{m}$

Equation (5.96), $\sigma_c^* = (1 - 300/800)(0.2)(3200 \text{ MPa}) + (1 - 0.2)(37.5 \text{ MPa})$
$= 430 \text{ MPa}$

(Adapted from McCrum, p. 282)

---

approaches: bounding methods, a special model called the composite cylinder assemblage, and semiempirical equations called the *Halpin–Tsai equations*. In all cases, only unidirectional, discontinuous fibers are considered—that is, short fibers that are preferentially aligned along a certain axis. Two directions are important in this analysis, as defined in Figure 5.98: the longitudinal direction, along the direction of the applied force, and the transverse direction, perpendicular to the applied force. Only the bounding methods and Halpin–Tsai approximations will be described here.

Upper and lower bounds on the elastic constants of transversely isotropic unidirectional composites involve only the elastic constants of the two phases and the fiber volume fraction, V_f. The following symbols and conventions are used in expressions for mechanical properties: Superscript plus and minus signs denote upper and lower bounds, and subscripts f and m indicate fiber and matrix properties, as previously. Upper and lower bounds on the composite axial tensile modulus, E_a, are given by the following expressions:

$$E_a^+ = V_f E_f + (1 - V_f)E_m + \frac{4V_f(1 - V_f)(v_f - v_m)^2}{V_f/K_m + (1 - V_f)/K_f + 1/G_f} \qquad (5.117)$$

$$E_a^- = V_f E_f + (1 - V_f)E_m + \frac{4V_f(1 - V_f)(v_f - v_m)^2}{V_f/K_m + (1 - V_f)/K_f + 1/G_m} \qquad (5.118)$$

Where v is Poisson's ratio, K is the bulk modulus, and G is the shear modulus. Similar relations exist for Poisson's ratio, bulk modulus, and shear moduli in both the axial

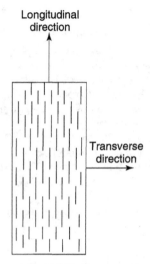

Figure 5.98   Schematic illustration of discontinuous, aligned fiber-reinforced composite.

and transverse directions [8]. Note that the final terms in both expressions provide the corrections to the simple rule-of-mixture approximation.

For most materials, the bounds on axial tensile modulus tend to be reasonably close together, and the rule-of-mixtures prediction given by Eq. (5.82) is generally accurate enough for practical purposes for both the elastic modulus

$$E_a = V_f E_f + (1 - V_f) E_m \qquad (5.119)$$

and, to a lesser extent, Poisson's ratio

$$\nu_a = V_f \nu_f + (1 - V_f) \nu_m \qquad (5.120)$$

The rule of mixtures generally provides a poor estimate of the bulk and shear moduli in both the axial and transverse directions, respectively, in unidirectional, discontinuous-fiber-reinforced composites.

The Halpin–Tsai equations represent a semiempirical approach to the problem of the significant separation between the upper and lower bounds of elastic properties observed when the fiber and matrix elastic constants differ significantly. The equations employ the rule-of-mixture approximations for axial elastic modulus and Poisson's ratio [Equations. (5.119) and (5.120), respectively]. The expressions for the transverse elastic modulus, E_t, and the axial and transverse shear moduli, G_a and G_t, are assumed to be of the general form

$$P = P_m \left(\frac{1 + V_f \xi \eta}{1 - V_f \eta} \right) \qquad (5.121)$$

where P is the composite property (modulus) of interest, P_m is the corresponding matrix property, V_f is the volume fraction of fibers, and ξ and η are correlations as follows:

$$\eta = \frac{\left(\dfrac{P_f}{P_m} - 1\right)}{\left(\dfrac{P_f}{P_m} + \xi\right)} \tag{5.122}$$

where P_f and P_m are again the fiber and matrix properties of interest, respectively, and

$$\xi = l/d \tag{5.123}$$

where l/d is the fiber aspect ratio.

5.4.2.4 Discontinuous-Fiber-Reinforced Polymer–Matrix Composites: Sheet Molding Compound*.

Of the parameters influencing the mechanical properties in short-fiber-reinforced polymer–matrix composites, fiber composition, matrix composition, fiber geometry, and manufacturing method will be elaborated upon here.

The composition and properties of some common reinforcing fibers were given earlier in Table 1.31, the most common of which is glass fiber. The compositions of some commercial glass fibers were given earlier in Table 1.32, and the most common polymer matrices were given in Table 1.28. For this section, we will concentrate on discontinuous-glass-fiber-reinforced polymer–matrix composites.

Glass fiber loadings from 10% to 40% by weight produce increases in strength and rigidity, along with a marked decrease in the coefficient of thermal expansion. Some of these properties develop gradually with increasing glass fiber content, while others change sharply. Both modulus and tensile strength are proportional to glass content. Figure 5.99 shows these relationships for short glass-fiber-reinforced polypropylene and styrene–acrylonitrile copolymer.

Glass fiber reinforcement often improves impact strength, but long-fiber reinforcements are more effective in this regard than short-fiber composites. The most significant effect of glass-fiber reinforcement in thermoplastics is the retention of impact strength down to very low temperatures, but the relationship between glass content and impact strength is not always linear. With low-modulus thermoplastics, optimum impact strength may or may not be achieved at less than the maximum glass content. Normally, room temperature impact strengths of low elastic modulus materials suffer by incorporating glass reinforcement. With rigid thermoplastics, marked improvement in room-temperature impact strengths usually occur with increasing glass content. In virtually all thermoplastics, impact strengths at low temperatures improve with increasing glass content.

The reinforcement can also cause increases in hardness and abrasion resistance, along with decreases in creep and dimensional changes with humidity. Undesirable effects brought on by the addition of glass reinforcement include opacity, low surface gloss, some difficulty in electroplating, and loss of mechanical flexibility.

The effect of fiber diameter on the tensile strength of a glass-fiber-reinforced polystyrene composite is shown in Figure 5.100. Some reinforcements also have a distribution of fiber diameters that can affect properties. Recall from the previous section that the fiber aspect ratio (length/diameter) is an important parameter in some mechanical property correlations.

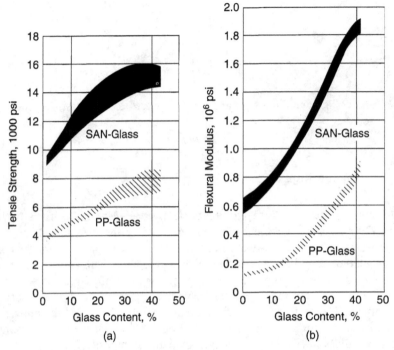

Figure 5.99 Relationship of glass fiber content with (a) tensile strength and (b) flexural modulus for styrene acrylonitrile (SAN) and polypropylene (PP). Reprinted, by permission, from G. Lubin, *Handbook of Fiberglass and Advanced Plastics Composites*, p. 130. Copyright © 1969 by Van Nostrand Reinhold.

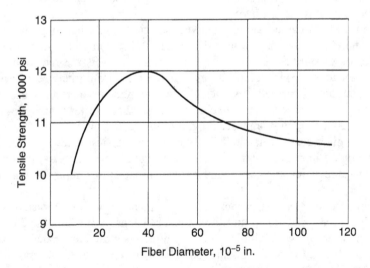

Figure 5.100 Effect of glass fiber diameter on strength of reinforced polystyrene. Reprinted, by permission, from G. Lubin, *Handbook of Fiberglass and Advanced Plastics Composites*, p. 130. Copyright © 1969 by Van Nostrand Reinhold.

A number of processing methods exist to produce glass-fiber/polymer–matrix composites, including reinforced reaction injection molding (RRIM), resin transfer molding (RTM), filament winding, and compression molding of *sheet molding compound* (SMC), to name a few. These processing techniques will be elaborated upon in Chapter 7. For this section, we concentrate on one processing technique that is commonly used to make useful components out of discontinuous-glass-fiber-reinforced polymer matrix composites—namely, compression molding of SMCs. This method comprises a wide range of thermosetting prepreg molding materials reinforced generally with short glass fiber strands about 25–50 mm long. Traditionally, unsaturated polyester resins have been used for SMC; however, compositions based on vinyl ester, epoxy, and urethane–polyester hybrids are currently in use. The most commonly used fiber in sheet molding compounds is E-glass (see Table 1.32), although other fibers such as S-2 glass, carbon, and Kevlar® 49 are also used.

Both short (discontinuous) and long (continuous) forms of fibers are used in SMC. Short fibers are commonly used in applications requiring isotropic properties. Long fibers, which will be described in the next section, provide high strength and modulus for structural applications. The conventional SMC material uses a continuous fiber *roving*, which is a collection of 25–40 untwisted fiber strands. Each fiber strand typically contains 200–400 E-glass filaments. Short fibers are obtained by chopping the rovings into desired lengths at the time of SMC sheet production. Sheet molding compounds are designated according to the form of fibers (see Figure 5.101). In both SMC-R and SMC-CR, the fiber content is shown by the weight percent at the end of each letter designation; for example, SMC-R40 contains 40 wt% of randomly oriented short fibers. For the remainder of this section, we will concentrate on randomly oriented short-fiber sheet molding compound (SMC-R) in order to illustrate some of the common mechanical properties of discontinuous-glass-fiber-reinforced polymer-matrix composites.

The tensile stress–strain diagram for two SMC-R compounds are shown in Figure 5.102. Unlike the unreinforced polymers, SMC composites do not exhibit any yielding and their strain to failure is low, usually less than 2%. However, at low fiber contents, the tensile stress–strain diagrams are nonlinear, which indicates the presence of microdamage, including crazing. At high fiber contents, the nonlinearity decreases, but the stress–strain diagram exhibits a bilinearity with a break point (knee) close to the matrix failure strain. At stress levels higher than the break point, the slope of the

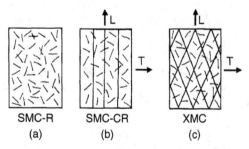

Figure 5.101 Various forms of sheet molding compounds: (a) random, SMC-R; (b) continuous/random, SMC-CR; and (c) criss-crossed continuous/random XMC; L and T refer to longitudinal and transverse directions, respectively. Reprinted, by permission, from *Composite Materials Technology*, P. K. Mallick and S. Newman, eds., p. 33. Copyright © 1990 by Carl Hanser Verlag.

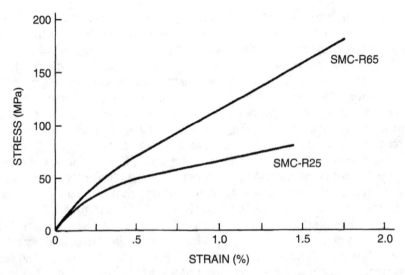

Figure 5.102 Tensile stress–strain diagrams for two SMC compounds: 25 wt% fiber, bottom, and 65 wt% fiber, top. Reprinted, by permission, from *Composite Materials Technology*, P. K. Mallick and S. Newman, eds., p. 45. Copyright © 1990 by Carl Hanser Verlag.

stress–strain diagram is lower than the initial slope. The tensile strengths of SMC-R composites increase significantly with increasing fiber content, but the modulus is affected only moderately. The following empirical relationships for tensile strength, σ_t, and tensile modulus, E, of SMC composites have been developed:

$$\sigma_t = 0.33 V_f \sigma_f + 0.31(1 - V_f)\sigma_m \tag{5.124}$$

and

$$E = 0.59 V_f E_f + 0.71(1 - V_f)E_m \tag{5.125}$$

where σ_f and σ_m are the ultimate tensile strengths of the fiber and matrix, respectively, E_f and E_m are their respectively tensile moduli, and V_f is the volume fraction of fibers. Note the similarity in form of these equations to the rule of mixtures given by Eq. (5.82). The tensile strength of SMC-R composites increases as the chopped fiber length is increased, but fiber length has virtually no effect on the tensile modulus. The use of S-glass instead of E-glass (see Table 1.32) has a significant effect on increasing the mechanical properties of the composite. The resin type can also influence the tensile properties, particularly at low fiber contents. The temperature dependence of both tensile strength and modulus for a SMC-R50 composite are shown in Figure 5.103. Similar effects are also seen for the flexural strength and modulus. As with unreinforced polymers, humidity and liquids environments can affect strength and modulus.

Flexural strengths tend to be higher than tensile strengths in SMC composites. Flexural load-deflection modulus values are nonlinear, which indicates the occurrence of microcracking even at low loading. In general, flexural properties follow the same trends as the tensile properties and are affected by fiber content, fiber lengths, type, and orientation.

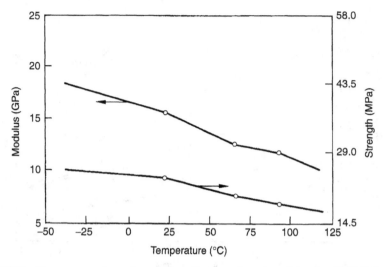

Figure 5.103   Temperature dependence of tensile strength and modulus for SMC-R50 composite. Reprinted, by permission, from P. K. Mallick and S. Newman, eds., *Composite Materials Technology*, p. 59. Copyright © 1990 by Carl Hanser Verlag.

Table 5.8   Selected Compressive and Shear Properties of Various SMC Composites

Property	SMC-R25	SMC-R50	SMC-R65
Compressive strength, MPa	183	225	241
Compressive modulus, GPa	11.7	15.9	17.9
In-plane shear strength, MPa	79	62	128
In-plane shear modulus, GPa	4.48	5.94	5.38
Interlaminar shear strength, MPa	30	25	45

Source: Composite Materials Technology, P. K. Mallick and S. Newman, ed. Copyright © 1990 by Carl Hanser Verlag.

As with tensile properties, both compressive strength and modulus depend on the fiber content and fiber orientation (see Table 5.8). The interlaminar shear strength reported in Table 5.8 is a measure of the shear strength in the thickness direction of the SMC sheet. It is determined by three-point flexural testing of beams with short span-to-depth ratios and is considered to be a quality-control test for molded composites.

Most of the data available on the fatigue properties of SMC composites are based on tension-tension cyclic loading of unnotched specimens. The S–N diagram (cf. Section 5.2.1.4) of an SMC-R65 composite at various temperatures is shown in Figure 5.104. In general, the fatigue strength of SMC-R composites increases with increasing fiber content. The microscopic fatigue damage appears in the forms of matrix cracking and fiber–matrix interfacial debonding. The matrix cracking is influenced by the stress concentrations at the chopped fiber ends, as described in Section 5.4.2.1. Matrix cracks are generally normal to the loading direction. However, in fiber-rich areas, with fibers oriented parallel to the loading direction, the matrix crack lengths are small and limited by the interfiber spacing.

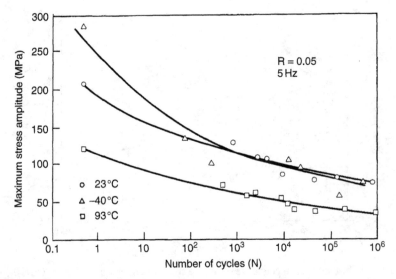

Figure 5.104 Fatigue $S-N$ diagram for SMC-R65 composite. Reprinted, by permission, from *Composite Materials Technology*, P. K. Mallick and S. Newman, eds., p. 52. Copyright © 1990 by Carl Hanser Verlag.

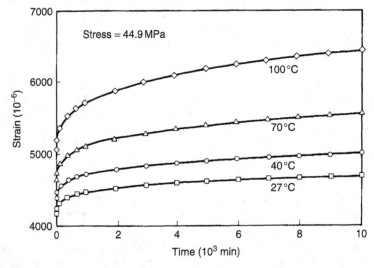

Figure 5.105 Creep response of SMC-R50 composite at various temperatures. Reprinted, by permission, from P. K. Mallick and S. Newman, eds., *Composite Materials Technology*, p. 62. Copyright © 1990 by Carl Hanser Verlag.

A power-law expression of the form of Eq. (5.115) adequately describes the creep properties of SMC composites. The creep response of the SMC-R50 composite at various temperatures shown in Figure 5.105 is representative of this behavior. Among the material variables, fiber content once again has the greatest influence on the creep strain. At a given temperature and stress level, creep strain is higher if the fiber content is reduced.

The mechanical behavior of other discontinuous-glass-fiber-reinforced polymer–matrix composites, while highly variable, exhibits many of the same functional dependences as SMC composites.

5.4.2.5 *High-Performance Polymer–Matrix Composites**. We continue our survey of important fiber–matrix composites in this section by extending the principles introduced in the previous section to continuous fibers. Certainly the continuous-fiber versions of sheet molding compounds (SMCs) of the previous section (see Figure 5.101) can be elaborated upon here, but we shall take a slightly different approach. Just as the primary consideration in the use of discontinuous fibers for SMCs was low cost, the primary purpose in utilizing continuous fibers over discontinuous fibers is to improve mechanical properties. It makes sense, then, to take this benefit to its extreme and describe continuous-fiber-reinforced composites with the ultimate in mechanical properties. Typically, these are not reinforced with glass fibers, but rather with so-called *high-performance fibers* such as Kevlar®, boron, and carbon. The specific mechanical properties of these fibers are compared in Figure 5.106. Notice the enormous improvement in both strength and modulus relative to E-glass, and even S-glass fibers. We will concentrate on carbon fibers for illustrative purposes, but also present data on Kevlar® and other types of continuous fibers where appropriate. Polymer–matrix composites will still be the focus of our descriptions, though, since it is the lightweight matrix that provides the added benefit of weight savings to the composite. The combination of

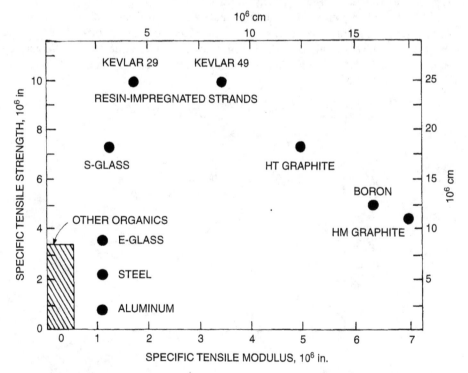

Figure 5.106 Specific tensile strength and specific tensile modulus of reinforcing fibers. Reprinted, by permission, from M. Schwartz, *Composite Materials Handbook*, 2nd ed., p. 2.70. Copyright © 1992 by McGraw-Hill, Inc.

mechanical properties and weight savings that carbon-fiber-reinforced polymer (CFRP) composites offer is phenomenal. For example, if a civilian aircraft were constructed, where possible, from CFRP instead of aluminum alloy, the total weight reduction would be approximately 40%. This creates enormous potential for increased fuel efficiency and increased payload.

Recall from Section 1.4.5.1 that there are two primary types of carbon fibers: polyacrylonitrile (PAN)-based and pitch-based. There are also different structural forms of these fibers, such as amorphous carbon and crystalline (graphite) fibers. Typically, PAN-based carbon fibers are 93–95% carbon, whereas graphite fibers are usually 99+%, although the terms carbon and graphite are often used interchangeably. We will not try to burden ourselves with too many distinctions here, since the point is to simply introduce the relative benefits of continuous-fiber composites over other types of composites, and not to investigate the minute differences between the various types of carbon-fiber-based composites. The interested reader is referred to the abundance of literature on carbon-fiber-reinforced composites to discern these differences.

The primary resin of interest is epoxy. Carbon-fiber–epoxy composites represent about 90% of CFRP production. The attractions of epoxy resins are that they polymerize without the generation of condensation products that can cause porosity, they exhibit little volumetric shrinkage during cure which reduces internal stresses, and they are resistant to most chemical environments. Other matrix resins of interest for carbon fibers include the thermosetting phenolics, polyimides, and polybismaleimides, as well as high-temperature thermoplastics such as polyether ether ketone (PEEK), polyethersulfone (PES), and polyphenylene sulfide.

Unlike short-fiber composites, modulus and strength in long-fiber composites are little affected by the interface. However, the interface does affect off-axis properties, such as compressive modulus, and the toughness and shear strength are affected. As shown in Figure 5.107, a reduction in the interfacial strength, such as through the

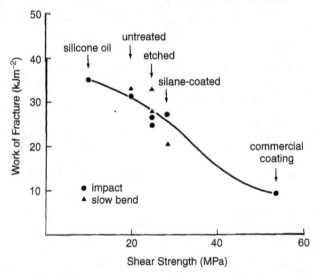

Figure 5.107 Variation in toughness and shear strength for various surface treatments of continuous carbon fiber-reinforced composites. Reprinted, by permission, from T. L. Vigo and B. J. Kinzig, ed., *Composite Applications*, p. 224. Copyright © 1992 by VCH Publishing, Inc.

treatment of the fiber surface with a coating, increases toughness, represented here as the work of fracture, while simultaneously reducing the shear strength. We will describe the mechanism behind this toughening effect in Section 5.4.2.7.

Due to the fact that the mechanical properties of unidirectional, continuous-fiber-reinforced composites are highly anisotropic, maximum effectiveness is often achieved by making laminate composites of multiple layers. This is particularly true of carbon and Kevlar®-reinforced polymers, which will be described in Section 5.4.3.

5.4.2.6 Whiskers and Metal–Matrix Composites*.
Metal–matrix composites (MMCs) are of interest because they offer a combination of properties not available through alloying or processing methods alone. Specifically, light-metal MMCs with aluminum, titanium, or magnesium matrices have very high specific strengths and specific moduli relative to alloys, owing to the high-tensile-strength and high-modulus fibers that are incorporated in them. The plastic deformation of the metal–matrix composite imparts excellent toughness to the MMC.

The common types of metals used in MMCs were first introduced in Table 1.29, and the types of reinforcing fibers were first listed in Table 1.31. Many of the principles previously discussed apply to MMCs: discontinuous versus continuous reinforcement, effect of fiber geometry (aspect ratio), the effect of interfacial bonding on composite properties, and variations due to processing methods. Just as we focused on glass fibers in discontinuous polymer–matrix composites, and carbon fibers for continuous polymer–matrix composites, we will select one system as representative of MMCs. Certainly, carbon fibers find widespread use in MMCs, as do boron fibers, alumina (Al_2O_3) fibers, and Kevlar®. However, the most widely varied usages employ SiC as a reinforcement in MMCs. The selection of SiC as the reinforcement of choice in this section also allows us to discuss a topic that has not yet been addressed: whisker technology.

Though the distinction between a *whisker* and a fiber is sometimes vague, whiskers are generally considered a subclass of fibers that have been grown in controlled conditions, often in the vapor phase, which leads to high-purity single crystals. The resultant highly ordered structure produces not only unusually high strengths, but also significant changes in electrical, optical, magnetic, and even superconductive properties. As the diameter of the whisker is reduced, its strength increases rapidly due to the increasing perfection in the fiber. The tensile strengths of whiskers are far above those of the more common, high-volume production reinforcements. This is illustrated in Figure 5.108, where the schematic stress–strain curves of whiskers and other reinforcements are compared. Notice that whiskers can be strained elastically as much as 3% without permanent deformation, compared with less than 0.1% for bulk ceramics.

Whiskers have been produced in a range of fiber sizes and in three forms: *grown wool*, *loose fiber*, and *felted paper*. The wool has a fiber diameter of 1–30 μm and aspect ratios of 500–5000. The wool bulk density is about 0.03 g/cm³, which is less than 1% solids. This is a very open structure, suitable for a vapor-deposition coating on the whiskers or direct use as an insulation sheet. Loose fibers are produced by processing the larger diameter fiber to yield lightly interlocked clusters of fibers with aspect ratios of 10–200. In felt or paper, the whiskers are randomly oriented in the plane of the felt and have fiber aspect ratios ranging from 250 to 2500. The felted paper has approximately 97% void volume and a density of 0.06–0.13 g/cm³.

Whiskers can be formed from a variety of materials, as illustrated in Table 5.9, but there are only two general methods for forming them: *basal growth* and *tip growth*.

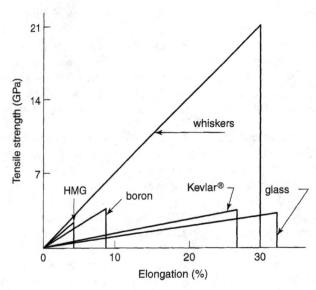

Figure 5.108 Schematic comparison of stress–strain diagrams for common reinforcing fibers (HMG = high modulus graphite) and whiskers. Reprinted, by permission, from A. Kelly, ed., *Concise Encyclopedia of Composite Materials*, revised edition, p. 312. Copyright © 1994 by Elsevier Science Publishers, Ltd.

Table 5.9   Selected Mechanical Properties of Some Common Whiskers

Material	Density (10^3 kg m^{-3})	Melting point (°C)	Tensile Strength (GPa)	Young's modulus (GPa)
Aluminum oxide	3.9	2082	14–28	550
Aluminum nitride	3.3	2198	14–21	335
Beryllium oxide	1.8	2549	14–21	700
Boron oxide	2.5	2449	7	450
Graphite	2.25	3593	21	980
Magnesium oxide	3.6	2799	7–14	310
Silicon carbide (α)	3.15	2316	7–35	485
Silicon carbide (β)	3.15	2316	7–35	620
Silicon nitride	3.2	1899	3–11	380

Source: *Concise Encyclopedia of Composite Materials*, A. Kelly, ed., revised edition. Copyright © 1994 by Elsevier Science Ltd.

In basal growth, the atoms of growth migrate to the base of the whisker and extrude the whisker from the substrate. Tip growth involves fabrication at high temperatures such that the vapor pressure of the whisker-forming material becomes significant to the point that the atoms attach themselves to the tip of the growing whisker. Silicon carbide whiskers are generally formed by the pyrolysis of chlorosilanes in hydrogen at temperatures above 1400°C. Selection of proper silanes, optimum reaction conditions, and suitable reaction chambers are important process parameters. In some cases, a coating is also produced on the whiskers to reduce reaction between the whisker and the metal matrix after composite formation.

Whiskers can be incorporated into the metallic matrix using a number of composite-processing techniques. *Melt infiltration* is a common technique used for the production of SiC whisker–aluminum matrix MMCs. In one version of the infiltration technique, the whiskers are blended with binders to form a thick slurry, which is poured into a cavity and vacuum-molded to form a pre-impregnation body, or *pre-preg*, of the desired shape. The cured slurry is then fired at elevated temperature to remove moisture and binders. After firing, the preform consists of a partially bonded collection of interlocked whiskers that have a very open structure that is ideal for molten metal penetration. The whisker preform is heated to promote easy metal flow, or infiltration, which is usually performed at low pressures. The infiltration process can be done in air, but is usually performed in vacuum.

The benefits of SiC whisker-reinforced aluminum can be seen in Figure 5.109. The room-temperature stiffness (modulus) of the composite is about 50% higher than that

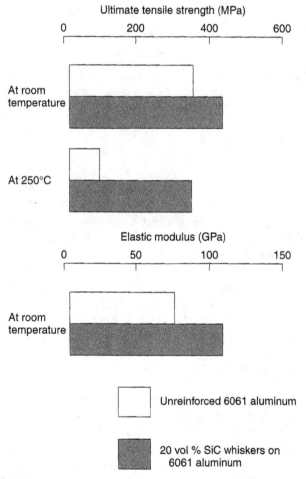

Figure 5.109 Variation of tensile strength and tensile modulus with temperature of SiC-reinforced aluminum. Reprinted, by permission, from A. Kelly, ed., *Concise Encyclopedia of Composite Materials*, revised edition, p. 189. Copyright © 1994 by Elsevier Science Publishers, Ltd.

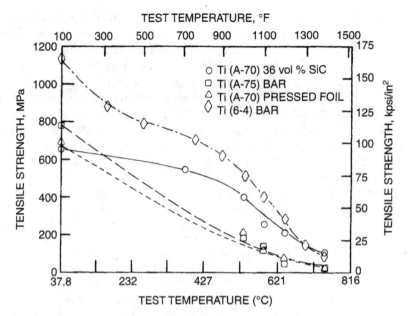

Figure 5.110 Variation of tensile strength with temperature for some titanium-based alloys and composites. Reprinted, by permission, from M. Schwartz, *Composite Materials Handbook*, 2nd ed., p. 2.111. Copyright © 1992 by McGraw-Hill, Inc.

of the aluminum alloy. When the lower density of aluminum is considered, the specific modulus is almost twice that of pure titanium. The primary benefit, however, is in the high-temperature strength retention of the composite relative to the unreinforced aluminum. This effect is also illustrated in Figure 5.110 for SiC fiber-reinforced titanium matrix composite relative to the base metal formed by different techniques. Unlike dispersion-strengthened metals (cf. Section 5.4.1.1), where the particles strengthen by blocking the movement of dislocations in the matrix, fiber reinforcements improve strength by a transference of load from the matrix to the fiber phase. Thus, the relationships for composite strength as a function of volume fraction fiber [Eqs. (5.114) and (5.116)] generally apply to metal-matrix composites, as does Figure 5.97. The primary difference is that the matrix strength at fiber failure, σ'_m, is subject to strain hardening in the metallic matrix. With regard to composite modulus, the contribution of the ductile metal matrix to the composite stiffness is complex. A complete description of the internal state of stress for every applied load is required, and it is being approached with analytical and finite-element models.

Non-whisker SiC fibers are also of importance in MMCs, and they are currently available in two commercial forms: Tyranno® and Nicalon®. As with the whisker form, the primary advantages of SiC fibers is their oxidation resistance and high-temperature mechanical property retention relative to other fibers. The high-temperature strength of three commercially available Nicalon® SiC fibers is shown in Figure 5.111.

In addition to aluminum, other types of metal matrices in MMCs include magnesium, which is relatively easy to fabricate due to its low melting point, lead for batteries, titanium for aircraft turbine engines, copper for magnetohydrodynamics, and iron, nickel, or cobalt alloys.

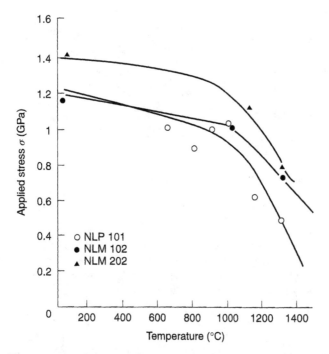

Figure 5.111 The strengths of three commercially available Nicalon ® fibers as a function of temperature. Reprinted, by permission, from A. Kelly, ed., *Concise Encyclopedia of Composite Materials*, p. 256, revised edition. Copyright © 1994 by Elsevier Science Publishers, Ltd.

5.4.2.7 Matrix Cracking in Ceramic-Matrix Composites*.

Silicon carbide fibers and whiskers also find common usage in ceramic-matrix composites (CMCs). However, unlike their use in matrices that undergo plastic deformation such as polymers and metals, in which a strong interfacial bond between the fiber and the matrix is desirable in order to facilitate load transfer, their use in brittle matrix composites such as CMCs is based upon the ability of the fiber to pull away from the matrix in a controlled manner in what is called *fiber pull-out* or *graceful failure*. The tensile modulus of a unidirectional continuous-fiber-reinforced ceramic-matrix composite is given to a good approximation by the rule of mixing [Eq. (5.119)]. The strength, however, is more complex than the rule of mixing in brittle-matrix composites. A successful continuous, unidirectional fiber-reinforced CMC will have fibers whose failure strain is significantly greater than that of the matrix and are not too strongly bonded to the matrix. The matrix thus fractures at a stress, σ_{mu}, that is lower than the ultimate failure stress of the composite, σ_{cu}. The typical stress–strain behavior for a unidirectional-reinforced CMC stressed in the axial fiber direction is shown in Figure 5.112. For a matrix material with a reasonable density of inherent flaws and zero fiber/matrix interfacial shear strength, τ_i, the curve would follow the path OACD, where the segment AC represents fiber pull-out as the matrix fails. In more realistic situations, the curve OAD is followed.

Matrix cracking and the ultimate composite strength can be analyzed using a number of approaches, but we will utilize the fracture mechanics approach so that we can compare the results with those we developed in Sections 5.2.1.2 and 5.2.3.1 for

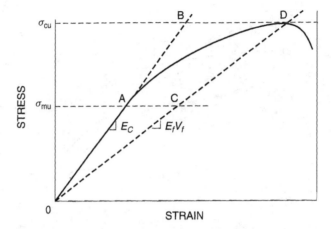

Figure 5.112 Idealized stress–strain curve for a tough ceramic-matrix composite. Reprinted, by permission, from R. W. Davidge and J. J. R. Davies, in *Mechanical Testing of Engineering Ceramics at High Temperatures*, B. F. Dyson, R. D. Lohr, and R. Morrell, eds., p. 251. Copyright © 1989 by Elsevier Science Publishers, Ltd.

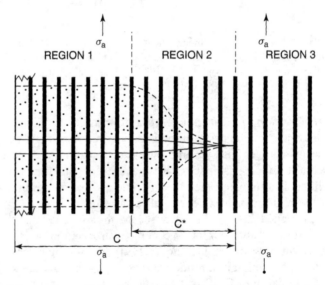

Figure 5.113 Microcrack parallel to fibers under applied stress, σ_a. Reprinted, by permission, from R. W. I. Davidge, and J. J. R. Davies, in *Mechanical Testing of Engineering Ceramics at High Temperatures*, B. F. Dyson, R. D. Lohr, and R. Morrell, eds., p. 253. Copyright © 1989 by Elsevier Science Publishers, Ltd.

fracture in monolithic ceramics. Consider the unidirectional, fiber-reinforced CMC in Figure 5.113. A crack of length *c* has passed some way through the material, creating three distinct regions. Region 3 is remote from the crack, and is unaffected by it. At the mouth of the crack (Region 1), the crack is fully established and the crack faces have relaxed to a constant crack opening displacement. Near the crack tip (Region 2) there is a transition between the outer two regions. It has been shown that the stress required to propagate the crack, σ_m, is given by

$$\sigma_m = \frac{12\tau\gamma_m}{r}\left[\frac{V_f^2 E_f(1-v^2)^2}{V_m E_c E_m^2}\right]^{1/3}\frac{E_c}{1-v^2} \tag{5.126}$$

where τ is the fiber–matrix interfacial frictional force, r is the fiber radius, γ_m is the surface energy of the matrix, v is Poisson's ratio of the matrix phase, V_f is the volume fraction of fibers, and E_c, E_f, and E_m are the elastic moduli of the composite, fiber, and matrix, respectively. This situation is to be contrasted with that for the unreinforced material where the stress at the crack tip increases with an increase in crack length [Eq. (5.45)]. For these composite systems, the stress at the crack tip is independent of the crack length once the crack is longer than the equilibrium length, c. Thus, matrix cracking is impossible below the threshold stress, σ_m, no matter how large the preexisting defect. Composites with intrinsic flaws less than c will be expected to have a matrix cracking stress greater than σ_m, as indicated in Figure 5.114. The threshold crack size is usually several times larger than the average fiber spacing, which is of the order of the inherent flaw size.

At stresses greater than σ_{mu}, the matrix will fracture into a series of parallel-sided blocks. The ultimate strength of the composite depends on the strength of the fibers σ_{fu}, such that

$$\sigma_{cu} = \sigma_{fu} V_f \tag{5.127}$$

Note that at σ_{cu}, some of the load is still carried by the matrix in the unbroken regions and that the maximum stress on the fibers is restricted to the bridging regions between single matrix cracks. Thus, σ_{fu} may be greater than the value obtained for unrestrained fibers.

The flexural and fracture properties of some common CMCs are given in Table 5.10, compared to their unreinforced matrix materials. Notice that the SiC fibers described

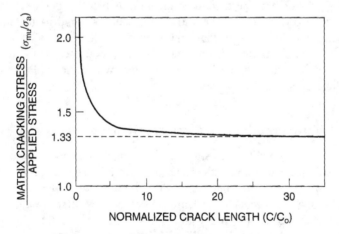

Figure 5.114 Matrix cracking as a function of normalized crack length in CMCs. Reprinted, by permission, from R. W. Davidge and J. J. R. Davies, in *Mechanical Testing of Engineering Ceramics at High Temperatures*, B. F. Dyson, R. D. Lohr, and R. Morrell, eds., p. 254. Copyright © 1989 by Elsevier Science Publishers, Ltd.

Table 5.10 Comparison of Mechanical Properties of CMCs and the Equivalent Unreinforced Ceramic

Material[a]	Bend Strength (MPa)	K_{1c} (MPam$^{1/2}$)
Unreinforced Al_2O_3	385–600	4–5
30 vol% SiC whisker-Al_2O_3	650–800	6–9.5
20 vol% SiC fibers (C coated)-Al_2O_3		8.3
20 vol% SiC fibers (SiO_2 coated)-Al_2O_3		7.2
Unreinforced LAS glass ceramic	200	2
50 vol% 1-D SiC fiber LAS	600	17
HPSN	400	3.4–8.2
30 vol% SiC-whisker HPSN	450	10.5
Unreinforced RBSN	200–400	1.5–2.8
20 vol% 1-D SiC fiber RBSN	650	54

[a] HP, hot pressed; RB, reaction bonded; LAS, lithium aluminosilicate.
Source: W. E. Lee, and W. M. Rainforth, *Ceramic Microstructures*. Copyright © 1994 by William E. Lee and W. Mark Rainforth.

in the previous section for use in MMCs are also commonly used for CMCs; also notice that just as for MMCs, processing methods figure prominently in dictating the ultimate properties of the CMC. One ceramic-matrix material worthy of further description is a glass ceramic (cf. Section 1.2.5), such as lithium aluminosilicate (LAS). In these CMCs, the major crystalline phase is β-spudomene, and the useful temperature range of the composite is 10–1200°C. The SiC–LAS composite can be formed by either of two common techniques: *filament winding* or *hot pressing*. As with all glass-ceramics, controlled crystallization of the matrix phase is necessary to obtain the glass-ceramic structure (see Figure 3.17). Unidirectional reinforcement is unlikely to be used because the strength normal to the fibers is extremely low (20 MPa; see curve 1D 90° in Figure 5.115). For most practical applications at least two-dimensional reinforcement is used. Some temperature-dependent mechanical property data are available, as presented in Figure 5.116 for the variation of bend strength with temperature for the one-dimensional, two-dimensional, and unreinforced LAS glass ceramic. An interesting characteristic of the two-dimensional CMC is that there is a significant increase in bending strength around 1000°C due to oxidation, which increases the fiber–matrix interfacial shear strength.

5.4.3 Mechanical Properties of Laminate Composites

At the end of Section 5.4.2.5, the statement was made that most continuous, unidirectional fiber-reinforced composites are used to produce layers that are subsequently assembled to form laminate composites. In this section, we expound upon this statement by examining the mechanics of laminate composites, first through a generalized description of their mechanics, then with some specific stress–strain behavior.

In the case of unidirectionally reinforced layers, or *laminae* (not to be confused with the similar-looking *lamellae*, cf. Section 1.3.6.1), a composite is formed by laminating the layers together. If all the layers were to have the same fiber orientation (see Figure 5.117a), the material would still be weak in the transverse direction. Therefore,

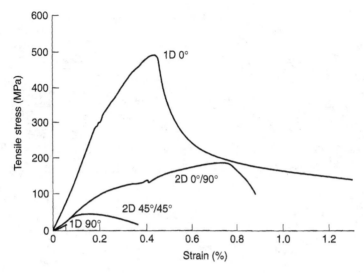

Figure 5.115 Stress–strain diagrams for lithiumaluminosilicate glass ceramic reinforced with 50% SiC fibers in various orientations. From *Ceramic Microstructures*, by W. E. Lee and W. M. Rainforth, p. 103. Copyright © 1994 by William E. Lee and W. Mark Rainforth, with kind permission of Kluwer Academic Publishers.

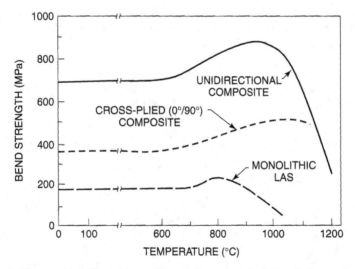

Figure 5.116 Temperature dependence of bending strength for SiC-reinforced lithium alumi-nosilicate (LAS) CMC. Reprinted, by permission, from R. W. Davidge and J. J. R. Davies, in *Mechanical Testing of Engineering Ceramics at High Temperatures*, B. F. Dyson, R. D. Lohr, and R. Morrell, eds., p. 264. Copyright © 1989 by Elsevier Science Publishers, Ltd.

angle-ply laminates are preferred for most applications. Typically, the laminae are stacked so that the fiber directions are symmetrical about the mid-plane of the com-posite, as in Figure 5.117b. A *symmetrical* stacking sequence ensures that in-plane stresses do not produce out-of-plane twisting or bending, as in Figure 5.118. The stacking sequence can be either 0°/90°, as in the bottom two plies of Figure 5.117b,

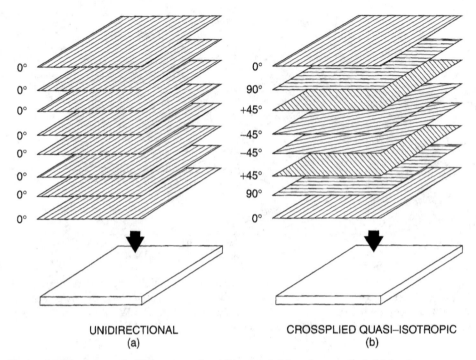

Figure 5.117 Schematic illustration of unidirectional (a) and crossplied (b) laminar composites. Reprinted, by permission, from M. Schwartz, *Composite Materials Handbook*, 2nd ed., p. 3.71. Copyright © 1992 by McGraw-Hill, Inc.

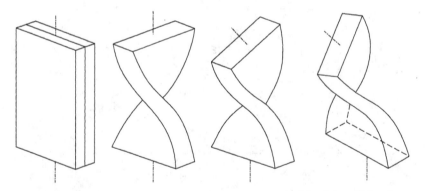

Figure 5.118 Twisting, bending, and shear of an unsymmetrical laminate under direct stress. Reprinted, by permission, from P. C. Powell and A. J. I. Housz, *Engineering with Polymers*, p. 201. Copyright © 1998 by Stanley Thorned Publishers.

termed *cross-ply*, −45°/+45° as in the third and fourth ply of Figure 5.117b, termed *angle-ply*, or a combination of the two. The layers can be of the same thickness, or different thickness, and multiple layers may be oriented in the same direction before changing the orientation of the next multiple layers—for example, three layers at 0°, three layers at +45°, three layers at −45°, and so on. The outer layers of the composite may be selected to provide a protective coating, as described previously.

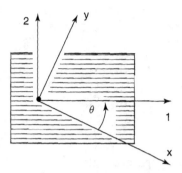

Figure 5.119 Cartesian coordinate system relative to principle fiber directions in unidirectional fiber composite. Reprinted, by permission, from P. C. Powell and A. J. I. Housz, *Engineering with Polymers*, p. 215. Copyright © 1998 by Stanley Thorned Publishers.

5.4.3.1 *Stress–Strain Behavior for Unidirectional Laminae**. In Section 5.1.1.2,

we developed a generalized view of elastic response (Hooke's Law) that included all stresses and strains within the framework of a matrix notation. Equation (5.6) gave the stress matrix, and Eq. (5.8) gave the matrix of modulus components, Q_{ij}. Let us relate the generalized Hooke's Law model to one lamina of continuous, unidirectional fiber-reinforced matrix, as first illustrated in Figure 5.85. The directions 1, 2, and 3, defined in terms of the fiber orientation direction, can be related to the Cartesian coordinates x, y, and z as in Figure 5.119, through the introduction of the orientation angle, θ. We will retain the coordinates 1, 2, and 3 to relate relationships developed in this section to those previously presented for unidirectional composites.

Equation (5.7) related the stress tensor to the strains through the modulus matrix, [Q], also known as the *reduced stiffness matrix*. Oftentimes, it is more convenient to relate the strains to the stresses through the inverse of the reduced stiffness matrix, $[Q]^{-1}$, called the *compliance matrix*, [S]. Recall from Section 5.3.1.3 that we introduced a quantity called the compliance, which was proportional to the inverse of the modulus. The matrix is the more generalized version of that compliance. The following relationship then holds between the strain and stress tensors:

$$\varepsilon = [S]\sigma \qquad (5.128)$$

Let us first consider the case of an isotropic material, then simplify it for the case of an *orthotropic* material (same properties in the two directions orthogonal to the fiber axis—in this case, directions 2 and 3), such as a unidirectionally reinforced composite lamina. Equation (5.128) can be written in terms of the strain and stress components, which are coupled due to the anisotropy of the material. In order to describe the behavior in a manageable way, it is customary to introduce a reduced set of nomenclature. Direct stresses and strains have two subscripts—for example, σ_{11}, ε_{22}, τ_{12}, and γ_{21}, depending on whether the stresses and strains are tensile (σ and ε) or shear (τ and γ) in nature. The moduli should therefore also have two subscripts: E_{11}, E_{22}, and G_{12}, and so on. By convention, engineers use a contracted form of notation, where possible, so that repeated subscripts are reduced to just one: σ_{11} becomes σ_1, E_{11} becomes E_1, but G_{12} stays the same. The convention is further extended for stresses and strains, such that distinctions between tensile and shear stresses and strains are dropped. So, the coupling between stresses and strains in an anisotropic material can now be expressed in the generalized Hooke's Law form:

$$
\begin{bmatrix} \varepsilon_1 \\ \varepsilon_2 \\ \varepsilon_3 \\ \gamma_{31} \\ \gamma_{32} \\ \gamma_{12} \end{bmatrix} =
\begin{bmatrix}
S_{11} & S_{12} & S_{13} & S_{14} & S_{15} & S_{16} \\
S_{21} & S_{22} & S_{23} & S_{24} & S_{25} & S_{26} \\
S_{31} & S_{32} & S_{33} & S_{34} & S_{35} & S_{36} \\
S_{41} & S_{42} & S_{43} & S_{44} & S_{45} & S_{46} \\
S_{51} & S_{52} & S_{53} & S_{54} & S_{55} & S_{56} \\
S_{61} & S_{62} & S_{63} & S_{64} & S_{65} & S_{66}
\end{bmatrix}
\begin{bmatrix} \sigma_1 \\ \sigma_2 \\ \sigma_3 \\ \tau_{31} \\ \tau_{32} \\ \tau_{12} \end{bmatrix}
\tag{5.129}
$$

where $[S]$ is the symmetric compliance matrix.

For an anisotropic lamina under in-plane loads, Eq. (5.129) reduces to

$$
\begin{bmatrix} \varepsilon_1 \\ \varepsilon_2 \\ \gamma_{12} \end{bmatrix} =
\begin{bmatrix}
S_{11} & S_{12} & S_{16} \\
S_{21} & S_{22} & S_{26} \\
S_{61} & S_{62} & S_{66}
\end{bmatrix}
\begin{bmatrix} \sigma_1 \\ \sigma_2 \\ \tau_{12} \end{bmatrix}
\tag{5.130}
$$

and for an orthotropic lamina, such as in Figure 5.119, under in-plane stresses in the principal directions, Eq. (5.130) further reduces to

$$
\begin{bmatrix} \varepsilon_1 \\ \varepsilon_2 \\ \gamma_{12} \end{bmatrix} =
\begin{bmatrix}
S_{11} & S_{12} & 0 \\
S_{21} & S_{22} & 0 \\
0 & 0 & S_{66}
\end{bmatrix}
\begin{bmatrix} \sigma_1 \\ \sigma_2 \\ \tau_{12} \end{bmatrix}
\tag{5.131}
$$

where the subscripts 1 and 2 refer to the fiber and transverse directions, respectively. The *elastic compliance constants*, S_{ij} (often called the *engineering constants*) can be expressed in terms of unidirectional lamina properties: $S_{11} = 1/E_1$; $S_{12} = S_{21} = -v_{12}/E_1 = -v_{21}/E_2$; $S_{22} = 1/E_2$, and $S_{66} = 1/G_{12}$, where v is Poisson's ratio.

Equation (5.131) can be inverted to give the reduced stress response to the in-plane stress:

$$
\begin{bmatrix} \sigma_1 \\ \sigma_2 \\ \tau_{12} \end{bmatrix} =
\begin{bmatrix}
Q_{11} & Q_{12} & 0 \\
Q_{21} & Q_{22} & 0 \\
0 & 0 & Q_{66}
\end{bmatrix}
\begin{bmatrix} \varepsilon_1 \\ \varepsilon_2 \\ \gamma_{12} \end{bmatrix}
\tag{5.132}
$$

where again $[Q] = [S]^{-1}$, and relationships for each Q_{ij} can be similarly expressed in terms of the lamina moduli; for example, $Q_{66} = G_{12}$.

The next level of complexity in the treatment is to orient the applied stress at an angle, θ, to the lamina fiber axis, as illustrated in Figure 5.119. A *transformation matrix*, $[T]$, must be introduced to relate the principal stresses, σ_1, σ_2, and τ_{12}, to the stresses in the new $x-y$ coordinate system, σ_x, σ_y, and τ_{xy}; and the *inverse transformation matrix*, $[T]^{-1}$, is used to convert the corresponding strains. The entire development will not be presented here. The results of this analysis are that the tensile moduli of the composite along the x and y axes, E_x, and E_y, which are parallel and transverse to the applied load, respectively, as well as the shear modulus, G_{xy}, can be related to the lamina tensile modulus along the fiber axis, E_1, the transverse tensile modulus, E_2, the lamina shear modulus, G_{12}, Poisson's ratio, v_{12}, and the angle of lamina orientation relative to the applied load, θ, as follows:

$$
\frac{1}{E_x} = \frac{\cos^4 \theta}{E_1} + \left(\frac{1}{G_{12}} - \frac{2v_{12}}{E_1} \right) \sin^2 \theta \cos^2 \theta + \frac{\sin^4 \theta}{E_2}
\tag{5.133}
$$

$$\frac{1}{E_y} = \frac{\sin^4 \theta}{E_1} + \left(\frac{1}{G_{12}} - \frac{2\nu_{12}}{E_1} \right) \sin^2 \theta \cos^2 \theta + \frac{\cos^4 \theta}{E_2} \tag{5.134}$$

and

$$\frac{1}{G_{xy}} = 2 \left(\frac{2}{E_1} + \frac{2}{E_2} + \frac{4\nu_{12}}{E_1} - \frac{1}{G_{12}} \right) \sin^2 \theta \cos^2 \theta + \frac{\sin^4 \theta + \cos^4 \theta}{G_{12}} \tag{5.135}$$

Poisson's ratio for the off-axis loaded lamina, ν_{xy}, can also be derived. The relative tensile modulus, E_x/E_1, shear modulus, G_{xy}/G_{12}, and Poisson's ratio, ν_{xy}, are plotted as a function of the angle of rotation, θ, for a glass-fiber-reinforced epoxy lamina and a graphite-fiber-reinforced epoxy lamina in Figures 5.120a and 5.120b, respectively.

The final level of complexity is to arrange F layers of unidirectional laminae to form the laminate composite, as shown in Figure 5.121. The lower surface of the fth lamina is assigned the coordinates h_f relative to the midplane, so the thickness of the fth lamina is $h_f - h_{f-1}$. The mathematics of stress analysis are an extension of those used in the foregoing description, and they include force and moment analysis on the bending of the laminate. The complete development will not be presented here, but only outlined. The interested reader is referred to two excellent descriptions of the results [9,10].

The laminate forces and moments are obtained by integrating the stresses of each lamina through the total laminate thickness to obtain a force vector, N, and a moment vector, M, the components of which can be related to the strains, ε^0 and γ^0, and curvatures, κ, of the laminae, where the superscript indicates that the strains are measured

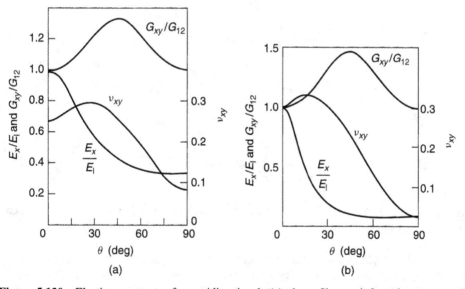

(a)   (b)

Figure 5.120 Elastic constants for unidirectional (a) glass fiber reinforced epoxy and (b) graphite fiber-reinforced epoxy laminae. Reprinted, by permission, from P. C. Powell and A. J. I. Housz, *Engineering with Polymers*, pp. 222, 223. Copyright © 1998 by Stanley Thorned Publishers.

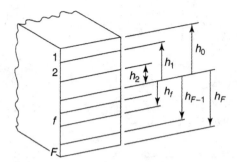

Figure 5.121   Thickness coordinates in a laminate composite. Reprinted, by permission, from P. C. Powell and A. J. I. Housz, *Engineering with Polymers*, p. 231. Copyright © 1998 by Stanley Thorne Publishers.

relative to the midplane of the laminate:

$$
\begin{bmatrix} N_x \\ N_y \\ N_{xy} \end{bmatrix} = [A] \begin{bmatrix} \varepsilon_x^0 \\ \varepsilon_y^0 \\ \gamma_{xy}^0 \end{bmatrix} + [B] \begin{bmatrix} \kappa_x \\ \kappa_y \\ \kappa_{xy} \end{bmatrix}
\tag{5.136}
$$

and

$$
\begin{bmatrix} M_x \\ M_y \\ M_{xy} \end{bmatrix} = [B] \begin{bmatrix} \varepsilon_x^0 \\ \varepsilon_y^0 \\ \gamma_{xy}^0 \end{bmatrix} + [D] \begin{bmatrix} \kappa_x \\ \kappa_y \\ \kappa_{xy} \end{bmatrix}
\tag{5.137}
$$

where the arrays $[A]$, $[B]$, and $[D]$ are given by

$$
A_{ij} = \sum_{f=1}^{F} (\overline{Q}_{ij})_f (h_f - h_{f-1})
\tag{5.138}
$$

$$
B_{ij} = \frac{1}{2} \sum_{f=1}^{F} (\overline{Q}_{ij})_f (h_f^2 - h_{f-1}^2)
\tag{5.139}
$$

and

$$
D_{ij} = \frac{1}{3} \sum_{f=1}^{F} (\overline{Q}_{ij})_f (h_f^3 - h_{f-1}^3)
\tag{5.140}
$$

Here, the array $[\overline{Q}]_f$ is the *transformed reduced stiffness matrix*, and each $\overline{Q}_{ij}$ term can be related to the Q_{ij} terms of the reduced stiffness matrix, $[Q]$, through the angle, θ.

5.4.3.2 *Strength of Laminate Composites.* The strength of angle-ply laminates can be calculated using the same type of elastic analysis as for modulus in the previous section. The strains produced in the laminate by a given set of applied stresses are first calculated, using the computed laminate moduli. Stresses corresponding to these strains are then calculated for each layer of the laminate. These stresses are then expressed in terms of stresses parallel and normal to the fibers, and the combination of stresses

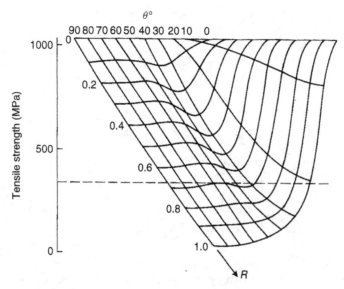

Figure 5.122 Strength chart for a $(0, \pm\theta)$ unidirectional carbon-fiber-reinforced epoxy laminate composite. Reprinted, by permission, from N. G. McCrum, C. P. Buckley, and C. B. Bucknall, *Principles of Polymer Engineering*, 2nd ed., p. 408. Copyright © 1997 by Oxford University Press.

is compared with strength criteria for the unidirectional material. Results for a three-direction carbon-fiber-reinforced epoxy composite $(0, \pm\theta)$ under tensile stress in the $\theta = 0°$ direction are shown in Figure 5.122. The composite contains 60 vol% fibers. Here, R is the fraction of plies which are angled at $\pm\theta$, and $1\text{-}R$ is the fraction at $\theta = 0$.

In cross-ply laminates, the stress–strain behavior is slightly nonlinear, as illustrated in Figure 5.123. The stress–strain behavior of a unidirectional lamina along the fiber axis is shown in the top curve, while the stress–strain behavior for transverse loading is illustrated in the bottom curve. The stress–strain curve of the cross-ply composite, in the middle, exhibits a knee, indicated by strength σ_k, which corresponds to the rupture of the fibers in the 90° ply. The 0° ply then bears the load, until it too ruptures at a composite fracture strength of σ_f.

One additional problem area for fiber-reinforced thermosetting laminates is *inter-laminar splitting*, which occurs particularly easily when the void content is high. Measurements of *interlaminar shear strength* (ISS) are made by subjecting short bar specimens to three-point bending. Deformation of laminated composites can also occur because of changes in temperature and the absorption of moisture. This is known as the *hygrothermal effect*, and it causes the polymer matrix to undergo both dimensional and property changes with temperature and humidity.

5.5 MECHANICS OF BIOLOGICS

Most of the terms, concepts, and equations for studying the mechanical properties of biological materials have already been developed in the previous sections of this

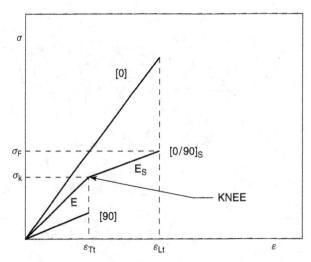

Figure 5.123 Schematic tensile stress–strain diagram for a symmetric $(0, \pm 90°)$ cross-plied laminate tested at $2 = 0°$. Reprinted, by permission, from P. K. Mallick, *Fiber-Reinforced Composites*, p. 187. Copyright © 1988 by Marcel Dekker, Inc.

chapter. What remains is to relate the unique structural characteristics of biologics to their mechanical properties. As in Chapter 1, we will divide the biologics into soft and hard classifications, since their structure and mechanical responses are uniquely different. These differences are summarized in Figure 5.124, in which the stress–strain curves of some soft and hard biologics are compared. The types of materials used as substitutes for these two classifications are also different. In each case, we will begin by examining the structure and mechanical properties of true biological materials, then briefly describe materials that can be used to replace them.

5.5.1 Mechanical Properties of Soft Biologics

Because most biological structures are composed of α helices, β sheets (cf. Section 1.5.1.1), and collagen triple helixes, it is important to understand the mechanical properties of these basic structures. For example, the mechanical properties of tendon collagen fibers are believed to reflect the mechanical properties of the collagen triple helix which serves as the building block for the collagen protein fiber (see Figure 5.125, cf. Figure 1.93). Ultimate tensile strength values as high as 40 MPa, strains up to 10%, and high strain moduli in excess of 500 MPa have been reported for isolated collagen fibers with diameters from 50 to 100 μm. These properties are related to the stiffness of the triple helix, which has been estimated to have an elastic modulus as high as 4 GPa.

Although the mechanical properties of connective tissues such as tendons and ligaments are dependent upon the properties of the collagen fibrils, their diameters, and the content of the collagen in the tissue, they are also dependent upon several other factors, including the orientation of the collagen fibers and the extent of tissue hydration. The latter is believed to occur due to the interaction of water with the hydrophobic regions of the collagen protein.

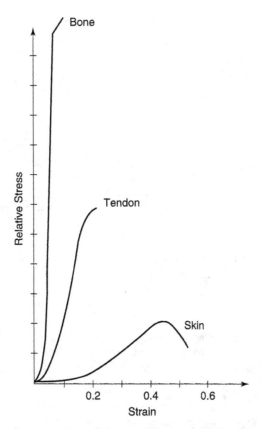

Figure 5.124 Schematic comparison of stress–strain curves for some soft and hard biological materials. Reprinted, by permission, from F. H. Silver and D. L. Christiansen, *Biomaterials Science and Biocompatibility*, p. 18. Copyright © 1999 by Springer-Verlag.

In tissues such as tendons that function to transmit loads along a single direction, the collagen fibers are oriented in only one direction. Ultimate tensile strengths as high as 120 MPa, along with strains of about 10%, have been reported for tendons. For ligaments, these values are 147 MPa and 70%, respectively (see Table 5.11). Other tissues, such as skin, bear loads within a plane and need to have collagen fibers oriented in the plane. In the next section, we look more closely at the structure–mechanical property relationships in skin.

5.5.1.1 *Skin*. Tissue such as tendons and ligaments are termed *oriented* networks because the collagen fibers are oriented in a preferred direction. Other collagen-containing tissues such as skin are termed *orientable*. Because skin normally bears loads within the plane of the skin, the collagen fibers become oriented with the direction of loading and reorient during mechanical deformation. Thus, skin is able to bear loads in any direction within its plane. For this reason, the collagen network in skin must be loose enough to be able to rearrange. Other types of soft tissue such as aorta and cartilage contain some orientable collagen networks.

The stress–strain curve for skin is shown in Figure 5.126. For strains up to about 30%, the collagen network offers little resistance to deformation, which is similar to

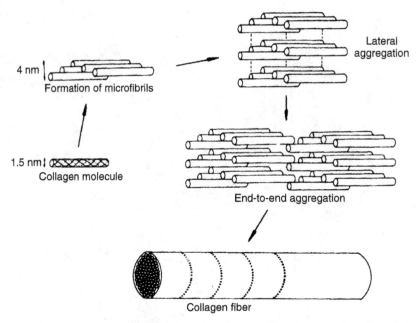

4 nm Formation of microfibrils

1.5 nm Collagen molecule

Lateral aggregation

End-to-end aggregation

Collagen fiber

Figure 5.125 Structure of collagen fibers based on collagen molecule, microfibrils and fibrillar aggregation. Reprinted, by permission, from I. V. Yannas, in *Biomaterials Science*, B. D. Ratner, A. S. Hoffman, F. J. Schoen, and J. E. Lemons, eds., p. 86. Copyright © 1996 by Academic Press.

Table 5.11 Selected Mechanical Properties of Some Soft Biologics

Tissue	Maximum Strength (MPa)	Maximum Strain (%)	Elastic Modulus (MPa)
Arterial wall	0.24–1.72	40–53	1.0
Tendons/ligaments	50–150	5–50	100–2500
Skin	2.5–15	50–200	10–40
Hyaline cartilage	1–18	10–120	0.4–19

Source: F. H. Silver and D. L. Christiansen, *Biomaterials Science and Biocompatibility*. Copyright © 1999 by Springer-Verlag.

tendon in this region. Collagen fibers are initially wavy and become straight upon loading. Between 30% and 60% strains, the collagen fibers become aligned and offer resistance to further extension. Over 60% elongation, the fibers begin to yield and fail. At high strains, the modulus is about 22.4 MPa, and the ultimate tensile strength is 14 MPa, based on the true cross-sectional area. The *elastic fraction* of skin is defined as the ratio of the total force divided by the force at equilibrium and is a measure of the viscoelastic nature of a material. For completely elastic materials, the elastic fraction is 1.0, whereas for viscoelastic materials, the value lies between 0 and 1.0. In skin, the elastic fraction increases from about 0.50 to 0.73 with increasing strain, as illustrated in Figure 5.127. This is in contrast to other types of collagen-containing soft tissue such as aorta.

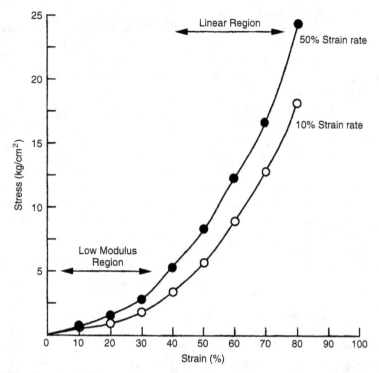

Figure 5.126   Stress–strain curves for wet back skin at various strain rates. Reprinted, by permission, from F. H. Silver and D. L. Christiansen, *Biomaterials Science and Biocompatibility*, p. 203. Copyright © 1999 by Springer-Verlag.

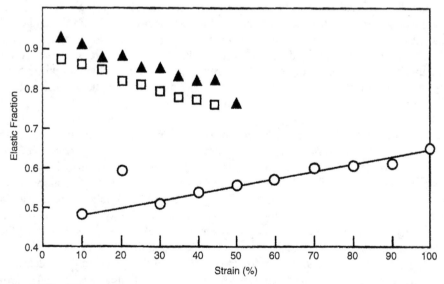

Figure 5.127   Effect of strain on elastic fraction of aorta in the circumferential (solid triangles) and axial (open squares) directions, as well as on skin (open circles). Reprinted, by permission, from F. H. Silver and D. L. Christiansen, *Biomaterials Science and Biocompatibility*, p. 202. Copyright © 1999 by Springer-Verlag.

The mechanical behavior of skin is time-dependent and strain-rate-dependent, and for this reason the elastic fraction is less than 1.0. The time dependence of the mechanical properties and the strain dependence of the elastic fraction imply an interesting structural transition that occurs in skin with strain. This behavior is attributable to the collagen network. At low strains, the elastic fraction is low and appears to involve the viscous realignment of the collagen fibers with the applied stress. At higher strains, the elastic fraction approaches 0.75, which is near the value for tendon material. In fact, several types of connective tissue exhibit an elastic fraction approaching 0.75 at high strains. This

suggests that the collagen network limits the high strain deformation in all types of connective tissues.

Although the proteins in skin are also composed of about 5% elastic fibers, they do not appear to affect the mechanical properties of the tissue. The elastic fibers are believed to contribute to the recoil of the skin, which gives it the ability to be wrinkle-free when external loads are removed. As humans age, the elastic fiber network of the skin is lost, and wrinkles begin to appear. The mechanical role of the elastic fibers is very different in vascular tissue, however.

Fatigue testing of skin in uniaxial tension shifts the stress–strain curve to the right within the linear region, and the elastic fraction approaches 0.85. Therefore, mechanical tests must be carried out carefully to eliminate any preconditioning effects. Further complicating testing is the fact that the mechanical properties of skin depend on the location from which the specimen is obtained. The contribution of the epithelial layer to the mechanical properties depends on its thickness. For most regions, the stratum corneum is thin and does not contribute significantly to the mechanical properties of skin; however, in places where it is thick, such as the palms of the hand, it needs to be considered.

5.5.1.2 Sutures. A *suture* is a biomedical product used to hold tissue together following separation by surgery or trauma. Sutures are broadly categorized according to the type of material (natural, synthetic) from which they are made, the permanence of the material (absorbable or nonabsorbable), and the construction process (braided, monofilament). As shown in Table 5.12, the most popular natural materials used for sutures are *silk* and *catgut* (animal intestine). The synthetic materials are exclusively polymeric, except for fine-sized stainless steel sutures. Approximately half of today's sutures are nonabsorbable and remain indefinitely intact when placed in the body. Polymers such as polypropylene, nylon, poly(ethylene terephthalate), and polyethylene, as well as their copolymers, are used as sutures. *Absorbable sutures* were commercially introduced with poly(glycolic acid) (PGA), and they were followed by copolymers of glycolide and lactide. A PGA suture will lose strength over

Table 5.12   Representative Mechanical Properties of Some Commercial Sutures

Suture Type	St. Pull (MPa)	Kt. Pull (MPa)	Elongation to Break (%)	Subjective Flexibility
Natural materials				
Catgut	370	160	25	Stiff
Silk	470	265	21	Very supple
Synthetic absorbable				
Poly(glycolic acid)	840	480	22	Supple
Poly(glycolide-co-lactide)	740	350	22	Supple
Poly(p-dioxanone)	505	290	34	Moderately stiff
Poly(glycolide-co-trimethylene carbonate)	575	380	32	Moderately stiff
Synthetic nonabsorbable				
Poly(butylene terephthalate)	520	340	20	Supple
Poly(ethylene terephthalate)	735	345	25	Supple
Poly[p(tetramethylene ether) terephthalate-co-tetramethylene terephthalate]	515	330	34	Supple
Polypropylene	435	300	43	Stiff
Nylon 66	585	315	41	Stiff
Steel	660	565	45	Rigid

Source: D. Goupil, in *Biomaterials Science*, B. D. Ratner, A. S. Hoffman, F. J. Schoen, and J. E. Lemons, ed., p. 356. Copyright © 1996 by Academic Press.

a 28-day period. More recently, novel absorbable polymers of polydioxanone and poly(glycolide-co-trimethylene carbonate) have been developed for surgical use.

Regardless of whether a suture is natural or synthetic, or if it is absorbable or permanent, it must meet the strength requirements necessary to close a wound under a given clinical circumstance. Suture strength is generally measured and reported in both *straight pull* and *knotted pull* tests. The former is essentially a traditional tensile test; the second is obtained after the suture has been knotted. As illustrated in Table 5.12, knotted pull strengths are approximately one-half to two-thirds of the straight pull strength values. There are a number of other characteristics a suture must have in addition to strength, including knot security, strength retention, resistance to infection, flexibility, and ease of tying.

5.5.1.3 *Artificial Soft Biologics.* In addition to sutures, polymers are used for a number of biomedical applications, as illustrated in Figure 5.128. Polymers used for hard structural applications such as dentures and bones are presented in this figure, but will be described in the next section. In this section, we will concentrate on polymers for soft biological material applications and will limit the description to mechanical properties as much as possible.

Artificial biologics, whether soft or hard, can be categorized as either temporary (short term) or permanent (long term) in their intended application. Most, but certainly not all, polymers for biomedical applications are of the short-term type and include sutures, drug delivery devices, temporary vascular grafts and stents, tissue scaffolds,

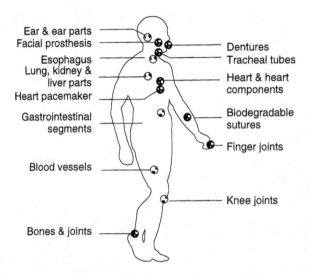

Ear & ear parts
Facial prosthesis
Esophagus
Lung, kidney & liver parts
Heart pacemaker
Gastrointestinal segments
Blood vessels
Bones & joints

Dentures
Tracheal tubes
Heart & heart components
Biodegradable sutures
Finger joints
Knee joints

Ear & ear parts: acrylic, polyethylene, silicone, poly(vinyl chloride) (PVC)
Dentures: acrylic, ultrahigh molecular weight polyethylene (UHMWPE), epoxy
Facial prosthesis: acrylic, PVC, polyurethane (PUR)
Tracheal tubes: acrylic, silicone, nylon
Heart & heart components: polyester, silicone, PVC
Heart pacemaker: polyethylene, acetal
Lung, kidney & liver parts: polyester, polyaldehyde, PVC
Esophagus segments: polyethylene, polypropylene (PP), PVC
Blood vessels: PVC, polyester
Biodegradable sutures: PUR
Gastrointestinal segments: silicones, PVC, nylon
Finger joints: silicone, UHMWPE
Bones & joints: acrylic, nylon, silicone, PUR, PP, UHMWPE
Knee joints: polyethylene

Figure 5.128 Common applications of polymers as soft biologics. Reprinted, by permission, from S. A. Visser, R. W. Hergenrother, and S. L. Cooper, in *Biomaterials Science*, B. D. Ratner, A. S. Hoffman, F. J. Schoen, and J. E. Lemons, eds., p. 57. Copyright © 1996 by Academic Press.

and orthopedic fixation devices. Short-term devices can either be removable or degradable. We will concentrate for the moment on biodegradable polymers and will make no distinction between often-confused terms such as biodegradable, bioresorbable, bioerodable, and bioabsorbable. In all cases, we describe a polymer that is converted under physiological conditions into water-soluble materials.

Some of the common biodegradable polymers used in short-term biomedical applications, and their corresponding mechanical properties, are listed in Table 5.13. Polyhydroxybutyrate (PHB), polyhydroxyvalerate (PHV), and their copolymers are derived from microorganisms and are used in such applications as controlled drug release, sutures, and artificial skin. Polycaprolactone is a semicrystalline polymer that is used in some applications as a degradable staple. Polyanhydrides are very reactive and hydrolytically unstable, which is both an advantage and a limitation to their use. A wide variety of drugs and proteins, including insulin, bovine growth factors, heparin, cortisone, and enzymes, have been incorporated into polyanhydride matrices. Poly(ortho esters) have also been used for controlled drug delivery. Poly(amino acids) are of

Table 5.13   Selected Mechanical Properties of Some Biodegradable Polymers

Polymer	Glass Transition (°C)	Melting Temperature (°C)	Tensile Strength (MPa)	Tensile Modulus (MPa)	Flexural Modulus (MPa)	Elongation Yield (%)	Elongation Break (%)
Poly(glycolic acid) (MW: 50,000)	35	210	n/a	n/a	n/a	n/a	n/a
Poly(lactic acids)							
L-PLA (MW: 50,000)	54	170	28	1200	1400	3.7	6.0
L-PLA (MW: 100,000)	58	159	50	2700	3000	2.6	3.3
L-PLA (MW: 300,000)	59	178	48	3000	3250	1.8	2.0
D, L-PLA (MW: 20,000)	50	—	n/a	n/a	n/a	n/a	n/a
D, L-PLA (MW: 107,000)	51	—	29	1900	1950	4.0	6.0
D, L-PLA (MW: 550,000)	53	—	35	2400	2350	3.5	5.0
Poly(β-hydroxybutyrate) (MW: 422,000)	1	171	36	2500	2850	2.2	2.5
Poly(ε-caprolactone) (MW: 44,000)	−62	57	16	400	500	7.0	80
Polyanhydrides							
Poly(SA-HDA anhydride) (MW: 142,000)	n/a	49	4	45	n/a	14	85
Poly(ortho esters)							
DETOSU: t-CDM: 1,6-HD (MW: 99,700)	55	—	20	820	950	4.1	220
Polyiminocarbonates							
Poly(BPA iminocarbonate) (MW: 105,000)	69	—	50	2150	2400	3.5	4.0
Poly(DTH iminocarbonate) (MW: 103,000)	55	—	40	1630	n/a	3.5	7.0

Source: J. Kohn, and R. Langer, in *Biomaterials Science*, B. D. Ratner, A. S. Hoffman, F. J. Schoen, and J. E. Lemons, ed., p. 64. Copyright © 1996 by Academic Press.

interest due to the presence of amino acid groups that can potentially be modified, but have found few biomedical applications because they are highly insoluble and difficult to process. Many of these biodegradable polymers, as well as nonbiodegradable fibers, are fabricated into fabrics for a variety of biomedical applications.

An example of how these materials can be assembled into a useful artificial soft biologic is *artificial skin*. The use of artificial skin allows for prompt closure of the largest of burn wounds, thereby limiting wound contracture and providing permanent coverage with a readily available, easily stored skin replacement material. Key characteristics of artificial skin are biocompatibility and resistance to infection.

Several types of synthetic skin grafts exist, but are all essentially *bilaminate membranes*, as shown in Figure 5.129. A small pore diameter is required in the outer layer to prevent the wound from drying out. However, a large pore diameter is required in the lower layer to promote the ingrowth of fibroblasts and long-term adhesion to the wound. One such bilaminate artificial skin consists of an inner layer of collagen-chondroitin-6-sulfate and an outer layer of silastic. The collagen binds to the wound bed and is invaded by neovasculature and fibroblasts. The collagen chondroitin 6-sulfate very significantly inhibits wound contracture, and with time the collagen is resorbed and is completely replaced by remodeled dermis. The pore size of the collagen layer is typically 50 μm. The silicone layer is 0.1 mm thick and provides a bacterial barrier in addition to the required water flux rate of 1 to 10 mg/cm^2·hr. The silicone layer

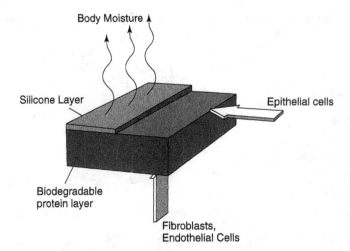

Figure 5.129 Schematic diagram of bilaminar artificial skin. Reprinted, by permission, from J. B. Kane, R. G. Tompkins, M. L. Yarmush, and J. F. Burke, in *Biomaterials Science*, B. D. Ratner, A. S. Hoffman, F. J. Schoen, and J. E. Lemons, eds., p. 367. Copyright © 1996 by Academic Press.

also provides the mechanical rigidity needed to suture the graft in place, thus preventing movement of the material during wound healing. The collagen layer adheres to the wound rapidly, as evidenced by the peel strength of the graft, which is 9 N/m at 24 hours and 45 N/m at 10 days. Histological cross sections of wounds closed with this artificial skin show complete replacement of the bovine collagen at seven weeks with remodeled human dermis.

5.5.2 Mechanical Properties of Hard Biologics

In contrast to soft biologics, whose mechanical properties primarily depend upon the orientation of collagen fibers, the mechanical properties of mineralized tissues, or hard biologics, are more complicated. Factors such as density, mineral content, fat content, water content, and sample preservation and preparation play important roles in mechanical property determination. Specimen orientation also plays a key role, since most hard biologics such as bone are composite structures. For the most part, we will concentrate on the average properties of these materials and will relate these values to those of important, man-made replacement materials.

5.5.2.1 Bone. The structure of bone was described in Section 1.5.2. Recall that bone is a composite material, composed primarily of a calcium phosphate form called hydroxyapatite (HA). The major support bones consist of an outer load-bearing shell of *cortical* (or compact) bone with a medullary cavity containing *cancellous* (or spongy) bone toward the ends.

The stress–strain curves for cortical bones at various strain rates are shown in Figure 5.130. The mechanical behavior is as expected from a composite of linear elastic ceramic reinforcement (HA) and a compliant, ductile polymer matrix (collagen). In fact, the tensile modulus values for bone can be modeled to within a factor of two by a rule-of-mixtures calculation on the basis of a 0.5 volume fraction HA-reinforced

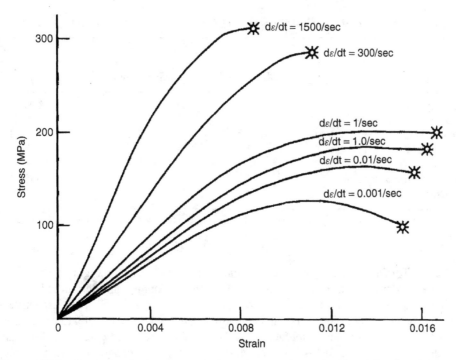

Figure 5.130 Stress–strain curves at various strain rates for cortical bone. Reprinted, by permission, from F. H. Silver and D. L. Christiansen, *Biomaterials Science and Biocompatibility*, p. 208. Copyright © 1999 by Springer-Verlag.

collagen composite. The ultimate tensile strength for cortical bone varies from 100 to 300 MPa, the modulus varies up to 30 GPa, and the strain at failure is typically only 1% or 2%. Bone has either isotropic symmetry, requiring five stiffness coefficients (see Section 5.4.3.1 on generalized Hooke's Law), or orthotropic symmetry, which requires nine such coefficients. These coefficients are particularly important for the modeling of bone resorption and implant performance.

Figure 5.130 shows that at higher strain rates, the modulus increases and the bone behaves in a more brittle manner. Values for the critical stress intensity factor, K_{1c}, of $2-12$ MN/m$^{3/2}$ and critical strain energy of $500-6000$ J/m^2 have been determined. Conversely, bone is much more viscoelastic in its response at very low strain rates. Recent experiments have shown that bone is a complex viscoelastic material; that is, time–temperature superposition cannot be used to obtain its properties. However, the nonlinearity in its mechanical behavior tends to occur outside the range of physiological interest, such that linear models are often useful for creep and stress relaxation behavior. Mineralization tends to increase modulus and ultimate tensile strength, as well as reducing the strain at failure. Mineralization does not affect the viscoelastic behavior of bone, however, and the increasing elasticity with decreasing strain rate observed in Figure 5.130 is followed, even with increased mineralization.

The mechanical properties of cancellous bone are dependent upon the bone density and porosity, and the strength and modulus are therefore much lower than those for cortical bone. The axial and compressive strength are proportional to the square of the bone density, and moduli can range from 1 to 3 GPa.

Table 5.14 Selected Mechanical Properties of Human Dentin and Enamel

Property	Dentin Mineralized	Dentin Demineralized	Enamel
Density, g/cm^3	2.14–2.18	—	2.95–2.97
Elastic modulus, GPa	14.7	0.26	84–130
Tensile strength, MPa	105.5	29.6	10.3
Compressive strength, MPa	297	—	384
Shear strength, MPa	138	—	90.2
Knoop hardness, kg/mm^2	68	—	355–431
Poisson's ratio	0.31	—	0.33

Source: Biomaterials Properties Database, W. J. O'Brien, director, www.lib.umich.edu/dentlib/Dental_tables/toc.html

5.5.2.2 *Teeth*. The structure of a human tooth was given in Figure 1.91. Recall that the two most prominent components of teeth are enamel and dentin and that the primary components of each are hydroxyapatite (HA) and HA/collagen, respectively. The mechanical properties of dentin and enamel—and, where available, their demineralized forms—are presented in Table 5.14. Notice the severe decrease in modulus and tensile strength as dentin loses its mineral content. A knowledge and understanding of these properties are important for developing suitably compatible dental amalgams, adhesives, and prosthetics.

5.5.2.3 *Artificial Hard Biologics*. A summary of polymeric materials used in biomedical applications was given in Figure 5.128. Recall that many of these application were for replacement of soft tissues, such as skin and tendons. In this section, we concentrate not only on those polymeric materials used for hard biologics, but also on all classes of materials that find use in the bonding, replacing, and growth-scaffolding of hard biologics. A schematic summary of some of these applications that involve metallic, ceramic, polymeric and composite components is given in Figure 5.131. We will approach each of these materials classes in order and will describe only a few examples of the biomedical applications for each. In most cases, these applications will be as structural replacements, and in all cases we will concentrate on the specific mechanical properties that make each material suitable for the particular application. Topics such as biocompatibility, longevity, and infection resistance are beyond the scope of this text.

Metals were some of the first materials to be used for the replacement of hard biologics. This was a reasonable selection, since the principal function of the long bones of the lower body is to act as load-bearing members. The *hip prosthesis* has been the most active area of joint replacement research. Both stainless steel, such as 316L, and Co–Cr alloys became the early materials of choice, since their stiffness, rigidity, and strength considerably exceed those of natural bone (see Table 5.15). Titanium and the so-called *aviation alloy* Ti–6Al–4V are now also used for the femoral portion of hip prostheses. In addition to strength and modulus, corrosion resistance and fatigue life are of prime importance. Also, the metals must be nontoxic, mutagenic, or carcinogenic within the body, either of themselves or due to the release of chemical components as a result of interactions with various body fluids. In fact, these are requirements for

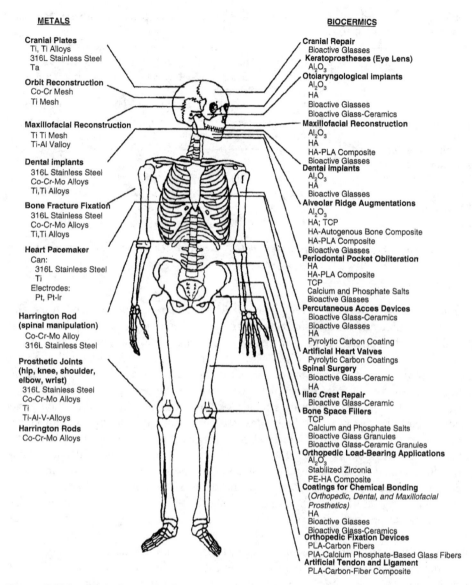

METALS

Cranial Plates
Ti, Ti Alloys
316L Stainless Steel
Ta

Orbit Reconstruction
Co-Cr Mesh
Ti Mesh

Maxillofacial Reconstruction
Ti Ti Mesh
Ti-Al Valloy

Dental implants
316L Stainless Steel
Co-Cr-Mo Alloys
Ti,Ti Alloys

Bone Fracture Fixation
316L Stainless Steel
Co-Cr-Mo Alloys
Ti,Ti Alloys

Heart Pacemaker
Can:
  316L Stainless Steel
  Ti
Electrodes:
  Pt, Pt-Ir

Harrington Rod
(spinal manipulation)
Co-Cr-Mo Alloy
316L Stainless Steel

Prosthetic Joints
(hip, knee, shoulder,
elbow, wrist)
316L Stainless Steel
Co-Cr-Mo Alloys
Ti
Ti-Al-V-Alloys

Harrington Rods
Co-Cr-Mo Alloys

BIOCERMICS

Cranial Repair
Bioactive Glasses
Keratoprostheses (Eye Lens)
Al_2O_3
Otolaryngological implants
Al_2O_3
HA
Bioactive Glasses
Bioactive Glass-Ceramics
Maxillofacial Reconstruction
Al_2O_3
HA
HA-PLA Composite
Bioactive Glasses
Dental Implants
Al_2O_3
HA
Bioactive Glasses
Alveolar Ridge Augmentations
Al_2O_3
HA; TCP
HA-Autogenous Bone Composite
HA-PLA Composite
Bioactive Glasses
Periodontal Pocket Obliteration
HA
HA-PLA Composite
TCP
Calcium and Phosphate Salts
Bioactive Glasses
Percutaneous Acces Devices
Bioactive Glass-Ceramics
Bioactive Glasses
HA
Pyrolytic Carbon Coating
Artificial Heart Valves
Pyrolytic Carbon Coatings
Spinal Surgery
Bioactive Glass-Ceramic
HA
Iliac Crest Repair
Bioactive Glass-Ceramic
Bone Space Fillers
TCP
Calcium and Phosphate Salts
Bioactive Glass Granules
Bioactive Glass-Ceramic Granules
Orthopedic Load-Bearing Applications
Al_2O_3
Stabilized Zirconia
PE-HA Composite
Coatings for Chemical Bonding
(Orthopedic, Dental, and Maxillofacial
Prosthetics)
HA
Bioactive Glasses
Bioactive Glass-Ceramics
Orthopedic Fixation Devices
PLA-Carbon Fibers
PlA-Calcium Phosphate-Based Glass Fibers
Artificial Tendon and Ligament
PLA-Carbon-Fiber Composite

Figure 5.131   Clinical uses of inorganic biomaterials. Reprinted, by permission, from L. Hench, in *Materials Chemistry*, L. V. Interrante, L. A. Casper, and A. B. Ellis, eds., p. 525. Copyright © 1995 by American Chemical Society.

all classes of materials in these types of biomedical applications. In most prosthetic applications, the metal or alloy is coated with, and held in place by, polymers.

Polymers have found two specific applications in orthopedics. First, they are used for one of the articulating surface components in joint prostheses. Thus, they must have a low coefficient of friction and low wear rate when they are in contact with the opposing surface, which is usually made of metal or ceramic. Initially, poly(tetrafluoroethylene), PTFE, was used in this type of application, but its accelerated creep rate and poor stress corrosion caused it to fail *in vivo*. Ultimately, this material was replaced with

Table 5.15 Comparison of Some Mechanical Properties of Current Implant Materials with Those of Cortical Bone

Materials	Young's Modulus (GPa)	Ultimate Tensile Strength (MPa)	Critical Stress Intensity Factor K_{1C} (MN m$^{-3/2}$)	Critical Strain Energy Release Rate, G_{1C} (J m^{-2})
Alumina	365	6–55	~3	~40
Cobalt–chromium alloys	230	900–1540	~100	~50,000
Austenitic stainless steel	200	540–1000	~100	~50,000
Ti–6wt% Al–4wt% V	106	900	~80	~10,000
Cortical bone	7–30	50–150	2–12	~600–5000
PMMA bone cement	3.5	70	1.5	~400
Polyethylene	1	30		~8000

Source: *Concise Encyclopedia of Composite Materials*, A. Kelly, ed., revised edition. Copyright © 1994 by Elsevier Science Ltd.

ultrahigh-molecular-weight polyethylene (UHMWPE). The second-place polymers find biomedical application is as a structural interface between the implant and bone tissue. In this case, the appropriate mechanical properties of the polymer are of major importance. The first type of use was with PMMA as a grouting material to fix both the stem of the femoral component and the acetabular component in place and to distribute the loads more uniformly from the implants to the bone. Since high interfacial stresses result from the accommodation of a high-modulus prosthesis within the much lower modulus bone, the use of a lower-modulus interpositional material has been introduced as an alternative to PMMA fixation. Thus, polymers such as polysulfone have been used as porous coatings on the implant's metallic core to permit mechanical interlocking through bone and/or soft tissue ingrowth into the pores. This requires that the polymers have surfaces that resist creep under the stresses found in clinical situations and have high enough yield strength to minimize plastic deformation. As indicated earlier, the mechanical properties of concern in polymer applications are yield stress, creep resistance, and wear rate. These factors are controlled by such polymer parameters as molecular chain structure, molecular weight, and degree of branching.

Ceramics and glasses have played an increasingly important role in implants in recent years. This has occurred because of two quite disparate uses. First, there is the use associated with improved properties such as resistance to further oxidation, high stiffness, and low friction and wear in articulating surfaces. This requires the use of full-density, controlled, small, uniform-grain-size (usually less than 5 μm) materials. The small grain size and full density are important since these are the two principal bulk parameters controlling the ceramic's mechanical properties. Clearly, any void within the ceramic's body will increase stress, thereby degrading the mechanical properties. Thus, full-density, small-grain-size aluminum oxide, Al_2O_3, has proved quite successful as the material in matched pairs of femoral head and acetabular components in total hip arthoplasty. Other ceramics for this use include TiO_2 and Si_3N_4. The second application takes advantage of the *osteophilic surface* of certain ceramics and glass ceramics. These materials provide an interface of such biological compatibility with bone-forming cells called *osteoblasts* that these cells lay down bone in direct apposition to the material with a direct chemicophysical bond. Special compositions of glass ceramics, termed *bioglasses*, were originally used for orthopedic implants. The bioglasses have nominal compositions in the ranges 40–50% SiO_2, 5–31% Na_2O, 12–35% CaO, 0–15% P_2O_5, plus in some cases MgO, B_2O_3, or CaF_2. The model proposed for the "chemical bond" formed between the glass and bone is that the former undergoes a controlled surface degradation, producing an SiO_2-rich layer and a Ca, P-rich layer at the interface. This latter layer eventually crystallizes as a mixture of hydroxy-carbonate apatite structurally integrated with collagen, which permits subsequent bonding by newly formed mineralized tissues.

There is an entirely different series of ceramic-based compounds that have been shown to be osteophilic. These include hydroxyapatite (HA), which you will recall is the form of the naturally occurring inorganic component of calcified tissues, and calcite, $CaCO_3$, and its Mg analog, dolomite, among others. The most extensive applications in both orthopedics and dentistry have involved HA. The elastic properties of HA and related compounds such as fluorapatite (FAp), chlorapatite (ClAp), and cobalt apatite (Co_3Ap) are compared with those of bone, dentin, and enamel in Table 5.16. The use of both HA and the glass ceramics as claddings on the metallic stems of prostheses is still another method of providing fixation instead of using PMMA. The fracture toughness properties of these ceramic-based systems are compared to natural bone in Figure 5.132.

Composites provide an attractive alternative to the various metal-, polymer- and ceramic-based biomaterials, which all have some mismatch with natural bone properties. A comparison of modulus and fracture toughness values for natural bone provide a basis for the approximate mechanical compatibility required for artificial bone in an exact structural replacement, or to stabilize a bone–implant interface. A precise matching requires a comparison of all the elastic stiffness coefficients (see the generalized Hooke's Law in Section 5.4.3.1). From Table 5.15 it can be seen that a possible approach to the development of a mechanically compatible artificial bone material

Table 5.16 Selected Mechanical Properties of Various Calcium-Bearing Minerals Compared to Bone and Teeth

Material	Bulk Modulus (10^{10}N m^{-2})	Shear Modulus (10^{10}N m^{-2})	Elastic Modulus (10^{10}N m^{-2})	Poisson's Ratio
FAp (mineral)	9.40	4.64	12.0	0.26
HAp (mineral)	8.90	4.45	11.4	0.27
HAp (synthetic)	8.80	4.55	11.7	0.28
ClAp (synthetic	6.85	3.71	9.43	0.27
Co$_3$Ap (synthetic, type B)	8.17	4.26	10.9	0.28
CaF$_2$	7.74	4.70	11.8	0.26
Dicalcium phosphate dihydrate	5.50	2.40	6.33	0.31
Bone (human ferour, powdered)	2.01	0.800	2.11	0.33
Dentine (bovine, powdered)	3.22	1.12	3.02	0.35
Enamel (bovine, powdered)	6.31	2.93	7.69	0.32

Source: L. J. Katz, in *Biomaterials Science*, B. D. Ratner, A. S. Hoffman, F. J. Schoen, and J. E. Lemons, ed., p. 335. Copyright © 1996 by Academic Press.

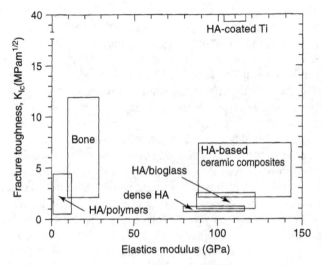

Figure 5.132 Fracture toughness and modulus of some biomaterials for bone replacement. Reprinted, by permission, from W. Suchanek, and M. Yoshimura, *J. Mater. Res.*, **13**(1), 94(1998). Copyright © 1998 by Materials Research Society.

is to stiffen one of the polymers already used as an implant with a suitable second phase. The general effect has been well demonstrated for polymers in that a progressive increase in the volume fraction of the second phase will correspondingly increase the elastic modulus but decrease the fracture toughness. Hence, starting with a ductile polymer, an eventual transition into brittle behavior is observed. Because PMMA is brittle, whereas PE is ductile, with an elongation to fracture of about 100%, PE provides the more suitable starting materials and has comparable mechanical behavior to collagen. Various materials could be considered as additions to PE that could stiffen the materials at the expense of fracture toughness, but the selection of HA has proved particularly appropriate.

Starting with polyethylene granules of average molecular weight 400,000, along with hydroxyapatite particles of 0.5–20 μm in size, composites of HA-reinforced PE are prepared to various volume fractions by a compounding and molding technique that minimizes polymer degradation. The composites are pore-free, with a homogeneous distribution of HA particles of volume fractions from 0.1 to 0.5. The rule of mixtures applies to the prediction of tensile modulus in these composites, and values increase with increasing HA content up to about 9 GPa at the 50% volume level. The increase in modulus is accompanied by a decrease in the elongation-to-fracture. However, the composite remains unusually ductile and maintains its fracture toughness, up to about 40% volume HA. From 40% to 50% volume fraction, the elongation to fracture decreases rapidly, with brittle fracture observed at $V_f = 0.45$. At these volume fractions, a fracture toughness of 3–8 MN/m$^{3/2}$ is observed (cf. Figure 5.132). Hence, in terms of cortical bone replacement, an HA–PE composite with 50% volume HA has a tensile modulus within the lower band of the values for bone and a comparable fracture toughness.

Another example of composites in biomedical applications is graphite-fiber-reinforced bone cement. Self-curing poly(methyl methacrylate), PMMA, is used extensively as a bone cement in orthopedic surgery for fixation of endoprostheses

---

Cooperative Learning Exercise 5.13

A biomedical composite is formed by incorporating 1.05 volume % short carbon fibers into a PMMA matrix. The carbon fibers have the following properties: $\sigma_f = 2.241$ GPa, $l_c = 2.54 \times 10^{-4}$ m, $l = 1.27 \times 10^{-2}$ m, $E_f = 248.3$ GPa, and aspect ratio = 2500. The PMMA has the following properties: $\sigma_m = 34.48$ MPa, $E_m = 2$ GPa. Recall that $n = 0.33$ for a random fiber composite.

Person 1: Use the rule of mixtures to estimate the tensile strength of the composite.

Person 2: Use the appropriate equation from Section 5.4.2.2 to estimate the tensile strength of the composite.

Compare your answers. Are they the same? Why or why not? How would you go about estimating the modulus for this composite?

Answers: $\sigma_c^* = (0.0105)2241 + (0.9895)34.48 = 57.6$ MPa; Eqn. 5.114 for low-volume-fraction fibers: $\sigma_c^* = [2.54 \times 10^{-4}/2(1.27 \times 10^{-2})](0.0105)(2241) + (0.9895)(34.34) = 34.34$ MPa;

The corrections for fiber length of Eq. (5.114) introduce a significant and necessary correction. The experimental value is 38.3 MPa. The modulus could be determined from Eq. (5.110), but the product $na = 825$, which according to Figure 5.95 means $\beta = 1.0$, such that a simple rule of mixing can be used for the modulus.

---

and the repair of bone defects. However, such bone cement is weak in tension and is also significantly weaker than compact bone. The reinforcement of bone cement with graphite fibers has been shown to increase its tensile, compressive, and shear strengths, as well as increase fatigue life and reduce creep. The tensile modulus of fiber-reinforced PMMA also increases with increasing fiber content when the fiber volume percentage

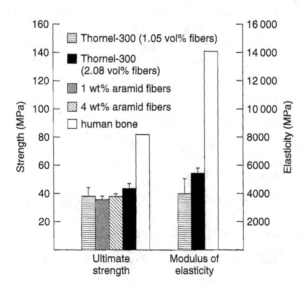

Figure 5.133 Comparison of tensile properties of fiber-reinforced bone cement (PMMA) and human compact bone. Reprinted, by permission, from *Concise Encyclopedia of Composite Materials*, A. Kelly, ed., p. 270, revised edition. Copyright © 1994 by Elsevier Science Publishers, Ltd.

is kept small, as illustrated in Figure 5.133, where Thornel is an ultrahigh-strength graphite fiber, and the aramid fiber in use is Kevlar® -29. Incorporation of 2% Thornel-300 fibers improves the tensile strength of PMMA to 50%, and its modulus to about 40%, of that of compact human bone. As described in Section 5.4.2.7, the energy absorption capacity of the composite is increased due to fiber pull-out.

REFERENCES

Cited References

1. Li, X., and B. Bhushan, A review of nanoindentation continuous stiffness measurement technique and its applications, *Mater. Char.*, **48**, 11–36 (2002).
2. Jenkins, C. H. M., The strength of Cd–Zn and Sn–Pb alloy solder, *J. Inst. Metals*, **40**, 21–39 (1928).
3. Pearson, C. E., The viscous properties of extruded eutectic alloys of lead-tin and bismuth-tin, *J. Ins. Metals*, **54**, 111–124 (1934).
4. Bochavar, A. A., and Z. A. Sviderskaya, Superplasticity in Zn–Al alloys, *Bull. Acad. Sci.* (USSR) **9**, 821–824 (1945).
5. Underwood, E. E., A review of superplasticity, *JOM*, 914–919 (1962).
6. Holt, D. L., and W. A. Backofen, Superplasticity in the Al–Cu eutectic alloy, *Trans. ASM*, **59**, 755–767 (1966).
7. Ashby, M. F., *Materials Selection in Mechanical Design*, 2nd ed., Butterworth Heinemann, Oxford, 1992.
8. *Concise Encyclopedia of Composite Materials*, A. Kelly, ed., Pergamon, Elmsford, NY, 1994, p. 100.
9. *Concise Encyclopedia of Composite Materials*, A. Kelly, ed., Pergamon, Elmsfords NY, 1994, p. 161.
10. Powell, P. C., and A. J. Ingen Housz, *Engineering with Polymers*, Stanley Thornes, Cheltenham, UK, 1998.
11. Suchanek, W., and M. Yoshimura, Processing and properties of hydroxyapatite-based biomaterials for use as hard tissue replacement implants, *J. Mater. Res.*, **13**(1), 94 (1998).

Mechanics of Materials

Knott, J., and P. Withey, *Fracture Mechanics: Worked Examples*, Institute of Materials, London, 1993.

Mechanical Properties of Metals

Metals Handbook, Desk Edition, American Society for Metals, 1985.

Mechanical Properties of Ceramics and Glasses

Wachtman, J. B., *Mechanical Properties of Ceramics*, John & Sons, Wiley, New York, 1996.

Mechanical Properties of Polymers

McCrum, N. G., C. P. Buckley, and C. B. Bucknall, *Principles of Polymer Engineering*, Oxford Science, Oxford, 1997.

Mechanical Properties of Composites

Kelly, A., *Composite Materials*, American Elsevier, New York, 1966.

Composite Materials Technology, P. K. Mallick and S. Newman, eds., Hanser Publishers, Munich, 1990.

Mechanical Testing of Engineering Ceramics at High Temperatures, B. F. Dyson, R. D. Lohr, and R. Morrell, ed., Elsevier Applied Science, London, 1989.

Lee, W. E., and W. M. Rainforth, *Ceramic Microstructures: Property Control by Processing*, Chapman & Hall, London, 1994.

Composite Applications: The Role of Matrix, Fiber and Interface, T. L. Vigo and B. J. Kinzig, eds., VCH Publishing, New York, 1992.

Mechanical Properties of Biologics

The mechanical properties of biological materials, *Symposium, Society of Experimental Biologists*, Cambridge University Press, New York, 1980.

Biomaterials Properties Database, University of Michigan, Dr. William J. O'Brien, Director
http://www.lib.umich.edu/dentlib/Dental_tables/toc.html

Evans, F. E., *Mechanical Properties of Bone*, C. Charles, Springfield, IL, 1973.

PROBLEMS

Level I

5.I.1 The following three stress–strain data points are provided for a titanium alloy: $\varepsilon = 0.002778$ at $\sigma = 300$ MPa; $\varepsilon = 0.005556$ at $\sigma = 600$ MPa; $\varepsilon = 0.009897$ at $\sigma = 900$ MPa. Calculate the modulus of elasticity for this alloy.

5.I.2 Experimentally, it has been observed for single crystals of a number of metals that the critical resolved shear stress is a function of the dislocation density, ρ_D:

$$\tau_{cr} = \tau_0 + A(\rho_D)^{0.5}$$

where τ_0 and A are constants. For copper, the critical resolved shear stress is 2.10 MPa at a dislocation density of 10^5 mm^{-2}. If it is known that the value of A for copper is 6.35×10^{-3} MPa·mm, compute the critical resolved shear stress at a dislocation density of 10^7 mm^{-2}. Callister, p. 182, Problem 7.29.

5.I.3 Calculate the composite modulus for epoxy reinforced with 70 vol% boron filaments under isotropic strain conditions.

5.I.4 A single crystal of aluminum is oriented for a tensile test such that its slip plane normal makes an angle of 28.1° with the tensile axis. Three possible slip directions make angles of 51.2°, 36.0°, and 40.6° with the same tensile axis. Which of these three slip directions is most favored? Adapted from Callister, P. 180, Problem 7.12 (numerical values changed, editorial changes).

5.I.5 According to the Kelvin (Voigt) model of viscoelasticity, what is the viscosity (in Pa·s) of a material that exhibits a shear stress of 9.32×10^9 Pa at a shear strain of 0.5 cm/cm over a duration of 100 seconds? The shear modulus of this material is 5×10^9 Pa.

5.I.6 A high-strength steel has a yield strength of 1460 MPa and a K_{1c} of 98 MPa·m$^{\frac{1}{2}}$. Calculate the size of a surface crack that will lead to catastrophic failure at an applied stress of $\frac{1}{2}\sigma_y$.

5.I.7 An alloy is evaluated for potential creep deformation in a short-term laboratory experiment. The creep rate is found to be 1% per hour at 800°C and 5.5 × 10^{-2}% per hour at 700°C. (a) Calculate the activation energy for creep in this temperature range. (b) Estimate the creep rate to be expected at a service temperature of 500°C. (c) What important assumption underlies the validity of your answer to part b?

5.I.8 A glass plate contains an atomic-scale surface crack. Assume that the crack tip radius is approximately equal to the diameter of an O^{2-} ion. Given that the crack is 1 μm long and the theoretical strength of the defect-free glass is 7.0 GPa, calculate the breaking strength of the glass plate.

Level II

5.II.1 In normal motion, the load exerted on the hip joint is 2.5 times body weight. (a) Estimate the corresponding stress (in MPa) on an artificial hip implant with a cross-sectional area of 5.64 cm^2 in an average-sized patient, and (b) calculate the corresponding strain if the implant is made of Ti–6Al–4V alloy. Clearly state your assumptions and sources of data.

5.II.2 In the absence of stress, the center-to-center atomic separation distance of two Fe atoms is 0.2480 nm along the [111] direction. Under a tensile stress of 1000 MPa along this direction, the atomic separation distance increases to 0.2489 nm. (a) Calculate the modulus of elasticity along the [111] direction. (b) How does this compare with the average modulus for polycrystalline iron with random grain orientation? (c) Calculate the center-to-center separation distance of two Fe atoms along the [100] direction in unstressed α-iron. (d) Calculate the separation distance along the [100] direction under a tensile stress of 1000 MPa, if the modulus in this direction is 125 GPa.

5.II.3 It is sometimes convenient to express the degree of plastic deformation in metals as the percent cold work, CW%, rather than as strain:

$$\%CW = [(A_0 - A_f)/A_0] \times 100$$

where A_0 and A_f are the initial and final cross-sectional areas, respectively. Show that for a tensile test

$$\%CW = [\varepsilon/(\varepsilon + 1)] \times 100$$

Look up the stress–strain diagram for naval brass and determine the CW% where a stress of 400 MPa is applied. Cite the source of your information. Adapted from Callister, problem 7.24, p. 181.

Level III

5.III.1 In a recent article [Shinzato, S., et al., A new bioactive bone cement: Effect of glass bead filler content on mechanical and biological properties, *J. Biomed.*

Mater. Res., **54**(4), 491 (2000)] the mechanical properties of glass-bead-filled poly(methylmethacrylate) (PMMA) bone cement are described. Four plots from this article are reproduced below. "GBC" refers to the glass bead composite, and the number refers to the weight percent glass beads; for example, GBC50 has 50 wt% glass beads in the PMMA matrix. The data from all tests were obtained from cylindrical samples with a gauge length of 20 mm. (a) Calculate the radius (in mm) of the cylindrical samples used for the compressive tests. Note that the compressive strengths of samples GBC30, GBC40, and GBC50 shown in Figure (a) were taken as the inflection point in the stress–strain diagrams shown in Figure (d), rather than the break point, as in samples GBC60 and GBC70. (b) Estimate Poisson's ratio for this material. (c) What is the force (in terms of weight, kg) required to break the GBC samples in a three-point bending test? (d) Sketch the stress–strain diagrams that result from the three-point bend test for the following three materials: glass beads (assume they are a generic silicate glass), PMMA, and GBC50. You should have three separate curves, each representing the relative behaviors of the three materials.

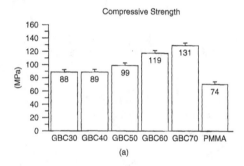

(a)

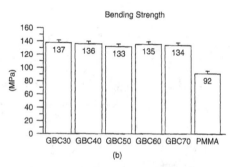

(b)

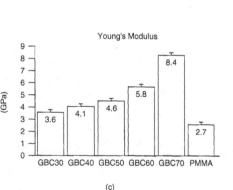

(c)

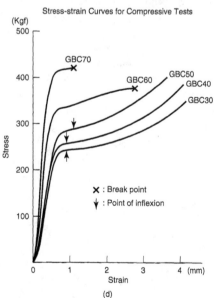

(d)

5.III.2   In another article on bone cements [Vallo, C, Influence of filler content on static properties of glass-reinforced bone cement, *J. Biomed. Mater. Res.*, **53**(6), 717 (2000)], the following data for the critical intensity factor in these same glass-bead-filled PMMA composites as was used in Problem 5.III.1 are given:

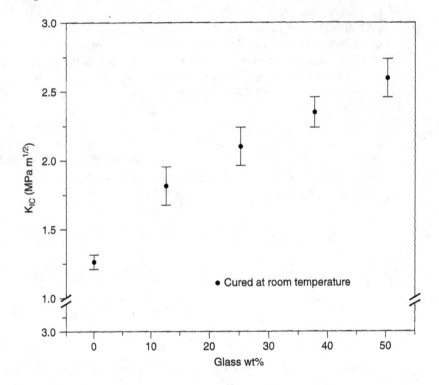

Using the values of the compressive strength from Problem 5.III.1 as representative of the stress at the fracture instability (critical stress for crack propagation), estimate the crack length at the fracture instability, *a* (in mm), for the GBC material. Clearly indicate which composition you are using in your calculation and any assumptions you must make.

CHAPTER 6

Electrical, Magnetic, and Optical Properties of Materials

This chapter is organized much like Chapter 4. We will describe three related properties of materials, and we will subdivide each section into the five materials classes. Just as in Chapter 4, there are some fundamental equations that relate important fluxes to driving forces and, in doing so, create proportionality constants that are material properties. And, just as in Chapter 4, we will be more concerned with the constants—material properties—than with the driving forces. In fact, we will be even less concerned with the driving forces and fluxes of electricity and magnetism in this chapter than we were with momentum, heat, and mass transport topics, since the latter have more to do with materials processing, which we will touch upon in Chapter 7, whereas the former are more the realm of solid-state physics. This is not to say that neither the origins of the driving forces nor their resultant fluxes are unimportant to the engineer. On the contrary, these are the design parameters that must often be met. The point is that the derivations of many of these fundamental equations will be dispensed with in favor of concentration upon their application. The third topic of this chapter is related to the first two, although no fundamental equation of the driving-force form exists. It has to do with how electromagnetic radiation interacts with materials, and it is a logical extension of the description of electrical and magnetic phenomena.

The first topic will deal with *electrical conductivity*. The driving force for the flow of electrons that leads to electrical conduction is the *electric field*, $\mathscr{E}$, and the resulting flow of electrons is characterized by the *current density*, $\mathbf{J}$. The flow and driving force are related through *Ohm's Law*:

$$\mathbf{J} = \sigma \mathscr{E} \tag{6.1}$$

where $\mathbf{J}$ is the current density in units of amps/m^2, $\mathscr{E}$ is the electric field strength in units of volts/m, and σ is the *electrical conductivity* in units of ohms^{-1}/m or $(\Omega \cdot \text{m})^{-1}$. It is indeed unfortunate that the symbol for conductivity is also the symbol we utilized for the strength of a material, but it is rare that the two are used in the same context, so that confusion should be easily avoided.

The second section will address the topic of *magnetism*. The magnetic driving force is the *magnetic field*, $\mathbf{H}$, and is related to a quantity known as the magnetic flux density, or *magnetic induction*, $\mathbf{B}$, through the magnetic analogue to Ohm's Law:

$$\mathbf{B} = \mu \mathbf{H} \tag{6.2}$$

An Introduction to Materials Engineering and Science: For Chemical and Materials Engineers,
by Brian S. Mitchell
ISBN 0-471-43623-2 Copyright © 2004 John Wiley & Sons, Inc.

where **B** has units of teslas (webers/m^2), and **H** has units of amps/m. The proportionality constant, μ, is called the *magnetic permeability*. We will have much more to say about the units of permeability in Section 6.2, since there is some variability in its definition. It is once again unfortunate that the symbol for magnetic permeability is identical to that of another important engineering parameter, Newtonian viscosity, but there is little opportunity for confusion in real applications.

The third topic has to do with how materials interact with radiation that has both an electrical and a magnetic component—that is, electromagnetic radiation. This topic is often generically referred to as "optical properties" of materials, even though the full range of electromagnetic radiation, of which the optical portion is only a small segment, can be considered. There is no unifying, driving-force-containing equation to describe the optical properties of materials. However, there is one useful relationship that we will use to build most of the descriptions of Section 6.3 upon which relates the intensity of incident electromagnetic radiation, I_0, to the transmitted, absorbed, and reflected intensities, I_T, I_A, and I_R, respectively:

$$I_0 = I_T + I_A + I_R \tag{6.3}$$

By the end of this chapter, you should be able to:

- Relate conductivity, resistivity, resistance, and conductance.
- Calculate conductivity from charge carrier concentration, charge, and mobility.
- Differentiate between a conductor, insulator, semiconductor, and superconductor.
- Differentiate between an intrinsic and an extrinsic semiconductor.
- Differentiate between a *p*- and *n*-type semiconductor.
- Identify different types of electrical polarizability, and determine if they are relevant to a given chemical structure.
- Describe how a *p*–*n* semiconductor junction works.
- Differentiate between diamagnetism, paramagnetism, ferromagnetism, ferrimagnetism, and antiferromagnetism.
- Describe the temperature dependence of magnetic susceptibility.
- Define the Curie temperature for a substance.
- Define the Neél temperature for a substance.
- Determine coercivity and remnant induction from a hysteresis loop.
- Relate transmissivity, absorptivity, and reflectivity.
- Define refractive index.
- Calculate reflectivity from refractive index differences at an interface.
- Describe the interactions of light with materials that result in color.

6.1 ELECTRICAL PROPERTIES OF MATERIALS

Without going into a complete derivation, let us examine Eq. (6.1) more closely to see where the driving force, electrical current, and electrical conductivity come from.

The electric field, $\mathscr{E}$, is a vector quantity, as indicated by the boldfaced type. For the time-being, we will limit the description to an electric field in one direction only and

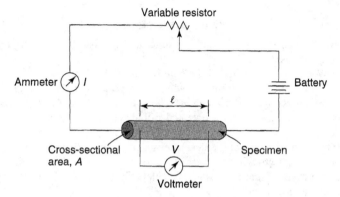

Figure 6.1 Schematic diagram of conductivity measurement in a material. Reprinted, by permission, from W. Callister, *Materials Science and Engineering: An Introduction*, 5th ed., p. 607. Copyright © 2000 by John Wiley & Sons, Inc.

will consider only the magnitude of the electric field, $\mathscr{E}$. An electric field is generated by applying a voltage, V, across a distance, l, as illustrated in Figure 6.1. The electric field strength is then given by

$$\mathscr{E} = V/l \tag{6.4}$$

When a conductive material is placed within the electric field, current begins to flow, as characterized by the current density, **J**, of Eq. (6.1). The current density is also a vector quantity, but since our field is in one dimension only, current will similarly flow only in one direction, so that we will use only the scalar quantity from here on, J. The current density is simply the *current, I*, per unit area in the specimen, A:

$$J = \frac{I}{A} \tag{6.5}$$

where I is in units of ampere (amps, A, for short) and A is in units of m^2. The current, in turn, is defined as the rate of charge passage per unit time:

$$I = \frac{dq}{dt} \tag{6.6}$$

where q is the charge in units of coulombs, so that an ampere is equal to a coulomb/s. Substitution of Eqs. (6.4) and (6.5) into Eq. (6.1) and isolation of V gives

$$V = I\left(\frac{l}{\sigma A}\right) \tag{6.7}$$

We define the quantity in parentheses as the *resistance, R*, to obtain the more familiar form of Ohm's Law:

$$V = I \times R \tag{6.8}$$

Cooperative Learning Exercise 6.1

Consider a wire, 3 mm in diameter and 2 m long. Use the data in Appendix 8 to answer the following questions.

Person 1: What is the resistance in the wire if it is made of copper?

Person 2: What is the resistance in the wire if it is made of stainless steel?

Exchange the results of your calculations, check them, and use them to make the following calculations, assuming that a potential difference of 0.05 V is placed across the ends of the wire.

Person 1: If the wire is made of stainless steel, calculate the current flow, current density, and magnitude of the electric field.

Person 2: If the wire is made of copper, calculate the current flow, current density, and magnitude of the electric field.

Answers: $R_{Cu} = 4.84 \times 10^{-3}$ Ω; $R_{ss} = 0.202$ Ω; $J_{ss} = 1.5 \times 10^6$ A/m², $\mathcal{E}_{ss} = 35{,}000$ A/m²; $J_{Cu} = 0.25$ A; $I_{ss} = 10.3$ A; $\mathcal{E}_{Cu} = 1.5 \times 10^6$ A/m², $\mathcal{E}_{Cu} = 0.025$ V/m.

The relationship between conductivity and resistance is an important one, and it can be further generalized through the introduction of two additional quantities called the *conductance*, G, and the *resistivity*, ρ. The resistivity is simply the resistance per unit length times the cross-sectional area, $\rho = RA/l$, and as such is a geometry-independent material property, as is the conductivity, σ. Similarly, the conductance is the conductivity per unit area times length, $G = \sigma l/A$. It is important to remember that the geometry of the sample must be known to calculate its resistance and/or conductance, whereas the resistivity and conductivity are inherent material properties that can vary with temperature and composition, but are generally independent of sample geometry. Moreover, it is useful to keep in mind that conductivity and resistivity are inverses of one another, and if we know one, we can find the other.

Conductivity requires a *charge carrier*. There are two types of charge carriers we will consider: electrons and ions. The structural descriptions of Chapter 1 will be helpful in determining the primary type of charge carrier within a material, if any. In subsequent sections, we explore the molecular origins of each type of conductivity, investigate the important parameters that cause conductivity to vary in materials, and describe additional electrical conduction phenomena that have revolutionized our lives.

6.1.1 Electrical Properties of Metals and Alloys

Electrical conduction in metals and alloys occurs by the motion of electrons. It can be shown that the conductivity is proportional to the number of electrons per unit volume, n_e, the charge per electron, q_e, and the *electron mobility*, μ_e:

$$\sigma = n_e q_e \mu_e \tag{6.9}$$

where the units of n_e are m^{-3}, q_e is the magnitude of charge on an electron (1.602×10^{-19} coulombs), and μ_e is in $m^2/V \cdot s$. Any of these three quantities can, and do, affect the overall conductivity of a material, leading to an astounding 27 orders of magnitude in the variation of electrical conductivity values. This immense variation in a physical property causes us to further classify materials according to their ability to conduct electricity. Those materials for which the value of σ is low, ranging from 10^{-20} to $10^{-10} (\Omega \cdot m)^{-1}$, are termed *insulators*, while those with intermediate values on the order of 10^{-6} to $10^{4} (\Omega \cdot m)^{-1}$ are called *semiconductors*. Materials with high electrical conductivities above $10^{7} (\Omega \cdot m)^{-1}$, such as most metals and alloys, are called *conductors*. Notice that the range of conductivities for these classes are still very large and that there are significant gaps between them for which absolute classification is difficult. In fact, a given material can cross over from one classification to another, depending on such factors as temperature. As a result, it is important to understand the origins of electrical conduction in these materials.

6.1.1.1 *The Molecular Origins of Electrical Conduction.* Recall from Section 1.0.4.3 that the bonds in metals tend to form bands of electrons. The electrons involved in bonding are found in the valence band, inner electrons not involved in bonding are found in the core band, and the remaining unfilled orbitals of the outer bands form the conduction band. It is the conduction band that gives metals and alloys the ability to freely conduct electrons. As shown in Figure 6.2, the "distance" between the conduction and valence bands in a solid, called the *band gap*, E_g, can be used to characterize the electrical behavior of a material. Insulators have large band gaps of several electron volts (eV), whereas conductors have small band gaps, or even overlapping valence and conduction bands, as illustrated in Figure 6.2b. Metallic solids are good electrical conductors because their valence band is not completely filled or because it may overlap with a second empty valence band, thereby causing the total energy band to be only partially filled.

According to the band theory, the motion of electrons under an external electric field is possible only for electrons in partially filled energy bands. The *probability function,*

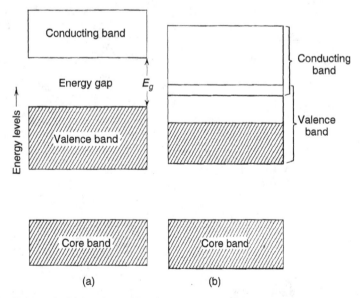

Figure 6.2 Schematic illustration of band gaps in (a) an insulator and (b) conductor. From Z. Jastrzebski, *The Nature and Properties of Engineering Materials*, 2nd ed. Copyright © 1976 by John Wiley & Sons, Inc. This material is used by permission of John Wiley & Sons, Inc.

$f(E)$, that describes the probability that an energy level E is occupied by an electron is given by Fermi as

$$f(E) = \frac{1}{\exp[(E - E_f)/k_B T] + 1} \tag{6.10}$$

where E_f is called the Fermi energy, k_B is Boltzmann's constant, and T is the absolute temperature. The Fermi energy in a metal represents the kinetic energy of the electrons having the maximum energy level that some electrons can attain. At absolute zero, the Fermi–Dirac distribution law requires that the Fermi function, $f(E)$, must have a value of 1 or 0 (see Figure 6.3a). Thus, at $T = 0$ K we have $f(E) = 1$ and $E < E_f$. At temperatures greater than 0 K, $f(E)$ changes from 1 to 0 over an energy range of about one unit of $k_B T$. At $E = E_f$ we obtain $f(E) = \frac{1}{2}$ (see Figure 6.3b). If $E - E_f >> k_B T$, the exponential term in the denominator of Eq. (6.10) becomes very large relative to unity, and $f(E)$ becomes

$$f(E) = \exp\left[\frac{-(E - E_f)}{k_B T}\right] \tag{6.11}$$

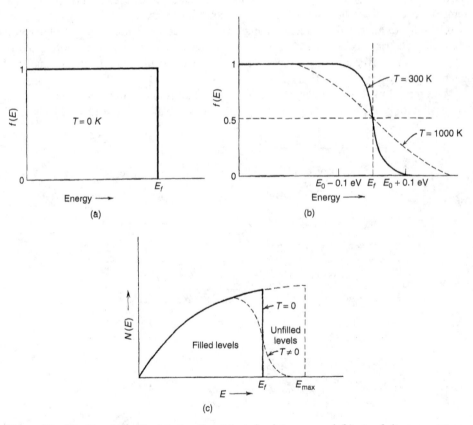

Figure 6.3 The Fermi distribution function (a) at absolute zero and (b) at a finite temperature. (c) The population density of electrons in a metal as a function of energy. From Z. Jastrzebski, *The Nature and Properties of Engineering Materials*, 2nd ed. Copyright © 1976 by John Wiley & Sons, Inc. This material is used by permission of John Wiley & Sons, Inc.

For electron energies, E, below the Fermi level, $E_f - E \gg k_B T$, and Eq. (6.11) can be approximated as

$$f(E) \approx 1 - \exp\left[\frac{E - E_f}{k_B T}\right] \qquad (6.12)$$

indicating that the probability of occupation of the energy band is nearly unity.

The state densities are not uniform across the energy band, and their *population density*, $N(E)$, is the greatest in the center of the band. The number of electrons in the band, n_e, can then be evaluated and used in Eq. (6.9) by integrating the product of the density of state $N(E)$ and the probability of their occupation, $f(E)$, over the band energy range (see Figure 6.3c). Thus, for metals,

$$n_e = \int_0^{E_f} N(E) \cdot f(E) \, dE \qquad (6.13)$$

The energy distribution in the conduction zone extends over several electron volts and is about 5 eV for some metals at absolute zero. Thus, only a small fraction of the electrons in the energy band can be excited above the Fermi level, and only those within an energy range of the order $k_B T$ can be excited thermally.

In addition to the number of electrons, the other factor in Eq. (6.9) that affects conductivity is the electron mobility, μ_e. The mobility is the average charge carrier velocity, or *drift velocity*, $\bar{v}$, divided by the electric field strength, $\mathscr{E}$:

$$\mu_e = \bar{v}/\mathscr{E} \qquad (6.14)$$

In a typical metal, $\mu_e = 5 \times 10^{-3}$ m^2/V · s, which gives a drift velocity of 5×10^{-3} m/s for an electric field of 1 V/m. The average electron velocity, in turn, is related to an important structural parameter, the mean free path, l, which we utilized in Section 4.2.1.4 to describe thermal conductivity:

$$l = \tau \bar{v} \qquad (6.15)$$

where τ is a quantity known as the *relaxation time* and has units of sec^{-1}. The relaxation time is closely related to the mean time of flight between electron collisions and also to the mean free path of the conduction electrons, as indicated above.

As the mobilities are likely to depend on temperature only as a simple power law over an appropriate region, the temperature dependence on conductivity will be dominated by the exponential dependence of the carrier concentration. We will have more to say about carrier mobility in the section on semiconductors.

In summary, metals are good electrical conductors because thermal energy is sufficient to promote electrons above the Fermi level to otherwise unoccupied energy levels. At these levels ($E > E_f$), the accessibility of unoccupied levels in adjacent atoms yields high mobility of conduction electrons known as *free electrons* through the solid.

6.1.1.2 Resistivity in Metals and Alloys.

Resistivity is the reciprocal of conductivity. We will find it useful to momentarily concentrate on resistivity instead of conductivity to explain some important electrical phenomena. If the metal crystal lattice were perfect and there were no lattice vibrations, the electrons would pass through the lattice unscattered, encountering no resistance. But, as we saw in the description of thermal conductivity (cf. Section 4.2.2.1), lattice vibrations and phonon scattering play a role in disrupting the mean free path of electrons. Working backwards through the previous equations, then, shows that a reduction in mean free path, l [Eq. (6.15)], due to phonon scattering and lattice vibrations decreases the electron mobility [Eq. (6.14)], which in turn can reduced electrical conductivity [Eq. (6.9)] or increase resistivity. This effect leads to the temperature-dependent portion of resistivity, ρ_t. Since the thermal lattice vibrations increase directly with increasing temperature, the electrical resistivity of metals will increase with temperature in a linear manner. At high temperatures approaching the melting point, the resistivity usually rises somewhat more rapidly than linearly with temperature. Only at very low temperatures can lattice defects and impurity atoms affect conductivity and resistivity, rather than thermal lattice vibrations. This type of resistivity is known as the *residual resistivity*, ρ_r, and is relatively temperature-independent. The two types of resistivity are additive, such that the total resistivity is given by

$$\rho = \rho_t + \rho_r \qquad (6.16)$$

This relationship is shown in Figure 6.4 for a series of copper alloys. The variation of resistivity with temperature is approximately linear over a wide temperature range, and near absolute zero the alloys have a relatively temperature-independent residual resistivity. Notice that as the level of nickel impurities increases, so do both the residual

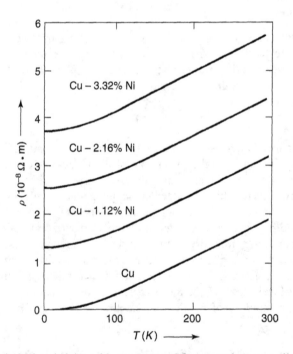

Figure 6.4 Variation of resistivity with temperature for several copper alloys. From K. M. Ralls, T. H. Courtney, and J. Wulff, *Introduction to Materials Science and Engineering*. Copyright © 1976 by John Wiley & Sons, Inc. This material is used by permission John Wiley & Sons, Inc.

and thermal vibration resistivities. This is directly related to a reduction in the electron mean free path.

Oftentimes the variation of resistivity with temperature in the linear region is approximated with an empirical expression of the form

$$\rho_t = \rho_0 + aT \qquad (6.17)$$

where ρ_0 and a are experimentally determined constants for a particular metal. Do not confuse the empirical parameter ρ_0 with the residual resistivity, ρ_r - the former is an equation-fitting parameter only for the linear region of the resistivity-temperature plot, whereas the latter is a fundamental property of a material. The resistivities of some metals and alloys are presented in Appendix 8.

The variation of resistivity with composition is also expressed in an empirical fashion. For two-phase alloys consisting of phases α and β, the rule of mixtures can be used to approximate the alloy resistivity from the individual metal resistivities:

$$\rho = V_\alpha \rho_\alpha + (1 - V_\alpha)\rho_\beta \qquad (6.18)$$

where V_α is the volume fraction of phase α.

Other factors such as cold deformation and processing conditions can affect resistivity. The cold deformation effect is not as pronounced as the addition of impurity or alloying elements, but the effect of processing on resistivity can be large, as illustrated in Figure 6.5 for a Cu_3Au compound. The effect on resistivity is again related to structure. The rapidly quenched compound maintains its disordered structure, which

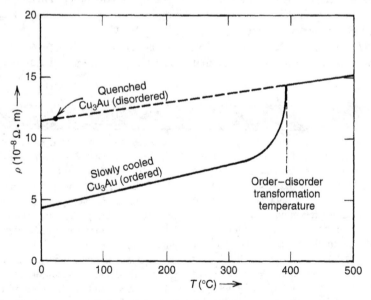

Figure 6.5 Temperature variation of the resistivity for Cu_3Au under two different processing conditions. From K. M. Ralls, T. H. Courtney, and J. Wulff, *Introduction to Materials Science and Engineering*. Copyright © 1976 by John Wiley & Sons, Inc. This material is used by permission John Wiley & Sons, Inc.

causes an increase in the resistivity relative to the slowly cooled, ordered structure. In the ordered structure, gold atoms occupy the corner positions and copper atoms occupy the face-centered positions of the FCC unit cell, which results in less electron scattering than the disordered structure in which copper and gold atoms occupy FCC atomic sites at random within each unit cell. Only at the order–disorder transition temperature do the two resistivity curves become coincident, corresponding to a transformation from the ordered to the disordered FCC structure as a result of increased thermal motion.

For metals in general, any mechanical or chemical action that alters the crystalline perfection will raise the residual resistivity and, therefore, the total resistivity, according to Eq. (6.16). Thus, vacancies in metals, in contrast to those in ionic solids, increase the resistivity. The reason for this lies in the inherent differences between conduction mechanisms in these two classes of materials. The differences between ionic and electronic conduction will be elaborated upon in Section 6.1.2.

6.1.1.3 Superconduction.
You may have noticed in Figure 6.3 that the resistivity of pure copper approaches zero at absolute zero temperature; that is, the residual resistivity is zero. An expanded scale shows that this is not really the case. Figure 6.6 shows that the residual resistivity in pure copper is about 10^{-10} $\Omega \cdot m$. This is the "normal" behavior of many metals. Some metals, however, such as Pb, lose all resistivity abruptly and completely at some low temperature. This phenomenon is called *superconductivity*, and the materials that exhibit it are called *superconductors*. The temperature at which the resistivity vanishes is called the *critical transition temperature*, T_c. For example, lead has a $T_c = 7.19$ K, and it has a resistivity that is less than 10^{-25} $\Omega \cdot m$ at 4.2 K. Many elemental metals (see Figure 6.7), solid–solution alloys, and intermetallic compounds exhibit superconductivity. Often, such materials have relatively high electrical resistivities above T_c, compared to other nonsuperconducting metals such as Ag and Cu. Observed T_c values range from a few millidegrees Kelvin to over 21 K for intermetallic compounds such as Nb_3Ge.

In 1957 the research team of Bardeen, Cooper, and Schrieffer produced a theory, now known as the *BCS theory*, that managed to explain all the major properties of

---

HISTORICAL HIGHLIGHT

The discovery of superconductivity was not very dramatic. When Dutch physicist Heike Kamerlingh Onnes succeeded in liquefying helium in 1908 he looked around for something worth measuring at that temperature range. His choice fell upon the resistivity of metals. He tried platinum first and found that its resistivity continued to decline at lower temperatures, tending to some small but finite value as the temperature approached absolute zero. He could have tried a large number of other metals with similar prosaic results. But he was in luck. His second metal, mercury, showed quite unorthodox behavior, and in 1911 he showed that its resistivity suddenly decreased to such a small value that he was unable to measure it—and no one has succeeded in measuring it ever since.

In the subsequent years, many more superconductors were found at these very low temperatures. By the 1960s, certain alloys of niobium were made that became superconductors at 10–23 K. It was generally believed on theoretical grounds that there would be no superconductors above 30 K. See Section 6.1.2.4 to see if this belief is still held today.

Source: Adapted from Solymar and Walsh [1] and Sheahan [2].

10^{-7}
10^{-8}
10^{-9}
10^{-10}
10^{-11}
$\rho\ (\Omega \cdot m)$
Pb
(a superconductor)
Cu
(a normal metal)
10^{0}
T_c
10^{1}
10^{2}
T (K)

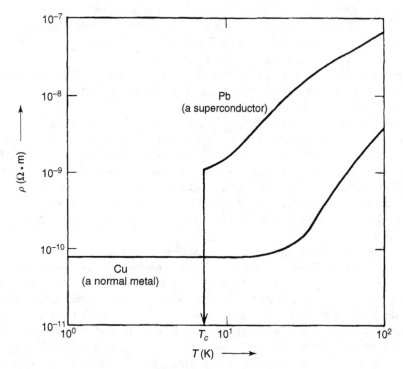

Figure 6.6 Variation of resistance at low temperatures for a nonsuperconductor and a superconductor. From K. M. Ralls, T. H. Courtney, and J. Wulff, *Introduction to Materials Science and Engineering*. Copyright © 1976 by John Wiley & Sons, Inc. This material is used by permission John Wiley & Sons, Inc.

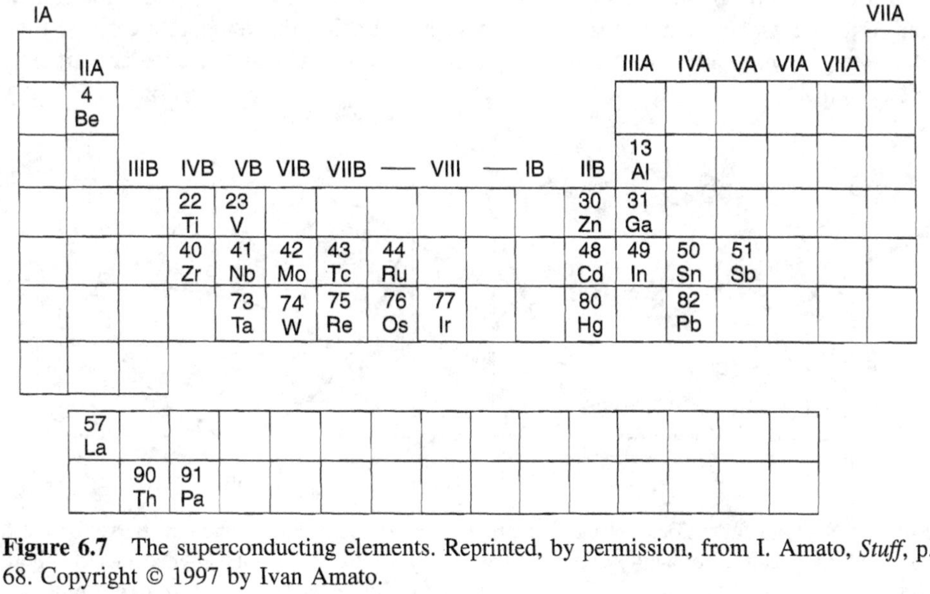

IA
VIIA
IIA
IIIA IVA VA VIA VIIA
4 Be
IIIB IVB VB VIB VIIB — VIII — IB IIB
13 Al
22 Ti
23 V
30 Zn
31 Ga
40 Zr
41 Nb
42 Mo
43 Tc
44 Ru
48 Cd
49 In
50 Sn
51 Sb
73 Ta
74 W
75 Re
76 Os
77 Ir
80 Hg
82 Pb
57 La
90 Th
91 Pa

Figure 6.7 The superconducting elements. Reprinted, by permission, from I. Amato, *Stuff*, p. 68. Copyright © 1997 by Ivan Amato.

superconductivity. The essence of the theory is that superconductivity is caused by electron–lattice interaction and that the superconducting electrons consist of paired ordinary electrons called a *Cooper pair*. We give here a qualitative description of how Cooper pairs form and how they lead to superconductivity.

Figure 6.8a shows the energy–momentum curve of an ordinary conductor with seven electrons in their discrete energy levels. In the absence of an electric field, the current from electrons moving to the right is exactly balanced by the electrons moving to the left, and the net current is zero. When an electric field is applied, all the electrons acquire some extra momentum, and this is equivalent to shifting the whole distribution in the direction of the electric field, as shown in Figure 6.8b. When the electric field is once again removed, the faster electrons will be scattered into lower energy states due to collisions with the vibrating lattice or impurity atoms, and the original distribution in Figure 6.8a is reestablished. For the model of Figure 6.8b, it means that the electron is scattered from energy level *a* to energy level *b* in order to reestablish the original distribution.

In the case of a superconductor, it becomes energetically more favorable for the electrons to pair up. Those of opposite momenta pair up to form a new particle called a superconducting electron or Cooper pair. This link between two electrons is shown in Figure 6.9a by an imaginary spring. The velocity of the entire center of mass is zero. According to the *de Broglie relationship* ($\lambda = h/p$, where λ is the wavelength, h is Planck's constant, and p is the momentum), this means that the wavelength associated with the new particle, λ, is infinitely long. This is valid for all the Cooper pairs. So, there is a large number of identical particles all with infinite wavelength—that is, a quantum phenomenon on a macroscopic scale. An applied electric field will displace all the particles again as shown in Figure 6.9b, but this time, when the electric field disappears, there is no return to the previous state. Scattering from energy level *a* to energy level *b* is no longer possible because then the electrons at both *b* and *c* would become pairless, which is energetically unfavorable. So, the asymmetrical distribution remains and there will be more electrons going to the right than to the left, and the current persists, in theory, forever. As the temperature rises, thermal energy eventually overcomes the pairing interaction, and the superconductivity effect vanishes.

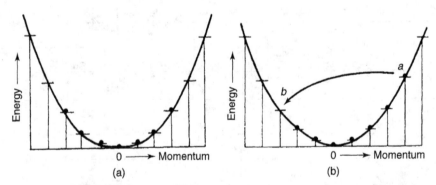

Figure 6.8 One dimensional schematic representation of the energy momentum curve for seven electrons in a conductor with (a) no applied electric field and (b) an applied electric field. Reprinted, by permission, from L. Solymar, and D. Walsh, *Lectures on the Electrical Properties of Materials*, 5th ed., p. 427. Copyright © 1993 by Oxford University Press.

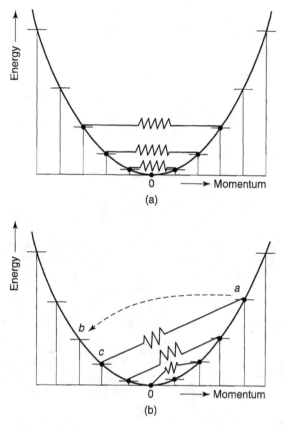

Figure 6.9   The energy–momentum curve for seven electrons in a superconductor with (a) no applied electric field and (b) an applied electric field. Reprinted, by permission, from L. Solymar and D. Walsh, *Lectures on the Electrical Properties of Materials*, 5th ed., p. 428. Copyright © 1993 by Oxford University Press.

One technologically significant application for superconductivity is in the use of electromagnets. It would be beneficial to have high magnetic fields without any power dissipation. Early experimenters, however, found that above certain magnetic fields, the superconductor became normal, regardless of the temperature. Thus, in order to have zero resistance, a superconductor must be held not only below the critical transition temperature, T_c, but also below a *critical magnetic field*, H_c. The interrelation between H_c and T_c is well described by the formula

$$H_c = H_0 \left[1 - \left(\frac{T}{T_c} \right)^2 \right] \tag{6.19}$$

where H_0 is the critical magnetic field at absolute zero. This relationship is graphically illustrated in Figure 6.10, where the material is a "normal" conductor above the curve and a superconductor below the curve. The values of H_0 and T_c for a number of elemental superconductors are given in Table 6.1. We will have more to say about superconductors in Section 6.1.2.4.

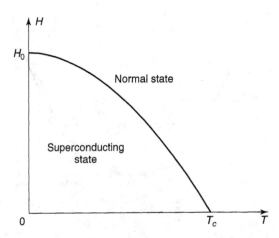

Figure 6.10 Temperature dependence of critical magnetic field for a superconductor. Reprinted, by permission, from L. Solymar and D. Walsh, *Lectures on the Electrical Properties of Materials*, 5th ed., p. 429. Copyright © 1993 by Oxford University Press.

Table 6.1   Critical Temperature and Critical Magnetic Field of a Number of Supercon-ducting Elements

Element	$T_c(K)$	$H_0 \times 10^{-4}(A/m)$	Element	$T_c(K)$	$H_0 \times 10^{-4}(A/m)$
Al	1.19	0.8	Pb	7.18	6.5
Ga	1.09	0.4	Sn	3.72	2.5
Hg α	4.15	3.3	Ta	4.48	6.7
Hg β	3.95	2.7	Th	1.37	1.3
In	3.41	2.3	V	5.30	10.5
Nb	9.46	15.6	Zn	0.92	0.4

Source: L. Solymar, and D. Walsh, *Lectures on the Electrical Properties of Materials*, 5th ed. Copyright © 1993 by Oxford University Press.

6.1.1.4   *Intrinsic Semiconduction.* We return now to our description of band gaps from Section 6.1.1.1 in order to elaborate upon those materials with conductivities in the range 10^{-6} to $10^4 (\Omega \cdot m)^{-1}$. At first, it may not seem important to study materials that neither readily conduct electricity nor protect against it, as do insulators. How-ever, the development of semiconducting materials has, and continues to, revolutionize our lives.

Recall from Figure 6.2 that the gap between the valence and conduction bands called the band gap, E_g, can be used to classify materials as conductors, insulators, or semiconductors; also recall that for semiconductors the value of E_g is typically on the order of 1–2 eV. The magnitude of the band gap is characteristic of the lattice alone and varies widely for different crystals. In semiconductors, the valence and conduction bands do not overlap as in metals, but there are enough electrons in the valence band that can be "promoted" to the conduction band at a certain temperature to allow for limited electrical conduction. For example, in silicon, the energies of the valence electrons that bind the crystal together lie in the valence band. All four

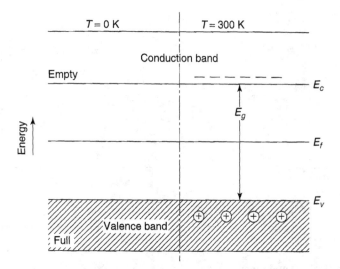

Figure 6.11   Energy bands of an intrinsic semiconductor: E_f is the Fermi energy level; E_c is the lower edge of the conduction band; E_v is the upper edge of the valence band; and E_g is the band gap. From Z. Jastrzebski, *The Nature and Properties of Engineering Materials*, 2nd ed. Copyright © 1976 by John Wiley & Sons, Inc. This material is used by permission of John Wiley & Sons, Inc.

valence electrons for each atom are tied in place, forming covalent bonds. There are no electrons in the conduction band at absolute zero, and the material is effectively an insulator (see Figure 6.11). At an elevated temperature, enough electrons have been thermally promoted to the conduction band to permit limited conduction. Promotion of electrons leaves behind positively charged "holes" in the valence band which maintain charge neutrality. The term *hole* denotes a mobile vacancy in the electronic valence structure of a semiconductor that is produced by removing one electron from a valence band. Holes, in almost all respects, can be regarded as moving positive charges through the crystal with a charge magnitude and mass the same as that of an electron.

Conduction that arises from thermally or optically excited electrons is called *intrinsic semiconduction*. The conduction of intrinsic semiconductors usually takes place at elevated temperatures, since sufficient thermal agitation is necessary to transfer a reasonable number of electrons from the valence band to the conduction band. The elements that are capable of intrinsic semiconduction are relatively limited and are shown in Figure 6.12. The most important of these are silicon and germanium.

In intrinsic semiconductors, the number of holes equals the number of mobile electrons. The resulting electrical conductivity is the sum of the conductivities of the valence band and conduction band charge carriers, which are holes and electrons, respectively. In this case, the conductivity can be expressed by modifying Eq. (6.9) to account for both charge carriers:

$$\sigma = n_e q_e \mu_e + n_h q_h \mu_h \tag{6.20}$$

where n_h, q_h, and μ_h are now the number, charge, and mobility of the holes, respectively, and contribute to the overall conductivity in addition to the corresponding

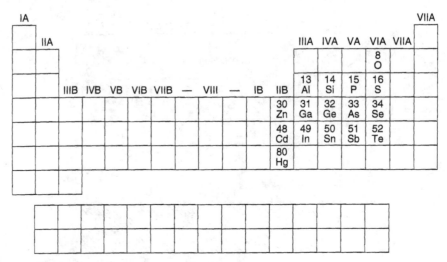

Figure 6.12   The intrinsic semiconducting elements. Reprinted, by permission, from I. Amato, *Stuff*, p. 68. Copyright © 1997 by Ivan Amato.

quantities for the electrons, indicated by the subscript e. The charge of the holes has the same numerical value as that of the electron, and since one hole is created for every electron that is promoted to the conduction band, $n_e = n_h$, such that Eq. (6.20) simplifies to

$$\sigma = n_e q_e (\mu_e + \mu_h) \tag{6.21}$$

The mobilities of holes are always less than those of electrons; that is $\mu_h < \mu_e$. In silicon and germanium, the ratio μ_e / μ_h is approximately three and two, respectively (see Table 6.2). Since the mobilities change only slightly as compared to the change of the charge carrier densities with temperature, the temperature variation of conductivity for an intrinsic semiconductor is similar to that of charge carrier density.

It can be shown from quantum mechanics that the effective density of states for the conduction and valence bands, N_c and N_v, respectively, are

$$N_c = 2 \left(\frac{2\pi m_e^* k_B T}{h^2} \right)^{3/2} \tag{6.22}$$

Table 6.2   Some Room-Temperature Electronic Properties of Group IV Elements

Element	E_g (eV)	σ (ohm$^{-1} \cdot$ m^{-1})	μ_e (m^2/V $\cdot$ s)	μ_h (m^2/V $\cdot$ s)
C (diamond)	~7	10^{-14}	0.18	0.14
Silicon	1.21	4.3×10^{-4}	0.14	0.05
Germanium	0.785	2.2	0.39	0.19
Gray tin	0.09	3×10^5	0.25	0.24
White tin	0	10^7	—	—
Lead	0	5×10^6	—	—

Source: K. M. Ralls, T. H. Courtney, and J. Wulff, *Introduction to Materials Science and Engineering*. Copyright © 1976 by John Wiley & Sons, Inc.

and

$$N_v = 2\left(\frac{2\pi m_h^* k_B T}{h^2}\right)^{3/2} \tag{6.23}$$

where m_e^* and m_h^* are the effective mass of electrons and holes, respectively, k_B is Boltzmann's constant, T is absolute temperature, and h is Planck's constant. If we multiply these densities of state by the distribution of Eq. (6.11), in which we replace the energy, E, with the respective energy of the conduction and valence bands (E_v is lower in energy than the Fermi energy, E_f, and E_c is higher), we can obtain the number of mobile charge carriers in each band:

$$n_e = N_c \exp\left[\frac{-(E_c - E_f)}{k_B T}\right] \tag{6.24}$$

and

$$n_h = N_v \exp\left[\frac{-(E_f - E_v)}{k_B T}\right] \tag{6.25}$$

The product of the mobile positive and negative charge carriers is then obtained by multiplication of Eqs. (6.24) and (6.25), and recalling that $n_e = n_h$ we obtain

$$n_h \cdot n_e = n_e^2 = N_c N_v \exp\left[\frac{-(E_c - E_v)}{k_B T}\right] \tag{6.26}$$

and since $E_c - E_v = E_g$, the band gap energy is

$$n_e = (N_v N_c)^{1/2} \exp\left[\frac{-E_g}{2k_B T}\right] \tag{6.27}$$

Equation (6.27) illustrates that the product of charge carriers, $n_e n_h$, is independent of the position of the Fermi level and depends on the energy gap. It is a further result of this relationship that

$$E_f = \frac{E_c + E_v}{2} \tag{6.28}$$

Thus, the Fermi level is halfway between the conduction and valence bands.

The term $(N_v N_c)^{1/2}$ in Eq. (6.27) depends on the band structure of the semiconductor and is usually a constant for a specific material, outside of its temperature dependence. Thus, we can remove the temperature dependence of N_v and N_c from Eqs. (6.22) and (6.23) and simplify Eq. (6.27) to

$$n_e = CT^{3/2} \exp\left[\frac{-E_g}{2k_B T}\right] \tag{6.29}$$

where C is a constant. Substitution of Eq. (6.29) into Eq. (6.21) gives

$$\sigma = CT^{3/2} q_e (\mu_e + \mu_h) \exp\left[\frac{-E_g}{2k_B T}\right] \tag{6.30}$$

Neglecting the variation of the $T^{3/2}$ term, which is negligible compared to the variation with temperature in the exponential term, and recalling that the mobilities are less sensitive to temperature than are the charge carrier densities, Eq. (6.30) can be rewritten as

$$\sigma = \sigma_0 \exp\left(\frac{-E_g}{2k_B T}\right) \tag{6.31}$$

where σ_0 is the overall constant.

Equation (6.31) indicates that the conductivity of intrinsic semiconductors drops nearly exponentially with increasing temperature. At still higher temperatures, the concentration of thermally excited electrons in the conduction band may become so high that the semiconductor behaves more like a metal.

6.1.1.5 Extrinsic Semiconduction. The charge carrier density can also be increased through impurities of either higher or lower valence. For example, if pentavalent substitutional atoms such as P, As, or Sb are placed into a covalently bonded tetravalent material such as Si or Ge, in a process known as *doping*, only four of their five valence electrons are required to participate in covalent bonding. Since the fifth electron remains weakly bound to the impurity or *donor* atom, it is not entirely free in the crystal. Nevertheless, the binding energy, which is of the order 0.01 eV, is much less than that of a covalently bonded electron. This "extra" electron can be easily detached from the impurity or donor atom. The energy state of this electron is indicated by E_d, since it is the donor level, and the energy required to excite it to the conduction band can be represented by the band model shown in Figure 6.13. Notice that the energy gap between the donor and conducting bands is much smaller than that between the valence and conduction bands, or the normal band gap, E_g. Thermal agitation, even at room temperature, is sufficient to transfer this electron to the conduction band, leaving behind a positively charged hole in the donor

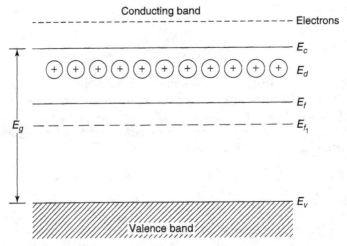

Figure 6.13   Schematic illustration of energy bands in an *n*-type, extrinsic semiconductor. From Z. Jastrzebski, *The Nature and Properties of Engineering Materials*, 2nd ed. Copyright © 1976 by John Wiley & Sons, Inc. This material is used by permission of John Wiley & Sons, Inc.

band. The conductivity in this case is due to motion of electrons in the conduction band, and this type of material is called an *n-type semiconductor*. This is one type of *extrinsic semiconductor*, or *impurity semiconductor*, in which an insulating or intrinsic semiconducting material is made to increase its semiconduction capabilities through the addition of impurity atoms. Notice also, in Figure 6.13, that the Fermi level of the *n*-type extrinsic semiconductor, E_f, is higher than that of the undoped, intrinsic semiconductor, E_{f1}.

From a conductivity standpoint, the other possibility in the addition of impurity atoms is to substitute with an impurity of lower valence, such as a trivalent element in a tetravalent lattice. Examples would be the addition of boron, aluminum, gallium, indium, and thallium to silicon or germanium. In this case, a *p-type*, extrinsic semiconductor is created. Since the substitutional atoms are deficient in bonding electrons, one of their bonding orbitals will contain a hole that is capable of "accepting" an electron from elsewhere in the crystal. As shown in Figure 6.14, the binding energy is small, and the promotion of an electron from the valence band to the acceptor band, E_a, leaves a hole in the valence band that can act as a charge carrier. In summary, then, for an *n*-type semiconductor, electrons in the conduction band are charge carriers and holes participate in bonding, whereas for a *p*-type semiconductor, the electrons participate in bonding, and the holes are the charge carriers.

In both types of extrinsic semiconductors, the doping elements are chosen so that both acceptor and donor levels are located closer to the corresponding energy bands and only a small energy gap is involved when exciting electrons. Intrinsic semiconductors are prepared routinely with initial total impurity contents less than one part per million (ppm), along with unwanted donor or acceptor atoms at less than one part in 10^7. To such highly purified materials, controlled amounts of substitutional donor or acceptor impurities are intentionally added on the order of 1–1000 ppm to produce extrinsic semiconductors having specific room temperature conductivities.

Unlike intrinsic semiconductors, in which the conductivity is dominated by the exponential temperature and band-gap expression of Eq. (6.31), the conductivity of extrinsic semiconductors is governed by competing forces: charge carrier density and charge carrier mobility. At low temperatures, the number of charge carriers initially

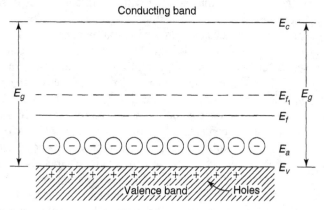

Figure 6.14   Schematic illustration of energy bands in a *p*-type, extrinsic semiconductor. From Z. Jastrzebski; *The Nature and Properties of Engineering Materials*, 2nd ed. Copyright © 1976 by John Wiley & Sons, Inc. This material is used by permission of John Wiley & Sons, Inc.

Cooperative Learning Exercise 6.2

The following electrical conductivity characteristics for both the intrinsic and extrinsic forms of a semiconductor have been determined at room temperature. Recall that $q_e = q_h = 1.602 \times 10^{-19}$ coulombs.

	$\sigma(\Omega \cdot m)^{-1}$	$n_e(m^{-3})$	$n_h(m^{-3})$
Intrinsic	2.5×10^{-6}	3.0×10^{13}	3.0×10^{13}
Extrinsic	3.6×10^{-5}	4.5×10^{14}	2.0×10^{12}

Person 1: Use the data to write a relationship for the intrinsic conductivity of this substance as a function of the electron and hole mobilities.

Person 2: Use the data to write a relationship for the extrinsic conductivity of this substance as a function of the electron and hole mobilities.

Assuming that the mobilities are the same in both the intrinsic and extrinsic states, combine your information to solve for the electron and hole mobilities of this substance. Can you tell whether the substance is an *n*-type of *p*-type extrinsic semiconductor by looking at the data?

Adapted from Callister, p. 654, problem 19.33.

Answers: Equation (6.20) can be used for both: $0.52 = \mu_e + \mu_h$; $112.4 = 225\mu_e + \mu_h$. Solve simultaneously to get $\mu_e = 0.50$ m^2/V-s; $\mu_h = 0.02$ m^2/V-s. This is an *n*-type semiconductor, since $n_e \gg n_h$.

increases with temperature, as shown in Figure 6.15 for an *n*-type semiconductor, because thermal activation promotes donor electrons to the conduction band. At intermediate temperatures, most of the donor electrons have been promoted and the number of charge carriers is nearly independent of temperature in what is known as the *exhaustion range*. At higher temperatures, the number of valence (bonding) electrons excited to the conduction band greatly exceeds the total number of electrons from substitutional donor atoms, and the extrinsic semiconductor behaves like an intrinsic semiconductor (dashed line in Figure 6.15).

On the basis of charge carrier density (conduction electrons) alone, then, the conductivity of an extrinsic semiconductor should vary with temperature as shown in Figure 6.16. However, charge carrier mobility has a temperature-dependent effect. As mentioned in the previous section, the temperature dependence of carrier mobility has little effect on intrinsic conductivity. However, when extrinsic conductivity predominates and donor levels are close the conduction band, the carrier mobility can determine the temperature dependence of conductivity. This occurs in the exhaustion range where mobility typically decreases with increasing temperature because of increased phonon scattering. Thus, in this region, characterized by an approximately constant number of charge carriers and a decreasing carrier mobility with temperature, the temperature dependence of conductivity of an extrinsic semiconductor is similar to that of a pure metal (cf. resistivity plot for pure copper in Figure 6.4). In summary, at low temperature the conductivity varies with temperature as the charge carrier concentration, at higher temperatures the charge carrier mobility dominates and the conductivity decreases with

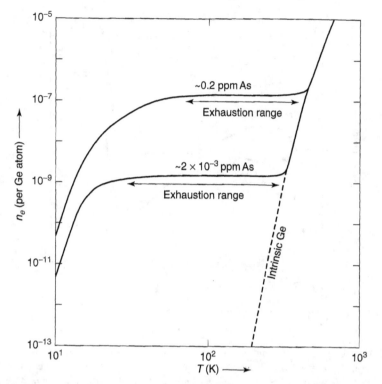

Figure 6.15 Temperature dependence of charge carrier concentration of *n*-type extrinsic germanium with two different As impurity levels. From K. M. Ralls, T. H. Courtney, and J. Wulff, *Introduction to Materials Science and Engineering.* Copyright © 1976 by John Wiley & Sons, Inc. This material is used by permission John Wiley & Sons, Inc.

temperature, and at still higher temperatures the extrinsic semiconductor behaves as an intrinsic semiconductor. This behavior is summarized in Figure 6.16.

6.1.1.6 Semiconductor Junctions and Devices*.

Semiconductor crystals can be made in such a way that an *n*-type region can be made adjacent to a *p*-type region. This is normally accomplished by selective doping during the growth of a single crystal. The boundary between such regions within a single crystal is called the *pn junction*, and it has important applications to the development of transistors, rectifiers, and solar energy cells. Let us examine some of these devices.

The *rectifier*, or *diode*, is an electronic device that allows current to flow in only one direction. There is low resistance to current flow in one direction, called the *forward bias*, and a high resistance to current flow in the opposite direction, known as the *reverse bias*. The operation of a *pn* rectifying junction is shown in Figure 6.17. If initially there is no electric field across the junction, no net current flows across the junction under thermal equilibrium conditions (Figure 6.17a). Holes are the dominant carriers on the *p*-side, and electrons predominate on the *n*-side. This is a dynamic equilibrium: Holes and conduction electrons are being formed due to thermal agitation. When a hole and an electron meet at the interface, they recombine with the simultaneous emission of radiation photons. This causes a small flow of holes from the *p*-region

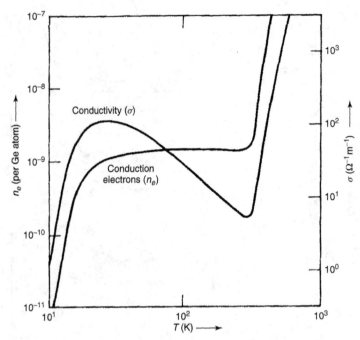

Figure 6.16   Temperature dependence of charge carrier density and conductivity of extrinsic semiconductor Ge doped with 2 ppb As. From K. M. Ralls, T. H. Courtney, and J. Wulff, *Introduction to Materials Science and Engineering.* Copyright © 1976 by John Wiley & Sons, Inc. This material is used by permission John Wiley & Sons, Inc.

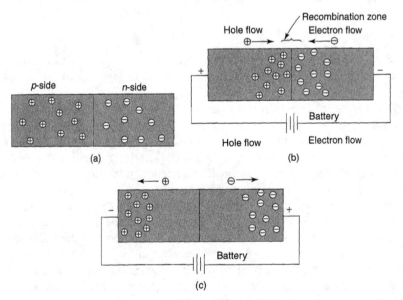

Figure 6.17   Schematic illustration of electron and hole distribution for a *pn*-rectifying junction for (a) no applied electrical potential, (b) forward bias, and (c) reverse bias. Reprinted, by permission, from W. Callister, *Materials Science and Engineering: An Introduction*, 5th ed., p. 631. Copyright © 2000 by John Wiley & Sons, Inc.

to the *n*-region and a flow of electrons from the *n*-region to the *p*-region. New holes and electrons continue to form to balance this flow, and an equilibrium is established with no external current flowing through the system.

If the *n*-region is connected to the negative terminal and the *p*-region to the positive terminal of a battery (Figure 6.17b), holes will move from left to right while electrons will move from right to left. These meet each other and recombine as before in the *recombination zone*. This produces an abundant number of current carriers flowing across the junction; hence the resistance in the forward direction is low. Only a small voltage is necessary to maintain a large current flow.

For reverse bias (Figure 6.17c), holes are attracted to the negative terminal, and electrons are attracted to the positive terminal. Thus, electrons and holes move away from each other and the junction region will be relatively free from charge carriers, in which case it is called the *barrier region*. Recombination will not occur to any appreciable extent, so that the junction is now highly insulative. There may be, however, a small flow of current as the result of the thermal generation of holes and electrons in the semiconductor, termed *leakage current*. The current–voltage behavior of the *pn* junction under both forward and reverse biases is shown in Figure 6.18. If the voltage is allowed to vary sinusoidally (Figure 6.19a), as in an alternating current, the current also varies, but the maximum current for the reverse bias, I_R, is extremely small compared to that for the forward bias, I_F.

The *pn* junction can be fabricated so that the leakage current in the reverse bias is extremely small, and potentials of up to several hundred volts can be applied before *breakdown* will occur. Breakdown occurs at high reverse voltages (over several hundred volts), when large numbers of charge carriers are generated, giving rise to an abrupt increase in current. Atoms may actually be ionized during breakdown due to the acceleration of electrons and holes in a phenomenon known as the *avalanche effect*. Some devices, called *Zener diodes*, are designed to utilize breakdown behavior

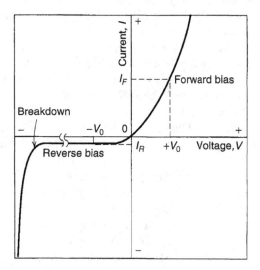

Figure 6.18 The current–voltage characteristics of a *pn*-junction. for forward and reverse biases. Reprinted, by permission, from W. Callister, *Materials Science and Engineering: An Introduction*, 5th ed., p. 632. Copyright © 2000 by John Wiley & Sons, Inc.

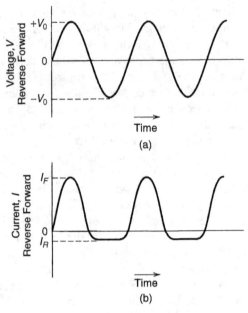

Figure 6.19   (a) Voltage–time and (b) current–time for a *pn*-rectifying junction. Reprinted, by permission, from W. Callister, *Materials Science and Engineering: An Introduction*, 5th ed., p. 632. Copyright © 2000 by John Wiley & Sons, Inc.

as voltage regulators. The voltage breakdown can be varied by controlling the dopant level; generally, the higher the impurity concentration, the lower the breakdown voltage. In silicon semiconductors, for example, the breakdown voltage can be varied from about 10 V to about 1000 V by regulating the impurity concentration.

Another type of semiconductor junction device is a transistor, which is capable of two primary types of functions. The first is to amplify an electrical signal. The second is to serve as a switching device for the processing and storage of information. We will briefly describe the *junction transistor* here, and we will defer the description of another type of transistor called a *MOSFET* to Section 6.1.2.6.

The junction transistor is composed of two *pn* junctions arranged back to back in either the *p-n-p* or *n-p-n* configurations. The latter configuration is shown in Figure 6.20. It is composed of a thin *n*-type *base* region, sandwiched between a *p*-type *emitter* and a *p*-type *collector*. The first *pn* junction is forward-biased, and the second *np* junction is reverse-biased. Since the emitter is *p*-type and Junction 1 is forward biased, a large number of holes enter the *n*-type base region. These holes combine with electrons in the base. However, in a properly prepared junction transistor, the base is thin enough to allow most of the holes to pass through the base without recombination and into the *p*-type collector. A small increase in input voltage within the emitter-base circuit produces a large increase in current across Junction 2 due to the passage of holes. The large increase in collector current also results in a large increase in voltage across the load resistor. Thus, a voltage signal that passes through a junction transistor experiences *amplification*, as illustrated by the input and output voltage plots in Figure 6.20. A similar effect is observed in an *n-p-n* junction transistor, except that electrons instead of holes are injected across the base to the collector.

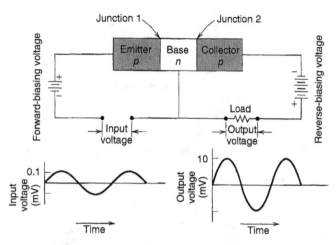

Figure 6.20 Schematic diagram of a *pnp*-junction transistor and associated input and output voltage characteristics. Reprinted, by permission, from W. Callister, *Materials Science and Engineering: An Introduction*, 5th ed., p. 633. Copyright © 2000 by John Wiley & Sons, Inc.

In addition to their ability to amplify electrical signals, transistors and diodes may act as switching and information storage devices. Computer information is stored in terms of binary code—for example, zeros and ones. Transistors and diodes can operate as switches that have two states: on and off, or conducting and nonconducting. Thus, a conducting circuit can represent a zero while a nonconducting circuit can represent a one. In this way, information and logical operations can be stored and retrieved from a series of semiconducting circuits known as *integrated circuits*. Inexpensive integrated circuits are mass produced by forming layers of high-purity silicon in a precisely detailed pattern and introducing specific dopants into specific areas by diffusion or ion implantation. The details of some of these processes will be described in Chapter 7.

6.1.2 Electrical Properties of Ceramics and Glasses

Many of the fundamental relationships and concepts governing the electrical properties of materials have been introduced in the previous section. In this section, we elaborate upon those topics that are more prevalent or technologically relevant in ceramics and glasses than in metals, such as electrical insulation and superconductivity, and introduce some topics that were omitted in Section 6.1.1, such as dielectric properties.

6.1.2.1 Electrical Resistance.
At room temperature, many ceramics (and polymers, for that matter) have electrical resistivities that are approximately 20 orders of magnitude higher than those of metals (see Figure 6.21). As a result, they are often used as electrical insulators. With rising temperature, however, insulating materials experience an increase in electrical conductivity, which may ultimately be greater than that for semiconductors. In fact, some oxide ceramics such as ReO_3 can achieve metallic levels of electrical conductivity at high temperatures. In all cases, the electrical conductivity can be interpreted in terms of carrier concentrations and carrier mobilities as previously described, though we shall see in Section 6.1.2.3 that the carriers may be different for ceramics and glasses than for metals.

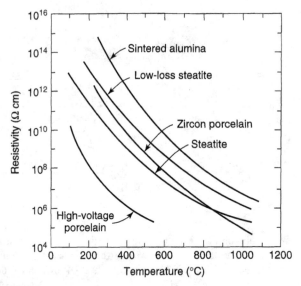

Figure 6.21   Temperature dependence of resistivity for some typical oxide ceramics. From W. D. Kingery, H. K. Bowen, and D. R. Uhlmann, *Introduction to Ceramics.* Copyright © 1976 by John Wiley & Sons, Inc. This material is used by permission of John Wiley & Sons, Inc.

Despite the potential for high electrical conductivities, we will concentrate on the more common high-resistivity aspects of ceramics in this section and will describe some of the factors that affect the use of ceramics and glasses as electrical insulators. For ceramics and glasses having electrical resistivities above about $10^5 \, \Omega \cdot cm$, the concentration–mobility product of the charge carriers is small; as a result, minor variations in composition, impurities, heat treatment, stoichiometry, porosity, oxygen partial pressure, and other variables can have a significant effect on the resistivity. The effect of porosity is similar to that described for thermal conductivity in Section 4.2. As porosity increases, the electrical resistivity increases proportionally for small values of porosity. At higher porosities, the effect is more substantial. The effect of grain boundaries in polycrystalline materials is related to the mean free path of the ions, as described in Section 6.1.1.1. Except for very thin films or extremely fine-grained materials (less than 0.1 μm), the effects of grain boundary scattering are small compared to lattice scattering. Consequently, the grain size in uniform-composition materials has little effect on resistivity. However, substantial effects can result from impurity concentrations at grain boundaries. Particularly for oxide materials, there is a tendency to form a glassy silicate phase at the boundary between particles. Ceramic compositions with the highest glassy phase have the lower resistivities at moderate temperatures. In glasses, the alkali content should be kept to a minimum when high resistivities are required. The reasons for this are related to the electrical conduction via ions, which will be described in Section 6.1.2.3. For now, we maintain our emphasis on the electrical insulating properties of ceramics and glasses in order to describe a unique application known as dielectrics.

6.1.2.2 Dielectrics, Ferroelectricity, and Piezoelectricity.   A *dielectric* is a material separating two charged bodies, as illustrated in Figure 6.22b. For a substance

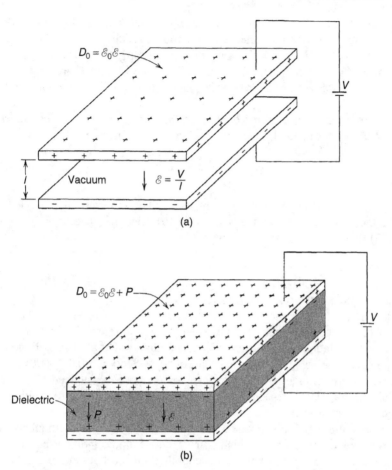

Figure 6.22   A parallel-plate capacitor (a) in vacuum and (b) with a dielectric material between the plates. Reprinted, by permission, from W. Callister, *Materials Science and Engineering: An Introduction*, 5th ed., p. 640. Copyright © 2000 by John Wiley & Sons, Inc.

to be a good dielectric, it must be an electrical insulator. As a result, any electrical insulator is also called a dielectric.

The interaction of a dielectric with an electric field is conveniently described in terms of a parallel plate capacitor. When a voltage is applied between two parallel plates in a vacuum (Figure 6.22a), one plate becomes positively charged and the other becomes negatively charged. If an electrical conductor were to be placed in contact with the plates, current would flow. In a vacuum, however, no current flows, and the electric field strength is a function of the applied voltage, V, and the separation distance between the plates, l, as given earlier by Eq. (6.4):

$$\mathscr{E} = V/l \tag{6.4}$$

The magnitude of the charge per unit area on either plate, called the *electric displacement* or *flux density*, $\mathbf{D}_0$, is directly proportional to the field:

$$\mathbf{D}_0 = \varepsilon_0 \mathscr{E} \tag{6.32}$$

where ε_0 is a constant called *electric permittivity* and has a value of 8.854×10^{-12} C/V · m (or farad/m, F/m, where a farad equals a coulomb/volt). The electric displacement therefore has units of C/m^2. Notice the similarity in form of Eq. (6.32) to Eq. (6.2) for the magnetic flux in a magnetic field. As before, both $\mathbf{D}_0$ and $\mathscr{E}$ are vector quantities, but for our purposes a description of their magnitudes is sufficient, so the boldface will be dropped.

Since D_0 is the surface charge density on the plate in vacuum, it can be related to the capacitance of a parallel plate capacitor in vacuum, C_0, which is defined as

$$C_0 = q/V \qquad (6.33)$$

where q is the magnitude of charge on each plate and V is the applied voltage. With A the plate area and l the separation between plates, along with $q = D_0 A$, Eq. (6.33) becomes

$$C_0 = \varepsilon_0 \frac{A}{l} \qquad (6.34)$$

when a vacuum exists between the plates.

If we now place a dielectric (insulator) material between the plates, as in Figure 6.22b, a displacement of charge within the material is created through a progressive orientation of permanent or induced dipoles. Recall that a dipole is simply any type of charge separation, as first described in Section 1.0.3.2 and illustrated in Figure 6.23, and that it may be permanent or temporarily induced. The magnitude of the electric dipole, p_e, is simply proportional to the charge, q, and the separation distance, a, as $p_e = qa$.

The interaction between permanent or induced electric dipoles with an applied electric field is called *polarization*, which is the induced dipole moment per unit volume. Polarization causes positive charge to accumulate on the bottom surface next to the

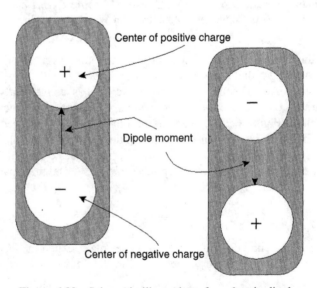

Figure 6.23   Schematic illustration of an electric dipole.

negatively charged plate and causes negative charge to accumulate toward the positively charged plate on the top. This tends to decrease the effective surface charge density on either plate. The expected decrease in effective charge corresponds to the polarization, P, of the material, which equals the induced dipole moment per unit volume of polarizable material. The electric field is now found from Eq. (6.32) by accounting for the reduction in displacement due to polarization:

$$\mathscr{E} = (D - P)/\varepsilon_0 \tag{6.35}$$

The displacement can now be found by rearranging Eq. (6.2) to

$$D = \varepsilon_0 \mathscr{E} + P \tag{6.36}$$

Equation (6.36) is often written

$$D = \varepsilon \mathscr{E} \tag{6.37}$$

where ε is called the *permittivity*.

The ratio of permittivity with the dielectric to the permittivity in vacuum, $\varepsilon/\varepsilon_0$, is called the *relative permittivity*, ε_r, or *dielectric constant*. The dielectric constant is a material property. Some values of dielectric constants for common ceramic and glass insulators are given in Table 6.3. Since a polarizable material causes an increase in charge per unit area on the plates of a capacitor, the capacitance also increases, and it can be shown that the dielectric constant is related to the capacitance and displacement in vacuum and with the dielectric material as follows:

$$\varepsilon_r = \frac{\varepsilon}{\varepsilon_0} = \frac{D}{D_0} = \frac{C}{C_0} \tag{6.38}$$

Cooperative Learning Exercise 6.3

A voltage of 10 V is applied across a parallel-plate capacitor with a plate separation of 2×10^{-3} m and plate area of 6.45×10^{-4} m^2, both in vacuum and then with a dielectric material placed between the plates. The polarization, P, due to the presence of the dielectric material is 2.22×10^{-7} C/m^2.

Person 1: Calculate the displacement in vacuum, D_0, and the capacitance in vacuum, C_0.

Person 2: Calculate the displacement, D, due to the presence of the dielectric.

Combine your information to calculate the dielectric constant and the capacitance in the presence of the dielectric.

Answers: Eq. (6.32) $D_0 = 8.854 \times 10^{-12}(10/0.002) = 4.427 \times 10^{-8}$ C/m^2.
Eq. (6.34) $C_0 = 8.854 \times 10^{-12}(6.45 \times 10^{-4}/0.002) = 2.85 \times 10^{-12}$ F.
Eq. (6.36) $D = 8.854 \times 10^{-12}(10/0.002) + 2.22 \times 10^{-7} = 2.66 \times 10^{-7}$ C/m^2.
Eq. $\varepsilon_r = D/D_0 = 6.0$; $C = \varepsilon_r C_0 = 1.71 \times 10^{-11}$ F.

Table 6.3   Dielectric Constants of Selected Ceramics and Glasses at 25°C and 10° Hz

Glasses	ε_r	Inorganic Crystalline Materials	ε_r
Silica glass	3.8	Barium oxide	3.4
Vycor glasses	3.8–3.9	Mica	3.6
Pyrex glasses	4.0–6.0	Potassium chloride	4.75
Soda–lime–silica glass	6.9	Potassium bromide	4.9
High-lead glass	19.0	Cordierite ceramics (based on $2MgO \cdot 2Al_2O_3 \cdot 3SiO_2$)	4.5–5.4
		Diamond	5.5
		Potassium iodide	5.6
		Forsterite (Mg_2SiO_4)	6.22
		Mullite ($3Al_2O_3 \cdot 2SiO_2$)	6.6
		Lithium fluoride	9.0
		Magnesium oxide	9.65

Source: K. M. Ralls, T. H. Courtney, and J. Wulff, *Introduction to Materials Science and Engineering.* Copyright © 1976 by John Wiley & Sons, Inc.

We can further describe the polarization, P, according to the different types of dipoles that either already exist or are induced in the dielectric material. The polarization of a dielectric material may be caused by four major types of polarization: electronic polarization, ionic (atomic) polarization, orientation polarization, and space-charge (interfacial) polarization. Each type of polarization is shown schematically in Figure 6.24 and will be described in succession. In these descriptions, it will be useful to introduce a new term called the *polarizability*, α, which is simply a measure of the ability of a material to undergo the specific type of polarization.

Electronic polarization arises because the center of the electron cloud around a nucleus is displaced under an applied electric field, as shown in Figure 6.24. The resulting polarization, or dipole per unit volume, is

$$P_e = N\alpha_e \mathscr{E}_{\text{loc}} \tag{6.39}$$

where N is the number of atoms or molecules, α_e is the electronic polarizability, and $\mathscr{E}_{\text{loc}}$ is the local electric field that an atom or molecule experiences. The local electric field is greater than the average field, $\mathscr{E}$, because of the polarization of other surrounding atoms.

Ionic polarization occurs in ionic materials because an applied field acts to displace cations in the direction of the applied field while displacing anions in a direction opposite to the applied field. This gives rise to a net dipole moment per formula unit. For an ionic solid, the atomic polarization is given by

$$P_i = N\alpha_i \mathscr{E}_{\text{loc}} \tag{6.40}$$

where N is the number of formula units per unit volume, and α_i is the ionic polarizability. Both ionic and electronic polarization contribute to the formation of induced dipoles.

No field   Field applied

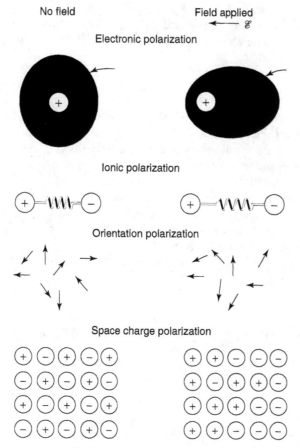

Figure 6.24 Schematic illustration of different types of polarization. From W. D. Kingery, H. K. Bowen and D. R. Uhlmann, *Introduction to Ceramics*. Copyright © 1976 by John Wiley & Sons, Inc. This material is used by permission of John Wiley & Sons, Inc.

Orientation polarization can occur in materials composed of molecules that have permanent electric dipole moments. The permanent dipoles tend to become aligned with the applied electric field, but entropy and thermal effects tend to counter this alignment. Thus, orientation polarization is highly temperature-dependent, unlike the forms of induced polarization which are nearly temperature-independent. In electric fields of moderate intensity, the orientation polarization is proportional to the local electric field, as for the other forms of polarization

$$P_o = N\alpha_o \mathscr{E}_{\text{loc}} \qquad (6.41)$$

but in strong electric fields this proportionality is not maintained since saturation must occur when all the permanent dipoles are aligned. Orientation polarization occurs mostly in gases and liquids where molecules are free to rotate. In some solids the rotation of polar molecules may be restricted by lattice forces, which can reduce orientation polarization. For example, the dielectric constant of water is about 82, while that of ice is about 10. The orientation polarizability, α_o, can be further elaborated upon, since the permanent electric dipole moment per molecule, p_e, is a measurable quantity. The two quantities are related by a relationship similar to the Curie Law, which we shall elaborate upon in Section 6.2:

$$\alpha_o = \frac{p_e{}^2}{3k_B T} \tag{6.42}$$

where k_B is Boltzmann's constant and T is absolute temperature. A list of permanent dipole moments for selected polar molecules is given in Table 6.4.

The final type of polarization is *space-charge polarization*, sometimes called *interfacial polarization*, and results from the accumulation of charge at structural interfaces in heterogeneous materials. Such polarization occurs when one of the phases has a much higher resistivity than the other, and it is found in a variety of ceramic materials, especially at elevated temperatures. The space-charge polarization, P_{sc}, has a corresponding space-charge polarizability, α_{sc}. The two are related via a relationship of the form for the other types of polarization.

The total polarization, then, is the sum of all the contributions from the different types of polarization:

$$P = P_e + P_i + P_o + P_{sc} \tag{6.43}$$

Table 6.4 Permanent Dipole Moments of Some Polar Molecules in the Gaseous Phase

Molecule	$10^{30} \times$ Dipole Moment(C $\cdot$ m)
AgCl	19.1
LiCl	23.78
NaCl	30.0
KCl	34.26
CsCl	34.76
LiF	21.1
NaF	27.2
KF	28.7
RbF	27.5
CsF	26.3
HF	6.07
HCl	3.60
HBr	2.74
HI	1.47
HNO_3	7.24
H_2O	6.17
NH_3	4.90
CH_3Cl	6.24
CH_2Cl_2	5.34
$CHCl_3$	3.37
C_2H_5OH	5.64
$C_6H_5NO_2$	14.1

Source: K. M. Ralls, T. H. Courtney, and J. Wulff, *Introduction to Materials Science and Engineering.* Copyright © 1976 by John Wiley & Sons, Inc.

With appropriate substitution of the expressions for the various types of polarization, it is easy to show that the total polarizability, α, is also the sum of the various types of polarizability:

$$\alpha = \alpha_e + \alpha_i + \alpha_o + \alpha_{sc} \tag{6.44}$$

The total polarization, P, and hence the dielectric constant, ε_r, subjected to an alternating electric field depend upon the ease with which the permanent or induced dipoles can reverse their alignment with each reversal of the applied field. The time required for dipole reversal is called the *relaxation time*, and its reciprocal is called the *relaxation frequency*. When the time duration per cycle of the applied field is much less than the relaxation time of a particular polarization process, the dipoles cannot change their orientation rapidly enough to remain oriented in the applied field, and they "freeze" their reorientation process. This particular polarization process will not contribute to the total polarization. The order in which the polarization processes cease to contribute to the total polarization is related to their relative size scales. Obviously, electronic polarization is the smallest since it deals with electron density. Ionic polarization is next in size scale, since it deals with the separation of ions, followed by orientation polarization, which is on the order of molecular dimensions and can sometimes be large. Finally, space-charge polarization, if it exists at all, is on the scale of interfacial dimensions. The larger the size scale, the more difficult it is for that type of polarization to respond to an alternating electric field. Thus, we would expect space charge to "freeze" up first with increasing electric field, followed by orientation, ionic, and electronic polarization, respectively. This process is shown in a plot of polarizability (which recall is proportional to polarization) versus the logarithm of the alternating electric field frequency in Figure 6.25a.

As the frequency of the applied field approaches the relaxation frequency, the polarization response increasingly lags behind the applied field. The reorientation of each type of dipole is opposed by internal friction, which leads to heating in the sample and power loss. The *power loss* depends on the degree to which the polarization lags behind the electric field. This lag can be quantified by measurement of the displacement–field relationship in a sinusoidally varying field. It is found that if the alternating field is given by

$$\mathscr{E} = \mathscr{E}_0 \sin(\omega t) \tag{6.45}$$

then the polarizability, P, is given by

$$P = P_0 \sin(\omega t - \delta) \tag{6.46}$$

where δ is the phase angle that the polarization lags behind the field, and ω is the angular frequency of the field. It can be shown that the power loss is related to the phase angle and the applied field:

$$\text{Power loss} = 2\pi \mathscr{E}^2 \, \omega \, \varepsilon_0 \tan \delta \tag{6.47}$$

where $\tan \delta$ is called the *loss tangent*. For small phase angles, $\tan \delta \approx \delta$. As shown in Figure 6.25b, the power loss for a particular polarization type increases as that process becomes slower to respond to frequency changes, reaches a maximum as the electric field frequency reaches the relaxation frequency, and decreases after that polarization type has reached its relaxation frequency.

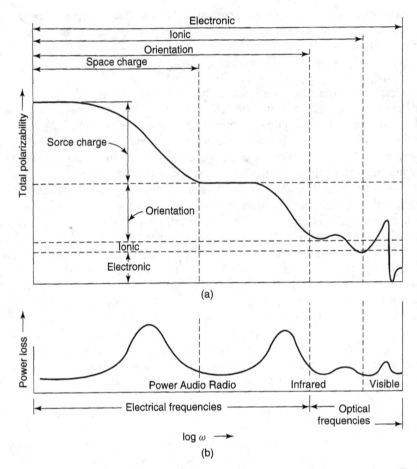

Figure 6.25 Frequency dependence of (a) total polarizability and (b) power loss. From K. M. Ralls, T. H. Courtney, and J. Wulff, *Introduction to Materials Science and Engineering*. Copyright © 1976 by John Wiley & Sons, Inc. This material is used by permission John Wiley & Sons, Inc.

The loss tangent is an important quantity. The product of the loss tangent and the dielectric constant, $\varepsilon_r \tan \delta$, is called the *loss factor* and is the primary criterion for judging the usefulness of a dielectric as an insulator material. For this purpose, it is desirable to have a low dielectric constant and a small loss angle. For applications in which it is desirable to obtain a high capacitance in the smallest physical space, a high dielectric constant must be used, and a low value of the loss angle is needed.

At high enough frequencies, the dielectric will experience electrical breakdown and the insulator will become a conductor. As in a semiconductor, breakdown is initiated by the field-induced excitation of a number of electrons to the conduction band with sufficient kinetic energy to knock other electrons into the conduction band. The result is the avalanche effect described earlier. The magnitude of the electric field required to cause dielectric breakdown is called the *dielectric strength*, or *breakdown strength*. The dielectric strength for oxide ceramic insulators lies in the range 1–20 kV/mm, with the value of 9 kV/mm for Al_2O_3 being typical.

The temperature dependence of the dielectric constant is highly dependent upon molecular structure, as well as the predominant polarization mechanism. For large molecules, orientation polarization is difficult below the material melting point, whereas orientation polarization can persist below the melting point in smaller molecules. Some substances undergo structural transformations in the solid state from higher- to lower-ordered structures which can restrict molecular rotation. The effects of temperature and frequency on the dielectric constant are not entirely independent. For example, the frequency effect on electronic and ionic polarization is negligible at frequencies up to about 10^{10} Hz, which is the limit of normal use. Similarly, the effect of temperature on electronic and ionic polarization is small. At higher temperatures, however, there is an increasing contribution resulting from ion mobility and crystal imperfection mobility. The combined effect is to give a sharp rise in the apparent dielectric constant at low frequencies with increasing temperature, as illustrated for alumina porcelain, whose primary component is Al_2O_3, in Figure 6.26. The loss tangent is also shown

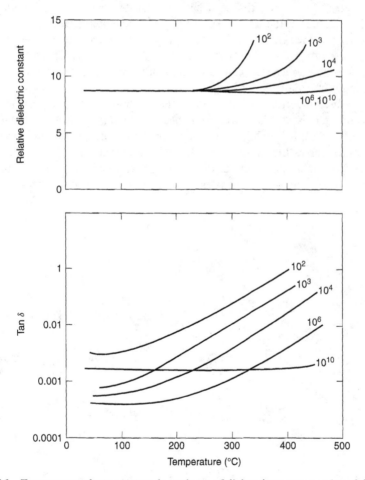

Figure 6.26   Frequency and temperature dependence of dielectric constant and tan δ for alumina porcelain. From W. D. Kingery, H. K. Bowen, and D. R. Uhlmann, *Introduction to Ceramics*. Copyright © 1976 by John Wiley & Sons, Inc. This material is used by permission of John Wiley & Sons, Inc.

in Figure 6.26 for reference. Alumina is an important component of most ceramic dielectrics. The other base components are silica, SiO_2, and magnesia, MgO.

Two phenomena related to the electric dipoles in a material are *ferroelectricity* and *piezoelectricity*. Ferroelectricity is defined as the spontaneous alignment of electric dipoles by their mutual interaction in the absence of an applied electric field. The source of ferroelectricity arises from the fact that the local field increases in proportion to the polarization. Thus, ferroelectric materials must possess permanent dipoles. One of the most important ferroelectric materials is barium titanate, $BaTiO_3$. The spontaneous polarization of $BaTiO_3$ is related directly to the positions of the ions within the tetragonal unit cell, as shown in Figure 6.27a. Divalent barium ions occupy each corner position, with the tetravalent titanium ion near, but just above, the center of the cell. The oxygen ions are located below the center of the (001) plane and just below the centers of the (100) and (010) planes, as illustrated in Figure 6.27b. Consequently, per unit cell, the centers of positive and negative charge do not coincide, and there is spontaneous polarization. At a certain temperature called the *ferroelectric Curie temperature* (120°C for barium titanate), barium titanate becomes cubic with Ti^{4+} ions located at the center of the unit cell and O^{2-} ions occupying face-centered positions, and it loses its spontaneous polarization. Ferroelectric materials have extremely high dielectric constants at relatively low applied field frequencies. Consequently, capacitors made from these materials can be significantly smaller than capacitors made from other dielectric materials.

Closely related to ferroelectricity is *piezoelectricity* in which polarization is induced and an electric field is established across a specimen by the application of external force (see Figure 6.28a,b). Reversing the direction of the external force, as from tension to compression, reverses the direction of the field. Alternatively, the application of an external electric field alters the net dipole length and causes a dimensional change, as in Figure 6.28c. Piezoelectric materials can be used as transducers—devices that

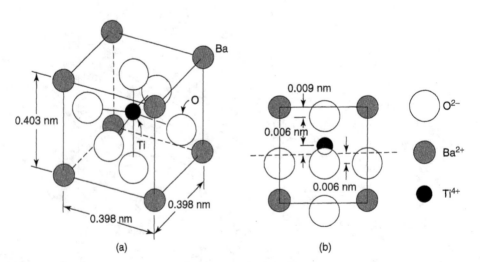

Figure 6.27   View of (a) the tetragonal unit cell of $BaTiO_3$ and (b) the ionic positions projected in a [100] direction onto a (100) face. From K. M. Ralls, T. H. Courtney, and J. Wulff, *Introduction to Materials Science and Engineering*. Copyright © 1976 by John Wiley & Sons, Inc. This material is used by permission John Wiley & Sons, Inc.

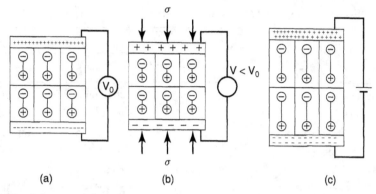

Figure 6.28 Schematic illustration of the piezoelectric effect that occurs in (a) an unstressed and (b) stressed piezoelectric material. Mechanical deformation can also occur when (c) a voltage is applied to a piezoelectric material. From K. M. Ralls, T. H. Courtney, and J. Wulff, *Introduction to Materials Science and Engineering.* Copyright © 1976 by John Wiley & Sons, Inc. This material is used by permission John Wiley & Sons, Inc.

convert mechanical stress into electrical energy and vice versa. Such devices are used in microphones and mechanical strain gauges. Barium titanate is also a piezoelectric material.

Before proceeding, it may be useful to summarize some of the myriad of terms and symbols that have been used to this point. Such a summary is presented in Table 6.5. With the exception of a few more terms dealing with superconductivity, we have introduced all the variables we will need to describe the electrical properties of materials.

6.1.2.3 Ion-Conducting Ceramics and Glasses.

In addition to the conduction of charge via electrons, charge can be conducted via ions. Ions are present in most crystalline ceramic materials such as oxides and halides. This process is termed *ionic conduction* and may occur either in conjunction with or separately from electronic conduction. As a result, we must expand our definition of conductivity to include both types of conduction:

$$\sigma = \sigma_{electronic} + \sigma_{ionic} \tag{6.48}$$

In a manner analogous to conduction via electrons and holes, the ionic conductivity can be related to the charge carrier density, n_i, charge carrier mobility, μ_i, and carrier charge, q_e, as originally given in Eq. (6.9) for electronic conduction, except that the charge carriers are now ions, which have multiple charges, or valences, z_i, so that Eq. (6.9) must be modified slightly:

$$\sigma_{ionic} = f n_i (z_i q_e) \mu_i \tag{6.49}$$

An additional factor, f, has been included to account for the number of equivalent sites to which an ion may transfer in a particular crystal structure; for example $f = 4$ for an ion vacancy in the cubic crystal structure.

For an ion to move through the lattice under an electric field driving force, it must have sufficient thermal energy to pass over an energy barrier, which in this case is

Table 6.5   Electrical Symbols, Terms, and SI Units

Quantity	Symbol	SI Units Primary	SI Units Derived
Electric charge	q	coulomb, C	(none)
Electric potential	V	$kg \cdot m^2/s^2 \cdot C$	volt, V
Electric current	I	C/s	ampere, amp, A
Electric field strength	ε	$kg \cdot m/s^2 \cdot C$	V/m
Resistance	R	$kg \cdot m^2/s \cdot C^2$	ohm, Ω
Resistivity	ρ	$\mathbf{kg \cdot m^3/s \cdot C^2}$	$\mathbf{\Omega \cdot m}$
Conductance	G	$s \cdot C^2/ kg \cdot m^2$	Ω^{-1}
Conductivity	σ	$\mathbf{s \cdot C^2/ kg \cdot m^3}$	$\mathbf{(\Omega \cdot m)^{-1}}$
Electrical capacitance	C	$s^2 \cdot C^2/ kg \cdot m^2$	farad, F
Permittivity	ε	$s^2 \cdot C^2/ kg \cdot m^3$	F/m
Dielectric Constant (Relative Permittivity)	ε_r	**(dimensionless)**	**(dimensionless)**
Dielectric displacement	D	C/m^2	$F \cdot V/m^2$
Electric polarization	P	C/m^2	$F \cdot V/m^2$
Polarizability	α	$\mathbf{s^2 \cdot C^2/ kg}$	$\mathbf{F \cdot m^2}$
Permanent Dipole Moment	p	$\mathbf{C \cdot m}$	**debeye, D**
Dielectric loss angle	δ	(none)	(none)
Charge carrier mobility	μ	$s \cdot C/kg$	$m^2/V \cdot s$
Charge carrier density	n	m^{-3}	(none)
Magnetic field strength	H	$C/m \cdot s$	A/m

Items in boldface are material properties.

an intermediate position between lattice sites. This activation energy barrier to motion has already been described in Section 4.3.2.1 for diffusion in crystalline solids, so it will not be described here. It makes sense, though, that the ion mobility, μ_i, should be related to the ion diffusivity (diffusion coefficient), D_i, and can be derived from Eq. (6.14) presented earlier:

$$\mu_i = \frac{z_i q_e D_i}{k_B T} \tag{6.50}$$

where z_i is the valence of the ion, q_e is the electronic charge, and k_B and T are Boltzmann's constant and absolute temperature, as before. Substitution of Eq. (6.50) into Eq. (6.49) gives a relationship for the ionic conductivity:

$$\sigma_{ionic} = \frac{f n_i (z_i q_e)^2 D_i}{k_B T} \tag{6.51}$$

Typically, there is a predominate charge carrier in ionic solids. For example, in sodium chloride, the mobility of sodium ions is much larger than the mobility of chloride ions. This phenomenon can be temperature-dependent, however. As shown in Table 6.6, the fraction of total conductivity attributable to positive ions (K^+) in

Table 6.6   Ionic and Electronic Contributions to Total Electrical Conductivity for Some Selected Compounds and Glasses

Compound	Temperature ($^\circ$C)	Cationic Conductivity ($\sigma^+_{ionic}/\sigma_{total}$)	Anionic Conductivity ($\sigma^-_{ionic}/\sigma_{total}$)	Electronic Conductivity ($\sigma_{e,h}/\sigma_{total}$)
NaCl	400	1.00	—	—
	600	0.95	0.05	—
KCl	435	0.96	0.04	—
	600	0.88	0.12	—
KCl + 0.02% $CaCl_2$	430	0.99	0.01	—
	600	0.99	0.01	—
AgCl	20–350	1.00	—	—
AgBr	20–300	1.00	—	—
BaF_2	500	—	1.00	—
PbF_2	200	—	1.00	—
CuCl	20	—	—	1.00
	366	1.00	—	—
ZrO_2 + 7% CaO	>700	—	1.00	10^{-4}
$Na_2O \cdot 11Al_2O_3$	<800	1.00(Na^+)	—	$< 10^{-6}$
FeO	800	—	—	1.00
ZrO_2 + 18% CeO_2	1500	—	0.52	0.48
ZrO_2 + 50% CeO_2	1500	—	0.15	0.85
$Na_2O \cdot CaO \cdot SiO_2$ glass	—	1.00(Na^+)	—	—
15%($FeO \cdot Fe_2O_3$) $\cdot$ CaO$\cdot$ $SiO_2 \cdot Al_2O_3$ glass	1500	0.1(Ca^+)	—	0.9

Source: W. D. Kingery, H. K. Bowen, and D. R. Uhlmann, *Introduction to Ceramics*. Copyright © 1976 by John Wiley & Sons, Inc.

KCl drops from 0.96 at 435°C to 0.88 at 600°C. Similarly, the conductivity of CuCl is wholly attributable to electronic conduction at room temperature, but completely due to ionic conduction at 366°C. This temperature dependence is illustrated further for NaCl in Figure 6.29. At high temperatures (to the left in Figure 6.29), the number of sodium ion vacancies [n_i in Eq. (6.51)] is a thermodynamic property, and the conductivity varies with temperature as the product of the vacancy concentration and the diffusion coefficient, each of which has an exponential temperature dependence. This is the intrinsic range. At lower temperatures (to the right in Figure 6.29), the concentration of sodium ions is not in thermal equilibrium but is determined by minor solutes. As a result, this is the extrinsic region, and the temperature dependence of conductivity depends only on the diffusion coefficient. The activation energy for mobility can be determined from the data in Figure 6.29 for this region, but the slope of the curve in the intrinsic region gives the sum of the activation energy for mobility and lattice-defect formation.

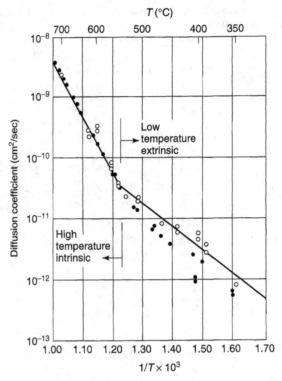

Figure 6.29   Diffusion coefficient measured directly (open circles) and calculated from electrical conductivity data (closed circles) for Na^+ in sodium chloride. From W. D. Kingery, H. K. Bowen, and D. R. Uhlmann, *Introduction to Ceramics*. Copyright © 1976 by John Wiley & Sons, Inc. This material is used by permission of John Wiley & Sons, Inc.

Several types of compounds show exceptionally high ionic conductivity. Such phases fall into three broad categories: halides and *chalcogenides* of silver and copper, in which the metal atom is disordered over several alternative sites; oxides with the β-alumina structure, in which a monovalent cation is mobile; and oxides of fluorite structure, with large concentrations of defects caused either by a variable valence cation or solid solution with a second cation of lower valence. Chalcogenide are chiefly based on the elements of Group VI—that is, S, Se, and Te. The silver and copper halides and chalcogenides often have simple arrays of anions. The cations occur in disorder in the interstices among the anions. The number of available sites is larger than the number of cations. In the highly conductive phases, the energy barrier between neighboring sites is very small, and channels are provided along which cations are free to move. The β-aluminas are hexagonal structures with approximate composition $AM_{11}O_{17}$. The mobile ion A is a monovalent species such as Na, K, Rb, Ag, Te, or Li, and M is a trivalent ion, Al, Fe, or Ga. The conductivities for several β-aluminas are plotted as a function of temperature in Figure 6.30. The crystal structure consists of planes of atoms parallel to the basal plane. Four planes of oxygens in a cubic close-packed sequence comprise a slab within which aluminum atoms occupy octahedral and tetrahedral sites as in spinel. The spinel blocks are bound together by a rather open layer of the monovalent ion and oxygen. This loosely bound layer is thought to be disordered

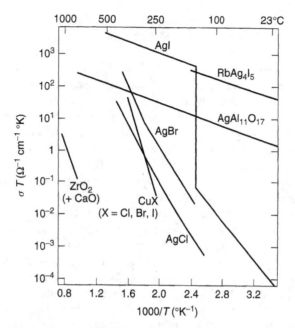

Figure 6.30   Conductivity of some highly conducting solid electrolytes. From W. D. Kingery, H. K. Bowen, and D. R. Uhlmann, *Introduction to Ceramics*. Copyright © 1976 by John Wiley & Sons, Inc. This material is used by permission of John Wiley & Sons, Inc.

and provide a two-dimensional channel for atom motion. Finally, the high dopant levels in the fluorite-type solid solutions leads to large defect concentrations and vacancy ordering. In such materials rapid oxygen migration occurs. This is believed to be due to the high concentration of vacancies. The effect of structure on the ionic conductivity of ceramics leads us naturally into one of the most exciting technological developments of recent decades: high-temperature superconductors.

6.1.2.4 *Oxide Superconductors.*

In Section 6.1.1.3, we learned that some metals can lose all of their residual resistivity near absolute zero and become what are called superconductors. We also learned that for many years, there was believed to be an upper temperature limit of 30 K for the critical transition temperature, T_c, below which a substance begins to superconduct. In 1986, however, the first indication of superconductivity above 30 K was found in barium-doped La_2CuO_4. Then researchers found that high pressure could increase T_c to 50 K. Work began to find combinations of atoms of different sizes and valences that could simulate the pressure effect through chemical substitution. One of the variations was to substitute yttrium into the perovskite structure of $BaCuO_3$. Of these so-called YBC compounds, $YBa_2Cu_3O_7$ (termed 1–2–3 superconductor after the chemical formula) was found to superconduct at 92 K. Why all the fuss over a few tens of degrees? The answer is stated simply with the boiling points of three common gases: T_b(helium) = 4.2 K; T_b(nitrogen) = 77 K; T_b(oxygen) = 90 K. In order to cool the superconductors below T_c, a liquefied gas must be used. It is far less costly to liquefy nitrogen and oxygen (the major components of air) than helium. For any application in which liquid nitrogen can replace liquid helium, the refrigeration cost is about 1000 times less. In this way, high-temperature superconductors (HTSC), with $T_c > 77$ K, are of great technological interest.

Another striking difference the HTSC compounds exhibited was the anomalous behavior when placed in a magnetic field. Recall from Section 6.1.1.3 that there is also a critical magnetic field, H_c, above which the superconductor loses its superconductivity, and that this critical field is a function of temperature. In the metal and compound-based superconductors of Section 6.1.1.3, the transition from superconducting to nonsuperconducting at a specified temperature is a relatively sharp one as the critical field is reached. This relationship was illustrated in Figure 6.10 and Eq. (6.19). When compounds belonging to the new class of HTSC are placed in a magnetic field, however, the temperature region for the superconducting–nonsuperconducting transition broadens as the field is increased, as illustrated for $YBa_2Cu_3O_7$ in Figure 6.31. At low fields the resistivity drops sharply over a very narrow temperature range, whereas at high fields the transition range is quite large—nearly 15 K. Compare this resistivity–temperature behavior with that of Pb in Figure 6.6. Thus, two classes of superconductors are defined: Type I, also called "hard" superconductors, which exhibit a sharp transition; and Type II, also known as "soft" superconductors, which have a range of transition values. In Type II superconductors, the magnetic field starts penetrating into the material at a *lower critical field*, H_{c1}. Penetration increases until at the *upper critical field*, H_{c2}, the material is fully penetrated and the normal state is restored.

In addition to a critical temperature and critical field, all superconductors have a *critical current density*, J_c, above which they will no longer superconduct. This limitation has important consequences. A logical application of superconductors is as current-carrying media. However, there is a limit, often a low one, to how much current they can carry before losing their superconducting capabilities. The relationship between J_c, H_c, and T_c for a Type II superconductor is shown in Figure 6.32. Notice that the H_c–T_c portion of this plot has already been presented in Figure 6.10 for a Type I superconductor.

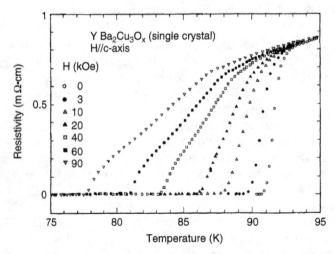

Figure 6.31 Variation of superconducting transition range in a magnetic field. Reprinted, by permission, from T. P. Sheahan, *Introduction to High-Temperature Superconductivity*, p. 121. Copyright © 1994 by Plenum Press.

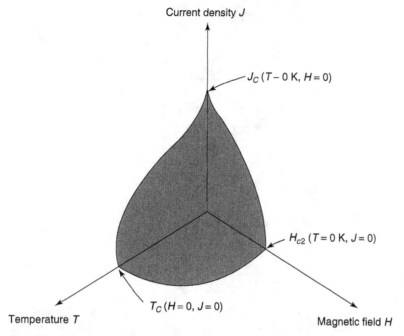

Figure 6.32 The relationship between temperature, magnetic field and current density in a Type II superconductor. Reprinted, by permission, from W. Callister, *Materials Science and Engineering: An Introduction*, 5th ed., p. 699. Copyright © 2000 by John Wiley & Sons, Inc.

Table 6.7   Some HTSC Materials Based on the Copper Oxide Lattice

Compound	T_c(K)
$YBa_2Cu_3O_7$	92
$(BiPb)_2Sr_2Ca_2Cu_3O_x$	105
$TlBa_2Ca_2Cu_3O_y$	115
$HgBa_2Ca_2Cu_3O_y$	135

Most of the oxide Type II HTSC are based upon the copper oxide lattice. Some examples and their transition temperatures are given in Table 6.7. In all cases, the structure is essentially a laminar one, with planes of copper oxide in the center (see Figure 6.33), through which the superconducting current flows. These materials are thus highly anisotropic—very little current can flow perpendicular to the copper oxide planes. As a result, great pains are taken to produce these materials in a highly oriented manner and to use them such that current flows parallel to the copper oxide planes. The role of the elements other than copper and oxygen is secondary. In YBCO, yttrium is only a spacer and a contributor of charge carriers. Nearly any of the rare earth elements (holmium, erbium, dysprosium, etc.) can be substituted for yttrium without significantly changing T_c.

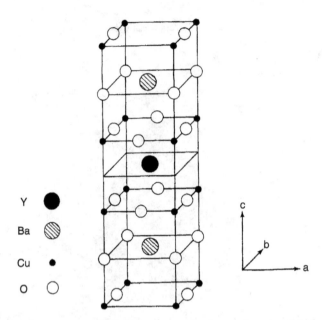

Figure 6.33   Unit cell structure of $YBa_2Cu_3O_7$ superconductor. Reprinted, by permission, from Sheahan, T. P., *Introduction to High-Temperature Superconductivity*, p. 5. Copyright © 1993 by Plenum Press.

It is beyond the scope of this text to describe the mechanism for charge transfer in the high-temperature superconductors. While there is still a great deal of discussion on the applicability of BCS theory (cf. Section 6.1.1.3) to HTSC materials, it is safe to say that many of the principles still apply—for example, density of states at the Fermi level. The interested reader should refer to existing literature [3,4] for more information on the structure and theory of copper oxide superconductors.

Finally, it would be misleading to suggest that these materials are the technological revolution they were first imagined to be. Like most other oxide ceramics, these materials are very brittle, and attempts to fabricate them in useful forms, such as wire, have proved problematic. The additional difficulty is with current density, as described earlier. For large-scale applications such as particle accelerator magnets, current densities on the order of 10^{10} A/m^2 are required. A high critical magnetic field would also be beneficial for some applications, where magnetic fields of 30 tesla (30,000 gauss) are required. Oxide superconductors meet neither of these requirements. As a result, the metallic superconductors are still the materials of choice for these high-end applications, since they are more easily formed into wires, and have higher J_c and H_c values, even though their T_c is low. Metallic superconductors of the niobium family are particularly useful in these regards and include NbTi, Nb$_3$Sn, NbZr, Nb$_3$Al, and Nb$_3$Ge.

6.1.2.5 *Compound Semiconductors.* The niobium-based superconducting compounds lead us naturally into another use for intermetallics—namely, semiconductors. This topic, too, was introduced earlier in this chapter (cf. Section 6.1.1.4 and 6.1.1.5), and we shall build upon those principles here to describe the semiconducting properties of compounds, ceramics, and glasses. The classification of intermetallics as ceramics may be confusing at this point, but we do so due to the strong covalent nature of their bonds that are more similar to oxide ceramics than to the metallic bond of elemental semiconductors. It is this bond structure that influences the electrical properties; hence, we discuss them together in this section.

Recall that elemental semiconductors such as silicon, germanium, tin, and diamond belong to group IV in the periodic table. They are characterized by four valency electrons, resulting in a tetrahedral arrangement of atoms in a diamond-like structure (cf. Figure 1.40). A large number of compounds possess structural and electrical characteristics that are similar to those of elemental semiconductors, many of which are compounds formed between elements of groups III and V of the periodic table, termed III–V compounds. In these cases, slight deviations from stoichiometry can lead to extrinsic conduction. One of the more successful examples is gallium arsenide, GaAs. Gallium is a group III element and arsenic a group V element, and the resulting average valency in a 50–50 compound is four, just as for a group IV element. However, since the elemental substances have different electronegativities (cf. Table 1.4), the bond will be partly ionic in nature and the band gap will be higher than in the corresponding elemental semiconductor, Ge (see Table 6.8; some values may differ slightly from Table 6.2). These larger band gaps extend the useful range of conduction since intrinsic conduction becomes important only at correspondingly higher temperatures. The general decrease in band gap with increasing atomic number of the II–V compounds is a result of an increasing tendency toward metallic bonding.

Some semiconducting compounds can be of the II–VI type, which also has an average valence of four, but these have much more ionic character than III–V compounds. Their band gaps are thus larger, and in some cases they may even be viewed as insulators. For example, ZnS, with a band-gap energy of 3.6 eV, is an insulator, whereas ZnSe has an band gap of 2.8 eV, which is closer to a semiconductor. A wide variety of

Table 6.8   Electronic Properties of Some Elemental and Compound Semiconductors at Room Temperature

Material	Energy Gap (eV)	Mobility, μ $\dfrac{m^2}{V \cdot s}$		Effective Mass, m^*/m_0		Dielectric Constant
		Electrons	Holes	Electrons	Holes	
Silicon (Si)	1.1	0.135	0.048	0.97	0.5	11.7
Ge	0.67	0.390	0.190	1.60	0.3	16.3
Sn (gray)	0.08	0.200	0.100			
GaAs	1.40	0.800	0.025	0.072	0.5	12.5
GaSb	0.77	0.400	0.140	0.047	0.5	15.0
GaP	2.24	0.050	0.002			10.0
InSb	0.16	7.800	0.075	0.013	0.60	17.0
InAs	0.33	3.300	0.040	0.02	0.41	14.5
InP	1.29	0.460	0.015	0.07	0.40	14.0
PbS	0.40	0.060	0.020			

Source: Z. Jastrzebski, *The Nature and Properties of Engineering Materials*, 2nd ed. Copyright © 1976 by John Wiley & Sons, Inc.

compounds between transition metals and nonmetals also exhibit semiconductivity, but these materials are used infrequently because impurity control is much more difficult to achieve.

For nonstoichiometric compounds, the general rule is that when there is an excess of cations or a deficiency of anions, the compound is an n-type semiconductor. Conversely, an excess of anions or deficiency of cations creates a p-type semiconductor. There are some compounds that may exhibit either p- or n-type behavior, depending on what kind of ions are in excess. Lead sulfide, PbS, is an example. An excess of Pb^{2+} ions creates an n-type semiconductor, whereas an excess of S^{2-} ion creates a p-type semiconductor. Similarly, many binary oxide ceramics owe their electronic conductivity to deviations from stoichiometric compositions. For example, Cu_2O is a well-known p-type semiconductor, whereas ZnO with an excess of cations as interstitial atoms is an n-type semiconductor. A partial list of some impurity-controlled compound semiconductors is given in Table 6.9.

The conductivity of these materials can be controlled by the number of defects. In a p-type semiconductor such as Cu_2O, in which vacancies are formed in the cation lattice when the oxygen partial pressure is increased, we can develop relationships between conductivity and oxygen partial pressure. The overall reaction for the formation of vacancies and electron holes can be written in Kroger–Vink notation (cf. Section 1.2.6.1) as

$$\tfrac{1}{2}O_2(g) \rightleftharpoons O_O + 2\,V'_{Cu} + 2h^\bullet \tag{6.52}$$

From this reaction, we can then write a mass-action relation [cf. Eq. (3.1)] that is a function of temperature:

$$K(T) = \frac{[V'_{Cu}]^2[h^\bullet]^2}{P_{O_2}^{1/2}} \tag{6.53}$$

Table 6.9   Some Impurity Compound Semiconductors

			n-*Type*		
TiO_2	Nb_2O_5	CdS	Cs_2 Se	$BaTiO_3$	Hg_2S
V_2O_5	MoO_2	CdSe	BaO	$PbCrO_4$	ZnF_2
U_3O_8	CdO	SnO_2	Ta_2O_5	Fe_3O_4	
ZnO	Ag_2S	Cs_2S	WO_3		
			p-*Type*		
Ag_2O	CoO	Cu_2O	SnS	Bi_2Te_3	MoO_2
Cr_2O_3	SnO	Cu_2S	Sb_2S_3		Hg_2O
MnO	NiO	Pr_2O_3	CuI		
			Amphoteric		
Al_2O_3	SiC	PbTe		Ti_2S	
Mn_3O_4	PbS	UO_2			
Co_3O_4	PbSe	IrO_2			

If the concentration of vacancies is largely determined by reaction with the atmosphere, then the conductivity should be proportional to the concentration of holes [which is equivalent to the concentration of vacancies according to Eq. (6.52)], and we obtain a relationship:

$$\sigma \propto [h^{\bullet}] = K(T)^{1/4} P_{O_2}^{1/8} \tag{6.54}$$

Experimentally, as illustrated in Figure 6.34, the electrical conductivity of Cu_2O is found to be proportional to $P_{O_2}^{1/7}$, which is in reasonable agreement with the prediction of Eq. (6.54). The variation of another oxide semiconductor, CdO, with temperature is also shown in Figure 6.34 for comparison.

As a final example of oxide-based semiconductors, recall from Section 6.1.2.3 that some chalcogenides exhibit high ionic conductivities. Chalcogenides also play a role in the high conductivities found in some glasses (see Figure 6.35). In addition to the group VI elements, chalcogenide glasses sometimes contain elements of Group V, such as P, As, Sb, and Bi, as well as other ingredients like Ge or the halides. These conductivities are electronic, and not ionic, in nature. Their melting points are low, sometimes below 100°C, and they are quite sensitive to corrosion. The conductivity in these glasses can be explained in terms of a special type of glass structure defects around which pairs of positive and negative centers are present, and with which electrons couple. In addition to their semiconductivity, these glasses have a unique *switching behavior*. If a voltage is applied at the normal high-resistivity state, called the "off state," a critical current–voltage pair of values is eventually achieved at which point the glass suddenly switches into a low-resistivity state, called the "on state." Each state is stable, and the process is reversible. This is a type of storage, or memory, effect.

6.1.2.6 *MOSFETs.* A type of semiconductor device that utilizes oxide ceramics is a *metal-oxide-semiconductor field-effect transistor*, abbreviated as MOSFET. Just like the semiconductor junction devices of Section 6.1.1.6, the MOSFET is composed of *n*- and *p*-type semiconductor regions within a single device, as illustrated in Figure 6.36.

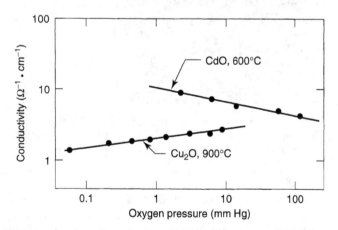

Figure 6.34 Electrical conductivity of two oxide semiconductors as a function of oxygen partial pressure. From W. D. Kingery H. K. Bowen, and D. R. Uhlmann, *Introduction to Ceramics.* Copyright © 1976 by John Wiley & Sons, Inc. This material is used by permission of John Wiley & Sons, Inc.

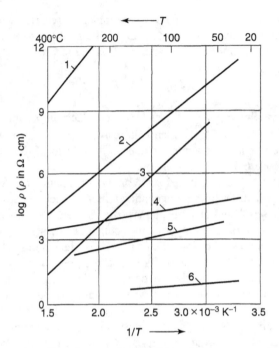

Figure 6.35 Temperature dependence of resistivity for some semiconducting glasses: (1) vitreous silica; (2) soda–lime silicate glass; (3) AsSeTe chalcogenide glass; (4) silicate glass with 18 mol% (Fe_3O_4 + MnO); and (6) borosilicate glass with 45 mol% (Fe_3O_4 + MnO). Reprinted, by permission, from H. Scholze, *Glass*, p. 312. Copyright © 1991 by Springer-Verlag.

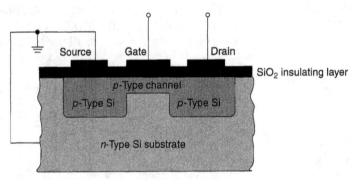

Figure 6.36 Schematic cross-sectional view of a MOSFET transistor. Reprinted, by permission, from W. Callister, *Materials Science and Engineering: An Introduction*, 5th ed., p. 634. Copyright © 2000 by John Wiley & Sons, Inc.

One such device consists of two small islands of *p*-type semiconductor with and *n*-type silicon substrate. The islands are joined by a narrow *p*-type channel. The oxide portion of the MOSFET is an insulating layer of silicon dioxide that is formed by surface oxidation of the silicon. Gate, drain, and source connectors are attached. The MOSFET differs from the junction transistor in that a single type of charge carrier, either an electron or a hole, is utilized, instead of both. The conductivity of the channel is varied by the presence of an electric field imposed on the gate. For example, a positive field on the gate will drive holes out of the channel, thereby reducing the electrical conductivity. Thus, a small alteration in the field at the gate will produce a relatively large variation in current between the source and the drain. In this respect, the MOSFET is similar to the junction transistor, insofar as both create a signal amplification. The primary difference is that the gate current is exceedingly small in comparison to the base current of the junction transistor. As a result, MOSFETs are used where signal sources to be amplified cannot sustain an appreciable current.

6.1.3 Electrical Properties of Polymers

In this section, we consider the insulating properties of polymers as dielectrics, and we also describe their electrical- and ion-conducting capabilities. Given the highly insulative nature of most polymers, we begin with dielectric properties and then describe special types of polymers that conduct either electrons or ions.

6.1.3.1 Dielectric Polymers. Nonpolar polymers generally have dielectric constants in the 2–3 range while polar polymers can have values up to 7. Typical dielectric strengths of polymers are in the range of 20–50 kV/mm, which are approximately 2–10 times higher than ceramics and glasses and are hundreds of times higher than conducting metals and alloys.

Data from mechanical and dielectric measurements can be related in a qualitative, if not quantitative, manner. Formally, the dielectric constant can be regarded as the equivalent of the mechanical compliance. This highlights the fact that the dynamic mechanical techniques of Section 5.3 measure the ability of a polymer to resist movement, whereas the dielectric technique measures the ability of the system to move, at least the dipolar portions. Interestingly, the dielectric loss appears to match the loss modulus more closely than the loss compliance when the data are compared for the same system. Both techniques respond in a similar fashion to a change in the frequency of the measurement. When the frequency is increased, the transitions and relaxations that are observed in a sample appear at higher temperatures. We expect maximum losses at transition temperatures for both electrical and mechanical deformation, as illustrated for poly(chlorotrifluoroethylene) in Figure 6.37. Just as mechanical fatigue after many alternating cycles can cause a polymer to fail at a lower stress, so dielectric insulation can fatigue under electrical stresses.

6.1.3.2 Electrically Conducting Polymers. High values of resistivity are common for organic polymers, with 10^{12} to 10^{18} $\Omega \cdot$ cm being typical. Electrical resistance can be lowered markedly by the addition of conductive fillers, such as carbon black or metal fillers, but the conductivity is due to the filler in this case. When we consider electronic conduction solely in homogeneous polymers, band theory is not totally suitable because the atoms are covalently bonded to one another, forming polymeric chains that experience weak intermolecular interactions. Thus, macroscopic conduction will require electron movement, not only along chains but also from one chain to another.

Polymers that display electronic conductivity are usually insulators in the pure state but, when reacted with an oxidizing or reducing agent, can be converted into polymer salts with electrical conductivities comparable to metals. Some of these polymers are listed in Figure 6.38, along with the conductivities of metals and ceramics for

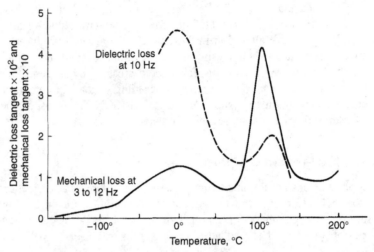

Figure 6.37 Mechanical and dielectric loss tangents for poly (chlorotrifluoroethylene). Reprinted, by permission, from F. Rodriguez, *Principles of Polymer Systems*, 2nd ed., p. 271. Copyright © 1982 by Hemisphere Publishing Corporation.

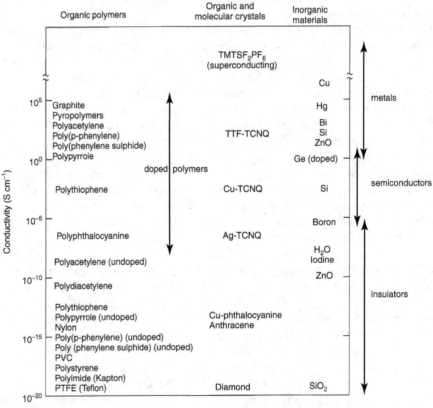

Figure 6.38 Conductivity ranges for doped and undoped polymers, inorganic materials and molecular crystals. One siemen (S) $= 1\ \Omega^{-1}$. Reprinted, by permission, from J. M. G. Cowie, *Polymers: Chemistry & Physics of Modern Materials*, 2nd ed., p. 411. Copyright © 1991 by J.M.G. Cowie.

comparison. The preparation and mechanism of conductivity in a few of these polymers will be elaborated upon.

One of the first polymers to exhibit electrical conductivity was polyacetylene. Normally a very poor conductor in the pure state, it was found that a highly conductive polymer could be formed by reacting it with I_2. The result is a dramatic increase of over 10^{10} in conductivity. Polyacetylene is a *polyconjugated polymer*, the structure of which is shown in Table 6.10. Polyconjugation refers to the existence of multiple C–C bonds in the polymer, typically along the backbone. In a polyconjugated system, the π orbitals (cf. Section 1.0.4.2) are assumed to overlap, and they form a valance and a conduction band as predicted by band theory. If all the bond lengths were equal—that is, delocalization that led to each bond having equal double bond character—then the bands would overlap and the polymer would behave as a quasi-one-dimensional metal having good conductivity. Experimental evidence has shown that this is an unstable system that will undergo lattice distortion. The alternative is to have single and double bonds alternate along the backbone, which leads to an energy gap between the valence and conduction bands. The *trans* structure of polyacetylene has such an alternating single-bond–double-bond backbone. The *trans* structure of polyacetylene is also unique, because it has a twofold degenerate ground state in which sections A and B in Figure 6.39 are mirror images, and the single and double bonds can be interchanged without changing the energy. Thus, if the *cis* configuration begins to isomerize to the *trans* structure from different locations in a single chain, an

Table 6.10   Structures and Conductivities of Doped Conjugated Polymers

Polymer	Structure	Typical Methods of Doping	Typical Conductivity (S cm)$^{-1}$
Polyacetylene		Electrochemical, chemical (AsF$_5$, I$_2$, Li, K)	500–1.5 × 10^5
Polyphenylene		Chemical (AsF$_5$, Li, K)	500
Poly(phenylene sulphide)		Chemical (AsF$_5$)	1
Polypyrrole		Electrochemical	600
Polythiophene		Electrochemical	100
Poly(phenyl-quinoline)		Electrochemical, chemical (sodium naphthalide)	50

One siemen equals 1 Ω^{-1}. *Source*: J. M. G. Cowie, *Polymers: Chemistry & Physics of Modern Materials*, 2nd ed. Copyright © 1991 by J.M.G. Cowie.

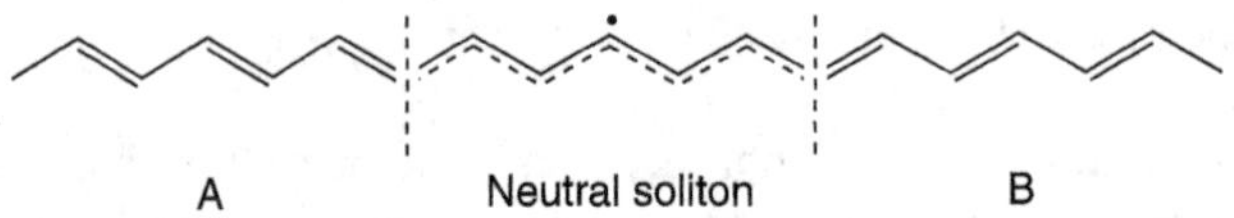

A   Neutral soliton   B

Figure 6.39   Isomerization of *cis* (A) and *trans* (B) sequences in polyacetylene that meet to form a soliton. Reprinted, by permission, from J. M. G. Cowie, *Polymers: Chemistry & Physics of Modern Materials*, 2nd ed., p. 417. Copyright © 1991 by J. M. G. Cowie.

A sequence may form and eventually meet a B sequence, as shown, but in doing so, a free radical, called a *soliton*, is produced. The soliton is a relatively stable electron with an unpaired spin and is located in a nonbonding state in the energy gap, midway between the conduction and valence bands. It is the presence of these neutral solitons which gives *trans*-polyacetylene the characteristics of an intrinsic semiconductor with conductivities of 10^{-7} to $10^{-8} (\Omega \cdot cm)^{-1}$.

The conductivity of polyacetylene can be magnified by doping. Exposure of a polyacetylene film to dry ammonia gas leads to a dramatic increase in conductivity of $10^3 (\Omega \cdot cm)^{-1}$. Controlled addition of an acceptor, or *p*-doping, agent such as AsF_5, I_2, or $HClO_4$ removes an electron and creates a positive soliton (or a neutral one if the electron removed is not the free electron). Similarly, a negative soliton can be formed by treating the polymer with a donor, or *n*-doping, agent that adds an electron to the mid-gap energy level. This can be done by an electrochemical method. At high doping levels, the soliton regions tend to overlap and create new mid-gap energy bands that may merge with the valence and conduction bands, allowing freedom for extensive electron flow. Thus, in polyacetylene the charged solitons are responsible for making the polymer a conductor.

Poly(*p*-phenylene) has all the structural characteristics required of a potential polymer conductor. It is an insulator in the pure state, but can be doped using methods similar to those of polyacetylene to form both *n*- and *p*-type semiconductors. However, because poly(*p*-phenylene) has a higher ionization potential, it is more stable to oxidation and requires strong *p*-dopants. Examination of the structure shows that the soliton defect cannot be supported in poly(*p*-phenylene) because there is no degenerate ground state. Instead, it is assumed that conduction occurs because the mean free path of charge carriers extends over a large number of lattice sites and the residence time on any one site is small compared with the time it would take for a carrier to become localized. If, however, a carrier is trapped, it tends to polarize the local environment, which relaxes into a new equilibrium position. This deformed section of the lattice and the charge carrier then form a species called a *polaron*. Unlike the soliton, the polaron cannot move without first overcoming an energy barrier so movement is by a hopping motion. In poly(*p*-phenylene) the solitons are trapped by the changes in polymer structure because of the differences in energy and so a polaron is created which is an isolated charge carrier. A pair of these charges is called a *bipolaron*. In poly(*p*-phenylene) and most other polyconjugated conducting polymers the conduction occurs via the polaron or bipolaron (see Figure 6.40).

There are other types of semiconducting polymers as well, some of the more important of which are listed in Table 6.10. The conduction mechanism in most of these polymers is the polaron model described above. Applications for these polymers are growing and include batteries, electromagnetic screening materials, and electronic devices.

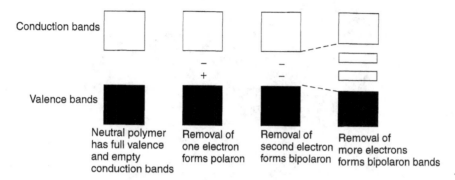

Figure 6.40 Proposed band structure for an oxidized *p*-type semiconducting polymer. Reprinted, by permission, from J. M. G. Cowie, *Polymers: Chemistry & Physics of Modern Materials*, 2nd ed., p. 419. Copyright © 1991 by J. M. G. Cowie.

6.1.3.3 Ion-Conducting Polymers.

Recall that electrical conductivity can be the result of electronic conductivity, in which electrons are the charge carriers, ionic conductivity, in which ions are the charges carriers, or a combination of both. In the previous section, we saw how some conjugated polymers can conduct electrons. In this section we concentrate on ion-conducting polymers. There are generally two types of polymers in this category: those that conduct primarily hydrogen ions and those that conduct primarily larger anions. In both cases, the term *polymer electrolyte* is used to describe the type of ion-transporting polymeric material, although it is more commonly used to describe the latter. This should not be confused with the term *ionomer*, which describes polymers that contain ionic groups primarily involved in creating ionic bonds between chains. This is not to say that ionomers are incapable of ionic conductivity—they may well be. It is rather a matter of terminology usage. Ionomers typically have applications other than as ion conductors.

We begin by examining the structure and properties of polymer electrolytes. Initial investigations of ionically conductive polymers were principally focused on poly(ethylene oxide), PEO, when it was discovered that PEO can solvate a wide variety of salts, even at very high concentrations. The solvation occurs through the association of the metallic cations with the oxygen atoms in the polyether backbone (see Figure 6.41). The solvated ions can be quite mobile, and thus give rise to significant bulk ionic conductivities. Pure PEO is a semicrystalline polymer possessing both an amorphous and a crystalline phase at room temperature. Significant ionic transport occurs within the amorphous phase. This explains the dramatic decrease in ionic conductivity seen in many PEO-based systems for temperatures below the melting point of pure crystalline PEO($T_m = 66°C$), as shown in Figure 6.42. The crystalline PEO regions are nonconductive and serve to hinder bulk ionic transport.

The mobility of the ions in polymer electrolytes is linked to the local segmental mobility of the polymer chains. Significant ionic conductivity in these systems will occur only above the glass transition temperature of the amorphous phase, T_g. Therefore, one of the requirements for the polymeric solvent is a low glass-transition temperature; for example, $T_g = -67°C$ for PEO.

One problem associated with polymer electrolytes arises from the opposing effects of increasing the salt concentration. Higher salt concentrations generally imply a higher density of charge carriers, which increases conductivity. However, through their

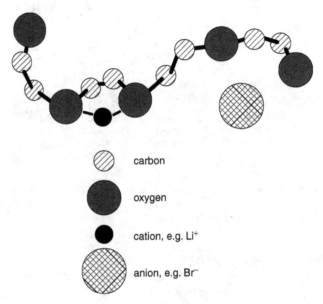

Figure 6.41   Schematic illustration of salt solvation by a PEO chain. Reprinted, by permission, from L. V. Interrante, L. A. Casper, and A. B. Ellis, eds. *Materials Chemistry*, p. 109. Copyright © 1995 by the American Chemical Society.

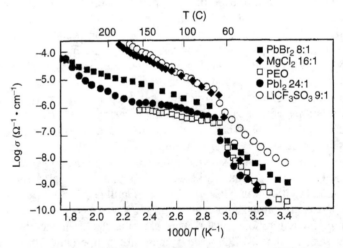

Figure 6.42   Temperature dependence of conductivity for various PEO-based polymer electrolytes containing divalent cations. Reprinted, by permission, from L. V. Interrante, L. A. Casper, and A. B. Ellis, eds., *Materials Chemistry*, p. 110. Copyright © 1995 by the American Chemical Society.

interactions with the polyether chains, inorganic salts increase the T_g as their concentrations increases. This increase in T_g tends to lower the mobility of the ions and thus decreases ionic conductivity. The balance of these two effects leads to a maximum in the conductivity at a specific salt concentration, which varies from system to system.

One method of reducing crystallinity in PEO-based systems is to synthesize polymers in which the lengths of the oxyethylene sequences are relatively short, such as through copolymerization. The most notable linear copolymer of this type is oxymethylene-linked poly(oxyethylene), commonly called amorphous PEO, or aPEO for short. Other notable polymer electrolytes are based upon polysiloxanes and polyphosphazenes. Polymer blends have also been used for these applications, such as PEO and poly(methyl methacrylate), PMMA. The general performance characteristics of the polymer electrolytes are to have ionic conductivities in the range of $10^{-3}(\Omega \cdot cm)^{-1}$ or (S/cm).

Still higher total conductivities are achievable in a special class of hydrogen-ion (proton)-conducting polymers. These polymers are particularly useful for fuel-cell and chlor-alkali processing applications in which the efficient transfer of protons is critical. These polymers are of produced in membrane form and are therefore referred to as *proton-exchange membranes* (PEM).

The most common polymeric proton-exchange membrane is Nafion®, which is manufactured in various forms by Du Pont. The generalized chemical structure of Nafion® is shown in Figure 6.43. It is a perfluorinated polymer, similar to Teflon®, with small amounts of sulfonic (SO_3^-) or carboxylic functional groups. The microstructure of Nafion® is actually quite complex. In general, the models, supported primarily by small-angle X-ray scattering (SAXS) results, predict that the ionic portions of Nafion® tend to aggregate due to electrostatic interactions to form ionic clusters. It has been determined that the clusters are on the order of 30–50 Å in size, resulting in (1) a phase-separated polymer with discrete hydrophobic regions formed from the polymer backbone and (2) hydrophilic regions formed from the ionic clusters. These distinct regions are functionally important. Nafion® is believed to conduct protons due to extensive hydration that takes place in the ionic clusters. Waters of hydration are necessary, since according to one theory the protons are transported through the polymer in the form of hydronium ions, H_3O^+ [5].

The proton conductivity of Nafion® is orders of magnitudes higher than most other proton-conducting polymers (see Table 6.11). The conductivity, however, is

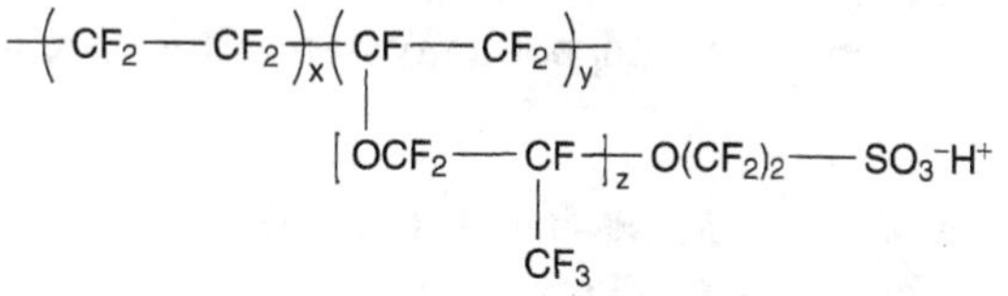

$$\left(CF_2 - CF_2 \right)_x \left(CF - CF_2 \right)_y$$
$$\mid OCF_2 - CF \mid_z O(CF_2)_2 - SO_3^-H^+$$
$$CF_3$$

Figure 6.43   The generalized chemical structure of Nafion®.

Table 6.11   Proton Conductivities of Selected Polymers

Polymer	Conductivity $(\Omega \cdot cm)^{-1}$
Nafion®	2.2×10^{-1}
Polybenzimidazole (PBI)	2×10^{-4}–8×10^{-4}
Poly(vinylidine)fluoride	10^{-2}

highly dependent upon (a) the relative humidity in the membrane and (b) the ability of the membrane to stay hydrated during operation. One particular drawback to all tetrafluoroethylene-based polymers is the inherent difficulty in processing due to limited solubility in most solvents. As a result, only certain types of geometries, such as thick films, are commercially available.

6.1.4 Electrical Properties of Composites

Because electronic and ionic conduction are so structure-sensitive, the simple rule-of-mixtures approach to estimating the conductivity and resistivity of composites is not normally of use. As a result, the conductivity of specific composites for specific applications must be experimentally determined. In the next two sections, we examine two examples of how composites can be used in electrical applications, and we describe the influence of each component on the electrical properties. The first example involves the electrical insulating properties of polymers, and the second one involves enhancing the electrically conducting properties of polymers.

6.1.4.1 Dielectric Properties of Glass-Fiber-Reinforced Polymers. As described in several of the previous sections on composites, the most common type of glass fiber used for reinforcement purposes is E-glass, which was originally developed for electrical insulation purposes, hence its name ("E" stands for "electrical grade"). It is not surprising, then, that when combined with organic resins such as epoxy, a class of materials results which can be readily molded into complex shapes that possess excellent insulting properties and high dielectric strength.

Some electrical properties of reinforcing fibers, composite resins, and the resulting composites are given in Tables 6.12, 6.13, and 6.14, respectively. These values should be taken as approximate only, especially for the composites, since fiber orientation, content, and field strength have an enormous impact on the dielectric properties of these materials. Some of the most widespread electrical applications for glass-fiber-reinforced epoxy systems are in printed circuit boards and electrical housing such as junction boxes.

One interesting application of the dielectric properties of the resins in these composite systems is that it provides an opportunity to monitor the cure characteristics of

Table 6.12 Some Electrical Properties of Selected Reinforcing Glass Fibers

Property	E-Glass	S-Glass	D-Glass
Dielectric constant, 10^6 Hz	5.80	4.53	3.56
Dielectric constant, 10^{10} Hz	6.13	5.21	4.00
Loss tangent, 10^6 Hz	0.001	0.002	0.0005
Loss tangent, 10^{10} Hz	0.0039	0.0068	0.0026

Source: Handbook of Fiberglass and Advanced Plastics Composites, G. Lubin, ed. Copyright © 1969 by Reinhold Book Corporation.

Table 6.13 Some Electrical Properties of Selected Resins

Property	Phenolic	Epoxy	Polyester	Silicone
Resistivity, $\Omega \cdot$ cm	$10^{12}-10^{13}$	$10^{16}-10^{17}$	10^{14}	$10^{11}-10^{13}$
Dielectric strength, V/μm	14–16	16–20	15–20	7.3
Dielectric constant, 60 Hz	6.5–7.5	3.8	3.0–4.4	4.0–5.0

Source: Handbook of Fiberglass and Advanced Plastics Composites, G. Lubin, ed. Copyright © 1969 by Reinhold Book Corporation.

Table 6.14 Some Electrical Properties of Selected Glass-Fiber-Reinforced Resins

Property	GFR Nylon	GFR Polypropylene	GFR Phenolic	GFR (Laminate)	GFR Polyester	GFR Silicone (Laminate)
Resistivity, $\Omega \cdot$ cm	$2.6-5.5 \times 10^{15}$	1.7×10^{16}	$10^{10}-10^{11}$	$6.6 \times 10^{7}-10^{9}$	$10^{12}-10^{13}$	$2-5 \times 10^{14}$
Dielectric strength, V/μm	16–18	12–19	15–17	26–30	8–16	28.5
Dielectric constant, 60 Hz	4.6–5.6	2.3–2.5	7.1–7.2	4.4	4.6–5.2	3.9–4.2

Source: Shackelford, J. and W. Alexander, *The CRC Materials Science and Engineering Handbook*. Copyright © 1992 by CRC Press.

Cooperative Learning Exercise 6.4

Assume that the conductivity of a undirectional, continuous fiber-reinforced composite is a summation effect just like elastic modulus and tensile strength; that is, an equation analogous to Eq. (5.88) can be used to describe the conductivity in the axial direction, and one analogous to (5.92) can be used for the transverse direction, where the modulus is replaced with the corresponding conductivity of the fiber and matrix phase. Perform the following calculations for an aluminum matrix composite reinforced with 40 vol% continuous, unidirectional Al_2O_3 fibers. Use average conductivity values from Appendix 8.

  Person 1: Calculate the conductivity of the composite in the axial direction.
  Person 2: Calculate the conductivity of the composite in the transverse direction.

  Compare your answers. How many orders of magnitude difference is there in the two conductivities?

Answers: $\sigma_{axial} = (0.4)(10^{-11}) + (0.6)(3.45 \times 10^{7}) = 20.7 \times 10^{6}$ $(\Omega$-m$)^{-1}$;
$\sigma_{trans} = (3.45 \times 10^{7})(10^{-11})/[(0.4)(3.45 \times 10^{7}) + (0.6)(10^{-11})] = 2.5 \times 10^{-11}$ $(\Omega$-m$)^{-1}$.
17 orders.

the resin in the composite. In-process cure monitoring ensures that each molded part has properly cured before the mold is opened. The dielectric loss [see Eq. (6.47)] has been shown to increase rapidly in the beginning of the cure cycle (see Figure 6.44), attain a peak, then reduce to a constant value as the curing reactions are completed. At the beginning of the cure cycle, the resin viscosity in the uncured prepreg is relatively high so that the dipole and ion mobilities are restricted. This results in low loss factors just after the uncured prepreg is placed in the mold. As the temperature of the prepreg increases in the mold, the resin viscosity is reduced and the loss factor increases owing to greater dipole and ion mobilities (see Figure 6.25). As soon as the gel point is reached, the resin viscosity increases rapidly and the loss factor decreases. At a full degree of cure, the loss factor levels off to a constant value.

The technique for monitoring the dielectric loss factor is relatively simple. Two metal electrodes are placed opposite each other at critical locations on opposite sides of the mold. When the sheet molding compound (SMC), is placed between the electrodes, a capacitor is formed. The dielectric power loss is monitored continually throughout the molding cycle, as outlined in Section 6.1.2.2.

6.1.4.2   *Metal Oxide-Polymer Thermistors.*   The variation of electrical properties with temperature heretofore described can be used to tremendous advantage. These so-called thermoelectric effects are commonly used in the operation of electronic temperature measuring devices such as *thermocouples, thermistors*, and resistance-temperature detectors (*RTD*s). A thermocouple consists of two dissimilar metals joined at one end. As one end of the thermocouple is heated or cooled, electrons diffuse toward

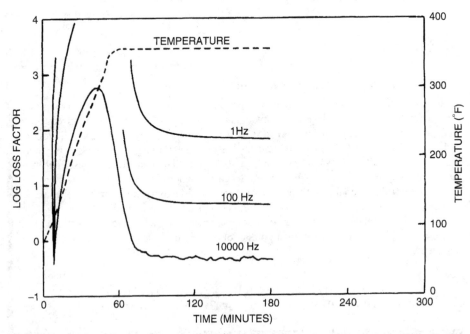

Figure 6.44   Dielectric loss factor as a function of cure time and frequency of the oscillating electric field in a fiber-reinforced polymer. Reprinted, by permission, from P. K. Mallick, *Fiber-Reinforced Composites*, p. 365. Copyright © 1988 by Marcel Dekker, Inc.

the cold end, and a *Seebeck voltage* is produced in each wire. Owing to the intentional dissimilarity between the Seebeck voltage in the two wires that compose the thermocouple, a voltage difference is established between the two wires, which can be measured and calibrated to temperature. A thermistor, on the other hand, directly correlates the change in resistivity of a substance to temperature. Although thermocouples are more widely utilized due to their inherent stability and larger operating temperature range, the thermistor concepts allows us to illustrate a thermoelectric effect that utilizes composite materials.

Thermistors can exhibit two primary resistance effects (see Figure 6.45): the *negative temperature coefficient* (NTC) effect, and the *positive temperature coefficient* (PTC) effect. In composites, these transitions arise from an individual component, such as a semiconductor–metal transition exhibited by V_2O_3 metallic filler, or result from a product property of the composite. Traditionally, PTC thermistors have been prepared from doped-$BaTiO_3$, such that the grains are semiconducting and the grain boundaries are insulating. The PTC effect occurs during heating through the Curie point, at which point the crystal structure changes from tetragonal to cubic, accompanied by a ferroelectric transition (see Section 6.1.2.2). In the ferroelectric state, the spontaneous polarization compensates for the insulating grain boundaries, resulting in a low resistivity. Although $BaTiO_3$ thermistors exhibit relatively large (six orders of magnitude) changes in resistivity with temperature, they have high room-temperature resistivities and are expensive to produce.

Composite-based PTC thermistors are potentially more economical. These devices are based on a combination of a conductor in a semicrystalline polymer—for example, carbon black in polyethylene. Other fillers include copper, iron, and silver. Important filler parameters in addition to conductivity include particle size, distribution, morphology, surface energy, oxidation state, and thermal expansion coefficient. Important polymer matrix characteristics in addition to conductivity include the glass transition temperature, T_g, and thermal expansion coefficient. Interfacial effects are extremely important in these materials and can influence the ultimate electrical properties of the composite.

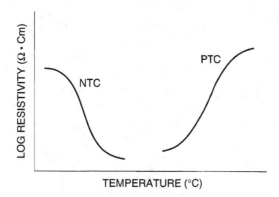

Figure 6.45   Schematic illustration of negative temperature coefficient (NTC) and positive temperature coefficient (PTC) effects. Reprinted, by permission, from D. M. Moffatt, J. Runt, W. Huebner, S. Yoshikawa, and R. Nenham, in *Composite Applications*, T. L. Vigo and B. J. Kinzig, eds., p. 52. Copyright © 1992 by VCH Publishing, Inc.

There is a relatively sharp increase in resistivity near the melting temperature of the polymer, T_m, at which point a discontinuous change in specific volume occurs. The PTC phenomenon in these composites is believed to be the result of a separation of the conducting particles due to the thermal expansion of the polymer. Because the transition occurs at the melting point, one may control the PTC transition by utilizing a different polymer matrix. Composites prepared with crystalline polymers can also exhibit dramatic NTC effects just above the melting point, due to the relaxation of the polymer structure. This effect may be eliminated by crosslinking the polymer or adding a third filler to stabilize the matrix.

Although a majority of these composite thermistors are based upon carbon black as the conductive filler, it is difficult to control in terms of particle size, distribution, and morphology. One alternative is to use transition metal oxides such as TiO, VO_2, and V_2O_3 as the filler. An advantage of using a ceramic material is that it is possible to easily control critical parameters such as particle size and shape. Typical polymer matrix materials include poly(methyl methacrylate) PMMA, epoxy, silicone elastomer, polyurethane, polycarbonate, and polystyrene.

The resistivity–temperature behavior for a V_2O_3–PMMA composite is shown in Figure 6.46. When the NTC transition in the V_2O_3 is combined with the PTC effect in the composite, a resistivity–temperature curve with a square-well appearance results. As the amount of V_2O_3 is increased, the PTC effect begins to appear near 50 vol% V_2O_3. The optimum composition is that which yields the lowest room-temperature resistivity and the largest PTC effect. As the amount of conducting filler is increased, the temperature at which the PTC effect occurs moves to higher temperatures. The PTC exhibited in the composite is related to a phenomenon called *percolation*, which deals with the number and properties of neighboring entities and the existence of a sharp transition at which long-range connectivity in the system appears or disappears. At low

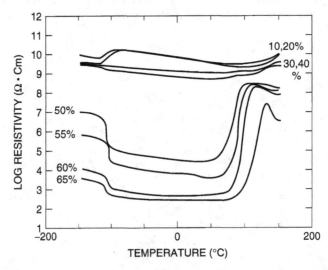

Figure 6.46 Temperature dependence of resistivity for a V_2O_3–PMMA composite of various volume fractions V_2O_3 at 1 kHz. Reprinted, by permission, from D. M. Moffatt, J. Runt, W. Huebner, S. Yoshikawa, and R. Nenham, in *Composite Applications*, T. L. Vigo and B. J. Kinzig, eds., p. 56. Copyright © 1992 by VCH Publishers, Inc.

filler concentrations, a high resistivity is due to the insulating phase. As the volume fraction is increased, conducting paths start to form, resulting in a decrease in resistivity. This region is known as the *percolation threshold*. At higher filler concentrations, a saturation region is reached wherein the resistivity is relatively constant due to extensive particle contacts and eventually approaches the value of the pure conductor.

6.1.5 Electrical Properties of Biologics

The principles introduced in the previous sections have important applications in the description of many biological processes. Both ionic and electron charge transfer play important roles in bioprocesses such as nerve and muscle cell function. While an understanding of these processes would be a worthwhile academic exercise, there is little new to be learned about the materials involved—that is, structure–property relationships for the biological materials. What would be useful to us is a description of the electrical properties of the biological materials—for example, the electrical conductivity of various tissues, or the ion-conducting capacity of membranes. Unfortunately, while some of this information may indeed exist, it is difficult to find all the necessary information, and it is even more difficult to make generalizations. In other words, the human body, and other biological units like it, are complex chemoelectric devices that are not yet fully understood from a materials standpoint. There are any number of ongoing scientific studies aimed at better understanding these biological processes and utilizing them for the development of biomaterials, such as the stimulation of nerve cell growth using electrically conducting polymers [6], the electrical properties of amphiphilic lipid membranes [7], and the electrical properties of hydroxyapatite thin films [8]. These topics, while fascinating, are much too specific for this introductory text. Instead, let us concentrate on two, more general, classes of artificial biomedical devices that may give us some insight into the electrical processes in biologics: biosensors and bioelectrodes.

6.1.5.1 Biosensors. The term *bioelectrode* is broadly used to denote a class of devices that transmit information into or out of the body as an electrical signal. Those bioelectrodes that transmit information out of the body are generally called *biosensors*. Any type of sensor falls into one of two general categories: physical or chemical. Physical parameters of biological interest include pressure, volume, flow, electrical potential, and temperature. Chemical sensing involves the determination of the concentration of chemical species in a volume of gas, liquid, or tissue. The species can vary in size from a hydrogen ion to a live pathogen. When the species is complex, such as a parasite or a tumor, an interaction with another biological entity is required to recognize it. In this section, we give two examples of chemical biosensors: ion sensors and glucose sensors.

Many simple ions such as K^+, Na^+, Cl^-, and Ca^{2+} are normally kept within a narrow range of concentrations in the body, and they must be monitored during critical care. Potentiometric sensors for ion, also called *ion-selective electrodes* or ISEs, utilize a membrane that is primarily semipermeable to one ionic species. The ionic species is used to generate a voltage that generally obeys the Nernst equation [cf. Section 3.1.3.2 and Eq. (3.24)]

$$E = C + \frac{RT}{zF} \ln[a_i + k_{ij}(a_j)^{z/y}] \tag{6.55}$$

where E is the potential in response to an ion, i, of activity a_i and charge z; C is a constant; k_{ij} is the selectivity coefficient; and j is any interfering ion of charge y and activity a_j.

Glasses exist that function as selective electrodes for many different monovalent and some divalent cations. Alternatively, a hydrophobic membrane can be made semipermeable if a hydrophobic molecule called an *ionophore* that selectively binds an ion is dissolved in it. The selectivity of the membrane is determined by the structure of the ionophore. Some ionophores are natural products, such as gramicidin, which is highly specific for K^+, whereas others such as crown ethers and cryptands are synthetic. Ions such as S^{2-}, I^-, Br^-, and NO_3^- can be detected using quaternary ammonium cationic surfactants as a lipid-soluble counterion. ISEs are generally sensitive in the 10^{-1} to 10^{-5} M range, but are not perfectly selective. The most typical membrane material used in ISEs is polyvinyl chloride plasticized with dialkylsebacate or other hydrophobic chemicals.

Biosensors are also available for glucose, lactate, alcohol, sucrose, galactose, uric acid, alpha amylase, choline, and L-lysine. All are amperometric sensors based on O_2 consumption or H_2O_2 production in conjunction with the turnover of an enzyme in the presence of substrate. In the case of glucose oxidase reaction, the normal biological reaction is:

$$\text{Glucose} + O_2 + H_2O \rightleftharpoons \text{Gluconic acid} + H_2O_2 \tag{6.56}$$

Under many circumstances, the concentration of oxygen is rate-limiting, so the sensor often measures not glucose, but the rate at which oxygen diffuses to the enzyme to reoxidize its cofactor. There are two electrochemical ways to couple the reaction to electrodes: (a) monitor depletion of oxygen by reducing what is left at an electrode or (b) monitoring buildup of H_2O_2 by oxidizing it to O_2 and protons:

$$H_2O_2 \rightleftharpoons O_2 + 2H^+ + 2e^- \tag{6.57}$$

An alternative method is to use electrochemical mediators that are at a higher concentration that O_2 and can therefore be shuttled back and forth between the protein and the electrode faster than the enzyme is reduced, so that the arrival of the glucose is always rate-limiting. A typical chemical that works in this way is ferrocene, which is an iron cation between two cyclopentadienyl anions, as shown in Figure 6.47. It exists in neutral and $+1$ oxidation state that are readily interconvertible at metal or carbon electrodes.

Most of the sensors are macroscopic and are employed in the controlled environment of a clinical chemistry analyzer. However, complete glucose sensors are available that

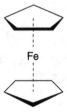

Figure 6.47 The structure of ferrocene.

contain disposable glucose oxidase-based electrodes, power supply, electronics, and readout in a housing the size of a ballpoint pen. The ultimate goal is an implantable glucose sensor, toward which much progress is being made.

6.1.5.2 Electrical Stimulation Devices.

Bioelectrodes that transmit electrical signals into the body are generally known as *electrical stimulation* devices, examples of which include cardiac pacemakers, transcutaneous electronic nerve stimulators (TENs) for pain suppression, and neural prostheses such as auditory stimulation systems for the deaf and phrenic nerve stimulators for artificial respiratory control. In these, and other similar devices, electrodes transmit current to appropriate areas of the body for direct control of, or indirect influence over, target cells.

Electrically excitable cells, such as those of the nervous system or heart, possess a potential difference across their membranes of approximately 60–90 mV. The inside of the cell is negative with respect to the outside. Such cells are capable of transmitting electrical signals, called *action potentials*, along their lengths. During an action potential, the cellular membrane (cf. Figure 1.89) changes polarity for a duration of about 1 ms, so that the inside of the cell becomes positive with respect to the outside. Excitable cells can be artificially stimulated by electrodes that introduce a transient electric field of proper magnitude and distribution. In general, the field generated near a cathodically driven electrode can be used to depolarize most efficiently an adjacent excitable cell's axonal process above a threshold value at which an artificially generated action potential results. Cathodes will tend to draw current outward through nearby cell membranes. In their passive state, such membranes can be electrically modeled as parallel arrangements of resistors and capacitors. Outward current therefore elicits depolarizing resistive potential drops and capacitive charging. Depending upon the specific application and electrode properties, a single action potential in a nerve cell might be elicited using currents on the order of a few microamps to a few milliamps, for durations of micro- to milliseconds.

The materials used in bioelectrodes for electrical stimulation vary, depending upon the specific application. For example, intramuscular stimulation requires an electrode with excellent long-term strength and flexibility, but not very high stimulation charge densities. Stimulation of the visual cortex through placement of electrodes on the cortical surface, however, requires little mechanical flexibility in an electrode material with relatively moderate charge needs. Intracortical stimulation electrodes must be fabricated on a very small scale and must be capable of injecting very high charge densities. In all instances, the device must be biocompatible. In most neural control devices, the surface area of the exposed material is relatively small (< 1 cm^2) so that problems associated with passive biocompatibility are minor. The noble metals such as platinum, iridium, rhodium, gold, and palladium are generally preferred for the conductor material due to their low chemical reactivity and high resistance to corrosion. The insulating materials around the conductive electrodes, however, have a much higher volume and surface area, and they represent the largest problems associated with biocompatibility. Thin layers of medical-grade silicone rubber, poly(tetrafluoroethylene), polyimide, or epoxy are typically used for this purpose. Other devices are fabricated using lithography and thin-film technology that are typically insulated with ceramics such as SiO_2 and Si_3N_4. These insulting materials must be carefully selected to provide pinhole-free coatings, good adhesion to the conductors, and biocompatibility with the tissue.

Once inserted, charge transfer between the metal electrode and biological tissue requires a change in charge carriers from electrons in the metal to ions in the tissue fluid. This change in charge carriers occurs by one of two mechanisms. The first is a capacitive mechanism involving only the alignment of charged species at the electrode–tissue interface, in which charging and discharging occurs at the electrode double layer. This mechanism is preferred because no chemical change occurs in either the tissue or the electrode. The amount of charge that can be transferred solely by capacitive charging is only about 20 μC/cm^2 of "real" electrode surface area, which is the actual area of the electrode–tissue interface. The second mechanism involves the exchange of electrons across the electrode–tissue interface and therefore requires that some chemical species be oxidized or reduced. Metal electrodes almost always inject charge by this so-called *faradaic charge transfer process*, because the amount of charge required greatly exceeds that available from the capacitive mechanism alone. The types of reactions that occur at the electrode–tissue interface include monolayer oxide formation and reduction, the electrolysis of water, or the oxidation of chloride ions.

Capacitor electrodes are considered the ideal type of stimulation electrode because the introduction of a dielectric at the electrode–tissue interface allows charge flow completely by charging and discharging of the dielectric film. The oxide film that is produced on tantalum by anodic polarization withstands substantial voltage without significant current leakage and permits charge flow without the risk of faradaic charge transfer reactions. Anodized titanium has higher dielectric strength than anodized tantalum, but its current leakage is too high for electrode applications. Capacitor electrodes based on tantalum–tantalum pentoxide have a charge storage capacity of about 100–150 μC/cm^2. Their use is limited to neural prosthesis applications having electrodes about 0.05 cm^2 or larger in surface area, such as peripheral nerve stimulators. It is not possible to obtain adequate charge storage capacity in electrodes smaller then 10^{-3} cm^2.

6.2   MAGNETIC PROPERTIES OF MATERIALS

Just as materials have a response when placed in an electric field, they can have a response when placed in a magnetic field. We will see in this section that many of the concepts of permanent dipoles and dipole alignment in response to an applied field that were described in the context of electrical fields apply to magnetic fields as well. There are a few differences, however, and we will also see that there are fewer materials with specialized magnetic properties than there were with specialized electrical and electronic properties. The magnetic properties of materials are nonetheless important, and they are applied in a number of technologically important areas.

6.2.1   Magnetic Properties of Metals and Alloys

6.2.1.1   The Molecular Origins of Magnetism. Recall from Section 1.0.3.2 that charge separation in a molecule leads to an electric dipole, and also recall that this dipole can be either permanent (intrinsic) or induced. And just as the magnitude of the electric dipole moment, p_e, was proportional to the charge and separation distance, so is the magnitude of the *magnetic dipole moment*, p_m, proportional to the pole strength, m, and the separation distance, a:

$$p_m = m \cdot a \tag{6.58}$$

Intrinsic magnetic dipole moments arise from two sources: (a) the orbital motion of electrons about the nucleus and (b) the net spin of unpaired electrons. In the first case, the intrinsic magnetic dipole is not observed until acted on by an external field. In the presence of an external magnetic field, the component of the magnetic moment that is parallel to the applied field is given by $m_l(eh/4\pi m_e)$, where m_l is the magnetic quantum number, e and m_e are the magnitude of the electronic charge and mass, respectively, and h is Planck's constant. Any filled shell or subshell contributes nothing to the orbital magnetic moment of an atom, but occupied outer electronic orbitals contribute to a net atomic magnetic moment (parallel to the applied field) if the summation of magnetic quantum numbers is not equal to zero. The quantity $eh/4\pi m_e$ is a fundamental quantity in magnetism called the *Bohr magneton* and is assigned the symbol μ_B. The Bohr magneton has a value of $9.27 \times 10^{-24} A \cdot m^2$. The quantity $h/2\pi$ is the angular momentum of the electron. The second type of intrinsic magnetic dipole moment is associated with the spin of an electron. When the spin quantum number, m_s, is $+\frac{1}{2}$ (spin up), the magnetic moment is $+\mu_B$ (parallel to the magnetic field), and when $m_s = -\frac{1}{2}$ (spin down), the magnetic moment is $-\mu_B$ (antiparallel to the magnetic field). For free atoms like Na and Mg, no permanent magnetic moment exists because there are no unpaired electrons and the individual orbital momentums sum to zero. Free atoms like Na and molecules like O_2, each of which have unpaired electrons (see Figure 1.7), possess a permanent magnetic dipole moment. When both spin and orbital components of the magnetic dipole moment exist, they are additive, though not in a simple arithmetic manner.

Recall from Section 1.0.4.2 that the pairing of electrons in molecular orbitals of molecules results in two types of magnetism. Diamagnetism is the result of all paired electrons in the molecular orbitals, whereas paramagnetism results from unpaired electrons in the molecular orbitals. Let us now examine how these two types of magnetism behave in the presence of an imposed magnetic field. This can be done by comparing the *magnetic induction, B,* or internal magnetic field strength, within a substance in the presence of an external magnetic field, H, to the induction produced by the field alone in a vacuum, B_0.

First, consider the case of a field in a vacuum. The field can be created by passing current, I, through a coil of N turns (called a *solenoid*) of total length, l, as shown in Figure 6.48. The field is simply given by:

$$H = \frac{N \cdot I}{l} \tag{6.59}$$

Recall from Eq. (6.2) that the induction and field are related as follows:

$$B_0 = \mu_0 H \tag{6.60}$$

where the subscript "0" has been added to indicate a vacuum, and boldface has been dropped to indicate that the inductance and field—normally vectors—are treated as scalar quantities only. The constant μ_0 is the *magnetic permeability in vacuum* and has a value of $4\pi \times 10^{-7}$ Wb/A · m.

If a substance is now placed inside the solenoid, as shown in Figure 6.49, the magnetic field remains the same according to Eq. (6.59) as long as the current remains

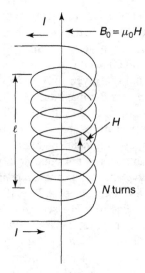

Figure 6.48   Schematic diagram of a solenoid in a vacuum. From K. M. Ralls, T. H. Courtney, and J. Wulff, *Introduction to Materials Science and Engineering*. Copyright © 1976 by John Wiley & Sons, Inc. This material is used by permission John Wiley & Sons, Inc.

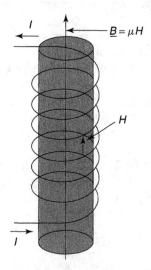

Figure 6.49   Schematic diagram of a substance within a solenoid. From K. M. Ralls, T. H. Courtney, and J. Wulff, *Introduction to Materials Science and Engineering*. Copyright © 1976 by John Wiley & Sons, Inc. This material is used by permission John Wiley & Sons, Inc.

the same, but due to interactions between the field and the substance, the inductance is different (lower or higher), and Eq. (6.60) becomes

$$B = \mu_0 H + \mu_0 M \qquad (6.61)$$

where the second term contains the *magnetization, M*, of the substance. Magnetization represents the net magnetic dipole moment per unit volume that is aligned parallel to the external field. For many substances, a direct proportionality exists between the magnetization and the magnetic field:

$$M = \chi H \tag{6.62}$$

with the dimensionless proportionality constant, χ, called the *magnetic susceptibility*. Do not confuse the magnetic susceptibility with the electronegativity of an element, which is represented by the same variable. The benefit of introducing magnetic susceptibility is that it is a material property, independent of quantity. Even if a direct proportionality does not exist between the magnetization and the magnetic field, Eq. (6.62) still defines the instantaneous magnetic susceptibility and can be substituted into Eq. (6.61) to yield

$$B = \mu_0(1 + \chi)H \tag{6.63}$$

The quantity $(1 + \chi)$ is called the *relative magnetic permeability* (also dimensionless), which is also a material property.

Cooperative Learning Exercise 6.5

A coil of wire 0.20 m long and having 200 turns carries a current of 10 A.

Person 1: Calculate the magnetic field strength in vacuum.

Person 2: Look up the magnetic susceptibility for titanium.

A bar of titanium is now placed in the coil. Combine your information to make the following calculations.

Person 1: Calculate the flux density (magnetic induction) with the titanium bar in the coil.

Person 2: Calculate the flux density (magnetic induction) in vacuum (without the titanium bar).

Compare your answers. Does the titanium bar affect the magnetic induction significantly? Why or why not?

Person 1: What is the relative magnetic permeability of titanium?

Person 2: What is the magnetization of titanium inside the coil?

Answers: Eq. (6.59) $H = (200)(10)/(0.2) = 10,000$ A/m; from Table 6.15 $\chi = 0.182 \times 10^{-3}$. Eq. (6.63) $B = 4\pi \times 10^{-7}(1 + 0.182 \times 10^{-3})(10,000) = 0.0126$ Wb/m^2; $B_0 = 4\pi \times 10^{-7}(10,000) = 0.0125$ Wb/m^2; they are essentially the same due to the small susceptibility of titanium. $(1 + \chi) = 1.000182$; Eq. (6.62) $M = (0.000182)(10,000) = 1.82$ A/m.

With this background on magnetic field production and susceptibility in hand, we can return to our description of diamagnetic and paramagnetic behavior in an applied magnetic field. Substances consisting of atoms or molecules that do not possess permanent magnetic moments exhibit diamagnetism in a magnetic field. That is, they display a negative, but generally quite small, magnetic susceptibility because of the weak, induced opposing magnetic dipole moments. An induced diamagnetic component of magnetization is inherent in all substances, but is often masked by larger intrinsic moments that reinforce an applied field. Inert gases and solids like Bi, Cu, MgO, and diamond are examples of diamagnetic materials and have typical magnetic susceptibilities on the order of -10^{-5}. The diamagnetic component of susceptibility is independent of temperature and magnetic field. For the most part, diamagnetic materials are of little engineering significance, with the notable exception of superconductors. Type I superconductors (cf. Section 6.1.2.4) can be perfectly diamagnetic ($\chi = -1$ or $B = 0$) up to the critical magnetic field because induced shielding currents persist indefinitely. Type II superconductors exhibit perfect diamagnetism or partial paramagnetism depending on the applied field (cf. Figure 6.31). The strong diamagnetism of superconductors makes it possible to use such materials for magnetic field shielding purposes.

Substances having magnetic susceptibilities in the range from about 10^{-6} to 10^{-2} are called paramagnetic. As described above, paramagnetism results from net intrinsic magnetic moments associated with electronic orbitals and electron spins. For paramagnetic materials, the interactions between atomic moments and an applied field can be described in terms of thermodynamics. When the individual atomic moments are randomly arrayed, the entropy of the material is high. Countering this tendency is the magnetic interaction energy, which is decreased when the moments align with the field. The final degree of moment alignment, therefore, depends on the competing factors of energy and entropy. Thus, the susceptibility, which reflects the degree of alignment, tends to decrease as temperature increases and entropy becomes more dominant. The results of thermodynamic calculations that minimize the free energy as a function of temperature, and they show that the susceptibility decreases with temperature and at intermediate temperatures is given approximately by

$$\chi = \frac{\mu_0 N p_m^2}{3k_B T} \tag{6.64}$$

where p_m is the magnetic dipole moment per atom as given in Eq. (6.58), N is the number of atoms per unit volume, k_B is Boltzmann's constant, T is absolute temperature, and μ_0 is the magnetic permeability in vacuum constant. The dependence of magnetic susceptibility on reciprocal absolute temperature is called the *Curie law*, and it holds for most paramagnetic nonmetallic materials above room temperature. As shown in Figure 6.50, the Curie Law does not hold at low temperatures, and the susceptibility approaches a constant value due to saturation.

In metals, conduction electrons provide both paramagnetic and diamagnetic moments, with a net paramagnetic moment, in addition to the induced diamagnetic moment resulting from core electrons. The paramagnetic component arising from the conduction electrons usually predominates, and most metals have positive susceptibilities. The origin of this paramagnetism can be explained in terms of the conduction electron energy band viewed as two subbands, one containing electrons with spin up and the other containing valence electrons with spin down. With no applied magnetic field, the subbands have equal populations of electrons and there is no net spin. Application of a magnetic field lowers the potential energy for the subband having spins parallel to the field and raises the potential energy of the other subband. Since the Fermi energy must be at the same level in both subbands when equilibrium exists (i.e., the electronic free energy is constant), a greater number of conduction electrons reinforce the field than oppose it, as illustrated in Figure 6.51. Those electrons with moments aligned parallel to the field ($m_s = +\frac{1}{2}$) have a magnetic potential energy $2\mu_0 \mu_B H$ lower than those electrons with moments antiparallel ($m_s = -\frac{1}{2}$).

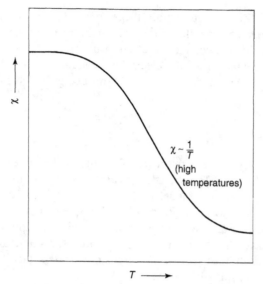

Figure 6.50 Temperature dependence of magnetic susceptibility for a paramagnetic material. From K. M. Ralls, T. H. Courtney, and J. Wulff, *Introduction to Materials Science and Engineering*. Copyright © 1976 by John Wiley & Sons, Inc. This material is used by permission John Wiley & Sons, Inc.

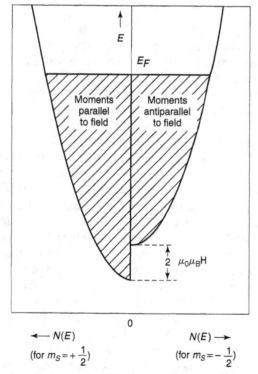

Figure 6.51 Distribution of electrons within the conduction band of a metal in the presence of an applied magnetic field. From K. M. Ralls, T. H. Courtney, and J. Wulff, *Introduction to Materials Science and Engineering*. Copyright © 1976 by John Wiley & Sons, Inc. This material is used by permission John Wiley & Sons, Inc.

Equalization of the Fermi energy within these subbands leads to a greater population of electrons with moments parallel to the field and a resulting net magnetic moment.

It can be shown that the conduction electron net spin susceptibility is proportional to the temperature coefficient of the electronic heat capacity [cf. Eq. (4.42)] and, for free electrons in a single band, having the Fermi energy much lower than any band gap, is given by

$$\chi = \frac{\mu_0 N_v \mu_B^2}{E_F} \tag{6.65}$$

where E_F is the energy of the Fermi level, N_v is the total number of valence electrons per unit volume [see Eq. (6.23)], μ_0 is the magnetic permeability in vacuum constant, and μ_B is the Bohr magneton. Figure 6.51 shows us that only those electrons having energies near the Fermi level are responsible for spin paramagnetism, and this has been accounted for in Eq. (6.65). In general, the spin susceptibility is proportional to the density of states at the Fermi energy, so that transition metals exhibit larger paramagnetic susceptibilities than do nontransition metals (see Table 6.15), just as they exhibit larger electronic heat capacities.

Some metals are diamagnetic because the conduction electron spin susceptibility is smaller than the induced diamagnetic susceptibility component. On the other hand, various rare earth metals display very strong paramagnetism because of unpaired f electrons that remain associated with individual atoms rather than entering into energy bands.

6.2.1.2 Types of Magnetism.

In addition to diamagnetic and paramagnetic behavior, there are other types of magnetism which are important. These include *ferromagnetism, antiferromagnetism,* and *ferrimagnetism.* The various types of magnetism will be summarized and compared here, and then described in more detail under the appropriate material-related section. In most cases, the different types of magnetism can be differentiated between by their variation of magnetic susceptibility with temperature.

The elements Fe, Co, Ni, Gd, Tb, Dy, Ho, and Tm, as well as a number of alloys and compounds, exhibit very large positive susceptibilities below a critical temperature, θ_c, known as the ferromagnetic *Curie temperature.* (We use the Greek lowercase theta and subscript "c" to designate the Curie temperature to differentiate it from the critical transition temperatures of superconductors, which are designated as T_c.) Such substances display spontaneous magnetization, the origin of which arises from an internal interaction that causes a permanent offset in the electronic subbands. Only certain metals and other solids having partially filled d subshells or partially filled f subshells are capable of ferromagnetism (see Table 6.16). In addition to this necessary, but not sufficient, condition for spontaneous permanent magnetization, the interatomic spacing must be such that an internal interaction, termed the *exchange interaction,* occurs. As shown in Table 6.16, the elements Fe, Co, and Ni are magnetized spontaneously, whereas Cu does not. This is because in Fe, Co, and Ni, the average number of electrons in the d band having spin up is not the same as that having spin down, and the difference between the two gives a net magnetic moment. With increasing temperature, the lowering of internal energy because of the exchange interaction is opposed by a greater entropy that results from the disorder of randomized spins. Ferromagnetic materials become paramagnetic above θ_c; susceptibilities can be on the order of 250 to 100,000 or more and continue to increase with decreasing temperature

Table 6.15 Magnetic Susceptibilities (×10³) of Common Paramagnetic Metals at 20°C

IA	IIA	IIIA	IVA	VA	VIA	VIIA	VIII	VIII	VIII	IB	IIB	IIIB	IVB	VB
Li 0.014	Be Dia-magnetic													
Na 0.008	Mg 0.012	Al 0.021												
K 0.006	Ca 0.019	Sc 0.264	Ti 0.182	V 0.375	Cr 0.313	Mn 0.871	Fe Ferro-magnetic	Co Ferro-magnetic	Ni Ferro-magnetic	Cu Dia-magnetic	Zn Dia-magnetic	Ga Dia-magnetic		
Rb 0.004	Sr 0.034	Y 0.114	Zr 0.109	Nb 0.226	Mo 0.119	Tc 0.395	Ru 0.066	Rh 0.169	Pd 0.802	Ag Dia-magnetic	Cd Dia-magnetic	In Dia-magnetic	Sn 0.002	
Cs 0.005	Ba 0.007	Rare Earths 0.007	Hf 0.070	Ta 0.178	W 0.078	Re 0.094	Os 0.015	Ir 0.038	Pt 0.279	Au Dia-magnetic	Hg Dia-magnetic	Tl Dia-magnetic	Pb Dia-magnetic	Bi Dia-magnetic

Source: K. M. Ralls, T. H. Courtney, and J. Wulff, *Introduction to Materials Science and Engineering.* Copyright © 1976 by John Wiley & Sons, Inc.

(see Figure 6.52a). Typically, susceptibilities at $0.9\theta_c$ are approximately half those at absolute zero.

In ferromagnetic materials, the exchange interaction gives rise to a parallel alignment of spins. In some metals like Cr, this exchange interaction, which is extremely sensitive to interatomic spacing and to atomic positions, results in an antiparallel alignment of spins. When the strength of antiparallel spin magnetic moments are equal, no net spin moment exists and the resulting susceptibilities are quite small. Such materials are called antiferromagnetic. The most noticeable characteristic of an antiferromagnetic material is that the susceptibility attains a maximum at a critical temperature, θ_N, called the *Néel temperature*.

The maximum in susceptibility as a function of temperature can be explained using thermodynamic reasoning. At absolute zero, the spin moments are aligned as antiparallel as possible, yielding a small susceptibility. With increasing temperature, the tendency for disorder or more random alignment of spin moments become greater,

Table 6.16   Distribution of Electrons and Electronic Magnetic Moments in Some Transition Metals

Element	Fe	Co	Ni	Cu
Total number $(s + d)$ valence electrons	8	9	10	11
Average number of electrons in s band	0.2	0.7	0.6	1.0
Average number of electrons in d band	7.8	8.3	9.4	10.0
Number of d electrons with moments up	5.0	5.0	5.0	5.0
Number of d electrons with moments down	2.8	3.3	4.4	5.0
Net moment per atom (units of Bohr magnetons, μ_B)	2.2	1.7	0.6	0

Source: K. M. Ralls, T. H. Courtney, and J. Wulff, *Introduction to Materials Science and Engineering.* Copyright © 1976 by John Wiley & Sons, Inc.

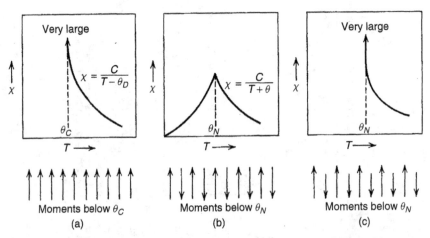

Figure 6.52   The magnetic spin alignments and variation of susceptibility with temperature for (a) ferromagnetic, (b) antiferromagnetic, and (c) ferrimagnetic materials. From K. M. Ralls, T. H. Courtney, and J. Wulff, *Introduction to Materials Science and Engineering.* Copyright © 1976 by John Wiley & Sons, Inc. This material is used by permission John Wiley & Sons, Inc.

and a higher net moment results in the presence of an applied magnetic field. This same process accounts for the decrease in net moment of ferromagnetic materials between absolute zero and θ_c. For antiferromagnetic materials, the spin moments become essentially random at θ_N, and above this temperature, paramagnetic behavior with decreasing χ is observed (see Figure 6.52b). Thus, the Néel temperature of an antiferromagnetic temperature is analogous to the Curie temperature of a ferromagnetic material. We will discuss antiferromagnetic materials in further detail in a later section.

Ferrimagnetism is similar to antiferromagnetism in that the spins of different atoms or ions tend to line up antiparallel. In ferrimagnetic materials, however, the spins do not cancel each other out, and a net spin moment exists. Below the Néel temperature, therefore, ferrimagnetic materials behave very much like ferromagnetic materials and are paramagnetic above θ_N (see Figure 6.52c). Ferrites, a particularly important class of ferrimagnetic materials, will be described later.

The different types of magnetism described so far are summarized in terms of the sign and magnitude of their susceptibilities in Table 6.17. A summary of the terms, symbols, and units used in magnetism are provided in Table 6.18, in a manner analogous to the electrical terms in Table 6.5.

6.2.1.3 Magnetic Domains and Hysteresis.
In addition to susceptibility differences, the different types of magnetism can be distinguished by the structure of the magnetic dipoles in regions called *magnetic domains*. As shown in Figure 6.53, each domain consists of magnetic moments that are aligned, giving rise to a permanent net magnetic moment per domain. The domains are separated by domain boundaries (not the same as the grain boundaries), or domain walls, across which the direction of the spin alignment gradually changes. Typically, the domain wall has a thickness of about 100 nm. This type of arrangement represents the lowest free energy in the absence of an externally applied magnetic field. When the bulk material is unmagnetized, the net magnetization of these domains is zero, because adjacent domains may be oriented randomly in any number of directions, effectively canceling each other out.

Table 6.17 Summary of the Various Types of Magnetism

Type of Magnetic Behavior	Characteristics of Magnetic Susceptibility		Typical Materials
	Sign	Magnitude	
Diamagnetism	Negative	Small[a] χ = constant	Organic materials, superconducting metals, and other metals (e.g., Bi)
Paramagnetism	Positive	Small χ = constant	Alkali and transition metals, rare earth elements
Ferromagnetism	Positive	Large $\chi = f(H)$	Some transition metals (Fe, Ni, Co) and rare earth metals (Gd)
Antiferromagnetism	Positive	Small χ = constant	Salts of transition elements (MnO)
Ferrimagnetism	Positive	Large $\chi = f(H)$	Ferrites ($MnFe_2O_4$, $ZnFe_2O_4$) and chromites

[a]Diamagnetic susceptibilities for superconducting metals are large.

Table 6.18 Magnetic Symbols, Terms, and SI Units

Quantity	Symbol	SI Units Primary	SI Units Derived
Magnetic field strength	H	$C/m \cdot s$	$A/m = 10^{-3}$ oersted (Oe)[a]
Magnetic induction (magnetic flux density)	B	$kg/s \cdot C$	$V \cdot s/m^2 =$ Weber (Wb)$/m^2 =$ tesla (T) $= 10^4$ gauss (G)[a]
Magnetic susceptibility	χ	**(dimensionless)**	**(dimensionless)**
Magnetization	M	$C/m \cdot s$	A/m
Magnetic dipole moment	p_m	$C \cdot m^2/s$	$A \cdot m^2$
Remnant induction	B_r	$kg/s \cdot C$	$V \cdot s/m^2 =$ weber$/m^2 =$ tesla (T) $= 10^4$ gauss(G)[*]
Coercivity (coercive force)	H_c	$C/m \cdot s$	$A/m = 10^{-3}$ oersted(Oe)[*]
Magnetic permeability in vacuum	μ_0	$kg \cdot m/C^2$	Henry/m $=$ Wb/A $\cdot$ m
Saturation magnetization	M_s	$C/m \cdot s$	A/m

[a] cgs unit; items in boldface are material properties.

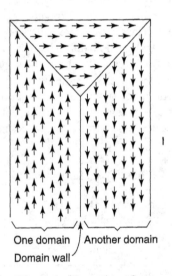

Figure 6.53 Schematic illustration of magnetic domains.

One domain / Another domain
Domain wall

The average magnetic induction of a ferromagnetic material is intimately related to the domain structure. When a magnetic field is applied, the domains most nearly parallel to the direction of the applied field grow in size at the expense of the others. This is called *boundary displacement of domains* or *domain growth*. This process continues until only the most favorably oriented domains remain. The spontaneous moments are aligned naturally with respect to certain crystallographic directions; in the case of iron, they are parallel to the [100] direction. When domain growth is completed, a further increase in the magnetic field causes the domains to rotate and align parallel to the applied field. The material reaches the point of saturation magnetization, and no further change will take place on increasing the strength of the magnetic field.

This change in magnetization with applied magnetic field is shown schematically in Figure 6.54, in a diagram known as a *hysteresis loop*. Hysteresis is defined as the lag in magnetization change that follows variations in the applied magnetic field. If an unmagnetized specimen, such as iron, is placed in a magnetic field, H, a magnetic induction, B, results, as described earlier. When the field is increased, the magnetic induction will also increase, as shown in curve (*a*) of Figure 6.54. This is the domain growth region. At sufficiently high magnetic fields, all of the magnetic dipoles become aligned with the external field, and the material becomes magnetically saturated; that is, the magnetization becomes constant and is called the *saturation magnetization*, M_s. There is a corresponding *saturation flux density*, B_s, as the induction reaches a maximum value. Once magnetic saturation has been achieved, a decrease in the applied field back to zero, as illustrated in curve (*b*) of Figure 6.54, results in a macroscopically permanent induction, called the *remnant induction*, B_r, under which the material is in the magnetized condition, even at zero applied field. At this point, spin orientations within domains have readily rotated back to their favorable crystallographic positions, but the original random domain arrangement is not achieved because domain wall motion is limited and the domain growth process is not entirely reversible. To bring the induction back to zero, a field of magnitude H_c, termed the *coercivity*, must be

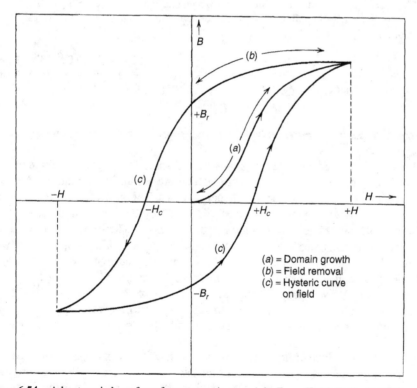

Figure 6.54 A hysteresis loop for a ferromagnetic materials. From K. M. Ralls, T. H. Courtney, and J. Wulff, *Introduction to Materials Science and Engineering*. Copyright © 1976 by John Wiley & Sons, Inc. This material is used by permission John Wiley & Sons, Inc.

applied antiparallel to the original magnetic field, shown as curve (c) in Figure 6.54. A further increase in the magnetic field in the negative direction results in a maximum induction and saturation magnetization, but in the opposite direction of the previous maximum achieved under an applied field in the forward direction. The field can once again be reversed, and the field-induction loop can be closed. The area within the hysteresis loop represents the energy loss per unit volume of material for one cycle and, unless suitable cooling is provided, the material heats up. Smaller hysteresis loops result for cycling between applied fields that do not cause saturation magnetization.

Cooperative Learning Exercise 6.6

Recall that the saturation magnetization, M_s, is the maximum possible magnetization in the material, and is simply the product of the net magnetic moment per atom, P_m, and the number of atoms per unit volume, N. The net magnetic moment, in turn, is related to the electronic structure (paired or unpaired electrons), although a number of other factors come into play. Use this information to calculate the saturation magnetization for nickel.

Person 1: Look up the net magnetic moment of nickel in Table 6.16 and convert it from units of Bohr magnetons to A · m².

Person 2: Look up the density and atomic weight of nickel in Tables 1.11 and 1.3, respectively, and calculate the number of atoms per cubic meter for nickel. (Don't forget Avogadro's number.)

Combine your information to calculate M_s for nickel in units of A/m.

Since the magnetic susceptibilities of ferromagnetic materials are so high, $H \ll M$, and Eq. (6.61) can be simplified. Use this information and your value of M_s to calculate the saturation flux density, B_s.

Answers: $P_m = 0.60$ Bohr magnetons $= 0.60(9.27) \times 10^{-24}$ A · m²
$= 5.56 \times 10^{-24}$ A · m²; $N = P_{Ni}N_A/(A.W.) = \rho_{Ni}N_A/(A.W.) = (8.91)(10^6)(6.02 \times 10^{23})/(58.71)$
$= 9.136 \times 10^{28}$ atoms/m³. $M_s = 5.1 \times 10^5$ A/m. Equation (6.61) simplifies to
$B_s = \mu_0 M_s = 0.638$ Wb/m².

When applied magnetic fields cyclically vary at some finite speed, *eddy currents* are formed. Eddy currents are electrical currents that are set up in the material due to the applied field and that lead to additional energy losses. The magnitude of these currents depends on the frequency and flux density imposed by the magnetic field, as well as on the specific resistance and geometry of the material. The effect of eddy currents is to increase the magnetic field, H, for the same induction, causing a much wider hysteresis loop. The sum of the hysteresis loss and eddy current loss is known as the *total core loss*, and it is usually expressed in watts per pound of a standard sample thickness and at a given magnetic induction.

Magnetic materials can be divided into soft and hard magnets, depending on their behavior in the presence of a magnetic field. This is the topic of the next two sections.

6.2.1.4 Soft and Hard Magnets.

Soft magnets, sometimes called *permeable magnets*, have high magnetic permeabilities and low coercive forces, H_c; they are easily magnetized and demagnetized. In devices subjected to alternating magnetic fields, such as transformer cores, not only is a high magnetic permeability and high saturation induction desirable, but it is also necessary that the material dissipate as little energy per cycle as possible. In other words, the core material must have a hysteresis loop of small area, as characterized by a low coercive force, as illustrated in the inner loop of Figure 6.55.

In ferromagnetic materials, the saturation magnetization is primarily a function of composition, but the susceptibility, coercivity, and shape of the hysteresis curve are very sensitive to structure. A low coercivity coincides with easy motion of domain walls. Domain wall motion is restricted to structural defects such as nonmagnetic inclusions, voids, and precipitates of a nonmagnetic phase. Domain walls are pinned by these defects because the wall energy is lowered. Consequently, the number of such defects must be minimized in a soft magnetic material. Impurities and dislocation structures resulting from cold deformation also lead to higher hysteresis.

Energy losses in soft magnetic materials arise due to both hysteresis and eddy currents, as described in the previous section. Eddy current losses can be reduced by increasing the electrical resistivity of the magnetic material. This is one reason why solid-solution iron–silicon alloys (~4% Si) are used at power frequencies of around 60 Hz and why iron–nickel alloys are used at audio frequencies. Some magnetically soft ferrites (see Section 6.2.2.1) are very nearly electrical insulators and are thus immune to eddy current losses. Some common soft magnetic materials and their properties are listed in Table 6.19. Soft magnetic alloys are described further in Section 6.2.1.6.

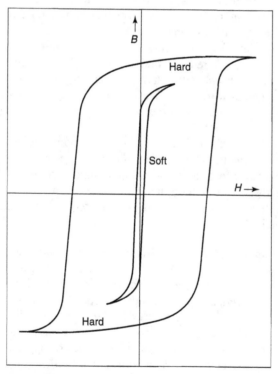

Figure 6.55 Comparison of hysteresis loops for soft and hard magnetic materials. From K. M. Ralls, T. H. Courtney, and J. Wulff, *Introduction to Materials Science and Engineering*. Copyright © 1976 by John Wiley & Sons, Inc. This material is used by permission John Wiley & Sons, Inc.

Table 6.19 Selected Properties of Some Soft Magnetic Materials

Material	Initial Susceptibility (χ at $H{\sim}0$)	Hysteresis Loss per Cycle (J/m^3)	Saturation induction (Wb/m^2)
Commercial iron ingot	250	500	2.16
Fe–4% Si, random	500	50–150	1.95
Fe–3% Si, oriented	15,000	35–140	2.0
45 Permalloy (45% Ni–55% Fe)	2,700	120	1.6
Mu metal			
(75% Ni–5% Cu–2% Cr–18% Fe)	30,000	20	0.8
Supermalloy			
(79% Ni–15% Fe–5% Mo–0.5% Mn)	100,000	2	0.79
Ferroxcube A (Mn, Zn)Fe_2O_4	1,200	~40	0.36
Ferroxcube B (Ni, Zn)Fe_2O_4	650	~35	0.29

Source: K. M. Ralls, T. H. Courtney, and J. Wulff, *Introduction to Materials Science and Engineering*. Copyright © 1976 by John Wiley & Sons, Inc.

6.2.1.5 *Hard Magnets.* Permanent magnets require materials with high remnant inductions and high coercivities. These materials, called *hard magnets*, also generally exhibit large hysteresis losses. A typical hysteresis loop for a hard magnetic material is shown in the outer loop of Figure 6.55, which is shown relative to a soft material for comparison. The term hard magnet, is derived from the mechanical properties of these materials, since increases in mechanical strength in ferromagnetic materials also tend to increase hardness. A relative measure of hardness in magnetic materials is the product of the remnant induction and the coercivity, $B_r \times H_c$, which is roughly twice the energy required to demagnetize a unit volume of material. That is, $(B_r \times H_c)/2$ is roughly twice the area under the B versus H curve between $H = 0$ and $H = -H_c$.

Hard magnetic behavior is intimately related to microstructure. Some permanent magnetic materials, including carbon steel and alloy steels containing combinations of Cr, Co, or W, undergo martensitic transformations upon cooling, resulting in a fine-grained microstructure. Others undergo an order–disorder transformation that results in internal strains. Specific hard magnetic materials will be described further in Sections 6.2.1.6 and 6.2.2.2.

6.2.1.6 *Magnetism in Alloys.* Recall from Table 6.15 that the transition metal Ni exhibits spontaneous magnetization, whereas Cu does not. The difference in magnetic behavior between these two transition metals can be explained in terms of band theory—specifically, the overlap of the s conduction band with the d band immediately below in energy. In the transition metals, the d band is not normally filled entirely. One exception is copper, Cu, which has a $3d^{10}4s^1$ outer shell configuration (cf. Table 1.3). The filled $3d$ band of copper has two separate subbands of opposite electron spin orientation, each holding five electrons. With both subbands filled, the net spin, and hence the net magnetization, of the d band are zero. Nickel, which has one less electron than copper, is ferromagnetic and has a saturation magnetic moment of $0.60\mu_B$ per atom at absolute zero. After a small correction for orbital electron motion, the result is that Ni at saturation has an excess of 0.54 electrons per atom with spin preferentially oriented

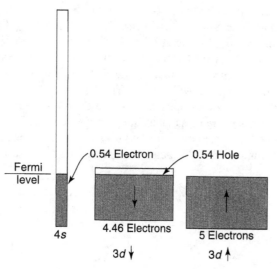

Figure 6.56   Schematic relationship of outer-shell energy bands in nickel at absolute zero.

in one direction. At absolute zero, the electrons are distributed as shown in Figure 6.56. We take one $3d$ subband as filled, and we take the other as filled except for a 0.54 hole. The 0.54 excess electron goes into the $4s$ band as shown, and it creates a net spin in the d-band due to the 0.54 hole in one of its subbands. The separation in energy between the $3d$ subbands is a result of a phenomenon called the *exchange interaction energy*, which is a function of temperature. When Ni is heated above θ_c(631 K), the exchange interaction is dominated by thermal effects, the 0.54 hole distributes itself equally in the two $3d$ subbands as shown in Figure 6.57, and the saturation magnetic moment disappears.

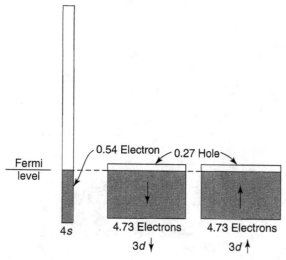

Figure 6.57   Schematic relationship of outer-shell energy bands in nickel over the Curie temperature.

When a series of solid solutions between Cu and Ni is formed, it is observed that the spontaneous magnetization decreases linearly with increasing Cu content until no spontaneous magnetization remains at 60 at % Cu (see Figure 6.58). At this composition, we have added about 0.54 electrons to the d band and about 0.06 electron to the s band. But the 0.54 electron added to the d band in Figure 6.57 will fill both d subbands, and the magnetization will be zero. The 10.6 total valence electrons per atom are now equally divided between spin up and spin down, as shown in Figure 6.59.

This simple band model predicts that an alloying metal with $10 + z$ valence electrons outside a filled p shell (rare gas configuration) is expected to decrease the magnetization of nickel by approximately z Bohr magnetons per solute atom (μ_B). As illustrated in

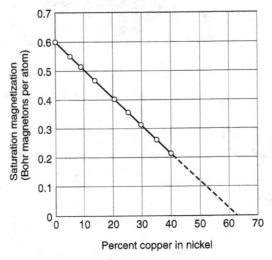

Figure 6.58 Variation in saturation magnetization for a copper–nickel alloy. Reprinted, by permission, from C. Kittel, *Introduction to Solid State Physics*, p. 333. Copyright © 1957 by John Wiley & Sons, Inc.

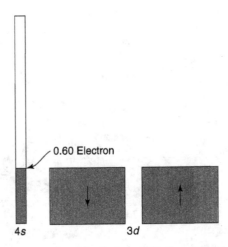

Figure 6.59 Schematic relationship of bands in a 60% Cu, 40% Ni alloy with zero magnetization.

Figure 6.60, this simple relationship is obeyed extraordinarily well for small amounts of such atoms as Sn(+4), Al(+3), Zn(+2), Cu(+1), Co(−1), Fe(−2), and Mn(−3).

It is believed that palladium may have a distribution of electrons similar to that of nickel. At room temperature the picture for palladium is like Figure 6.57 for Ni above the Curie temperature, except that we now have to deal with a $5s$ band instead of $4s$, and $4d$ instead of $3d$. The addition of hydrogen in solution in metallic palladium reduces the susceptibility. The hydrogen is ionized, with the free electrons joining the palladium bands just as the valence electron of copper joins the bands of nickel.

Widely used permanent magnetic materials include low-alloy steels, containing between 0.6% and 1.0% carbon, that are hardened by quenching or precipitation hardening. Another important group is iron−nickel−aluminum alloys with a certain amount of cobalt, known as *Alnico alloys*. The permanent-magnet materials tend to be brittle and difficult to shape, so that they must be cast and finished by grinding. More ductile alloys such as copper−nickel−iron and copper−nickel−cobalt are available that can be cold-worked and machined by ordinary methods. All three types of alloys are subject to precipitation hardening, which can be controlled in such a way as to produce a microscopically heterogeneous structure composed of acicular particles oriented in a required direction. The size of these particles is very small and approaches the dimensions of a magnetic domain (about 1 µm). The reversal of magnetization can then occur only by a complete rotation of the magnetic domain in a particle. This, however, would require a very high magnetizing force because of the crystal anisotropy. Consequently, the boundary displacement of the domains is suppressed, resulting in very high coercive forces.

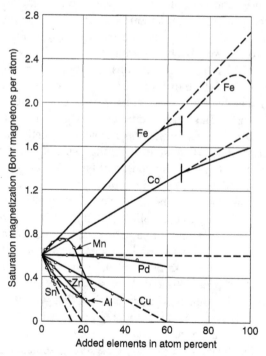

Figure 6.60 Saturation magnetization of some nickel alloys. Reprinted, by permission, from C. Kittel, *Introduction to Solid State Physics*, p. 335. Copyright © 1957 by John Wiley & Sons, Inc.

The same principle has been observed in the production of permanent magnets from powdered materials having particle sizes of colloidal dimensions (0.1 to 0.01 μm). A number of iron-based permanent magnet alloys made from these fine powders have very high coercive forces. For example, a piece of iron has a coercive force of 1 oersted (Oe = 1000A/4πm), whereas a compact made of a fine iron powder may have a coercive force as great as 500 Oe. Other compact powdered alloys, such as iron–cobalt, exhibit still higher coercive forces reaching thousands of oersteds.

As mentioned in the previous section, the magnetization behavior of a single crystal depends upon its crystallographic orientation relative to the applied magnetic field. For example, the easy magnetization directions are [100] for BCC iron and [111] for FCC nickel since these correspond to the natural spin moment alignments. A crystal oriented in the easy magnetization direction is easier to magnetize, or softer, than one oriented in another direction. Beneficial use of this characteristic is realized in the case of Fe–Si alloys that are formed by rolling into a sheet, with appropriate intermediate annealing treatments. When the thermomechanical processing is carefully controlled, a preferred crystallographic orientation develops within the polycrystalline sheet, such that the direction of easy magnetization is parallel to the rolling direction. Such a textured alloy is readily magnetized, and the hysteresis loop is small.

Magnetically soft Fe–Ni alloys can have their properties altered by heat treatment. The compound Ni_3Fe undergoes an order–disorder transformation at about 500°C. Since the susceptibility of the ordered phase is only about half that of the disordered phase, a higher susceptibility is realized when the alloy is quenched from 600°C, a process that retains the high-temperature, disordered structure. Heat treatment of Fe–Ni alloys in a magnetic field further enhances their magnetic characteristics (see Figure 6.61), and the square hysteresis loop of 65 Permalloy so processed is desirable in many applications. A related alloy called Supermalloy (see Table 6.19) can have an initial susceptibility of approximately one million.

Other alloys undergo an order–disorder transformation that results in internal strains. For example, Co −50 at % Pt transforms from a high-temperature, disordered FCC structure to an ordered body-centered tetragonal structure having Co atoms and Pt atoms on alternating (002) planes. The partially ordered alloy can have coercivities as large as 3.7×10^5 A/m. In addition, there are many permanent magnetic materials having two-phase microstructures produced by solid precipitation. Among these are Cu–Ni–Fe alloys (Cunife), Cu–Ni–Co alloys (Cunico), and Fe–Al–Ni–Co–Cu alloys (Alnico). Selected hard magnetic alloys are listed in Table 6.20.

The Alnico-type alloys are commercially very important. At high temperatures, these alloys exist as a homogeneous solid solution which, below a miscibility gap (cf. Section 2.3.1), decomposes into a fine two-phase mixture consisting of FeCo-rich particles within an NiAl-rich matrix. Both product phases tend to be ordered with the CsCl structure. A suitable phase decomposition heat treatment yields particles of the strongly magnetic FeCo-rich phase that are sufficiently small so that each particle contains a single magnetic domain. This, in combination with the fact that the NiAl-rich phase is only weakly magnetic, leads to a high coercivity. The magnetic properties of Alnico alloys can be enhanced by various processing techniques. For example, directional solidification leads to a preferred crystallographic orientation, and phase decomposition in the presence of an applied magnetic field causes particles of the strongly magnetic phase to be elongated in the field direction.

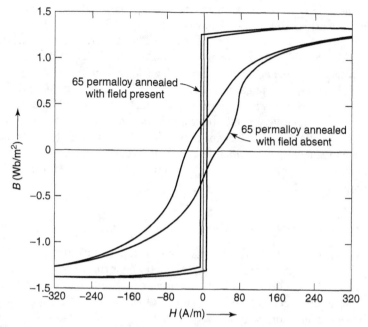

Figure 6.61 Hysteresis loop for Permalloy 65 heat treated with and without an applied magnetic field. From K. M. Ralls, T. H. Courtney, and J. Wulff, *Introduction to Materials Science and Engineering*. Copyright © 1976 by John Wiley & Sons, Inc. This material is used by permission John Wiley & Sons, Inc.

Table 6.20 **Some Hard Magnetic Alloys and Their Selected Properties**

Material	Composition (wt%)	B_r (Wb/m^2)	H_c (kA/m)	$B_r H_c$ (J/m^3)
Martensitic				
carbon steel	98.1 Fe–1 Mn–0.9 C	0.95	4.0	3.8
Tungsten steel	92.7 Fe–6 W–0.7 C–0.3 Cr–0.3 Mn	1.05	5.6	5.9
Chromium steel	95.3 Fe–3.5 Cr–0.9 C–0.3 Mn	0.97	5.2	5.0
Cobalt steel	71.8 Fe–17 Co–8 W–2.5 Cr–0.7 C	0.95	11.9	11.3
Cunife	60 Cu–20 Ni–20 Fe	0.54	43.8	23.7
Cunico	50 Cu–29 Co–21 Ni	0.34	52.5	17.9
Alni–a	59.5 Fe–24 Ni–13 Al–3.5 Cu	0.62	38.0	23.6
Alnico-I	63 Fe–20 Ni–12 Al–5 Co	0.72	35.0	25.2
Alnico-V	51 Fe–24 Co–14 Ni–8 Al–3 Cu	1.25	43.8	54.8
Alnico-XII	52 Fe–24.5 Co–13.5 Ni–8 Al–2 Nb	1.20	64.0	76.8
Co–Pt	77 Pt–23 Co	0.52	246.8	128.3
Ferroxdur	60.3 Fe–12.4 Ba–27.3 O (BaFe$_{12}$O$_{19}$)	0.20	120.0	24.0
Ferroxdur				
(oriented)		0.39	240.0	93.6

Source: K. M. Ralls, T. H. Courtney, and J. Wulff, *Introduction to Materials Science and Engineering*. Copyright © 1976 by John Wiley & Sons, Inc.

Rare-earth magnets have superior magnetic properties and are particularly suitable for use in compact and lightweight devices due to their coercive forces that are up to 10 times that of ordinary magnets. They are alloys of rare-earth elements, which include elements 57 through 71, so-named because of the inherent difficulty in extracting them from ores, not due to their rarity. The most common rare-earth alloys for magnetic applications are samarium–cobalt (Sm–Co) and neodymium–iron–boron (Nd–Fe–B). Sm–Co *rare-earth magnets*, of which the composition $SmCo_5$ is the most common, are relatively stable to changes in temperature. Nd–Fe–B rare-earth magnets, such as $Nd_2Fe_{14}B$, have magnetic properties superior to those of Sm–Co. However, the magnetic flux variation is very sensitive to changes in temperature, making it necessary to select a magnetic material with properties suited to meet its specific service conditions.

6.2.2 Magnetic Properties of Ceramics and Glasses

Most of the important magnetic ceramics are of the ferrimagnetic class. However, some ceramics do exhibit other types of magnetic behavior. These ceramic materials will be described first, followed by a more thorough description of an important class of ferrimagnetic ceramics called ferrites. Finally, a topic related to the magnetic properties of ceramic superconductors will be introduced.

6.2.2.1 *Diamagnetic, Paramagnetic and Antiferromagnetic Ceramics and Glasses.* In general, diamagnetism is associated with all ceramic materials in which the ions have closed electronic shells—that is, no unpaired electrons. This means that ceramics with no transition metal ions or rare-earth ions are generally diamagnetic. We will return to a special class of diamagnetic ceramics in Section 6.2.2.3.

Paramagnetism results from unpaired electrons. As a result, most compounds containing transition, rare-earth, and actinide elements, including oxides, nitrides, carbides, and borides, exhibit paramagnetism. Such ceramics are generally not of importance due to their paramagnetism alone, since they often exhibit other types of magnetism, as well.

Aside from ferrimagnetic ceramics, which is the subject of the next section, the only other type of magnetism exhibited by ceramics is antiferromagnetism. Several of the transition metal monoxides exhibit antiferromagnetic behavior, in which the exchange between the unpaired electrons causes antiparallel spin alignments. These oxides include MnO, FeO, NiO, and CoO, all of which have the rock salt structure. For example, in FeO, the O^{2-} ions possess no net magnetic moment, whereas the Fe^{2+} ions do. Since neutrons have a magnetic moment, neutron diffraction can be employed to determine the spin arrangement, as shown in Figure 6.62. The spins of the *d* electrons of adjacent iron ions are aligned in opposite directions. Furthermore, any ions an a (111) plane have parallel spins, but ions in adjacent (111) planes have antiparallel spins, as illustrated in Figure 6.62. The aligned moments of the ions in the two directions cancel, and the FeO crystal as a whole has no magnetic moment. The ordering of spins in antiferromagnetic substances often introduces a complexity with respect to viewing their crystal structures. Because of the spin ordering, the magnetic unit cell has twice the lattice parameter of the standard unit cell. Some antiferromagnetic ceramics and their magnetic properties are listed in Table 6.21.

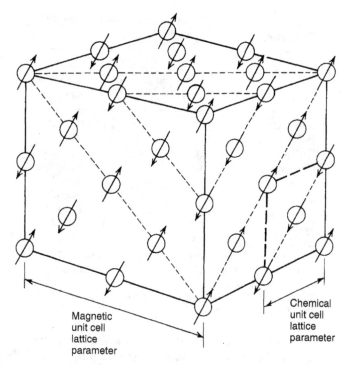

Figure 6.62   Schematic illustration of spin alignments in adjacent (111) planes of FeO. From K. M. Ralls, T. H. Courtney, and J. Wulff, *Introduction to Materials Science and Engineering.* Copyright © 1976 by John Wiley & Sons, Inc. This material is used by permission John Wiley & Sons, Inc.

Table 6.21   Selected Magnetic Properties of Some Antiferromagnetic Ceramics

Substance	Néel Temperature, θ_N, (K)	$\chi(0\text{ K})/\chi(\theta_N)$
MnO	116	0.67
MnS	160	0.82
MnF_2	67	0.76
FeF_2	79	0.72
FeO	198	0.80
CrSb	723	~0.25
$FeCO_3$	35	~0.25

Source: K. M. Ralls, T. H. Courtney, and J. Wulff, *Introduction to Materials Science and Engineering.* Copyright © 1976 by John Wiley & Sons, Inc.

   Before moving on to ferrites, we should at least mention that magnetic glasses do exist, although any further coverage of them is beyond the scope of this text. Not surprisingly, the magnetic glasses are found chiefly in the group of glasses with high contents of transition elements or rare earths, as well as in the corresponding fluoride glasses but especially in the metallic glasses.

Table 6.22   Some   Ferrites   and   their   Magnetic Properties

Composition	Saturation Magnetization (A/m)	Curie Temperature (°C)	Crystal Structure
$MnFe_2O_4$	4.13×10^5	300	Inverse spinel
$FeFe_2O_4$	4.77×10^5	585	Inverse spinel
$NiFe_2O_4$	2.71×10^5	590	Inverse spinel
$CoFe_2O_4$	3.98×10^5	520	Inverse spinel
$CuFe_2O_4$	1.35×10^5	455	Inverse spinel
$BaFe_{12}O_{19}$	3.18×10^5	450	Hexagonal
$SrFe_{12}O_{19}$	3.18×10^5	453	Hexagonal
$MgFe_2O_4$	1.11×10^5	440	Inverse spinel
$Li_{0.5}Fe_{2.5}O_4$	3.10×10^5	670	Inverse spinel
$\gamma - Fe_2O_3$	4.13×10^5	575	Inverse spinel
$Y_3Fe_5O_{12}$	1.35×10^5	287	Garnet

Source: S. Somiya, *Advanced Technical Ceramics*. Copyright © 1984 by Academic Press.

6.2.2.2   *Ferrites.* Ferrites are ceramic materials having the general formula MO · Fe_2O_3, where M is a bivalent element such as Fe, Cu, Mn, Zn, and Ni. The oldest known ferrite with magnetic properties is magnetite (Fe_3O_4), which may be considered a double oxide composed of ferrous and ferric oxides, FeO · Fe_2O_3. Mixed ferrites can also be fabricated in which the divalent cation may be a mixture of ions, such as $Mg_{1-x}Mn_xFe_2O_4$, so that a wide range of composition and magnetic properties are available. The major ferrite solid solutions are Mn–Zn, Ni–Zn, and Cu–Zn. These, along with the Ba and Sr ferrites, form the mainstream of magnetic materials for electronic applications. Table 6.22 presents the most important types of ferrites and their basic properties.

The ferrites generally have one of three crystal structures: inverse spinel, garnet, and hexagonal. The spinel structures (cf. Figure 1.41) have the oxygen ions in a nearly close-packed cubic array. The unit cell contains 32 oxygen ions, with 32 octahedral and 64 tetrahedral sites, of which 16 of the octahedral and 8 of the tetrahedral sites are filled. It is the position of these 24 cations within the unit cell that determines magnetic behavior. The distribution of cations in the sites is specific

to the type of cations, and it must be determined experimentally. There are two idealized spinel structures. In the *normal spinel*, all the divalent ions are on the tetrahedral sites, as in $ZnFe_2O_4$. In the *inverse spinel*, the 8 occupied tetrahedral sites are filled with trivalent ions and the 16 occupied octahedral sites are equally divided between di- and trivalent ions (see Figure 6.63). The prototypical inverse spinel ferrite is magnetite, whose structure consists of an FCC oxygen array with Fe^{2+} and Fe^{3+} ions in the interstices.

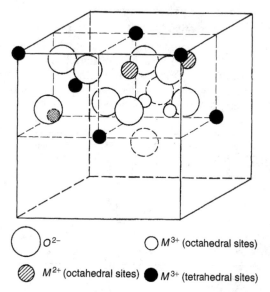

○ O^{2-} ○ M^{3+} (octahedral sites)

◍ M^{2+} (octahedral sites)   ● M^{3+} (tetrahedral sites)

Figure 6.63   A portion of the inverse spinel unit cell found in many ferrites. From K. M. Ralls, T. H. Courtney, and J. Wulff, *Introduction to Materials Science and Engineering.* Copyright © 1976 by John Wiley & Sons, Inc. This material is used by permission John Wiley & Sons, Inc.

Ferrite compounds with the inverse spinel structure are similar to magnetite, with different ions substituting for the iron atoms. As with FeO (cf. Figure 6.62), the oxygen ions have no permanent magnetic moment. Tetrahedral sites in the FCC oxygen array are occupied by half of the trivalent cations, and octahedral sites are occupied equally by divalent cations and the remaining trivalent cations.

The permanent moments of the cations in the octahedral and tetrahedral sites are antiparallel. Thus, the moments of the trivalent cations (Fe^{3+}) cancel out, and the net moment is due to the divalent cations. If the individual divalent cation moments are known, then it is possible to calculate the net moment per formula unit (see Cooperative Learning Exercise 6.7). In a formula unit of magnetite, there is one divalent iron ion that has a total moment of $4\mu_B$ resulting from four unpaired electrons, which is very close to the observed value. Similarly, in $NiFe_2O_4$, the moment per formula unit is $2\mu_B$ identical to that of the divalent nickel ion.

In some systems, and particularly at high temperatures, the cation distribution may be disordered and the cations nearly randomly distributed between tetrahedral and octahedral sites, but generally there is a tendency for individual ions to fit into particular sites so that either the normal or inverse arrangement is preferred. The extent to which inversion of the cations occurs depends on the heat treatment, but, in general, increasing the temperature of a normal spinel causes an excitation of the ions to the inverted position. For example, manganese ferrite is about 80% normal spinel, and this arrangement does not change greatly with heat treatment. Since the Mn^{2+} ion has a moment of $5\mu_B$, the inversion should not affect the net moment of $MnFe_2O_4$. Nickel ferrite has similar properties, and it is 80–90% normal spinel. Magnesium ferrite, however, transforms at high temperature to the normal spinel structure as the divalent magnesium ions are thermally excited onto the tetrahedral sites. In this case, the magnetization is strongly influenced by the cooling rate. Slow cooling allows the inverse spinel structure

---

Cooperative Learning Exercise 6.7

The number of Bohr magnetons contributed by various divalent transition metal ions are summarized below. Since the trivalent ions are equally distributed between half of the occupied octahedral sites (8) and all the occupied tetrahedral sites (8), their moments cancel out, and the net magnetic moment of a ferrite can be predicted from the moment of the divalent ions that occupy the remainder of the octahedral sites (8).

Ion	Magnetic Moment (Bohr magnetons, $\mu_B = 9.27 \times 10^{-24}$ A $\cdot$ m^2)
Mn^{2+}	5
Fe^{2+}	4
Co^{2+}	3
Ni^{2+}	2
Cu^{2+}	1

Person 1: Calculate the net magnetic moment per unit cell for copper ferrite. Remember that there is more than one formula unit (CuFe$_2$O$_4$) per unit cell in the inverse spinel structure.

Person 2: Calculate the net magnetic moment per unit cell for nickel ferrite. Remember that there is more than one formula unit (NiFe$_2$O$_4$) per unit cell in the inverse spinel structure.

Exchange information.

Person 1: Calculate the saturation magnetization, M_s, for nickel ferrite, which has a lattice parameter of $a = 0.833$ nm.

Person 2: Calculate the saturation magnetization, M_s, for copper ferrite, which has a lattice parameter of $a = 0.838$ nm.

$$Answers: p_m(\text{Cu}^{2+}) = 8\mu_B; \; p_m(\text{Ni}^{2+}) = 16\mu_B \cdot M_s(\text{NiFe}_2\text{O}_4) =$$
$$(16)(9.27 \times 10^{-24})/(0.833 \times 10^{-9})^3 = 2.6 \times 10^5 \text{ A/m}; \; M_s(\text{CuFe}_2\text{O}_4) =$$
$$(8)(9.27 \times 10^{-24})/(0.838 \times 10^{-9})^3 = 1.3 \times 10^5 \text{ A/m.}$$

---

to occur because enough thermal energy and time are available to allow the magnesium ions to migrate to the preferred octahedral sites. The saturation moment for a rapidly quenched sample is $2.23\mu_B$; a slow furnace cool results in $1.28\mu_B$. The Mg^{2+} ion has no net moment so that the inverse spinel should have zero magnetization and the normal spinel $10\mu_B$. Magnesium ferrite has high resistivity and low magnetic and dielectric losses that make it most suitable for microwave applications.

Some ferrimagnetic oxides have the garnet structure, which can accommodate large trivalent rare-earth ions with large magnetic moments. The chemical formula for ferrimagnetic garnets is M$_3$Fe$_5$O$_{12}$, where M is a rare-earth ion or a yttrium ion. The garnet Y$_3$Fe$_5$O$_{12}$, called yttrium–iron–garnet (YIG), has a high electrical resistivity and, correspondingly, a very low hysteresis loss even at microwave frequencies.

The hexagonal ferrites have a structure related to the spinel structure but with hexagonal close-packed oxygen ions and a unit cell made up of two formulae of MN$_{12}$O$_{19}$, where M is divalent (Ba, Sr, or Pb) and N is trivalent (Al, Ga, Cr, or Fe), which gives a molecular formula of $M^{2+}O \cdot N_2^{3+}O_3$. The best-known examples are magnetoplumbite, which has a formula $PbFe_{12}O_{19}$, and barium ferrite, $BaFe_{12}O_{19}$. As in the spinel structures, magnetization of these compounds is the result of unpaired electrons, usually in the Fe ions. Hexagonal ferrites are of interest because of their high magnetocrystalline anisotropy and high coercivity, which make them suitable for permanent magnets.

For all the ferrites, a number of other factors influence the magnetic properties beside structure. These include impurities, oxygen partial pressure, grain size, and porosity. For example, nickel ferrite thin films deposited by reaction sputtering at 0°C are noncrystalline, and the magnetic susceptibility shows the material to behave as a paramagnet. If the deposition temperature is raised to 400°C, a micropolycrystalline film results with grain sizes less than 15 nm, and the resulting film shows superparamagnetic behavior. *Superparamagnetism* is the result of clusters of paramagnetism. If there are n clusters, each with N spins, it can be shown that the magnetic susceptibility is now proportional to N^2/n, where in normal paramagnetic materials the Curie Law [Eq. 6.64] tells us that χ is proportional to N. This causes the magnetic moment to increase by orders of magnitude, provided that $N >> n$. If the size of the clusters get so big as to form magnetic domains, the advantage is lost.

Ferrite can be used as soft magnets, hard magnets, and a class of materials known as *semihard magnets*. Soft magnetic ferrite is used for core materials for coils and transformers. Hard magnetic ferrite is used in making permanent magnets. Semihard magnetic ferrites have properties between those of soft and hard magnetic ferrites. They are comparatively easy to magnetize, can retain magnetization, and can produce changes in the magnetic flux from this state, as needed. The typical application for semihard magnetic ferrites is in memory cores for computers. In general, the Mn–Mg ferrites and the lithium ferrites are useful for semihard applications. There is a growing list of applications for ferrites, including microwave applications that utilize the unique magnetic properties of these materials.

6.2.2.3 The Meissner Effect.

In Section 6.1.2.4, we describe the interrelationship between electric current density, magnetic field, and temperature for oxide superconductors, and although a great deal of description was given to upper and lower critical magnetic fields, one particularly important topic related to magnetism was deferred to this section. That topic has to do with an important phenomenon related to superconductivity called the *Meissner effect*. Like superconductivity itself, the Meissner effect is relatively old as far as scientific discoveries go. It was found in 1933 by W. Meissner and colleagues, who determined that a superconducting metal expels any magnetic field when it is cooled below the critical transition temperature, T_c, and becomes superconducting. By expelling the field and thus distorting nearby magnetic field lines, as shown in Figure 6.64b, a superconductor will create a strong enough force field to overcome gravity. This expulsion of the magnetic field is to be distinguished from simply preventing the field lines from penetrating—any metal with infinite conductivity would do the latter. If a magnetic field is already present, and a substance is cooled below T_c to become a superconductor, the magnetic field is expelled. The significance of this difference cannot be explained merely by infinite conductivity.

The ability of the superconductor to overcome gravitational forces has led to the famous demonstration, typically involving an oxide superconductor and a rare-earth magnet, in which the magnet is levitated above the superconductor when it is cooled

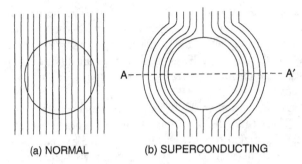

(a) NORMAL   (b) SUPERCONDUCTING

Figure 6.64 Schematic illustration of Meissner effect in which magnetic flux lines that (a) normally penetrate the material are (b) expelled in the superconducting state.

with liquid nitrogen. In this experiment, the magnetic field lines from the strong (but small) magnet penetrate the superconductor to a small depth called the *penetration depth*, λ, which sets up circulating currents—called *eddy currents*—in the supercon-ductor. Typically, the penetration depth is less than 0.05 μm, so that on a macroscopic scale, the superconductor is said to exclude the magnetic field. In actuality, these eddy currents create their own magnetic field lines that exactly cancel the applied magnetic field from the magnet, thus levitating it.

No superconductor can keep out very strong magnetic fields. As described previ-ously, Type I superconductors have a critical magnetic field, H_c, above which they no longer superconduct. Type II superconductors have a range of critical fields from the upper critical field, H_{c2}, to the lower critical field, H_{c1}. Below these critical fields, the superconductor is perfectly diamagnetic. For a perfect diamagnet, $M = -H$, so that according to Eq. (6.61) the inductance is $B = 0$. Thus, the earlier statement that diamagnetism is not important in ceramic materials must be strongly qualified for the oxide superconductors—diamagnetism is critical to their importance.

At first, the superconducting state was not thought to be a thermodynamic equilib-rium state. But, as we know from Chapter 2, any equilibrium state must be the result of a free energy minimization. It can be shown that the superconducting and normal states will be in equilibrium when their free energies are equal, and that the free energy difference between the normal state at zero field, $G_n(H = 0)$, and the superconducting state at zero field, $G_s(H = 0)$, is

$$G_n(H = 0) - G_s(H = 0) = \frac{\mu_0 H_c^2}{2} \tag{6.66}$$

The free energies for the respective states can be determined from heat capacity data, and the thermodynamic critical field can be determined from a plot of free energy versus applied field for the two states, as illustrated in Figure 6.65.

The Meissner effect is a very important characteristic of superconductors. Among the consequences of its linkage to the free energy are the following: (a) The super-conducting state is more ordered than the normal state; (b) only a small fraction of the electrons in a solid need participate in superconductivity; (c) the phase transition must be of second order; that is, there is no latent heat of transition in the absence of any applied magnetic field; and (d) superconductivity involves excitations across an energy gap.

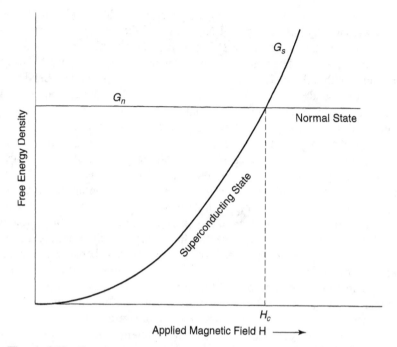

Figure 6.65 Free energy curves for both normal and superconducting states.

6.2.3 Magnetic Properties of Polymers

Much like ceramics, the electrical and magnetic properties of polymers are closely linked. Initially, this statement can be viewed with considerable disappointment, since, with only a few notable exceptions, polymers are poor electrical conductors, and what is known about their potential for magnetization is even less than is known about their electrical conductivity. Nonetheless, there is hope. Just as new classes of polymers are currently being developed with improved electrical conductivities, there is reason to believe that polymers can be developed that will have significant magnetic properties. What a potential benefit this could be for the development of lightweight magnets. We briefly describe here just one topic related to magnetic polymers that is still too "young" in its scientific development to warrant significant coverage in an introductory text: *molecular magnets*. The more common use of polymers in magnetic media, that as a matrix for the support of nonpolymeric magnetic particles, is more appropriately addressed in the section on composites.

6.2.3.1 Molecular Magnets*. Much of the information presented here on molecular magnets is drawn from some excellent, recent review articles [9–12] on this emerging field. Following a brief introduction to the general topic of molecular magnets, a description of organic molecular magnets will be given, with an emphasis on the development of polymeric molecular magnets.

A molecular magnet is a magnet that is molecule-based—that is, composed of molecular or molecular ion entities, and not atoms or atomic ions. This distinction is not new to us, though previously we have not elaborated upon it. Simply compare the crystal structures of such atom-based solids as $BaTiO_3$ in Figure 1.42, in which the positions of the atoms determine the crystal structure, with that of a molecule-based solid such as polyethylene in Figure 1.65, in which the positions of the molecules, in this case chains, determine the crystal structure. Think about this distinction for a moment, and you will realize that most molecular materials are organic in nature. The covalent bonds within the molecules are stronger than the noncovalent bonds that draw the organic compounds into a molecular crystal, such as hydrogen bonding and van der Waals interactions. This is different from metallic and ionic solids, in which any Na–Cl interaction, for example, is essentially the same anywhere in the crystal, so that ordering takes place on the atomic scale.

The magnetic materials described so far in this chapter are all atom-based: Their bulk magnetic properties are due to the incorporation of atoms such as Fe or Nd. In these materials, magnetic order arises from cooperative spin–spin interactions between unpaired electrons located in d or f orbitals of the metal atoms. Molecular-based materials, on the other hand, do not tend to exhibit magnetic properties for some very good reasons. Organic compounds generally have closed-shell structures and are thus diamagnetic. Paramagnetic organic molecules are very rare because the presence of unpaired electrons—which are required for paramagnetism—in the s and p orbitals gives instability to the molecules, which tend to spontaneously form a covalent bond with another molecule, thus pairing the spins. Even if one or more unpaired electrons are maintained stably in an organic molecule, stabilization of a triple state (parallel alignment of spins) requires that certain quantum mechanical conditions be met which are difficult to satisfy, such that they prefer to couple in an antiferromagnetic manner with no spontaneous magnetic moment. Thus, the development of ferro- or ferrimagnetic organic molecules has been elusive. However, recent advances in synthetic molecular chemistry have made possible the formation of organic molecules with sufficient magnetic function that can serve as building blocks for bulk magnetic behavior. Without delving into the theory of magnetism in these materials, we describe some of these building blocks that may one day be the precursors to polymer magnets.

The typical synthetic approach to designing molecular magnets consists of choosing molecular precursors, each of which contains an unpaired spin, as represented by the arrow in Figure 6.66. The precursors are then assembled in such a way that there is

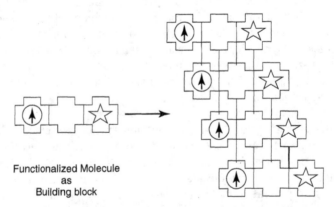

Functionalized Molecule
as
Building block

Figure 6.66   Schematic illustration of building blocks used to construct molecular magnets. An arrow indicates an unpaired electron, and a star represents some other desirable functionality. Reprinted, by permission, from J. V. Yakhmi, *Physica B*, **321**, 206. Copyright © 2002 by Elsevier Science B. V.

no compensation of the spins at the crystal lattice scale. In Figure 6.66, the precursors may contain other functional groups, represented by the star, that can assist in self-assembly, or impart some additional property such as chirality or ferroelectricity. The primary criteria for designing a molecular magnet, then, are that (a) all the molecules in the lattice have unpaired electrons, and (b) the unpaired electrons have spins aligned parallel along a given direction. Since magnetism is a cooperative effect, the spin–spin interaction must extend to all three dimensions in the lattice. These criteria are met in molecules such as the nitronyl nitroxide radical, NITR, shown in Figure 6.67a, where R is an alkyl group. Through different substitutions for the alkyl group, the spin delocalization effect can be fine-tuned. An example is when R is a nitrophenyl compound, which results in 4-nitrophenyl nitronyl nitroxide, or *p*-NPNN for short, as shown in Figure 6.67b. Ferromagnetism has been exhibited in this purely organic compound at around 0.6 K. Other similar, organic compounds have Curie temperatures of 16 K and above.

An alternative strategy is to utilize ferrimagnetic chains containing alternating spins of unequal magnitude and assemble them in such a way that there is a net spin. For example, two different spin carriers such as Mn ions, with $m_s = 5/2$, and Cu ions, with $m_s = 1/2$, can be incorporated within the same molecular precursor, as schematically illustrated in Figure 6.68. It is also possible to create different spins using organic radicals instead of metal ions. An example of such a building block is shown in Figure 6.69 for [Cu(opba)]$^{2-}$, where "opba" stands for ortho-phenylenebis(oxamato). This copper

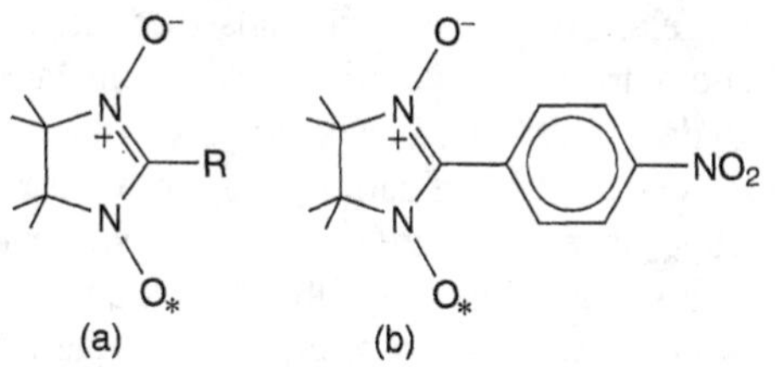

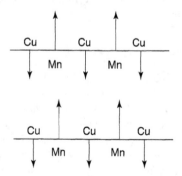

(a)   (b)

Figure 6.67   Chemical structures of (a) NITR and (b) *p*-NPNN used in molecular magnets. An asterisk, *, indicates an unpaired electron. Reprinted, by permission, from J. V. Yakhmi, *Physica B*, **321**, 206. Copyright © 2002 by Elsevier Science B. V.

Figure 6.68   Schematic illustration of spin assembly metal-based ferrimagnetic chains. Reprinted, by permission, from J. V. Yakhmi, *Physica B*, **321**, 206. Copyright © 2002 by Elsevier Science B. V.

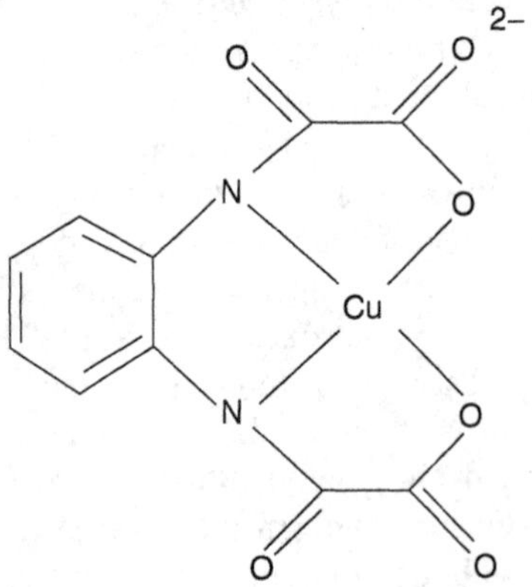

Figure 6.69 Structure of the copper dianion precursor $[Cu(opba)]^{2-}$. Reprinted, by permission, from J. V. Yakhmi, *Physica B*, **321**, 206. Copyright © 2002 by Elsevier Science B. V.

precursor can be reacted with a divalent ion such as Mn^{2+}, which results in an amorphous magnet with a spontaneous magnetization below 6.5 K. Similar precursors can be assembled into crosslinked chains with still higher Curie temperatures, as shown in Figure 6.70. Note that even though these structures contain transition metals such as Mn and Cu, it is the entire *molecule*, including the metal ion, that exhibits the spin, and that can be used as a building block for larger structures.

One final example is a class of related materials that undergo reversible magnetic behavior through a mild dehydration-rehydration process. Because these materials show a reversible crossover under dehydration to a polymerized long-range, magnetically ordered state with spontaneous magnetization, and transformation back to isolated molecular units under rehydration, they are called *molecular magnetic sponges*. Coercivity values for these sponges are high (see Table 6.23), and some even undergo color changes that coincide with the change in magnetic properties. In addition to the "opba" anion described above, these molecular magnetic sponges are composed of "obbz" and "obze" ligands, where obbz = oxamido bis(benzoato) and obze = oxamido-*N*-benzoato-*N'*-ethanoato.

As this field continues to develop, look for improvements not only in the magnetic properties of these organic polymers, but also in the theoretical descriptions of intrinsic ferro- and ferrimagnetism in polymers.

6.2.4 Magnetic Properties of Composites

In this section, we describe the properties of two magnetic composites that represent extremes in terms of applications for magnetic materials. The first, flexible magnets illustrate how the incorporation of less-expensive components into a material can retain the functionality of the magnetic material while significantly reducing cost. The second, magnetic storage media illustrates how a simple concept such as a reversible magnetic moment can be utilized to create an entirely new industry and revolutionize the manner in which we conduct our daily lives.

6.2.4.1 Flexible Magnets. Flexible magnets are made by embedding ferrimagnetic ceramics in a polymer matrix, such as barium ferrite or strontium ferrite in nylon or

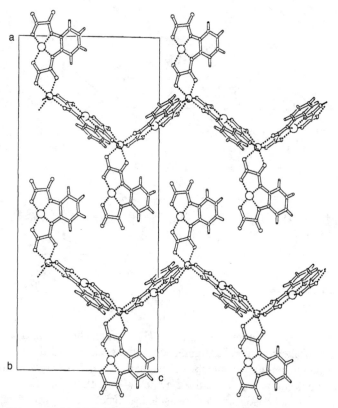

Figure 6.70 Structure of 1:2 Mn(II)Cu(II) compound consisting of a zigzag chain with terminal "opba" groups. Reprinted, by permission, from J. V. Yakhmi, *Physica B*, **321**, 207. Copyright © 2002 by Elsevier Science B. V.

Table 6.23 Selected Magnetic Properties of Some CoCu-Based Molecular Magnetic Sponges

Compound	T_c (K)	H_c (kOe)
$CoCu(pbaOH)(H_2O)_3 \cdot 2H_2O$	38	5.66
$CoCu(pba)(H_2O)_3 \cdot 2H_2O$	33	3
$CoCu(obbz)(H_2O)_4 \cdot H_2O$	25	1.3
$CoCu(obze)(H_2O)_4 \cdot 2H_2O$	20	1

Source: J. V. Yakhmi, *Physica B*, Vol. 321, pp. 204–212. Copyright © 2002 by Elsevier Science B.V.

polyphenylene sulfide. In an anisotropic magnet (one-sided) the ferrite and polymer components are mixed and processed, usually by extrusion, into a thin sheet and are then coated or printed on one side. Typical tape thicknesses are 1–2 mm. To obtain maximum alignment, the ferrite particles are physically rotated in a magnetic field during the molding process. In this way, energy products equivalent to those of cobalt steel are obtained.

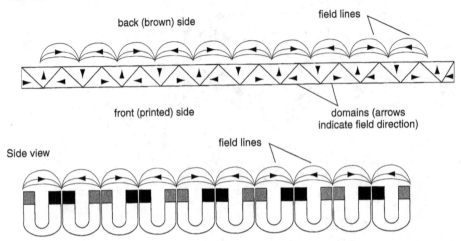

Figure 6.71 Side view of magnetic flux lines in an anisotropic flexible magnet. (mrsec.wisc.edu/edetc).

The magnetic field structure (see Figure 6.71) in an anisotropic magnet is made of a striped pattern of alternating pole alignments on the back side of the flexible magnetic sheet. This arrangement also channels the magnetic field to the back of the magnet so that maximum holding power is obtained, which is important for many applications. It is also possible to make isotropic, or two-sided, magnetic tapes and sheets, for which the field lines shown in Figure 6.71 are present on both sides of the sheet.

The magnetic properties of the flexible magnets depend upon the properties of the ferrite, of course, and the relative amounts of ferrite and binder, as well as processing conditions, but generally for strontium ferrite-based materials, residual inductions, B_r, range from 1.6 to 2.6 gauss, coercive forces, H_c, range from 1.3 to 2.3 kOe, and densities range from 3.5 to 3.8 g/cm^3. The magnetic properties degrade linearly with increasing temperature, and there is generally an upper-use temperature of around 100°C at which point the polymeric component begins to melt or flow.

6.2.4.2 Magnetic Storage Media. Magnetic materials are ideal candidates for information storage media because magnets can be used like switches. They can be made to point in one direction or another in the presence of a magnetic field. Reverse the direction of the field, and the magnet will switch to point in the other direction. This represents a simple way to store binary information. The magnet represents a "1" when pointing in one direction and represents a "0" when pointing in the other. A magnetic memory can be constructed from an entire array of magnets, in which the binary data can be encoded by applying an external magnetic field to each magnet in turn, the direction of which determines whether a "1" or a "0" is written. As we have seen, individual atoms can act as magnets. In a ferromagnetic material such as iron, for example, the magnetic moments in each atom tend to line up so that they all point in the same direction. Furthermore, recall that the magnetic moments tend to align in domains, separated by domain walls, the net magnetic moments of which are randomly oriented to create a bulk material with no net magnetic moment.

The ability to influence the alignment of magnetic moment by the application of a magnetic field forms the basis of magnetic recording. By applying a strong, localized magnetic field to one region of a thin layer of a ferromagnetic material such as magnetite (above the Curie temperature), a domain can be created in which all the magnetic moments are aligned. The magnetized domains, in general, are larger than the magnetic particles. Application of the field in the opposite direction in another part of the magnetic medium creates a domain of opposite alignment. One alignment direction represents the "1" in the binary code, and the opposite direction represents the "0." The alignment is generated by a *recording head*, which generates a magnetic field that changes the direction of alignment in the magnetic particles, as shown in Figure 6.72. The recording head contains an electromagnet in which the field is generated by flowing current across a small gap. When the recording head is magnetized, a magnetic field bridges the gap, and the flux lines spread out so that they pass through the magnetic medium lying below. The magnetic moments in the medium align themselves with the direction of the field across the recording head gap. Essentially the same process in reverse is used to retrieve information from the device.

Magnetic tape technology, while still important in limited applications, has been almost completely replaced by two-dimensional disk technology due to the limitations of the one-dimensionality of the tape. In order to retrieve information from one part of the tape, the entire preceding information must first be trawled through. Though the disk allows information to be retrieved from a specific site on the disk surface, the principles of magnetization and storage of binary information via alignment of magnetic particles is essentially the same. The primary principle leading to the development of new materials in this area is storage density. As illustrated in Figure 6.73, there must be a *transition region* between each oriented domain in the magnetic film. This buffer zone allows adjacent magnetic domains to retain their alignment without influencing each other. The resistance to alignment flipping due to an adjacent domain is related to the coercivity of the particles. The higher the coercivity, the more resistant the particle

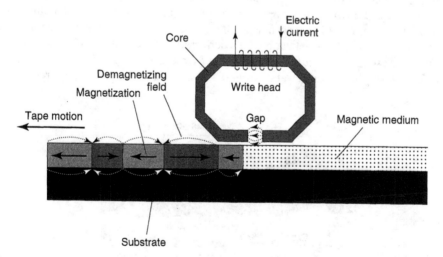

Figure 6.72 Schematic diagram of a magnetic recording device and magnetization in magnetic tape recording medium. Reprinted, by permission, from P. Ball, *Made to Measure*, p. 72. Copyright © 1997 by Princeton University Press.

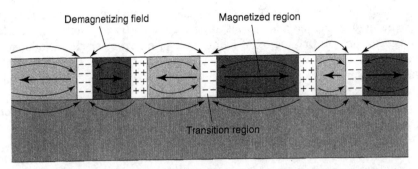

Demagnetizing field   Magnetized region

Transition region

Figure 6.73   Schematic illustration of transition region between magnetic domains in a magnetic information storage medium. Reprinted, by permission, from P. Ball, *Made to Measure*, p. 75. Copyright © 1997 by Princeton University Press.

is to flipping, and the smaller the transition zone between domains can be. Smaller transition zone and magnetic domains translate directly into increased storage density.

Composite magnetic recording media consist of submicroscopic, single-domain particles of magnetic oxides or metals immersed in a polymeric binder that separates the particles and binds them to the substrate. The substrates used in tapes, magnetic cards, and floppy disks are generally made from poly(ethylene terephthalate), while rigid disks are fabricated from an Al–Mg alloy. Particles of γ-Fe_2O_3 have been used in tapes for a long time, but as the bit length of recorded signals becomes shorter, further improvements in coercivity are required. Coercive fields have been raised from 100 to 500 A/m by impregnating the surface of the iron oxide particles with cobalt.

In addition to magnetic particle/polymer composites, laminar magnetic composite thin films are used for information storage. The magnetic film is typically either a CoPtCr or CoCrTa alloy of thickness 10 to 50 nm on a substrate of pure chromium or chromium alloy (see Figure 6.74). The carbon layer overcoat is to improve mechanical stability, reduce corrosion of the magnetic layer, and provide a hard, low-friction surface for the head to glide over. The magnetic thin film is polycrystalline, with an average grain size of 10–30 nm. Each grain is a single magnetic domain, with the direction of easy magnetization for each grain aligned in the direction of disk motion. In these materials, it is desirable to have a relatively large and square hysteresis loop. Saturation flux densities range from 0.4 to 0.6 tesla, and coercivities range from

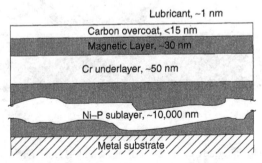

Lubricant, ~1 nm
Carbon overcoat, <15 nm
Magnetic Layer, ~30 nm
Cr underlayer, ~50 nm
Ni–P sublayer, ~10,000 nm
Metal substrate

Figure 6.74   Schematic illustration of layers in a composite thin film for magnetic information storage.

1.5×10^5 to 2.5×10^5 A/m (2 to 3 kOe). These films are formed by a process called chemical vapor deposition, which will be described in more detail in the next chapter.

Laminar composite technology is also used in the readout heads. Instead of relying solely on the change in magnetic field strength from the magnetic storage medium to change an electrical signal in the head (essentially the reverse of the writing situation presented in Figure 6.72), these heads use a phenomenon known as *magnetoresistance*. The electrical resistance of magnetoresistive materials varies markedly with magnetization. The magnetic field from the information storage medium creates a change in resistivity in the readout head, which can be registered as a change in voltage across the material under conditions of constant current. The advantage of this device is that the head is far more sensitive to changes in the magnetic flux of the storage medium. As a result, magnetoresistive heads can be made very thin and can actually be placed within the gap of the induction loop used to write on the medium (see Figure 6.72). The magnetoresistive heads are made by alternating layers of two different metals, such as iron and chromium. Cobalt and chromium are also used. The layers are again formed by chemical vapor deposition and are only a few nanometers thick. Stacks of these thin layers are known as *superlattices*. In this way, iron–chromium superlattices can be made with resistances in the presence of magnetic fields of just half that in the absence of a field, as shown in Figure 6.75. The magnetic coupling between the layers is like that in an antiferromagnetic material, and it leads to changes in the mobility of the charge carriers in the superlattice.

Additional information storage phenomena such as compact disks (CDs) and digital video disks (DVD) involve laser technology and will be described in the next section. The area of information storage is an ever-evolving one and will potentially involve phenomena other than alignment of magnetic domains. At present, however, magnetic storage media are an excellent example of how composites are utilized in areas other than for structural applications.

6.2.5 Magnetic Properties of Biologics*

Much as was the case in general with polymeric materials, biologics do not possess significant magnetic activity. There is one topic related to the interaction of biological

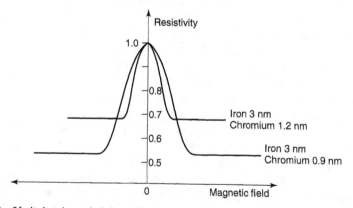

Figure 6.75 Variation in resistivity of two ironchromium superlattices of various layer thicknesses due to magnetoresistive effect. Reprinted, by permission, from P. Ball, *Made to Measure*, p. 77. Copyright © 1997 by Princeton University Press.

tissues with magnetic fields that is of such technological importance, and related enough to the chemical makeup of biologics, that it is worth describing here. *Nuclear Magnetic Resonance Imaging*, or MRI for short, has revolutionized diagnostic medicine. What is so intriguing about it from a scientific standpoint is that it is based upon the interaction of only one atom with a magnetic field—hydrogen.

6.2.5.1 *Nuclear Magnetic Resonance Imaging.*

Virtually all of the topics in this textbook deal in some way with the chemical aspects of atoms and molecules—that is, how electrons and the electronic structure dictate interactions between atoms and molecules, and ultimately, the bulk properties of an assembly of atoms or molecules. This is why such a great deal of attention was devoted to potential energy functions, molecular orbital structure, and bonding in the first chapter. And, as we have seen in this section on magnetic properties, simple electronic orbital diagrams can provide very clear explanations for why certain materials behave as diamagnets while others behave as paramagnets. Indeed, all of the types of magnetism described in this section are related in some way to the pairing of electrons.

What we describe in this subsection is not a bulk magnetic property of a material and is not related to any of the types of magnetism described heretofore. *Nuclear magnetic resonance*, as the name implies, is related to the interaction of an atom's nucleus with a magnetic field. Let us take a moment to review a bit of nuclear chemistry, then describe the general topic of nuclear magnetic resonance, or NMR, before completing this section with a description of the application of NMR to imaging of hydrogen in biological tissues—MRI.

Recall from general chemistry that an atom consists not only of orbiting electrons, but of a central nucleus, composed of *nucleons* called protons and neutrons. The protons are positively charged, and the neutrons have no charge associated with them. In addition to electrons orbiting the nucleus, there are two other kinds of spins associated with an atom: (a) the spin of an electron on its own axis and (b) the spin of a nucleus on its own axis. We have concentrated a great deal upon the motion of electrons and have completely ignored the spin of the nucleus to this point. We must no longer ignore nuclear spin, since it is the basis upon which MRI functions.

Particles that spin have an associated angular momentum; and as dictated by quantum chemistry, the angular momentum has values that are integral or half-integral multiples of $h/2\pi$, where h is Planck's constant. The maximum spin component for a particular nucleus is its *spin quantum number, I*, and a nucleus has $(2I + 1)$ discrete states. The component of angular momentum for these states in any chosen direction will have values of $I, I - 1, I - 2, \ldots, -I$, called the *magnetic quantum number, m*. In the absence of an external magnetic field, the various states have identical energies. The spin number for a proton is $1/2$; thus two spin states are possible: $I = +1/2$ and $I = -1/2$. Heavier nuclei have spin numbers that range from 0 to over 9/2. As shown in Table 6.24, the spin number of a nucleus is related to the relative number of protons and neutrons it contains. Since the nucleus contains a charge, its spin gives rise to a magnetic field, much like the field produced when electricity flows through a coil of wire. The resulting magnetic dipole is oriented along the axis of spin and has a value that is characteristic for each type of nucleus. For a hydrogen atom, whose nucleus contains only one proton, the spin quantum number is $I = 1/2$, and the magnetic quantum number can have values of $+1/2$ and $-1/2$. These two quantum energies correspond to the two possible orientations of the spin axis: spin up and spin down. Those nuclei

Table 6.24   Spin Quantum Number for Various Nuclei

Number of Protons	Number of Neutrons	Spin Quantum Number I	Examples
Even	Even	0	^{12}C, ^{16}O, ^{32}S
Odd	Even	$\frac{1}{2}$	^{1}H, ^{19}F, ^{31}P
		$\frac{3}{2}$	^{11}B, ^{79}Br
Even	Odd	$\frac{1}{2}$	^{13}C
		$\frac{3}{2}$	^{127}I
Odd	Odd	1	^{2}H, ^{14}N

with a spin-up orientation are termed low energy, and those with spin-down orientation are termed high energy. In the absence of a magnetic field, the energies of these two states are identical, despite their names. Consequently, a large assemblage of hydrogen atoms will contain an identical number of nuclei with $m = +\frac{1}{2}$ and $m = -\frac{1}{2}$.

When brought into the influence of an applied magnetic field, the spins tend to become oriented parallel to the field, termed spin up, and the lower-energy spin state ($m = +\frac{1}{2}$) is more highly populated. Some spins do populate the high-energy state, however. In fact, at room temperature there is only a minute excess (< 10 ppm) of nuclei in the lower-energy state due to thermal energies that are several orders of magnitude greater than the magnetic energy differences. This small difference in spin orientations is sufficient, however, to result in a net magnetic moment, as characterized by the *net magnetization vector*, NMV. The orientation of nuclear spins in a magnetic field is shown in Figure 6.76.

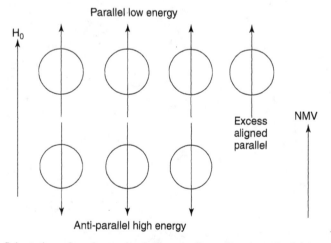

Figure 6.76   Orientation of nuclear spins in the presence of a magnetic field, and the resulting net magnetization vector (NMV) from excess low-energy spins aligned parallel to field. Adapted from C. Westbrook and C. Kaur, *MRI in Practice*, 2nd ed., p. 6. Copyright © 1998 by Blackwell Science Ltd.

The application of a magnetic field to the spinning nucleus creates an additional movement in the nucleus, called *precession*. Much like a compass needle that will swing back and forth indefinitely in the presence of an external magnet, a spinning particle will be forced to move in a circular path in the presence of a magnetic field. This is a gyroscopic effect similar to what occurs in a gyroscope when it is displaced from vertical by application of an external force. In this case, the external force is the applied magnetic field, which causes the spinning particle to move in a circular path, or precess, about the magnetic field, as illustrated in Figure 6.77. The *angular velocity of precession*, ω_0, is directly proportional to the applied force and inversely proportional to the angular momentum of the spinning body. The force on a spinning nucleus in a magnetic field is the product of the field strength, H_0, and the magnetic moment of the particle. When combined with the angular momentum, the magnetic moment creates a quantity called the *magnetogyric ratio*, γ, sometimes (inappropriately) called the gyromagnetic ratio. The result is the Larmor equation:

$$\omega_0 = \gamma H_0 \tag{6.67}$$

The magnetogyric ratio is then the ratio between the magnetic moment and the angular momentum of a rotating particle and has a characteristic value for each type of nucleus:

$$\gamma = \frac{\mu\beta}{I\,(h/2\pi)} \tag{6.68}$$

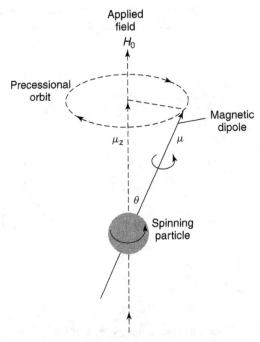

Figure 6.77 Precession of a spinning particle in a magnetic field. Reprinted, by permission, from D. A. Skoog and D. M. West, *Principles of Instrumental Analysis*, 2nd ed., p. 381. Copyright © 1980 by Saunders College.

where μ is the magnetic moment of the nucleus expressed in nuclear magnetons, β is a constant called the *nuclear magneton* (5.051×10^{-24} erg/G), I is the spin quantum number as described previously, and h is Planck's constant. For reference, the magnetic moment for the proton is 2.7927 nuclear magnetons. The angular velocity of precession given by the Larmor equation can be converted to a frequency of precession, ν_0, by simply dividing by 2π. Thus, as the field strength is increased, the precessional frequency increases.

We turn our attention now to nuclear magnetic resonance, NMR, which forms the basis for MRI. *Resonance* is a phenomenon that occurs when an object is exposed to an oscillating perturbation that has a frequency close to its own natural frequency of oscillation. The nucleus gains energy and resonates if the energy is delivered at exactly its precessional frequency. If energy is delivered at a frequency different from that of the Larmor frequency of the nucleus, resonance does not occur. When this energy, typically in the form of radio-frequency electromagnetic radiation, is absorbed by the nucleus, *excitation* occurs, and the angle of precession, θ, shown in Figure 6.77, must change. If the frequency of the radio wave is the same as the precessional frequency, absorption and flipping can occur. The first result of resonance is that the NMV moves out of alignment away from the direction of the applied field. The angle to which the NMV moves out of alignment is called the *flip angle*. The magnitude of the flip angle depends upon the amplitude and duration of the electromagnetic pulse. Usually, the flip angle is 90°; that is, the NMV is given enough energy by the pulse to move through 90° relative to H_0 (see Figure 6.78). With a flip angle of 90° the nuclei are given sufficient energy so that the longitudinal NMV is completely transferred into a transverse NMV. This transverse NMV rotates in the transverse plane at the Larmor frequency. We will see that the longitudinal and transverse components of the NMV are critical to NMR analysis.

The process is reversible, and the excited particle can thus return to the ground state by reemission of the radiation. This is known as the *relaxation process*. Relaxation following resonance is the basis for nuclear magnetic resonance spectroscopy. It turns

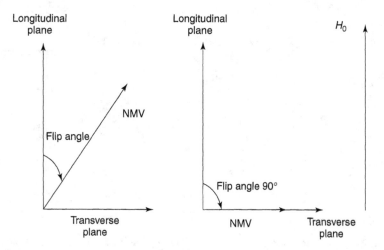

Figure 6.78 The flip angle with and without an applied field. Adapted from C. Westbrook and C. Kaur, *MRI in Practice*, 2nd ed., p. 9. Copyright © 1998 by Blackwell Science Ltd.

out that there is a low probability of emission of radiation of the exact frequency of resonance. Instead, relaxation processes usually involve radiationless mechanisms. In order to produce readily detectable signals, the relaxation processes should be as rapid as possible. Conversely, the lifetime of the excited state should be small. Due to line-broadening effects, which tend to counter the need for short lifetimes, the optimal half-life for an excited species is from about 0.1 to 1 s. There are two types of relaxation processes that are important in this regard: longitudinal, or *spin–lattice relaxation*, and transverse, or *spin–spin relaxation*. Each relaxation process is a first-order process that can be characterized by a time, T, which is the average lifetime of the nuclei in the higher energy state. Spin–lattice relaxation is characterized by a relaxation time T_1, and spin–spin relaxation is characterized by time T_2.

T_1 recovery is caused by the nuclei giving up their energy to the surrounding environment, or lattice. The rate of recovery is an exponential process, with a recovery time constant, T_1, which is defined as the time it takes 63% of the longitudinal magnetization to recover. In addition to depending upon the magnetogyric ratio and the absorbing nuclei, T_1 is strongly affected by the mobility of the lattice. In crystalline solids and viscous liquids, where mobilities are low, T_1 is large. As the mobility increases, the vibrational and rotation frequencies increase, and T_1 becomes shorter. At very high mobilities, however, the probability for a spin–lattice transition begins to decrease. Thus, there is a minimum in the relationship between T_1 and lattice mobility.

T_2 is caused by nuclei exchanging energy with neighboring nuclei. The energy exchange is caused by the magnetic fields of each nucleus interacting with its neighbor. The rate of decay is also an exponential process, and it represents the time it takes 63% of the transverse magnetization to be lost. Values for T_2 are generally so small for crystalline solids and viscous liquids (as low as 10^{-4} s) that these samples are not typically used for high-resolution NMR spectra.

The NMR process, then, essentially involves placing a sample in a magnetic field, applying a radio-frequency (RF) pulse, and determining the decrease in power (attenuation) of the radiation caused by the absorbing sample. This process is shown schematically in Figure 6.79.

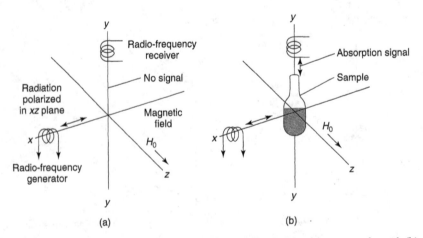

Figure 6.79 Schematic illustration of an NMR experiment (a) without sample and (b) with sample. Reprinted, by permission, from D. A. Skoog and D. M. West, *Principles of Instrumental Analysis*, 2nd ed., p. 285. Copyright © 1980 by Saunders College.

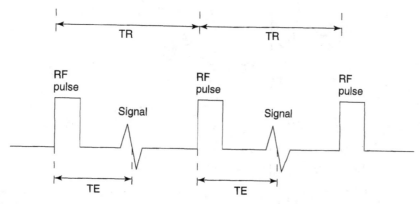

Figure 6.80 A basic NMR pulse sequence. Reprinted, by permission, from C. Westbrook and C. Kaur, *MRI in Practice*, 2nd ed., p. 15. Copyright © 1998 by Blackwell Science Ltd.

The manner in which the RF pulse is applied is critical to NMR analysis. A very simplified pulse sequence is a combination of RF pulses, signals, and intervening periods of recovery, as illustrated in Figure 6.80. The main components of the pulse sequence are the *repetition time*, TR, which is the time from the application of one RF pulse to the application of the next RF pulse (measured in milliseconds) and the *echo time*, TE. The repetition time determines the amount of relaxation that is allowed to occur between the end of one RF pulse and the application of the next. Therefore, the repetition time determines the amount of T_1 relaxation that has occurred. The echo time is the time from the application of the RF pulse to the peak of the signal induced in the coil (also measured in milliseconds). The TE determines how much decay of transverse magnetization is allowed to occur before the signal is read. Therefore, TE controls the amount of T_2 relaxation that has occurred.

Now that the basic principles of NMR have been described, we can describe how this phenomenon is adapted to the generation of images from biological tissues. Magnetic resonance imaging relies upon the nuclear magnet response of only one atom: hydrogen. Though there are certainly other magnetically susceptible atoms in the human body such as C^{13}, N^{15}, O^{17}, Na^{23}, and P^{31}, hydrogen is in such abundance and has such a relatively large magnetic moment owing to its single proton in the nucleus that its selection as the nucleus of choice for MRI is an easy one. As mentioned previously, the spin–spin relaxation time, T_2, for an atom is influenced by its surroundings. As a result, hydrogen atoms in different environments within the body, in principle, can be differentiated according to their relaxation times. Practically, this means that the NMV can be separated into the longitudinal and transverse components for differing tissues such as fat, cerebrospinal fluid, and muscle.

A tissue has a large NMR signal if it has a large transverse component of magnetization. If there is a large component of transverse magnetization, the amplitude of the signal received by the coil is large, resulting in a bright area on the MRI image. A tissue returns a low signal if it has a small transverse component of magnetization, resulting in a dark area on the image. In general, the two extremes of image contrast in MRI are the result of hydrogens found in fat (lipids) and water (see Figure 6.81).

The hydrogens in lipids are linked to carbon, whereas the hydrogen in water is linked to electron-withrawing oxygens. The removal of electrons from around the

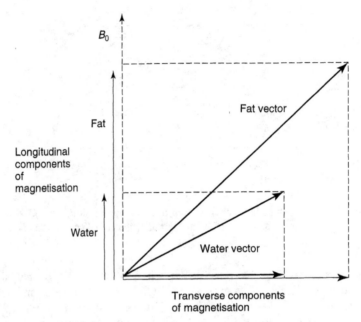

Figure 6.81 Magnitude of transverse and longitudinal magnetization for fat and water. Reprinted, by permission, from C. Westbrook and C. Kaur, *MRI in Practice*, 2nd ed., p. 18. Copyright © 1998 by Blackwell Science Ltd.

hydrogens in water causes its nucleus to be less shielded and to interact more strongly with the applied magnetic field. The carbon in lipids is less electron-withdrawing, so lipid hydrogens do not interact with the magnetic field as strongly. Therefore, the Larmor frequency of hydrogen in water is higher than that of hydrogen in lipids. Lipid hydrogens recover more rapidly along the longitudinal axis than water, and they lose transverse magnetization faster than water hydrogens. The result is different imaging characteristics in MRI.

Images obtain contrast mainly through T_1 and T_2 relaxation processes, as well as proton density. Let us briefly describe each of these effects in both fat and water. T_1 recovery occurs due to nuclei transferring their energy to the surroundings. The slow molecular tumbling in fat allows the recovery to proceed relatively rapidly. This means that the magnetic moments of hydrogens in fat are able to relax and regain their longitudinal magnetization quickly. The NMV of lipid hydrogens realigns rapidly with the applied field, and the T_1 for fat is short. In water, molecular mobility is high, resulting in less efficient T_1 recovery. The magnetic moments of water hydrogens take longer to relax and regain their longitudinal magnetization. The NMV of hydrogens in water take longer to realign with the applied field, so their T_1 is long. Because the T_1 time of hydrogens in fat is shorter than for those in water, its vector realigns with the applied field vector faster than that of water. The longitudinal component of magnetization of lipid hydrogens is therefore larger than water hydrogens. After a certain repetition time, TR, the next RF pulse is applied, flipping the longitudinal components of magnetization of both fat and water hydrogens into the transverse plane (assuming that a 90° pulse is applied). Since there is more longitudinal magnetization in the lipid hydrogens than before the RF pulse, there is more transverse magnetization after the RF pulse. Lipid hydrogens therefore have a high signal and appear bright on a T_1 contrast image. The opposite is true for water hydrogens, which appear as dark on a T_1 contrast image. These images are called T_1-*weighted images*.

Recall that T_2 occurs as a result of the nuclei exchanging energy with neighboring atoms. As energy exchange is more efficient for hydrogens in fat, the T_2 is short, approximately 80 ms. Energy exchange is less efficient in water than in fat, so the T_2 times for water hydrogens are longer, about 200 ms. As a result, the transverse component of magnetization decays faster for lipid hydrogens than water hydrogens. The magnitude of transverse magnetization in water is large, and it has a high signal, thus appearing bright on a T_2 contrast image. Similarly, fat appears dark on a T_2 *contrast image*.

The final effect on image contrast is proton density, or the relative number of protons per unit volume. To produce contrast due to proton density differences, the transverse

Table 6.25   Relative Repetition and Echo Times for Different MRI Image Contrasts

	T_1 Weighting	T_2 Weighting	Proton Density Weighting
TR	Short	Long	Long
TE	Short	Long	Short

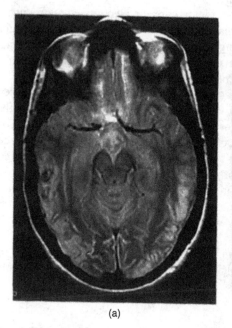 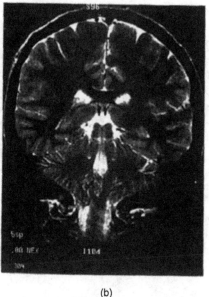

(a)   (b)

Figure 6.82   (a) Proton density weighted axial MRI image of the brain, TE = 20 ms, TR = 2700 ms. (b) T_2 weighted coronal image of the brain, TE = 90 ms, TR = 2700 ms. Reprinted, by permission, from C. Westbrook and C. Kaur, *MRI in Practice*, 2nd ed., p. 35. Copyright © 1998 by Blackwell Science Ltd.

component of magnetization must reflect these differences. Tissues with high proton density, such as brain tissue, have a large transverse component of magnetization and therefore have a high signal. A bright *proton density contrast image* results. Tissues with low proton density, such as cortical bone, have small transverse components of magnetization and have low signal strengths, and they appear dark on proton density contrast images.

To demonstrate either T_1, proton density or T_2 contrast, specific values of TR and TE are selected for a given pulse sequence (see Figure 6.80). The selection of appropriate TR and TE weights an image so that one contrast mechanism predominates over the other two (see Table 6.25). There are, of course, more topics and techniques associated with MRI imaging, such as spin echo and gradient pulse echo pulse sequences, but the basics of nuclear magnetic resonance and its application to MRI imaging have been presented. The result is astounding detailed images of biological tissue, like that of the human brain shown in Figure 6.82.

6.3 OPTICAL PROPERTIES OF MATERIALS

The final, general topic of this chapter deals, in a way, with a combination of the previous two properties: electricity and magnetism. The interaction of electromagnetic radiation with materials involves many of the same principles used to describe the electrical and magnetic properties of materials, such as magnetic moments, spins, and electronic configurations. However, we will see that both wave (frequency) and particle (photon) descriptions of light will be of benefit in this chapter. Thus, we will vacillate between the two with little fanfare or warning.

In this section, we are concerned with the interaction of electromagnetic radiation with a material. When electromagnetic radiation (which we will generally call "light" for ease of visualization, while recognizing that the visible spectrum is but a small portion of all electromagnetic radiation) strikes a surface, three types of processes may take place, either individually or in tandem. The light may be *absorbed, reflected*, or *transmitted*. Like many technical terms that find their way into common usage, these three terms are well known to us through their everyday occurrences. We know that black clothing absorbs energy, that a mirror reflects light, and that a pane of window glass transmits light. What is not always clear is that these three effects may occur simultaneously and that more than just the visible spectrum of light is involved. Since we are dealing primarily with "optical" properties of materials, this implies that we are primarily considering the visible portion of the electromagnetic spectrum. In some instances, radiation of other energies, such as infrared and ultraviolet radiation, will be examined. In these cases, more specificity will be exercised in the description of the type of "light."

In all instances, the total intensity of the incident light striking a surface, I_0, is the sum of the absorbed, reflected, and transmitted intensities, I_A, I_R, and I_T, respectively:

$$I_0 = I_A + I_R + I_T \tag{6.69}$$

The intensity is defined as the number of photons impinging on a surface per unit area per unit time. If we divide Eq. 6.69 through by the incident radiation, we end up with an equally useful relationship

$$1 = A + R + T \tag{6.70}$$

where A is the *absorptivity* (I_A/I_0), R is the *reflectivity* (I_R/I_0), and T is the *transmissivity* (I_T/I_0). The utility of Eq. (6.70) is that A, R, and T are material properties. Thus, it is not possible for a material to be simultaneously highly reflective, highly absorptive, and highly transmissive. We have some intuition as to which materials are absorptive, reflective, and transmissive, at least in terms of visible light. Let us utilize this knowledge to examine the fundamentals of how electromagnetic radiation interacts with materials from our four classes and how these interactions can be used in some important applications.

6.3.1 Optical Properties of Metals and Alloys

The "shininess" and inability to transmit visible light inherent in most metals indicates that absorptivity and reflectivity are high. In this section, we utilize our knowledge of electronic structure from previous sections to describe why this is so, and we investigate what happens to the energy that is absorbed in these materials.

Consider an incident beam of light interacting with a material as shown in Figure 6.83. Upon initial interaction, a certain fraction of the incident beam is reflected. According to Eq. (6.70), this fraction is R, or I_R/I_0. The associated intensity is I_R, or

---

Cooperative Learning Exercise 6.8

The reflectivity of silicon at 633 nm is 35% and the absorption coefficient is $3.8 \times 10^5 \text{ m}^{-1}$. Calculate the following quantities for a 10 μm-thick sample of silicon at this wavelength.

Person 1: Calculate the fraction of incident light that is absorbed.
Person 2: Calculate the fraction of incident light that is transmitted through the sample.
Combine your information to check that Eq. (6.70) is satisfied.

Answers: If Eq. (6.72) is the fraction of light reaching the back face, then that portion subtracted from the light entering, $I_0(1-R)$, is the fraction absorbed, so $I_A/I_0 = 0.35 - (1 - 0.35) \exp[(-3.8 \times 10^5)(1 \times 10^{-5})] = 0.6355$. Equation (6.73) gives $I_T/I_0 = (0.65)_2 \exp[(-3.8 \times 10^5)(1 \times 10^{-5})] = 0.009545$. $A + R + T =$ $0.6355 + 0.035 + 0.009545 \approx 1.0$.

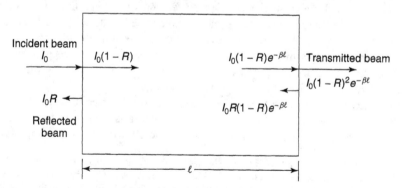

Figure 6.83 Reflection, absorption and transmission by a solid. From K. M. Ralls, T. H. Courtney, and J. Wulff, *Introduction to Materials Science and Engineering*. Copyright © 1976 by John Wiley & Sons, Inc. This material is used by permission John Wiley & Sons, Inc.

$I_0 R$. Metals and alloys typically have high values of R, as high as 1.0 in some cases. The light that is not reflected enters the material, the intensity of which is thus $I_0 - I_R$, or $I_0(1 - R)$. Some or all of this remaining light will be absorbed through electronic interactions with the solid, and the result is a continuously decreasing intensity as the light passes through the length of the solid, indicated by l. The fractional change in light intensity, dI/I, over a distance x is directly proportional to a material property called the *linear absorption coefficient*, β:

$$\frac{dI}{I} = -\beta \, dx \tag{6.71}$$

where the minus sign indicates a decrease in intensity with distance, since β is taken as a positive quantity. Integration of Eq. (6.71) over the distance l in Figure 6.83 results in

$$I = I_0(1 - R)\exp\{-\beta \cdot l\} \tag{6.72}$$

This is the intensity of light reaching the back surface in Figure 6.83. At the back surface, a fraction of the light is once again reflected internally, which is R times the amount given in Eq. (6.72), and the remainder passes out of the material or is transmitted. Thus, the fractional amount of transmitted light, T, is one minus the amount reflected at the back face, or

$$T = (1 - R)^2 \exp\{-\beta \cdot l\} \tag{6.73}$$

provided that the same medium, e.g., air is on both the entering and exiting side of the solid. (We will return to this condition when we discuss refractive index in Section 6.3.2.1). These relationships indicate that the linear absorption coefficient, β, and reflectivity, R, are the two important parameters in determining how light interacts with a material. Let us investigate these two quantities more carefully, first for metals and alloys.

6.3.1.1 *Reflectance and Color.* Recall that in metals and alloys, there are many empty electronic states just above the occupied levels in the conduction bands. When incident radiation strikes a metallic surface, energy can be absorbed by promoting electrons from the occupied to the unoccupied electronic states. This generally occurs over a very short distance in the metal surface, and only very thin films of metals (< 0.1 μm) exhibit any transmittance at all. Once the electrons have been excited, they decay back to the lower energy levels, and reemission of light from the metal surface occurs. It is this immediate reemission process that gives rise to reflectivity. As we will see, there are other types of reemission processes, but they differ in time scale and frequency of the emitted light from reflectivity.

The efficiency of the reflection process depends on the frequency of the incident light, ν. For example, as shown in Figure 6.84, silver is highly reflective over the entire visible range. As a result, it has a white metallic *color* and a bright *luster**.

---

*The technical definition of *luster* is a bit elusive. In its common use, it is defined as "the quality, condition, or fact of shining by reflected light; gloss; sheen." In fact, the term *metallic luster* is used in mineralogy as a distinguishing feature from *mineral luster*, which is less "shiny" than metallic luster. In this context, luster is defined as "the appearance of a mineral in reflected light." However, this does not help us differentiate between the luster of two metals. So, while luster is commonly used to describe both metals and minerals, there is no absolute or relative scale for luster that makes it a useful quantity for describing metals.

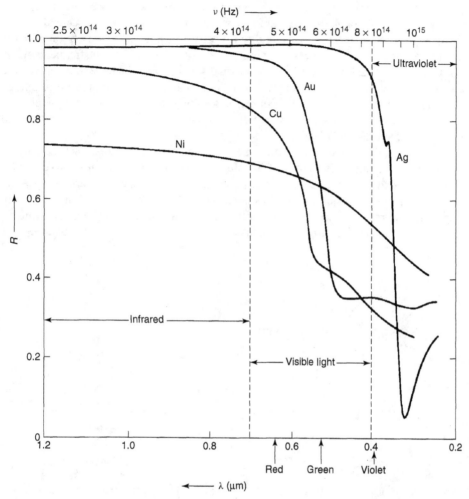

Figure 6.84   The frequency dependence of reflectivity for selected metals. From K. M. Ralls, T. H. Courtney, and J. Wulff, *Introduction to Materials Science and Engineering.* Copyright © 1976 by John Wiley & Sons, Inc. This material is used by permission John Wiley & Sons, Inc.

The white color of silver indicates that the reemitted photons cover much of the same frequencies and are comparable in number (intensity) to the incident photons. Metals like copper and gold exhibit characteristic red-orange and yellow colors, respectively, because incident photons above a certain frequency cause excitation of some electrons from a filled d electronic band to empty s electronic bands. Since this frequency falls in the visible region for Cu and Au (see Figure 6.84), strong absorption of light with higher frequencies occurs, and light with lower frequencies (red and yellow) is strongly reemitted. Many metals, like nickel, iron, and tungsten, have grayish colors because the density of electronic states both for occupied and unoccupied levels exhibit maxima and minima. This, in combination with the fact that the density of electronic states is very large in different parts of the d band, leads to relatively strong absorption and thus leads to a relatively low reflectivity in the visible region. We will return to a more quantitative description of color in the section on ceramics and glasses, where there is more variability in color than is commonly found in metals and alloys. Outside of the visible region, most metals become highly reflective for photons having frequencies in the infrared region. This is because unlike free atoms where discrete frequencies are absorbed and emitted, photons having a virtually continuous band of frequencies are absorbed and emitted by a metal because of the almost continuous variation of energy levels in the conduction band. Of course, some intensity is lost because of phonon generation resulting from collisions suffered by the excited electrons in a metal and because of electronic polarization.

The color and luster of a metal can be affected by other factors, such as surface roughness. As illustrated in Figure 6.85, reflection from a smooth or mirror-like surface is termed *specular reflectance*, whereas reflection from surfaces that are not parallel to the average surface due to roughness is termed *diffuse reflectance*. On a completely rough surface, reflectance can occur at all angles, θ, relative to the incident light, as shown in Figure 6.86. The relative intensity of reflection, I_θ, varies as the cosine of

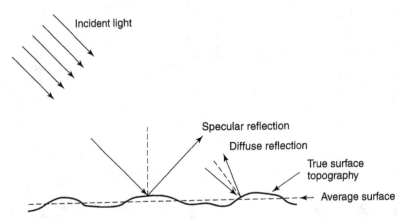

Figure 6.85   Schematic illustration of specular reflection from a flat surface and diffuse reflection from a rough surface, both relative to the average surface. Reprinted, by permission, from J. F. Shackelford, *Introduction to Materials Science for Engineers*, 5th ed., p. 597. Copyright © 2000 by Prentice-Hall, Inc.

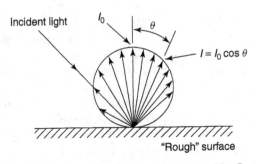

Figure 6.86   Polar diagram illustrating the directional intensity of reflection from a rough surface. Reprinted, by permission, from J. F. Shackelford, *Introduction to Materials Science for Engineers*, p. 598, 5th ed. Copyright © 2000 by Prentice-Hall, Inc.

the angle as

$$I_\theta = I_0 \cos\theta \tag{6.74}$$

where I_0 is the intensity of scattering at $\theta = 0°$. Since any area segment, A_θ, will be reduced when viewed at an angle, the luster of the diffuse surface will be a constant independent of viewing angle:

$$\text{luster} = \frac{I_\theta}{A_\theta} = \frac{I_0 \cos\theta}{A_0 \cos\theta} = \text{constant} \tag{6.75}$$

This, at least in part, helps us understand the concept of luster.

6.3.1.2 The Photoelectric Effect.
A phenomenon that is related to the bombardment of metallic surfaces with light (photons) is the *photoelectric effect*, which is the release of electrons due to the absorption of light energy. The photoelectric effect arises from the fact that the potential energy barrier for electrons is finite at the surface of the metal. This process is illustrated in Figure 6.87. The finite energy barrier at the surface is represented by W. In order for an electron to escape from the metal, it must possess a kinetic energy equal to or greater than W. Since the free electrons in the metal have kinetic energies ranging from nearly zero at the bottom of the potential energy well to E_f at the Fermi level, the electrons must absorb energy between W and $(W - E_f)$ to be ejected from the metal. The work that must be supplied to eject a free electron is called the *work function* and is denoted by the Greek lowercase phi, φ, where:

$$\varphi = W - E_f \tag{6.76}$$

The critical energy required for a photon to remove an electron is then $h\nu_c = \varphi$, where ν_c is the critical frequency of the photon and h is Planck's constant. When the frequency of the incident radiation is less than the critical frequency, electrons will not be ejected. Similarly, when the wavelength of the incident radiation is greater than the *critical wavelength*, λ_c, electrons will also not be emitted (recall that $\nu = c/\lambda$, where c is the speed of light in vacuum). This relationship between photon energy (in terms of wavelength) and electron emission for several alkali metals is shown in Figure 6.88. The decrease in number of emitted electrons at short wavelengths (high energy) is due to a decrease in the efficiency of the electronic excitation. In this way, the work function for each element, in principle, can be determined and can be used to identify a metallic sample.

We will see that a number of important optical phenomena are functions of the wavelength of light. This is probably a good time to review the various categories

> **HISTORICAL HIGHLIGHT**
>
> It is somewhat surprising that despite all of his contributions to science and engineering, including his work on Brownian motion and the theory of relativity, Albert Einstein (1879–1955) won his only Nobel Prize in 1921 "for his services to Theoretical Physics, and especially for his discovery of the law of the photoelectric effect."
>
> Source: www.nobel.se/physics/laureates/1921

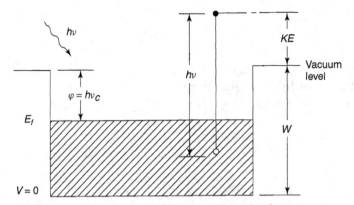

Figure 6.87   Illustration of potential energy well for the surface of a metal, and the energy required by an incident photon to remove an electron from it.

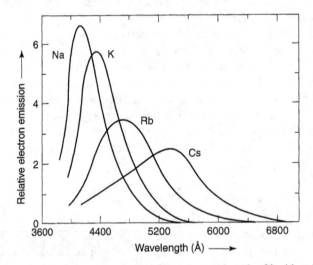

Figure 6.88   Dependence of photoelectric emission on wavelength of incident light for various alkali metals. Reprinted, by permission, from C. R. Barrett, W. D. Nix, and A. S. Tetelman, *The Principles of Engineering Materials*, p. 378. Copyright © 1973 by Prentice-Hall, Inc.

and associated wavelengths (and frequencies) of electromagnetic radiation through the summary in Figure 6.89.

6.3.2 Optical Properties of Ceramics and Glasses

6.3.2.1 Refractive Index and Dispersion. The velocity of light varies depending upon the density of the medium in which it is propagating. In a vacuum, the speed of light is a constant, c, which has a value of 3.08×10^8 m/s. In any other medium, such as a gas, liquid, or solid, the velocity is given by the variable, v. The ratio between these two velocities determines the *index of refraction, n,* sometimes called the *refractive index*:

$$n = \frac{c}{v} \tag{6.77}$$

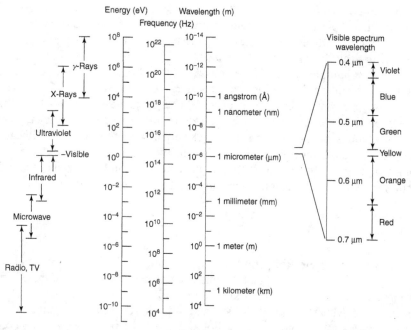

Figure 6.89 Energies, wavelengths, and frequencies of various categories of electromagnetic radiation. Reprinted, by permission, from W. Callister, *Materials Science and Engineering: An Introduction*, 5th ed., p. 709. Copyright © 2000 by John Wiley & Sons, Inc.

Hence, the refractive index is a dimensionless quantity. The velocity of light in a vacuum can be related to the electric permittivity in a vacuum, ε_0 [cf. Eq. (6.32)], and the magnetic permeability in a vacuum, μ_0 [cf. Equation (6.60)]:

$$c = \frac{1}{\sqrt{\varepsilon_0 \mu_0}} \tag{6.78}$$

Similarly, the velocity of light in a medium is related to the electric permittivity and magnetic permeabilities in the medium, ε and μ, respectively:

$$v = \frac{1}{\sqrt{\varepsilon \cdot \mu}} \tag{6.79}$$

Thus, we see the initial connection between optical properties and the electrical and magnetic properties from the two previous sections. Substitution of Eqs. (6.78) and (6.79) into (6.77) shows that the refractive index can be expressed in terms of the relative electric permittivity (dielectric constant), ε_r (cf. Table 6.5), and relative magnetic permeability of the medium, $(1 + \chi)$ [cf. Eq. (6.63)], where χ is the magnetic susceptibility:

$$n = \sqrt{\frac{\varepsilon_r}{(1 + \chi)}} \tag{6.80}$$

Since most ceramic substances (and most nonmetallic substances, for that matter) possess small magnetic susceptibilities, the quantity $(1 + \chi)$ is approximately unity, and the refractive index can be approximated as the square root of the dielectric constant:

$$n \approx \sqrt{\varepsilon_r} \tag{6.81}$$

Refractive index values vary from 1.0003 for air to over 2.7 for some solid oxide ceramics. Silicate glasses have a much narrower range of values, from about 1.5 to 1.9. The refractive indices (or indexes) of other materials can be found in Appendix 9.

The relationship between refractive index and the dielectric constant is a logical one, and brings into play many of the considerations of electronic structure utilized in Section 6.1.2.2 to describe the dielectric constant. A dielectric material interacts with electromagnetic radiation because it contains charge carriers that can be displaced. Light waves are retarded in a dielectric; that is, the velocity is decreased because of the interactions of the electromagnetic radiation with the electrons. Recall from Section 6.1.2.2 that the polarizability, α, is a measure of the average dipole moment per unit field strength [cf. Eq. (6.39)]. The polarizability can be related to the index of refraction for a simple, monatomic gas through the Lorentz–Lorenz equation

$$\alpha = \frac{3\varepsilon_0 M (n^2 - 1)}{N_A \rho (n^2 + 2)} \tag{6.82}$$

where ε_0 is again the electric permittivity in vacuum, M is the atomic weight of the gas, ρ is its density, N_A is Avogadro's number, and n is the refractive index.

In ionic solids, we expect the ionic polarizability to dominate. Ionic polarization should increase with the size of the ion and with the degree of negative charge on isoelectric ions. Since the index of refraction increases with polarizability in accordance with Eq. (6.82), large refractive indices arise with large ions (e.g., $n_{PbS} = 3.912$) and low refractive indices occur with small ions (e.g., $n_{SiCl_4} = 1.412$). However, as we know, the electronic environment of the ions also affects the polarizability and, in turn, the refractive index. Only in glasses and in cubic crystals is the refractive index independent of direction. In other crystal systems, the refractive index is larger in directions that are close-packed. This also follows directly from Eq. (6.82). Similarly, the more open structures of high-temperature polymorphic forms have lower refractive indices than do crystals of the same composition; for example, for SiO_2, $n_{glass} = 1.46$, $n_{tridymite} = 1.47$, $n_{cristobalite} = 1.49$, and $n_{quartz} = 1.55$. The refractive index of a typical soda-lime glass is about 1.5. The addition of large ions such as lead and barium have the effect of increasing the index of refraction. For example, glasses containing 90 wt % PbO have an index of refraction of around 2.1, which is about the upper limit of the obtainable values for practical optical glasses.

In the same way that close-packed directions in a crystal have larger refractive indices, so too can the application of a tensile stress to an isotropic glass increase the index of refraction normal to the direction of the applied stress. Uniaxial compression has the reverse effect. The resulting variation in refractive index with direction is called *birefringence*, which can be used as a method of measuring stress.

The refractive index is a function of the frequency of the light, and it normally decreases with increasing wavelength, as illustrated in Figure 6.90 for some typical glasses. The variation in refractive index with wavelength, λ, is called *dispersion*, and it is given in one definition by

$$\text{Dispersion} = \frac{dn}{d\lambda} \tag{6.83}$$

As can be seen in Figure 6.90, the slope of the refractive index versus wavelength—the dispersion—varies with wavelength. However, most practical measurements are made

Cooperative Learning Exercise 6.9

It can be shown that for some ionic crystals, such as LiF, both ionic and electronic polarization can contribute to the overall dielectric constant, ε_r. In such cases, Eq. (6.81) is not entirely correct, and the electronic contribution to the polarizability, α_e, is given by Eq. (6.82), since the refractive index affects only the frequencies in the electronic range, and the number of ions per unit cell, in this case two, must be included in the denominator. The total polarizability, $\alpha = \alpha_e + \alpha_i$, is then given by

$$\alpha = \frac{3\varepsilon_0 M}{N_A \rho}\left(\frac{\varepsilon_r - 1}{\varepsilon_r + 2}\right),$$ where ε_r is the dielectric constant as given in Table 6.3 and the

remaining quantities are all the same as in Eq. (6.82). The index of refraction for LiF is 1.395, its density is 2.635×10^3 kg/m³, and its molecular weight is 26×10^{-3} kg/mol. Recall that $\varepsilon_0 = 8.854 \times 10^{-12}$ C/V-m.

Person 1: Calculate the total polarizability for LiF.

Person 2: Calculate the electronic contribution to the total polarizability.

Combine your information to calculate the ionic polarizability, α_i.

Answers: $\alpha = [3(8.854 \times 10^{-12})(26 \times 10^{-3})/(6.02)(2.635 \times 10^{23})][(9-1)/(9+2)] = 3.166 \times 10^{-40}$ F-m²; $\alpha_e = [3(8.854 \times 10^{-12})(26 \times 10^{-3})/(2)(6.02)(2.635 \times 10^{23})][(1.95-1)/(1.95+2)] = 5.2 \times 10^{-41}$ F-m². $\alpha_i = \alpha - \alpha_e = 3.16 \times 10^{-40} - 5.2 \times 10^{-41} = 2.64 \times 10^{-40}$ F-m².

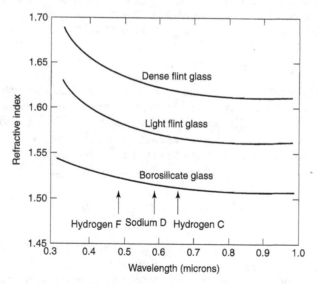

Figure 6.90 Effect of light wavelength on refractive index for some common glasses. From W. D. Kingery, H. K. Bowen, and D. R. Uhlmann, *Introduction to Ceramics*. Copyright © 1976 by John Wiley & Sons, Inc. This material is used by permission of John Wiley & Sons, Inc.

by determining the dispersion at specific wavelengths, such as the hydrogen F line ($\lambda = 486.1$ nm), sodium D line ($\lambda = 589.3$ nm), and hydrogen C line ($\lambda = 656.3$ nm), as indicated in Figure 6.90. The complete dispersion curves for some common inorganic materials are shown in Figure 6.91.

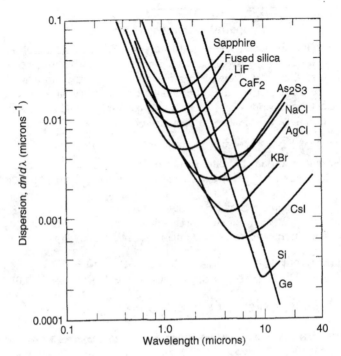

Figure 6.91   Effect of light wavelength on dispersion for some common inorganic materials. From W. D. Kingery, H. K. Bowen, and D. R. Uhlmann, *Introduction to Ceramics*. Copyright © 1976 by John Wiley & Sons, Inc. This material is used by permission of John Wiley & Sons, Inc.

The data shown in Figures 6.90 and 6.91 are a subset of the more general behavior, which is composed of two types of dispersion: *normal dispersion* and *anomalous dispersion*. Normal dispersion arises in the visible range of light, and the index of refraction increases with decreasing wavelength (increasing energy), as in Figure 6.90. In this region, the refractive index can be correlated with wavelength according to the following empirical expression, known as the Cauchy formula:

$$n = A + \frac{B}{\lambda^2} + \frac{C}{\lambda^4}$$ (6.84)

where A, B, and C are constants. In the region where refractive index decreases with decreasing wavelength (typically in the ultraviolet), anomalous dispersion occurs. Anomalous dispersion is the result of interactions between the light and the electronic oscillators in the material. At the so-called *resonant frequencies* of the electronic oscillators, absorption of the electromagnetic radiation can occur, resulting in reinforcement of the natural electronic oscillations. This resonance tends to decrease the refractive index.

6.3.2.2   *Reflection and Refraction.* With the principle of refractive index now in mind, we can return momentarily to the concept of reflection, which was first introduced in Section 6.3.1. Recall that the velocity of light changes as it passes from

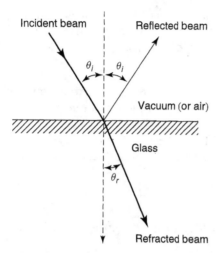

Figure 6.92 Reflection and refraction at a solid surface. Reprinted, by permission, from J. F. Shackelford, *Introduction to Materials Science for Engineers*, 5th ed., p. 597. Copyright © 2000 by Prentice-Hall, Inc.

vacuum to a medium, such as air. The velocity also changes at any interface between two phases with different indices of refraction, such as at an air–solid interface as illustrated in Figure 6.92. The change in velocity causes the light to bend upon passing through the interface. The bending of transmitted light at an interface is more properly termed *refraction*—hence the term refractive index. We also know that some of the light is reflected. The ratio of reflected to refracted (transmitted) light at the interface is a function of (a) the refractive index difference between the two phases and (b) the angle of the incident light.

If the *angle of incidence* from a normal to the surface is θ_i (see Figure 6.92) and the *angle of refraction* is θ_r, the refractive index of the solid medium, n (provided that the incident light is coming from a phase of low refractive index such as vacuum or air), is given by

$$n = \frac{\sin \theta_i}{\sin \theta_r} \tag{6.85}$$

Recall from Figure 6.85 that the light that is reflected at the same angle of incidence, θ_i, is due to specular reflectance. We can now use the refractive index to show that the reflectivity, R [cf. Eq. (6.70)], at the interface in Figure 6.92 is given by

$$R = \frac{(n-1)^2}{(n+1)^2} \tag{6.86}$$

Equation (6.86) holds for a solid–gas interface where the refractive index of the gas (or vacuum) is approximately unity. It is a simplification of a more general expression for light propagating through a medium of refractive index n_1, striking perpendicular to a surface ($i = 0$) of refractive index n_2, which is presented here without derivation:

$$R = \frac{(n_2 - n_1)^2}{(n_2 + n_1)^2} \tag{6.87}$$

Thus, the higher the refractive index of the solid, the greater the reflectivity. For example, the reflectivity of a typical silicate glass, with $n = 1.5$, is about 0.04. Just as the refractive index varies with wavelength, so too does the reflectivity.

6.3.2.3 Absorbance and Color. Once the light that is not reflected enters the medium, it can either continue to be transmitted or be absorbed by the medium (cf. Figure 6.83). The absorption process, too, is a function of the energy (wavelength) of the light.

Recall from Eq. (6.71) that the fractional change in light intensity, dI/I, over a distance x within the absorbing medium is directly proportional to the linear absorption coefficient, β:

$$\frac{dI}{I} = -\beta \, dx \tag{6.71}$$

The absorption coefficient is a material property and is a function of the wavelength of light, λ:

$$\beta = \frac{4\pi k}{\lambda} \tag{6.88}$$

where k is called the *index of absorption*.

Light is absorbed by two basic mechanisms: electronic polarization and electronic excitation. Electronic polarization and its effect on refractive index were described in the previous section and will not be elaborated upon here. The process of electronic excitation is an important one, however, and has implications to a number of optical phenomena such as lasing and luminescence.

Absorption of a photon of light may occur by the promotion or excitation of an electron from the valence band to the conduction band. In doing so, the promoted electron leaves behind a hole in the valence band and is free to move in the conduction band. Absorption and excitation occur only if the photon has energy greater than that of the band gap, E_g:

$$h\nu \geq E_g \tag{6.89}$$

Cooperative Learning Exercise 6.10

Use the data in Figure 6.89 to perform the following calculations.

Person 1: Calculate the maximum band gap energy for which absorption of visible light is possible.

Person 2: Calculate the minimum band gap energy for which absorption of visible light is possible.

Compare your answers. What type of electrical properties will materials within these band gap energy limits possess?

Answers: $E_g(\text{max}) = hc/\lambda_{\text{min}} = (4.13 \times 10^{-15} \text{ eV} \cdot \text{s})(3 \times 10^8 \text{ m/s})/(4 \times 10^{-7} \text{ m}) = 3.1 \text{ eV}; E_g(\text{min}) = 1.8 \text{ eV}$

The magnitude of the band gap energy, then, will determine whether the material absorbs no light, and is transparent, or absorbs certain wavelengths of light, thus becoming *opaque*.

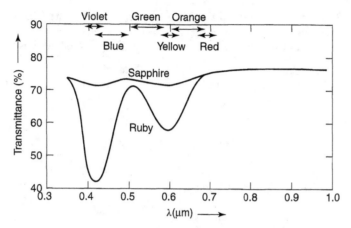

Figure 6.93   Transmission spectrum for sapphire and ruby. From K. M. Ralls, T. H. Courtney, and J. Wulff, *Introduction to Materials Science and Engineering.* Copyright © 1976 by John Wiley & Sons, Inc. This material is used by permission John Wiley & Sons, Inc.

The absorption of certain wavelengths of light in the visible spectrum results in *color*. As illustrated in Figure 6.93 for a ruby, the absorption of light centered at 0.42 and 0.6 μm (represented here as decreases in transmittance rather than increases in absorbance) correspond to absorption of blue-violet and yellow-green light, respectively. The result is that red is the most strongly transmitted visible light, giving ruby its characteristic red color. The transmission spectrum for sapphire, which has a slight bluish color, is also shown in Figure 6.93. The interesting fact is that both ruby and sapphire are primarily Al_2O_3, with only trace amounts of Ti^{3+} or Cr^{3+} substituted for Al^{3+} in sapphire and ruby, respectively. In the case of ruby, the Cr^{3+} ions are point defects with associated energy states that lie between the valence and conduction bands of Al_2O_3. The separation of levels is such that blue-violet light can be absorbed and red light emitted. A characteristic of such an absorption–emission process is that the emitted light cannot be of a higher energy than the absorbed light. We will further describe these radiative transitions when we get to the topics of luminescence and phosphorescence.

In general, the most commonly used ions for creating color in ceramics are transition elements with their incomplete *d* shells, such as V, Cr, Mn, Fe, Co, Ni, and Cu, and, to a lesser extent, the rare-earth elements with their incomplete *f* shells. In addition to the individual ion and its oxidation state, absorption phenomena are affected by their ionic environment. The introduction of impurity atoms to affect color is also utilized in glasses. In this case, a single ion such as Co^{2+} can give different colorings in different glasses, since it can have different coordination numbers associated with it (cf. Section 1.2.4.3). Some common metal ions and their effect on the color of silicate glasses are listed in Table 6.26.

6.3.2.4   *Transmittance, Transparency, and Translucency.*   The descriptions of absorption processes and their relationship to transmittance and color in the preceding section serve to underscore the relationship first presented by Eq. (6.70)—namely, that the fractions of light reflected, absorbed, and transmitted must sum to unity. In this

Table 6.26   Selected Metal Ions and their Effect on Color in Silicate Glasses

Ion	In Glass Network		In Modifier Position	
	Coordination Number	Color	Coordination Number	Color
Cr^{2+}				Blue
Cr^{3+}			6	Green
Cr^{6+}	4	Yellow		
Cu^{2+}	4		6	Blue-green
Cu^{+}			8	Colorless
Co^{2+}	4	Blue-purple	6–8	Pink
Ni^{2+}		Purple	6–8	Yellow-green
Mn^{2+}		Colorless	8	Weak orange
Mn^{3+}		Purple	6	
Fe^{2+}			6–8	Blue-green
Fe^{3+}		Deep brown	6	Weak yellow
U^{6+}		Orange	6–10	Weak yellow
V^{3+}			6	Green
V^{4+}			6	Blue
V^{5+}	4	Colorless		

Source: Shackelford, J.F., *Introduction to Materials Science for Engineers*, 5th ed. Copyright © 2000 by Prentice-Hall, Inc.

section, we complete our description of these three processes with some issues related to transmittance—namely, transparency and translucency.

A material is transparent if it transmits a clear image through it. Dielectrics tend to be transparent to visible radiation because they possess such large energy gaps that light cannot cause electronic excitation. Thus, materials such as diamond, sodium chloride, and silicate glasses readily transmit light. When the energy of the incident radiation is sufficiently high, as in ultraviolet radiation, electrons can be excited across the energy gap, and strong absorption occurs, as described in the previous section.

Many inherently transparent materials appear "milky" because of diffusive transmission, also known as *translucency*. Translucence results from multiple internal reflections. In noncubic polycrystalline materials that possess an anisotropic refractive index, light is reflected or scattered at grain boundaries where the index of refraction undergoes a discontinuous change. This effect is similar to that which results in reflection at a free surface. Translucency is slightly different from opacity, which was first described in the previous section, insofar as a diffuse image is transmitted through a translucent medium, whereas a total loss of image occurs in an opaque medium. In both cases, however, a certain fraction of some light of certain wavelengths is transmitted.

In addition to surface roughness and absorbing chemical species that affect transmittance and have already been described, such factors as porosity and particulate inclusions can affect translucence and opacity, mostly through a process called *scattering*. Figure 6.94 illustrates how scattering can occur at a single pore by refraction. The light is refracted as it exits a medium of higher refractive index and enters one of lower refractive index, such as air, inside the pore. This is essentially a miniaturized, reversed situation of the refraction process illustrated in Figure 6.92. A similar

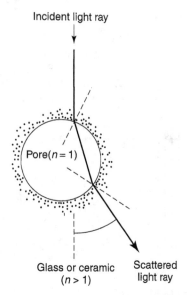

Incident light ray

Pore($n = 1$)

Glass or ceramic
($n > 1$)

Scattered
light ray

Figure 6.94   Schematic illustration of light scattering in a solid due to a pore. Reprinted, by permission, from J. F. Shackelford, *Introduction to Materials Science for Engineers*, 5th ed., p. 599. Copyright © 2000 by Prentice-Hall, Inc.

scattering result can be created with second-phase particles such as SnO_2 with a higher index of refraction ($n = 2.0$) than that of the medium, such as glass ($n = 1.5$). Such materials are called *opacifiers*. The degree of scattering and opacification created by the pores or particles depends on their average size and concentration, as well as the mismatch between the refractive indices. If individual pores or particles are significantly smaller than the wavelength of light, they are ineffective scattering centers. Conversely, particles or pores in the 400- to 700-nm size range maximize the scattering effect. This effect is what leads to transparent ceramics, such as glass ceramics (cf. Section 1.2.5) where the crystallite sizes are carefully controlled through nucleation and growth to be less than the wavelength of visible light (cf. Figure 1.52).

6.3.2.5 *Decay Processes: Luminescence and Lasers**. The absorption and reemission of electromagnetic radiation leads to some interesting phenomena that can be utilized for important applications. Both lasers and luminescence are built upon the principle of emission processes and will be described here in brief.

As we have already seen, the absorption of photons by a material can lead to the reemission of other photons of equal or lesser energy. When photon absorption is accompanied by emission of photons in the visible spectrum, the process is called *photoluminescence*, or simply *luminescence*. In fact, luminescence can be used to describe the emission of visible light as a result of absorption of any number of forms of energy, including thermal, mechanical, or chemical energy. The photon emission is the result of excited electrons returning to their ground state. The duration of this relaxation process is used to distinguish between the two main types of luminescence: *fluorescence* and *phosphorescence*. If the relaxation process occurs in less than about 10 ns (10^{-8} s), it is termed fluorescence. The fluorescent process also involves additional internal or

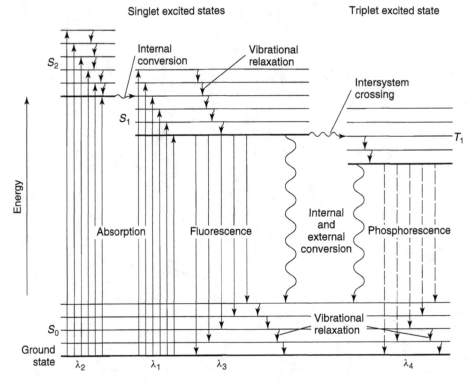

Figure 6.95   Schematic diagram of transitions between electronic energy levels involved in fluorescence and phosphorescence. Reprinted, by permission, from D. A. Skoog and D. M. West, *Principles of Instrumental Analysis*, 2nd ed., p. 282. Copyright © 1980 by Saunders College.

external energy conversion, as illustrated with the energy diagrams in Figure 6.95. In this way, the energy of the reemitted photons is less than the energy of the absorbed photons. If the reemission process takes longer than about 10 nanoseconds, it is termed phosphorescence.

Luminescence is generally produced through the addition of impurities to compounds. For example, zinc sulfide containing small amounts of excess zinc, or impurities such as gold or manganese, emits light in the visible spectrum after excitation with ultraviolet radiation (see Figure 6.96). Zinc sulfide and a similar compound, Mn-activated zinc orthosilicate, Zn_2SiO_4, are also examples of compounds used as *phosphors*. Phosphors are useful for detecting X-ray or γ-ray radiation because they emit visible light upon exposure. They are used for other, more mundane purposes as well, such as the luminous hands on watches and clocks.

Light amplification by stimulated emission of radiation, or *LASER*, is another application of radiative relaxation processes. Unlike most radiative processes, such as luminescence, which produce *incoherent* light (light waves are out of phase with each other), the light produced by laser emission is *coherent* (light waves are all in phase with each other). The principal benefit of coherent light is that the photons do not disperse as they do in an incoherent beam of light. Consequently, the laser beam does not spread out in the same way that an incoherent beam can. Additionally, lasers produce a beam of light that is monochromatic—that is, entirely of one wavelength. This is

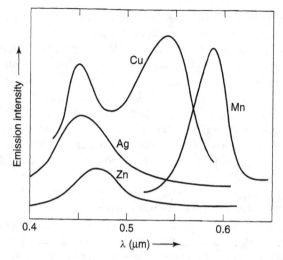

Figure 6.96 Luminescence spectra for Zn doped with Cu, Ag, Mn, and excess Zn. From W. D. Kingery, H. K. Bowen, and D. R. Uhlmann, *Introduction to Ceramics*. Copyright © 1976 by John Wiley & Sons, Inc. This material is used by permission of John Wiley & Sons, Inc.

in contrast to most light sources, even phosphors, that produce a range of wavelengths, or polychromatic light. Polychromatic light is not entirely a bad thing—we much prefer white light (all or most of the wavelengths in the visible spectrum) to monochromatic light for our daily and nightly operations. But the monochromatic, coherent, and collimated (photons directed along the same straight line) beam of a laser opens up a realm of possible applications. Let us see how this type of beam can be produced.

The radiative transitions of the previous descriptions have all been spontaneous: Relaxation from the excited state to the ground state and emission of photons occur without external aid. In contrast, a stimulated emission occurs when the half-life of the excited state is relatively long, and relaxation can occur only through the aid of a stimulating photon. In stimulated emission, the emitted photon has the same direction as, and is in phase with, the stimulating photon. The example of Cr^{3+}-doped Al_2O_3 that we utilized earlier for our description of the color of ruby works equally well for a description of stimulated emission. Recall that the presence of chromium in alumina alters the electronic structure, creating a metastable state between the valence and conduction bands. Absorption of a blue-violet photon results in the excitation of an electron from the ground state to the excited state, as illustrated in Figure 6.97. The excited electron relaxes to a metastable electronic state through a non-radiative process involving the transfer of energy to a phonon. *Spontaneous decay* results in the emission of a photon with wavelength in the red part of the visible spectrum ($\lambda = 694.3$ nm). However, most of the electrons reside in this metastable state for up to 3 ms before decaying to the ground state. This time span is sufficiently long that several electrons can occupy this metastable state simultaneously. This fact is utilized to create artificially large numbers of electrons in the metastable state through a process known as *optical pumping*, in which a xenon flash lamp excites a large number of electrons into the excited state, which decay rapidly (in the nonradiative process) to the metastable state. When spontaneous emission finally occurs due to a few electrons, these electrons stimulate the emission of the remaining electrons in the metastable state, and an avalanche of emission occurs. The photons are of the same energy and are in phase (coherent).

The beam is collimated through the use of a tube with silvered mirrors at each end, as illustrated in Figure 6.98. As stimulated emission occurs, only those photons traveling nearly parallel to the long axis of the ruby crystal are reflected from the fully silvered surface at the end of the tube. As these reflected photons travel back through the tube, they stimulate the emission of more photons. A fraction of these photons are re-reflected at the partially silvered face at the other end of the tube. Thus, photons travel up and down the length of the crystal, producing ever-increasing numbers of stimulated photons. When the beam is finally emitted through the partially silvered face, it is a high-energy, highly collimated, monochromatic beam of coherent light. Typical power densities for a ruby laser are on the order of a few watts.

In addition to ruby, lasers can be constructed from gases such as CO_2 or He–Ne and from semiconductors such as GaAs (cf. Section 6.1.2.5) and InGaAsP. Gas lasers generally produce lower intensities and powers, but are more suitable for continuous operation since solid-state lasers generate appreciable amounts of heat. The power in a solid-state laser can range from microwatts to 25 kW, and more than a megawatt in lasers for defense applications. Some common lasers and their associated powers are listed in Table 6.27. Lasers that can selectively produce coherent light of more than one wavelength are called *tunable lasers*.

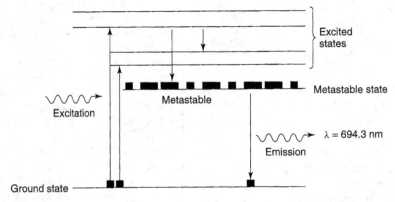

Figure 6.97   Schematic illustration of energy levels in ruby that are used to create a populated metastable electronic state, which can then be stimulated to emit monochromatic, coherent radiation for a laser. Reprinted, by permission, from J. F. Shackelford, *Introduction to Materials Science for Engineers*, 5th ed., p. 607. Copyright © 2000 by Prentice-Hall, Inc.

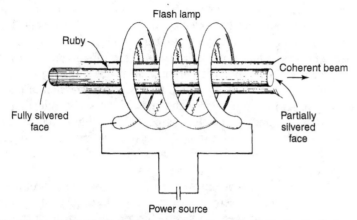

Figure 6.98 Schematic illustration of a ruby laser. From K. M. Ralls, T. H. Courtney, and J. Wulff, *Introduction to Materials Science and Engineering*. Copyright © 1976 by John Wiley & Sons, Inc. This material is used by permission of John Wiley & Sons, Inc.

Table 6.27   Characteristics and Applications of Some Common Lasers

Laser	Type	Common Wavelengths (μm)	Maximum Output Power (W)[a]	Applications
He-Ne	Gas	0.6328, 1.15, 3.39	0.0005–0.05 (CW)	Line-of sight communications, recording/playback of holograms
CO_2	Gas	9.6, 10.6	500–15,000 (CW)	Heat treating, welding, cutting, scribing, marking
Argon	Gas ion	0.488, 0.5145	0.005–20 (CW)	Surgery, distance measurements, holography
HeCd	Metal vapor	0.441, 0.325	0.05–0.1	Light shows, spectroscopy
Dye	Liquid	0.38–1.0	0.01 (CW) 1×10^6 (P)	Spectroscopy, pollution detection
Ruby	Solid state	0.694	(P)	Pulsed holography, hole piercing
Nd-YAG	Solid state	1.06	1000 (CW) 2×10^8 (P)	Welding, hole piercing, cutting
Nd-glass	Solid state	1.06	5×10^{14} (P)	Pulse welding, hole piercing
Diode	Semiconductor	0.33–40	0.6 (CW) 100 (P)	Bar-code reading, CDs and video disks, optical communications

[a]"CW" denotes continuous; "P" denotes pulsed.
Source: W. Callister, *Materials Science and Engineering: An Introduction*, 5th ed. Copyright © 2000 by John Wiley & Sons, Inc.

Semiconducting lasers have become increasingly important in information storage, such as in the use of compact discs (CDs) and digital video discs (DVDs). In these materials, a voltage is applied across a layered semiconductor (see Figure 6.99), which excites electrons into the conduction band, just as in the ruby laser. However, the holes created in the valence band play an important role in the emission process. It is the recombination of the electron–hole pairs that results in the emission of photons with a specific wavelength. As in the ruby laser, the ends of the semiconducting "sandwich" are fully and partially silvered, and a voltage is continuously applied to ensure that there is a steady source of holes and electrons. These devices are extremely temperature-sensitive, since the current required to cause lasing, the wavelength of emitted radiation, and the lifetime of the diode all depend on temperature.

The conventional III–V compound semiconductors such as GaAs are limited to infrared or red emitting diode lasers and are not capable of generating shorter wave-length light, which is a requirement for size reduction in information storage devices. Short-wavelength LEDs and laser diodes require a different class of semiconductor materials; principally, the group III nitrides. The growth of high-quality (AlGaIn)N single-crystalline layer sequences has been a challenge, and only recent developments in fabrication processes have produced blue lasers that can be used in commercially available devices such as DVD players. The wavelengths of some common compound semiconductors are listed in Table 6.28.

6.3.2.6 *Photoconductivity**. In Section 6.3.1.2 we saw how electricity can be generated from the surface of a metal when it is bombarded with photons. A similar process occurs when certain semiconductors absorb photons and create electron–hole pairs that can be used to generate current. This process is called *photoconductivity* and is slightly different from the photoelectric effect, since it is an electron–hole pair that is being generated instead of a free electron and since the energy of the electron–hole pair is related to the band gap energy, and not the energy of the Fermi level.

When a semiconductor is bombarded with photons equal to or greater in energy than the band gap, an electron–hole pair is formed. The current that results is a direct function of the incident light intensity. Photoconductive devices consist of a *p–n* junction called a *photodiode*, or a *p–i–n* junction commonly used in a photodetector, an

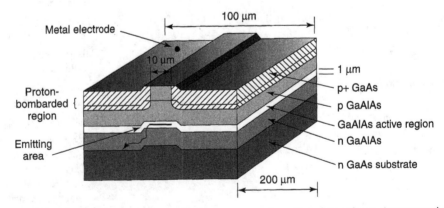

Figure 6.99 Schematic diagram of a layered compound semiconductor laser. (www.mtmi.vu. lt/pfk/funkc_dariniai/diod/led.htm).

Table 6.28   Some Common Compound Semiconductor Lasers and Their Characteristic Wavelengths

Compound	Wavelength (μm)	Compound	Wavelength (μm)
ZnS	0.33	GaAs	0.84–0
ZnO	0.37	InP	0.91
GaN	0.40	GaSb	1.55
ZnSe	0.46	InAs	3.1
CdS	0.49	Te	3.72
ZnTe	0.53	PbS	4.3
GaSe	0.59	InSb	5.2
CdSe	0.675	PbTe	6.5
CdTe	0.785	PbSe	8.5

Source: www.mtmi.vu.lt/pfk/funkc_dariniai/diod/led.htm

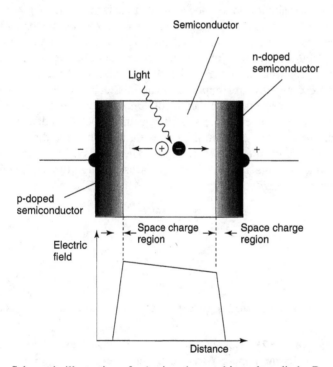

Figure 6.100   Schematic illustration of *p-i-n* junction used in a photodiode. Reprinted, by permission, from P. Ball, *Made to Measure*, p. 46. Copyright © 1997 by Princeton University Press.

example of which is schematically illustrated in Figure 6.100. When light is absorbed in the undoped region (*i* region), the electron–hole pair is created, and the two charge carriers are pulled in opposite directions by the applied electric field to the two poles, thus creating current. Silicon can be used to detect infrared wavelength light, and InGaAs is used for longer wavelengths, including visible light. However, optical detectors tend to be relatively slow and have low sensitivity. CdS is widely used for the detection of visible light, as in photographic light meters. Photoconductivity is also the underlying principle of the *photovoltaic cell*, commonly called the *solar cell*, used for the conversion of solar energy into electricity.

6.3.2.7 Optical Fiber*. One of the benefits that lasers have provided is a method for transmitting information in binary code (cf. Section 6.1.1.6). This is easily performed by assigning a "one" to a high-intensity pulse (on) and assigning a "zero" to a low intensity pulse (off). The key to utilizing this technology for telecommunications, however, is the ability to transmit this digital information over long distances. One method for transmitting optical information has been through the optical analog to copper wire, the optical fiber. The conversion from the transmission of information via analog signals (such as telephone conversations) on copper wires to the transmission of information via digital signals on fiber-optic cable has been a rapid one and is due primarily to the advantages of optical fiber over copper wire, some of which are listed in Table 6.29. In Table 6.29, the number of two-way transmissions is a measure of information capacity, and the repeater distance is the distance between amplifiers, indicative of the attenuation loss associated with each medium. It is ironic that even with the proliferation of wireless communications (e.g., cellular telephones), the use of land-based transmission media, such as optical fiber, has grown in a concurrent manner, since for every cellular transmission tower, there needs to be the associated land-based components that connect the tower to the communications "backbone." Hence, it is worthwhile to describe how this important materials technology works.

Optical fiber operates on the principle of *total internal reflectance*. Recall from Eq. (6.85) that when light encounters a change in refractive index, it is bent at an angle. Consider now the case where light goes through three media, each of differing refractive index, as illustrated in Figure 6.101. The angles of refraction are related to the refractive index change at each of the interfaces, through the more general form of Eq. (6.85) known as *Snell's Law*:

$$\frac{n_2}{n_1} = \frac{\sin \theta_1}{\sin \theta_2} \tag{6.90}$$

and

$$\frac{n_3}{n_2} = \frac{\sin \theta_2}{\sin \theta_3} \tag{6.91}$$

Note that Eq. (6.90) reduces to Eq. (6.85) for the case where medium 1 is air ($n_1 = 1$).

Table 6.29 Comparison of Transmission Characteristics for Copper Wire and Optical Fiber

Characteristic	Copper Wire	Optical Fiber
Diameter, mm	1.5	0.025
Density, g/cm^3	8.96	2–4
Repeater distance, km	2	30
No. of two-way transmissions	24	1344
Corrosion resistance	Low	High
Susceptibility to EM interference	High	Low

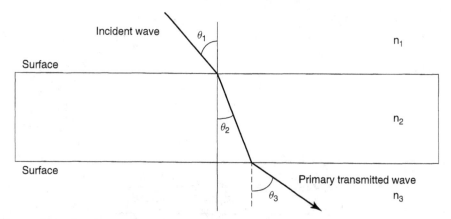

Figure 6.101 Schematic illustration of light propagating through three media of differing refractive indices.

At some angle of θ_2, the transmitted wave will travel parallel to the second interface (*i.e.*, $\theta_3 = 90°$), and all of the light is reflected back into medium 2. This value of θ_2 is called the *critical angle for total internal reflectance*, θ_c. For example, if medium 3 were air ($n_3 = 1$) and medium 2 were a glass of $n_2 = 1.5$, the condition for which $\theta_3 = 90°$ could be found from Eq. (6.91) to be $\theta_c = 42°$. You should be able to visualize, with the aid of Figure 6.101, that total internal reflectance can also be accomplished by selecting an index of refraction for medium 3 such that $\theta_3 = 90°$. This is called the *core and cladding* structure, where medium 1 is air at the entrance to the optical fiber, medium 2 is the core, and medium 3 is the cladding. The core–cladding structure is the principle behind fiber optics, and simply consists of a center core, through which the optical signal is transmitted, surrounded by a cladding of different refractive index. The indices of refraction are selected such that $n_{cladding} < n_{core}$. Once the light enters the core from the source, it is reflected internally and propagates along the length of the fiber. Typically, both the core and cladding are made of special types of glass with carefully controlled indices of refraction. A polymeric coating is usually applied to the outside of the fiber for protection only.

The sharp change of refractive index between the core and cladding results in what is called a *step-index optical fiber*, as illustrated in Figure 6.102a. It is also possible, through chemical manipulation of the glass components, to vary the index of refraction in a continuous, parabolic manner, resulting in a *graded-index optical fiber*, as shown in Figure 6.102b. This results in a helical path for the light rays, as opposed to a zigzag path in a step-index fiber. The digital pulse is less distorted in a graded-index fiber, which allows for a higher density of information transmission. Both step- and graded-index fibers are termed *multimode fibers*. A third type of optical fiber is called a *single-mode fiber*, as shown in Figure 6.102c, in which light travels largely parallel to the fiber axis with little distortion of the digital light pulse. These fibers are used for long transmission lines, in which all of the incoming signals are combined, or *multiplexed*, into a single signal.

Core and cladding materials are selected not only on the basis of their refractive index, but also for their processability, attenuation loss, mechanical properties, and dispersion properties. However, density, ρ, and refractive index, n, are critical and have been correlated for over 200 optical glasses, for which the following formula

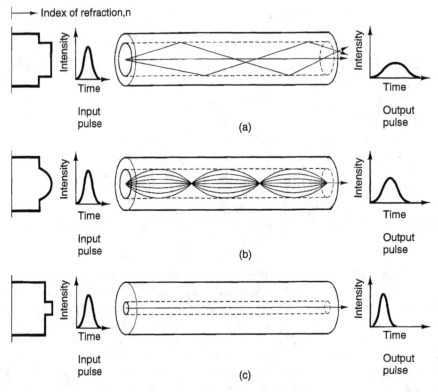

Figure 6.102   Schematic illustration of (a) step-index, (b) graded-index, and (c) single-mode optical fibers. Reprinted, by permission, from J. F. Shackelford, *Introduction to Materials Science for Engineers*, 5th ed., p. 611. Copyright © 2000 by Prentice-Hall, Inc.

applies with an accuracy of 2% or better:

$$n = \frac{\rho + 10.4}{8.6} \tag{6.92}$$

Refractive index can also be found from the weighted average of the refractive index of its components, to a first approximation.

High-purity silica-based glasses are used as the fiber material, with fiber diameters ranging between about 5 and 100 μm. The fibers are carefully fabricated to be virtually free of flaws and, as a result, are extremely strong and flexible. We will examine this unique fabrication process in more detail in the next chapter.

6.3.3   Optical Properties of Polymers

As with virtually every other material property, if a polymer possesses optical properties comparable to those of a glass or metal, it will be utilized simply on the basis of weight savings, all other factors being equal. As a result, much of the development in the optical properties of polymers is related to creating glass-like transmissivities or metal-like reflectivities. There are also a few unique optical characteristics of polymers that make them useful for applications for which there is no true glass or metal analogue. We investigate some of these optical characteristics of polymers in this section.

6.3.3.1 *Polymer Structure, Transmittance and Birefringence.*

There is a great deal of variability in the light transmission properties of polymers. Some of this variability is due to structural differences, and some is due to the presence of fillers and colorants that are purposely added to make otherwise transluscent polymers appear opaque. We will focus for the moment on structural differences in polymers that lead to differences in the transmission of light.

The primary structural characteristic affecting the transmission of light in polymers is crystallinity. As in ceramic materials, the presence of grain boundaries, pores and particles that are of the order in size of the wavelength of light will create scattering. Crystallites in polymers act as these scattering centers. An example is found in polyethylene. Low-density polyethylene is semicrystalline, with only a few crystallites to act as scattering centers, so that it is mostly transparent. However, high-density polyethylene, which is highly crystalline, is translucent. In fact, most polymer crystals are of the right size to cause scattering, so a rule of thumb is that crystalline polymers are translucent (or opaque), whereas noncrystalline polymers are transparent. Some examples of noncrystalline, transparent polymers are polycarbonate, acrylics such as poly(methyl methacrylate), and polystyrene. There are some transparent, crystalline polymers, such as polyethylene terephthalate (PET), in which the crystallites are smaller than the wavelength of light.

An interesting artifact of polymer crystallinity is the anisotropic nature of the refractive index, which leads to birefringence, as first introduced in Section 6.3.2.1. When the topic of polymer crystallinity was first introduced back in Chapter 1, we presented a photomicrograph of polyethylene spherulites that was produced under cross-polarized light (cf. Figure 1.62). This image is obtained by placing a sample of the polymer between two quarter-wave plate optical filters, or *polarizers*, which allow the light passing through to vibrate only along one plane. In this way, if the second filter, called the *analyzer*, is rotated 90° from the polarizer, no light will pass through the filter combination, since the plane of vibration allowed through the polarizer is extinguished in the analyzer. However, when a crystalline sample such as polyethylene is placed between the polarizer and analyzer, an interesting pattern is transmitted through the analyzer, which includes some dark regions (no transmission) and light regions (complete transmission). An idealized form of the resulting *Maltese cross* is shown in Figure 6.103, and an actual image of which was shown in Figure 1.62. This image is the result of birefringence, or refractive index differences in the oriented polymer chains. The index of refraction for most polymers is greater parallel to the chain than normal to the molecular axis. As a result, the refractive index in the tangential direction of the spherulite is greater (see Figure 1.61) than that along the spherulite radius. This birefringence, coupled with the spherical geometry of the spherulite, produces light extinction along the axis of each of the filters, hence the 90° angles in the Maltese cross.

Birefringence can also be used to analyze polymer samples after melt processing. As we will see in the next chapter, the shear produced in certain molding techniques, such as injection molding, can orient polymer chains in certain parts of the mold, especially near the mold walls, whereas the chains in low-shear regions, such as in the middle of the mold, are not as oriented. Figure 6.104 shows the variation in birefringence, as

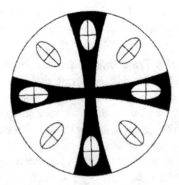

Figure 6.103   Schematic of Maltese cross produced by spherulites in crosspolarized filters. Reprinted, by permission, from Strobl, G., *The Physics of Polymers*, 2nd ed., p. 146. Copyright © 1997 Springer-Verlag.

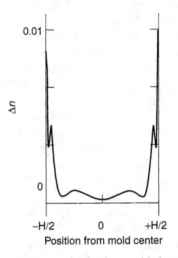

Figure 6.104   Variation in birefringence of injection molded polymer of width H. Adapted from P. C. Powell and A. J. Ingen Housz, *Engineering with Polymers*, p. 314. Copyright © 1998 by P. C. Powell and A. J. Ingen Housz.

measured by the refractive index difference along two directions, with position in the mold for an injection-molded polymer. The birefringence is largest at the periphery of the mold, where shear rates are highest and the polymer is most highly oriented.

6.3.3.2   *Electroluminescence**. In Section 6.3.2.5, we saw that some materials—in particular, semiconductors—can reemit radiation after the absorption of light in a process called photoluminescence. A related type of emission process, which is common in polymer-based semiconductors, called *electroluminescence*, results when the electronic excitation necessary for emission is brought about by the application of an electric field rather than by incident photons. The electric field injects electrons into the conduction band, and holes into the valence band, which upon recombination emit light.

As we saw in Section 6.1.3.2, most electrically conducting polymers contain poly-conjugated structures, such as extended double bonds along the polymer backbone. It is not surprising, then, that electrically conducting polymers such as polyacetylene can also be electroluminescent. The first visible-light-emitting polymer was made from poly(phenylene vinylene), or PPV, the structure for which is shown in Figure 6.105a. PPV exhibits luminescence in the yellow part of the spectrum, and it can be fabricated into a *light-emitting diode* (LED) by sandwiching a micron-thick layer of PPV between contacts made of indium-tin oxide (a transparent electrode material) on the bottom and cadmium metal on the top. These contact materials are selected to match their electronic bands to that of the polymer. When 12 volts are applied across the contacts, electrons and holes enter the polymer and recombined to generate photons of yellow light. Other polymers with different band gaps can be used to emit light of various wavelengths, such as polythiophene-based compounds (see Figure 6.105b) for the emission of blue light. The applications for polymer LEDs (PLEDs) continue to grow, including paper-thin computer and television displays, which will ultimately require a single polymer that is capable of emitting blue, red, or yellow light, depending upon the applied voltage. The requirements that must be met before PLEDs can become viable candidates for use in flat panel displays include sufficiently high brightness, good photoluminescence profiles, high efficiency (i.e., low operating voltage and current), good color saturation, and long lifetime. A good display device must have at least a brightness of 100 cd m^{-2} at an operating voltage of between 5 and 15 V and a lifetime of 10,000 h. Moreover, charge conduction in the PLEDs requires electric fields in the range of 2–5 MV cm^{-1}.

6.3.3.3 *Liquid Crystal Displays (LCDs)**. Liquid crystalline polymers, first introduced in Section 1.3.6.3, are utilized for a different type of computer and television display, the *liquid crystal display* (LCD). Most of today's laptop computers and hand-held devices utilize color flat panel displays where the light transmission from the

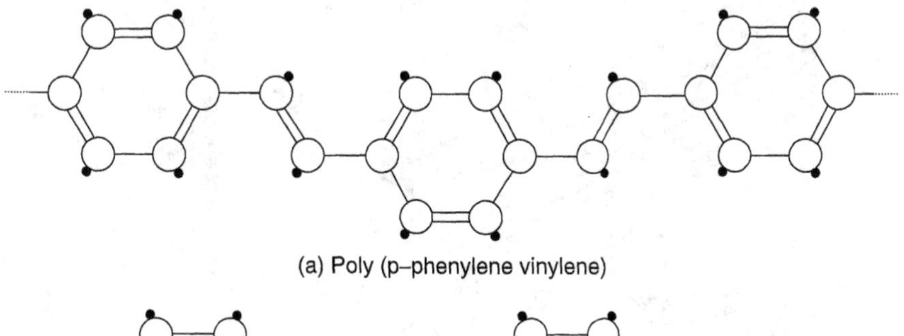

(a) Poly (p–phenylene vinylene)

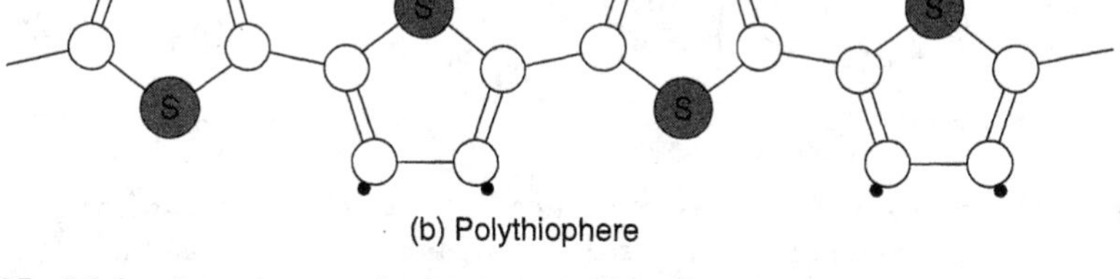

(b) Polythiophere

Figure 6.105 Molecular structure of two electroluminescent polymers. Reprinted, by permission, from P. Ball, *Made to Measure*, p. 381. Copyright © 1997 by Princeton University Press.

back to the front of the display is modulated by orientational changes in liquid crystal molecules. As illustrated in Figure 6.106, liquid crystalline polymers are sandwiched between two transparent polarizers. The types of liquid crystalline polymers can vary, but generally they are of the cholesteric or twisted nematic type. The LC polymers are initially oriented in the display manufacturing process using a "rubbing process." The mechanism of alignment is not well understood, however, and is an expensive step in the fabrication process. The transparent substrates between which the liquid crystals are deposited are usually polyimides.

Under an applied electric field, the liquid crystal molecules tend to align parallel to each other, resulting in optical anisotropy. The liquid crystal cell is thus a tiny shutter that is controlled by the applied electric field. When the area encompassed on the top, bottom, and sides by electrodes, called a *pixel*, is turned on, the liquid crystal molecules align, allowing only one plane of the electromagnetic wave (polarized light) to penetrate to the front panel, where it is absorbed in the second polarizer, effectively blocking the light (left image in Figure 6.106). When the field is turned off, the shutter filters the light by rotating its optical plane, as shown on the right in Figure 6.106. Changing bright pixels into colored ones requires yet another filter. As a result, LCDs

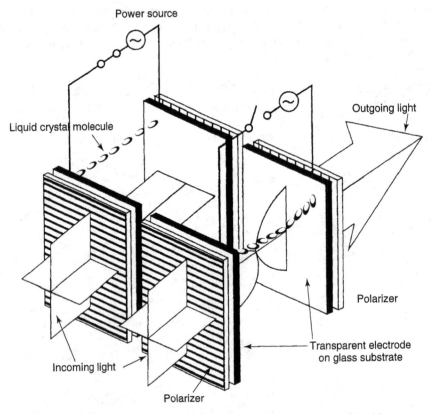

Figure 6.106 Schematic illustration of liquid crystalline polymers sandwiched between polarizers and effect on light in a liquid crystal display. From Jun-ichi Hanna and Isamu Shimizu, Materials in active-matrix liquid-crystal displays, *MRS Bulletin*, **21**(3), 35 (1996). Reproduced by permission of MRS Bulletin.

are very dim, and must be back-illuminated with a lamp, which is an energy-intensive process. LCDs also suffer from a limited viewing angle. These are the primary reasons behind the interest in the electroluminescent materials of the previous section for flat panel displays.

6.3.3.4 Nonlinear Optical Materials*.

We have seen several examples in the previous sections of how photons can be used to generate and transmit data. Though it is beyond the scope of this text, there are also devices for storing data in an optical fashion. This emerging area of using photons to replace what was previously performed by electrons is called *photonics*. The integration of photonic and electronic devices is called *optoelectronics*. In this section, we describe how the optical properties of materials can be used in yet another manner to create devices that can act as switches, multiplexers, and mirrors and even change the wavelength of light. The materials that exhibit these effects and that are used in these photonic and optoelectronic devices are generally referred to as *nonlinear optical materials* (NLO), because their outputs are not a linear response to some input. We are most familiar with linear responses of materials; for example, the intensity of photons emitted from a tungsten filament in a light bulb increases (more or less) linearly with the power we put into it. But even in this example, nonlinear behavior can occur in the form of *saturation*, where further increases in power do not result in more output, or in the form of *breakdown*, when the filament burns out. The diodes described in the preceding sections are nonlinear electronic components—their output currents remain negligibly low until the driving voltages reach a threshold value whereupon the output current increases sharply.

The best way to describe a nonlinear optical response is first through some examples, then through some mathematics. Shortly after the invention of lasers, it was observed that a ruby laser beam passing through a quartz crystal produced a faint beam at the laser's second harmonic—that is, at twice the fundamental frequency of the ruby laser (cf. Section 6.3.2.5). A much simpler-to-perform example of nonlinear optical behavior is found when the polarization of light passing through a crystal is modified upon application of an electric field to the crystal. The origin of both of these NLO effects is related to the change in refractive indices that result by an applied electric field and the modulation of light beams by these field-dependent indices. We have already seen the relationship between refractive index and polarizability, and this is where our mathematical description of NLO effects begins.

We can examine how induced polarization behaves as a function of an applied electric field, $\mathscr{E}$, by considering the induced electric dipole moment (cf. Section 6.1.2.2), p_e, as a Taylor series expansion in $\mathscr{E}$:

$$p_e = p_{e,0} + \mathscr{E}(\partial p_e/\partial \mathscr{E})_{\mathscr{E}\to 0} + \tfrac{1}{2}\mathscr{E}^2(\partial^2 p_e/\partial \mathscr{E}^2)_{\mathscr{E}\to 0} + \tfrac{1}{6}\mathscr{E}^3(\partial^3 p_e/\partial \mathscr{E}^3)_{\mathscr{E}\to 0} + \cdots \tag{6.93}$$

or, more simply,

$$p_e = p_{e,0} + \alpha\mathscr{E} + (\beta/2)\mathscr{E}^2 + (\gamma/6)\mathscr{E}^3 + \cdots \tag{6.94}$$

where $p_{e,0}$ is the static dipole of the molecule, α is the linear polarizability [previously called just "polarizability"; cf. Eq. (6.35)], and the higher-order terms β and γ are called the *first* and *second hyperpolarizabilities*, respectively. Note that once again, we consider only the magnitude of the electric field, which is actually a vector quantity.

The terms beyond $\alpha \mathscr{E}$ in Eq. (6.94) are not linear in $\mathscr{E}$, so they are referred to as the *nonlinear polarization* and give rise to NLO effects. You should now see why nonlinear polarization becomes more important at higher electric field strengths, since it scales with higher powers of the field. For most materials, $\alpha \mathscr{E} > (\beta/2)\mathscr{E}^2 > (\gamma/6)\mathscr{E}^3$, so the first few observations of NLO effects were made prior to the invention of lasers and their associated large electric fields. The bulk polarization, P, which you will recall is the dipole moment per unit volume, can then be expressed as

$$P = P_0 + \chi^{(1)}\mathscr{E} + \chi^{(2)}\mathscr{E}^2 + \chi^{(3)}\mathscr{E}^3 + \cdots \tag{6.95}$$

where the $\chi^{(n)}$ *susceptibility coefficients* (not to be confused with the magnetic susceptibility of the same symbol) are tensors of order $n + 1$, and P_0 is the intrinsic static dipole moment density of the material.

We now consider the polarization induced by an oscillating electric field, such as that found in electromagnetic radiation (light), which can be expressed as

$$\mathscr{E} = \mathscr{E}_0 \cos(\omega t) \tag{6.96}$$

where ω is the frequency of the electric field and t is time. Substitution of Eq. (6.96) into Eq. (6.95) yields

$$P = P_0 + \chi^{(1)}\mathscr{E}_0 \cos(\omega t) + \chi^{(2)}\mathscr{E}_0^2 \cos^2(\omega t) + \chi^{(3)}\mathscr{E}_0^3 \cos^3(\omega t) + \cdots \tag{6.97}$$

Because $\cos^2(\omega t) = \frac{1}{2} + \frac{1}{2}\cos(2\,\omega t)$, the first three terms of Eq. (6.97) become

$$P = (P_0 + \tfrac{1}{2}\chi^{(2)}\mathscr{E}_0^2) + \chi^{(1)}\mathscr{E}_0 \cos(\omega t) + \tfrac{1}{2}\chi^{(2)}\mathscr{E}_0^2 \cos(2\,\omega t) + \cdots \tag{6.98}$$

Equation (6.98) indicates that the polarization consists of a second-order, direct-current (not dependent upon ω) field contribution to the static polarization (the first term), a frequency component, ω, corresponding to the incident light frequency (the second term), and a new, frequency-doubled component, $2\,\omega$ (the third term). Thus, if an intense light beam passes through a second-order NLO material, light at twice the input frequency will be produced, as well as a static electric field. The process of frequency-doubling, which explains the first example of NLO behavior from the beginning of this section, is called *second harmonic generation* (SHG). The oscillating dipole reemits at all of its polarization frequencies, so light is observed at both ω and $2\,\omega$. The process of static field production is called *optical rectification*.

The general condition of second-order NLO effects involves the interaction of two distinct waves of frequencies ω_1 and ω_2 in an NLO material. In this case, polarization occurs at sum $(\omega_1 + \omega_2)$ and difference $(\omega_1 - \omega_2)$ frequencies. This electronic polarization will therefore reemit radiation at these frequencies, with contributions that depend on $\chi^{(2)}$, which is itself frequency-dependent. The combination of frequencies is called *sum* (or difference) *frequency generation* (SFG). SHG is a special case of SFG, in which the two frequencies are equal.

An applied electric field can also change a material's linear susceptibility, and thus its refractive index. This effect is known as the *linear electro-optic* (LEO) or *Pockel's effect*, and it can be used to modulate light by changing the voltage applied to a second-order NLO material. The applied voltage anisotropically distorts the electron density within the material, and the optical beam "sees" a different polarizability and anisotropy of the polarizability. As a result, a beam of light can have its polarization state changed by an amount related to the strength of the orientation of the applied voltage and can travel at a different speed, and possibly in a different direction. The change in refractive index as a function of the applied field is approximated by the general expression

$$1/N_{ij}^2 = 1/n_{ij}^2 + r_{ijk}\mathscr{E}_k + s_{ijkl}\mathscr{E}_k\mathscr{E}_l + \cdots \qquad (6.99)$$

where N_{ij} are the induced refractive indices, n_{ij} are the refractive indices in the absence of the electric field, r_{ijk} are the linear or Pockels coefficients, s_{ijkl} are the quadratic or Kerr coefficients, and $\mathscr{E}$ is the applied field. Subscripts indicate the orientation of the field with respect to a material coordinate system. Equation (6.99) indicates that light traveling through an electro-optic material can be phase- or polarization-modulated by refractive index changes induced by an applied electric field. Devices exploiting this effect include optical switches, modulators, and wavelength filters.

An example of an electro-optic switch based on NLO materials is shown in Figure 6.107. The switch is comprised of two parallel waveguides made of NLO materials. The waveguide channels have a different refractive index from the surrounding material. The light can be switched back and forth between the channels by applying and removing a voltage across the bottleneck. In the absence of an electric field, the light traveling through the lower waveguide interacts with the upper waveguide in a nonlinear manner at the bottleneck, causing the light to switch channels. Switching *does not* occur when an electric field is applied. The electric field polarizes the NLO material and alters the refractive indices of the two channels, such that the nonlinear interaction at the bottleneck is modified and the light stays in the lower waveguide.

Many current NLO devices are based upon crystalline materials (such as lithium niobate for the electro-optic switch) and nonlinear optical glasses, but there is intense

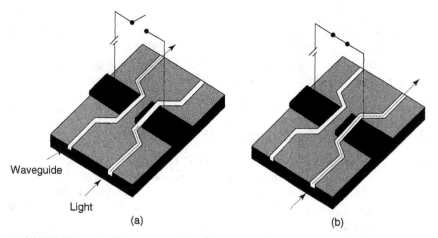

Figure 6.107   Schematic illustration of an electro-optic switch. (a) An open circuit causes light to switch channels, and (b) a closed switch keeps light in the same channel. Reprinted, by permission, from P. Ball, *Made to Measure*, p. 53. Copyright © 1997 by Princeton University Press.

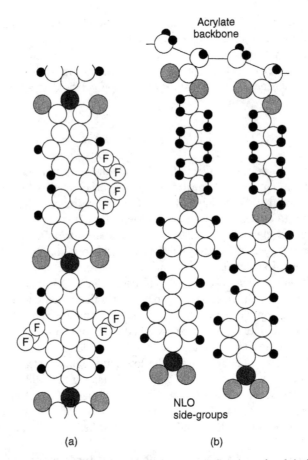

Figure 6.108   Examples of nonlinear optical polymers (a) fluorinated polyimide and (b) poly-acrylate with NLO side groups. Reprinted, by permission, from P. Ball, *Made to Measure*, p. 55. Copyright © 1997 by Princeton University Press.

interest in utilizing polymers for these applications, due to their inherently low densities and potential ease of processing. Some of the polymers that have been used for photonic devices are based upon architectures that have highly polarizable electronic structures. Several polyimides fall into this category (Figure 6.108a), as do polyacrylates and polyurethanes with optically responsive side groups (Figure 6.108b).

6.3.4   Optical Properties of Composites and Biologics*

Of the three physical properties covered in this chapter, optical properties have the least importance in composite and biological applications. This is not to say that there are no applications of optical properties in composites or biological materials. There are indeed, such as the use of birefringence in the analysis of stress distribution and fiber breakage in fiber–matrix composites [14] and in the development of materials for ophthalmic implants such as intraocular devices [15]. These topics are beyond the scope of this text, however, even as optional information, and introduce no new concepts from a material property standpoint. There are many interesting articles and books on these topics, however, and the interested reader is encouraged to consult them as these, and many more, new uses for the optical properties of materials are applied to composite and biological materials systems.

REFERENCES

Cited References

1. Solymar, L., and D. Walsh, *Lectures on the Electrical Properties of Materials*, 5th ed., Oxford University Press, Oxford, 1993, p. 425.

2. Sheahan, T., *Introduction to High-Temperature Superconductivity*, Plenum, New York, 1994, p. 3.

3. Cava, R. J., Oxide superconductors, *J. Am. Ceram. Soc.*, **83**(1), 5–28 (2000).

4. Sheahan, T., *Introduction to High-Temperature Superconductivity*, Plenum Press, New York, 1994, p. 223.

5. Zawodzinski, T. A., C. Derouin, S. Radzinski, R. J. Sherman, V. T. Smith, T. E. Springer, and S. Gottesfeld, *J. Electrochem. Soc.*, **140**, 1041 (1993).

6. Schmidt, C. E., V. R. Shastri, J. P. Vacanti, and R. Langer, Stimulation of neurite growth using an electrically-conducting polymer, *Proc. Natl. Acad. Sci. USA*, **94**, 8948–8953 (1997).

7. Sun, A., H. Xu, Z. Chen, L. Cui, and X. Hai, Research on electrical properties of amphiphilic lipid membranes by means of interdigital electrodes, *Mater. Sci. Eng. C*, **2**(3), 159–163 (1995).

8. Hontsu, S., T. Matsumoto, J. Ishii, M. Nakamori, H. Tabata, and T. Kawai, Electrical properties of hydroxyapatite thin films grown by pulsed laser deposition, *Thin Sol. Films*, **295**, 214–217 (1997).

9. Gatteschi, D., P. Carretta, and A. Lascialfari, Molecular magnets and magnetic nanoparticles: new opportunities fo μSR investigations, *Physica B*, **289–290**, 94–105 (2000).

10. Yakhmi, J. V., Magnetism as a functionality at the molecular level, *Physica B*, **321**, 204–212 (2002).

11. Chavan, S. A., J. V. Yakhmi, and I. K. Gopalakrishnan, Molecular ferromagnets—a review, *Mater. Sci. Eng. C*, **3**, 175–179 (1995).

12. Miller, J. S. and A. J. Epstein, Molecular magnets: An emerging area of materials chemistry, in *Materials Chemistry: An Emerging Discipline*, L. V. Interrante, L. A. Casper, and A. B. Ellis, eds., American Chemical Society, Washington, D.C., 1995, p. 161.

13. Hanna, J., and I. Shimizu, Materials in Active-Matrix Liquid Crystal Displays, *MRS Bulletin*, March, 35 (1996).

14. Schuster, D. M., and E. Scala, *Trans. AIME*, **230**, 1639 (1964).

15. Obstbaum, S. A., Ophthalmic Implantation, in *Biomaterials Science*, B. D. Ratner, A. S. Hoffman, F. J. Schoen, and J. E. Lemons, eds., Academic Press, San Diego, 1996, p. 435.

Electrical Properties of Materials

Electrical Properties of Bone and Cartilage, C. T. Brighton, J. Black, and S. Pollack, eds., Grune & Stratton, New York, 1979.

Solymar, L., and D. Walsh, *Lectures on the Electrical Properties of Materials*, 5th ed., Oxford University Press, New York, 1993.

Magnetic Properties of Materials

Moorjani, K., and J. M. D. Coey, *Magnetic Glasses*, Elsevier, Amsterdam, 1984.

Westbrook, C., and C. Kaut, *MRI in Practice*, Blackwell Science, Oxford, 1998.

Optical Properties of Materials

Campbell, S. A., *The Science and Engineering of Microelectronic Fabrication*, Oxford University Press, New York, 1996.

Wolf, S., and R. N. Tauber, *Silicon Processing for the VLSI Era*, Vol. 1, Lattice Press, Sunset Beach, CA, 1986.

Fox, M., *Optical Properties of Solids*, Oxford University Press, New York, 2001.

PROBLEMS

Level I

6.I.1 What will be the resistance of a copper wire 0.08 in. in diameter and 100 ft long if its resistivity is 1.7 $\mu\Omega \cdot$ cm?

6.I.2 A maximum resistance of 1 Ω is permitted in a copper wire 25 ft long. What is the smallest wire diameter that can be used?

6.I.3 What is the electrical conductivity of iron (a) at room temperature? (b) at 212°F?

6.I.4 Silicon has a density of 2.40 g/cm^3. (a) What is the concentration of the silicon atoms per cubic centimeter? (b) Phosphorus is added to the silicon to make it an *n*-type semiconductor with a conductivity of 1 mho/cm and an electron mobility of 1700 cm^2/V-s. What is the concentration of the conduction electrons per cubic centimeter?

6.I.5 (a) How many silicon atoms are there for each conduction electron in problem 6.I.4? (b) The lattice constant for silicon is 5.42 Å, and there are eight atoms per unit cell. What is the volume associated with each conduction electron?

6.I.6 Germanium used for transistors has a resistivity of 2 $\Omega \cdot$ cm and an electron "hole" concentration of 1.9×10^{15} holes/cm^3. (a) What is the mobility of the electron holes in the germanium? (b) What impurity element could be added to germanium to create electron holes?

6.I.7 Calculate the mobility of electrons in Cu. The resistivity of Cu is 1.72×10^{-8} $\Omega \cdot$ m at 25°C and its density is 8.9 g/cm^3. Assume each copper atom donates one valence electron to the conduction band.

6.I.8 A coil of wire 0.1 m long and having 15 turns carries a current of 1.0 A. (a) Compute the magnetic induction if the coil is within a vacuum. (b) A bar of molybdenum is now placed in the coil, and the current adjusted to maintain the same magnetic induction as in part (a). Calculate the magnetization. Adapted from Callister, pg. 703, problem 21.2.

6.I.9 Why is there a maximum power loss when the dielectric is changing most rapidly as a function of temperature and frequency?

6.I.10 Crown glass has a refractive index of 1.51 in the visible spectral region. Calculate the reflectivity of the air–glass interface, and determine the transmission of a typical glass window.

6.I.11 Look up the refractive indices for fused silica and dense flint glass, and calculate the ratio of their reflectivities. Cite the source of your information.

Level II

6.II.1 At room temperature, the temperature dependence of electron and hole mobilities for intrinsic germanium is found to be proportional to $T^{-3/2}$ for temperature in degrees Kelvin. Thus, a more appropriate form of Eq. (6.31) is

$$\sigma = C'' T^{-3/2} \exp\left(\frac{-E_g}{2k_B T}\right)$$

where C'' is a temperature-independent constant. (a) Calculate the intrinsic electrical conductivity using both Eq. (6.31) and the equation above for intrinsic germanium at 150°C, if the room temperature electrical conductivity is $2.2 \ (\Omega \cdot m)^{-1}$. (b) Compute the number of free electrons and holes for intrinsic germanium at 150°C assuming this $T^{-3/2}$ dependence of electron and hole mobilities. Adapted from Callister, problem 19.40, page 654.

6.II.2 Assume there exists some hypothetical metal that exhibits ferromagnetic behavior and that has a simple cubic structure, an atomic radius of 0.153 nm, and a saturation flux density of 0.76 tesla. Determine the number of Bohr magnetons per atom for this material. Callister, Problem 21.10, p. 704.

6.II.3 The formula for yttrium iron garnet ($Y_3Fe_5O_{12}$) may be written in the form $Y_3{}^cFe_2{}^aFe_3{}^dO_{12}$, where the superscripts a, c, and d represent different sites on which the Y^{3+} and Fe^{3+} ions are located. The spin magnetic moments for the Y^{3+} and Fe^{3+} ions positioned in the c and a sites are oriented parallel to one another and antiparallel to the Fe^{3+} ions in d sites. Compute the number of Bohr magnetons associated with each Y^{3+} ion, given the following information: (1) Each unit cell consists of eight formula ($Y_3Fe_5O_{12}$) units; (2) the unit cell is cubic with an edge length of 1.2376 nm; (3) the saturation magnetization for this material is 1.0×10^4 A/m, and (4) assume that there are 5 Bohr magnetons associated with each Fe^{3+} ion. Callister problem 21.17, p. 704.

6.II.4 The detectors used in optical fiber networks operating at 850 nm are usually made of silicon, which has an absorption coefficient of $1.3 \times 10^5 \ m^{-1}$ at this wavelength. The detectors have coating on the front face that make the reflectivity at the design wavelength negligibly small. Calculate the thickness of the active region of a photodiode designed to absorb 90% of the light.

Level III

6.III.1 The *magnetocaloric effect* (*Scientific American*, May 1998, p. 44) relies on the ability of a ferromagnetic material to heat up in the presence of a magnetic field and then cool down once the field is removed. When a ferromagnet is placed in a magnetic field, the magnetic moments of its atoms become aligned, making the material more ordered. But the amount of entropy must be conserved, so the atoms vibrate more rapidly, raising the material's temperature. When the material is taken out of the field, the material cools. Water, or some other heat transfer fluid, can be cooled down by running it through the ferromagnetic material as it cools. (a) The magnitude of the magnetocaloric effect reaches a maximum at the Curie temperature of the ferromagnet. Based upon the above description of the application and the Curie temperature, which of the following three materials will make the best ferromagnetic material for a typical household refrigeration unit? *Justify your selection.*

Material	Saturation Magnetization[a] at Room Temperature	Curie Temperature (K)	Melting Point (K)
Iron	1707	1043	1811
Gadolinium	1090	289	1585
Nickel	485	631	1728

[a]Maximum possible magnetization that results when all magnetic dipoles are aligned with the external field.

(b) The amount of refrigeration is also determined by the strength of the applied magnetic field. Superconducting magnets are used to generate massive magnetic fields that cause large magnetizations in the ferromagnetic inductor. Below are three candidate materials for the superconducting magnet. Based on what you know about superconductors, which of these three materials will make the best superconducting magnet for this application? *Justify your selection.*

Material	T_c, K	H_c, tesla
$Bi_2Sr_2Ca_2Cu_3O_{10}$	110	100
Gadolinium	5.9	0.0001
NbTi	9.5	10

6.III.2 A team of chemical engineers [*Proc. Natl. Acad. Sci. USA*, **94**, 8984 (1997)] has shown that oxidized polypyrrole supports the growth of structural support cells needed for nerve regeneration in rats. In particular, the team showed that rat nerve cells responded to electrical stimulus from the polymer by producing extensions called neurites that are twice as long as neurites produced in the absence of the stimulus. (a) Calculate the current generated in a 10-mm-long, 10.0×10^{-6}-m-diameter polypyrrole filament that is subjected to an electric field of 12 V. Assume that the conductivity of oxidized polypyrrole is $1.5 \times 10^7 \ \Omega^{-1}m^{-1}$. (b) One drawback for this application is that polypyrrole is very fragile and not biodegradable. Suggest a method for evaluating the long-term biocompatibility of polypyrrole fibers *in vivo*.

CHAPTER 7

Processing of Materials

7.0 INTRODUCTION

With an understanding of the structure and properties of engineering materials now firmly in place, we can discuss how these materials can be formed or fabricated into useful products and components. Most of the important processing methods are described here, with little or no distinction made between microscale and macroscale processes—for example, processes that form both integrated circuits and components for highway bridges are described here. The common thread is that all the chemical and physical phenomena needed to introduce these processing techniques have already been described in the previous chapters.

By the end of this chapter, you should be able to:

- Identify and describe different types of metal processing techniques.
- Calculate the force imparted on a workpiece in forging operations.
- Calculate the power and penetration depth of metals heated by induction heating.
- Identify a proper milling technique for the preparation of metallic and ceramic powders of a desired size.
- Calculate pressures and identify the neutral point in rolling operations.
- Identify and describe different types of ceramic slurry processing techniques.
- Differentiate between kinetic and transport rate limitation in the vapor phase processing of ceramics.
- Identify and describe different types of continuous and batch polymer processing techniques.
- Use equations of change to derive simple design equations for flow and pressure drop in polymer processing techniques.
- Differentiate between pultrusion, resin transfer molding, and filament winding methods for fiber-reinforced composite production.
- Describe the benefits of plasma processing to improvement of biocompatibility in biologics.

7.1 PROCESSING OF METALS AND ALLOYS

Despite the relatively recent development of polymers and composites as structural materials, metals continue to be the dominant group of engineering materials for many

An Introduction to Materials Engineering and Science: For Chemical and Materials Engineers,
by Brian S. Mitchell
ISBN 0-471-43623-2 Copyright © 2004 John Wiley & Sons, Inc.

applications, including the automotive, aerospace, and construction industries. This is due, in part, to their ease in component fabrication, or *formability*. In this section, we describe three of the most common and widely utilized metal-processing techniques: casting, wrought processing, and powder metallurgy. The topic of metal joining (e.g., welding), albeit an important one, is beyond the scope of this text.

7.1.1 Casting

In the context of metals processing, *casting* is the process of melting a metal and solidifying it in a cavity, called a *mold*, to produce an object whose shape is determined by mold configuration. Almost all metals are cast during some stage of the fabrication process. They can be cast either (a) directly into the shape of the component or (b) as ingots that can be subsequently shaped into a desired form using processes that are described later in this section. Casting offers several advantages over other methods of metal forming: It is adaptable to intricate shapes, to extremely large pieces, and to mass production; and it can provide parts with uniform physical and mechanical properties throughout. From an economic standpoint, it would be desirable to form most metal components directly from casting, since subsequent operations such as forming, extruding, annealing and joining add additional expense. However, since the growth rate of crystalline phases is much higher in the liquid state than in the solid state (cf. Sections 2.1.5 and 3.2.1), the microstructure of as-cast materials is much coarser than that of heat-treated or annealed metals. Consequently, as-cast parts are usually weaker than wrought or rolled parts. Casting is generally only used as the primary fabrication process when (1) the structural components will not be subjected to high stresses during service; (2) high strength can be obtained by subsequent heat treatment; (3) the component is too large to be easily worked into shape; or (4) the shape of the component is too complicated to be easily worked into shape. Let us more closely examine the important steps in the casting process and see how some properties can be related to the important processing parameters in each step.

7.1.1.1 Melting. The initial step in the casting operation is melting. Melting can be achieved through any number of well-established techniques, including resistance heating elements or open flame, but there are several alternative methods that offer distinct advantages in the melting of certain metals. For example, metals such as zirconium, titanium, and molybdenum are melted in water-cooled copper crucibles by *arc melting*, in which a direct or alternating current is applied across an electrode and the material in the crucible, called the *charge*. The electrode may be nonconsumable when made of carbon or tungsten, or consumable when made of the metal to be melted. An *arc* is formed between the electrode and charge, and the energy imparted to the charge causes it to melt. This process is carried out under a vacuum, or in an inert atmosphere such as helium or argon, to prevent contamination with oxygen or nitrogen.

In a process known as *induction melting*, the metal is placed in a container within an electrical coil that produces a high-frequency alternating magnetic field (see Figure 7.1). The alternating current flowing through the induction coil generates an alternating magnetic field that produces a current in the metal, called a *workpiece*. Heat is generated within the metal due to the passage of these *secondary currents* induced in the coils by electromagnetic induction (cf. Section 6.2.1.1). The electrical energy through the coil is thus converted to magnetic energy, which is in turn converted back to heat in the

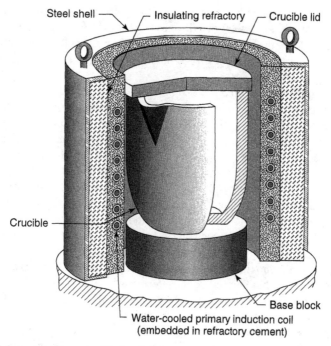

Steel shell — Insulating refractory — Crucible lid

Crucible —

Base block

Water-cooled primary induction coil
(embedded in refractory cement)

Figure 7.1 Schematic diagram of induction furnace and crucible used in the melting of copper. Reprinted, by permission, from *Metals Handbook*, pp. 23–60. Copyright © 1985 by the American Society for Metals.

workpiece due to the induced secondary electrical currents called *eddy currents*. The current density in the workpiece is dictated by the *skin effect*, in which most of the eddy currents occur on the workpiece surface. Toward the center the current density decreases exponentially. In the center of the workpiece the current flow is nearly zero. The heat develops in the workpiece itself; there is no need for a transmission medium, such as air, water, or a mechanical connection. As the outer portions of the metal melt, mixing begins and thermal conduction eventually results in the melting of the entire workpiece.

The power imparted to the workpiece or melt, P, is a function of the magnetic susceptibility of the metal, χ, and the frequency of the applied field, ω:

$$P = kI^2\sqrt{(1+\chi)\rho\omega} \qquad (7.1)$$

where P is in watts, k is a constant, I is the current in the induction coil (in amps), ρ is the specific resistivity of the metal in $\Omega \cdot mm^2/m$, and ω is in Hz. As mentioned above, the induced currents occur at the outer of portions of the metal, as given by the *penetration depth*, δ, of the currents (in mm)

$$\delta = 503\sqrt{\frac{\rho}{(1+\chi)\omega}} \qquad (7.2)$$

Obviously, the melt or workpiece must be larger in diameter than the penetration depth. The induction process, too, is often carried out under vacuum or inert gas atmosphere.

The advantage of induction melting is the ability to melt cleanly and quickly. It is applicable to virtually any metal or alloy, even if they are not inductively suscepting, since it is possible to use the crucible as a susceptor as well. Furthermore, once the metal is molten, the induction lines from the secondary currents stir the melted charge and improve melt homogeneity.

Other techniques such as *electron beam melting* and *plasma flame melting* are used for highly refractory metals, such as tantalum. These too are often performed in vacuum. The use of vacuum in melting operations is used not only to prevent reactions such as oxidation, but to prevent, and even remove, dissolved gases in the metals. Most metals dissolve considerable amounts of gases in their liquid state. The solubility of these gases is much less in the solid state than in the liquid state, and as the metal solidifies, gases can escape from the melt, forming *blowholes* and *pinholes* in the metal ingot. This results in porosity than can adversely affect properties of the metal or alloy.

Oxidation is not always to be avoided in the melting process. It is an important step in the making of steel, in which *pig iron*, the raw product of iron production, is re-melted in a blast of oxygen. The pig iron contains about 4 wt% carbon, which is converted to carbon monoxide in the presence of the oxygen. The furnace used to melt iron and form steel is thus termed a *blast furnace*, due to the blast of oxygen that is provided. The amount of oxygen is carefully controlled, so that not all of the carbon is removed, but just enough to form steel of the desired carbon content (cf. Figure 2.8 and Table 2.4). The oxygen also reacts to form oxides of highly reactive trace impurities in the pig iron, such as silicon, aluminum, and manganese. These oxides float to the top of the melt, since they are less dense than the molten metal, and form a product called *slag* that can be skimmed from the top of the melt. An additional benefit of the oxygen blast is that these reactions with carbon and trace impurities produce an enormous amount of energy that can be used to melt the iron. Many modifications have

---

HISTORICAL HIGHLIGHT

The physical limitations of cast-iron cannons in the Crimean War (1854–1856)—in which England, France, Turkey, and Sardinia defeated Russia for the domination of southeastern Europe—were a turning point in the history of steel. Those limitations caught the attention of Henry Bessemer, an Englishman who had already enjoyed considerable commercial success for inventing the "lead" of "lead pencils." Bessemer's breakthrough in 1855 was to drop the standard approach, which involved making steel by heating carbon-free wrought iron with carbon-bearing fuel. He tried to turn the process around by starting with carbon-rich pig iron and using oxygen in an air blast to get the excess carbon out.

Iron and steel became the stuff of big and visible things—ships, bridges, buildings,

even the Eiffel Tower. These metals, particularly steel, were the harbingers of skyscrapers and audacious urban skylines. In 1874 James Eads used steel to build a giant arched bridge across the Mississippi River in St. Louis. Some 50,000 tons of steel went into the building of a railway bridge over the Firth River in Firth, Scotland. For the enormous suspension cables that would hold up his Brooklyn Bridge, John Roebling in 1883 could choose steel rather than iron. In 1890 the second Rand-McNally Building in Chicago was the first building to be framed entirely in steel. Hundreds of ever higher buildings and skyscrapers followed. Steel also was showing up in hundreds of smaller everyday items like food cans, home appliances, automobiles, and road signs.

Adapted from *Stuff*, I. Amato, pp.42–46

since been made to this process, and may variations exist, including the formation of pig iron from iron ore (Fe_2O_3). The chemistry of the steel-making operation is quite complex, and the interested reader is referred to the metallurgy texts cited at the end of this chapter for further information.

7.1.1.2 *Forming Operations.*

In principle, the molten metal can then be poured into a mold of the desired shape and become solidified to form a component. As mentioned above, and as described further in the next section, solidification from casting operations usually leads to undesirable crystal structures, such that the most common operation after melting is the formation of a large metal "blank" called an *ingot*. Thus, there are two main categories of metal casting processes: *ingot casting* and *casting to shape*.

Ingot casting makes up the majority of all metal castings and can be separated into three categories: *static casting*, semicontinuous or *direct-chill casting*, and *continuous casting*. Static ingot casting simply involves pouring molten metal into a permanent, refractory mold with a removable bottom or plug in the bottom, as schematically illustrated in Figure 7.2. Ingots have a slight taper toward the bottom of the mold that facilitates their removal and aids in the solidification process. The plug in the bottom of the mold allows the metal to be drained from the mold upon remelting, if desired.

Semicontinuous ingot casting is primarily employed in the production of cast aluminum. In this process, the molten metal is transferred to a water-cooled permanent mold that has a movable base mounted on a long piston (see Figure 7.3a). As the metal solidifies, the piston is moved downward and more metal continues to fill the reservoir (see Figure 7.3b). The piston is allowed to move its entire length, at which point the process is stopped.

Continuous casting is used in the steel and copper industries. In this process, molten metal is delivered to a permanent mold in much the same way as semicontinuous

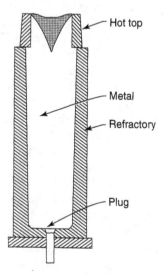

Figure 7.2 Schematic diagram of an ingot in a refractory mold. From Z. Jastrzebski, *The Nature and Properties of Engineering Materials*, 2nd ed. Copyright © 1976 by John Wiley & Sons, Inc. This material is used by permission of John Wiley & Sons, Inc.

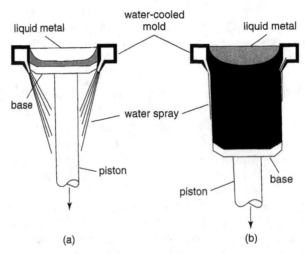

Figure 7.3 Schematic illustration of semicontinuous casting process in which (a) molten metal solidifies in a water-cooled mold with a movable base and (b) the base is moved downward so more metal can be poured into the reservoir. Reprinted, by permission, from the *McGraw-Hill Encyclopedia of Science and Technology*, Vol. 11, 8th ed., p. 31. Copyright © 1997 by McGraw-Hill.

casting. However, instead of the process ceasing after a certain length of time, the solidified ingot is continually cut into lengths and removed from the process as quickly as it is cast. This method has many economic advantages over the more conventional ingot casting techniques.

Casting to shape is generally classified according to the molding process, molding material, or method of feeding melt to the mold. There are four basic types of casting to shape processes: *sand casting, permanent mold casting, die casting*, and *centrifugal casting*. In sand casting, the mold consists of a mixture of sand grains, water, clay, and other materials, which are compacted around a pattern to 20–80% of their mixture bulk density. The two halves of the mold, called the *cope* and *drag*, are clamped together to form internal cavities, into which the metal is cast. When the molds are made of cast iron, steel, or bronze, the process is called *permanent mold casting* because the molds can be re-used. In this case, the mold cavity is formed by machining. This process is used to cast iron and nonferrous alloys, and it has advantages over sand casting such as smoother surface finish, closer tolerances, and higher production rates. A further development of the permanent molding process is die casting. Molten metal is forced through a plate with an opening of a desired geometry known as a *die*, under pressures of 0.7–700 MPa (100–100,000 psi). Two basic types of die-casting machines are *hot-chamber* and *cold-chamber*. In the hot-chamber machine, a portion of the molten metal is forced into a cavity at pressures up to 14 MPa (2000 psi). The process is used for casting low-melting-point alloys such as lead, zinc, and tin. In the cold-chamber process (see Figure 7.4), the molten metal is ladled into the injection cylinder and forced into the cavity under pressures that are about 10 times those in the hot-chamber process. High-melting-point alloys such as aluminum-, magnesium-, and copper-based alloys are used in this process. Die casting has the advantages of high production rates, high quality and strength, excellent surface finish, and close

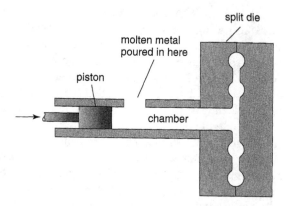

Figure 7.4   Schematic illustration of cold chamber die casting process. *McGraw-Hill Encyclopedia of Science and Technology*, Vol. 11, 8th ed., p. 33. Copyright © 1997 by McGraw-Hill.

tolerances. Finally, inertial forces of rotation distribute molten metal into the mold cavities during centrifugal casting. The rotational speed in centrifugal casting is selected to give 40 to 60 gravitational constants of acceleration. Dies may be made of forged steel or cast iron. Other types of casting processes include *evaporative casting* and *zero gravity casting*.

7.1.1.3   Solidification. When the ingot or casting solidifies, there are three main possible microstructures that form (see Figure 7.5). We will describe here only the final structures; the thermodynamics of the liquid–solid phase transformation have been described previously in Chapter 2. The outside layer of the ingot is called the *chill zone* and consists of a thin layer of equiaxed crystals with random orientation.

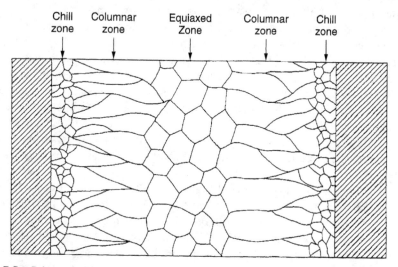

Figure 7.5   Schematic illustration of crystallization zones in a solidified metal ingot. From *The Nature and Properties of Engineering Materials*, 2nd ed., by Z. Jastrzebski. Copyright © 1976 by John Wiley & Sons, Inc. This material is used by permission of John Wiley & Sons, Inc.

This is due to the high degree of subcooling of the melt adjacent to the mold wall. Recall from Eq. (3.43) that a high degree of undercooling creates a driving force for a high nucleation rate. Many nuclei are formed at the mold wall, and a fine grain size results. The second zone, which constitutes the majority of the ingot, consists of elongated grains, often of dendritic shape, having great length and oriented parallel to the heat-flow direction in the ingot. This is called the *columnar zone*. Finally, the *central zone* of the ingot is the last to solidify, resulting in a much lower heat transfer rate and a coarser, equiaxed grain size.

The relative amounts of grains in each zone can be controlled by the rate of solidification (cooling) and the composition of the alloy. The cooling rate is determined by the thermal properties of the mold material, the size of the ingot, and the temperature of the melt. If the melt temperature is too high, the chill zone may completely disappear due to redissolution of the chilled grains. In pure metals, as well as in alloys with very little supercooling, the columnar structures may progress to the center of the ingot, where the grains advancing from the opposite wall meet. Thus, no equiaxed zone occurs. At very high cooling rates, the chill zone may extend throughout the ingot, producing a fine-grained, equiaxed microstructure. This usually occurs for small ingots and in cold metal molds. The formation of the central equiaxed zone is caused by a process called *constitutional supercooling*. This enhances nucleation in the melt at the center of the casting prior to the advance of the solid interface. The growth of the equiaxed zone can also be promoted by adding some nucleating agents. In general, columnar growth dominates in pure metals, while equiaxed crystallization prevails in solid–solution alloys. With increasing amounts of solute atoms, and with increasing freezing range of the solid alloy, the equiaxed crystal zone increases, and in some cases the columnar growth can be completely eliminated. In the extreme of cooling rates, nucleation and crystallization can be completely eliminated, leading to amorphous metals and alloys. This is the topic of the next section.

Some common phenomena that occur during solidification are *segregation* and *shrinkage*. During the solidification process, variations in the concentration can occur in certain regions of alloys that result in segregation. Two main types of segregation can occur: *macrosegregation* and *microsegregation*. In macrosegregation, compositional changes occur over relatively large dimensions as a result of gravity, normal, or inverse segregation. Gravity segregation is caused by density differences in melt components. In normal segregation, the solute is rejected at an advancing solid–liquid interface because of different diffusion rates in solid and in liquid. Inverse segregation is the result of outward movement of the impurity-enriched interdendritic liquid, thereby increasing the average impurity content at the head of the ingot. Microsegregation refers to compositional variations over relatively small distances. Insoluble foreign particles can be trapped during solidification, resulting in segregation at the grain boundaries.

Just as polymers undergo volume changes upon cooling (cf. Figure 1.67), so, too, do metals. The resultant shrinkage is equal to the difference in volume between the liquid at the casting temperature and the solid metal at room temperature. The total change in volume is the combined effect of the contraction of the liquid when cooled from its casting temperature to its freezing point, the contraction during freezing from liquid to solid, and the contraction of the solid metal when cooled from its freezing point to room temperature (cf. thermal expansion in Section 5.1.3.1). Shrinkage may cause the formation of cavities in the ingot and adversely affect the properties of the

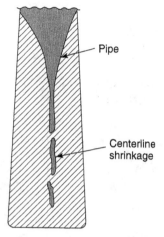

Figure 7.6   Schematic illustration of piping in a solidified metal ingot. From Z. Jastrzebski, *The Nature and Properties of Engineering Materials*, 2nd ed. Copyright © 1976 by John Wiley & Sons, Inc. This material is used by permission of John Wiley & Sons, Inc.

metal. Such cavities, particularly in the extreme case of piping (see Figure 7.6), can cause considerable damage to the ingot since air can penetrate into the cavity and cause oxidation. The presence of oxide layers inside the ingot will prevent proper welding of the cavities during subsequent rolling or forging operations. In order to minimize piping, molds with tapered walls as shown in Figure 7.2 are used, which permit the liquid metal to penetrate into any cavities or pores that may have already formed in the solidified parts of the ingot. Shrinkage can also be minimized in cast parts through use of a *riser*, which is a volume added to the casting to provide metal as shrinkage occurs.

The relationship between the processing history and the solidification event is illustrated in Figure 7.7 for a cast component. Solidification begins with the formation of dendrites on the mold walls, which grow with time into the liquid and eventually consume the entire liquid volume. In alloys, it is the growth of dendrites that leads to segregation. In addition, the dendrites provide opportunities for gases to become trapped and form porosity. Most or all of these defects can be eliminated by processing in a *microgravity* environment. Macrosegregation, for example, is driven by gravity effects in which solute-rich materials sink in response to density differences.

7.1.1.4 *Rapid Solidification.*

In Section 6.2.1.3, we saw the relationship between grain size, magnetic domain size, and magnetic properties in ferrous materials. In the previous section, we saw the effect of cooling rate on grain size in metals. You would expect, then, that there would be a strong relationship between solidification processes and magnetic properties in ferrous alloys, and this is indeed the case. We also saw that increased solidification rates at the walls of a mold can produce a fine-grained crystalline structure due to increased nucleation rates. But in the extreme, even nucleation rates drop off at sufficient subcooling (cf. Figure 3.15), such that it is possible, in theory, to produce *amorphous metals and alloys*, also known as *metallic glasses*, with high enough cooling rates. Such amorphous materials are formed by a class of techniques known collectively as *rapid solidification processing*, in which solidification rates, or *quench rates* as they are called for these processes, are on the

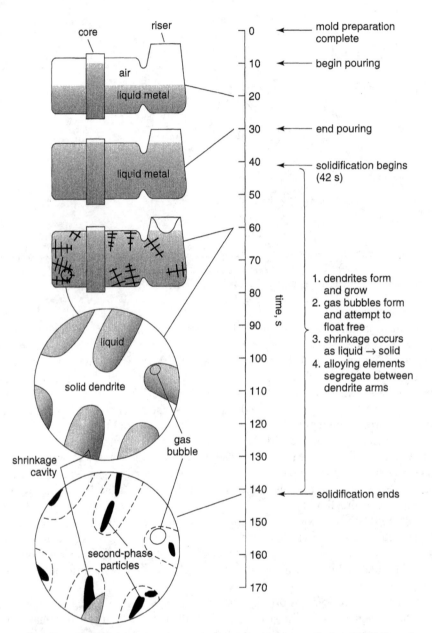

Figure 7.7   Solidification history for a cast-to-shape component. *McGraw-Hill Encyclopedia of Science and Technology*, Vol. 11, 8th ed., p. 35. Copyright © 1997 by McGraw-Hill.

order of 10^5 to 10^6 °C/s. It is not yet possible to produce amorphous pure metals, since the minimum cooling rates through the liquid–solid transition must be at least 10^{14} °C/s, so metallic glasses always contain at least two or more different kinds of atoms.

Although rapid solidification may not produce a truly amorphous (noncrystalline) material for some alloy compositions, crystallite sizes of rapidly solidified crystalline

Table 7.1   Composition of Some Amorphous Ferrous Alloys

Composition (wt%, balance Fe)					
B	Si	Cr	Ni	Mn	P
20	—	—	—	—	—
10	10	—	—	—	—
28	—	6	—	6	—
6	—	—	40	—	14

alloys are characteristically in the range of 500 nm or below, compared with 50 μm for traditionally processed alloys. The absence of grain boundaries in alloys such as iron–silicon improve their magnetization properties and help make them especially attractive for soft-magnet applications such as transformer cores. Boron is another common alloying element in ferrous alloys that assists in the formation of noncrystalline structures. The compositions of some common amorphous ferrous alloys produced by rapid solidification techniques are listed in Table 7.1. Other known glass-forming metallic systems include alloys of transition metals or noble metals that contain about 10–30% semimetals, such as the iron–boron systems above, platinum–phosphorus, and niobium–rhodium, alloys containing group II metals such as magnesium–zinc, and alloys of rare-earth metals and transition metals such as yttrium–iron and gadolinium–cobalt. With further solute substitution, the stability and glass-forming ability can be drastically enhanced. Ternary alloy glasses such as palladium–copper–silicon and platinum–nickel–phosphorus have been prepared as cylindrical rods of 2.5 mm in diameter at quench rates of 10^2 °C/s or less. The X-ray diffraction pattern (cf. Section 1.1.2) of an amorphous alloy is very similar to that of the same alloy in the liquid state. From the XRD pattern, it can be determined that the coordination number in metallic glasses is close to 12, just as in the close-packed crystalline structure.

The high quench rates necessary to achieve the noncrystalline structures are primarily achieved through contacting the molten alloys with a surface that has a high thermal conductivity and is maintained at a low temperature. In a process called *splat-quenching*, the molten metal is propelled by means of a shock tube onto copper substrates with high thermal conductivity. Ribbons, or tapes, of metallic glasses can be formed by feeding a continuous jet of liquid alloy on the outside rim of a rapidly rotating cylinder. The molten alloy solidifies rapidly into a thin ribbon, about 50 μm thick, that is ejected tangentially to the rotating cylinder at rates as high as 2000 m/min. The width of the ribbon is limited to a few millimeters. Ribbons up to 15 cm in width can be produced by using a planar flow casting method, as illustrated in Figure 7.8.

By injecting a stream of melt through a nozzle into a rotating drum containing water instead of onto a rotating cylinder, wires with circular cross sections can be formed. The centrifugal force ensures the formation of a water layer on the inner surface of the drum. The molten jet solidifies rapidly into a wire that is collected at the bottom of the drum. The wire has a uniform cross section with diameter ranging from 50 to 200 μm. The metallic glass wires show excellent mechanical and magnetic properties, comparable to or surpassing those of corresponding ribbons.

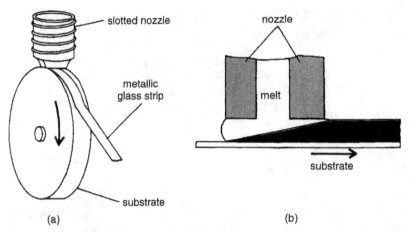

Figure 7.8   Schematic illustration of (a) nozzle-substrate orientation and (b) cross section of the nozzle substrate in planar flow casting. *McGraw-Hill Encyclopedia of Science and Technology*, Vol. 11, 8th ed., p. 57. Copyright © 1997 by McGraw-Hill.

7.1.2   Wrought Metals and Alloys

The ingots produced by casting can either be used as cast, or further processed into usable geometries such as bars, rods and sheets. Those metals and alloys that are manufactured by plastic deformation in the solid state are referred to as *wrought metals and alloys*.* The mechanical working of metals during wrought processing improves the properties of the wrought metals over that of the cast products. Plastic deformation, when carried out below the recrystallization temperature (cf. Section 5.1.2.3), results in work hardening, which improves the mechanical properties of the metal.

In all wrought processes, the flow of metal is caused by application of an external force or pressure that pushes or pulls a piece of metal or alloy through a metal die. The pressure required to produce plastic flow is determined primarily by the yield stress of the material (cf. Section 5.1.4.3) which, in turn, controls the load capacity of the machinery required to accomplish this desired change in shape. The pressure, P, used to overcome the yield stress and cause plastic deformation is given by

$$P = \sigma_y \varepsilon \tag{7.3}$$

where σ_y is the yield stress of the material and ε is the strain during the deformation. It is often useful to utilize the true strain in this relationship [see Eq. (5.31)] rather than the engineering strain. The pressure calculated in this way is generally much lower than that required in the actual process due to losses caused by friction, work hardening, and inhomogeneous plastic deformation. For example, the pressure predicted in Eq. (7.3) is only 30–55% of the working pressure required for extrusion, 50–70% of that required for drawing, and 80–90% of the pressure required for rolling. The actual work requirement is the sum of (a) the work due to plastic deformation and (b) the losses mentioned above. The mechanical working of metals can be accomplished by various processes, but the methods of forging, rolling, and extrusion will be described here.

---

*There is no participial form of the word "wrought" in common use that is analogous to the term "casting," so the modified gerund *wrought processing*, albeit awkward, is generally used.

7.1.2.1 Forging. *Forging* is the process of applying large forces, mainly compressive forces, to cause plastic deformation to occur in the metal. The use of compressive forces means that tensile necking and fracture are generally avoided. Forging can be used to produce objects of irregular shape. There are two main categories of forging: *pressing* and *hammering*.

Pressing is the relatively slow deformation of metal to the required shape between mating dies. This method is preferred for homogenizing large cast ingots because the deformation zone extends throughout the cross section. All presses are composed of some basic components. These include a *ram* for applying the compressive force, a *drive* to move the ram up and down, power to the drive, a *bed* to rest the workpiece upon, and a *frame* to hold the components. Most press forging is done on upright *hydraulic presses* or *mechanical presses*. There are a variety of frame geometries and types of ram drives that vary depending upon the pressing application. A press is rated on the basis of the force (typically in tons) it can deliver near the bottom of a stroke. Pressures (in tons/in.^{2} of projected area) have been found to be 5–20 for brass, 19–20 for aluminum, 15–30 for steel, and 20–40 for titanium. The force the press must deliver is equal to the unit pressure times the projected area. Presses in excess of 50,000 tons have been built for forging large and complex parts.

Hammering involves more rapid deformation than pressing and is mainly restricted to the surface region of the metal. Large forgings require very large forces and usually involve elevated temperatures. The most common forging hammers are steam or air operated. An example of a steam forging hammer is shown in Figure 7.9. A related operation is *drop forging*, in which a drop hammer forms parts with impression or cavity dies. Stock metal in the form of bars, slugs, or billets is placed in a cavity in the bottom half of a forging die on the anvil of a drop hammer. The upper half of the die is attached to the hammer, which falls on the stock, shaping it into the desired geometry. Generally, a finished forging cannot be formed in one blow, so most dies have several impressions, and the workpiece is transferred from one to the next between blows until the finished product is attained.

7.1.2.2 Rolling. In the process of *rolling*, the metal is continuously drawn between rotating rollers by the friction forces between the surfaces of the rolls and the metal, as illustrated in Figure 7.10, where $h_f < h_0$. The process can be run either hot or cold, and it is much more economical than forging since it is faster, consumes less power, and produces items of a uniform cross section in a continuous fashion. In *cold rolling*, it is possible to attain production speeds of over 1500 m/min for thin strips of metal. Temperatures in the *hot rolling* process are similar to those in forging, namely 400–450°C for aluminum alloys, 820°C for copper alloys, 930–1260°C for alloys steels, 760–980°C for titanium alloys, and 980–1650°C for refractory metals. A number of roll arrangements are possible, as illustrated in Figure 7.11. Unlike other wrought processes, rolling can be used without a die.

Deformations are restricted to a small volume at any given time, so the loads are relatively low. The metal is subjected to a longitudinal tension force that, combined with the normal force, produces a shear stress, τ, which causes deformation

$$\tau = f \times P \tag{7.4}$$

where P is the normal pressure applied by the roller and f is the coefficient of friction between the metal surface and the roller surface. In hot rolling, f may be as much as

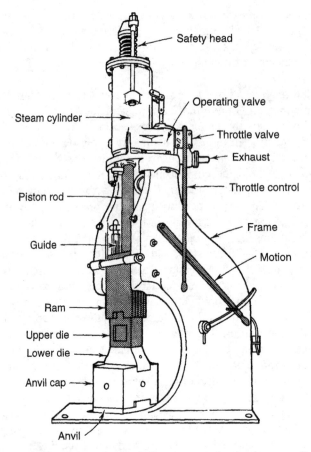

Figure 7.9 Schematic illustration of a single frame steam forging hammer. Reprinted, by permission, from Doyle, L.E., *Manufacturing Processes and Materials for Engineers*, 2nd ed., p. 254. Copyright © 1969 by Prentice-Hall, Inc.

0.7, whereas in cold rolling it generally ranges from 0.02 to 0.3. The normal pressures entering, P_e, and exiting, P_x, the neutral point can be determined at any point of contact along the arc formed between the roller and the sheet by

$$P_e = \sigma' \frac{h}{h_0} \exp[f(H_0 - H)] \tag{7.5}$$

and

$$P_x = \sigma' \frac{h}{h_f} \exp(fH) \tag{7.6}$$

where f is again the coefficient of friction, σ' is the plane strain flow stress at a particular point in the roll gap where the instantaneous thickness is h, h_0 is the entering thickness, h_f is the exiting thickness, and H is a parameter given by

$$H = 2\sqrt{\frac{R}{h_f}} \tan^{-1}\left(\theta\sqrt{\frac{R}{h_f}}\right) \tag{7.7}$$

Example Problem 7.1

The ability of a hammer to deform metal depends on the energy it is able to deliver on impact. Consider a steam hammer (see Figure 7.9) that has a falling weight of 2000 lb and a steam bore, d, equal to 12 in. Assume that the mean effective steam pressure, P, is 80 psi and that the stroke is 30 in. If the hammer travels 1/8 in. into the metal after striking it, determine the average force exerted on the workpiece. You should be able to do this without an equation, through the application of some dimensional analysis.

Answer: Steam force $= \left(\dfrac{\pi d^2}{4}\right) \times P = 9050$ lb$_f$

$$\text{Total downward force} = \text{Steam force} + \text{Hammer weight}$$

$$= 9050 + 2000 = 11{,}050 \text{ lb}_f$$

$$\text{Energy in a blow} = \text{Force} \times \text{Distance} = 11{,}050 \text{ lb}_f \times 30 \text{ in.}$$

$$= 331{,}500 \text{ in.-lb}_f$$

Average force exerted $= 331,500/0.125 = 2,652,000$ lb$_f$ (1326 tons)

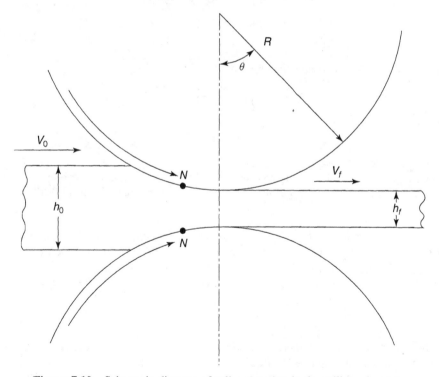

Figure 7.10   Schematic diagram of roll gap region in the rolling of metals.

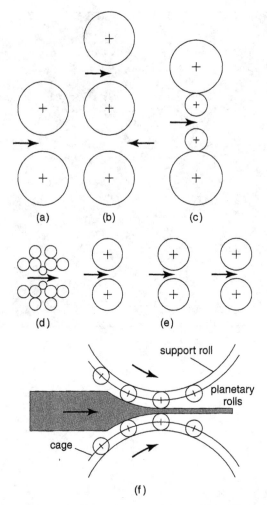

Figure 7.11 Some basic types of roll arrangements: (a) two-high, (b) three-high, (c) four-high, (d) cluster, (e) tandem rolling with three stands, and (f) planetary milling. *McGraw-Hill Encyclopedia of Science and Technology*, Vol. 11, 8th ed., p. 54. Copyright © 1997 by McGraw-Hill.

where R is the radius of the roller and θ is the angle subtended by the roll center by the vertical and the point of interest at the roll surface. H takes on a value of H_0 when $\theta = \theta_0$, the angle at the plane of entry, which can be approximated by

$$\theta_0 = \sqrt{\frac{h_0 - h_f}{R}} \tag{7.8}$$

The speed at which the metal moves through the rollers must change in order to keep the volume rate of flow constant through the roll gap. Hence, as the thickness decreases, the velocity increases. However, the surface speed of the roller is constant, so there is relative sliding between the roller and the metal. The direction of the relative velocity changes at a point along the contact area called the *neutral* or *no-slip point* (point N in Figure 7.10). At the neutral point, the roller and metal have the same

---

Cooperative Learning Exercise 7.1

The neutral plane (point N in Figure 7.10) is the point at which the subtended angle is $\theta = \theta_N$, and can be found in several ways. A precise method is to balance the entry and exit forces pressures; i.e., $P_e = P_x$ and solve for H at the neutral plane, H_N, which can then be substituted into Eq. (7.7) to solve for θ_N. Alternatively, there are empirical relations for θ_N. Consider the cold-rolling of an aluminum strip that is rolled from 4 mm to 3.3 mm in thickness with a roller 500 mm in diameter. The coefficient of friction is 0.06.

Person 1: Use Eq. (7.5) to (7.8) to determine the subtended angle at the neutral point. Hint: First, calculate θ_0 from Eq. (7.8), then calculate H_N by equating Eq. (7.5) and (7.6) at $H = H_N$. Then use Eq. (7.7) to find θ_N.

Person 2: Use the following empirical relationship to determine the subtended angle at the neutral point

$$\theta_N = \sqrt{\frac{h_0 - h_f}{4R}} - \frac{h_0 - h_f}{4fR}.$$

Then help your partner finish his or her calculations. Compare your answers. How similar or different are they?

Answers: $\theta_N = 0.0141$ radians; $\theta_N = 0.0148$ radians

---

velocity. Prior to the neutral point, the metal moves more slowly than the roller; to the right it moves faster ($V_f > V_0$ in Figure 7.10). The direction of frictional forces are thus opposite in these two regions. The net frictional force acting on the metal must be in the direction of exit to enable the rolling operation to take place.

It is sometimes necessary to use multiple rolling steps in sequence, interspersed with annealing steps to improve plastic deformation. The largest product in hot rolling is called a *bloom*. By successive hot- and cold-rolling operations the bloom is reduced to a *billet, slab, plate, sheet, strip*, and *foil*, in decreasing order of thickness and size. The initial breakdown of the ingot by rolling changes the course-grained, brittle, and porous structure into a wrought structure with greater ductility and finer grain size.

7.1.2.3 Extrusion.

Extrusion involves the pressing of a small piece of metal, called a *billet* or *slug*, through a die by the application of force. A billet or slug is differentiated from an ingot primarily by its smaller size. The die in extrusion processing is sometimes called an *orifice* or *nozzle*. The forces necessary to push the metal through the orifice are generated by a ram that pushes on the billet, generating both compressive and shear forces. Because no tensile forces are involved, fracture is avoided. When the billet is forced through the orifice by a ram, the process is called *direct extrusion* (see Figure 7.12). It is also possible to force the die down around the billet to cause the same change in cross-sectional geometry, in a process called *indirect extrusion*. In either case, the efficiency of the extrusion process is measured by the ratio of work necessary for deformation [Eq. (7.3)] to the actual work performed.

Extrusion is primarily used for lower-melting point, nonferrous metals (see Table 7.2); but with the aid of lubricants and improvements in presses, the extrusion of steel has become available. Most carbon steels and stainless steels are extruded at temperatures of about 1200°C, in which glass is used as a lubricant. Glass is inexpensive, easy

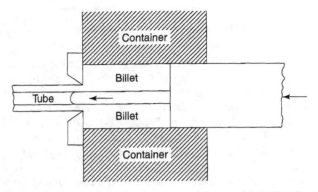

Figure 7.12   Schematic illustration of direct extrusion of a metallic billet. From Z. Jastrzebski, *The Nature and Properties of Engineering Materials*, 2nd ed. Copyright © 1976 by John Wiley & Sons, Inc. This material is used by permission of John Wiley & Sons, Inc.

Table 7.2   Some Typical Processing Parameters for the Extrusion of Nonferrous Metals

Materials	Minimum Section Thickness, (in.)	Extrusion Temperature (°F)	Extrusion Pressure (ksi)	Extrusion Exit Speed (fpm)	Extrusion Ratio $\ln(A_0/A_1)^a$
Aluminum alloys	0.04–0.30	550–1050	40–130	4–300	4–7
Copper alloys	0.05–0.30	1200–1650	30–130	80–1000	5–6.5
Magnesium alloys	0.04	570–800	100–130	4–100	4–5.3
Zinc alloys	0.06	400–660	90–110	75–100	4–5

[a] A_0 is the cross-sectional area of the billet; A_1 is the cross-sectional area of the extruded shape.
Source: T. Altan, S.-I. Oh, and H. L Gegel, *Metal Forming: Fundamentals and Applications*. Copyright © 1983 by the American Society for Metals.

to use, and chemically stable, and it assists in dissolving oxide layers on the metal surface. The extrusion process is commonly used to form tubes and other continuous shapes of complex cross section (see Figure 7.13). We will describe the extrusion process in more detail in Section 7.3 within the context of polymer processing.

7.1.3   Powder Metallurgy

The process of converting metallic powders into ingots or finished components via compaction and sintering is called *powder metallurgy*. Powder metallurgy is used whenever porous parts are needed, whenever the parts have intricate shapes, whenever the alloy or mixture of metals cannot be achieved in any other manner, or whenever the metals have very high melting points. The technique is suitable to virtually any metal, however. An organic binder is frequently added to the metallic powders during the mixing state to facilitate compaction, but volatilizes at low temperatures during the sintering process. The advantages of powder metallurgy included improved microstructures and improved production economies. Although the cost of the powders can be greater than that of ingots and mill products obtained by casting or wrought processing, the relative

Some product forms (not to scale)

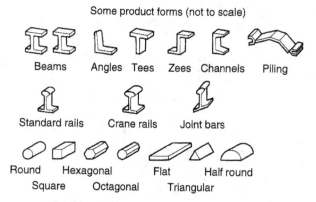

Beams   Angles Tees Zees Channels   Piling

Standard rails   Crane rails   Joint bars

Round   Hexagonal   Flat   Half round
     Square   Octagonal   Triangular

Figure 7.13 Some cross-sectional geometries possible from metal extrusion. Reprinted, by permission, from J. F. Shackelford, *Introduction to Materials Science for Engineers*, 5th ed., p. 421. Copyright © 2000 by Prentice-Hall, Inc.

ease of forming the powder into a final product more than offsets the labor costs and metal losses associated with the other two processes.

The powder metallurgy process consists of three basics steps: powder formation; powder compaction; and sintering. Each of the steps in powder metallurgy will be described in more detail.

7.1.3.1 *Powder Formation.* Metallic powders can be formed by any number of techniques, including the reduction of corresponding oxides and salts, the thermal dissociation of metal compounds, electrolysis, atomization, gas-phase synthesis or decomposition, or mechanical attrition. The *atomization* method is the one most commonly used, because it can produce powders from alloys as well as from pure metals. In the atomization process, a molten metal is forced through an orifice and the stream is broken up with a jet of water or gas. The molten metal forms droplets to minimize the surface area, which solidify very rapidly. Currently, iron–nickel–molybdenum alloys, stainless steels, tool steels, nickel alloys, titanium alloys, and aluminum alloys, as well as many pure metals, are manufactured by atomization processes.

Chemical powder formation methods include the reduction of oxides, reduction of ions in solution by gases, electrolytic reduction of ions in solution, and thermal decomposition of gaseous molecules containing metal atoms. *Sponge iron* powder, for example, is produced by reacting a mixture of magnetite (Fe_3O_4) ore, coke and limestone. The product is crushed to control the iron particle size, and the iron is magnetically separated. Iron powder can also be produced by hydrogen reduction of ground mill scale and the reaction of atomized high-carbon steel particles with iron oxide particles. Copper, tungsten and cobalt powders can be produced by hydrogen reduction of oxide powders. Nickel and copper powders can be produced by hydrogen reduction of their corresponding amine sulfates, and iron and nickel powders of high purity and fine particle size are produced by thermal decomposition of their gaseous carbonyls, $Fe(CO)_5$ and $Ni(CO)_4$, respectively.

Mechanical methods of powder production, also known as *mechanical attrition*, typically require a brittle material, or at least one that becomes brittle during processing. There are various types of equipment that perform particle size reduction,

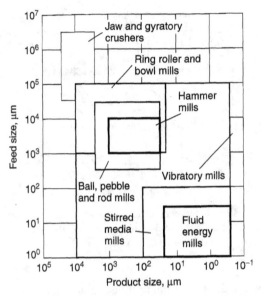

Figure 7.14 Classification of equipment used in mechanical attrition. Reprinted, by permission from *ASM Handbook*, Vol. 7, *Powder Metallurgy* (1984), ASM International, Materials Park, OH 44073-0002. p. 70.

termed *comminution*, but they are generally called *mills* and operate on the principles of grinding, crushing, and particle fracture (see Figure 7.14). In a device known as a *ball mill*, spherical milling media are used inside a vial that is agitated via any one of a number of mechanical means to cause ball–ball and ball–vial collisions. The metallic powder is trapped during these collisions. Figure 7.15 shows the process of trapping an incremental volume of metallic powder between two balls. Compaction begins with a powder mass that is characterized by large spaces between particles compared with the particle size. The first stage of compaction starts with the rearrangement and restacking of particles. Particles slide past one another with a minimum of deformation and fracture, producing some irregularly shaped particles. The second stage of compaction involves elastic and plastic deformation of particles. Cold welding may occur between particles in metallic systems during this stage. The third stage of compaction, involving particle fracture, results in further deformation and/or fragmentation of the particles. In the case of metals, fracture is the result of work hardening and embrittlement (cf. Section 5.1.2.2). For brittle materials, particle fracture is well described by Griffith theory (cf. Section 5.2.1.2). Cooling of the metal below the brittle–ductile transition, such as with liquid nitrogen in *cryogenic milling*, can aid in increasing brittle fracture during milling. As fragments decrease in size, the tendency to aggregate increases, and fracture resistance increases. Particle fineness approaches a limit as milling continues and maximum energy is expended. The major factors contributing to grind limit are increasing resistance to fracture, increasing cohesion between particles, excessive clearance between impacting surfaces, coating of the grinding media by fine particles, surface roughness of the grinding media, bridging of large particles to protect fine particles, and increasing apparent viscosity as the particle size decreases. We will see that these milling methods are also useful for the formation of powders and particles from brittle materials such as ceramics.

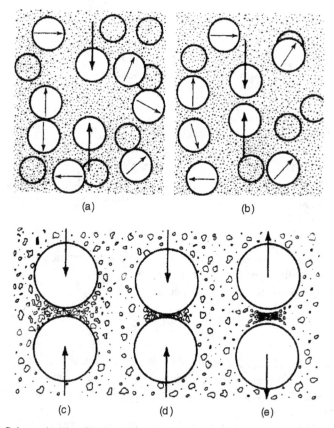

(a)   (b)

(c)   (d)   (e)

Figure 7.15 Schematic illustration of compaction and fracture processes in the formation of metallic powders via ball milling. *ASM Handbook*, Vol. 7, *Powder Metallurgy* (1984), ASM International, Materials Park, OH 44073-0002. p. 2.

7.1.3.2 Compaction and Hot Pressing.

The shaping of components from metallic powders is accomplished by compacting the particles under pressure, and sometimes at elevated temperatures. Typically, compaction temperatures are in excess of 75% of the absolute melting temperature, and compaction times are on the order of 10^3 to 10^4s. The compaction operation is performed primarily to provide form, structure, and strength, known as *green strength*, necessary to allow for handling before sintering. Compaction at room temperature involves pressures of 70–700 MPa, depending on the powder, and *green densities* (density prior to sintering) of 65–95% of theoretical (powder material) densities are achievable. Pressures for hot pressing are usually lower due to die or pressure chamber limitations.

The density–pressure relationship for powder compaction at room temperature typically increases from the apparent density at zero pressure to values that approach the theoretical density at high pressures, as illustrated in Figure 7.16. A compact with 100% theoretical density would indicate that it contains no porosity. Soft powders are more easily densified than hard powders at a given pressure, and irregularly shaped powders have lower densities than spherical powders in the low-pressure regime.

The mechanism of densification begins with individual particle motion at low pressure before interparticle bonding becomes extensive. At higher pressures, the compact

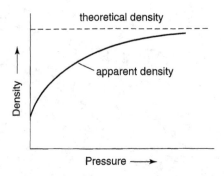

Figure 7.16   Pressure-density relationship in the compaction of metallic powders. Reprinted, by permission, from *Encyclopedia of Materials Science & Engineering*, Vol. 5, p. 3874. Copyright © 1986 by Pergamon Press.

acts as a coherent mass and densifies with decreased interparticle motion. At pressures of about 100–700 MPa, it has been shown that the slope of the density–pressure curve is generally proportional to the pore fraction, $1 - \rho_r$:

$$\frac{d\rho_r}{dP} \approx K(1 - \rho_r) \tag{7.9}$$

where ρ_r is the ratio of actual compact density to theoretical density (ρ/ρ_t), P is the pressure, and K is a proportionality constant that increases as the powder becomes softer. In general, K is proportional to the reciprocal of the nominal yield strength of the powder material.

Powders can be compacted at room temperature using a so-called *die and punch* design, in which the powders are compacted in a tool steel or cemented carbide die. The process can be automated, as illustrated in Figure 7.17, so that the die can be filled with powder, the upper and lower portions of the die, called *punches*, are brought together to compact the powder, and the compact is then ejected to prepare for the next cycle. The powder capacity of these presses can be as high as 1000 kg, and the maximum part size is limited to about 300 cm^2 in cross section. As the height-to-width ratio of the die and compact increases, it becomes more difficult to densify the powder in a uniform manner. Thus, the top of the compact may reach a higher density than the bottom.

Die compaction of simple shapes can be carried out at elevated temperature using carbide, superalloy, refractory metal, or graphite dies. An inert gas atmosphere or vacuum is often used to protect the die and/or the powder. For example, beryllium powder is compacted at about 1350°C in a graphite die under vacuum with pressures of 2–4 MPa.

The die-and-punch method of compact imparts an anisotropic pressure to the compact. It is possible to apply pressure equally in all directions using a process called *isostatic pressing*, in which the powder is compacted with a fluid, such as hydraulic fluid, water, or an inert gas. Isostatic pressing is carried out using formed, flexible molds, called *cans*, that contain the powder and allow the pressure from the fluid to be transferred to the powder. The flexible molds are then stripped off the compact after densification. Isostatic pressing eliminates the die-wall friction inherent in die compaction and allows for consolidation of large components. However, isostatic pressing

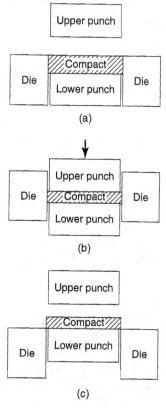

Figure 7.17   Schematic illustration of automated die-and-punch process in which (a) die is filled, (b) die is compacted, and (c) compact is ejected. Reprinted, by permission, from *Encyclopedia of Materials Science & Engineering*, Vol. 11, p. 3875. Copyright © 1986 by Pergamon Press.

cannot provide the dimensional tolerances and cycle rates of die compaction. Isostatic pressing is useful for forming intricate shapes and for forming parts that are near their final, desired dimensions prior to sintering, in what is termed *near net-shape processing*.

When the powder is isostatically compacted at elevated temperatures, the process is called *hot isostatic pressing* (HIP). In this case, the flexible dies are often made of thin metals, and high-pressure gases such as argon are used to heat the part rapidly and reduce thermal losses. Pressure up to 100 MPa and temperatures in excess of 2000°C are possible using HIP, and parts up to 600 kg can be fabricated. A schematic diagram of a typical HIP apparatus is shown in Figure 7.18. Metals that are processed commercially by HIP include various specialty steels, superalloys, hard metals, refractory alloys, and beryllium. We will see in Section 7.2 that HIP is also particularly useful for the densification of ceramic components.

7.1.3.3   *Sintering and Densification.* Recall from Section 2.2.2.4 that sintering is the consolidation of particles during heat treatment, and that it is controlled by the reduction in free energy of the system that accompanies surface area elimination during densification. Since the thermodynamics of sintering have already been described, we

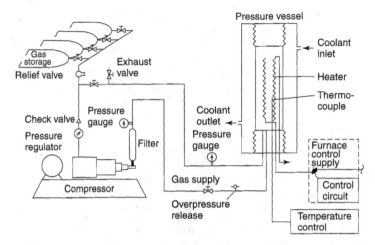

Figure 7.18 Schematic diagram of a hot isostatic pressing (HIP) operation. Reprinted, by permission, from *Encyclopedia of Materials Science & Engineering*, Vol. 3, p. 2188. Copyright © 1986 by Pergamon Press.

will concentrate here on the practical aspects of the sintering of metallic particles in powder metallurgy. A similar approach will be used in an upcoming section on the sintering of ceramic powders.

For sintering to take place, the temperature must generally be maintained above one-half the absolute melting point of the material. This condition results in atomic diffusion and neck formation between the powder particles in the solid state (cf. Figure 2.31). In the initial stages of sintering, interparticle bonds formed during compaction grow in size, resulting in a significant increase in strength. The pore channels in the compact then begin to round and become discontinuous. Further thermal processing forms rounded pores that can undergo coarsening. Typically, the ductility of the metal compacts increases as the pore morphology changes. Sometimes, density can decrease during sintering because of gas expansion or interdiffusion phenomena called *Kirkendall porosity formation*.

Various batch and continuous furnaces are employed for sintering operations. Composition and impurity control and oxidation protection are provided by the use of vacuum conditions or inert gas atmospheres. It is important to control the atmosphere to prevent undesired chemical reactions, such as oxidation or carburization. However, some reactions may be desirable during sintering, such as the control of carbon content in steels and oxide removal using a reducing furnace atmosphere. Hydrogen, either pure or diluted with nitrogen, and dissociated ammonia ($75\%H_2 - 25\%N_2$) are commonly used as reducing atmospheres. Exothermic gas formed by reacting air and natural gas mixtures (10:1 to 6:1) is inexpensive and reduces many materials. Endothermic gas formed by reacting air–natural gas mixtures (2.5:1 and 4:1) is both highly reducing and carburizing and is often used for sintering steel compacts. Additional considerations are involved in the sintering of ceramic components, and these are addressed in Section 7.2.

7.2   PROCESSING OF CERAMICS AND GLASSES

There are perhaps a wider variety of techniques that can be used to form and shape ceramics and glasses than for any of the other materials classes. Much of the variation in processing is due to the end use of the product. Certain techniques are used for structural ceramics, while others are used for semiconductor, superconductors, and magnetic materials. In general, there are three types of processes we will consider: powder-forming, melt processing, and chemical processing. The melt processing of ceramics and glasses is similar enough to the melt processing of polymers that its description will be delayed to Section 7.3. The latter category, which includes vapor phase and sol-gel processing, is considered optional in this chapter. Two topics related to powder processing, firing and sintering, will be described here.

The powder-forming processes are similar in many ways to those used for powder metallurgy described in the previous section. For example, pressing is a common method for processing ceramics; however, ceramic powders can be pressed in either dry or wet form. In wet form, they can also be extruded, just like metals, and cast in a variety of process variations. The nominal forming pressures and shear rates associated with some of these processing methods are summarized in Table 7.3. You may want to refer back to this table when each of the various processing techniques is described in more detail.

Before a description of powder processing is initiated, there is a consideration in ceramic-forming that is more prominent than in metal-forming which must be addressed: dimensional tolerance. Post-forming shrinkage is much higher in ceramics processing than in metals and polymer processing because of the large differential between the final density and the as-formed density. *Volume shrinkage* may occur during drying (Sv_d), during the removal of organic binders or reaction of bond phases producing consolidation (Sv_b), and during sintering (Sv_s). The total volume shrinkage Sv_{tot} is then the sum of these contributions:

$$Sv_{tot} = Sv_d + Sv_b + Sv_s \tag{7.10}$$

Table 7.3 Nominal Processing Pressures and Shear Rates in Some Ceramic Forming Processes

Forming Process	Pressure (MPa)	Shear Rate (1/s)
Pressing		
Roll/isostatic	>150	
Metal die	<100	
Plastic forming		
Injection molding	Varies	10–10,000
Extrusion	<40	10–1000 (die)
		100–10,000 (die-land)
Casting		
Slip (mold suction)	<0.2	<10 Pouring/ draining
Slip (slurry pressurized)	<10	<100 Pumping
Slip (vacuum on mold)	<0.7	<10 Filling
Gel	<0.1	<10 Filling
Tape	<0.1	10–2000

Source: J. S. Reed, *Principles of Ceramics Processing*, 2nd ed. Copyright © 1995 by John Wiley & Sons, Inc.

where each shrinkage is specified using an as-formed volume basis that is $(V_{formed} - V_{final})/V_{formed}$. Therefore, the final volume depends on the as-formed volume and the total shrinkage as follows:

$$V_{final} = (1 - Sv_{tot})V_{formed} \tag{7.11}$$

Precise control of the final product dimensions depends on control of the various shrinkages. For dry-pressed and injection-molded ceramics, Sv_d is in the range 3–12%. Binder elimination shrinkage is relatively high for injection-molded and tape-cast parts. The sintering shrinkage is commonly about 25–45% for products having high fired density and may range from 20% to 50%.

The final volume, V_{final}, is larger than the volume of the particulate material, V_{solids}, when the product has *residual porosity*. We define a ratio of these two volumes, D_r, that gives a measure of the residual porosity:

$$D_r = \frac{V_{solids}}{V_{final}} \tag{7.12}$$

When the densified component is pore-free, $V_{final} = V_{solids}$ and $D_r = 1.0$. Anything less than this value indicates the presence of porosity. Control of D_r requires control of both the microstructure of the as-formed part and the firing, because both affect the sintering shrinkage.

When shrinkage is isotropic, the *linear shrinkage* $(\Delta L/L_0)$ is calculated from the volume shrinkage as:

$$\frac{\Delta L}{L_0}(\%) = [1 - (1 - Sv_{tot})^{1/3}] \times 100 \tag{7.13}$$

Cooperative Learning Exercise 7.2

The same alumina ceramic product is made using two different hypothetic processes. In both cases, the ultimate density for the product is 3.96 g/cm³.

Person 1: In process A, the bulk density is 2.43 g/cm³ green and 3.84 fired. Calculate the volume shrinkage, linear shrinkage, and the densification ratio for this process.

Person 2: In process B, the bulk density is 2.74 g/cm³ green and 3.90 fired. Calculate the volume shrinkage, linear shrinkage, and the densification ratio for this process.

Compare your answers. Which process produces results in the greatest shrinkage?

Answers:

$Sv_{(A)} = 36.7\%; \Delta L/L_0 = 14.1\%; D_{r(A)} = 0.97$

$Sv_{(B)} = 29.7\%; \Delta L/L_0 = 11.1\%; D_{r(B)} = 0.98$

The linear shrinkage on drying is typically less than 2%, and on binder elimination and sintering less than 15%.

With these general principles of shrinkage and porosity in mind that apply to most ceramic processes, we proceed to the specific processing methods by examining some of the oldest materials processing techniques known, and we progress through time and technical difficulty to current techniques that produce complex components of incredible purity and dimensional tolerances.

7.2.1 Pressing

In much the same way that metal powders can be compacted and sintered to form a densified product, so, too, can many ceramic powders. Recall that pressing is the simultaneous compaction and shaping of a powder within a die or flexible mold. Just as we did for metals processing, we will look at each of the separate steps involved in the pressing operation: powder formation (called granulation for ceramics), die filling, and compaction. We reserve the final step, firing, to Section 7.2.3, since it applies not only to pressing operations, but to casting operations as well, which are the topic of Section 7.2.2.

7.2.1.1 Granulation. Powders of glass and ceramic materials may be formed in many of the same ways as those used to form metals particles. This is particularly true of milling operations, for which Figure 7.14 applies to both metals and ceramics. However, wet milling is much more common with ceramic materials, particularly when finer particle sizes are required. The combination of dry powders with a dispersant such as water is called a *slurry*. Unlike a solution or suspension, the solids contents of slurries are very high, typically 50 wt% solids and above. Depending on the type of milling equipment being used, the solids content can have an effect on the efficiency of the milling operation. In general, a high solids content is desirable, in order to develop a coating of adequate viscosity on the milling media (balls and vial) to prevent escape of the particles from the grinding zone (cf. Figure 7.15), but not completely dissipate the grinding stress. The effect of slurry solids content on milling efficiency, in terms of the percentage fines (smaller particles) produced, is illustrated in Figure 7.19 for the milling of an alumina slurry.

For the formation of fine particles necessary for most pressing and casting operations, ball-and vibratory-milling are the most common operations. Low-amplitude wet vibratory milling is generally limited to feed sizes smaller than 250 μm. Cylindrical milling media 1.3 cm in diameter are common, because smaller media do not produce a uniform distribution of vibration energy. The increase in specific surface area of the powders, called the *grinding rate*, is 10–50% greater in vibratory mills than in ball mills, as illustrated in Figure 7.20 for two types of ceramic materials. Power consumption and media wear are also lower for vibratory mills than for ball mills. The viscosity of the feed slurry is generally the maximum that can be discharged from the mill after attrition.

Feed material in ball mills is usually smaller than about 50 μm, and the solids contents of slurries range from 30% to 70%. The size of the spherical grinding media is in the range of 0.5–5 mm. Very rapid attrition is produced in ball mills by the intense combination of compression and shearing forces and the frequency of collisions, which is very high.

Milling efficiency is an important parameter in these operations. The total energy input used to produce a unit of milled product, U_T, is commonly used as a measure of

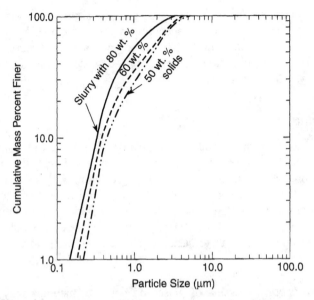

Figure 7.19 Effect of solids content in the milling of an alumina slurry on milling rate, expressed as cumulative mass percent fines. From J. S. Reed, *Principles of Ceramics Processing*, 2nd ed. Copyright © 1995 by John Wiley & Sons, Inc. This material is used by permission of John Wiley & Sons, Inc.

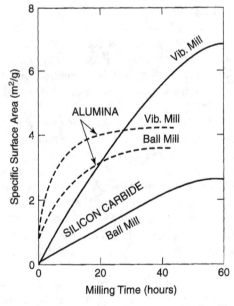

Figure 7.20 Comparison of grinding rates for vibratory and ball mills for two ceramics. From J. S. Reed, *Principles of Ceramics Processing*, 2nd ed. Copyright © 1995 by John Wiley & Sons, Inc. This material is used by permission of John Wiley & Sons, Inc.

efficiency. An empirical expression that is used to relate the dependence of U_T in kW to the characteristic mean particle size during milling, $\bar{a}$ (in μm), and the type of mill is

$$U_T = A_c \left(\frac{1}{\bar{a}^m} - \frac{1}{\bar{a}_0^m} \right) \tag{7.14}$$

where A_c is an efficiency constant for a particular milling system, $\bar{a}_0$ is the initial size of the material in μm, and m is a fracture constant for a particular material. Some representative values of m are presented in Table 7.4, and some values of A_c are presented in Table 7.5.

Powder-forming techniques are generally applied to reduce or eliminate porosity from raw materials and to create a known particle size distribution. However, dry bulky powders are not convenient feed materials for most pressing processes. They do not flow well, and they do not fill the molds or dies in a uniform manner. Semidry materials of controlled powder agglomerates called *granules* are much more satisfactory. Powder granules may be produced directly using pressing, extrusion, and spray granulation.

Spray granulation is the formation of granules when a liquid or binder solution is sprayed into a continually agitated powder. The formation of granules occurs in a nucleation and growth process, but on a larger size scale than the crystallization processes of Section 3.1.2.1. The nucleation and growth processes in granulation are

Table 7.4 Size-Dependence Index, *m*, for Several Industrial Raw Materials

Material	m
Fused alumina	1.1
Silicon carbide	1.3
Quartz	1.4
Tabular alumina	1.6
Bauxite	2.4
Calcined alumina (porous aggregate)	4.0
Calcined titania (porous aggregate)	4.4

Source: J. S. Reed, *Principles of Ceramics Processing*, 2nd ed. Copyright © 1995 by John Wiley & Sons, Inc.

Table 7.5 Values of *m* and A_c for Milling to a Submicron Mean Size for Various Types of Mills

Mill	Material	m	$A_c[\mathrm{KW}(\mu\mathrm{m})^m]$ (T)	$A_c[\mathrm{KW}(\mu\mathrm{m})^m]$ (m³)
Ball	Zircon	1.8	650	3350
Vibratory	Zircon	1.8	100	520
Attrition	Quartz	1.8	920	2680
Attrition	Limestone	1.8	500	1500

Multiply A_c by the appropriate milling temperature or mill capacity.
Source: J. S. Reed, *Principles of Ceramics Processing*, 2nd ed. Copyright © 1995 by John Wiley & Sons, Inc.

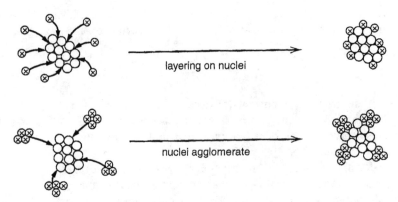

Figure 7.21 Schematic illustration of nucleation and growth process in spray granulation of $\cong 1$ μm alumina. From J. S. Reed, *Principles of Ceramics Processing*, 2nd ed. Copyright © 1995 by John Wiley & Sons, Inc. This material is used by permission of John Wiley & Sons, Inc.

illustrated in Figure 7.21. While the powder is stirred, particles roll and slide, and fines become airborne. An agglomerate nucleus forms when a droplet of liquid or binder hits and is absorbed on the surface of a group of particles. Capillary forces give the nuclei strength. Nuclei are more numerous when the liquid is introduced as a fine mist and the powder is vigorously agitated. The growth of granules by a process called *layering* occurs by the contact and adhesion of individual particles to the nuclei (see top illustration in Figure 7.21).

Alternatively, the agglomeration of small nuclei and fracture fragments can also produce a granule, as in the bottom illustration in Figure 7.21. The rate of each mechanism depends on the liquid feed rate, the adsorption of liquid into the agglomerate, and the mixing action. Rubbing between granules during tumbling may cause particle transfer at surfaces and result in surface smoothing. There is a critical range of liquid content for granulation. At low liquid binder contents, granules are eventually formed when the processing energy is sufficiently high. The addition of greater amounts of liquid generally increases the mean granule size, the size distribution, and the porosity of the granules, as illustrated in Figure 7.22 for alumina. The liquid requirement for granulation is greater when the specific surface area of the powder is higher. Common liquid requirements in the range of 20–36 vol% indicate that during granulation the pores in granules are incompletely saturated. Granules produced by spray granulation are more nearly spherical than granules produced by other methods such as compaction and extrusion. Spray-granulated materials are usually partially dried prior to use in pressing operations.

The process of *spray drying* is fundamentally different from spray granulation. In spray drying, a particle slurry is sprayed into a warm drying medium such that a portion of the binder is evaporated, resulting in nearly spherical powder granules that are more homogeneous than those formed in spray granulation. Spray drying is used for preparing granulated pressing feed from powders of ferrites, titanates, alumina, carbides, nitrides, and porcelains. A schematic illustration of one type of spray drying process is illustrated in Figure 7.23.

The granules from spray granulation or spray drying can then be used for pressing, though any one of a number of *additives* is typically used prior to forming. The liquid

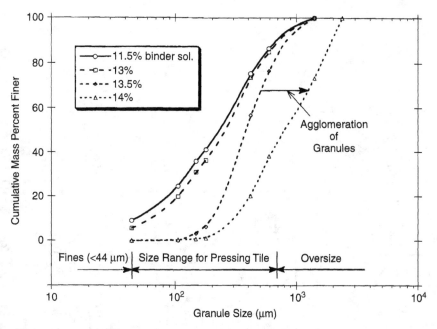

Figure 7.22 Effect of binder concentration on granule size in spray granulation of alumina. From J. S. Reed, *Principles of Ceramics Processing*, 2nd ed. Copyright © 1995 by John Wiley & Sons, Inc. This material is used by permission of John Wiley & Sons, Inc.

content of the ceramic and the composition of the processing additive will differ for each type of processing technique, but, in general, additives serve to stabilize the viscosity and provide a ceramic mixture suitable for consistent processing. A list of typical additives and their specific functions are given in Table 7.6. Some of these additives will be described in the course of the next several sections, as necessary.

7.2.1.2 Die Filling. Powder flow rates into the dies can be determined using traditional mass or volume per unit time techniques during processing. An alternative method for estimating the *flowability* of untested powders is to use the *angle of repose* that a powder forms when poured onto a flat surface (see Figure 7.24). The angle of repose is the greatest angle measured from the horizontal to the slope of a cone-shaped pile of free-flowing material. A highly flowable powder forms a relatively flat pile with a low angle of repose, whereas a poorly flowable powder has a higher angle of repose. The flow rate and angle of repose for the pressing of granular powders correlates well, as shown in Figure 7.25.

Dense, spherical particles with smooth surfaces and diameters of 20 μm or greater provide the optimal flow behavior. The presence of more than 5% *fines* (particles less than about 20 μm in diameter) may prohibit flow altogether. Binders and other additives can inhibit proper flow when the temperature exceeds the glass transition point of the additive, and relative humidity can affect the flow behavior of water-soluble binders.

The bulk density of the powder inside the die is called the *fill density*, D_{fill}. A high fill density reduces both the air content in the powder and the distance the punch needs to travel to further compress the powder during pressing. Poor powder flowability

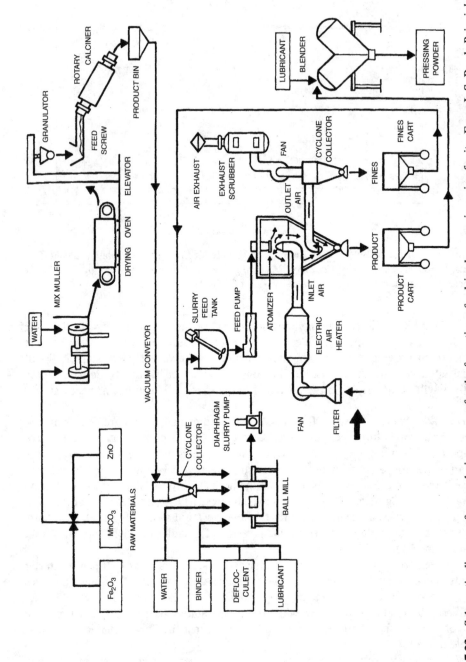

Figure 7.23 Schematic diagram of spray drying process for the formation of calcined magnesium zinc ferrite. From J. S. Reed, *Principles of Ceramics Processing*, 2nd ed. Copyright © 1995 by John Wiley and Sons, Inc. This material is used by permission of John Wiley & Sons, Inc.

Table 7.6   Some Typical Additives to Ceramic Powders Prior to Processing

Additive	Functions
Deflocculant	Particle charging, aid and maintain dispersion
Coagulant	Uniform agglomeration after dispersion
Binder/flocculant	Modify rheology, retain liquid under pressure, yield strength, green strength, adhesion
Plasticizer	Change viscoelastic properties of binder at forming temperature, reduce T_g of binder
Lubricant	Reduce die friction (external), reduce internal friction, mold release
Wetting agent	Improve particle wetting by liquid, aid dispersion
Antifoam	Eliminate foam
Foam stabilizer	Stabilize foam
Antistatic agent	Charge control of dry powder
Chelating agent/sequestering agent/precipitant	Inactivate undesirable ions
Antioxidant	Retard oxidative degradation of binder

Source: J. S. Reed, *Principles of Ceramics Processing*, 2nd ed. Copyright © 1995 by John Wiley & Sons, Inc.

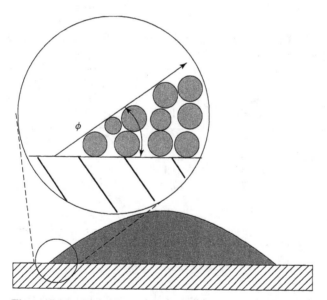

Figure 7.24   Definition of angle of repose for a powder pile.

can adversely affect the fill density and, in turn, the final product density. Mechanical vibration is often used to cause particle settling in the die, which further improves the fill density. A typical fill density for a granulated powder is in the range of 25–35%, as indicated for several representative ceramic powders in Table 7.7.

7.2.1.3 *Compaction.* *Compaction* is the process of granule densification via applied pressure to produce a cohesive part having a particular shape and microstructure. When

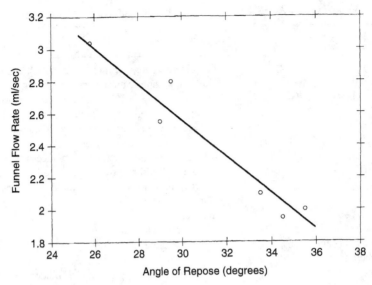

Figure 7.25 Relationship between angle of repose and flowability of a powder. From J. S. Reed, *Principles of Ceramics Processing*, 2nd ed. Copyright © 1995 by John Wiley & Sons, Inc. This material is used by permission of John Wiley & Sons, Inc.

Table 7.7 Particle Size and Bulk Density Characteristics of Some Spray-Dried Ceramic Pressing Powders

	Particle Size (μm)	Organics (vol%)	Granule Size (μm)	Granule Density (%)	Fill Density (%)
Alumina substrate	0.7	3.6	92	54	32
Spark plug alumina	2.0	13.3	186	55	34
Zirconia sensor	1.0	10.5	75	55	37
MnZn ferrite	0.7	10.0	53	55	32
MnZn ferrite	0.2	8.5	56	31	18
Silicon carbide	0.3	20.4	174	45	29

Source: J. S. Reed, *Principles of Ceramics Processing*, 2nd ed. Copyright © 1995 by John Wiley & Sons, Inc.

the pressure is first applied, it is transmitted throughout the powder, or compact, by means of intergranular contact. The rate of initial densification is high, but drops off rapidly as the particle compacts. As illustrated in Figure 7.26, relatively little densification occurs above about 50 MPa. In these initial stages, air from the pores is removed and exhausted between the punch and die. Most industrial presses operate below 100 MPa for high-performance technical ceramics and below 40 MPa for whiteware and tile compositions.

The final density of the compact is less than the maximum packing fraction of the particles, PF_{max} [cf. Eqs. (4.8) and (4.67)], due to frictional forces at particle contacts that retard particle sliding. The effectiveness of the compaction process is quantified

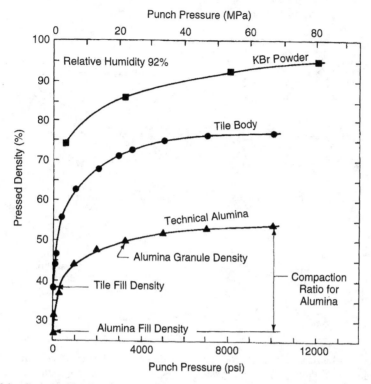

Figure 7.26 Compaction behavior of KBr powder and spray-dried granules of alumina and clay tile as a function of punch pressure. From J. S. Reed, *Principles of Ceramics Processing*, 2nd ed. Copyright © 1995 by John Wiley & Sons, Inc. This material is used by permission of John Wiley & Sons, Inc.

using the *compaction ratio, CR,* which is defined as the ratio of the pressed product density to the filled die density (an inverse ratio of volumes also works since the mass stays constant):

$$CR = \frac{V_{fill}}{V_{pressed}} = \frac{D_{pressed}}{D_{fill}} \qquad (7.15)$$

For most ceramic pressing, a $CR < 2.0$ is desired since it reduces both the punch displacement and the compressed air in the compact. As indicated in Eq. (7.15), a high fill density leads to a low CR. For comparison, the CR in metal powder pressing is typically much greater than 2.0 due to the ductility of the particles.

There are essentially three stages of densification in the compaction process. In Stage I, a small amount of sliding and granular rearrangement occurs at low pressure when the punch first contacts the powder. The density is only slightly above the fill density at this point. In Stage II, granular deformation into neighboring interstices begins when the pressure exceeds the apparent yield pressure of the granules, P_y. A reduction in the volume and size of the relatively large interstices occurs in Stage II, as illustrated in Figure 7.27. The apparent yield pressure of the granules is related to the packing fraction, the volume of the binder and particles, V_b and V_p, respectively,

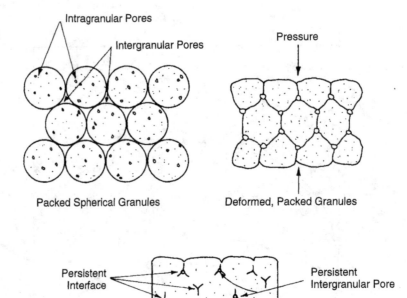

Figure 7.27 Schematic illustration of change in shape and change of pore size distribution during compaction. From J. S. Reed, *Principles of Ceramics Processing*, 2nd ed. Copyright © 1995 by John Wiley & Sons, Inc. This material is used by permission of John Wiley & Sons, Inc.

and the strength of the binder, σ_0, as follows:

$$P_y = C \left(\frac{PF}{1 - PF} \right) \left(\frac{V_b}{V_p} \right) \sigma_0 \tag{7.16}$$

where C is an experimentally-determined proportionality constant.

Most of the densification occurs in Stage II. Granule deformation is the predominant mechanism of densification. The *compact density*, $D_{compact}$, may be approximated in this stage as

$$D_{compact} = D_{fill} + m \log(P_a / P_y) \tag{7.17}$$

where P_a is the applied pressure, P_y is the apparent yield pressure of the granules, D_{fill} is the fill density, and m is a *compaction constant* that depends upon the deformability and densification of the granules. Powders containing high-density granules may produce a high compact density with a relatively low value of m, but the elimination of all intergranular pores is favored when powders contain granules of somewhat lower density, for which m has intermediate values of 0.07 to 0.10. For each powder, there is an optimum granule density, fill density, and value of m that produces the best compromise of compaction ratio and the elimination of large pores.

Finally, in Stage III of the compaction densification process, densification continues by the sliding and rearrangement of particles into a slightly denser packing configuration. In powders containing granules of varying yield strength and density, the transition between stages does not occur uniformly throughout the compact. Ideally, the large pores between deformed granules have disappeared in Stage II, and interfaces between granules no longer exist. The high pressures can cause particle fracture that blocks further densification. Interstices between large granules and large pores within doughnut-shaped granules may persist into Stage III. These defects are not eliminated during sintering and reduce the mechanical reliability of the product.

As with metals, compaction can be effected in an isostatic manner using inert gas pressurization. Isostatic pressing is used for products with one elongated dimension, complex shapes, or large volumes that cannot be easily compacted using dry pressing. Roll pressing can also be applied to ceramic powders, and it is used to produce substrates up to 30 cm in width with a uniform thickness in the range of 0.5–1.5 mm. Once the ceramic powder has been compacted by any of these means, it is ejected from the mold or die and is transferred to the final processing steps, which include sintering and surface finishing.

7.2.2 Casting

The casting of ceramics is similar to the casting of metals insofar as both involve the pouring of liquid-like material into a mold, followed by the development of a solidified product via a transformation that causes the development of yield strength. Unlike metal casting, which involves a molten metal, the pourable substance in ceramic casting is typically a slurry. And unlike metal casting, in which the transformation required to achieve sufficient yield strength in the cast part is invariably a physical transformation (solidification), the transformation in ceramic casting can be either a physical, chemical, or thermal transformation, or a combination thereof. Since the fundamental steps of casting have been previously described, let us proceed by examining some of the differences between the more common types of ceramic casting.

7.2.2.1 Slip Casting.
An aqueous slurry of fine clay is called a *slip*. *Slip casting* is then the casting of a slip, typically using a porous gypsum mold. As capillary suction of the water from the slurry into the mold proceeds, the slurry particles coagulate near the mold surface, and the cast is formed. Variations in the slip casting process include the application of pressure (*pressure casting*), vacuum (*vacuum-assisted casting*), or centrifugal force (*centrifugal casting*), all of which serve to increase the casting rate. There are two main types of slip casting as illustrated in Figure 7.28: *drain casting* and *solid casting*. In drain casting, the molds are first filled with the slurry. After some initial solidification has taken place, the excess slurry is drained from the molds. This technique is suitable for the formation of objects with hollow centers, such as crucibles or tubes. In solid casting, the entire amount of slurry in the mold is used to form a dense, finished shape. The important steps of the slip casting process that will be elaborated upon here include slurry preparation, mold filling, draining, and partial drying while in the mold. Other steps that will not be described here include (a) separation of the slip from the mold after drying and (b) post-mold trimming and surface finishing prior to the final firing or sintering step.

Slurries are prepared from a mixture of liquid (typically water), milled powders or granules, and additives (see Table 7.6). The compositions of two common ceramic slips are given in Table 7.8. Casting slurries are typically formulated to be shear thinning (cf. Section 4.1.2.2) with an apparent viscosity below 2000 cpoise (2000 mPa · s) at

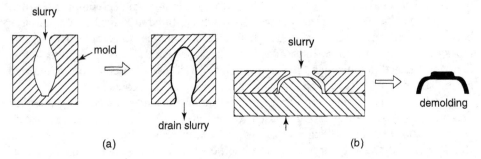

Figure 7.28 Schematic illustration of (a) drain and (b) solid types of slip-casting processes. Reprinted, by permission, from H. Yanagida, K. Koumoto, and M. Miyayama, *The Chemistry of Ceramics*, p. 160. Copyright © 1996 by John Wiley & Sons, Inc.

Table 7.8   Composition of Some Common Slip Casting Slurries

	Concentration (vol%)	
Material	Porcelain	Alumina
Alumina		40–50
Silica	10–15	
Feldspar	10–15	
Clay	15–25	
Water	45–60	50–60
	100 vol%	100 vol%
Processing additives (wt%[a])		
Deflocculant		
Na silicate	<0.5	
NH$_4$ polyacrylate		0.5–2
Na citrate		0.0–0.5
Coagulant		
CaCO$_3$	<0.1	
BaCO$_3$	<0.1	
Binder		
Na carboxymethylcellulose		0.0–0.5

[a] Based on weight of solids in slurry.
Source: Reprinted, by permission, from J. S. Reed, *Principles of Ceramics Processing*, 2nd ed. Copyright © 1995 by John Wiley & Sons, Inc.

a mold-filling shear rate of between 1 and 10 s^{-1} (see Table 7.3). A lower viscosity (<1000 cpoise) is desirable when the slurry is screened prior to mold filling, or when air bubbles must be quickly removed from the slurry. The density, apparent viscosity, and yield stress of the slurry should be high enough to minimize particle settling, called *deflocculation*, during casting. The yield stress of the slurry, τ_y, can be related to the slurry density, D_{slurry}, and the diameter and apparent density of the suspended particle,

a and D_p, respectively:

$$\tau_y = (2/3)a(D_p - D_{slurry})g \tag{7.18}$$

where g is the gravitational constant. When the slurry does not have a yield stress, a high viscosity at the low shear rates during casting can also retard settling of the particles.

The drying process is caused by the removal of water from the slurry through capillaries in the gypsum mold, as illustrated in Figure 7.29. The permeable gypsum molds typically have a porosity of 40–50% and an effective pore size of 1–5 μm. As water is removed from the slip due to capillary action in the mold, a compact layer of slip particles (clay in Figure 7.29) forms at the mold–slip interface. The rate of the process is determined by the transport of water out of the slip and into the capillaries; that is, the rate-limiting step is the flow of water through the compact layer. As the thickness of the compact layer, x, increases, the overall rate of water removal, $J^*_{H_2O}$, decreases. We can determine the molar flux of water using a form of Fick's Law [Eq. (4.4)], in which we replace the concentration gradient driving force with a pressure gradient driving force, dP/dx:

$$J^*_x = -K\frac{dP}{dx} \tag{7.19}$$

where the diffusivity, D, in Eq. (4.4) has been replaced by a related quantity, the *permeation coefficient*, K, to make the units work out correctly. Until the capillaries become filled with water, the pressure difference is effectively constant, $dP = \Delta P = P_s - P_m$, where P_s is the pressure in the slip, which is essentially atmospheric pressure,

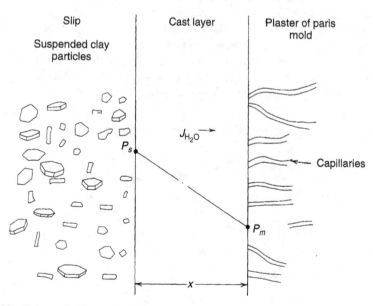

Figure 7.29 Schematic illustration of the formation of a slip-cast layer formed in the extraction of water by capillary action from a mold. From *Introduction to Ceramics*, by W. D. Kingery, H. K. Bowen and D. R. Uhlmann, Copyright © 1976 by John Wiley & Sons, Inc. This material is used by permission of John Wiley & Sons, Inc.

and P_m is the pressure in the mold. Recall from Section 2.2.2.2 that the pressure drop across a curved surface, such as at a capillary opening, is given by Eq. (2.69), such that

$$\Delta P = \frac{2\gamma}{R} \tag{2.69}$$

where γ is the surface tension of the slip, and R is the capillary radius in the mold. The surface tension will vary with the type of binders and deflocculants added to the slip, as well as with temperature and the chemical composition of the ceramic particles. To a first approximation, then, the molar flux of water is proportional to the change in thickness of the compact layer with time, dx/dt, and the change in thickness of the compact layer, dx, becomes x, which is now a function of time, so Eq. (7.19) becomes

$$J_x^* \propto \frac{dx}{dt} = -K' \frac{2\gamma}{Rx} \tag{7.20}$$

The permeation coefficient, K', has been modified slightly to include some other terms such as the density of the slip and a factor for converting the volume of water removed to the volume of clay particles deposited. The important point, however, is that Eq. (7.20) can now be integrated to give a relationship between the compact layer thickness and time:

$$x = (K't)^{0.5} \tag{7.21}$$

Equation (7.21) predicts that the compact layer should increase linearly with the square root of time. This functionality is illustrated in Figure 7.30 for the slip casting of porcelain in a gypsum mold. The rate of water removal can be affected in several ways. A higher solids content in the slip will increase the water flux, as can heating the slurry. The pressure gradient may also be increased to increase the flux, through application of an external pressure to the slip. This process is known as *pressure casting*.

In drain casting, the excess slurry is drained from the mold once a cast has been formed. The cast must possess adequate yield strength, however, to maintain its integrity through the draining process. In some systems, coagulation forces in the cast contribute significantly to the strength. This coagulation is commonly called *gelation*. For

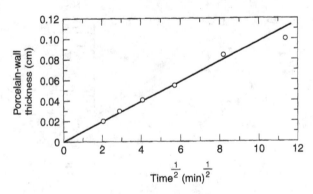

Figure 7.30 Dependence of compact layer thickness with time for a porcelain slip in a gypsum mold. From *Introduction to Ceramics*, by W. D. Kingery, H. K. Bowen and D. R. Uhlmann, Copyright © 1976 by John Wiley & Sons, Inc. This material is used by permission of John Wiley & Sons, Inc.

engineering ceramics, gelation may occur from the flocculation of an organic binder in the cast. The anionic polymer binder sodium carboxymethylcellulose (cf. Table 4.4) has been shown to act beneficially both as a deflocculant and as a binder in aqueous slip castings. Of course, in solid casting, the entire slip remains in the mold and must be dried prior to removal of the cast from the mold. Once the cast is removed from the mold, it is subjected to post-casting processes such as trimming and drying. The process of firing is described further in Section 7.2.3.

7.2.2.2 Tape Casting.

As with slip casting, a tape casting slurry is prepared that contains deflocculated powders or granules, binders and/or plasticizers, and a solvent such as water. Unlike slip casting, however, the slurry can be processed into thin, flexible sheets called *green sheets*, which are then cut and dried. The term "green" refers to the fact that it is a pre-fired or pre-sintered product and has nothing to do with color. The dried tape is rubbery and flexible due to the high binder content, and has a very smooth surface. Tape-cast ceramics are used primarily as substrates for electronic conductors, resistors, capacitors, photovoltaic cells, electrical sensors, and solid electrolytes for batteries. The electronic materials can be printed on the tape surface and they become an integral part of the ceramic component upon firing. As with slip casting, we will briefly investigate the topics of slurry composition, casting, and drying as they relate to this important processing technique.

Ceramic powders used for tape casting typically have a maximum particle size of 1–5 μm and a specific surface area of 2–5 m^2/g. Finer powders result in a smoother tape surface. Because the tapes are often used for electronic component fabrication, chemical homogeneity is critical. Particle size and dispersion also affect dimensional stability during the drying and firing processes. Therefore, the liquid–solvent system must dissolve the additives yet permit their adsorption on the ceramic particles. A solvent blend is thus typically used. Organic solvents have a low viscosity, low boiling point, low heat of vaporization, and high vapor pressure, all of which promote short drying times. Some common nonaqueous slurry compositions are listed in Table 7.9. Longer drying times and higher drying temperatures are required when an aqueous system is used.

Example Problem 7.2

Tape is cast at a velocity of 1200 mm/min and the blade height is 0.4 mm. Estimate the shear rate (s^{-1}) and compare the value to that in Table 7.3.

Answer: The shear rate for flow in the x direction is calculated as dv/dy, which for a planar geometry at steady state is the linear velocity divided by the separation distance between the moving and static surfaces (y direction):

$$\dot{\gamma} = \frac{dv}{dy} = \frac{v}{h_0} = \frac{(1200 \text{ mm/min})(1 \text{ min}/60 \text{ s})}{0.4 \text{ mm/min}} = 50 \text{ s}^{-1}$$

This shear rate is not high when compared to the range 10–2000 s^{-1} given in Table 7.3

As mentioned in the previous paragraph, binder and plasticizer concentrations for tape casting are much higher than for slip casting and for most other types of slurry

Table 7.9   Composition of Some Nonaqueous Tape Casting Slurries

Function	Composition	Vol%	Composition	Vol%
Powder	Alumina	27	Titanate	2.8
Solvent	Trichloroethylene	42	Methylethyl ketone	3.3
	Ethyl alcohol	16	Ethyl alcohol	16
Deflocculant	Menhaden oil	1.8	Menhaden oil	1.7
Binder	Polyvinyl butyral	4.4	Acrylic emulsion	6.7
Plasticizer	Polyethylene glycol	4.8	Polyethylene glycol	6.7
	Octyl phthalate	4.0	Butylbenzylphthalate	6.7
Wetting agent			Cyclohexanone	1.2

Source: J. S. Reed, *Principles of Ceramics Processing*, 2nd ed. Copyright © 1995 by John Wiley & Sons, Inc.

Table 7.10   Some Common Binders and Plasticizers Used in Tape Casting Slurries

	Binder	Plasticizer
Nonaqueous	Polyvinyl butyral	Dioctyl phthalate
	Polymethylmethacrylate	Dibutyl phthalate
	Polyvinyl alcohol	Benzyl butyl phthalate
	Polyethylene	Polyethylene glycol
Aqueous	Acrylics	Glycerine
	Methyl cellulose	Polyethylene glycol
	Polyvinyl alcohol	Dibutyl phthalate

Source: J. S. Reed, *Principles of Ceramics Processing*, 2nd ed. Copyright © 1995 by John Wiley & Sons, Inc.

processing such as extrusion and pressing. More flexible molecules such as acrylics and vinyls are common (see Table 7.10). These macromolecules orient during casting and impart toughness and strength to the casting. The combination of binder and plasticizer must be carefully controlled to provide the required mechanical properties but permit a high concentration of ceramic particles in the slurry. An excess of binder can increase the particle separation and reduce the PF of the particles in the tape. Other additives include a *wetting agent* to promote spreading of the slurry on the substrate, a *homogenizer* which enhances surface quality, and *antifoaming agents*. As for slip casting, the tape casting slurry should be shear thinning and have a viscosity greater than about 4000 cpoise at the tape casting shear rate (refer back to Table 7.3).

In the tape casting process, a schematic diagram for which is presented in Figure 7.31, the slurry is cast on a smooth, clean, impervious surface called the *carrier*, which is typically made of Teflon® or cellulose acetate. Casting machines range in size up to 25 m in length, are several meters in width, and have casting speeds up to 1500 mm/min. Tape thicknesses are in the range of 25–150 μm. The tape is formed when the slurry flows beneath a blade, forming a film on the carrier. The thickness of the tape varies directly with the height of the blade or gate above the carrier, the speed of the carrier, and the drying shrinkage. A flow model that gives the dependence of the tape thickness, H, as a

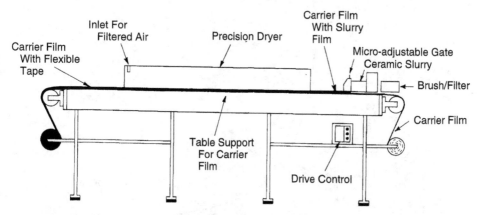

Figure 7.31 Schematic diagram of a typical tape-casting process. From *Principles of Ceramics Processing*, 2nd edition, by J. S. Reed; Copyright © 1995 by John Wiley & Sons, Inc. This material is used by permission of John Wiley & Sons, Inc.

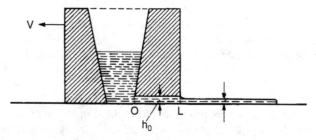

Figure 7.32 Tape casting flow model used for the development of Eq. (7.22). From J. S. Reed, *Principles of Ceramics Processing*, 2nd ed. Copyright © 1995 by John Wiley & Sons, Inc. This material is used by permission of John Wiley & Sons, Inc.

function of the carrier velocity, v, can be obtained using the parameters in Figure 7.32:

$$H = AD_r h_0 \left(1 + \frac{h_0^2 \Delta P}{6\eta_s v L}\right) \tag{7.22}$$

where A is a constant that depends on the amount of side flow, D_r is the ratio of the density of the slurry to the density of the as-dried tape, h_0 is the cast thickness at the blade, ΔP is the pressure that causes the slurry to flow, η_s is the viscosity of the slurry, and L is the length of the casting. Equation (7.22) indicates that the tape thickness is nearly independent of variations in the carrier speed when the ratio h_0/η_s is small. A high slurry viscosity and a carrier velocity exceeding about 0.5 cm cm/s are desirable for thickness uniformity, as illustrated in Figure 7.33.

In the final step, a viscoelastic tape is formed as the casting moves through a drying tunnel and the solvent vaporizes. The drying rate varies directly with the solvent concentration on the surface and the temperature and solvent content in the drying medium, which is typically air. Capillary forces transport the solvent to the drying surface. Initial vaporization is relatively slow because the counterflowing air contains

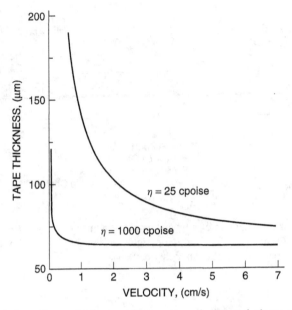

Figure 7.33   Variation of tape thickness with carrier velocity and slurry viscosity. Adapted from J. S. Reed, *Principles of Ceramics Processing*, 2nd ed. Copyright © 1995 by John Wiley & Sons, Inc. This material is used by permission of John Wiley & Sons, Inc.

much solvent and is cooler. Shrinkage occurs as solvent is lost and the particles move closer together. Shrinkage occurs primarily along the dimension of tape thickness, and the final thickness is typically about one-half the initial blade height. Some migration of the plasticizer to the drying surface occurs, but migration of higher-molecular-weight binders is minimal. The liquid surface is interrupted, and drying by vapor transport begins when the degree of pore saturation, DPS, is <1 (cf. Section 4.1.2.2). The viscoelastic tape becomes more elastic as drying continues. The packing fraction, PF, of the particles in the dried tape is about 55–60%, and the remaining volume is about 35 vol% organics and 15 vol% porosity. The dried tape may go directly to a post-processing step or may be stored on reels.

7.2.2.3 *Other Casting Techniques.* As alluded to in Section 7.2.2.1, there are several modifications to the slip casting technique that give rise to other types of casting processes. Most of these methods have to do with increasing the pressure, ΔP, which causes liquid migration from the slip into the mold, as indicated in Eq. (7.19). For a typical gypsum mold, ΔP is less than about 200 kPa. In pressure casting, the cast thickness can be increased by a factor of 2–3 through application of external pressures on the slip of up to 1.5 MPa. Furthermore, pressure casting can reduce the residual water content in the cast from 1% to 3.5%, but shrinkage anisotropy also tends to increase. Increasing the casting pressure reduces the cycle time for a cast, but the higher pressure increases the equipment requirements. High-pressure slip casting at up to 4 MPa is used for the small volume production of small shapes. One disadvantage of pressure casting is the larger differential pressure produced across the cast. A large gradient in the effective stress is produced, which produces a density gradient in the cast. The density gradient can cause differential shrinkage and shape distortion upon drying. A variation on pressure casting is used to form hard ferrite magnets (Section 6.2.2.2), which have a highly oriented microstructure. In this process, a die set with pistons that are magnetically permeable is drilled to permit flow of liquid. The die set is charged with a well-dispersed slurry of the ferrite particles. A magnetic field is applied to orient the particles, and the pistons are forced into the die to form a dense cast with oriented particles.

Vacuum casting is widely used for forming very porous refractory insulation having a complex shape. The slurry typically contains partially deflocculated ceramic powder and chopped refractory fiber. The fiber increases the viscosity and liquid requirement and increases the porosity of the cast. An expendable, permeable preform is coupled to a vacuum line and then submerged in the slurry. After the desired wall thickness has been cast, the preform is withdrawn and removed for drying. Vacuum casting may also be used to describe the process of applying a vacuum to the external surface of the gypsum mold during slip casting to increase the casting rate.

Centrifugal casting is used for forming advanced ceramics having a complex shape where a very high cast density is required. Centrifuging increases the drive force for the settling of the particles, and liquid is displaced to the top of a centrifuge cell. For a relatively dilute suspension, the settling rate is described by Stoke's Law [cf. Eq. (4.70)], and particles having a greater size or density settle at a faster rate and a structural gradient is produced. The casting rate increases with centrifugal speed and the flow rate of the liquid away from the deposition surface. Casts of complex shapes having a uniform density and sedimentation rates of 0.6 mm/min can be obtained with this process.

There are additional variations to the casting process, and additional post-processing steps such as shaping, trimming, laminating, and printing that are beyond the scope of this text. The interested reader is referred to the textbook by Reed on ceramic processing for more information.

7.2.3 Firing

Once the green ceramic component has been cast by any of the methods described in the previous section, dried, and surface finished, it is generally subjected to a process known as firing. Firing involves heat treatment of the green ceramic in a high-temperature furnace, known as a *kiln*, to develop the desired microstructure and properties. Kiln design varies widely, depending upon the ceramic to be fired, but they are generally designed to operate in either an intermittent or continuous fashion. Heating can be supplied by heating elements such as silicon carbide (to 1150°C), molybdenum disilicide (to 1700°C), refractory metals such as molybdenum or tungsten (>1700°C), graphite (>1700°C in inert gases), plasma heating sources, microwave heating, or induction (cf. Section 7.1.1.1). Combustion heating is also commonly utilized, in which methane, usually in the form of natural gas, is combusted with air to produce temperatures in excess of 1700°C.

Several processes occur during firing, including further drying beyond that performed in the casting process, binder burn-out, and, most importantly, sintering. We will briefly touch on each of these topics.

7.2.3.1 *Pre-sintering Processes.* Material changes that occur upon firing prior to sintering may include drying, the decomposition of organic additives, the vaporization

of hydrolyzed water from the surfaces of particles, the pyrolysis of particulate organic materials, changes in the oxidation state of some of the transition metal and rare-earth ions, and the decomposition of carbonates and sulfates, termed *calcination*, that are introduced either as additives or as raw component constituents.

The decomposition of organic additives, such as binders, called *thermolysis*, is an important step prior to densification on sintering. Incomplete binder removal and uncontrolled thermolysis may reduce yields or impair performance of the product. Ideally, the green ceramic should survive thermolysis without deformation, distortion, the formation of cracks, or expanded pores. Binder burnout is dependent upon the composition of the binder and the composition and flow rate of the kiln atmosphere. It is also very dependent upon the microstructure of the organic, powder, and porosity phases and dynamic changes in the microstructure as the binder is eliminated. Some of the chemical and physical changes that take place during thermolysis are summarized in Table 7.11. Most cast ceramic components have sufficient pore channel sizes to allow the transport of vapors and gases from the binder burn-out zone to the product surface. The time for thermolysis is controlled by the diffusion length for vapor phase transport rather than by the characteristic dimension of the binder phase. Decomposition and vaporization of the organics will cause an internal gas pressure that will depend on the gas evolution rate, gas permeability, and the compact size. The permeability varies with the binder loading, and the pore structure varies with pore radius and volume fraction. Thus, the gas permeation rate is much lower in dense compacts of very fine particles.

As an example of thermolysis, consider the burnout of polyvinyl alcohol from a barium titanate compact in air. Initially, there is loss of hydroxyl and hydrogen side groups, leaving a conjugated hydrocarbon:

$$\underset{\substack{|\\ OH\ \ H}}{\overset{\substack{H\ \ H\\ |\ \ |}}{H_2C-C-C-CH_2}} \xrightarrow{\text{150--250 °C}} \underset{\substack{|\\ OH}}{\overset{\substack{H\ \ H\\ |\ \ |}}{H_2C-C=C-CH}} + H_2O\ (g) \tag{7.23}$$

The conjugated hydrocarbon then degrades in an exothermic manner at higher temperatures. The volume of the gaseous products is several hundred times larger than

Table 7.11 Some Chemical and Physical Changes that Occur During Thermolysis

Chemical reactions	Decomposition
	Depolymerization
	Carubrization
	Oxidation
Physical changes	Melting/softening (solid–liquid)
	Sublimation (solid–gas)
	Evaporation (liquid–gas)
Reaction zone	External
	Internal
	Uniform displacement
	Nonuniform displacement
Transport	Gas diffusion
	Liquid flow

the volume of the compact. Similar results are observed on heating vinyl binders with other common side groups and cellulose binders, except that the burnout range may be narrower.

Other additives such as waxes and polyethylene glycols melt at relatively low temperatures and vaporize over a more narrow temperature range. Oxidation of the binder causes the temperature to increase more rapidly. For the thermolysis of polyethylene glycol (PEG) in air, decomposition occurs by the mechanism of chain scission and oxidative degradation.

Thermolysis of an organic binder in an inert atmosphere such as nitrogen, as is used when firing silicon nitride and aluminum nitride, proceeds differently. Some initial oxidation of the binder may occur from adsorbed oxygen. Vaporization is more predominant and in general the thermolysis is endothermic and shifted to a higher temperature, as indicated by the thermogravimetric analysis (TGA) plot of methylcellulose in Figure 7.34, which follows the weight loss of the ceramic sample as a function of increasing temperature. For PEG binders the onset temperature in nitrogen is about 300°C. Some residual carbon may be expected in the absence of oxygen. The residue after burnout of synthetic polymerized glycols and acrylic binders is relatively small, as indicated in Table 7.12. When firing nitride and carbide ceramics in an inert atmosphere, carbon is again the common residue, which may serve as a sintering aid.

Some ceramics, such as kaolin, (cf. Section 2.2.1.2), contain waters of crystallization that are eliminated between 450°C and 700°C:

$$Al_2Si_2O_5(OH)_4 \longrightarrow Al_2Si_2O_6 + H_2O(g) \qquad (7.24)$$

The gas pressure may produce a volume expansion, and the heating rate must be carefully controlled in bodies containing a high clay content. Other dehydration reactions producing gas are the dehydration of aluminum hydrates in the range 320–560°C and

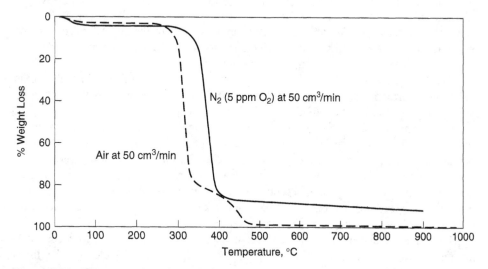

Figure 7.34 Thermogravimetric analysis of methylcellulose during thermolysis in both air and nitrogen environments. From J. S. Reed, *Principles of Ceramics Processing*, 2nd ed. Copyright © 1995 by John Wiley & Sons, Inc. This material is used by permission of John Wiley & Sons, Inc.

Table 7.12 Thermolysis Residue of Some Common Binders in Air

Binder Type	Residue (wt%)
Gums, waxes	>1.5
Methyl cellulose	>1.0
Polyvinyl alcohols	0.6–1.0
Hydroxyethyl cellulose	0.8
Polyvinyl butyral	0.4
Acrylic	0.14
Polyethylene glycols	0.05–0.07

Source: J. S. Reed, *Principles of Ceramics Processing*, 2nd ed. Copyright © 1995 by John Wiley & Sons, Inc.

talc at 900–1000°C. Carbon dioxide is produced upon the decomposition of magnesium carbonate at 700°C and dolomite at 830–920°C. Magnesium sulfate decomposes as low as 970°C, but the decomposition of calcium sulfate occurs in excess of 1050°C. Fine organic matter in clays is commonly oxidized between 200°C and 700°C, but coarse particulate carbon in the material may not be completely oxidized at 1000°C even when firing with excess air. Residual carbon discolors the microstructure and can lead to the production of carbon monoxide in the pores of the product. A chemical dopant such as a transition metal oxide in its higher oxidation state, such as Mn_2O_3, is sometimes added as an internal oxidizing agent, in which the following reaction occurs:

$$Mn_2O_3 + CO(g) \longrightarrow 2MnO + CO_2(g) \tag{7.25}$$

Carbon monoxide in the product is not inert and may change the stoichiometry and properties of transition metal oxide pigment or magnetic ferrites, such as Fe_2O_3.

7.2.3.2 Sintering of Ceramics. The thermodynamics of the sintering process were presented in Section 2.2.2.4. We describe here some of the more practical aspects of sintering as they relate to the firing of ceramics and glasses.

Recall from Section 2.2.2.4 that there are three primary stages of sintering: initial sintering, intermediate sintering, and final-stage sintering. These three stages, and the accompanying densification that occurs in each stage, can be observed for the sintering of two 0.8- (finer) and 1.3-μm (coarser), magnesia-doped alumina powders in Figure 7.35. For comparison, the densification process for hot pressing of the same powder is shown. Notice that most of the densification takes place in the intermediate stage. The mass transport mechanisms that lead to densification include grain boundary diffusion, lattice diffusion, viscous flow, and plastic flow. Such mass transport processes as surface diffusion and evaporation take place during sintering, but do not contribute to densification.

The shrinkage that often accompanies densification is controlled by different mechanisms in the three stages of sintering. In the initial stage of sintering, the linear shrinkage, $\Delta L/L_0$, is described by

$$\frac{\Delta L}{L_0} = \left(\frac{K D_v \gamma_s V_v t}{d^n k_B T} \right)^m \tag{7.26}$$

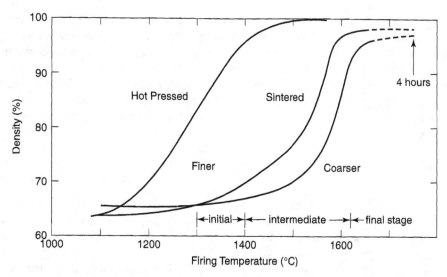

Figure 7.35 Densification behavior of compacts of two alumina powders (0.8- and $1.3 - \mu$m average particle size), with the sintering stages indicated for the coarser powder. A hot-pressed powder is shown for comparison. From J. S. Reed, *Principles of Ceramics Processing*, 2nd ed. Copyright © 1995 by John Wiley & Sons, Inc. This material is used by permission of John Wiley & Sons, Inc.

where D_v is the apparent diffusivity of the vacancies of volume V_v, d is the grain diameter, K is a geometry-dependent constant, γ_s is the solid surface energy [cf. Eq. (2.75)], t is time, k_B is Boltzmann's constant, T is absolute temperature, and m and n are constants depending on the mechanism of mass transport. The value of n is approximately 3, which implies that surface diffusion predominates, and m is in the range of 0.3–0.5. The shrinkage is very temperature-dependent, since in addition to the direct temperature dependence in Eq. (7.26), the diffusivity is also temperature-dependent [cf. Eq. (4.71)].

In the intermediate stage, Eq. (7.26) can still apply, but the situation is complicated by grain growth (i.e., d is no longer constant) and a change in the pore geometry. More than one mass transport mechanism may be contributing to the changes in microstructure. In the final stages of sintering, densification is very dependent upon the association of pores with grain boundaries and the rate and mode of grain growth. Diffusion of atoms across the grain boundary causes the grain boundary to be displaced. The grain boundary mobility, M_b, can also be described by an Arrhenius-type expression, so it is temperature-dependent. When M_b is roughly constant, it can be used to determine the rate of grain size growth with time, t:

$$d_t^n - d_0^n = 2AM_b\gamma_{gb}t \tag{7.27}$$

where n is the dimensionality parameter of Eq. (7.26) with a value of 2–3, d_0 is the original mean grain size, d_t is the mean grain size at time t, A is a constant that depends on geometry, and γ_{gb} is the grain boundary interfacial energy. A log-normal grain size distribution is often observed during regular grain growth.

Pores can also have an effect on grain growth and shrinkage, typically leading to exaggerated grain sizes. The tendency for exaggerated grains is higher when powder aggregates or dense, extremely coarse particles are present and the pore distribution is inhomogeneous. A high sintered density with minimal grain growth occurs when the material contains densely packed, homogeneously distributed, fine particles. A common *grain growth inhibitor* is magnesium oxide, MgO, which inhibits exaggerated grain growth by decreasing surface diffusion. Surface diffusion and pore mobility, M_p, are related by

$$M_p = \frac{K D_s}{T r_p^4} \qquad (7.28)$$

where K is a geometry-dependent constant, D_s is the surface diffusivity, T is absolute temperature, and r_p is the pore radius. Thus, M_p is larger for smaller pores. Pore growth can also occur during sintering, either by the coalescence of pores at grain boundaries or by the diffusion of vacancies from smaller pores to larger pores rather than to grain boundaries. This process of pore growth during heating is referred to as *Ostwald ripening*. A rapid heating schedule may produce densification with a smaller concomitant grain size, owing to the different activation energies for the two processes (see Figure 7.36). Surface diffusion, which predominates at low temperatures, can cause grain coarsening. Fast heating can increase the vacancy diffusivity (D_v) and densification at a faster rate than diffusion that causes grain coarsening.

The atmosphere is also important in sintering. Gas trapped in closed pores will limit pore shrinkage unless the gas is soluble in the grain boundary and can diffuse from the pore. Alumina doped with MgO can be sintered to essentially zero porosity in hydrogen or oxygen atmospheres, which are soluble, but not in air, which contains insoluble nitrogen. The density of oxides sintered in air is commonly less than 98% and often only 92–96%. The sintering atmosphere is also important in that it may influence the sublimation or the stoichiometry of the principal particles or dopants.

Finally, sintering can also be used to densify glass particles. In the initial stage, viscous flow produced by the driving force of surface tension causes neck growth. For two glass spheres of uniform diameter, d, the initial shrinkage, $\Delta L/L_0$, is given by

$$\frac{\Delta L}{L_0} = \frac{3\gamma_s t}{2\mu d} \qquad (7.29)$$

where γ_s is the solid surface energy of the glass, μ is the viscosity, and t is the isothermal sintering time. In the intermediate sintering stage, it can be assumed that gas diffuses rapidly through the interconnected pores and that the rate of pore shrinkage is proportional to γ_s/μ, which is highly temperature-dependent. Sintering occurs very rapidly when the glass is heated to its softening point of $10^{9.6}$ cpoise (cf. Figure 4.2), but gravity-induced flow causes slumping at this temperature. Closed pores form in the glass upon entering the final stage of sintering. The densification rate, dD/dt, of uniform spherical pores dispersed in an isotropic, incompressible viscous medium is described by the Mackenzie–Shuttleworth equation:

$$\frac{dD}{dt} = \frac{K\gamma_s n^{1/3}(1 - D)^{2/3} D^{1/3}}{\mu} \qquad (7.30)$$

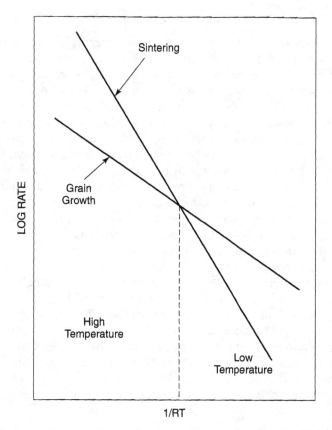

Figure 7.36 Temperature dependence of grain growth and sintering (densification) rates. The slope of each curve represents the activation energy. Adapted from J. S. Reed, *Principles of Ceramics Processing*, 2nd ed. Copyright © 1995 by John Wiley & Sons, Inc. This material is used by permission of John Wiley & Sons, Inc.

where D is the bulk density fraction of the porous glass containing n pores per unit volume of glass, K is a geometry-dependent constant, γ_s is the glass solid surface energy, and μ is its Newtonian viscosity.

The sintering rate is higher than that predicted by Eq. (7.29) when sintering a crushed glass powder because the effective radius of curvature is smaller than the equivalent sphere radius $(d/2)$. Water vapor in the sintering atmosphere may reduce the viscosity and increase the apparent densification rate. In the final stage, gas solubility and its diffusion in the glass and pore coalescence affect the final pore shrinkage. Finer pores have a larger driving force for shrinkage. An increase in the gas pressure due to an increase in temperature or from gas evolved from within the glass, or a reduction of the driving force due to pore coalescence, can cause pore enlargement and a volume expansion. Some glasses may be crystallized after being densified by sintering, to produce a glass ceramic.

There are many other aspects of ceramic and glass sintering that are beyond the scope of this text, including vitrification, sintering in the presence of liquid, the sintering of glazes and thick films, and sintering during hot pressing. The interested reader is directed to one of the many fine references on this topic at the end of the chapter.

7.2.4 Vapor Phase Synthesis and Processing*

In this section we investigate a class of processing techniques that are used to form a variety of components for the electronics, communications, and advanced ceramics industries. There are many common elements to these techniques, but the unifying theme is that they all involve the chemical reaction of gaseous, or vapor phase, components to form a solid compound, typically a solid ceramic. The formation of a nonvolatile solid, typically but not exclusively in film form upon a substrate, by the reaction of vapor phase chemical reactants is called *chemical vapor deposition* (CVD). What makes CVD processing so unique, and of great interest to chemical and materials engineers, is that it involves many of the topics we have covered in the previous chapters: bond strength, defect structure, thermodynamics, kinetics, diffusivity, and electrical and electronic properties. Perhaps more so than any other materials processing technique, all of the fundamental concepts of materials engineering must be brought to bear upon this topic. As a result, concepts from previous chapters will be introduced with relatively little cross-referencing, and the reader is advised to review pertinent concepts from earlier in this text when necessary.

We begin by describing the general topic of the vapor phase processing of structural ceramics such as carbides and nitrides, since the synthesis of these components does not involve other significant processing steps. We then look specifically at the vapor phase processing of semiconductor devices. This description will take us through some non-CVD processes such as the growth of single crystals from the melt and the patterning of electronic materials using optical techniques, but it is better that these topics remain as closely linked as possible. Finally, we look at how the CVD process can be modified and coupled with glass fiber drawing techniques to form optical fiber that is used for the communications industry.

7.2.4.1 *Fundamentals of Vapor Phase Synthesis.*

In this section we will concentrate on the vapor phase synthesis of some structural ceramics, such as carbides and nitrides. The principles described here apply equally well to the production of oxide ceramics, but we reserve some of this description for later sections, particularly with respect to the formation of optical fibers.

In general, carbides, nitrides, and borides are manufactured in the vapor phase in order to form high-purity powders. This procedure is fundamentally different than a strict CVD process, since in powder synthesis reactors, deposition on seed particles may be desirable, but deposition on the reactor walls represents a loss of product material. As we will see, in CVD, heterogeneous deposition on a surface will be sought. Aside from this issue of deposition, many of the thermodynamic and kinetic considerations regarding gas phase reactions are similar.

Once the particles have been produced by gas-phase synthesis, they can be used for fabrication into components via methods described in previous sections, such as dry pressing (Section 7.2.1) and slip casting (Section 7.2.2). These components are used in structural and electronic applications where wear resistance (e.g., WC), high-temperature oxidation resistance (e.g., SiC), high hardness (e.g., B_4C), high ceramic toughness (e.g., Si_3N_4), or resistance to chemical attack (e.g., AlN) is required. Although many of these compounds can be formed by either carbothermal reduction of oxides (e.g., the reaction of TiO_2 with carbon to form TiC) or combustion synthesis of metals (e.g., $Ti + C \rightarrow TiC$), we will concentrate here on the use of gas-phase precursors, such as halides and hydrides, to form the compound of interest (e.g., the gas-phase reaction of $TiCl_4$ and CH_4

to form TiC). Some examples of precursor gases, resultant ceramic phases, processing temperatures, and particle characteristics are presented in Table 7.13.

The ceramic particles can be synthesized in the gas phase via a number of techniques, including *thermal aerosol, laser processing,* or *plasma generation.* All of these methods employ the same general principle of gas-phase reactions, but differ slightly in how energy is supplied to the reaction and how particles are removed from the system. For simplicity, we will concentrate on the thermal aerosol process. Depending on the phase of the raw materials, aerosol processes can be classified into two main categories: *gas-to-particle conversion* or *particle-to-particle conversion.* Since the precursors we are interested in are gas phase, we will consider only the gas-to-particle synthesis route.

A schematic illustration of the gas-to-particle conversion route is shown in Figure 7.37. Precursor vapors react at high temperatures to form molecules of intermediate

Table 7.13 Some Structural Ceramics Formed in Gas-Phase Synthesis

	Precursors	Temperature [K (°C)]	Aggregate/ Primary Particle Diameter (μm)	Specific Surface area (m^2/g)	Purity/Phase Composition
AlN	Al, N$_2$, NH$_3$	900–2273 (627–2000)	0.05–1	1–40	0.5 < O$_2$ < 8 wt%
AlN	AlCl$_3$, NH$_3$	800–3073 (527–2800)	0.1–0.7	—	Wurtzite lattice, 97.5%
AlN	Al(C$_2$H$_5$)$_3$, NH$_3$	873–1373	0.013–0.025	20–145	Amorphous, 0.9 wt% O$_2$
	Al(i-C$_4$H$_9$)$_3$, NH$_3$	(600–1100)	0.011	173	
	Al(CH$_3$)$_3$, NH$_3$		0.019	95	18.8 wt% O$_2$
Si$_3$N$_4$	SiCl$_4$, NH$_3$	873–2773 (600–2500)	0.05–0.2	—	80% α- and 20% β-Si$_3$N$_4$
Si$_3$N$_4$	SiH$_4$, NH$_3$	823–1193 (550–920)	0.1	8–24	Amorphous
BN	BCl$_3$, NH$_3$	3073 (2800)	0.1	—	Hexagonal BN above 1373 K (1100°C)
SiC	SiCl$_4$, CH$_4$	1810 (1537)	—	—	—
SiC	SiH$_4$, CH$_4$, H$_2$	1473–1673 (1200–1400)	0.05	—	β-SiC, excess Si
SiC	SiO$_2$, C, N$_2$	1023–1773, 2073 (750–1500, 1800)	0.02–0.1	10–15	β-SiC
B$_4$C	B$_2$O$_3$, C	2073–2573 (1800–2300)	1–1.5	12–32	Crystalline
TiC	TiCl$_4$, CH$_4$	973–2273 (700–2000)	—	10.2	Cubic TiC
WC	WCl$_6$, CH$_4$	673–1773 (400–1500)	0.02–0.06	—	WC, W$_2$C
TiB$_2$	TiCl$_4$, BCl$_3$	1173 (900)	0.1	13.7	Titanium diboride

Source: *Carbide, Nitride and Boride Materials Synthesis and Processing*, A. W. Weimer, ed. Copyright © 1997 by Chapman & Hall.

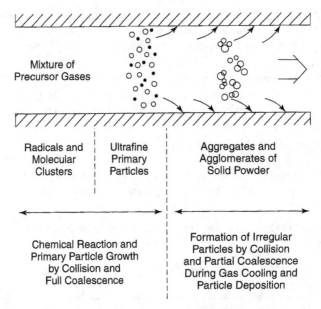

Mixture of
Precursor Gases

Radicals and Molecular Clusters	Ultrafine Primary Particles	Aggregates and Agglomerates of Solid Powder

Chemical Reaction and Primary Particle Growth by Collision and Full Coalescence

Formation of Irregular Particles by Collision and Partial Coalescence During Gas Cooling and Particle Deposition

Figure 7.37 Schematic illustration of gas-to-particle synthesis via thermal aerosol processing. From *Carbide, Nitride, and Boride Materials Synthesis and Processing*, A. W. Weimer, ed., p. 308. Copyright © 1997 by Chapman & Hall, London, with kind permission of Kluwer Academic Publishers.

and, eventually, product species. These clusters result in particles either by nucleation or by uninhibited coagulation. As the gas stream leaves the high-temperature zone and is cooled, particles grow by surface reactions and/or coagulation. During transport, particles can be removed from the process stream by diffusion and thermophoresis to the reactor walls. Eventually, the particles are collected by filtration or electrostatic precipitation. The residence time and process temperature play key roles in determining the extent of particle sintering, morphology, and crystallinity. Let us examine each of these process steps in more detail.

Thermodynamic analyses of various chemical pathways are useful in selecting appropriate CVD gaseous precursors. Thermodynamic equilibrium models provide a clear indication of the driving forces involved in gas-phase synthesis and help estimate reverse reaction rates when limited kinetic information is available. The thermodynamic predictions can also be used to assess the likelihood that a particular chemical route will be possible, as well as to identify phases that are likely to be produced. Equilibrium calculations can be performed in two different ways: (a) using equilibrium constants for individual reactions or (b) through global Gibbs free energy minimization. The former is most useful in examining a single chemical reaction—for example, for assessing the feasibility of a particular chemical route. The latter method is used for more complex systems and involves minimization of the Gibbs free energy subject to constraints on the numbers of moles of the various elements in the system, enthalpy or temperature, and pressure. We will briefly review the first method as it is applied to the gas-phase synthesis of silicon nitride, Si_3N_4.

Recall from Section 2.1 that chemical equilibrium is the thermodynamically stable state of a system; recall from Section 3.0 that the kinetics dictate how rapidly this process might occur. An example of a highly favorable reaction that is too slow to be of use in ceramic synthesis at modest temperatures is the reaction of silane, SiH_4, with molecular nitrogen to produce silicon nitride:

$$SiH_4(g) + 2N_2(g) \rightleftharpoons Si_3N_4(s) + 6H_2(g) \tag{7.31}$$

where the double arrow is included to remind us that the reaction is reversible and that an equilibrium exists between products and reactants. The enthalpy of formation for this reaction at 298 K is $\Delta H_R = -807$ kJ/mol. In fact, the Gibbs free energy change for this reaction, ΔG, is highly negative at all temperatures, indicating a spontaneous reaction (cf. Table 2.1). However, the reaction does not proceed rapidly because of the difficulty in breaking the nitrogen–nitrogen bond, for which the bond energy is about 950 kJ/mol. A more practical synthetic route is the reaction of silane with ammonia:

$$3SiH_4(g) + 4NH_3(g) \rightleftharpoons Si_3N_4(s) + 12H_2(g) \tag{7.32}$$

for which the free energy change is still highly negative. However, initiation of the reaction is more favorable than with nitrogen due to the lower energy required to break the N–H bond, for which the bond energy is only about 400 kJ/mol (see Appendix 1). Once some NH_2 radicals have been produced, the energy barrier is reduced even further. A thermodynamic comparison of some of the gas-phase reactions used to form Si_3N_4 is given in Table 7.14. Once the most desirable precursors have been identified, a system that will supply them to the reactor must be designed. For precursors that are gases under normal conditions, this process is simple. Vapors are readily obtained from liquids by bubbling an inert gas carrier such as Ar through the liquid.

From a kinetic standpoint, the overall formation reaction of Eq. (7.32) gives us little indication as to how the reaction actually proceeds. Unlike most of the condensed-phase reactions of Chapter 3, gas-phase reactions can only take place when molecules come close enough together that sufficient energy can be directly exchanged, resulting in bond breakage. For the small molecules of interest for high-temperature synthesis of ceramics, centers of mass of the molecules must approach one another within a few angstroms. At low pressures, such close interactions occur as discrete, short-lived collisions between molecules, on the order of 10^{-12} to 10^{-13} s. At the temperatures and

Table 7.14 Overall Reactions and Thermodynamic Data for the Gas-Phase Synthesis of Si_3N_4 from Selected Precursors

Reaction	ΔH_R (298 K) (kJ/mol)	ΔG_R (298 K) (kJ/mol)	T_G (K)	$\Delta H_R(T_G)$ (kJ/mol)
$3SiCl_4 + 2N_2 + 6H_2 \rightleftharpoons Si_3N_4(s) + 12HCl$	112.9	54.6	618	93.4
$3SiCl_4 + 4\ NH_3 \rightleftharpoons Si_3N_4(s) + 12HCl$	296.5	120.1	498	299.9
$3H_2SiCl_2 + 4\ NH_3 \rightleftharpoons Si_3N_4(s) + 6HCl + 6H_2$	−110.9	−226.2	$\Delta G < 0$	—
$3SiH_4 + 4\ NH_3 \rightleftharpoons Si_3N_4(s) + 12H_2$	−652.8	−741.4	$\Delta G < 0$	—
$3SiH_4 + 2N_2 \rightleftharpoons Si_3N_4 + 6H_2$	−836.4	−806.9	$\Delta G < 0$	—

Source: Carbide, Nitride and Boride Materials Synthesis and Processing, A. W. Weimer, ed. Copyright © 1997 by Chapman & Hall.

pressure of interest, the times between collisions are much longer than the interaction times, on the order of nanoseconds. Collision between molecules is a necessary, but not sufficient, condition for a chemical reaction to proceed. The molecules must possess sufficient energy to react when they collide. Expressed in terms of the energy per mole, the fraction of the collisions that involve energy greater than the required activation energy, E_a, is given by $\exp(-E_a/RT)$, which results in the rate constant, k, of Eq. (3.12). The overall reaction rate is then dictated by the reaction stoichiometry (cf. Section 3.0.1) and consists of both a forward and a reverse reaction.

A complete kinetic description of the gas phase reactions leading to the formation of a ceramic material is a set of microscopic reactions and the corresponding rate coefficients. The net rate of formation of species j, r_j by chemical reactions is the sum of the contributions of the various reactions in the set of elemental steps called the *mechanism*:

$$r_j = \frac{d[j]}{dt} = \sum_i^{reactions} v_{ij} r_i \tag{7.33}$$

where v_{ij} are the stoichiometric coefficients of the individual reactions and r_i are the corresponding rates (cf. Section 3.0.1). Surface reactions, which must include both adsorption and desorption of reactants and products, introduce an additional level of complexity to the overall rate expression. Comprehensive modeling, therefore, requires treatment of the gas-phase convection and diffusion in addition to the description of the effects of chemical reactions. In this case, equations of conservation for both mass and momentum must be included.

The chemical reactions describe only part of the synthesis of ceramic powders from gas-phase reactions. The process generally begins with gaseous precursors and forms the condensed phase by homogeneous nucleation (i.e., by the aggregation of vapor molecules to form a thermodynamically stable particle) or by *coagulation* (i.e., particle–particle collisions). The principles of particle nucleation have been introduced in Sections 3.1.2.1 and 3.2.1. A common criterion for determining the importance of nucleation and coagulation is the thermodynamically critical radius of the product species. If the radius is smaller than the radius of a single molecule of the product species, then coagulation determines particle formation. Nonoxide ceramics have low equilibrium vapor pressures; hence, coagulation tends to control particle formation. In general, homogeneous nucleation is of limited importance for ceramic particles by gas-phase chemical reactions. Nucleation is the dominant mechanism for particle formation by cooling of superheated vapors.

Once particles have been formed, they grow by the combined effect of vapor deposition and continued coagulation. The transport of vapor species to the surfaces of the aerosol particles depends on the size of the aerosol particle relative to the mean free path of the gas molecules, λ; that is, it depends on the Knudsen number characteristic of vapor diffusion to the particle, Kn:

$$Kn = \frac{\lambda_{AB}}{d_p} \tag{7.34}$$

where λ_{AB} is the mean free path of the vapor molecule A in a background of B, and d_p is the characteristic size of the particle. Recall that there are three important regimes of the Knudsen number (Section 4.3.2.3). For particles much smaller than the mean free path, Kn $\gg$ 1, and the particles are in the free molecular regime in which the flux of vapor to the particle surface is the product of the particle surface area and the effusion flux. When Kn $\ll$ 1, the particle is said to be in the continuum size regime, and mass transfer to the particle surface is described by the diffusion equation [Eq. (4.4)]. At intermediate Knudsen numbers, mass transfer to spherical particles cannot be rigorously derived, and empirical correction factors must be used.

When d_p is much larger than the mean free path of the gas (Kn $\ll$ 1, continuum regime), the mean free path is given by

$$\lambda_{AB} = \frac{\mu}{0.499\rho\bar{c}} \tag{7.35}$$

where μ is the gas viscosity, ρ is the gas density, and $\bar{c}$ is the average velocity of the gas molecules:

$$\bar{c} = \left(\frac{8k_B T}{\pi m}\right)^{1/2} \tag{7.36}$$

where m is the molecular mass of the gas, k_B is Boltzmann's constant, and T is absolute temperature. In this instance, the particle diameter is given by

$$d_p = d_{p,0}(1 + N_0 k_c t)^{1/3} \tag{7.37}$$

where t is the residence time, N_0 and $d_{p,0}$ are the initial particle concentration and diameter, respectively, and k_c is a constant for coagulation in the continuum regime:

$$k_c = \frac{4k_B T}{3\mu} \tag{7.38}$$

where the variables are the same as in Eqs. (7.35) and (7.36).

When d_p is much smaller than λ(Kn $\gg$ 1, free molecular regime), it is given by

$$d_p^{5/2} = d_{p,0}^{5/2} + \frac{10\ V_{vol}t}{\pi}\left(\frac{6k_B T}{\rho_p}\right)^{1/2} \tag{7.39}$$

where V_{vol} is the total volume of powder per unit volume of gas, ρ_p is the density of the powder, and all other variables are as in Eq. (7.37). Increasing the residence time gives an approximately $t^{1/3}$ increase in diameter because when two particles collide, their volume doubles, but diameter goes up by a factor of $2^{1/3}$. Increasing the temperature results in a $T^{1/6}$ and $T^{1/3}$ increase in particle diameter in the free molecular and continuum regimes, respectively.

The deposition of particles on macroscopic surface is the primary goal in CVD processes, but reduces the efficiency of vapor phase particle synthesis. Particles can deposit by Brownian motion, but in high-temperature reactors, *thermophoretic deposition* often dominates. Thermophoresis is the migration of small aerosol particles as a result of a temperature gradient. It causes particles carried in a hot gas to deposit on a cool surface. For small particles, Kn $\ll$ 1, a dimensionless group can be created to describe thermophoresis, Th:

$$Th = \frac{v_T \rho T}{\mu(dT/dx)} \tag{7.40}$$

in which v_T is the thermophoretic velocity (essentially a constant), T is temperature, μ is the gas viscosity, and ρ is the gas density. Theoretical predictions suggest a value of Th between 0.42 and 1.5, and experimental measurements indicate that Th $\approx$ 0.5. In laminar flow aerosol reactors, thermophoresis can cause substantial losses of product particles.

Powders made by the gas-to-particle conversion route are usually agglomerates of fine, nonporous, primary particles. This route is most suitable for the synthesis of single-component, high-purity powders of small particle size, high specific surface area, and controlled particle size distribution. Temperature, reactor residence time, and chemical additives affect the particle size, extent of agglomeration, and, consequently, powder morphology. Short reactor residence times and rapid quenching of the product stream lead to agglomerates of fine primary particles. Sintering additives can also be effective in altering the morphology of the powders. Particle morphology may affect the rate of particle growth by coagulation. In the free molecular limit, agglomerates collide much faster than spheres of the same mass since they have more surface area for collisions. In the continuum limit, the enhanced collisional surface of agglomerates does not play a major role because agglomerates experience enhanced drag by the medium.

In addition to single-component powders, gas-phase synthesis can be used to produce composite powders, such as SiC and BN. An example of an apparatus used for this particular synthesis, which is representative of the apparatuses used for most gas-phase synthesis of particles, is shown in Figure 7.38.

7.2.4.2 *Semiconductor Device Fabrication.* In this section we investigate the gas-phase synthesis of compounds, primarily semiconductors, that are preferentially deposited on a surface to form a layer called a *thin film*. This technology can be used to form structural and protective layers of materials described in the previous section, but is used primarily to form thin films from semiconductor materials for electronic devices. Recall from Figure 6.99, for example, that most semiconductor devices are made from layers of appropriately doped compounds. In order to fabricate these devices at ever smaller scales, the layers must be formed in a carefully controlled manner. This

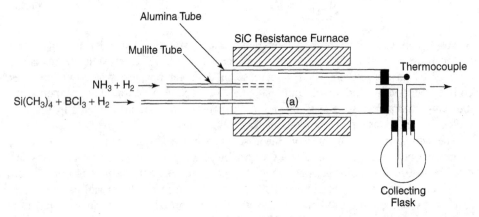

Figure 7.38 Schematic diagram of experimental apparatus used to form SiC/BN composite particles by gas-phase synthesis on a substrate (a). From *Carbide, Nitride, and Boride Materials Synthesis and Processing*, A. W. Weimer, ed., p. 336. Copyright © 1997 by Chapman & Hall, London, with kind permission of Kluwer Academic Publishers.

is most effectively performed using CVD processes. Most CVD processes are carried out in order to form layers of varying composition on an appropriate platform called a *substrate*. In addition to the GaAs substrate shown in Figure 6.99, substrates can be made of a variety of materials, but are typically composed of elemental silicon, Si. We will briefly describe the formation of Si wafers since they are such an integral component of the CVD process, even though the are formed from the melt rather than the vapor phase. We will then concentrate on the fundamental CVD processes that constitute layer formation upon the substrate, using specific examples. The examples will be selected as representative for the formation of semiconductor thin films. This will allow us to illustrate the important processing parameters and engineering principles involved in the formation of these components, while recognizing that specific material chemistries can, and inevitably will, change over time.

Silicon wafers are merely thin slices of a large, single crystal of silicon that is formed by a process known as *Czochralski growth* (see Figure 7.39). The material used in single-crystal silicon growth is *electronic grade polycrystalline silicon* (EGS), sometimes called *polysilicon*, which has been refined from quartzite, SiO_2, until it contains less than 1 ppb (part per billion) impurities. Given that the quartzite contains about eight orders of magnitude more contaminants than is required, the purification process is quite remarkable. In short, the formation of polysilicon involves four primary steps: reduction of quartzite to a *metallurgical grade silicon* (MGS) with a purity of about 98%; conversion of MGS to trichlorosilane, $SiHCl_3$; purification of $SiHCl_3$ by distillation; and CVD of Si from the purified $SiHCl_3$.

In the first step, quartzite is reacted with carbon in the form of coal, coke, or wood chips, to form silicon and carbon monoxide:

$$SiO_2(s) + 2C(s) \longrightarrow Si(s) + 2CO(g) \tag{7.41}$$

In the second step, trichlorosilane is formed by the reaction of the silicon with anhydrous hydrogen chloride, HCl. The reaction of the solid silicon and gaseous HCl occurs at 300°C in the presence of a catalyst. This step is important to purification, since chlorides of the aluminum and boron impurities also are formed. The resulting

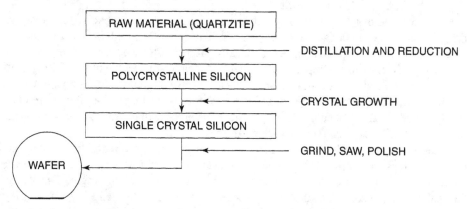

Figure 7.39 Process sequence used to form silicon wafer electronic substrates. Reprinted, by permission, from S. Wolf and R. N. Tauber, *Silicon Processing for the VLSI Era*, Vol. 1, p. 6. Copyright © 1986 by Lattice Press.

silicon trichloride is liquid at room temperature; thus it can be purified by distillation in the third step. In the final step, the purified silicon trichloride is converted back to electrical-grade polycrystalline silicon (EGS) by CVD in the presence of hydrogen. This process, called the *Siemens process*, is primarily comprised of the following reaction:

$$2\mathrm{SiHCl}_3(g) + 2\mathrm{H}_2(g) \longrightarrow 2\mathrm{Si}(s) + 6\mathrm{HCl}(g) \tag{7.42}$$

The substrate for this CVD process is a thin silicon rod, called a *slim rod*, which serves as a nucleation site for the depositing silicon. After deposition, the EGS, or polysilicon, is processed in the Czochralski (CZ) growth process.

In the CZ process, a fused silica crucible is loaded with a charge of EGS, and the growth chamber is evacuated (see Figure 7.40). The chamber is then refilled with inert gas, such as argon, and the charge is melted at 1421°C. Then, a small seed crystal of silicon, about 5 mm in diameter and 100–300 mm long, is lowered into the molten silicon. The seed crystal is then withdrawn at a controlled rate. Both the seed crystal and the crucible are rotated during the pulling process, but in opposite directions. The initial pull rate is relatively rapid, so that a thin neck is formed, which assists in forming a dislocation-free ingot. At this point the melt temperature is reduced and stabilized so that the desired ingot diameter can be formed. As the crystal solidifies and is pulled from the melt, the latent heat of fusion, ΔH_f, is transferred to the crystal along the solid–liquid interface. Heat is lost from the crystal surface due to radiation and convection. The maximum pull rate, V_{max}, is achieved when the maximum heat that can be lost by the ingot is being transferred to the ingot by the solidifying silicon, and it can be estimated from the following relation:

$$V_{max} = \frac{k_s}{\Delta H_f \rho} \left(\frac{dT}{dx} \right) \tag{7.43}$$

where ρ is the density of silicon, k_s is the thermal conductivity of silicon, and dT/dx is the temperature gradient along the axis of the ingot. The actual pull rate is 30–50% less than the maximum pull rate. Impurities are incorporated into the CZ crystal as it solidifies in order to create specific resistivities in the wafers. Common dopants include Al, As, B, C, Cu, Fe, O, P, and Sb.

Upon the completed growth of a silicon ingot, a variety of analysis steps are conducted before the ingot is sawn into wafers. The ingot is weighed and diameter is measured along its length. The ingot is visually inspected for evidence of any twinning or other gross crystalline imperfections. The dislocation-free attribute of CZ crystals refers to macroscopic (edge and screw) dislocations. Small dislocation loops, however, can be formed during the growth of the main crystal. After inspection, the entire crystal is etched in an acid bath and is ground to a precise diameter. The resistivity is measured at various points along its growth axis.

After a single-crystal ingot has been grown, a complex sequence of shaping and polishing steps are performed to create the wafers. A critical step in this *wafering* process is the grinding of one or more flats along the length of the ingot. The largest flat, called the *primary flat*, identifies a specific crystal direction and is located using diffracted X rays. Smaller flats, called *secondary flats*, are utilized to identify the orientation and conductivity type of the wafer, as illustrated in Figure 7.41. The wafers are then ground and polished.

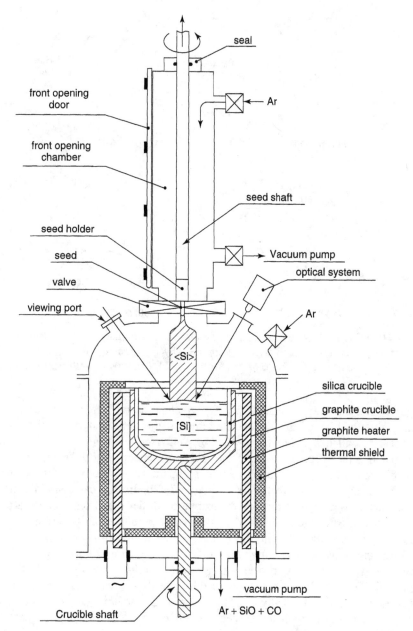

Figure 7.40 Schematic diagram of CZ silicon crystal puller. Reprinted, by permission, from W. Zuhlehner and D. Huber, Czochralski Grown Silicon, in *Crystals 8*. Copyright © 1982 by Springer-Verlag.

The orientation of the atomic planes in the single-crystal wafer is important for the formation of subsequent layers that are necessary to form the thin, multilayer structures illustrated in Figure 6.99. Typically, these layers are from 0.5 to 20 μm in thickness, and strict compositional control is necessary during their formation. The process of depositing these thin layers of single-crystal materials on the surface of

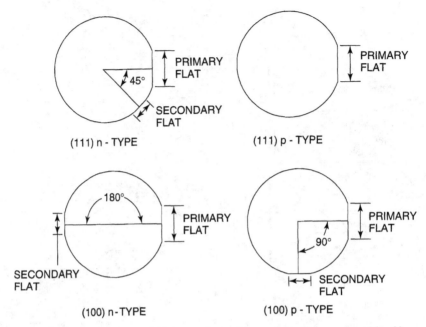

Figure 7.41 Schematic illustration of primary and secondary flats used to identify specific crystallographic planes in silicon wafers. Reprinted, by permission, from S. Wolf and R. N. Tauber, *Silicon Processing for the VLSI Era*, Vol. 1, p. 23. Copyright © 1986 by Lattice Press.

a single crystal substrate is called *epitaxy*, which is derived from two Greek words meaning "arranged upon." If the deposited layer is the same material as the substrate, the process is called *homoepitaxy*, and the result is two layers of exactly matched lattice spacings (see Figure 7.42a). If the layers are chemically different, the process is termed *heteroepitaxy*, and the lattice spacing of the two layers will not match, even if the film is laid down perfectly, as in Figure 7.42b. In this case, strain caused by lattice mismatch is alleviated through the formation of defects. These defects affect the electrical properties of the layers, but are necessary to reduce lattice strain.

The two most common techniques for making these thin films are *vapor-phase epi-taxy* (VPE) and *molecular-beam epitaxy* (MBE). In keeping with our general theme of chemical vapor deposition processing, we will describe only VPE here, since it is a form of CVD. In VPE, the substrate, in this case the Si wafer, is placed on induction suscepting support (typically graphite) under high vacuum. The wafers are heated through thermal contact with the suscepting support which is inductively heated (cf. Section 7.1.1.1). The precursor vapors are introduced through mass flow controllers (see Figure 7.43). This type of CVD reactor is known as a *cold-wall reactor*, because the substrates are heated directly and the reactor walls stay relatively cool. The advantage of such systems is that there is little or no deposition on the reactor walls, which allows for high precursor efficiency and low reactant depletion. A *hot-wall reactor* is simpler to construct, typically consisting of a resistance heating coil around the entire reactor, but deposition rates on the reactor walls are much higher. Before going on to the specific reactions and process conditions involved in VPE, it is important to recognize that the various layers in an integrated circuit can be, and are, formed with thin film techniques other than VPE, some of the more important of which are physical

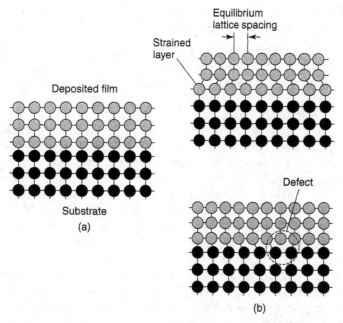

Figure 7.42   Illustration of (a) matched and (b) mismatched lattice spacings in epitaxial layers. Reprinted, by permission, from P. Ball, *Made to Measure*, p. 36. Copyright © 1997 by Princeton University Press.

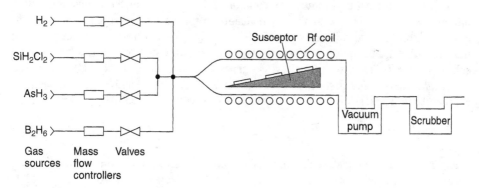

Figure 7.43   Schematic diagram of a typical vapor phase epitaxy apparatus. Reprinted, by permission, from S. A. Campbell, *The Science and Engineering of Microelectronic Fabrication*, p. 341. Copyright © 1996 by Oxford University Press.

deposition processes such as *evaporation* and *sputtering*. We will not describe these physical deposition techniques here, instead concentrating upon chemical deposition techniques such as VPE.

As illustrated previously in Figure 3.32, there are seven primary steps in the VPE–CVD process:

1. Gas-phase decomposition of the precursors
2. Transport of reactants to the substrate surface

3. Adsorption of the reactants to the substrate surface
4. Diffusion of reactants along the surface
5. Reaction/decomposition to form products
6. Desorption of products from the surface
7. Transport of products away from the surface

As pointed out earlier, CVD is a steady-state, but rarely equilibrium, process. It can thus be rate-limited by either mass transport (steps 2, 4, and 7) or chemical kinetics (steps 1 and 5; also steps 3 and 6, which can be described with kinetic-like expressions). What we seek from this model is an expression for the deposition rate, or growth rate of the thin film, on the substrate. The ideal deposition expression would be derived via analysis of all possible sequential and competing reactions in the reaction mechanism. This is typically not possible, however, due to the lack of activation or adsorption energies and preexponential factors. The most practical approach is to obtain deposition rate data as a function of deposition conditions such as temperature, concentration, and flow rate and fit these to suspected rate-limiting reactions.

An example of this approach was presented earlier in Figure 3.34, which contains Arrhenius plots (rate vs. $1/T$; cf. Section 3.0.2) at different total pressures. Figure 3.34 clearly shows the two types of deposition rate behavior. At low temperatures (higher $1/T$) the reaction kinetics are slow compared to mass transport, and the deposition rate is low. At higher temperatures (lower $1/T$) chemical kinetic processes are rapid compared to mass transport, resulting in a distinct change in slope and a higher deposition rate.

With this information in mind, we can construct a model for the deposition rate. In the simplest case, the rate of flux of reactants to the surface (step 2) is equal to the rate at which the reactants are consumed at steady state (step 5). All other processes (decomposition, adsorption, surface diffusion, desorption, and transport away from the substrate) are assumed to be rapid. It is generally assumed that most CVD reactions are heterogeneous and first order with respect to the major reactant species, such that a general rate expression of the form of Eq. (3.2) would reduce to

$$r = k_s[A]_s \tag{7.44}$$

where k_s is the rate constant for the surface reaction, and $[A]_s$ is the concentration of species A on the substrate surface. We will neglect the reverse reaction rate [Eq. (3.3)] for this analysis by assuming that it is negligible relative to the forward reaction. The rate of reactant transport to the substrate surface is given by the flux of component A, J_A^*, as given by Fick's Law in Eq. (4.4). [This is sometimes done by using a mass transfer coefficient, as in Eq. (4.79) instead of a diffusivity, but the principle is the same.] Here the concentration gradient can be approximated by $([A]_g - [A]_s)/l$, where l is the diffusion length and $[A]_g$ is the gas-phase concentration of species A. If we equate the reaction rate and mass transport rate expressions (steady-state assumption) and appropriately represent gas-phase concentrations as partial pressures (e.g., $[A]_g = p_A/T$ from the ideal gas law), it can be shown that the deposition rate, j [see Eq. (7.33)], is given by:

$$j = \left[\frac{1}{(1/D_A) + (1/k_s)}\right]\left(\frac{p_A}{T}\right) \tag{7.45}$$

Equation (7.45) clearly illustrates how the mass transfer and kinetic rates relate to each other (denominator in brackets); that is, when diffusion is the rate-limiting step (D_A is large and k_s is small), $1/k_s$ can be neglected and $j = D_A p_A / T$. The converse is true for rapid diffusion: $j = k_s p_A / T$. Of equal importance is the prediction that the deposition rate is proportional to the reactant gas concentration, represented as the partial pressure. This prediction is borne out by experiment for low concentrations, as illustrated in Figure 7.44 for the deposition of silicon from the reduction of $SiCl_4$ by H_2. At high concentrations, the deviations from linearity are extreme, and the reactant eventually begins to etch the substrate (negative growth rate).

Nearly all silicon epitaxy for integrated circuits is done by the reduction of chlorosilanes, $SiH_x Cl_{4-x}$ that are heavily diluted in hydrogen gas. The smaller the number of chlorine atoms in the precursor molecule, the lower the temperature required for the same growth rate. This temperature dependence on composition is particularly important in light of Eq. (7.45), since at high temperatures, mass transfer tends to be the rate-limiting step, and the growth rate becomes less temperature-sensitive as a result. The chlorosilanes all follow similar reaction paths. Figure 7.45 shows the equilibrium partial pressures for various species from silicon tetrachloride at one atmosphere for temperatures up to 1500 K. These values come from energy minimization calculations.

In order to control the conductivity type and carrier concentration of the epitaxial layers, dopants are added to the reactant gases (cf. Section 6.1.1.5). Dopants are typically introduced in the form of hydrides (e.g., B_2H_6, PH_3, and AsH_3) in low concentrations (10–1000 ppm). There are no simple rules to relate the incorporation of dopant atoms from the gas phase into the Si film, since the incorporation depends on many factors, including substrate temperature, deposition rate, dopant molar volume, and reactor geometry. Thus, the dopant/Si ratio in the film is different from that in the

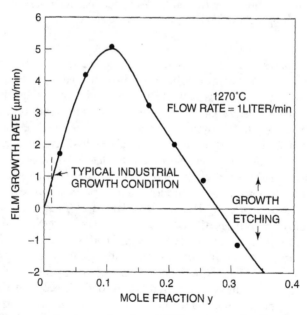

Figure 7.44   Silicon growth rate as a function of $SiCl_4$ concentration. Reprinted, by permission, from S. Wolf, and R. N. Tauber, *Silicon Processing for the VLSI Era*, Vol. 1, p. 127. Copyright © 1986 by Lattice Press.

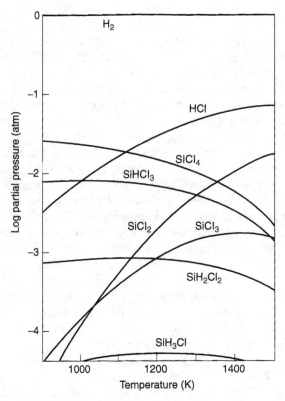

Figure 7.45   Equilibrium partial pressures in the Si–H–Cl system at 1 atm and Cl/H = 0.06. Reprinted, by permission, from S. A. Campbell, *The Science and Engineering of Microelectronic Fabrication*, p. 347. Copyright © 1996 by Oxford University Press.

gas phase. Nonetheless, epitaxial films with well-controlled dopant concentrations in the range of 10^{14}–10^{20} atoms/cm^3 can be routinely achieved.

Once layers of semiconductors have been deposited on the substrate, they can be patterned to form the necessary junctions for the integrated circuit to function properly. This is typically accomplished by an etching process, in which certain areas of the film are protected from etching by a crosslinked polymer called a *resist*. Since the polymer is often crosslinked by visible or ultraviolet (UV) light, the resist is often called a *photoresist*, and the process of imprinting a pattern onto a multilayer wafer in order to create an integrated circuit pattern is called *optical lithography*. As integrated circuits continue to decrease in size, the wavelength of light becomes increasingly important, and only electromagnetic radiation of very short wavelengths can reproduce the fine lines necessary. As a result, electron beam and X-ray lithography are growing in use and popularity. For our purposes, optical lithography adequately illustrates the concepts of circuit patterning. We will concentrate on the materials-related aspects of this process; the principles of circuit design are beyond the scope of this text, but are based upon the concepts of semiconductor devices similar to those illustrated previously in Figures 6.20 and 6.36.

The lithographic process is illustrated in Figure 7.46. The substrate, consisting of multiple layers of semiconducting compounds and metals, is covered with a crosslinkable

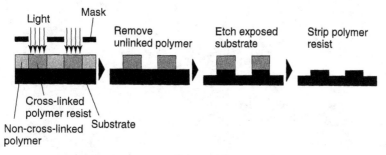

Figure 7.46 Schematic illustration of lithographic process. Reprinted, by permission, from P. Ball, *Made to Measure*, p. 39. Copyright © 1997 by Princeton University Press.

polymer. A negative pattern of the circuit, called a *mask*, is placed over the polymer, which is then exposed to light that causes the polymer to crosslink. The noncrosslinked portions of the polymer are removed, exposing certain portions of the semiconductor layers below. The exposed layers are removed by etching, while the crosslinked polymer protects the remaining areas of the circuit. Once the pattern has been etched, the crosslinked polymer is removed, and a patterned circuit remains.

Resists are categorized according to their polarity, or how the respond to the light during exposure. The example shown in Figure 7.46 is a *negative photoresist*; that is, the exposed regions of the polymer remain. *Positive photoresists* are also used, and in fact are preferred because they give better image resolution. In a positive photoresist, the exposure to light increases the solubility of the polymer, and the unexposed regions remain. Regardless of polarity, photoresists consist of three components: a resin or *base material*, a *photoactive compound* (PAC), and a solvent that controls the viscosity of the base, keeping it in the liquid state. In positive resists, the PAC acts as an inhibitor before exposure, slowing the rate at which the resist will dissolve when placed in the developing solution, or *developer*. Upon exposure to light, the inhibitor becomes a sensitizer, increasing the dissolution rate of the resist.

The most popular positive resists are referred to as *DQN*, corresponding to their photoactive compound (DQ) and base material (N), respectively. DQ stands for diazoquinone, and N stands for novolac resin. The repeat unit of a novolac resin is shown in Figure 7.47. Novolacs normally dissolve in aqueous solutions, but combinations of

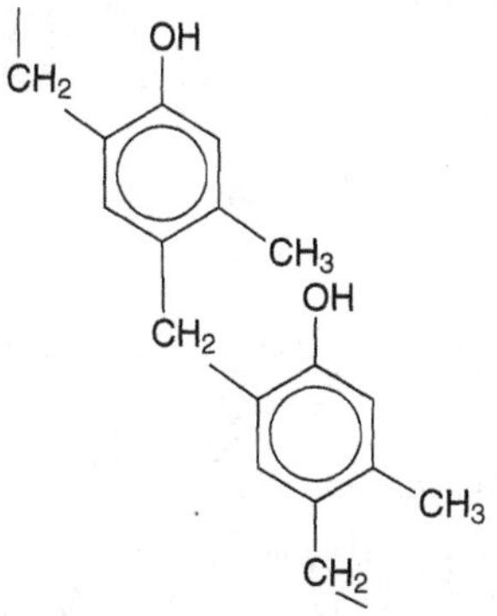

Figure 7.47 Chemical structure of a novolac polymer.

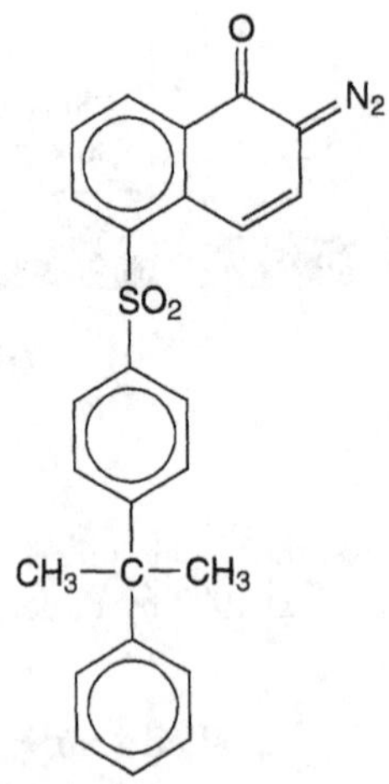

Figure 7.48 Chemical structure of diazoquinone.

aromatic compounds such as xylene and acetates are typically used as solvents. The structure of DQ is shown in Figure 7.48. The part of the molecule below the SO_2 group plays only a secondary role in the exposure process and can vary, so we will represent it with the generic group R. The PAC acts as an inhibitor, reducing the dissolution rate of the base in the developer by a factor of 10 or more. This occurs by chemical bonding of the PAC to the novolac at the surface of the resist where it is exposed to the developer, although the exact mechanism is not fully understood. The nitrogen group in the PAC is weakly bonded. As shown in Figure 7.49, the addition of UV light will free the nitrogen molecule from the carbon ring, leaving behind a highly reactive carbon site. One way to stabilize the structure is to move one of the

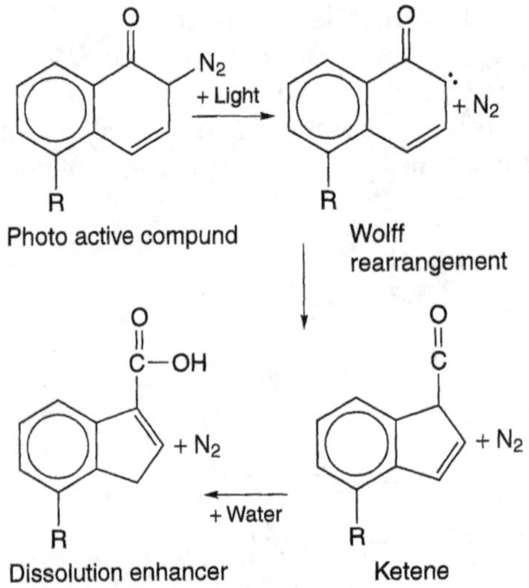

Figure 7.49 Photolysis and subsequent reactions of DQ upon UV exposure. Reprinted, by permission, from S. A. Campbell, *The Science and Engineering of Microelectronic Fabrication*, p. 185. Copyright © 1996 by Oxford University Press.

carbons outside the ring. The oxygen atom is then covalently bonded to this external carbon atom. This process is known as a *Wolff rearrangement*. The resultant molecule shown in Figure 7.49 is called a *ketene*. In the presence of water, a final rearrangement occurs in which the double bond to the external carbon atom is replaced with a single bond and an –OH group. This final product is a carboxylic acid. This process works as a PAC because the starting material will not dissolve in a basic solution. If the PAC is added to the resin in a 1:1 mixture, the photoresist is almost insoluble in basic solutions. Only light, water, and the ability to remove the nitrogen gas are required to drive this process. Typical developer solutions are KOH or NaOH diluted with water.

One of the great advantages of DQN photoresists is that the unexposed areas are essentially unchanged by the presence of the developer, since it does not penetrate the resist. Thus, a pattern of small lines on a clear field that are imaged onto a positive resist keep their linewidth and shape. Most negative resists work by crosslinking reactions, such as an azide-sensitized polyisoprene. Negative photoresists have very high photospeeds and adhere well to the wafer without pretreatment, but suffer from swelling, which broadens linewidths. As a result, negative resists are generally not suited for features less than 2.0 μm.

The mask is made of various types of fused silica. Each photomask contains an image of one layer of the process and may be the same size as the finished chip, or an integral factor of that size. The most important properties for the masks include a high degree of optical transparency, a small thermal expansion coefficient, and a flat, highly polished surface that reduces light scattering. On one surface of this glass is a patterned opaque layer, which on most masks is made of chromium. The imprinting of the pattern on the mask is a critical step, since any defect in the mask large enough to be reproduced will be on every wafer made from the mask.

Once the device is patterned, its elements must be electrically connected. This is usually done with a highly conductive metal layer, which is applied over an additional insulating layer. Thus, the photolithographic and VPE processes are alternated, with other deposition techniques, to produce patterned layers. A simple resistor may require four layers, which uses three photolithographic steps, three etch steps, and four thin-film deposition steps. A similar set of steps is required for a capacitor, and transistors require still more steps. However, the benefit of this painstaking process is that multiple chips, each containing tens of millions of transistors, can be fabricated on one wafer in a highly reproducible manner. The chips are then cut from the wafers. This process continues to decrease in size scale and increase in complexity and efficiency, but the principles of photolithography and vapor-phase epitaxy still apply.

7.2.4.3 Modified CVD Processes for Optical Fiber Production.
The principles of chemical vapor deposition can also be applied to the production of optical fibers. The fundamentals of fiber optics were described in Section 6.3.2.7. We concentrate here on the application of CVD to their production.

The development of optical fiber is a fascinating story, especially with regard to the introduction of CVD as the primary method of fabrication, but the details are beyond the scope of this text. Suffice to say that traditional glass-forming technologies, which were based on melt processing, did not provide fibers free from defects such as bubbles or with sufficient purity to prevent enormous attenuation losses. The lowest losses of optical quality glasses were on the order of 1000 dB/km, whereas optical fibers today are produced using CVD technology with less than 0.2 dB/km. Let us investigate how this is possible.

Recall that optical fibers function on the basis of refractive-index differences between a core and a cladding material that results in the phenomenon of total internal reflectance. The refractive index can change abruptly between the core and cladding, as in a step-index optical fiber or, gradually, as in a graded-index optical fiber. Regardless of the manner in which the refractive index changes across the fiber cross section, there is a need to change the refractive index of the glass from the center to the perimeter of the fiber. This is done by doping. As we saw with the production of semiconductors in the previous sections, chlorosilane precursors are easily doped to controlled levels with other gases. However, instead of reacting the chlorosilanes with hydrogen to reduce the gases to pure silicon, the precursor gases are oxidized to form SiO_2, which is the primary component of the optical glasses. The dopant is typically GeO_2, which can be derived from germanium-containing analogues to the chlorosilanes, such as $GeCl_4$.

Though there are multiple methods and process variations for the production of GeO_2–SiO_2 optical fibers using vapor deposition, we will concentrate on what is termed *inside processing*. Inside processing utilizes a quartz (SiO_2) tube, into which high-purity precursor gases are flowed. The tube is mounted on a lathe, which rotates the tube, while it is heated by a traversing oxygen–hydrogen torch. The precursor gases react by a homogeneous gas-phase reaction to form amorphous particles that are deposited downstream from the *hot zone*, as illustrated in Figure 7.50. The heat from the moving torch sinters the material to form a pure, glassy coating on the inside of the tube. Typical deposition temperatures are sufficiently high to sinter the deposited material, but not so high as to deform the substrate tube. The torch repeatedly traverses the length of the tube, simultaneously moving the hot zone, thereby building up layer upon layer of deposited material. The composition of the layer can be varied during each traversal to form the desired structure. Typically, 30–100 layers are deposited to make either single-mode or graded index multimode fiber.

This CVD procedure is somewhat different from that used to deposit semiconductor layers. In the latter process, the primary reaction occurs on the substrate surface, following gas-phase decomposition (if necessary), transport, and adsorption. In the fiber optic process, the reaction takes place in the gas phase. As a result, the process is termed *modified chemical vapor deposition* (MCVD). The need for gas-phase particle synthesis is necessitated by the slow deposition rates of surface reactions. Early attempts to increase deposition rates of surface-controlled reactions resulted in gas-phase silica particles that acted as scattering centers in the deposited layers, leading to attenuation loss. With the MCVD process, the precursor gas flow rates are increased to nearly 10 times those used in traditional CVD processes, in order to produce GeO_2–SiO_2 particles that collect on the tube wall and are vitrified (densified) by the torch flame.

In the final step, the tube is removed from the lathe after sufficient layers have been deposited, and the entire preform is heated to the softening point of the quartz tube. The tube collapses and is drawn into fiber. In this process, the inner core that remains after MCVD layering is removed. Typical preforms are on the order of 1 m in length, from which hundreds of kilometers of continuous optical fiber can be fabricated.

The chemistry of the $SiCl_4$ and $GeCl_4$ gas-phase reactions warrants further description. The oxidation reactions under consideration are as follows:

$$SiCl_4(g) + O_2(g) \longrightarrow SiO_2(g) + 2Cl_2(g) \qquad (7.46)$$

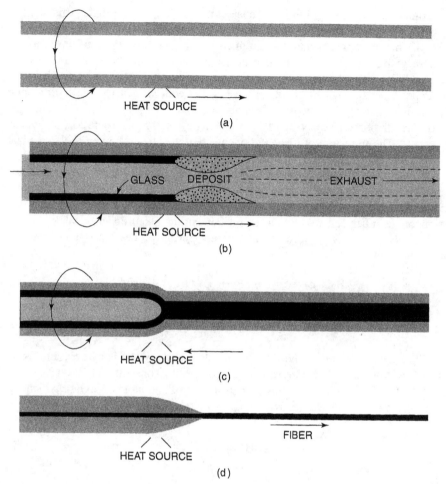

Figure 7.50   Schematic illustration of MCVD processing of optical fiber, including (a) tube setup, (b) deposition, (c) collapse, and (d) fiber drawing. Reprinted, by permission, from J. B. MacChesney, *J. Mater. Ed.*, **11**(14), 343 (1989). Copyright © 1989 by Materials Research Laboratory.

and

$$GeCl_4(g) + O_2(g) \longrightarrow GeO_2(g) + 2Cl_2(g) \qquad (7.47)$$

Equilibrium constants for these reactions can be written in accordance with Eq. (3.4). At low temperatures ($<1325°C$) the extent of the reactions for $SiCl_4$ and $GeCl_4$ are controlled by the reaction kinetics, while at higher temperatures thermodynamic equilibrium dominates. Experimental equilibrium constants have been determined for reactions (7.46) and (7.47), and oxide formation is strongly favored for Eq. (7.46), as indicated by the absence of detectable $SiCl_4$ concentrations in the effluent gases. On the other hand, $GeCl_4$ oxidation has an equilibrium constant less than 1.0 at temperatures above $1125°C$, which means that at equilibrium, only part of the germanium is present as GeO_2. The presence of significant Cl_2 concentrations from reaction (7.46) will further shift the equilibrium to the left in reaction (7.47).

Turning now from equilibrium considerations, let us investigate the deposition of gas-phase particles, called *soot*, on the tube walls. The SiO_2 soot particles are 0.02–0.1 μm in diameter and are entrained in the gas flow. Without the proper temperature gradient, they would remain in the gas stream and be exhausted. However, the traveling hot zone produces a temperature gradient so that the particles drift toward and deposit on the wall by thermophoresis (cf. Section 7.2.4.1). The thermophoretic deposition efficiency for this process is about 60%.

Another important aspect of MCVD chemistry is the incorporation of hydroxide impurities. The reduction of OH^- in optical fibers has been the key to the realization of low attenuation in the 1.3- to 1.5-μm wavelength region. Hydroxide contamination comes from three sources: diffusion of OH^- from the substrate tube during processing; impurities in the precursor gases, including the oxygen carrier gas; and contamination from leaks in the chemical delivery system. Diffusion of OH^- into the active optical region of the fiber can be eliminated by deposition of sufficiently thick, low OH^- cladding. The remaining hydroxide that enters during processing is governed by the following reactions:

$$H_2O(g) + Cl_2(g) \longrightarrow \tfrac{1}{2}O_2(g) + 2HCl(g) \qquad (7.48)$$

and

$$H_2O(g) + [\text{Si–O–Si}](s) \longrightarrow 2[\text{Si–OH}](s) \qquad (7.49)$$

where the square brackets in Eq. (7.49) indicate that the water is hydrolyzing the Si–O network in the silica. The corresponding equilibrium constants [cf. Eq. (3.4)] can be used to solve for the concentration of hydrolyzed silicon atoms, [SiOH], which, while technically a solid with unit activity, has a small concentration:

$$[\text{SiOH}] = \frac{P_{H_2O}\, P_{O_2}^{1/4}}{P_{Cl_2}^{1/2}} \qquad (7.50)$$

where p_i is the partial pressure of component i. Equation (7.50) tells us that one way to reduce the hydrolyzed silicon concentration is to increase the free chlorine partial pressure. There is typically 3–10% Cl_2 present in the MCVD process due to the reactions in Eqs. (7.46) and (7.47), which is sufficient to reduce the [SiOH] by a factor of about 4000 to a level of around 10^{-8}. However, during collapse of the tube, chlorine is typically not present, and significant amounts of OH^- can be incorporated. This dependence is shown in Figure 7.51 for an initial H_2O partial pressure of 10 ppm.

There are numerous variations to this MCVD process, including outside deposition on a mandrel, plasma CVD, and even fiber derived from sol–gel-processing, which will be described in the next section.

7.2.5 Sol–Gel Synthesis and Processing*

In addition to the formation of ceramic particles in the gas phase, particles can be formed in the liquid phase and consolidated via solvent evaporation to form useful products. Unlike slurry-based processes in which no liquid-phase reaction occurs, the processing of ceramics via the *sol–gel* method involves several important reactions. And like gas-phase reactions, the ceramics are formed from precursors that contain the component ions for the ceramic.

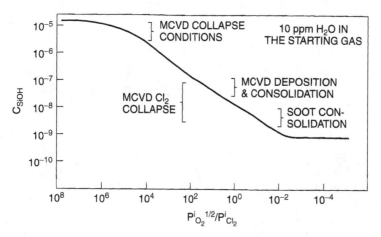

Figure 7.51 Relationship between oxygen and chlorine partial pressures and hydroxide concentration in optical glass. Reprinted, by permission, from J. B. MacChesney, *J. Mater. Ed.*, **11**(14), 325–356 (1989). Copyright © 1989 by Materials Research Laboratory.

The concept of a gel was briefly introduced in Section 1.2.4. There we saw that a silica gel was amorphous in a manner similar to a melt-derived silica. Unlike the melt processing route, however, the sol–gel method involves only moderate temperatures. In sol–gel processing, colloidal particles or molecules in a suspension are mixed with a liquid that causes them to join together in a continuous network. The definition of a *sol* is somewhat ambiguous, but, in general, the term is used to refer to the aforementioned colloidal suspension of particles that is sufficiently fluid to cause flow, but is stable for a long period of time. The rigid solid formed from the evaporation of solvent after the sol has been formed into a continuous network is called a *gel*. The gel can be dried, calcined, and milled to form a powder, or formed into a desired shape in the sol state. Some examples of how the sol–gel process can be used to form films or dense ceramics are given in Figure 7.52. Let us examine the sol-to-gel transition in more detail.

Sol–gel processing almost exclusively involves *metal alkoxides* as precursor chemicals. An alkoxide is an alkane with an oxygen interposed between at least one of the carbon atoms and the metal—for example, $(OC_2H_5)_4Si$, known as tetraethoxysilane (TEOS, a.k.a. tetratethyl orthosilicate). Other common alkoxide precursors are listed in Table 7.15. TEOS is used to produce the sol–gel-derived silica that generated the X-ray diffraction pattern in Figure 1.50. Let us use silica as an example to study the steps in the sol–gel process.

Silica gel is produced by the *hydrolysis, dehydration,* and *polymerization* (gelation) of TEOS in solution according to the following reactions:

$$(OC_2H_5)_4Si + H_2O \xrightarrow{\ hydrolysis\ } (OC_2H_5)_3SiOH + C_2H_5OH \tag{7.51}$$

$$(OC_2H_5)_3SiOH + OH^- \xrightarrow{\ dehydration\ } (OC_2H_5)_3SiO^- + H_2O \tag{7.52}$$

$$(OC_2H_5)_3SiO^- + (OC_2H_5)_3SiOH \xrightarrow{\ gelling\ } (OC_2H_5)_3Si\text{–}O\text{–}Si(OC_2H_5)_3 + OH^- \tag{7.53}$$

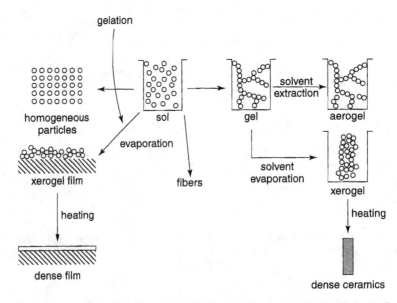

Figure 7.52   Some examples of sol-gel processing to form green ceramic products. Reprinted, by permission, from H. Yanagida, K. Koumoto, and M. Miyayama, *The Chemistry of Ceramics*, p. 147. Copyright © 1996 by John Wiley & Sons, Inc.

When a weak acidic solution of HCl and water is added to TEOS, the TEOS hydrates and forms a stable sol which contains short molecular chains called *oligomers* that contain a small number of OH^- groups [Eq. (7.51)]. Further polymerization leads to an Si–O–Si network called a *siloxane* (see Figure 7.53a). Thin films of SiO_2 can be formed by heating substrates coated with the sol by the dipping method. By increasing the polymerization of the sol at room temperature to 80°C, the sol can be made more viscous and can be drawn into fibers. By heating these threads to 400–800°C, SiO_2 fibers can be obtained through dehydration and gelation reactions [Eqs. (7.52) and (7.53)]. Silica glass can also be obtained by heating the sol to 800°C. When TEOS is hydrated with a large amount of an acidic solution, it tends to form a polymer network rather than polymer chains. A similar result occurs when TEOS is hydrated with a basic solution. Here, the reaction proceeds almost completely to $Si(OH)_4$, which can be further polymerized to form a three-dimensional network of the polymer, as illustrated in Figure 7.53b.

When multicomponent alkoxide solutions, or a single alkoxide and a soluble inorganic salt, are mixed, a multicomponent alkoxide may result. In this way, such complex oxides such as the YBCO superconductor (cf. Section 6.1.2.4) can be formed. Sol–gel processing can also be used to coat fibers for composites and to form ceramics with very fine pore sizes called *xerogels*. A xerogel commonly contains 50–70% porosity, a pore size of 1–50 nm, and a specific surface area exceeding 100 m^2/g.

7.3   PROCESSING OF POLYMERS

Just as was the case for a metal, a polymer must have formability in order to be shaped. Although an exact definition of formability is as difficult to give for polymers as it was for metals, it involves such polymer physical properties as viscosity, crystallinity, and

Table 7.15   Some Metal Alkoxides Used for Sol–gel Processing of Ceramics

		Alkoxide
Single cation alkoxides		
I A (1) group	Li, Na	$LiOCH_3$ (s), $NaOCH_3$ (s)
I B (11) grp.	Cu	$Cu(OCH_3)_2$ (s)
II A (2) grp.	Ca, Sr, Ba	$Ca(OCH_3)_2$ (s) $Sr(OC_2H_5)_2$, $Ba(OC_2H_5)_2$ (s)
II B (12) grp.	Zn	$Zn(OC_2H_5)_2$ (s)
III A (3) grp.	B, Al, Ga	$B(OCH_3)_3$ (l) $Al(i-OC_3H_7)_3$ (s) $Ga(OC_2H_5)_3$ (s)
III B (13) grp.	Y	$Y(OC_4H_9)_3$
IV A (4) grp.	Si, Ge	$Si(OC_2H_5)_4$ (l) $Ge(OC_2H_5)_4$ (l)
IV B (14) grp.	Pb	$Pb(OC_4H_9)_4$ (s)
V A (5) grp.	P, Sb	$P(OCH_3)_3$ (l) $Sb(OC_2H_5)_3$ (l)
V B (15) grp.	V, Ta	$VO(OC_2H_5)_3$ (l) $Ta(OC_3H_7)_5$ (l)
VI B (16) grp.	W	$W(OC_2H_5)_6$ (s)
lanthanide	La, Nd	$La(OC_3H_7)_3$ (s) $Nd(OC_2H_5)_3$ (s)
Alkoxides with various alkoxyl groups		
	Si	$Si(OCH_3)_4$ (l) $Si(OC_2H_5)_4$ (l) $Si(i-OC_3H_7)_4$ (l) $Si(t-OC_4H_9)_4$
	Ti	$Ti(OCH_3)_4$ (s) $Ti(OC_2H_5)_4$ (l) $Ti(i-OC_3H_7)_4$ (l) $Ti(OC_4H_9)_4$ (l)
	Zr	$Zr(OCH_3)_4$ (s) $Zr(OC_2H_5)_4$ (s) $Zr(OC_3H_7)_4$ (s) $Zr(OC_4H_9)_4$ (s)
	Al	$Al(OCH_3)_3$ (s) $Al(OC_2H_5)_3$ (s) $Al(i-OC_3H_7)_3$ (s) $Al(OC_4H_9)_3$ (s)
Double cation alkoxides		
	La-Al	$La[Al(i-OC_3H_7)_4]_3$
	Mg-Al	$Mg[Al(i-OC_3H_7)_4]_2$, $Mg[Al(s-OC_4H_9)_4]_2$
	Ni-Al	$Ni[Al(i-OC_3H_7)_4]_2$
	Zr-Al	$(C_3H_7O)_2Zr[Al(OC_3H_7)_4]_2$
	Ba-Zr	$Ba[Zr_2(OC_2H_5)_9]_2$

Source: H. Yanagida, K. Koumoto, and M. Miyayama, *The Chemistry of Ceramics*. Copyright © 1996 by John Wiley & Sons, Inc.

thermal conductivity and such processing parameters as temperature and pressure. Most pure polymers are difficult to process, due in part to their high molecular weights or their cohesive forces. As a result, additives, oftentimes low-molecular-weight species, are sometimes added to polymers to improve their formability (or for other reasons as we will see), to form *plastics*. Up to this point, we have made little distinction between polymers and plastics, and we will continue to do so. But be aware of the relationship between the two, and recognize that when the term "polymer" is used as an adjective to describe a certain type of processing, it may well be that the term "plastics" is more appropriate. Nonetheless, the principles of each process are fundamentally the same, whether a pure polymer or a plastic is being formed, such that the distinction is not necessary for our purposes.

There are three primary steps to any polymer forming procedure: *pre-shaping; shaping*; and *post-shaping*. Pre-shaping involves the melting, mixing and conveying of the material to next step. The thermomechanical history in this step can affect properties,

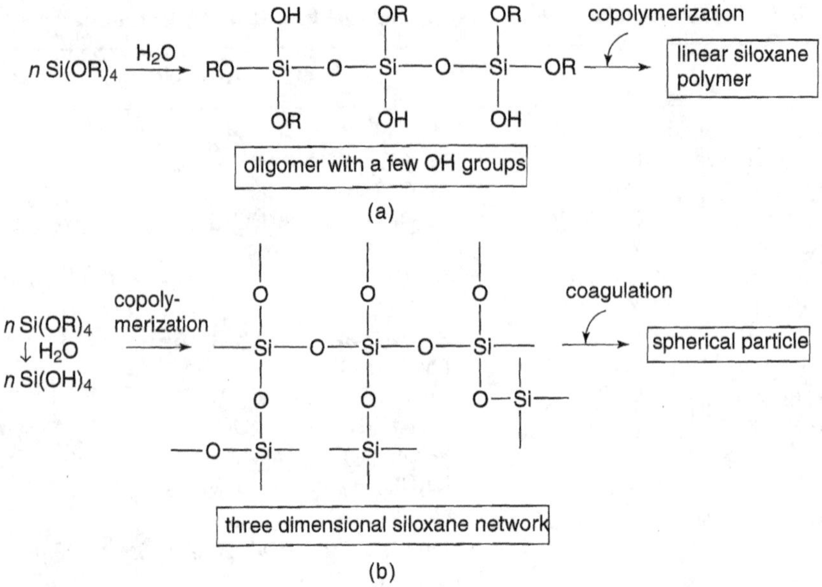

Figure 7.53 Hydrolysis and polymerization of a generic alkoxide $Si(OR)_4$ involving both (a) acid and (b) basic routes. Reprinted, by permission, from H. Yanagida, K. Koumoto, and M. Miyayama, *The Chemistry of Ceramics*, p. 148. Copyright © 1996 by John Wiley & Sons, Inc.

and time/temperature effects are important. In the shaping operation, deformation of the mass after pre-shaping forms the product into the desired shape. In this step, anisotropic properties can result. Finally, post-shaping consists primarily of solidification, although secondary operations such as printing, painting, fastening and electroplating can occur here, as well. Heat transfer is the primary consideration in solidification, and tolerance of dimensions are dictated by this step. We will concentrate primarily on the shaping step, since it is the one that involves the most engineering principles, but we will also introduce aspects of the other two steps where appropriate. Let us begin by looking at some of the more important shaping methods.

There are two primary polymer processing methods, *continuous processing* and *cyclic processing*. *Die forming* and *calendering* constitute the continuous processes; whereas *molding* and *mold casting* are cyclic processes. We will discuss the methods within each of these categories, as well as *secondary shaping* methods such as *thermoforming*. Finally, we will discuss some of the more specialized types of polymer processing.

7.3.1 Continuous Processing

The two primary types of continuous polymer processing techniques are die forming (extrusion) and calendering. The advantages these methods offer are those of any continuous process: high throughput and product uniformity. The product geometries that can be made with these methods are limited, however. As with metals, the products that can be formed in a continuous fashion are primarily sheets and rods. In addition to extrusion and calendering, we will describe a specialized type of extrusion process known as fiber-spinning, which is used to make synthetic fibers for a number of important applications. Where appropriate, we will describe the engineering principles that can affect polymer structure and performance in these continuous operations.

7.3.1.1 Die Forming and Extrusion.

Just as in the extrusion of metals, polymer extrusion is the pushing of a polymer melt across a metal die that continuously shapes the melt into a desired form. Recall that a die is a metal flow channel or restriction that serves the purpose of imparting a specific cross-sectional shape to a material—in this case, a stream of polymer melt. Dies are primarily used in extrusion and are used to form tubes, films, sheets, fibers, and complex profiles. They are positioned at the end of the melt generator or conveying equipment and consist of three general regions: the *manifold*, the *approach channel*, and the *die lip* (see Figure 7.54). The manifold distributes the incoming melt stream over a cross-sectional area similar in shape to the desired final product. The approach channel streamlines melt into the final die opening, and the die lip, or final die opening area, gives the proper cross-sectional shape to the product.

The objective of the die is to achieve the desired shape within set limits of dimensional uniformity at the highest possible production rate. However, nonuniformities can arise in the final part and are classified according to their geometry. The first type of nonuniformity is in the machine direction, or direction of flow [z direction, part (a) in Figure 7.55]. These arise due to time variations of inlet temperature, pressure and composition of the melt. The second type of part nonuniformity is in the cross-machine direction [perpendicular to the z direction, part (b) in Figure 7.55]. These are generally due to improper die design.

A common phenomenon in die processing of polymers that influences the geometry of the final product is called *extrudate swelling*, or *die swelling*. Extrudate swelling refers to the phenomenon observed with polymer melts and solutions that, when extruded, emerge with cross-sectional dimensions appreciably larger than those of the flow conduit. For the cylindrical extrudate geometry, the ratio of the final diameter to that of the die capillary, D/D_0, varies from 1.12 at low shear rates to 0.87 at high shear rates for Newtonian fluids. Non-Newtonian fluids such as polymer melts have similar D/D_0 values at low shear rates, but they swell 2–4 times the extrudate diameter at high shear rates (see Figure 7.56). Experimentally, D/D_0 depends on the following

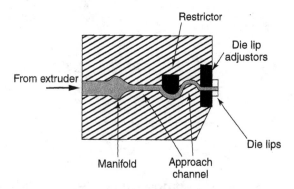

Figure 7.54 Schematic diagram of a polymer die. From Z. Tadmor and C. G. Gogos, *Principles of Polymer Processing*, Copyright © 1979 by John Wiley & Sons, Inc. This material is used by permission of John Wiley & Sons, Inc.

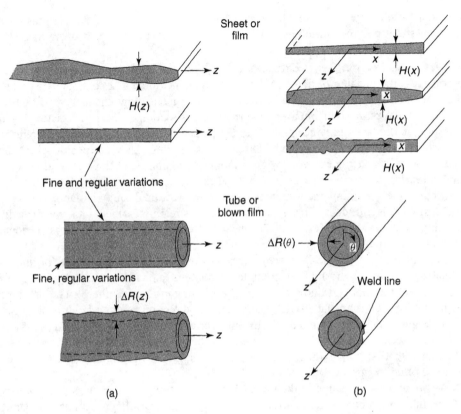

Figure 7.55 Die-formed product nonuniformities (a) in the machine direction and (b) in the cross-machine direction. From Z. Tadmor and C. G. Gogos, *Principles of Polymer Processing*. Copyright © 1979 by John Wiley & Sons, Inc. This material is used by permission of John Wiley & Sons, Inc.

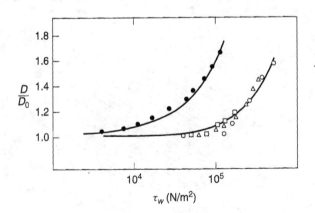

Figure 7.56 Extrudate swelling in polystyrene melts for (●) broad molecular weight distribution and (○, △) narrow molecular weight distribution samples. From Z. Tadmor and C. G. Gogos, *Principles of Polymer Processing*, Copyright © 1979 by John Wiley & Sons, Inc. This material is used by permission of John Wiley & Sons, Inc.

variables: the shear stress at the die wall, τ_w, the molecular weight distribution, and the length-to-diameter ratio of the die capillary. Extrudate swelling is related to the ability of polymer melts and solutions to undergo delayed elastic strain recovery (cf. Section 5.3.1.2). The more strained and the more entangled the melt is at the die exit, the more it will swell.

How the melt is conveyed to the die is of equal, if not greater, interest when compared to the die forming process, since much of the polymer flow characteristics through the die are determined in the conveying section of the extruder. A typical extrusion apparatus is shown in Figure 7.57. As with all extruders, there is a rotating screw, called an *extrusion screw*, that is used to convey the polymer from the feed, called a *hopper*, to the die. The polymer is typically placed in pellet form into the hopper. From the hopper, the material falls through a hole in the top of the extruder (feed throat) onto the extrusion screw. This screw, which turns inside the *extrusion barrel*, conveys the polymer forward into a heated region of the barrel where a combination of external heating and frictional heating melts the polymer. The screw moves the molten polymer forward until it exits through the die. The extruded polymer is immediately cooled and solidified, usually in a water tank. The continuous output from the extruder is called the *extrudate*. Auxiliary equipment will cut, form, or coil the extrudate for secondary shaping or processing.

In the simplest type of extruder, a single screw is used to convey the polymer from the feed to the die. Every screw has certain geometric characteristics that can be summarized in Figure 7.58. The *barrel diameter*, D_b, is the diameter of the screw barrel in which the screw resides (not shown). There is a small clearance between the tip of the screw, called the *flight*, and the barrel wall, δ_f, on each side of the screw. The *screw diameter* at the tip of the flight, D_s, is then related to the barrel diameter as $D_b - 2\delta_f$. The *channel depth*, H, is the distance between the barrel wall and surface of the main portion of the screw, called the *screw root*. The channel depth varies along the length of the screw, because the screw root diameter increases from the feed to the die. This is because the screw is conveying solids near the hopper, but molten polymer near the die. The ratio of the channel depth at the feed to the channel depth in near the die, called the *metering section*, is called the *compression ratio*. The axial distance of one full turn of the flight is called the *lead*, L_s, and e is the width of a flight. The

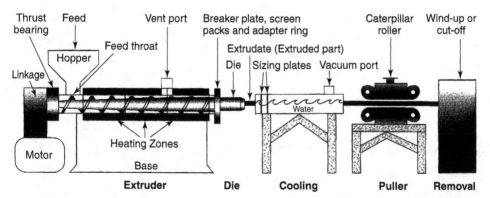

Figure 7.57 Schematic illustration of a typical extrusion apparatus. Reprinted, by permission, from A. B. Strong, *Plastics Materials and Processing*, 2nd ed., p. 352. Copyright © 2000 by Prentice Hall, Inc.

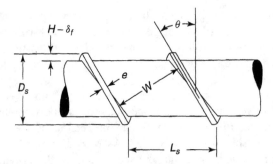

Figure 7.58 Characteristic dimensions of an extrusion screw. From Z. Tadmor and C. G. Gogos, *Principles of Polymer Processing*. Copyright © 1979 by John Wiley & Sons, Inc. This material is used by permission of John Wiley & Sons, Inc.

actual width of the flow channel, W, is related to the lead and flight width:

$$W = L_s \cos \theta - e \tag{7.54}$$

where θ is called the *helix angle*, or *screw pitch*, and is given by

$$\tan \theta = \frac{L_s}{\pi D_s} \tag{7.55}$$

For square pitched screws, $L_s = D_s$, and the flights are all perpendicular to the screw axis. Extruder screws range from 25 to 150 mm in diameter, with screw lengths typically of 25 to 30 screw diameters. They rotate at speeds from 50 to 150 rpm, and deliver melt at rates from 10 to 1000 kg/h with pressures up to 40 MN/m^2.

It is important to understand the flow characteristics of the polymer inside the screw. If we "unwind" the screw channel, as illustrated in Figure 7.59, we see that the polymer is conveyed to the die through nothing more than a square channel formed between the screw and the barrel surface. For the purposes of this development, we will assume that the channel has a constant channel depth, H (at least in the metering

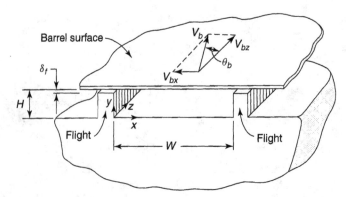

Figure 7.59 Schematic diagram of an "unwound" screw channel. From Z. Tadmor and C. G. Gogos, *Principles of Polymer Processing*. Copyright © 1979 by John Wiley & Sons, Inc. This material is used by permission of John Wiley & Sons, Inc.

region). Furthermore, since the screw channel moves relative to the barrel surface, we can model the flow either as in a moving channel against a fixed barrel wall or as a fixed channel with a moving barrel wall. Both give the same result, and we will find it easier to use the latter model. So, assuming that the barrel surface moves at constant velocity, V_b, the surface velocity of the barrel surface can be resolved into components as illustrated in Figure 7.59: the *down channel velocity*, or *drag velocity*, $V_{bz} = V_b \cos\theta$ and the *cross-channel velocity* $V_{bx} = V_b \sin\theta$. The components of the barrel surface motion lead to circulatory flow of the polymer in the channel.

To obtain the velocity profile of the polymer in down channel direction, v_z, we start with the appropriately simplified equation of motion in rectangular coordinates

$$\frac{\partial P}{\partial z} = \mu \left(\frac{\partial^2 v_z}{\partial x^2} + \frac{\partial^2 v_z}{\partial y^2} \right) \tag{7.56}$$

and apply the following boundary (no slip) conditions:

$$
\begin{aligned}
v_z &= 0 & @x = x, y = 0 \\
v_z &= V_{bz} & @x = x, y = H \\
v_z &= 0 & @x = 0, y = y \\
v_z &= 0 & @x = W, y = 0
\end{aligned}
$$

It can be shown that the pressure gradient is a constant. We then integrate v_z across the channel and perform similar analysis for v_x to get the volumetric flow rate, Q,

$$Q = \frac{V_{bz} W H}{2} F_d + \frac{W H^3}{12\mu} \left(-\frac{\partial P}{\partial z} \right) F_p \tag{7.57}$$

where F_d and F_p are *shape factors* for the drag and pressure flows, respectively. They have values less than 1.0 and depend only on the H/W ratio, as illustrated in Figure 7.60. The shape factors represent the reducing effect of the flights on flow rate between infinite parallel plates, as in our model.

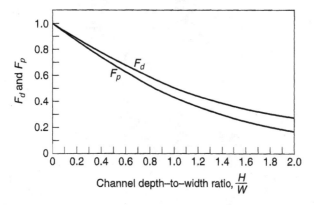

Figure 7.60 Shape factors for polymer extrusion. From Z. Tadmor and C. G. Gogos, *Principles of Polymer Processing*. Copyright © 1979 by John Wiley & Sons, Inc. This material is used by permission of John Wiley & Sons, Inc.

The first term in Eq. (7.57) represents the contribution of *drag flow*, Q_d, and the second term represents the contribution of *pressure flow*, Q_p. The two terms are independent of one another and represent the two ways in which polymer moves down the channel. The ratio of pressure to drag flow, Q_p/Q_d, is a measure of the relative amount of pressure to drag flow and indicates the velocity profile and movement of fluid elements in the flow channel, among other things. For pure drag flow, $\partial P/\partial z = 0$ (no pressure flow) and $Q_p/Q_d = 0$. The polymer flows by drag only. A value of -1 for this ratio indicates a situation of zero net flow rate, or $Q_p = -Q_d$. An increase in the positive pressure gradient will increase the pressure backflow. This decreases the net flow rate, despite an unchanging drag flow. Eventually, the drag and pressure flows are equal, but opposite, and discharge ceases. As the throughput increases from zero, the fluid elements travel along a sort of flattened helical path through the channel, as illustrated in Figure 7.61 for the two extremes cited above and for an intermediate case of $Q_p/Q_d = -0.5$. The closer the flow to pure drag flow, the more open the loops of the helix will be. Thus polymer particles travel in a path that is a flat helix within a helical channel.

In addition to conveying the polymer toward the die, the screw helps perform two more important functions. The first is melting. In addition to the heat supplied by resistance heating elements along the barrel length, the screw provides frictional heat to the melt to assist in melting. The last function is mixing. In addition to the mixing provided by a combination of drag flow and pressure flow, mixing elements can be added to the screw that accentuate dispersive or distributive mixing action.

In addition to single screw extruders, there are twin and multiscrew extruders that perform essentially the same functions, but with additional benefits. Among these, the *intermeshing twin screw* extruders are the most important ones. They are used primarily for heat-sensitive resins (such as PVC), which are difficult to process. The intermeshing screws create a relative motion of one flight in another, such that it acts

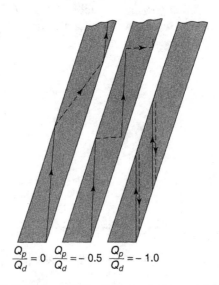

$$\frac{Q_p}{Q_d} = 0 \qquad \frac{Q_p}{Q_d} = -0.5 \qquad \frac{Q_p}{Q_d} = -1.0$$

Figure 7.61 The path of fluid elements in the extruder screw channel for various ratios of Q_p/Q_d. From Z. Tadmor and C. G. Gogos, *Principles of Polymer Processing*. Copyright © 1979 by John Wiley & Sons, Inc. This material is used by permission of John Wiley & Sons, Inc.

like a paddle to push the material from screw to screw and from flight to flight. Twin screw extruders have a more positive pumping action than single-screw extruders and can therefore be used more effectively in high-output situations.

The screws can rotate either in the same direction, known as *corotating screws*, or in opposite directions, known as *counterrotating screws*, as shown in Figure 7.62. In the corotating screws system the material is passed from one screw to another and follows a path over and under the screws. In a counterrotating screws system the material is brought to the junction of the screws and builds up what is called a *material bank* on the top of the junction. This buildup of material is conveyed along the length of the screws, high shear is created, but shear elsewhere is very low. Only a small amount of material passes between the screws. Therefore, total shear is lower than in single-screw extruders and in corotating twin-screw extruders. A comparison between some characteristics of single, corotating and counterrotating twin screws is given in Table 7.16.

7.3.1.2 Calendering. *Calendering* is used for the continuous manufacture of polymer sheet or film. As illustrated in Figure 7.63, the calendar consists of four horizontal, counterrotating steel rolls, called *bowls* or *nip rollers*. The top two bowls look just like a two-roll mill used in the rolling of metals (cf. Figure 7.10) and are charged with a uniform polymer melt. The melt may come from a melt-conveying operation such as extrusion. The gap between the two rolls is small, and the polymer builds up a small pool on the top of the nip area called the *bank*. The polymer is drawn through the first nip and, adhering to one bowl, is transferred to the next nip where it is

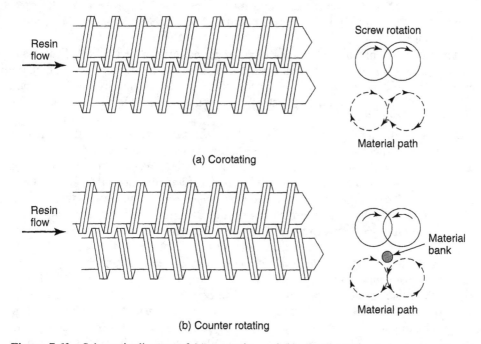

Figure 7.62 Schematic diagram of (a) corotating and (b) counterrotating twin screws. Reprinted, by permission, from A. B. Strong, *Plastics Materials and Processing*, 2nd ed., p. 365. Copyright © 2000 by Prentice-Hall, Inc.

Table 7.16 Comparison of Single- and Twin-Screw Extruder Screw Characteristics

| Type | Single Screw | Corotating Screw | | Counterrotating Twin Screw |
		Low-Speed Type	High-Speed Type	
Principle	Friction Between Cylinder and Materials and the Same Between Material and Screw	Mainly Depend on the Frictional Action as in the Case of Single Screw Extruder		Forced Mechanical Conveyance Based on Gear Pump Principle
Conveying efficiency	Low	Medium		High
Mixing efficiency	Low	Medium/High		High
Shearing action	High	Medium	High	Low
Self-cleaning effect	Slight	Medium/High	High	Low
Energy efficiency	Low	Medium/High		High
Heat generation	High	Medium	High	Low
Temp distribution	Wide	Medium	Narrow	Narrow
Max. revolving speed (rpm)	100–300	25–35	250–300	35–45
Max. effective length of screw L/D	30–32	7–18	30–40	10–21

Source: Z. Tadmor and C. G. Gogos, *Principles of Polymer Processing*. Copyright © 1979 by John Wiley & Sons, Inc.

reduced in thickness again. The gaps between subsequent rolls get smaller and the temperature of the rolls lower to finally set the dimensions of the part. The thickness reduction in the third nip provides fine control of the sheet and confers the required surface finish.

The advantage of calendering over direct sheet extrusion is that a complicated and expensive die is not required. Some additional mixing can be done as part of the calendering process, as well, and a surface finish can be applied by the final roller, if desired. Sheet up to a few millimeters thick and a meter wide can be made using calendering.

Just as we did for extrusion, we can derive useful expressions from the equations of change for such quantities as the pressure profile, velocity profile, shear stress and strain, and roller force for the calendering operation. We will briefly outline the procedure here. Consider the top two rollers, as illustrated in a more detailed fashion in Figure 7.64. The rolls have been rotated to a horizontal position for ease of the analysis. The two identical rolls of radii R rotate in opposite directions with frequency of rotation N. The minimum gap between the rolls is $2H_0$. The polymer is uniformly distributed laterally over the roll width, W. At a certain upstream axial position, $x = X_2$ ($X_2 < 0$), the rolls start biting into the polymer. The melt contacts both rolls at this position. At a certain downstream axial position, $x = X_1$ ($X_1 > 0$), the polymer detaches itself from one of the rolls. Pressure, which is assumed to be atmospheric at X_2, rises along the x direction as the melt proceeds through the nip area, and reaches a maximum before the minimum gap clearance, then drops back to atmospheric at X_1. As a result of this pressure profile, there is a force acting on the rolls that tends to increase the clearance between them, and even distort them. The location of points X_2 and X_1 depends on the roll geometry, gap clearance, and the total volume of polymer in the bank.

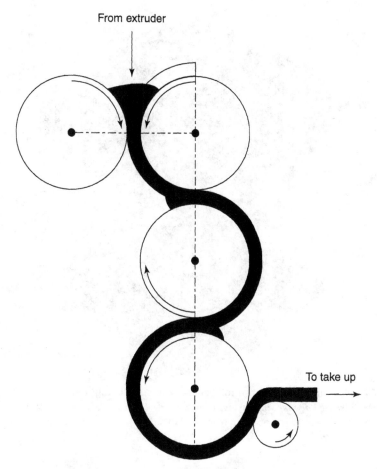

Figure 7.63   Schematic diagram of calendering system.

It is possible to derive an expression for the pressure profile in the x direction using a simple model. We assume that the flow is steady, laminar, and isothermal; the fluid is incompressible and Newtonian; there is no slip at the walls; gravity forces are neglected, and the polymer melt is uniformly distributed on the rolls. With these assumptions, there is only one component to the velocity, $v_x(y)$, so the equations of continuity and motion, respectively, reduce to

$$\frac{dv_x}{dx} = 0 \tag{7.58}$$

and

$$\frac{\partial P}{\partial x} = -\frac{\partial \tau_{yx}}{\partial y} = \mu \frac{\partial^2 v_x}{\partial y^2} \tag{7.59}$$

The equation of motion, Eq. (7.59), can be integrated twice with no difficulty because the pressure, P, is a function of x only. The boundary conditions are $v_x(\pm h) = U$,

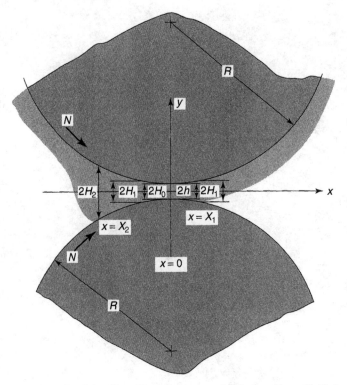

Figure 7.64 The nip area of the two roll geometry in calendering. From Z. Tadmor and C. G. Gogos, *Principles of Polymer Processing*, Copyright © 1979 by John Wiley & Sons, Inc. This material is used by permission of John Wiley & Sons, Inc.

where U is the tangential velocity of the roll surfaces

$$U = 2\pi N R \tag{7.60}$$

and the resulting velocity profile is

$$v_x = U + \frac{y^2 - h^2}{2\mu}\left(\frac{dP}{dx}\right) \tag{7.61}$$

We can also solve for the pressure profile, which we will do momentarily, and use this, along with a reduced velocity, $u_x = v_x/U$, and reduced gap parameter, $\xi = y/H$, to obtain the following form of the velocity profile:

$$u_x = 1 + \frac{3(1 - \xi^2)(\lambda^2 - \rho^2)}{2(1 + \rho^2)} \tag{7.62}$$

where

$$\lambda^2 = \frac{X_1^2}{2RH_0} \tag{7.63}$$

and

$$\rho^2 = \frac{x^2}{2RH_0} \tag{7.64}$$

The velocity profile, Eq. (7.62), can be plotted as a function of position in the gap along the x direction, now given by the reduced parameter, ρ, of Eq. (7.64). This is done schematically in Figure 7.65. Notice that at $\rho = \pm\lambda$ the velocity profiles are flat; that is, the flow is plug flow, because the pressure gradient vanishes at these locations.

As promised, the pressure profile can also be solved for, for which the result is

$$P = \frac{3\mu U}{4H_0}\sqrt{\frac{R}{2H_0}}\left\{\left[\frac{\rho^2 - 1 - 5\lambda^2 - 3\lambda^2\rho^2}{(1+\rho^2)^2}\right]\rho + (1-3\lambda^2)\tan^{-1}\rho + C(\lambda)\right\} \tag{7.65}$$

where all parameters are as described previously, and $C(\lambda)$ is a constant of integration given by

$$C(\lambda) = \frac{\lambda(1+3\lambda^2)}{(1+\lambda^2)} - (1-3\lambda^2)\tan^{-1}\lambda \tag{7.66}$$

The pressure profile between the rolls in the x direction, represented as ρ, is given for several values of λ in Figure 7.66 in the form of the reduced pressure, $P/P_{\max}$.

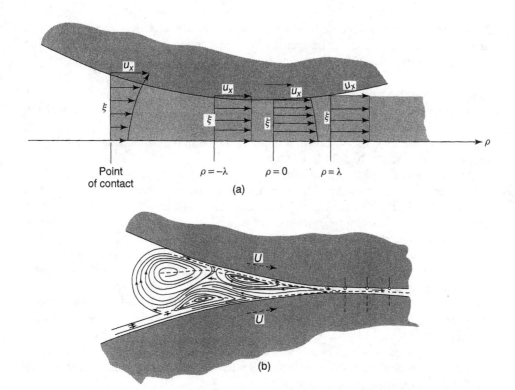

Figure 7.65 Velocity profiles between rolls for $\lambda^2 = 0.1$ from Eq. (7.32). From Z. Tadmor and C. G. Gogos, *Principles of Polymer Processing*. Copyright © 1979 by John Wiley & Sons, Inc. This material is used by permission of John Wiley & Sons, Inc.

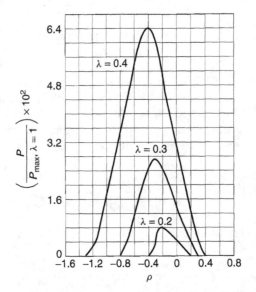

Figure 7.66   Pressure profile between rolls in calendering. From Z. Tadmor and C. G. Gogos, *Principles of Polymer Processing*. Copyright © 1979 by John Wiley & Sons, Inc. This material is used by permission of John Wiley & Sons, Inc.

Notice that the pressure profile is as described previously; that is, it rises along the x direction as the melt proceeds through the nip area, and it reaches a maximum before the minimum gap clearance, then drops back to atmospheric pressure at the exit.

7.3.1.3 Fiber Spinning.

The production of polymer fibers is important for a number of applications, including woven fabrics for such things as composite reinforcement and clothing. There are three primary methods of spinning polymeric fibers: *melt spinning; dry solution spinning;* and *wet solution spinning*. A fourth type of fiber spinning, known as *gel spinning*, is a hybrid form of dry-wet spinning.

As the name implies, melt spinning involves the drawing of fibers directly from a polymer melt. A schematic illustration of the apparatus is shown in Figure 7.67a. The polymer is fed and melted in much the same manner as it is for other melt processing techniques (e.g., extrusion), but instead of being forced through a single die, it is pressed through a multiple-orifice plate called a *spinneret*. The spinneret consists of 50–60 holes, each 0.12 mm in diameter. The final diameter of the fiber, or *denier*, is determined by pumping and winding rate—not hole size. Production rates of about 2500 feet per minute are typical for this operation. The multiple fibers are gathered together on a roller in a bundle known as a *tow*. To help keep the group of fibers together, the tow can be twisted slightly, at which point it is called a *yarn*. Melt spinning is used to make nylon, polyester, PET, and olefin fibers, among others. The desired range of viscosity at the spinneret is between 1000 and 2000 poise. If the polymer melt is too viscous and would require a dangerously high temperature to reach this viscosity range, it is spun in solution.

Although the term "dry solution spinning" may seem like an oxymoron, it simply describes a process by which a polymer is dissolved in an appropriate solvent, and the solution is pumped through a spinneret in much the same manner as in melt

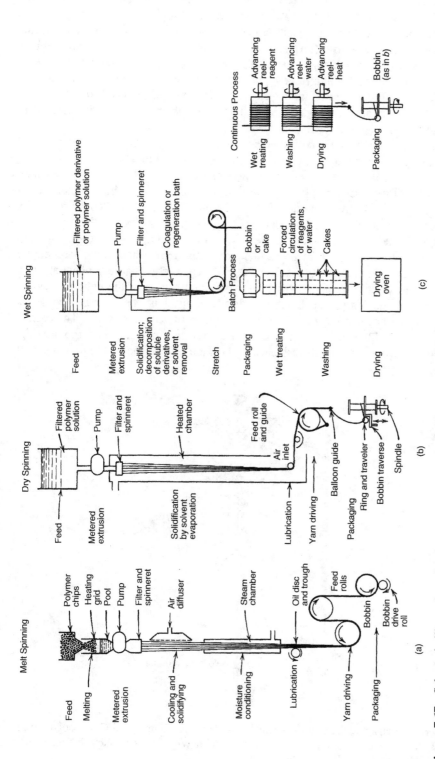

Figure 7.67 Schematic illustration of three types of polymer fiber spinning operations: (a) melt spinning, (b) dry spinning, and (c) wet spinning. From F. W. Billmeyer, *Textbook of Polymer Science*, 3rd ed. Copyright © 1984 by John Wiley & Sons, Inc. This material is used by permission of John Wiley & Sons, Inc.

769

spinning (see Figure 7.67b). In this case, however, the fibers enter air when they exit the spinneret and solidify because the solvent is evaporated and recycled. This process is used for the formation of such filaments as cellulose acetate, acrylic, spandex, triacetate, vinyon, and poly acrylonitrile (PAN) fibers, which are the precursors to carbon fibers (cf. Section 1.4.5.1). In wet solvent spinning, or wet spinning, the polymeric fiber is formed in an analogous fashion to dry spinning, except that the solvent is leached out by another liquid, as illustrated in Figure 7.67c. This process is used to make rayon (cellulosic), acrylic, aramid, modacrylic, and spandex fibers. Gel spinning is used to form some high-strength polyethylene fibers.

We can derive some useful design equations for fiber spinning, just as we did for extrusion and calendering, by using a capillary flow model, as illustrated in Figure 7.68. Though the principles apply to all spinning methods, we will concentrate on melt spinning, since it involves a single phase system. In capillary flow, the polymer melt is contained in a reservoir. In the entrance region, the melt is forced into a converging flow pattern and undergoes large axial accelerations—that is, stretching. As the flow rate is increased, the axial acceleration also increases, and the polymer becomes much more elastic. It is possible for melt fracture to occur in this regime due to the viscoelastic nature of the polymer (cf. Section 5.3.1). Barring any such instability phenomena, a fully developed velocity profile is reached a few diameters after the entrance region. The flow in the capillary, which for pseudoplastic fluid is characterized by a flat velocity profile, imparts a shear strain on the melt near the capillary walls. The core of the melt, if the capillary L/R is large and the flow rate is small, can undergo a partial strain recovery process during its residence in the capillary. At the exit region, the melt

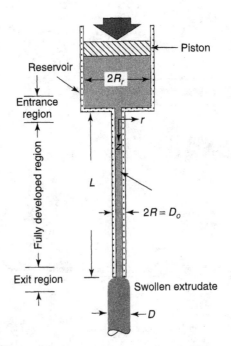

Figure 7.68   Schematic diagram for capillary flow. From Z. Tadmor and C. G. Gogos, *Principles of Polymer Processing*. Copyright © 1979 by John Wiley & Sons, Inc. This material is used by permission of John Wiley & Sons, Inc.

is under the influence of no externally applied stresses. It can thus undergo delayed strain recovery, which, together with the velocity profile rearrangement to one that is pluglike, results in the phenomenon of die swelling that was introduced earlier in this section. Let us develop some more quantitative descriptions of these phenomena.

The velocity profile for a Newtonian fluid in a capillary is well-described in most introductory transport texts by the Hagen–Poiseuille equation

$$v_z = \frac{\Delta P R^2}{4 \mu L} \left[1 - \left(\frac{r}{R} \right)^2 \right] \tag{7.67}$$

with the implicit assumption that the pressure gradient, dP/dz, is constant and given by $\Delta P/L$, where $\Delta P = P_0 - P_L$, and L is the length of the capillary. This assumption is not correct near the capillary entrance, where the converging flow causes extra velocities (e.g., v_r), and velocity gradients to be present. An example of the entrance flow patterns in a molten polymer is given in Figure 7.69. Although the flow pattern is not known precisely, we know that a higher pressure drop is needed to support the additional velocity gradients for any viscous or viscoelastic fluid. Schematically, then, the pressure profile along the length of the capillary can be represented as in curve a (bottom curve) in Figure 7.70. Thus,

$$\frac{-dP}{dz} = \frac{\Delta P}{L^*} = \frac{\Delta P}{(L + N D_0)} \tag{7.68}$$

where $N > 0$ is the *entrance loss correction factor*, sometimes called the *Bagley correction*, and is applied to increase the effective length of the capillary by a multiple of capillary diameters, D_0. This has the same effect as reducing the pressure drop by a value of ΔP_{ent}, the *entrance pressure drop*. It turns out that the value of the entrance pressure drop becomes larger than the total capillary pressure drop at higher shear rates. It follows then that in polymer processing, where the length-to-opening ratios are small and shear rates are high, entrance pressure drops must be included in calculations of the die pressure in die design equations. This is done through *Bagley plots*, an example

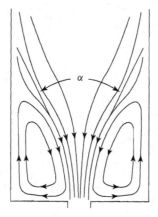

Figure 7.69 Entrance flow patterns in molten polymers. From Z. Tadmor and C. G. Gogos, *Principles of Polymer Processing*. Copyright © 1979 by John Wiley & Sons, Inc. This material is used by permission of John Wiley & Sons, Inc.

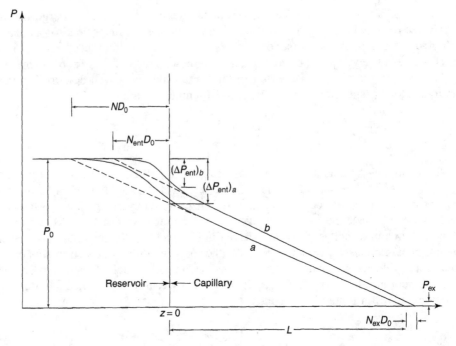

Figure 7.70 Schematic representation of the capillary pressure along its axis: curve *a*, without exit effects; curve *b* with exit effects. From Z. Tadmor and C. G. Gogos, *Principles of Polymer Processing*. Copyright © 1979 by John Wiley & Sons, Inc. This material is used by permission of John Wiley & Sons, Inc.

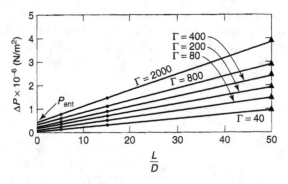

Figure 7.71 Bagley plots for a polystyrene melt at 200°C. From Z. Tadmor and C. G. Gogos, *Principles of Polymer Processing*. Copyright © 1979 by John Wiley & Sons, Inc. This material is used by permission of John Wiley & Sons, Inc.

of which is shown in Figure 7.71. If the L/D_0 of the capillary and the volumetric flow rate, Q, are known the pressure drop can be found from Figure 7.71, where ΔP_{ent} is the value of the pressure drop at $L/D_0 = 0$. The values of Γ are given by

$$\Gamma = \frac{32Q}{\pi D_0^3} \tag{7.69}$$

Cooperative Learning Exercise 7.3

At right is shown data for the ratio of entrance pressure drop to shear stress at the capillary wall, τ_w, as a function of Newtonian wall shear rate, Γ, for a number of common polymers. Consider the melt spinning of PS at a volumetric flow rate of 4.06×10^{-1} cm^3/s through a spinneret that contains 100 identical holes of radius 1.73×10^{-2} cm and length 3.46×10^{-2} cm. Assume that the molecular weight distribution is broad.

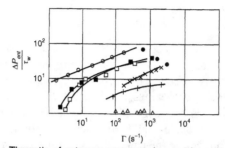

The ratio of entrance pressure drop to shear stress at the capillary wall versus Newtonian wall shear rate, Γ. ■, PP; □, PS; ○, LDPE; +, HDPE; ●, 2.5% PIB in mineral oil; ×, 10% PIB in decalin; △, NBS-OB oil. Reprinted, by permission, from Z. Tadmor and C. G. Gogos, *Principles of Polymer Processing*, p. 537. Copyright © 1979 by John Wiley & Sons, Inc.

Person 1: Estimate the entrance region pressure drop, ΔP_{ent}, using the Bagley correction. What can you say about the percentage of the entrance pressure drop relative to the total pressure drop?

Person 2: Estimate the ratio of entrance region pressure drop to shear stress at the capillary wall using the plot above.

Combine your results to obtain a value for the shear stress at the capillary wall, and use that value to estimate the extrudate swell, D/D_0 for polystyrene using Figure 7.56. What effect does molecular weight have on die swell at these shear levels?

Answer: $Q = 10^{-3}$ cm^3/s; $\Gamma = 10^3$ s^{-1}; $L/D_0 = 1$; $\Delta P_{ent} \approx 0.4 \times 10^6$ N/m^2; the close proximity of the entrance pressure drop to the overall pressure drop in the Bagley plot indicates that most of the pressure drop is due to entrance effects. $\Delta P_{ent}/\tau_w \approx 40$; $\tau_w \approx 10^4$ N/m^2; $D/D_0 \approx 1.1$. At this stress level, molecular weight does not have a large effect.

Similarly, one must worry about exit pressure effects, or *end corrections*. Experimentally, it has been observed that there is a nonzero gauge pressure at the capillary exit, P_{ex} (curve *b* of Figure 7.70). It has been found that the ratio of exit to entrance effects, $P_{ex}/\Delta P_{ent}$, is between 0.15 and 0.20 and that although ΔP_{ent} does not depend on L/D_0, P_{ex} decreases up to $L/D_0 = 10$ and then remains constant. With these observations in mind, we can rewrite Eq. (7.68) to include both entrance and exit effects:

$$\frac{\Delta P}{L^*} = \frac{\Delta P}{L + N_{ent} D_0 + N_{ex} D_0} \tag{7.70}$$

where N_{ent} and N_{ex} are Bagley corrections for entrance and exit effects, respectively. If $P_{ex}/\Delta P_{ent} \ll 1$, then $N = N_{ent}$. If P_{ex} is included, it is found that $N_{ent} < N$.

Other flow effects include extrudate swelling and *melt fracture*. The phenomenon of extrudate swell was elaborated upon in the context of polymer extrusion (cf. Figure 7.56).

We will not describe the phenomenon of melt fracture here. The interested reader is referred to the excellent text by Tadmor and Gogos for more information on this topic.

7.3.2 Cyclic Processing

In contrast to extrusion, calendering, and fiber spinning, which are continuous processes, molding operations are by their very nature cyclical, or batch, processes. In this section we describe five types of cyclic molding operations involving polymers: *injection molding, compression molding, blow molding, thermoforming*, and a special type of injection molding called *reaction injection molding*. These techniques all have in common the use of molds to shape the polymer into a desired geometry. Although molds are utilized in the batch processing of other materials, they are most widely used in the shaping of polymers, so we will briefly elaborate on the various types of molds before continuing on to the molding operations.

Molds can be of three basic types. Open molds are used to define one surface of a product. They can be of the female (indenting) or male (projecting) type. Two matched female half-molds can be used to make hollow items, and a matched female and male pair is used to form a cavity that is to be completely filled. Shaping of the polymer inside the mold is usually achieved by applying pressure to cause the fluid to conform to one or both mold surfaces. The variation in types of molding is primarily due to the type of mold used and the method of melt delivery to the mold.

The choice of process and of mold type depends greatly on the number of pieces to be made, bearing in mind the lowest possible price per product, and on the tolerances to be attained. In turn, the choice of the process will affect the choice of material and the shape of the product. For example, the price of producing a simple plastic tray is compared in Figure 7.72 for the various molding techniques described in this chapter. The price is determined by such things as raw material, mold, machine, energy, and personnel costs. There are, of course, considerations other than total cost, but all things being equal, it is clear that injection molding provides the lowest cost per part for the production of a large number of components. So, it is fitting that we begin our review of cyclic polymer processing techniques with injection molding.

7.3.2.1 Injection Molding. Injection molding is the process of producing identical articles from a hollow mold. Because of their high viscosity, polymers cannot be poured into a mold, or cast, in the same way that metals are, because gravitational forces are not sufficient to produce appreciable flow rates. Thus, the melt must be injected into the mold cavity by the application of large forces from a plunger. Moreover, once the mold is filled with melt and solidification starts, an additional amount of melt must be packed into the mold to offset polymer shrinkage during solidification.

Injection molding involves two distinct processes. The first is melt generation, mixing, and pressurization and flow, which is carried out in the injection unit of the molding machine. The second is product shaping, which takes place in the mold cavity. Injection molders, therefore, have two distinct parts: the injection unit and the mold/clamping unit. The function of the injection unit is to melt the polymer and inject it into the mold; the function of the clamping unit is to hold the mold, open and close the mold, and eject the finished product.

Two systems have been used in injection molding machines to melt and inject the polymer. The most commonly used types use a *reciprocating screw*, which has many

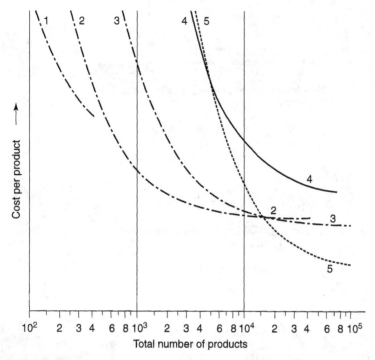

Figure 7.72 Comparison of production costs per product for a polymer tray formed using (1) thermoforming in a wooden mold, (2) thermoforming in an aluminum mold, (3) blow molding with a single mold, (4) compression molding, and (5) injection molding. Reprinted, by permission, from P. C. Powell and A. J. I. Housz, *Engineering with Polymers*, p. 58. Copyright © 1998 by Stanley Thorned Publishers.

similarities to an extruder screw, but with a unique reciprocating (back and forth) action. The other type of injector system is the *ram injector*. We will concentrate on the reciprocating screw since it offers many advantages over ram injectors, including more uniform melting, improved mixing, lower injection pressures, the formation of larger parts, fewer stresses in parts, and faster cycle time.

The reciprocating screw used for injection molding is similar to an extruder screw, but is generally much shorter. Typical length-to-diameter ratios for reciprocating screws are 12:1 to 20:1. The compression ratios are in the range of 2:1 to 5:1, which means that there is less mechanical action added during melting and more thermal energy is needed. As with an extrusion screw, the reciprocating screw has three sections: feed, compression, and metering. The reciprocating action of the screw is illustrated in Figure 7.73. The screw acts like an extrusion screw to melt and meter the polymer toward the mold, forming a pool at the mold entrance. It then moves forward like a piston to inject the polymer into the mold. After mold filling, the screw retracts, and the part is rejected. Let us look more closely at the entire injection molding cycle.

The injection molding cycle is depicting in Figure 7.74. We can begin the cycle at any point we wish, but let us start at the point the screw moves forward and fills the mold with polymer melt. The screw moves forward and fills the mold with melt and maintains the injected melt under pressure, during what is called the *hold time*. To ensure that polymer does not flow backward, a check valve is attached to the end

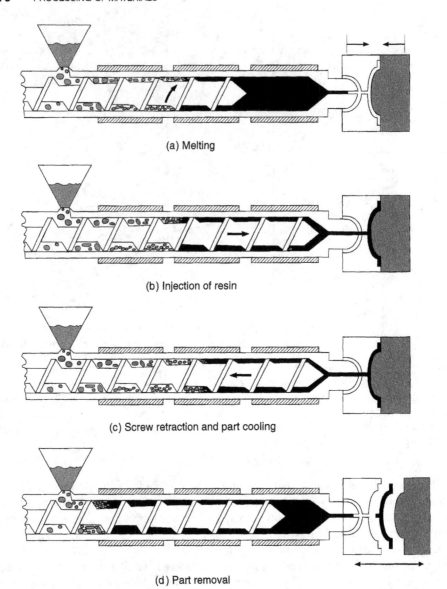

(a) Melting

(b) Injection of resin

(c) Screw retraction and part cooling

(d) Part removal

Figure 7.73   Schematic illustration of reciprocating action of injection molding screw. Reprinted, by permission, from A. B. Strong, *Plastics Materials and Processing*, 2nd ed., p. 424. Copyright © 2000 by Prentice-Hall, Inc.

of the screw. During the hold time, additional melt is injected, offsetting contraction due to cooling and solidification. As the melt cools and solidifies, the pressure drops. Once the part is cooled, screw rotation commences and the screw moves backward, conveying molten polymer to the front as it retracts. After sufficient melt generation for the next injection, called the *shot*, screw rotation ceases. The polymer on the stationary screw continues to melt by heat conduction from the hot barrel. Meanwhile, the mold opens, and solidified part is ejected from the mold. The mold then closes, and is ready to accept the next shot, thus initiating the cycle once again. All of these operations are

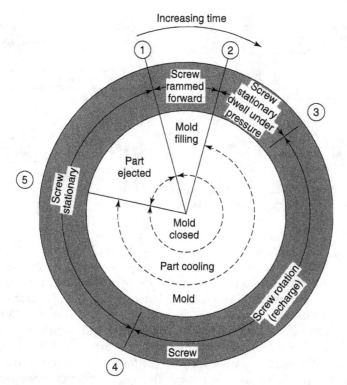

Figure 7.74 The injection molding cycle. From Z. Tadmor and C. G. Gogos, *Principles of Polymer Processing*, Copyright © 1979 by John Wiley & Sons, Inc. This material is used by permission of John Wiley & Sons, Inc.

automated and computer controlled. Perhaps the most important measure of the size of an injection molding machine is the amount of polymer that can be injected, or shot size. Typical shot sizes range from 20 g to 20 kg. Because the shot size is dependent upon the density of the polymer being injected, polystyrene has been chosen as the standard for the rating of machines.

A typical injection mold is made of at least two parts, one of which is movable so that it can open and close during different parts of the molding cycle. The entire mold is kept at a constant temperature below T_g or T_m of the polymer. The various components of an injection mold are illustrated in Figure 7.75. The *sprue* is the part of mold that joins the mold cavity to the machine nozzle. In Figure 7.75 the polymer, then, would be coming into the page from the injection screw. The *runners* are channels cut in the mold through which molten polymer flows after entering the sprue, and on to the *gates*, which are narrow constrictions between the runner and each individual cavity. The gates are purposely designed to be as narrow as possible to allow the molded part to be easily removed from the runner and sprue while still allowing polymer to flow easily into each mold. It is the material within the gate that solidifies first, due to the more rapid heat transfer from the small part. Once the gate is solidified, the parts at the end of the runners are isolated from the injection pressure, and the part can be safely ejected.

The function of the runner system is to transmit the molten polymer to the cavities with a minimum of material and pressure drop. Therefore, the runner length must be

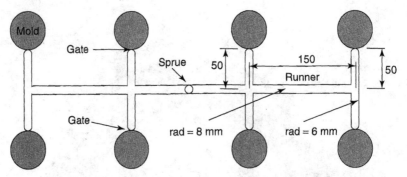

Figure 7.75 Schematic diagram of eight-cavity mold for injection molding operations. The molded parts of interest are indicated by the shaded regions.

kept to a minimum, and the cross section should be optimally set for low pressure drop, low material waste, and relatively slow cooling. Slow cooling is desirable to prevent premature solidification of the polymer in the molds, which causes incomplete filling called *short shot*. The runner dimension is about 1.5 times the characteristic thickness of the molded part, and it is of circular cross section to minimize heat loss.

Short shot can be avoided by proper mold design and control of polymer melt conditions—namely, temperature and injection pressure. This relationship is shown in Figure 7.76. Within the area bounded by the four curves, the specific polymer is moldable in the specific cavity. If the pressure and/or temperature are too low, short shot will result. If the temperature is too high, thermal degradation of the polymer can occur. If the temperature is too low, the polymer will not be molten. If the pressure is too high or the polymer is too fluid, the melt can flow into the gaps of the mold, creating thin webs of polymer attached to the molded article in an undesirable part

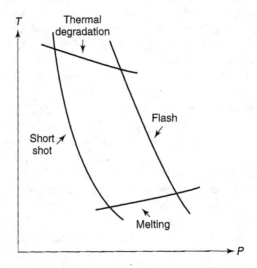

Figure 7.76 Schematic illustration of molding area for a given polymer/mold combination. From Z. Tadmor and C. G. Gogos, *Principles of Polymer Processing*. Copyright © 1979 by John Wiley & Sons, Inc. This material is used by permission of John Wiley & Sons, Inc.

defect called *flash*. Obviously, mold design is critical, as is the proper matching of mold design to process conditions. The most common material for injection molds is tool steel, although beryllium–copper alloys are used when high heat transfer rates are required. Aluminum is also used for molds.

Let us turn our attention now to developing some useful flow equations for injection molding, just as we have done for the other polymer processing techniques. Consider a steady, isothermal, laminar, fully developed pressure flow of an incompressible fluid in a horizontal tube, such as might be found in the runner of an injection mold. For a Newtonian fluid, the analysis is essentially the same as for capillary flow, and the velocity profile can be given by the Hagen–Poiseuille equation, Eq. (7.67). Let us take this opportunity, however, to add an additional level of complexity that is common in polymer processing. As introduced in Section 4.1.2.2 for ceramic slurries, and later applied to polymer melts and solutions, non-Newtonian behavior can occur when the viscosity, η, is a function of the shear rate, $\dot{\gamma}$. One model that is widely used for non-Newtonian fluids is the power law model [cf. Eqs. (4.15) and (4.17)]. We will dispense with the derivation here, but starting from the continuity equation and the equations of motion in cylindrical coordinates as before, the following velocity profile, $v_z(r)$, for a power law fluid in a horizontal tube can be obtained:

$$v_z(r) = \left(\frac{R}{1 + 1/n} \right) \left[\frac{-R}{2K} \left(\frac{dP}{dz} \right) \right]^{1/n} \left[1 - \left(\frac{r}{R} \right)^{1+1/n} \right] \tag{7.71}$$

where R is the tube radius, dP/dz is the pressure drop, and n and K are the power law exponent and consistency index, respectively, as defined in Eq. (4.15). The corresponding volumetric flow rate, Q, assuming a constant pressure drop over the tube length, L, can be obtained from the velocity profile:

$$Q = \frac{\pi R^3}{3 + 1/n} \left(\frac{-R\Delta P}{2KL} \right)^{1/n} \tag{7.72}$$

Notice that Eq. (7.72) reduces to the Hagen–Poiseuille Law for $n = 1$, where $K = \mu$, the Newtonian viscosity.

Obviously, molds are much too complicated to be adequately represented by simple flow in a tube, even if a power law model is employed. Other important factors to consider include heat transfer rates and shrinkage, which are dependent upon both the material of construction for the mold and the polymer being molded. One of the advantages of injection molding is that parts with very complex shapes can be made. Even hollow parts can be made. However, sharp bends in the part are to be avoided. These are difficult for the viscous resin to fill and can be the origin of stresses in the part. The minimum thickness of the part depends upon the type of resin used. Surface finish in the part is imparted by the surface of the mold. Part imperfections include (a) surface indentations where the ejector pins hit the part and (b) rough edges from mold lines or remaining after the part is separated from the secondary runner.

The most serious imperfection that must be accounted for is *mold shrinkage*, which is the difference in the part dimensions and the dimensions of the mold. The amount of shrinkage is dependent upon several factors, including the temperature of the polymer, the type of polymer, the flow field of the polymer in the mold, injection pressure and hold time, and the presence of additives. Characteristic shrinkages for several common

Example Problem 7.3

Consider a straight tubular runner of length L. A melt following the power-law model is injected at constant pressure into the runner. The melt front progresses along the runner until it reaches the gate located at its end. Calculate the melt front position, $Z(t)$, and the instantaneous flow rate, $Q(t)$, as a function of time. Assume an incompressible fluid and an isothermal and fully developed flow, and make use of the pseudo-steady-state approximation. For a polymer melt with $K = 2.18 \times 10^4 \text{ N} \cdot \text{s}^n/\text{m}^2$ and $n = 0.39$, calculate $Z(t)$ and $Q(t)$ for an applied pressure $P_0 = 20.6 \text{ MN/m}^2$ in a runner of dimensions $R = 2.54 \text{ mm}$ and $L = 25.4 \text{ cm}$.

Answer: We begin with the relationship for flow rate of a power law fluid, as given by Eq. (7.72), and recognize that the length the fluid has traveled, L, is now replaced by the melt front position, $Z(t)$, such that

$$Q(t) = \frac{\pi R^3}{3 + \frac{1}{n}} \left[\frac{R P_0}{2 K Z(t)} \right]^{1/n} \tag{E7.1}$$

The position expression, $Z(t)$, is obtained from a mass balance:

$$Z(t) = \frac{1}{\pi R^2} \int_0^t Q(t) \, dt \tag{E7.2}$$

Differentiation of Eq. (E7.2) with respect to time gives

$$\frac{dZ(t)}{dt} = \frac{Q(t)}{\pi R^2} \tag{E7.3}$$

Substitution of Eq. (E7.1) into (E7.3) gives

$$Z(t) = R \left[\frac{t(1+n)}{1 + 3n} \right]^{\frac{n}{n+1}} \left(\frac{P_0}{2K} \right)^{\frac{1}{n+1}} \tag{E7.4}$$

Substitution of Eq. (E7.4) into Eq. (E7.1) gives the flow rate expression:

$$Q(t) = \pi R^3 \left(\frac{1+n}{1+3n} \right)^{n/(1+n)} \left(\frac{n}{1+n} \right) \left(\frac{P_0}{2K} \right)^{1/(1+n)} \left(\frac{1}{t^{1/(1+n)}} \right) \tag{E7.5}$$

Insertion of the appropriate values for R, n, and K gives the following simplified expressions:

$$Z(t) = 0.188 t^{0.281} \quad \text{and} \quad Q(t) = 1.07 \times 10^{-6} t^{-0.719}$$

Values for the flow front and flow rate at various times are listed in the table below. The results clearly indicate that we should expect a very high flow rate and quick runner filling initially, followed by a rapid drop in Q and long filling times for the remainder of the long runner. The first 50% of the runner is filled in 10% of the total runner time. This is the situation with a constant applied pressure and decreasing flow rate with time. In practice, the initial part of the mold filling cycle is one of increasing applied pressure and almost constant flow rate.

$t(s)$	$Z(m)$	$Q(m^3/s)$
0	0	∞
0.5	0.155	1.76×10^{-6}
1	0.188	1.06×10^{-6}
1.5	0.211	8×10^{-7}
2	0.228	6.5×10^{-7}
2.88	0.253	5×10^{-7}

polymers are given in Table 7.17. These values are used by calculating the dimensions of the mold cavity from the following equation:

$$\text{Mold dimension} = \text{Part dimension} \times (1 + \text{Shrinkage value}) \qquad (7.73)$$

Notice the strong dependence of shrinkage on the presence of a reinforcement such as fiberglass. Polymers without reinforcement have a substantially greater shrinkage than those with reinforcement. In general, crystalline polymers tend to have higher characteristic shrinkages than do amorphous polymers. To a first approximation, shrinkage is isotropic, although nonuniform part thicknesses can result in substantial orientation effects on shrinkage.

7.3.2.2 Compression Molding. Whereas the previous polymer processing techniques are used primarily for thermoplastic polymers, *compression molding*, also called *matched die molding*, is used almost exclusively for molding thermoset polymers. In

Table 7.17 Characteristic Shrinkages for Various Plastics

Shrinkage (mm/mm)	Type of Material
0–0.002	Polyester (thermoset) BMC, SMC
0.001–0.004	Polycarbonate, 20% fiberglass
0.002–0.008	Acrylic
0.002–0.003	PVC
0.004–0.007	ABS
0.004–0.006	Polystyrene
0.005–0.007	Polycarbonate
0.005–0.008	Polyphenylene oxide
0.008–0.015	Nylon (6/6)
0.010–0.020	Polypropylene
0.018–0.023	Acetal
0.007–0.025	LDPE
0.020–0.040	HDPE

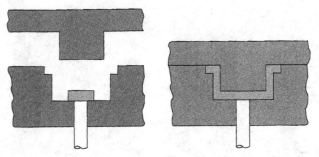

Figure 7.77 Schematic illustration of compression molding process with mold open (*left*) and mold closed (*right*). From Z. Tadmor and C. G. Gogos, *Principles of Polymer Processing*. Copyright © 1979 by John Wiley & Sons, Inc. This material is used by permission of John Wiley & Sons, Inc.

compression molding, a charge of thermosetting resin is placed in the cavity of a matched mold, as illustrated in Figure 7.77. The mold is closed by bringing the male and female halves together, and pressure is applied to squeeze the resin so that it uniformly fills the mold cavity. While under pressure, the material is heated, causing it to flow, crosslink, and harden. When the material is hard, the mold is opened and the part is removed. As with injection molding, then, this is a cyclic, or batch, operation.

The advantages of compression molding over other molding operations are the simplicity of the molds, which makes them less expensive to manufacture and maintain, and the relatively small amount of waste produced. Disadvantages include relatively slow cycle times and geometrical limitations to the moldable parts. Both compression molding and injection molding are used to produce discrete parts of widely differing geometries, but the difference in resin type commonly used between the two processes dictates very different mold conditions. In compression molding, the mold is heated so that the thermoset polymer will cure. In injection molding, the mold is cold so that the molten polymer will freeze. Since curing usually takes longer than freezing, the molding cycle for compression molding is longer than for injection molding, typically 1–2 minutes versus 20–60 seconds, respectively. Another consequence of making thermoset polymer parts by compression molding is that reject parts cannot be reprocessed. To overcome some of these limitations, *transfer molding* was developed. Transfer molding is a hybrid between compression molding and injection molding, in which the charge is melted in a separate pot, which is part of the heated mold, and then transferred under pressure by a ram, through runners and gates, into the mold cavity. Transfer molding cycles tend to be shorter than compression molding cycles. We will describe transfer molding in more detail in the context of composites processing (Section 7.4), for which it is most widely utilized.

Figure 7.78 represents the various stages of the compression molding cycle from the point of view of the plunger force needed to close the mold at a constant rate. In the first region, the force increases rapidly up to the fill time, t_f, as the preform is squeezed and heated. At t_f, the polymer is in the molten state and is forced to flow and fill the cavity. Filling terminates at t_c, when compression of the polymer melt begins, to compensate for the volume contraction that occurs during polymerization. The bulk of the curing process takes place after t_c. In order to derive relationships for the flow and pressure characteristics, then, we simply need to consider the time period $t_f < t < t_c$, since prior to t_f the polymer is solid and after t_c it is curing.

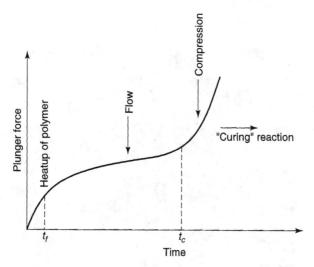

Figure 7.78 Schematic representation of the plunger force applied during compression molding. From Z. Tadmor and C.G. Gogos, *Principles of Polymer Processing.* Copyright © 1979 by John Wiley & Sons, Inc. This material is used by permission of John Wiley & Sons, Inc.

The flow characteristics can be determined using the model shown in Figure 7.79. During the time interval under consideration, it is assumed that the polymer has been heated to a uniform temperature, T_w, that is equivalent to the mold wall temperature. As long as the preform radius, R, is less than the radius of the outer wall of the mold cavity, R_o, we can treat the problem as an isothermal radial flow of an incompressible power law fluid flowing between two disks that approach each other at a constant rate, $\dot{h}$. In this way, the velocity field, $v_r(z, r, t)$, the pressure distribution in the mold, $P(z, r, t)$, and the plunger force, $F_N(z, r, t)$ can be obtained as follows:

$$v_r = \frac{nh^{\frac{n+1}{n}}}{n+1}\left(\frac{-\partial P}{K \partial r}\right)^{1/n}\left[1 - \left(\frac{z}{h}\right)^{\frac{n+1}{n}}\right] \tag{7.74}$$

$$P = P_a + \frac{-\dot{h}K R^{n+1}\left(\frac{2n+1}{n}\right)^n}{2^n(n+1)h^{2n+1}}\left[1 - \left(\frac{r}{R}\right)^{n+1}\right] \tag{7.75}$$

$$F_N = \frac{K\pi\left(\frac{2n+1}{n}\right)^n(-\dot{h})^n R^{n+3}}{2^n(n+3)h^{2n+1}} \tag{7.76}$$

where K and n are the power law parameters, and all other geometric variables are as defined in Figure 7.79. When the radius of the flowing preform reaches the mold wall (i.e., $R = R_o$), the fluid is forced to flow in the annular space $R_o - R_i$. Eqs. (7.74) through (7.76) can be modified to account for this flow. The pressure, in particular, rises substantially during annular flow. The interested reader is referred to the text by Tadmor and Gogos for more details on the addition of annular flow to these equations.

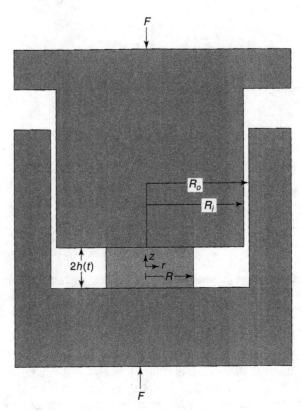

Figure 7.79 Model of compression molding process. From Z. Tadmor and C. G. Gogos, *Principles of Polymer Processing*. Copyright © 1979 by John Wiley & Sons, Inc. This material is used by permission of John Wiley & Sons, Inc.

After filling $(t > t_c)$, curing takes place, and the analysis switches from one of modeling flow to modeling heat transfer and kinetics. More details on the reaction kinetics will be given in Section 7.3.2.5 when we describe reaction injection molding, but for now we simply model the curing process as the reaction of trifunctional A groups, A_3, with bifunctional B groups, B_2, as follows:

$$A_3 + B_2 \longrightarrow \text{Crosslinked polymer} \tag{7.77}$$

The result is that the extent of reaction, p (cf. Section 3.3.1.1), can be related to the weight average molecular weight, $\overline{M}_w$, for a polyurethane system under equal stoichiometry:

$$\overline{M}_w = \frac{\frac{2}{3}(1 + p^2)M_{A_3}^2 + (1 + 2p^2)M_{B_2}^2 + 4pM_{A_3}M_{B_2}}{(\frac{2}{3}M_{A_3} + M_{B_2})(1 - 2p^2)} \tag{7.78}$$

where M_{A_3} and M_{B_2} are the molecular weights of species A_3 and B_2, respectively. The *gel point* is defined to be the condition when $\overline{M}_w$ goes to infinity, which occurs at $p = (\frac{1}{2})^{\frac{1}{2}}$. It turns out that the center of the curing polymer slab will gel faster than the skin. A knowledge of the temperature, conversion, and molecular weight distribution as a function of thickness and reaction time is essential in determining the required compression mold cycle or the time and temperature in the postcuring step.

7.3.2.3   *Blow Molding.* Blow molding is used to form hollow articles such as bottles. The process involves first forming and melting a preshaped sleeve called a *parison*, which is usually produced by extrusion. Hence, the term *extrusion blow molding* is sometimes used for this process. The parison is engaged between two mold halves, into which air is blown, causing the parison to take on the shape of the mold. The polymer quickly solidifies upon contacting the cold mold, and the finished hollow article is ejected. This procedure is schematically illustrated in Figure 7.80.

Blow molding is not the only process from which polymeric hollow articles can be made, but it does offer distinct advantages over other methods such as injection molding

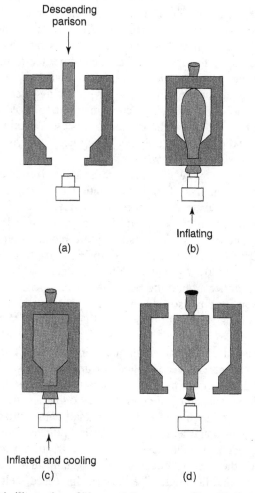

Figure 7.80   Schematic illustration of blow molding process. From Z. Tadmor and C. G. Gogos, *Principles of Polymer Processing*, Copyright © 1979 by John Wiley & Sons, Inc. This material is used by permission of John Wiley & Sons, Inc.

and *rotational molding* that makes it ideal for medium-sized hollow articles—for example, parts ranging from a few milliliters to over 500 liters in volume. Blow molding can create parts with much lower mold costs than injection molding and can create parts with narrow openings and wide bodies. Inflation pressures are quite modest, typically about 0.2 MN/m^2, so that the clamping forces on the mold are relatively small. The process is suitable for high production rates, and it is typically integrated with automatic polymer feeding systems and part removal. Cycles tend to be very short, so that equipment costs can be borne over many fabrication cycles. There are many variations to the blow molding process, primarily involving different methods of melt delivery and parison production. We will concentrate on continuous extrusion blow molding to illustrate the concepts of the blow molding process.

In extrusion blow molding, the parison is formed from an extrusion die. (This process is not to be confused with *blown film extrusion*, which is similar, but is used to form only films.) The molten material flows from the extruder through an adapter that changes the direction of the flow from horizontal to vertical, thus allowing gravity to act uniformly on the polymer melt. The material then enters the die and flows around a mandrel to create a cylindrical extrudate. The die may have a hole down the center of the mandrel for injecting the blown air, or the air is introduced through an inlet in the bottom of the mold. This procedure creates some interesting conflicts, namely, that the output from an extrusion operation is continuous, whereas the blow molding process is a cyclical one. There are methods for producing intermittent polymer flow, such as with the use of a reciprocating screw in injection molding, but the trend in extrusion blow molding is to keep the extrusion process a continuous one and match the cyclical molding process to the polymer flow rate. This type of blow molding is called *continuous extrusion blow molding*.

In continuous extrusion blow molding, the extruder is run continuously and its output matched through the use of multiple molds which seal and blow the parison. For example, if the mold cycle is no more than twice as long as the time to create a parison, a two-mold system can be used. Typically, multiple molds are required to match the extrusion rate of the parisons, so they are mounted on a rotating wheel, as illustrated in Figure 7.81. While one mold is closing to capture the parison, the mold ahead is in position for blowing of the part, and other molds are closed while the part is cooling. Still farther around the wheel the mold opens for part ejection and is prepared to capture another parison. In this system the rotational speed of the wheel and the number of molds mounted on the wheel are matched to the extrusion rate of the parisons.

From both the product design and economic points of view, the shape and thickness of the parison must be carefully controlled. In general, parison thicknesses are not uniform in the axial direction due to flow rate variations and gravitational effects. These parameters are neither predictable nor controllable, so that empirical relationships must be found. One such method used to quantitatively analyze the parison formation process and relate it to fundamental rheological properties utilizes four parison properties: parison diameter, product weight, severity of melt fracture, and pleating. *Pleating* involves buckling of the parison under its own weight, and it is a function of the parison wall thickness and the angle of extrusion. Increased melt strength and higher extrusion rates alleviate pleating. In this method, response surfaces are experimentally obtained, and by specifying minimum acceptable levels for each property, operating lines can be generated. An example of this procedure is shown in Figure 7.82 for

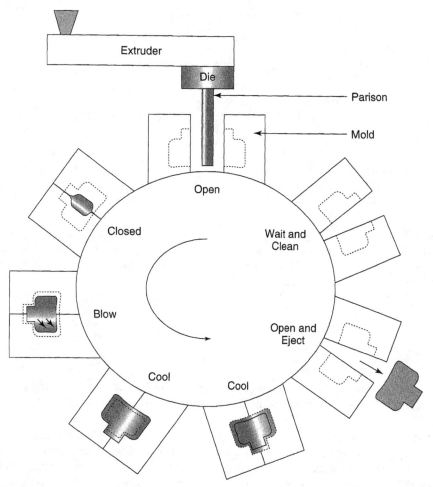

Figure 7.81 Schematic diagram of rotating mold system used in continuous extrusion blow molding. Reprinted, by permission, from A. B. Strong, *Plastics Materials and Processing*, 2nd ed., p. 489. Copyright © 2000 by Prentice Hall, Inc.

high-density polyethylene (HDPE). The heavy line represents the acceptable range of shot pressure and die gap values to produce the specified product. It should be noted that these results apply to HDPE only, and a new curve with new operating lines should be experimentally generated for each polymer under consideration.

Parison inflation is less difficult to model. In general, the parison is inflated very rapidly, and at a predetermined rate such that it does not burst while expanding. An approximate description of the blowing of a cylindrical parison of uniform radius R_i and thickness h_i to that of R_o and h_o can be obtained by assuming that the flow is planar extension, that the flow is isothermal, and that $h/R \ll 1$. In this instance, the parison inflation time, t, at constant inflation pressure, P, for a power-law polymer melt is

$$t = \frac{n}{2CP^{1/n}} \left[\left(\frac{1}{R_i} \right)^{2/n} - \left(\frac{1}{R_o} \right)^{2/n} \right] \tag{7.79}$$

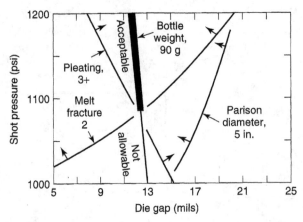

Figure 7.82 Operating diagram for the blow molding of HDPE. Arrows indicate regions of acceptable operation. From Z. Tadmor and C. G. Gogos, *Principles of Polymer Processing*, Copyright © 1979 by John Wiley & Sons, Inc. This material is used by permission of John Wiley & Sons, Inc.

where

$$C = \left(\frac{2\pi L}{VK}\right)^{1/n} \tag{7.80}$$

In Eqs. (7.79) and (7.80), K and n are the power-law parameters, L is the length of the parison, and V is the volume of the blown part, which is a function of both $R(t)$ and $h(t)$.

Parison cooling significantly impacts the cycle time only when the final parison thickness is large. In thin blown articles the mold is opened when the pinched-off parts have solidified so that they can be easily stripped off; thus they are the rate-controlling element in the cooling process. For fast blow molding of even very thin articles, the crystallization rate must be fast. For this reason, HDPE, which crystallizes rapidly, is ideally suited for blow molding, as are amorphous polymers that do not crystallize at all.

7.3.2.4 Thermoforming. *Thermoforming* is a process used to shape polymeric sheets and films into discrete parts. (Although there is a distinction between sheets and films based upon thickness, we will use the generic term "sheet" in this section to reflect the forming of both.) As with the other techniques in this section, thermoforming is a cyclic process. It differs from the others, however, insofar as it is technically a secondary shaping operation; that is, it utilizes sheets that must be formed by some primary shaping method such as extrusion or calendering. Other than that, many of the same processing principles apply to thermoforming as they do to other melt processing techniques, even though the polymer in thermoforming is only softened, instead of being melted.

In the thermoforming process, the sheet is heated to slightly above T_m or T_g and is placed in a clamp frame that clamps it along the part perimeter. The deformable sheet is then forced to conform to the shape of a mold by means of vacuum, air or gas pressure, or a plunger. There are thus three distinct categories of thermoforming:

vacuum forming, pressure forming, and *plug forming.* Schematic illustrations of the vacuum- and plug-forming processes are shown in Figure 7.83. Once the object has been shaped, cooling occurs by conductive heat transfer to the mold wall(s).

The main advantage of thermoforming is the low cost of the shaping equipment. The main disadvantage is that thermoforming is limited to simple shapes. Trimming operations are also often required after shaping, which increases cost and the amount of scrap. Cycle times can be long due to the slower heating and cooling rates for thick sheets.

We can model the thermoforming process by examining a preheated sheet that is vacuum-formed directly onto the walls of a cavity. The softened sheet undergoes a

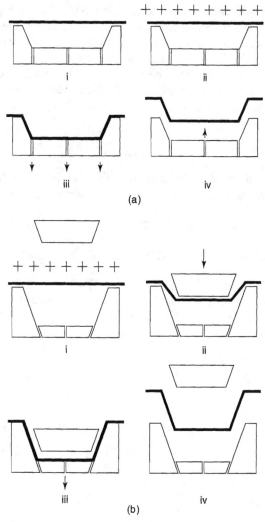

Figure 7.83   Schematic illustration of (a) vacuum-forming and (b) plug-forming variations of thermoforming. From Z. Tadmor and C. G. Gogos, *Principles of Polymer Processing.* Copyright © 1979 by John Wiley & Sons, Inc. This material is used by permission of John Wiley & Sons, Inc.

biaxial extension and thickness reduction, called *free blowing*, until it touches the cold mold surfaces. Further sheet expansion occurs as dictated by the mold in order for the rolling sheet to conform to the cavity. Thus, only the remaining free portion of the sheet continues to deform and decrease in thickness. The sheet cooling process commences upon contact with the cold mold surface. The analysis of thickness distribution in the part can be described through an example.

Consider the vacuum forming of a polymer sheet into a conical mold as shown in Figure 7.84. We want to derive an expression for the thickness distribution of the final, conical-shaped product. The sheet has an initial uniform thickness of h_0 and is isothermal. It is assumed that the polymer is incompressible, and it deforms as an elastic solid (rather than a viscous liquid as in previous analyses); the free bubble is uniform in thickness and has a spherical shape; the free bubble remains isothermal, but the sheet solidifies upon contact with the mold wall; there is no slip on the walls, and the bubble thickness is very small compared to its size. The present analysis holds for thermoforming processes when the free bubble is less than hemispherical, since beyond this point the thickness cannot be assumed as constant.

In Figure 7.84, we note that after a certain time the free bubble contacts the mold at height z and has a spherical shape of radius R. The radius is determined by the mold geometry and bubble position and is given by

$$R = \frac{H - z_k \sin \beta}{\sin \beta \tan \beta} \tag{7.81}$$

where H is the overall height of the cone, z_k is the slant height at the point of contact, and β is the angle indicated in Figure 7.84. The surface area of the bubble, A, is given by

$$A = 2\pi R^2 (1 - \cos \beta) \tag{7.82}$$

The thickness distribution of the wall is obtained by making a differential volume balance:

$$2\pi R^2 (1 - \cos \beta) h|_{z_k} - 2\pi R^2 (1 - \cos \beta) h|_{z_k + \Delta z_k} = 2\pi r h \Delta z_k \tag{7.83}$$

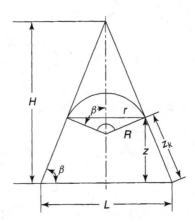

Figure 7.84 Schematic representation of thermoforming of a polymer sheet into a conical mold. From Z. Tadmor and C. G. Gogos, *Principles of Polymer Processing*. Copyright © 1979 by John Wiley & Sons, Inc. This material is used by permission of John Wiley & Sons, Inc.

where h is the thickness at z_k. The resulting differential equation for h, after substitution of $r = R \sin \beta$, is

$$\frac{-d}{dz_k}(R^2 h) = \frac{Rh \sin \beta}{1 - \cos \beta} \tag{7.84}$$

Differentiation of the expression for R from Eq. (7.81) gives

$$\frac{dR}{dz_k} = \frac{-1}{\tan \beta} \tag{7.85}$$

With Eqs. (7.85) and (7.81), Eq. (7.84) becomes

$$\frac{dh}{h} = \left(2 - \frac{\tan \beta \sin \beta}{1 - \cos \beta}\right) \sin \beta \frac{dz_k}{H - z_k \sin \beta} \tag{7.86}$$

Integration of Eq. (7.86) with the initial condition $h(0) = h_1$, where h_1 is the initial thickness of the bubble tangent to the cone at $z_k = 0$, gives

$$\frac{h}{h_1} = \left(1 - \frac{z_k \sin \beta}{H}\right)^{\sec \beta - 1} \tag{7.87}$$

Finally, the initial thickness h_1 can be related to the original sheet thickness h_0 with Eq. (7.82) where $R = L/2 \sin \beta$:

$$\frac{\pi L^2 h_0}{4} = \frac{\pi L^2 h_1 (1 - \cos \beta)}{2 \sin^2 \beta} \tag{7.88}$$

The thickness distribution, therefore, is given by

$$\frac{h}{h_0} = \frac{1 + \cos \beta}{2} \left(1 - \frac{z_k \sin \beta}{H}\right)^{\sec \beta - 1} \tag{7.89}$$

In a similar manner, thickness distributions can be derived in other relatively simple but more realistic molds, such as truncated cones. In such cases, the above model holds until the bubble comes in contact with the bottom of the mold at its center. From that point on, new balance equations must be derived.

7.3.2.5 Reaction Injection Molding*.
Reaction injection molding (RIM) is a process used to make polymeric articles by direct polymerization of two reactive monomeric species in a closed mold. It was developed to circumvent two major problems associated with injection molding: (a) the generation of sufficient, homogenized melt in the injection portion of the machine and (b) maintaining sufficient clamping pressure to keep the mold closed during the filling and packing stages of the injection molding cycle. Both of these issues become problematic as the size of the molded article increases. Thus, RIM is used to produced large molded parts such as automotive

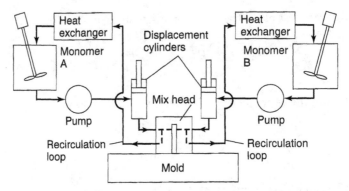

Figure 7.85   Schematic diagram of the reaction injection molding (RIM) process.

bodies. In the RIM process, two (or more) low-viscosity liquid streams are mixed prior to injection into the mold. This procedure is illustrated in Figure 7.85.

In many cases, the components are a polyol, which contain one or more –OH groups, and a diisocyanate, which contains two –NCO groups. These react to form a polyurethane (see Appendix 2). Other materials are also suitable for RIM, such as epoxies and nylons. The choice of monomers and additives yields a polymer with properties very much tailored to the function of the product, from flexible elastomers to rigid foams. Cycle times are in the range of 2–4 minutes, provided that the polymerization reaction is sufficiently fast; otherwise there is little advantage offered over injection molding. Foaming agents are often added to (a) ensure that the polymer completely fills the mold and (b) provide porous products with improved thermal insulating properties, for example.

The advantages of RIM, in addition to those cited above, are that the injection pressures are relatively small ($1-10$ MN/m^2), and homogenization of the polymer is readily accomplished. The primary disadvantages are that there are (a) nonisothermal and transient flow with chemical reaction during the filling stage and (b) conductive heat transfer with heat generation during the polymerization reaction, both of which are difficult to model. We address both of these issues with a model of the polymerization process.

Consider a long, thin mold being fed at constant temperature with two bifunctional monomers, AA and BB. The feed has a molecular weight of M_0, and the polymerization reaction, which is assumed to be reversible, proceeds by the reaction of A and B functional groups in an idealized step polymerization reaction (cf. Section 3.3.1.1):

$$AA + BB \underset{k_r}{\overset{k_f}{\rightleftharpoons}} AA - BB \tag{7.90}$$

For simplicity, we assume that the small molecule (such as water) typical of these reactions is not formed. The reaction under consideration, then, Eq. (7.90), is second order in A or B functional groups and first order in A–B bonds. The reaction rate is given by the rate of disappearance of AA or BB, or the rate of formation of AA–BB. This analysis was performed previously in Section 3.3.1.1 for the case of the forward reaction only. If we include the reverse reaction and assume that the concentrations of the two reactants are equal at all times, [AA] = [BB], the rate expression becomes

$$\frac{d[\text{AA}]}{dt} = -k_f[\text{AA}]^2 + k_r([\text{AA}]_0 - [\text{AA}]) \tag{7.91}$$

where the substitution [AA–BB] = [AA]$_0$ − [AA] has been employed. The forward and reverse reaction rate constants, k_f and k_r, follow the Arrhenius expression, Eq. (3.12).

Recall also from Section 3.3.1.1 that we introduced the extent of reaction, p, which is related to the degree of polymerization, $\bar{x}_n$. From Eq. (7.91), we can derive the following expressions for the number- and weight-average molecular weights in terms of the extent of reaction:

$$\overline{M}_n = \frac{M_0}{1 - p} \tag{7.92}$$

and

$$\overline{M}_w = M_0 \left(\frac{1 + p}{1 - p}\right) \tag{7.93}$$

To solve the filling flow and heat transfer problem with this reacting system, we need to specify the x-direction momentum and energy balances. The x-direction momentum equation during filling is

$$\rho \frac{\partial v_x}{\partial t} = \frac{-\partial P}{\partial x} + \eta \frac{\partial^2 v_x}{\partial y^2} + \frac{\partial \eta}{\partial y} \frac{\partial v_x}{\partial y} \tag{7.94}$$

where ρ is the density of the reacting fluid, P is the pressure, v_x is the velocity in the x direction, and η is the non-Newtonian viscosity, which can be described by a relationship such as the Ellis model (see Cooperative Learning Exercise 4.4) or the Carreau model:

$$\eta = \frac{\eta_0}{[1 + (\lambda \dot{\gamma})^2]^{(1-n)/2}} \tag{7.95}$$

where n is the power-law exponent, η_0 is the zero-shear rate viscosity, $\dot{\gamma}$ is the shear rate, and

$$\lambda = \frac{\lambda_0 \eta_0 \overline{M}_w^{0.75}}{\rho T} \tag{7.96}$$

where λ_0 is a curve-fitting parameter, T is temperature, and all other parameters are defined above. The zero-shear rate viscosity is a function of molecular weight, as well, and can incorporate variables to account for entanglements (cf. Section 4.1.3.2).

The energy equation for the filling stage is for constant density and thermal conductivity as follows:

$$\rho C_p \left(\frac{\partial T}{\partial t} + v_x \frac{\partial T}{\partial x}\right) = k \frac{\partial^2 T}{\partial y^2} + \eta \left(\frac{\partial v_x}{\partial y}\right)^2 - \Delta H \left(\frac{\partial p}{\partial t}\right) \tag{7.97}$$

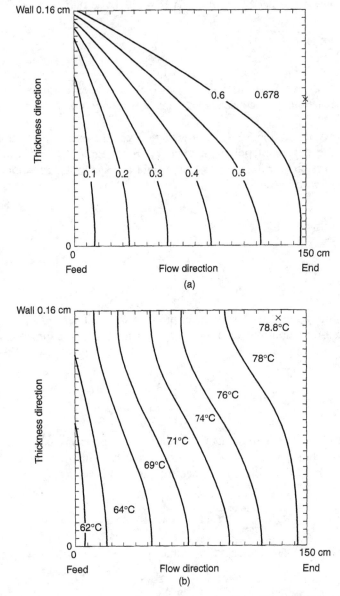

Figure 7.86 Simulation results of the RIM process for $T_0 = T_w = 60°C$, $k_f = 0.51/\text{mol} \cdot \text{s}$, and $t_{fill} = 2.4$ s. Contours in (a) are for conversion; those in (b) are for temperature. From Z. Tadmor and C. G. Gogos, *Principles of Polymer Processing*. Copyright © 1979 by John Wiley & Sons, Inc. This material is used by permission of John Wiley & Sons, Inc.

where ρ is the fluid density, C_p is its heat capacity, k is its thermal conductivity, η is its viscosity, T is its temperature, t is time, v_x is the velocity in the x direction, ΔH is the heat of the polymerization reaction [Eq. (7.97)], and p is the extent of the polymerization reaction.

Equations (7.96) and (7.97), along with the rate expression, Equation (7.91), can be used to analyze the conditions during the filling and polymerization portions of the RIM process. This is normally done by a *finite element analysis*, and not through analytical solution to the equations. We will not present the details of this analysis here, but rather show some representative results. Figure 7.86 shows some results of the finite element simulation for the filling and polymerization reaction conversion in a typical RIM process with stepwise polymerization. Both temperature and conversion increase with increasing flow direction, which are the result of larger residence times. Drastically different conversion and fill profiles are obtained when the fill time and forward reaction rate constant are changed.

At the end of the filling stage, only heat transfer with chemical reaction occurs, which can be described by the appropriate analogues of the species and energy balance equations, Eqs. (7.91) and (7.97), respectively. We will not present the solution to these equations here, but simply comment that the postfill cure process continues until the thickness-average tensile modulus is high enough at every position along the x direction that the part can be removed.

7.4 PROCESSING OF COMPOSITES

In this section we see how some of the processing methods described in previous sections can be modified to produce composite materials, and how new techniques are being developed to create composites with unique structures and properties. We will concentrate almost exclusively on fiber-reinforced matrix composites, especially those composites based on glass-fiber-reinforced polymers. There are three common methods for forming glass-fiber-reinforced polymer matrix composites: pultrusion, resin transfer molding, and filament winding. In the final section, we look at some vapor-phase-based synthetic methods for forming nonpolymeric matrix composites. Techniques not covered in this section include wet layup (also known as hand layup), autoclave molding, and mandrel wrapping.

7.4.1 Pultrusion

Pultrusion is derived from the terms "pulling" and "extrusion" and reflects a process that involves the pulling of continuous reinforcement fibers through a molten polymer (sometimes called a *resin*), through a die, and into a curing chamber. A schematic illustration of one simple variation on the pultrusion process is shown in Figure 7.87. Unlike extrusion, in which the material is pushed through a die, the material in pultrusion is pulled through the die. This is possible due to the continuous nature of the reinforcing fibers.

Pultrusion is a continuous process that is ideal for high throughput of constant-cross-section products. As long as the fiber volume passing through the die is constant, excess resin is squeezed out and returned to the resin bath. There is very little waste of the fiber reinforcement, and virtually any length of product is possible. Both solid and hollow cross-sectional profiles are possible, and inserts made of various materials can be encapsulated. Depending on the part complexity, resin viscosity, and cure schedule, the pultrusion line speed may range from 0.05 to 5 m/min. Unlike other molding processes, no external pressure is applied in the pultrusion process, other than the pulling force on the fiber tows.

The major reinforcing phase is typically continuous glass roving, which is a bundle composed of 1000 or more individual filaments. The most common glass roving used

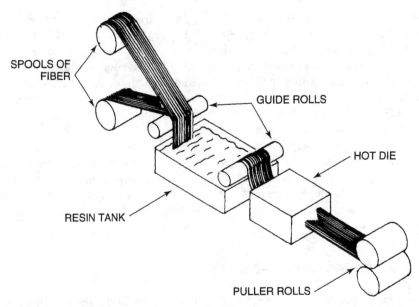

Figure 7.87   Schematic diagram of a typical pultrusion process. Reprinted, by permission, from M. Schwartz, *Composite Materials Handbook*, 2nd ed., p. 4.46. Copyright © 1992 McGraw-Hill.

in pultrusion is E-glass, although A-glass and graphite filaments are also common (cf. Tables 1.31–1.33). The total fiber content in a pultruded member is typically very high, over 70% by weight in some cases. In nearly all cases, the fibers are oriented along the axis of the pultrusion member, although mats or woven rings can also be incorporated to improve transverse strength and modulus.

In most applications, polyester and vinyl ester resins are used as the matrix materials. Epoxies are also used, although they require longer cure times and do not release easily from the pultrusion dies. Hence, thermosetting resins are most commonly used with pultrusion, although some high-performance thermoplastics such as PEEK and polysulfone can also be accommodated. In addition to the resin, the resin bath may contain a curing agent (initiator, cf. Section 3.3.1.2), colorants, ultraviolet stabilizer, and fire retardant.

The viscosity of the liquid resin, residence time of the fibers in the resin bath, and mechanical action of the fibers in the resin bath are adjusted to ensure complete wetting of the fibers by the resin, called *wet-out*. The fiber–resin stream is pulled first through a series of preformers and then through a long preheated die. The preformers distribute the fiber bundles evenly, squeeze out excess resin, and bring the material to its final configuration. Final shaping, compaction, and curing take place in the die, which has a gradually tapering section along its length. The entrance section of the die is usually water-cooled to prevent premature gelling, and the rest of the die is heated in a controlled manner by either oil or electric heaters. Infrared and microwave heating can also be employed to speed the curing process.

The most important factor controlling the mechanical performance of the pultruded members is the fiber wet-out. For a given resin viscosity, the degree of wet-out is improved as the residence time is prolonged by using slower line speeds or longer baths, the resin bath temperature is increased, and/or the degree of mechanical working is increased. The resin viscosity of commercial pultrusion lines may range from 400 to 5000 centipoise. Higher resin viscosities may result in poor fiber wet-out and slower line speed. On the other hand, very low resin viscosities may cause excessive resin draining from the fiber–resin stream after it leaves the resin bath.

Resin penetration takes place through capillary action as well as lateral squeezing between the bundles. The fiber and resin surface energies (see Appendix 4) are also important parameters in improving the amount of resin coating on fiber rovings. Thus, Kevlar 49 fibers, by virtue of their high surface energies, pick up more resin in the resin bath than either E-glass or carbon fibers under similar conditions.

As the fiber resin stream enters the heated die, the resin viscosity first decreases, which aids in the continued wet-out of uncoated fibers. Curing reactions commence a short distance from the die entrance, and soon the resin viscosity rapidly increases, as illustrated in Figure 7.88. The curing reaction continues at an increasing rate as the fiber–resin stream moves toward the die exit. Heat generated by the exothermic curing reaction raises the temperature in the fiber–resin stream. The location of the exothermic peak depends on the pulling speed, as illustrated in Figure 7.89. As the curing reactions finish, the exotherm temperature decreases and cooling begins. The rate of heat transfer from the cured material into the die walls is increased owing to a lower die temperature near the exit zone.

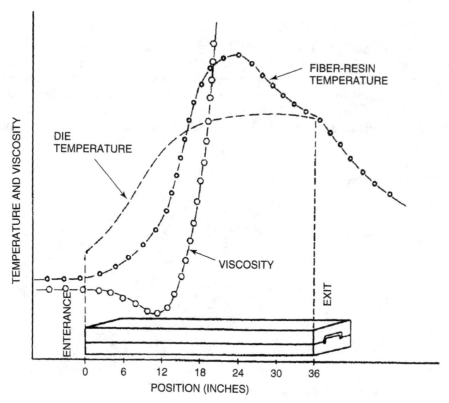

Figure 7.88 Viscosity change of a thermosetting resin in a pultrusion die. Reprinted, by permission, from P. K. Mallick, *Fiber-Reinforced Composites*, p. 349. Copyright © 1988 by Marcel-Dekker, Inc.

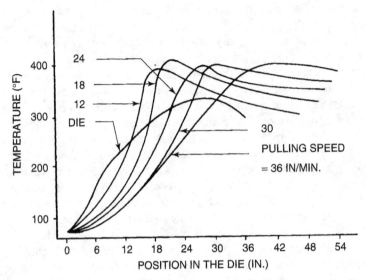

Figure 7.89 Temperature distribution along the length of a pultrusion die. Reprinted, by permission, from P. K. Mallick, *Fiber-Reinforced Composites*, p. 350. Copyright © 1988 by McGraw-Hill.

7.4.2 Resin Transfer Molding

Resin transfer molding (RTM) is a hybrid of two polymer processing techniques we have already described: transfer molding and reaction injection molding (RIM). The key difference is that a reinforcement is incorporated during molding to create a composite.

Several layers of dry, continuous-strand mat, woven roving, or cloth called *preform* are placed in the bottom half of a two-part mold, either by a hand-layup or by an automated technique (see Figure 7.90). The mold is closed, and a catalyzed liquid

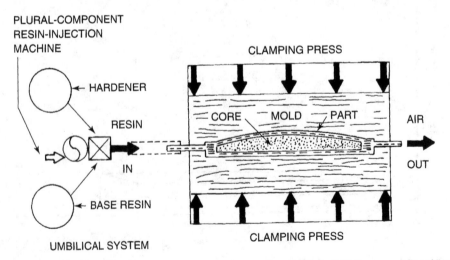

Figure 7.90 Schematic diagram of a typical resin transfer molding (RTM) process. Reprinted, by permission, M. Schwartz, *Composite Materials Handbook*, 2nd ed., p. 4.56. Copyright © 1992 by McGraw-Hill.

resin is injected into the mold via a centrally located sprue. The resin injection point is usually the lowest point of the mold cavity. The injection pressure is in the range of 70–700 kPa. Care must be taken to inject slowly enough that the fibers in the preform are not moved significantly as the resin fills the mold. As the resin spreads throughout the mold, it displaces the entrapped air through the air vents and impregnates the fibers. The mold can be held under vacuum to assist in air removal and allow inward resin flow. Alternatively, the edges of the mold are vented to allow air to escape as the resin enters. The local speed of the resin will be determined by the local permeability of the fiber material, the viscosity of the resin, and the local pressure decay. The permeability of the fiber material is low if its density is high, so a high-fiber volume fraction can only be realized at the cost of low production rate. By summation, these factors determine the total filling time. When excess resin begins to flow from the vent areas of the mold, the resin flow is stopped and the molded part begins to cure. When cure is completed, which can take from several minutes to several hours, it is removed from the mold and the process can begin again. As with other molding techniques, it is often necessary to trim the part at the outer edges once it is removed from the mold.

The reinforcements amenable to RTM are similar to those used for pultrusion, except that they need not be continuous in nature. Thus, E-glass, S-glass, aramid, and carbon fibers are commonly used, as are discontinuous filaments such as wood fiber and polyesters. Even metal and ceramic fibers can be used in this technique. In one method, the preform is fabricated by spraying 12- to 75-mm-long chopped fiber rovings onto a preshaped screen. A binder sprayed with the fibers keeps them in place and holds the preform shape, which is then placed in the mold.

Although polyurethanes or other reactive resins can be used in RTM, the most common resins are polyesters and epoxies. The two pumping reservoirs in Figure 7.90, then, contain polyester resin and initiator, or epoxy resin and hardener, respectively. Epoxies require relatively long cycle times. When cycle time is critical, low viscosity vinyl ester, acrylamate, or urethanes can be injected very rapidly into the mold. Even the use of thermoplastics is possible.

Compared with compression molding, RTM has very low tooling costs and simple mold clamping requirements. Another advantage of RTM is its ability to encapsulate metal ribs, stiffeners, and inserts within the molded product. Size capability is among the major benefits of RTM, and although RTM is well-suited to molding large objects, part complexity may restrict the use of the process.

A related process is called *structural reaction injection molding* (SRIM). Preform and mold preparation are similar to the RTM process, with changes in mold release and reinforcement sizing made to optimize their chemical characteristics for the SRIM chemistry. Once the mold has been closed, the SRIM resin is rapidly introduced into the mold and reacts quickly to cure fully within a few seconds. Therefore, wet-out and air displacement occur rapidly. No postcure is normally needed. RTM resins are typically two-component low-viscosity liquids in the range of 100–1000 cpoise, require preinjection mixing ratios in the range of 100:1 and can be mixed at low pressure using a static mixer. SRIM resins, on the other hand, are typically two-part, low-viscosity liquids being in the viscosity range of 10–100 cpoise at room temperature. They are highly reactive in comparison to RTM resins and require very fast, high-pressure impingement mixing to achieve thorough mixing before entering the mold. Mix ratios of typical systems are near 1:1, which is desirable for rapid impingement mixing. Preforms for RTM and SRIM are similar in most respects.

7.4.3 Filament Winding

In the *filament winding* process, a band of continuous resin-impregnated rovings or monofilaments is wrapped around a rotating mandrel and cured to produce axisymmetric hollow parts. Among the applications of filament winding are automotive drive shafts, helicopter blades, oxygen tanks, pipes, spherical pressure vessels, and large underground storage tanks. Figure 7.91 shows a schematic diagram of a basic filament winding process. A large number of fiber rovings are pulled from a series of rolls called *creels* into a liquid resin bath containing liquid resin, catalyst, colorants, and UV stabilizers, just as in the pultrusion process. Fiber tension is controlled by fiber guides or scissor bars located between each creel and the resin bath. Just before entering the resin bath, the rovings are usually gathered into a band by passing them through a textile thread board or a stainless steel comb.

At the end of the resin tank, the resin-impregnated rovings are pulled through a wiping device that removes the excess resin from the rovings and controls the resin coating thickness around each roving. Once the rovings have been impregnated and wiped, they are gathered together in a flat band and positioned on the mandrel. The band former is usually located on a carriage, which traverses back and forth parallel to the mandrel, like a tool stock in a lathe machine. The traversing speed of the carriage and the winding speeds of the mandrel are controlled to create the desired winding angle patterns. Typical winding speeds are on the order of 100 linear m/min. After winding a number of layers to generate the desired reinforcement thickness, the filament-wound part is generally cured on the mandrel. The mandrel is then extracted from the cured part.

There are a number of material options available for filament winding. Typical fibers are fiberglass, carbon and aramid. Fiberglass is still the most commonly used

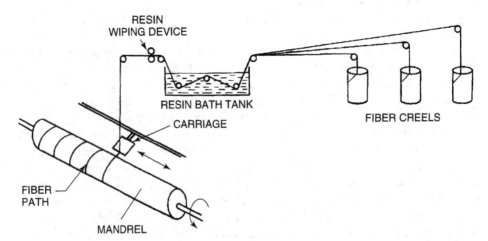

Figure 7.91 Schematic diagram of the filament winding process. Reprinted, by permission, from P. K. Mallick, *Fiber-Reinforced Composites*, p. 352. Copyright © 1988 by Marcel Dekker, Inc.

reinforcing fiber, and the choices are similar to those found in the other composite-forming techniques, with E-glass and S-glass rovings being the most common.

The typical resin systems include thermoset polyesters, vinyl esters, epoxies, polymidies, bismaleimide, and phenolics. Thermoplastics are also finding their way into filament winding. Wet thermoset filament winding requires a resin with viscosity in the range of 1000–3000 cpoise. Resin components are chosen on the basis of pot life, winding temperature, viscosity, gel time, and cure time.

Filament winding patterns are designated in one of two primary ways: helical and polar. In *helical winding*, fiber is fed from a horizontal translating delivery head to a rotating mandrel. The angle of the roving band with respect to the mandrel axis is called the *wind angle* or *wrap angle*, as illustrated in Figure 7.92. The wind angle is specified by the ratio of the two relative motions. For a circular mandrel of radius r, rotating with a constant rotational speed of N revolutions per minute and a constant carriage feed of V, the wind angle, θ, is given by

$$\theta = \frac{2\pi N r}{V} \tag{7.98}$$

Wind angles vary from 20° to 90°, with a 90° wind angle called a *hoop* or *circumferential winding* pattern. In *polar winding*, a delivery unit races around a slowly indexing mandrel, as illustrated in Figure 7.93. Technically speaking, polar winding is a helical winding pattern with a wind angle near 0°, typically between 5° and 15°. The principal advantage of polar winding is that it is a simple and rapid winding technique for short geometries with length-to-diameter ratios less than two. The winding pattern is a function of the desired winding angle, the polar port openings, the cylinder length, and the dwell time as the fiber direction is reversed at the end of the dome. The pattern can contain any number of circuits, but usually ranges from 5 to 15. The number of patterns to complete a layer requires adjustment; otherwise a gap or overlap will occur between the first and last pattern.

The mechanical properties of the filament wound composite are a complex combination of (a) winding pattern and thickness and (b) resin and filament properties. The mechanical properties of the dome regions, in particular, require specialized treatment. The interested reader is referred to the text by Lubin for more information.

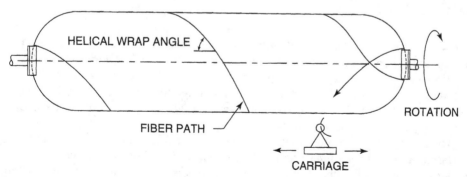

Figure 7.92 A helical winding pattern. Reprinted, by permission, from P. K. Mallick, *Composites Engineering Handbook*, p. 184. Copyright © 1997 by Marcel Dekker, Inc.

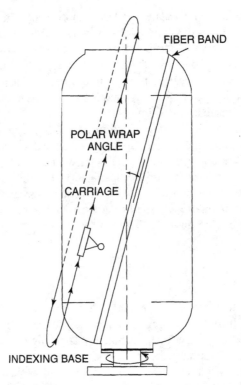

Figure 7.93 A polar winding pattern. Reprinted, by permission, from P. K. Mallick, *Composites Engineering Handbook*, p. 184. Copyright © 1997 by Marcel Dekker, Inc.

7.4.4 Infiltration Processing of Composites

In Section 3.4.2, we introduced the concept of chemical vapor infiltration, CVI, in which a chemical vapor deposition process is carried out in a porous preform to create a reinforced matrix material. In that section we also described the relative competition between the kinetic and transport processes in this processing technique. In this section we elaborate upon some of the common materials used in CVI processing, and we briefly describe two related processing techniques: *sol infiltration* and *polymer infiltration*.

7.4.4.1 Chemical Vapor Infiltration (CVI). Recall from Section 3.4.2 that CVI is primarily used to create ceramic matrix composites, CMCs. Fabrication of CMCs by CVI involves a sequence of steps, the first of which is to prepare a preform of the desired shape and fiber architecture. This is commonly accomplished by layup onto a shaped form of layers from multifilament fibers using some of the techniques previously described, such as filament winding.

The second step is to deposit a thin layer of a suitable interface material onto the surface of each filament in the preform. The purpose of this layer is twofold: to provide a barrier to protect the fiber from the CVI atmosphere and to prevent strong bonding between the fiber and the matrix so that high toughness is achieved be debonding, bridging, and fiber pull-out (cf. Section 5.4.2.7).

The third step is to heat the preform in a sealed chamber and pass a mixture of gases into the chamber that will react when they contact the hot fibers to form and deposit the desired chemical constituents of the matrix. The deposition rate is very slow and becomes even slower as the thickness of the deposit increases and the permeability of the preform decreases. To achieve high levels of densification, the partially densified part is removed from the CVI chamber, the surface is machined to reopen pore channels, and the part is returned to the chamber for further infiltration. This procedure is typically repeated a number of times to achieve a composite density of over 80% (<20% porosity).

The primary advantage of CVI over hot pressing is that complex shapes and larger size parts can be fabricated (cf. Section 7.1.3.2). Another advantage of CVI is the relatively low temperature of matrix formation. A variety of matrix materials can be deposited by CVI at temperatures between 900°C and 1200°C. This is substantially lower than the temperatures required for hot pressing. The disadvantage, as with most infiltration processes, is the high porosity of the final composite, which is typically 15% or higher. Furthermore, the reaction times required to reach even these high levels of porosity are large. Days or weeks in a standard reactor can be required, depending on the thickness, fiber volume fraction, and target density.

As noted earlier, CVI is used primarily to form ceramic-fiber-reinforced ceramic matrix composites. The most common of these combinations is SiC fiber/SiC matrix composites. One commercially available product has a two-dimensional 0/90 layup of plain weave fabric and fiber volume fraction of about 40%. This same composite can be fabricated with unidirectional fibers and with ±45° architectures. The most commonly used SiC fiber for the preforms is Nicalon®, the mechanical properties for which were provided earlier in Section 5.4.2.7. A number of other carbide and nitride fibers are also available, including Si_3N_4, BN, and TiC. Preform geometries can be tailored to the application in order to maximize strength and toughness in the direction of maximum stresses. The reactions used to form the matrix are similar to those used in CVD processes (cf. Section 7.2.4) and those described previously in Eq. (3.105).

The largest commercial production facilities employ the isothermal–isobaric CVI process to produce mostly SiC/SiC or C/SiC composites. The hot zones of these reactors measure 3 m or larger and can accommodate hundreds of parts.

7.4.4.2 Sol Infiltration.
Sol infiltration involves the use of sol–gel techniques, which were described earlier in Section 7.2.5. To fabricate a CMC by a sol–gel approach, the fiber preform is vacuum-infiltrated with the sol. The sol provides a thin coating on the fibers. The sol is converted to a gel and then dried and heat-treated to yield the polycrystalline ceramic matrix. The challenge is that sols have very low ceramic components, typically under 5% by volume. This requires many infiltration cycles. In one instance, 30 infiltration cycles and heat treatments are required to achieve over 80% theoretical density of the matrix. Some of the typical sol–gel approaches used for CMC matrices are listed in Table 7.18. Some of the more common fiber–matrix combinations include Al_2O_3/Al_2O_3, $SiZrO_4$/zirconium titanate, SiC/BN, SiC/SiO_2–GeO_2, and Al_2O_3/ZrO_2.

7.4.4.3 Polymer Infiltration.
Polymer infiltration is similar to sol infiltration. The polymer is introduced in liquid form under vacuum and coats the fibers. The polymer is then dried and pyrolyzed to yield the desired composition. Multiple infiltration cycles

Table 7.18   Examples of Sol–gel Approaches For Preparing Oxide-Matrix Composites

Matrix	Used Precursor	Gel Formation Process
SiO_2	Tetraethoxysilane (TEOS)	HCl catalysis in ethanolic solution
TiO_2–SiO_2	TEOS + Ti$(OC_2H_5)_4$	Simultaneous hydrolysis HCl catalyzed in ethanolic solution
	TEOS + Ti(iso-$OC_3H_7)_4$	
GeO_2–SiO_2	TEOS + tetraethoxygermane (TEOG)	HCl catalysis in ethanolic solution
	TEOS + $GeCl_4$	Ethanolic solution without additional catalyst
B_2O_3–SiO_2	TEOS + B_2O_3 hydr.	HCl catalyst in ethanolic solution
Al_2O_3–SiO_2	TEOS + Al (sec-OC_4H_9)	Aqueous ethanolic solution
Al_2O_3	Al (iso-$C_3H_7)_3$	HCl catalysis in ethanolic solution
	Al (sec-$OC_4H_9)_{13}$	Two-step (NH_4OH/HCl) catalysis
ZrO_2–SiO_2	TEOS + $ZrCl_4$	Ethanolic solution without catalyst

Source: Mallick, P.K., *Composites Engineering Handbook*, Marcel Dekker, Inc., 1997.

are required, but usually not as many as for a sol. Most of the polymers have a higher ceramic yield than sols, so more matrix is deposited during each cycle. However, the polymers also have a higher viscosity than sols, so infiltration is more difficult and may require pressure and/or elevated temperature of dilution of the polymer to achieve uniform infiltration. Dilution reduces the ceramic yield and increases the number of cycles necessary to achieve the target density. The process is analogous to fabrication of polymer matrix composites, except that a pyrolysis step is added to decompose the polymer to the ceramic composition, and multiple infiltration cycles are required.

Polymer infiltration has the same limitations as CVI and sol infiltration. It is difficult to achieve densifications over about 85%, and multiple cycles present a challenge to achieving cost. In addition, some of the polymers require higher temperatures to fully pyrolyze and crystallize the resulting ceramic composition, in some instances over $1400°C$. This increases the difficulty of finding an interface coating that will protect the fibers during processing and prevent strong bonding between the fibers and matrix.

Development has been conducted to fabricate SiC-based matrix/Si-C-based fiber composites by polymer infiltration. Nicalon fibers are used, and various polymers such as polysilazanes, polysiloxanes, and polycarbosilanes are used to yield matrices of SiCO, SiNC, and SiC. As with CVI of SiC/SiC composites, carbon acts as an interface layer but does not result in stability at high temperatures.

7.5   PROCESSING OF BIOLOGICS

The processing of biologics on a scale comparable to that of the previous material classes is as yet unknown, at least with regard to biomaterials for use in the human body. Most artificial biologics, both hard and soft, serve as scaffolds for the *in situ* manufacture of tissues; that is, the human body is actually the processing center in which the raw materials it provides are used to fabricate some portion of the component (e.g., bone or skin). Even when the component is fabricated almost entirely out of artificial materials, such as in an artificial hip, the device fabrication process is performed almost entirely by hand in order to meet case-dependent specifications. Furthermore, most of the artificial materials have already been processed into useful shapes using any of the previously described processing techniques, such as casting of a titanium rod or extrusion of a polymer sheet.

Despite these observations, there are some common processing techniques that are emerging in the area of processing of biologics, as self-contained biomedical devices become more widely utilized. We examine two such areas in this section: collagen processing and surface modification.

7.5.1 Processing of Collagen

As described in Section 1.5.3, collagen is one of the most useful and abundant substances in the human body. Both skin and tendon are composed of collagen, and in its purified form, collagen can be used for a wide variety of tissue repair applications. Let us briefly examine the processing of collagen.

There are two distinct ways in which collagen can be isolated and purified. One is a molecular-based technology, and the other is fibrillar technology. The isolation and purification of soluble collagen molecules from a collagenous tissue is achieved by using a proteolytic enzyme such as pepsin to cleave the natural crosslinking sites of collagen called *telopeptides*. This process renders the collagen soluble in aqueous solution. The pepsin-solubilized collagen can be purified by repetitive precipitation with a neutral salt. The collagen molecules may be reconstituted into fibrils of various polymorphisms. However, the reconstitution of the pepsin-solubilized collagen into fibrils of native molecular packing is not as efficient as the intact molecules.

In the second method, collagen fibers are isolated and purified upon the removal of noncollagenous materials from the collagenous tissue. Salt extraction removes the newly synthesized collagen molecules that have not been covalently incorporated into the collagen fibrils. Salt also removes the noncollagenous materials that are soluble in aqueous conditions and are bound to collagen fibrils by nonspecific interactions. Lipids are removed by low-molecular-weight organic solvents such as ethers and alcohols. Acid extraction facilitates the removal of acidic proteins and glycosaminoglycans due to weakening of the interactions between the acidic proteins and collagen fibrils. Alkaline extraction weakens the interaction between the basic proteins and collagen fibrils and thus facilitates the removal of basic proteins. Purified collagen fibers can be obtained through these sequential extractions and enzymatic digestions from the collagen-rich tissues.

The purified collagen obtained from either of these techniques is subjected to additional processing to fabricate the materials into useful devices for specific medical applications. Some of these matrices and their medical applications are shown in Table 7.19 and will be briefly described.

Collagen membranes can be produced by drying a collagen solution or a fibrillar collagen dispersion cast on a nonadhesive surface. The thickness of the membrane is dictated by the concentration and the initial thickness of the casting. In general, membrane thicknesses of up to 0.5 mm can be obtained in this way. Additional chemical crosslinking is required to stabilize the membrane from dissolution. The casting process does not permit manipulation of the pore structure, and the cast membrane is dense and amorphous with minimal permeability to macromolecules.

Table 7.19   Some Examples of Collagen Matrices and Their Medical Applications

Matrix Form	Medical Application
Membrane (sheet, film)	Oral tissue repair; wound dressings; dura repair; patches
Porous (sponge, felt)	Hemostats; wound dressings; cartilage repair; soft tissue augmentation
Gel	Drug delivery; soft- and hard-tissue augmentation
Solution	Drug delivery; soft- and hard-tissue augmentation
Filament	Tendon and ligament repair; sutures
Tubular	Nerve repair; vascular repair
Collagen/synthetic polymer	Vascular repair; skin repair; wound dressings
Collagen/biological polymer	Soft-tissue augmentation; skin repair
Collagen/ceramic	Hard-tissue repair

Source: *The Biomedical Engineering Handbook*, 2nd edition, J.D. Bronzino, editor. Copyright © 2000 by CRC Press.

Porous membranes may be obtained by freeze-drying a casting or by partially compressing a preformed matrix to a predetermined density and pore structure. The freeze-dried porous matrix also requires chemical crosslinking to stabilize the structure. A convenient way to do this is to use a vapor of a volatile crosslinking agent such as formaldehyde or glutaraldehyde. The pore structure depends on the concentration of the collagen in the casting. Other factors that contribute to the pore structure include the rate of freezing, the size of fibers in the dispersion, and the presence or absence of other macromolecules. Apparent densities from 0.05 to 0.3 g/cm^3 can be obtained. The pores in these matrices range in size from 50 to 15,000 μm in diameter.

A *gel matrix* is defined as a homogeneous phase between a liquid and a solid. As such, it may vary from a simple viscous liquid to a highly concentrated slurry. Collagen gels may be formed by shifting the pH of a dispersion away from its isoelectric point. Alternatively, the collagen material may be subjected to a chemical modification to change its charge profile to a net positively charged or net negatively charged protein before hydrating the material.

A *collagen solution* is obtained by dissolving the collagen molecules in an aqueous solution. These solutions are prepared as previously described using pepsin. The solubility of collagen depends on the pH, temperature, ionic strength, and molecular weight. Collagen molecules aggregate into fibrils when the temperature of the solution increases to body temperature. Collagen is more soluble at a pH away from the isoelectric point of the protein and is less soluble at higher ionic strength of the solution. The solubility of collagen decreases with increasing size of molecular aggregates. Thus, collagen becomes increasingly less soluble with increasing crosslink density.

Collagen filaments can be produced by extrusion techniques, in which a collagen solution or dispersion in the concentration range 0.5–1.5% weight to volume is first prepared. Collagen is extruded into a coacervation bath containing a high concentration of a salt or into an aqueous solution at a pH of the isoelectric point of the collagen. Tensile strengths of 30 MPa have been obtained for the reconstituted filaments. *Tubular matrices* may be formed by extrusion through a coaxial cylinder or by coating collagen onto a mandrel. Different properties of the tubular membranes can be obtained by controlling the drying properties.

Finally, collagen can form a variety of *collagen composites* with other water-soluble materials. Ions, peptides, proteins, and polysaccharides can all be uniformly incorporated into a collagen matrix. The methods of composite formation include ionic and covalent bonding, entrapment, entanglement, and co-precipitation. A two-phase composite can be formed between collagen, ceramics, and synthetic polymers for specific biomedical applications.

7.5.2 Biologic Surface Modification

The increasing demand for synthetic biomaterials, especially polymers, is mainly due to their availability in a wide variety of chemical compositions and physical properties, their ease of fabrication into complex shapes and structures, and their easily tailored surface chemistries. Although the physical and mechanical performance of most synthetic biomaterials can meet or even exceed that of natural tissue (see Table 5.15), they are often rejected by a number of adverse effects, including the promotion of thrombosis, inflammation, and infection. As described in Section 5.5, biocompatibility is believed to be strongly influenced, if not dictated, by a layer of host proteins and cells spontaneously adsorbed to the surfaces upon their implantation. Thus, surface properties of biomaterials, such as chemistry, wettability, domain structure, and morphology, play an important role in the success of their applications.

Several methodolgies have been developed for modifying the surfaces of polymeric materials to manipulate their interactions with biological environments, as summarized in Table 7.20. Biomaterial surfaces can be simply modified by physical deposition or adsorption of amphiphilic molecules, polymers, or proteins. However, such physical coatings are normally unstable. To overcome this limitation, chemical modifications to the surface through liquid- or gas-phase derivatization are usually favored. For example, strong oxidizing acids have been used to introduce carbonyl and carboxylic acid groups to polyolefinic surfaces. The treatment of polyolefins with ozone yields surfaces with hydroxyl, carbonyl, and carboxylic acid groups. Flame and thermal treatments induce radicals and chains scission on polymer surfaces. Also, ion beam treatments are utilized to induce chemical and morphological changes of the surface and can be used to incorporate various elements into the polymer surface. Laser treatments, which offer a combination of thermal and ultraviolet effects, have been used to induce chemical and morphological changes, as well. These surface modification methods can also

Table 7.20   Some Surface Modification Methods for Biomaterials

Physical Modification Methods	Chemical Modification Methods	Radiation Modification Methods
Physical adsorption	Oxidation by strong acids	Glow-discharge plasma
Langmuir-Blodgett film	Ozone treatment	Corona discharge
	Chemisorption	UV photoactivation
	Flame treatment	Laser
	Ion beam	Electron beam
		Gamma irradiation

Source: Sheu, M.S., D.M. Hudson, and I.H. Loh, in *Encyclopedic Handbook of Biomaterials and Bioengineering*, Part A: Materials, Vol. 1. Copyright © 1995 by Marcel Dekker, Inc.

be used to induce further surface functionalization via grafting methods. Radicals on the surfaces, once created, can act as surface-bound initiators for grafting reactions to monomers. For example, acrylic acid has been grafted to poly(ethylene terephthalate) by radiation or plasmas and has been grafted to polyethylene by electron beam, radiation, or plasma treatment.

Let us briefly examine plasma treatment in more detail as an example of how the surface of these biologics can be modified to improve biocompatibility. Recall that a plasma is a partially ionized complex gas composed of electrons, ions of both charges, free radicals, photons of various energies, and gas atoms and molecules in both the ground and excited electronic states. There are two general types of plasma: a *cold plasma*, also known as a low-temperature plasma, and a *hot plasma*. A hot plasma is generated by electrical arcs produced at atmospheric pressure, and it is a highly ionized discharge in thermal equilibrium. Cold plasmas are less ionized and are usually produced by low-pressure glow discharges. They are also called *glow-discharge plasmas*. A typical plasma system is illustrated in Figure 7.94, and consists of a reaction chamber, a vacuum system, a radio-frequency (RF) generator, a gas flow controller, and a computer-controlled process controller. A plasma system is sometimes equipped with an optical emission spectrometer to analyze active gas species present in the reactor. Plasma surface treatment is usually performed in a batch process, although continuous on-line treatment of fibers, tubing, and films is also common.

The plasmas are in a nonequilibrium state in which the Boltzmann temperature of the electrons is roughly two orders of magnitude higher than that of the gas ions and molecules. This lack of thermal equilibrium makes it possible to obtain a plasma in which the gas temperature is near ambient conditions, while the electrons

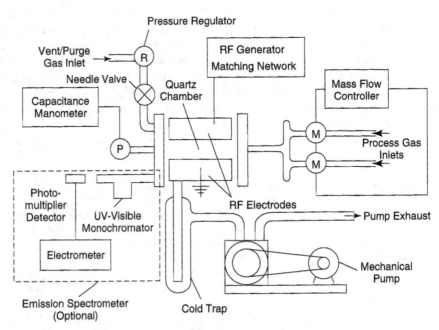

Figure 7.94   Schematic diagram of a typical plasma reactor system. Reprinted, by permission, from M. S. Sheu, D. M. Hudson, and I. H. Loh, in *Encyclopedic Handbook of Biomaterials and Bioengineering*, Part A: *Materials*, Vol. 1, p. 873. Copyright © 1995 by Marcel Dekker, Inc.

are sufficiently energetic to cause rupture of the molecular bonds at the material surface. It is this characteristic that makes glow-discharge plasmas well-suited for the promotion of chemical reactions involving thermally sensitive materials. Although the glow discharge is in a thermodynamically nonequilibrium state, the plasma reaction is a kinetically steady-state reaction. In other words, the rate of active species generated is equivalent to the rate of active species consumed on the plasma–solid interface. It is this steady-state condition that makes the plasma surface treatment so valuable.

A broad range of chemical and physical reactions occurs in a plasma. The following are some of the more significant electron–gas and gas–gas reactions in a plasma of a two-atom gas.

Excitation	$A_2 + e^- \longrightarrow A_2^* + e^-$	(7.99a)
Dissociation	$A_2 + e^- \longrightarrow 2A + e^-$	(7.99b)
Electron attachment	$A_2 + e^- \longrightarrow A_2^-$	(7.99c)
Dissociative attachment	$A_2 + e^- \longrightarrow A + A^-$	(7.99d)
Ionization	$A_2 + e^- \longrightarrow A_2^+ + 2e^-$	(7.99e)
Photoemission	$A_2^* \longrightarrow A_2 + h\nu$	(7.99f)
Abstraction	$A + B_2 \longrightarrow AB + B$	(7.99g)

During the plasma surface reaction, the plasma and the solid are in physical contact, but electrically isolated. Surfaces in contact with the plasma are bombarded by free radicals, electrons, ions, and photons, as generated by the reactions listed above. The energy transferred to the solid is dissipated within the solid by a variety of chemical and physical processes, as illustrated in Figure 7.95. These processes can change surface wettability (cf. Sections 1.4.6 and 2.2.2.3), alter molecular weight of polymer surfaces or create reactive sites on polymers. These effects are summarized in Table 7.21.

The wettability changes of polymers after plasma treatment are possibly due to the oxidation, unsaturation effects, electrostatic charging, and surface morphology changes. Improvement of surface wettability is one of the major interests in biomaterials surface modifications due to the fact that most common polymeric biologics are hydrophobic in nature, such as PE, PP, PET, PMMA, PS, PTFE, PVC, polyurethane, and silicone rubber. Hydrophilic surfaces can be obtained by treating polymers with oxygen, nitrogen, or water plasmas.

Molecular weight changes are due to such surface interactions as scission, branching, and crosslinking. In an inert gas plasma, molecular weight changes are due only to crosslinking or degradation. In addition to molecular weight changes, minor surface oxidation also can be observed on the inert-gas-plasma-treated surface when the surface is exposed to air after the treatment. This oxidation may affect the surface wettability.

Most polymeric biomaterials surfaces are chemically inert. In order to extend their biological performance, there is always great interest in functionalizing their surfaces for further immobilization of bioactive molecules such as enzymes, antibodies, ligands (cf. Section 2.5.2), and drugs. Hydroxyl, carboxylic, and primary amine groups are some of the commonly-desired reactive centers that can be generated by plasma treatment, as listed in Table 7.21. However, functional groups generated on the plasma-modified surface are usually not homogeneous due to the presence of undesired side

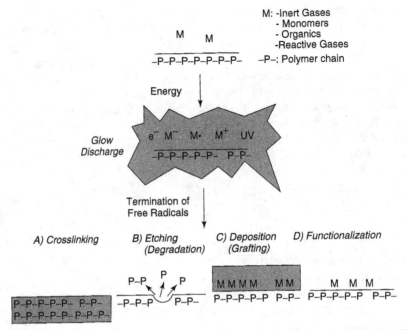

Figure 7.95 Schematic illustration of interactions of glow-discharge plasma with solid surfaces. Reprinted, by permission, from M. S. Sheu, D. M. Hudson, and I. H. Loh, in *Encyclopedic Handbook of Biomaterials and Bioengineering*, Part A: *Materials*, Vol. 1, p. 869. Copyright © 1995 by Marcel Dekker, Inc.

Table 7.21 Effects of Plasma Treatment on Polymer Surface Modification

Effect	Surface Change	Plasma Gas
Wettability	Oxidation, electrostatic unsaturation	O_2, N_2, H_2O, air (inert gases), NH_3, CO_2, inert gases
Molecular weight	Crosslinking	Inert gases (He, Ar), H_2, N_2
	Degradation	O_2, N_2
	Etching	Ar, CF_4
Functionalization (reactive sites)	$-OH$	O_2, H_2O, H_2O/H_2O_2
	$-C(O)OH$	CO_2
	$-C-O-O-$	Ar (quenching in O_2 or air)
	$-NH_2$	$N_2 + H_2$, NH_3
	$C=C$	Inert gases

Source: Sheu, M.S., D.M. Hudson, and I.H. Loh, in *Encyclopedic Handbook of Biomaterials and Bioengineering, Part A: Materials*. Vol. 1, Marcel Dekker, Inc., New York, 1995.

products from various gas fragments in the plasma. Reactive sites, especially peroxide groups, can also be used for surface grafting with other functional polymers, such as hydrogels. Surface peroxides can be obtained using an argon plasma treatment followed by quenching in air or oxygen.

A wide variety of parameters can directly affect the chemical and physical characteristics of a plasma, which in turn affect the surface chemistry obtained by the plasma modification. Some of the more important parameters include electrode geometry, gas type, radio frequency ($0-10^{10}$ Hz), pressure, gas flow rate, power, substrate temperature, and treatment time. The materials and plasmas used for specific biomedical applications are beyond the scope of this text, but the applications include surface modification for cardiovascular, ophthalmological, orthopedic, pharmaceutical, tissue culturing, biosensor, bioseparation, and dental applications.

REFERENCES

Processing of Metals

Handbook of Metal Forming, K. Lange, ed., McGraw-Hill, New York, 1985.

Hosford, W. F., and R. M. Caddell, *Metal Forming: Mechanics and Metallurgy*, Prentice-Hall, Englewood Cliffs, NJ, 1983.

Altan, T., S. I. Oh, and H. L. Gegel, *Metal Forming: Fundamentals and Applications*, American Society for Metals, Metals Park, OH, 1983.

Richardson, G. J., D. N. Hawkins, and C. M. Sellars, *Worked Examples in Metalworking*, The Institute of Metals, London, 1985.

Processing of Ceramics and Glasses

Campbell, S. A., *The Science and Engineering of Microelectronic Fabrication*, Oxford University Press, New York, 1996.

Carbide, Nitride, and Boride Materials Synthesis and Processing, A. W. Weimer, ed., Chapman & Hall, London, 1997.

Reed, J. S., *Principles of Ceramics Processing*, 2nd ed., Wiley-Interscience, New York, 1995.

Yanagida, H., K. Koumoto, and M. Miyayama, *The Chemistry of Ceramics*, John Wiley & Sons, New York, 1996.

Wolf, S., and R. N. Tauber, *Silicon Processing*, Volume 1, Lattice Press, Sunset Beach, CA, 1986.

Processing of Polymers

Tadmor, Z. and C. G. Gogos, *Principles of Polymer Processing*, John Wiley & Sons, New York, 1979.

Middleman, S., *Fundamentals of Polymer Processing*, McGraw-Hill, New York, 1977.

Powell, P. C., and A. J. Ingen Housz, *Engineering with Polymers*, S. Thornes, Cheltenham, England, 1998.

Crawford, R. J., *Plastics Engineering*, Butterworth-Heinemann, Boston, 1997.

Modern Plastics Encyclopedia, McGraw-Hill, New York, 1977.

Processing of Composites

Composite Materials Technology: Processes and Properties, P. K. Mallick and S. Newman, ed., Hanser, Munich, 1990.

Lubin, G., *Handbook of Fiberglass and Advanced Plastics Composites*, Van Nostrand Reinhold, New York, 1969.

Processing of Biologics

The Biomedical Engineering Handbook, 2nd ed., J. D. Bronzino, ed., CRC Press, 2000.

Encyclopedic Handbook of Biomaterials and Bioengineering, Parts A and B, D. L. Wise et al., ed., Marcel Dekker, New York, 1995.

PROBLEMS

Level I

7.I.1 A 1-in. × 36-in. (thickness × width) slab enters a hot-rolling mill at 500 feet per minute. It passes through seven stands and emerges as a strip 0.25 in. × 36 in. What is the exit speed of the strip from the last set of rollers?

7.I.2 In polymer extrusion, show that a square-pitched screw has a helix angle of 17.65°.

7.I.3 In polymer extrusion, show that, for a single-flighted screw, the channel width is related to the inner barrel diameter by $W = \pi D_s \sin \theta$. Clearly state any assumptions you make.

7.I.4 Estimate the shear rate during ceramic tape casting when the substrate velocity is 3 cm/s and the blade height is 50 μm.

Level II

7.II.1 The actual falling weight of a steam hammer is 1200 pounds, the cylinder is 10 inches in diameter, and the stroke is 27 in. The mean average steam pressure is 80 psi. (a) What is the energy of the hammer blow? (b) What is the average force exerted by the hammer if it travels 1/8 in. after striking the workpiece?

7.II.2 Refer to Figure 7.26. (a) Calculate the compact ratio for the granulated tile composition. (b) Estimate the *PF* of the alumina granules in the compact. Assume that the granules pack with $PF = 0.50$ and the particle density is 3.98 Mg/m³.

Level III

7.III.1 The following data for a polymer have been collected using a capillary viscometer. The capillary over which the pressure drop has been measured was 1 mm in diameter and 30 mm in length.

Q(cm³/s)	ΔP(MPa)
0.0196	48.0
0.0441	59.4
0.127	93.6
0.588	144
1.47	198
3.92	288

(a) Using the Hagen–Poiseuille equation, calculate the shear stress at the wall and the apparent shear rate at the wall. Use the ratio of these two values to calculate the apparent viscosity. (b) Calculate the power-law index for the linear portion of the data when plotted on a log–log plot. (c) Does the polymer exhibit power-law behavior over the entire range of shear rates? (d) Look up the equation for something called the Rabinowitsch correction factor in a polymer science or polymer processing reference, and calculate it for this polymer. Cite the source of your information.

CHAPTER 8

Case Studies in Materials Selection

8.0 INTRODUCTION AND OBJECTIVES

In this chapter we present some case studies that will allow us to examine how the concepts we have described in the previous chapters can be applied to solving real, materials-related problems. We begin with a description of some basic design principles, which will lead into how they can be adapted to materials selection.

By the end of this chapter, you should be able to:

- Identify the five basic steps in an iterative design process.
- Differentiate between iterative and concurrent design processes.
- Describe how materials selection fits into the mechanical design process.
- Utilize your knowledge of the structure and properties of materials engineering and of materials processing techniques to carry out a materials selection design problem.

8.0.1 The Design Process and Factors

Design is the process by which an idea or need is identified and brought to a useable end product or process. In this way, there are many types of design—even many types of engineering design. Chemical engineers are familiar with *chemical process design* as it relates to the design of industrial-scale chemical process facilities through the application of economic analysis, process flowsheets, and optimization of kinetic and transport processes. Many types of engineers and architects use *computer-aided design* (CAD) to assist them in the construction of devices ranging in size from microelectronic circuits to skyscrapers. Materials scientists, including chemists and physicists who work in this area, are more familiar with *materials design* as it applies to the "ground up" approach to developing a new material, such as in the production of carbon nanotubes or molecular magnets. Materials design is beyond our scope of expertise here. Perhaps when you get involved in materials research in graduate school or in industry, you will be able to use design principles for the development of truly novel materials based only upon a perceived need and the fundamentals of materials engineering you have been introduced to here. But for now, what we are really interested in is learning about some of the fundamentals that go into *mechanical design*, which deals with the physical principles, function, and production of mechanical parts or systems. The principles we

An Introduction to Materials Engineering and Science: For Chemical and Materials Engineers,
by Brian S. Mitchell
ISBN 0-471-43623-2 Copyright © 2004 John Wiley & Sons, Inc.

will describe here can also be used in many kinds of other design processes, such as the design of electrical components. Ultimately, we can use these design principles to better understand the more important process (for us) of *materials selection*, which is described in the next section.

8.0.1.1 The Iterative Design Process.

The most fundamental of design processes is the *iterative design process*, in which a series of design steps are repeatedly implemented until an appropriate product is made that meets or exceeds all of the *design specifications*. Most iterative design processes involve five steps, which can vary by name and in scope, depending on the specific type of product or process to be designed. These five steps are involved in almost all problem-solving techniques, and include (see Figure 8.1) defining the problem, gathering information, generating multiple solutions, analyzing each solution, and testing and implementing the solution. At any point, a preceding step may be returned to if a design specification is not properly met.

A modified form of the five-step iterative approach can be applied to the problem of mechanical design. A typical iterative design process flow chart used for mechanical design is shown in Figure 8.2. By way of illustration, the iteration loop is shown to return to the concept stage, but it could return to any of the previous stages depending upon the severity of the design difficulty. Each of the steps between the problem definition (which in this case is identification of a market need) and the product for testing is in itself a design process. In other words, *conceptual design, embodiment design,* and *detail design* all may involve iterative processes before moving to the next step.

We will not describe the iterative design process in any more detail than this. There are many fine books on the subject, and the interested reader is referred to the list at the end of the chapter for further information. After a brief diversion into alternative design strategies in the next section, we will return to the iterative design process to see how the concepts of materials selection fit into mechanical design.

8.0.1.2 Concurrent Engineering.

Concurrent engineering, sometimes called *simultaneous engineering*, utilizes a parallel, rather than series, approach to the design process. As described in the previous section, the iterative design process utilizes a step-by-step (series) approach in which each stage must be completed before the next stage starts. In concurrent engineering (see Figure 8.3), options are

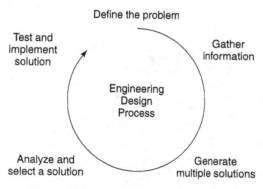

Figure 8.1 The iterative design process.

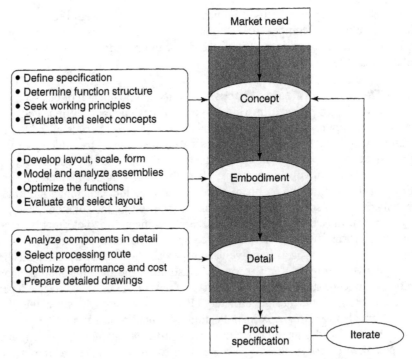

Figure 8.2 The iterative design process applied to mechanical design of components. Reprinted, by permission, from M. F. Ashby, *Materials Selection in Mechanical Design*, 2nd ed., p. 9. Copyright © 1999 by Michael F. Ashby.

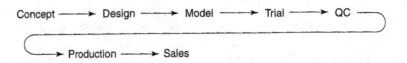

(*a*) **The traditional method**

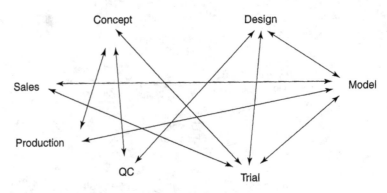

(*b*) **Concurrent method**

Figure 8.3   Comparison of (a) traditional and (b) concurrent engineering processes.

Table 8.1   Some Concurrent Engineering Techniques

Concurrent Engineering Technique	Reference
Knowledge-based engineering (KBE)	*www.daasolutions.com/faq/faq3.asp*
Total quality management (TQM)	*www.isixsigma.com/me/tqm/*
Taguchi (robust engineering)	*www.isixsigma.com/me/taguchi/*
Design for manufacturing	*www.design4manufacturability.com*
Just in time (JIT)	*domino.watson.ibm.com/cambridge/research.nsf/pages/papers.html*
Poka-yoke	*www.campbell.berry.edu/faculty/jgrout/pokayoke.shtml*

simultaneously considered (parallel) for each aspect of the design process before the final part is assembled and tested. The objective of concurrent engineering is to rapidly provide high-quality components, regardless of complexity. Inherent in the concurrent engineering process is the use of multidisciplinary teams to deal with each aspect of the design process, such as modeling, quality control (QC), testing, and manufacturing. Some specific types of concurrent engineering techniques are listed in Table 8.1 with references for more detailed information.

8.0.2   The Materials Selection Process

To examine the materials selection process in mechanical design, we will utilize the iterative process design approach illustrated in Figure 8.2. It should be clear that the principles described here can be used for any of the design processes, either series or parallel in nature, that are used to design a component, and that these principles need not be limited to mechanical design only. They can be used to design electrical, optical, biomedical, and chemical sensing devices as well.

As illustrated in Figure 8.4, there are different levels of materials selection that are involved in the three inner design loops (conceptual, embodiment, and detailed design). In the conceptual design stage, all materials are considered. A subset of these materials are then used in the embodiment design stage, followed by a (hopefully) small and manageable subset of materials for the final selection process. All of these materials selection processes involve essentially two important components: design tools and data. Some of the design tools are listed on the left of Figure 8.4 and involve some of the techniques you may have studied in other courses, such as numerical modeling, sensitivity analysis, statistical design of experiments, computer-aided design, economic analysis, and optimization. We will not describe these techniques here, except to add that many of these techniques now can be performed by commercially available computer programs.

Of more importance to us in the materials selection processes is the availability of data upon which to base our decisions. The nature of the data needed in the early stages differs greatly in its level of precision and breadth from that needed in later

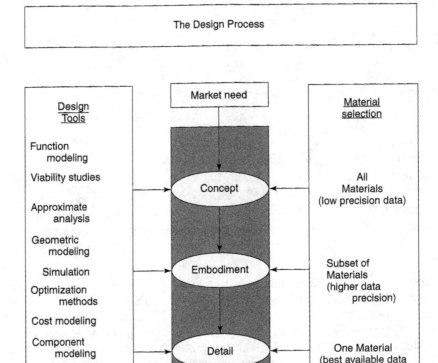

Figure 8.4   Incorporation of design tools and materials data into the iterative design process to assist in materials selection. Reprinted, by permission, from M. F. Ashby, *Materials Selection in Mechanical Design*, 2nd ed., p. 12. Copyright © 1999 by Michael F. Ashby.

stages. At the conceptual stage, the designer requires approximate property values for the widest possible range of materials. It is at this stage that the materials selection charts of Ashby [1] are particularly useful. An example of an Ashby chart is shown in Figure 8.5. Here, tensile modulus (Young's modulus) is plotted as a function of density for most classes of engineering materials. Such a plot assists the designer in selecting an appropriate class of material for consideration. The specific plot to use depends upon the design criteria; for example, modulus versus density may be important for one application, whereas thermal conductivity versus tensile strength is important for another. We will give specific examples of how these charts can be utilized in subsequent case studies. There are also computer programs that can assist in this stage of the materials selection process, and a wealth of data is available on the Internet, though its accuracy should always be questioned and websites change almost daily. A list of some Internet materials databases is given in Table 8.2.

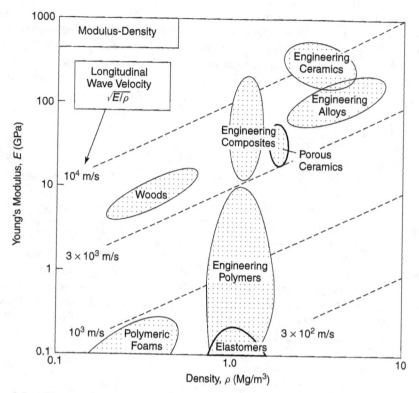

Figure 8.5 Example of a materials selection design chart for modulus vs. density. Reprinted, by permission, from M. F. Ashby, *Materials Selection in Mechanical Design*, 2nd ed., p. 34. Copyright © 1999 by Michael F. Ashby.

Table 8.2   Some Internet Material Property Databases

Website Title	Location
WebElements™	*www.webelements.com*
Biomaterials Properties	*www.lib.umich.edu/dentlib/Dental_tables/toc.html*
Materials X-Sight™	*www.centor.com/solutions*
Chemscope—Medical Materials	*chemscope.com*
Corrosion databases	*corrosionsource.com/links.htm*
Electronic Products from ASM International	*www.asminternational.org*
F*A*C*T—Compound Web	*www.crct.polymtl.ca/fact*
Granta Design Limited	*www.grantadesign.com*
STN Database Catalog	*www.cas.org/stn.html*
Japanese material database directory	*inaba.nims.go.jp/netnavi/link_F0831.htm*
IDEMAT, Environmental Materials Database	*www.io.tudelft.nl/research/dfs/index.html*

Data for the embodiment design stage are for a smaller subset of materials, but at a higher level of precision and detail than the conceptual stage. These types of data are found in specialized handbooks and software which usually deal with a single class of materials—for example, a metals handbook that covers most alloy compositions. The final stage of detailed design requires a still higher level of precision and detail, but for an even smaller subset of materials. Often this is one or just a few specific materials, for which the relevant data come from the specific manufacturer. For example, polyethylene has a range of properties that can depend upon molecular weight and manufacturing method. It is imperative that representative data on a specific type of polyethylene be used in order to properly evaluate that material. It may even be necessary to conduct in-house tests to measure critical properties at this stage of the design process.

In most materials selection processes, it is virtually impossible to make materials choices independent of the product shape. This includes not only the macroscopic, or bulk, shape of the object such as hammer or pressure relief valve, but also the internal or microscopic shape, such as a honeycomb structure or a continuous-fiber-reinforced composite. Shape is so important because in order to achieve it, the material must be subjected to a specific processing step. In Chapter 7, we saw how even simple objects made from a single-phase metal alloy could be formed by multiple processes such as casting or forging, and how these processing steps can affect the ultimate properties of the material. As illustrated in Figure 8.6, function dictates the choice of

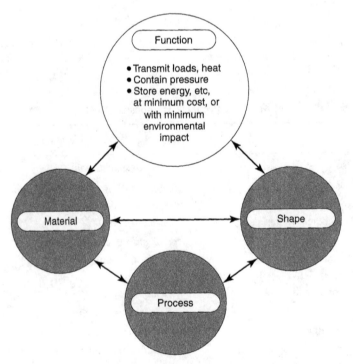

Figure 8.6 Schematic illustration of the interaction between function, material, processing, and shape in materials selection for mechanical design. Reprinted, by permission, from M. F. Ashby, *Materials Selection in Mechanical Design*, 2nd ed., p. 13. Copyright © 1999 by Michael F. Ashby.

both material and shape. Process is influenced by the material—for example, by its formability, machinability, and weldability. Process also determines the shape, the size, the precision, and the cost of the product. The interactions are two-way: Specification of shape restricts the choice of material and process, just as the specification of process limits the materials one can use and the shape they will take. The more sophisticated the design, the tighter the specifications and the greater these interactions become. The interaction between function, material, shape, and process lies at the heart of the material selection process. Let us examine these relationships with some case studies.

8.1 SELECTION OF METALS FOR A COMPRESSED AIR TANK

(Adapted from Lewis and Ashby)

8.1.1 Problem Statement and Design Criteria

You are asked to design a vessel to contain compressed air. Economics and weight are obviously issues to consider, but the primary consideration is safety; that is, the tank must not rupture. Specifically, the tank must meet three design criteria:

1. The maximum stress in the vessel must be below the yield strength of the material used.
2. The vessel must not fail by fast fracture.
3. The vessel must not fail by fatigue.

8.1.2 Problem Analysis

The vessel is subjected to an internal pressure from the compressed air, which we shall designate as p. The internal pressure is uniformly distributed over the internal surfaces of the vessel, giving rise to both circumferential stress, σ_H, also known as *hoop stress*, and longitudinal stress, σ_L (see Figure 8.7). We will examine each of these stresses independently before we begin the material selection process. In our development, we will make the following assumptions:

- The radial stresses in the cylinder wall are negligible.
- There are no longitudinal supports in the cylinder.
- The stresses are uniformly distributed throughout the section wall.

The hoop stress can be obtained by considering an elemental portion of the cylinder wall, δx, subtending an angle $\delta\theta$, at the center and at an angle θ with XX (see Figure 8.8). The internal radius of the cylinder is r and the length of the unit is L. Since the pressure of the fluid always acts perpendicular to the surface of contact, the total pressure, P, normal to the elemental section is given by

$$P = p \times (\text{area of elemental section}) = p\delta x L = pr\delta\theta L \tag{8.1}$$

The vertical component of the pressure is δP_v:

$$\delta P_v = pr\delta\theta L \sin\theta \tag{8.2}$$

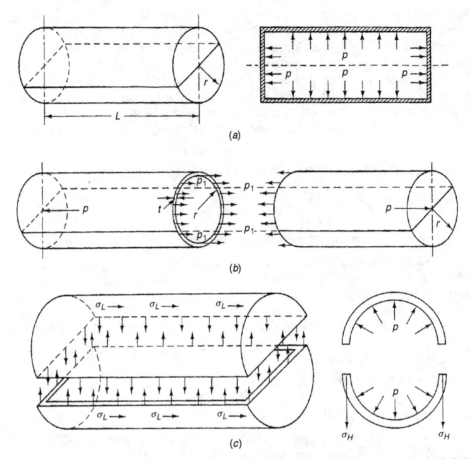

Figure 8.7 Stresses in a thin cylinder subjected to an internal pressure, P: (a) cylindrical shell under internal fluid pressure; (b) longitudinal stress development; (c) hoop stress development. Reprinted, by permission, from G. Lewis, *Selection of Engineering Materials*, p. 139. Copyright © 1990 by Prentice-Hill, Inc.

The total upward pressure on the semicircular portion of the cylindrical shell above the diametral plane XX is obtained by integrating δP_v over the entire angle, θ:

$$P_v = \int_0^\pi pr L \sin\theta\,\delta\theta = 2prL \tag{8.3}$$

Similarly, the total downward pressure on the semicircular portion of the cylindrical shell below the diametral plane XX is also $2prL$. These two equal and opposite pressures act to burst the cylinder longitudinally at the plane XX. The resisting force comes from the hoop stress. Thus

$$2prL = 2\sigma_H t L \tag{8.4}$$

where t is the thickness of the wall. We can then solve Eq. (8.4) for the hoop stress:

$$\sigma_H = \frac{pr}{t} \tag{8.5}$$

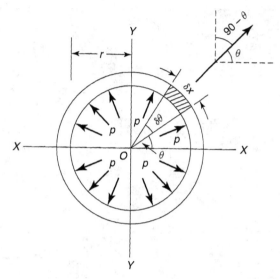

Figure 8.8 Element of the wall of a thin cylinder subjected to internal pressure P. Reprinted, by permission, from G. Lewis, *Selection of Engineering Materials*, p. 140. Copyright © 1990 by Prentice-Hill, Inc.

(This is the hoop stress for thin-walled vessels, i.e., $t < r/4$. For thick-walled vessels, see Ashby, p. 396 [1]) The longitudinal stress can be obtained from a similar shell balance (see Figure 8.7)

$$\sigma_L = \frac{pr}{2t} \tag{8.6}$$

By comparing Eqs. (8.5) and (8.6), we see that the maximum stress on the vessel is given by the hoop stress and is equal to pr/t. Let us now turn our attention to the design criteria.

8.1.2.1 *Yield Strength of Candidate Metals.* To avoid yield of the tank, the hoop stress must be less than the yield stress of the material used, σ_y:

$$\sigma_H = \frac{pr}{t} < \sigma_y \tag{8.7}$$

The higher the value of σ_y, the higher the hoop stress that can be tolerated in the vessel.

If we consult the strength versus density diagram of Ashby (Fig. 8.9), we see that the classes of materials with the highest strengths are engineering ceramics, engineering composites, and engineering alloys. Of the engineering alloys, the titanium, steel, and nickel alloys provide the highest ranges of strengths. With these material classes in mind, let us continue to the other criteria.

8.1.2.2 *Fracture Properties of Candidate Metals.* Recall that the opening mode stress intensity factor for the case of a component containing a single edge crack in tension, K_{1c}, is given by Eq. (5.47), where $Y = 1.12$ for the geometry under consideration here:

$$K_{1c} = 1.12\sigma\sqrt{\pi a} \tag{8.8}$$

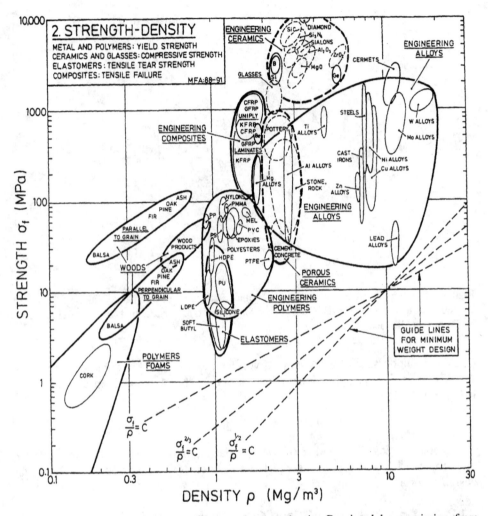

Figure 8.9   Materials selection chart of strength versus density. Reprinted, by permission, from M. F. Ashby, *Materials Selection in Mechanical Design*, p. 39, 2nd ed. Copyright © 1999 by Michael F. Ashby.

where σ is the applied stress and a is the length of the edge crack in the vessel. (See Ashby, p. 394 [1] for formulae for internal cracks or two cracks.) To avoid failure by fast fracture, the following condition must be met:

$$K_{1c} \geq 1.12\sigma_H\sqrt{\pi a_{cr}} \tag{8.9}$$

where σ_H is the maximum applied stress, a_{cr} is the critical flaw size, and K_{1C} is the fracture toughness of the vessel material. From Eq. (8.9),

$$a_{cr} = \frac{K_{1c}^2}{1.25\pi\sigma_H^2} \tag{8.10}$$

That is, a_{cr} is directly proportional to $(K_{1c}/\sigma_y)^2$ since σ_H is a fraction of σ_y. Thus, the larger the value of a_{cr}, the more attractive is the material, since cracks can be easily detected without the use of sophisticated equipment. The Ashby plot of fracture toughness versus density (Figure 8.10) indicates that of the three classes of materials selected with Criterion 1, only the engineering composites and engineering alloys provide suitable possibilities for Criterion 2. Again, of the alloys, titanium, steel, nickel, and copper alloys are the best here.

The real power of the Ashby diagrams comes when we realize that we can combine Figures 8.9 and 8.10 to yield one, more useful diagram (Figure 8.11), namely a plot of fracture toughness versus strength. This plot shows unequivocally that the steel, nickel, and titanium alloys are the best classes of materials to select for this application. We will use Criterion 3 to narrow this field even further.

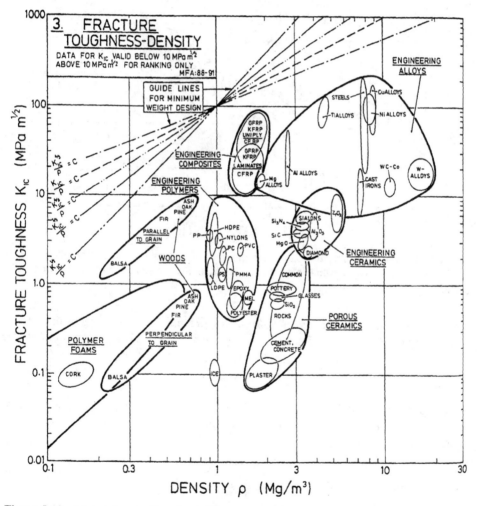

Figure 8.10 Materials selection chart of fracture toughness versus density. Reprinted, by permission, from M. F. Ashby, *Materials Selection in Mechanical Design*, 2nd ed., p. 41. Copyright © 1999 by Michael F. Ashby.

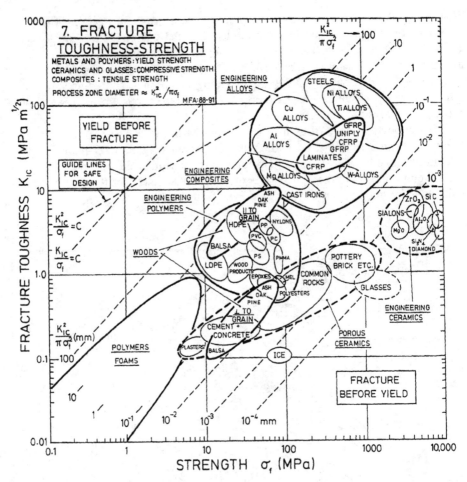

Figure 8.11   Materials selection chart of fracture toughness versus strength. Reprinted, by permission, from M. F. Ashby, *Materials Selection in Mechanical Design*, 2nd ed., p. 44. Copyright © 1999 by Michael F. Ashby.

8.1.2.3   *Fatigue Properties of Candidate Metals.*

We assume that the fatigue crack growth law in the vessel material is described by the Paris equation [Eq. (5.48)]:

$$\frac{da}{dN} = A(\Delta K)^m \tag{8.11}$$

where da/dN is the fatigue crack growth rate, ΔK is the range of stress intensity factors, and A and m are material constants. The number of cycles to failure, N_f, of the component is given by

$$N_f = \frac{2[a_i^{1-(m/2)} - a_{cr}^{1-(m/2)}]}{(m-2)AY^m(\sigma)^m\pi^{m/2}} \tag{8.12}$$

when $m \neq 2$. In the present case, $Y = 1.12$ [see Eq. (8.8)]. So, for the crack to grow from an initial size a_i to a critical size a_{cr}, the rate of growth is dependent upon K_{1c}. We also want small m and A to get a large number of cycles to failure, N_f.

Table 8.3   Comparison of Three Materials for Compressed Air Tank[a]

Material	σ_y (MPa)	K_{1c} (MPa $\cdot$ m$^{1/2}$)	A	m	a_{cr} (mm)	N_f
18% Ni maraging steel	**1340**	100	1.32×10^{-10}	**2.32**	5.65	1955
Al–Zn–Mg alloy	490	25	6.34×10^{-11}	3.14	2.64	2205
Ti–6Al–4V	830	**120**	9.60×10^{-13}	3.80	**21.21**	**4513**

[a] The most favorable value in each category is denoted by boldface type.

8.1.3   Material Selection

Though there are many possibilities of the engineering alloys, let us consider three common alloys from different classes: a steel, an aluminum alloy, and a titanium alloy. The three alloys and their appropriate design properties are listed in Table 8.3. The values that are the most favorable in each category are listed in bold typeface. On the basis of Criterion 1, the best material is maraging steel, but from the viewpoints of Criteria 2 and 3 the titanium alloy is obviously superior. Cost is an additional factor that could influence the final selection.

Cooperative Learning Exercise 8.1

(Extended Study)

Form groups of three. Each person should select a material from the three categories under consideration for this application (steel, aluminum alloy, and titanium alloy) other than the three listed in Table 8.3 and should perform a similar analysis—that is, calculate or look up yield strength, fracture toughness, critical crack size, number of cycles to failure, and the constants A and m in the Paris equation. Combine your results and compare your answers. Do you obtain a result similar to that in Table 8.3?

8.2   SELECTION OF CERAMIC PIPING FOR COAL SLURRIES IN A COAL LIQUEFACTION PLANT

(Adapted from Lewis)

In this case study, we see how the data needed to perform the selection process are not always readily available, and how material property principles may come into play that we have not covered in this text. Hence, it may be necessary to develop in-house analytical techniques and to do a little outside reading on a particular topic. We use the results of a completed study to illustrate how this can be accomplished.

8.2.1   Problem Statement and Design Criteria

Coal liquefaction is the process by which coal is crushed and dried, mixed with water to form a slurry, and gasified to form a series of hydrocarbon fuels such as fuel oil, liquid propane, and naphtha. One of the constraints in the development of coal liquefaction technology is the erosive wear of such components as pumps, valves, and piping used in processing and conveyance of flowing coal slurries. The problem is notoriously acute in valves used to feed the process stream from the high-pressure side of the valve to the low-pressure side. In these valves, which are designed to maintain the pressure drop, the slurry stream velocity is quite high. This speed is further enhanced by the expansion of the dissolved gases in the slurry stream. The consequence of this is erosive wear in various valve components. Nominal wear life as short as 14 days has been reported for letdown valves in some liquefaction plants. The problem is equally severe for the piping.

There is only one simply stated, yet poorly defined, design criterion for this problem: Propose some materials that have improved wear resistance for fabricating parts of high-pressure pumps and piping to be used in the pumping of coal slurries.

8.2.2 Problem Analysis

The selection of material for the pump components and piping should be based primarily on resistance to erosion by solid particles. We have not described *erosive wear* in this textbook, but this is a good example of how you can utilize your fundamental knowledge in materials engineering and science to do some background reading and devise a testing scheme based upon your research. The background data and analysis presented here are based upon literature sources that can be found at the end of this chapter [2–6].

Figure 8.12 illustrates a solid particle impinging on a surface. It has been found that the erosive wear rate depends upon the *impingement angle*, α, the particle velocity, v_0, and the size and density of the particle, as well as the properties of the surface material. It has also been found that there is a difference in erosive wear properties of brittle and ductile materials. The maximum erosive wear of ductile materials occurs at $\alpha = 20°$, whereas the maximum erosive wear for brittle materials occurs near $\alpha = 90°$. Since the impingement angle is probably lower than 90° for these type of flow situations, we might consider only brittle materials, such as ceramics for this application. Let us examine brittle erosive wear in a little more detail first.

The dependence of *erosive wear rate*, $\dot{W}$, on particle velocity, v_0, has been observed to follow the general form

$$\dot{W} = f(v_0^b) \tag{8.13}$$

Detailed studies have shown some further dependence of the wear rate on the attack angle, temperature of erosion, particle size, particle concentration, and particle hardness;

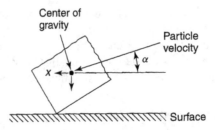

Figure 8.12   Schematic illustration of erosive wear due to a particle impacting a solid surface. Reprinted, by permission, from G. Lewis, *Selection of Engineering Materials*, p. 171. Copyright © 1990 by Prentice-Hill, Inc.

however, there are three specific models for erosive wear in brittle materials of the form of Eq. (8.13) that we will consider.

The Sheldon–Finnie model assumes that erosive wear occurs as a result of contact stresses during impact that cause cracks to grow from preexisting flaws on the target surface. The loads at which crack propagation occur is related to the distribution of surface flaws through Weibull statistics. The erosive wear rate, in terms of weight lost from the surface material per unit weight of impacting particle, is dependent upon particle size, r, and velocity, v_0, as follows:

$$\dot{W} = k_1 r^a v_0^b \tag{8.14}$$

where a and b are constants related to Weibull parameter, m, depending on the particle shape. For round particles

$$a = 3\left(\frac{m - 0.67}{m - 2}\right) \tag{8.15}$$

for angular particles

$$a = 3.6\left(\frac{m - 0.67}{m - 2}\right) \tag{8.16}$$

and for all types of particles

$$b = 2.4\left(\frac{m - 0.67}{m - 2}\right) \tag{8.17}$$

The constant k_1 in Eq. (8.14) is given by

$$k_1 = E^{\frac{0.8(m+1)}{m-2}} \rho^{\frac{1.2(m-0.67)}{m-2}} \sigma_0^{\frac{-2}{m-2}} \tag{8.18}$$

where E is the modulus of elasticity of the surface material, ρ is the density of the particle material, and σ_0 is the other Weibull parameter.

The theory of Evans includes lateral crack formation and leads to an expression for the volume lost from the surface material per volume of the impacting particle, S, as follows:

$$S = C' v_0^{19/6} r^{11/3} \rho^{19/12} K_{1c}^{4/3} H^{-1/4} \tag{8.19}$$

where C' is a constant, K_{1c} is the fracture toughness of the surface material, and H is the hardness of the surface material.

The elastic–plastic theory of Weiderhorn and Lawn can also be applied here, which leads to the following expression for S that is similar to Eq. (8.19):

$$S = C'' v_0^{22/9} r^{11/13} \rho^{11/9} K_{1c}^{-4/3} H^{1/9} \tag{8.20}$$

where C'' is a constant.

8.2.3 Materials Selection

From the preceding descriptions of erosive wear, we see that fracture toughness and hardness are important parameters in the materials selection process. There is some additional information we can use to help initiate our material selection process, namely *dry sliding friction* between two surfaces. We do not have dry surfaces in this case study, but at least we can use this information to initially select some candidate materials, then return to fracture toughness and hardness to finish the process.

For dry sliding friction, the wear rate constant, k_s, is given by

$$k_s = \frac{\dot{W}_s}{F_n} \tag{8.21}$$

where $\dot{W}_s$ is the sliding friction wear rate, defined as the volume of material removed from the contact surface per unit distance slid by the surface, and F_n is the normal load force on the surface. Thus, this wear rate constant has units of $(m^2/N)^{-1}$.

Figure 8.13 shows a material selection chart of the wear rate constant for dry sliding friction in Eq. (8.21) plotted as a function of hardness, H, in MPa. The most favorable materials will have a high hardness and a low wear rate constant. Diamond would be the best choice from this diagram, but may be prohibitive from a cost standpoint, even when used only as a liner, if it is even possible to produce such components.

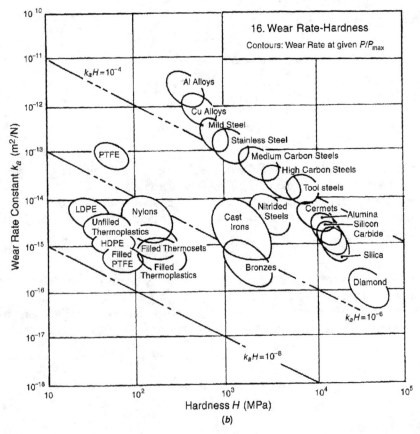

Figure 8.13   Materials selection chart of dry sliding wear rate constant [Eq. (8.21)] versus hardness. Reprinted, by permission, from M. F. Ashby, *Materials Selection in Mechanical Design*, 2nd ed., p. 60. Copyright © 1999 by Michael F. Ashby.

Table 8.4   Wear Rate of Candidate Piping and Pump Component Materials by Coal Slurries

Material	H^a (GPa)	K_{1c} (MPa·m$^{1/2}$)	$1/H$ (GPa^{-1})	$K_{1c}^{4/3}H^{1/4}$ (MPa·m)$^{4/3}$(GPa)$^{1/4}$	Mean Erosive Wear Rate^b (μm/h)
Cemented WC	16.3	10.50	0.06	45.95	36
SiC	32.5	3.75	0.03	13.90	10
Si_3N_4	14.0	3.75	0.07	11.26	241
Al_2O_3	18.2	3.10	0.05	9.33	152

^{a}Vicker's hardness, with indenter load of 2.5 kg.
^{b}Test temperature $= 343°C$, slurry jet velocity $= 100$ m/s, angle of attack of slurry stream $= 50°$.

More reasonable choices include silicon carbide and alumina. In addition, our intuition tells us that similar materials like silicon nitride and tungsten carbide may fit into this category, even though they are not specifically listed in the chart.

At this point, experiments must be performed. Experimental results for the erosive wear of the selected candidate ceramic materials in coal slurries are presented in Table 8.4. Notice that the wear rate has a very rough inverse correlation with $K_{1c}^{4/3}H^{1/4}$, which is consistent with some of the descriptions of erosive wear from the previous section. Any of these ceramic materials is suitable for the piping and pump components based solely on wear rate, with the lowest wear rate for SiC being the most attractive. Formability and economic criteria can be applied to assist in the final material selection.

Cooperative Learning Exercise 8.2

(Extended Study)

Form groups of three. Each person should select one of the following three materials to perform an analysis similar to that presented in Table 8.4. Note that although some of the materials are the same as in Table 8.4, their mechanical properties are different.

Cemented WC $H = 17.5$ GPa; $K_{1c} = 9.5$ MPa·m$^{1/2}$

SiC $H = 26.1$ GPa; $K_{1c} = 3.90$ MPa·m$^{1/2}$

Cemented WC $H = 19.0$ GPa; $K_{1c} = 4.00$ MPa·m$^{1/2}$

Each person should find the remaining parameters and physical property data for this material required to solve the three models [Eqs. (8.14), (8.19), and (8.20)] for the erosive wear of a coal slurry; that is, each person will have three calculations to do and three erosion rates as a result. Assume that the test temperature is 343°C, the slurry velocity is 100 m/s, and the angle of attack is 50°.

Clearly state any additional assumptions and approximations, and make sure that all three of you are making the same ones. Compare your answers from each of the three models for your material, with the three values from the other two people from their models, and with the experimental values listed in Table 8.4 for the other materials.

8.3   SELECTION OF POLYMERS FOR PACKAGING

In this example, we see how selection criteria need not be entirely quantitative. Such properties as resistance to solvents and printability are important to the selection of polymers for certain applications, yet difficult to quantify.

8.3.1   Problem Statement and Design Criteria

Justify the use of the following polymers as packing in the following applications:

- Polyethylene terephthalate (PET) for beverage containers.
- Oriented polypropylene for easy-open bags and wraps.
- Low-density polyethylene for bread bags.
- Nylon for flexible packaging for meats and cheeses.

The following criteria are important to one or more of these applications:

1. Strength (tensile, tear and impact) properties
2. Diffusion barrier properties
3. Packaging-contents interaction
4. Aesthetic properties such as color, transparency, printability, scuff resistance

8.3.2   Problem Analysis and Materials Selection

In this section we combine the problem analysis with the materials selection process, since the materials have already been specified and we seek only to justify their selection.

8.3.2.1   *Strength of Candidate Polymers.*   The materials selection chart for tensile strength versus density has been previously presented (Figure 8.9) and is a good starting point for the determination of candidate polymers for the various packaging applications. However, similar charts do not exist for tear strength and impact strength, since these tend to be polymer-specific properties. It is more useful to compile these data in tabular form, as in Table 8.5, where the three types of strength values are listed on a relative basis for some common polymers.

Notice that there is variation in values not only between polymer types, but within polymer type due to processing differences. Tensile strength is of most importance to the soda bottle application, since the contents are under pressure, and factors such as hoop stress (see Section 8.1) come into play. PET is obviously an excellent choice for this application from a tensile strength standpoint, as would be oriented polypropylene (OPP).

Tear strength often varies inversely with tensile strength, so the polyethylenes demonstrate high resistance to tearing. PET is an exception to this guideline, because it can have both relatively high tensile and tear strengths. This combination of properties makes it an excellent choice for grocery bags. The tear strength of poly(vinyl chloride), PVC, is excluded in Table 8.5 due to its widely varying tear strength values. For easy-open bags and wraps, a low tear strength is desirable, which justifies the selection of oriented polypropylene for this application.

Table 8.5   Tensile, Tear, and Impact Strengths of Some Packaging Polymers[a]

Polymer	Tensile Strength Ratio (vs. LDPE)	Tear Strength Ratio (vs. OPP)	Impact Strength Ratio (vs. Nylon or HDPE)
Low-density polyethylene (LDPE)	1.0–3.5	25.0–100.0	1.8–3.5
High-density polyethylene (HDPE)	3.0–9.0	10.5–75.0	1.0–1.7
Linear low-density polyethylene (LLDPE)	3.5–9.0	30.0–150.0	2.2–3.2
Polyvinyl chloride (PVC)	2.2–18.0	—	3.4–4.8
Nylon	6.0–19.0	4.5–18.0	1.0–1.7
Polyvinylidene chloride (PVDC)	8.0–20.0	2.5–4.5	3.0–4.0
Oriented polypropylene (OPP)	27.0–37.0	1.0–1.8	1.2–4.0
Polyethylene terephthalate (PET)	27.0–40.0	3.5–25.0	6.0–7.8

[a] Spread in ratios reflects the influence of grade of resin and orientation of production (in or normal to machine direction).

Source: G. Lewis, *Selection of Engineering Materials*. Copyright © 1990 by Prentice-Hill, Inc.

There is a much smaller variation in impact strength of polymers, since it is not directly related to either tensile or tear strength. Once again PET has a high impact strength relative to the other polymers, which is important for soda bottles (have you ever dropped a full, 2-liter bottle of soda?). PVC also has excellent impact strength.

8.3.2.2 Diffusion Barrier Properties of Candidate Polymers.

The second design criterion involves barrier properties, which are a measure of the resistance to permeation by gases and vapors, such as oxygen, water vapor, and carbon dioxide, that can modify the taste and/or lifetime of the package contents. The oxygen and water permeabilities of some common polymers are listed in Table 8.6. The permeability of carbon dioxide roughly follows the same trend as that of oxygen.

Polyvinylidene chloride (PVDC), polyethylene (LDPE and HDPE), polypropylene (PP), PVC, and polystyrene all have excellent resistance to oxygen permeation. Polyethylene also has relatively good resistance to the permeation of water, which makes it an excellent choice for bread bags.

8.3.2.3 Packaging–Contents Interaction of Candidate Polymers.

The acidity and oil content of packaged foods vary widely, such that it is difficult to make generalizations regarding the suitability of a particular polymer for food packaging from an interaction standpoint. Some qualitative descriptions of the resistance of some common packaging polymers to various food contents are listed in Table 8.7. Nylon, for example, has excellent resistance to oils and grease and is thus well-suited for packaging meats and cheeses. However, it is not appropriate for packaging acidic foods, such as tomato-based products. Here, polyethylene and polypropylene provide better compatibility.

8.3.2.4 Aesthetic Properties of Candidate Polymers.

Other desirable properties belong to the category of aesthetics and consumer acceptance. These properties include

Table 8.6 Barrier Properties of Some Polymers

Polymer	Permeability of Oxygen at 25°C, 65% RH (cc, mil/100 in^2/24 h)	Permeability of Water at 40°C, 90% RH (cc, mil/100 in^2/24 h)
Ethylene vinyl alcohol	0.05–0.18	1.4–5.4
Nitrile-barrier resin	0.80	5.0
High-barrier PVDC	0.15	0.1
Good-barrier PVDC	0.90	0.2
Moderate-barrier PVDC	5.0	0.2
Oriented PET	2.60	1.2
Oriented nylon	2.10	10.2
Low-density polyethylene	420	1.0–1.5
High-density polyethylene	150	0.3–0.4
Polypropylene	150	0.69
Rigid PVC	5–20	0.9–5.1
Polystyrene	350	7–10

Source: A. B. Strong, *Plastics Materials and Processing*, 2nd ed. Copyright © 2000 by Prentice-Hall, Inc.

Table 8.7 Qualitative Rating of the Resistance of Selected Packaging Polymers to Various Corrosive Contents

Resin	Content			
	Oils and Greases	Acids	Alkalis	Solvents
HDPE	Excellent	Acceptable	Good	Good
LDPE	Acceptable	Acceptable	Good	Good
Nylon	Excellent	Unacceptable	Good	Good
PET	Excellent	Good	Good	Good
Polypropylene	Excellent	Acceptable	Good	Good
Polystyrene	Acceptable	Acceptable	Good	Good
PVC	Good	Good	Good	Acceptable
PVDC	Excellent	Good	Good	Acceptable

Source: G. Lewis, *Selection of Engineering Materials*. Copyright © 1990 by Prentice-Hill, Inc.

Table 8.8 Number System for Polymer Recycling

Plastic Type	Recycling Symbol	Recycling Number
Polyethylene terephthalate	PETE	1
High-density polyethylene	HDPE	2
Polyvinyl chloride	V	3
Low-density polyethylene	LDPE	4
Polypropylene	PP	5
Polystyrene	PS	6
Other	OTHER	7

Source: A. B. Strong, *Plastics Materials and Processing*, 2nd ed. Copyright © 2000 by Prentice-Hall, Inc.

printability, color retention, scuff resistance, transparency (clarity), and recyclability. These properties can vary widely and are subject to surface treatment and processing conditions. The optical properties of polymers were described in Section 6.3.3.1. In general, PET, PVC, PVDC, and crystalline forms of polypropylene are acceptably transparent for packaging applications. For reference, the recycling identification scheme for most commodity polymers is listed in Table 8.8.

Cooperative Learning Exercise 8.3

(Extended Study)

Work in groups of two to four. Collect information on the recycling programs in your home towns—that is, the types of materials that are collected, whether it is mandatory or voluntary, the amount of material processed per year, and what happens to it. Pay particular attention to plastics recycling. Compare your answers and prepare a short presentation to your class.

8.4 SELECTION OF A COMPOSITE FOR AN AUTOMOTIVE DRIVE SHAFT

(Adapted from Lewis, Mallick)

The use of composites in mechanical design is often brought about by a need for increased mechanical strength or a need for reduced weight, or both. Only infrequently are composites selected due to something other than their excellent specific mechanical properties—for example, for electrical, magnetic, or oxidation resistance reasons. As a result, we will concentrate on the use of composites in mechanical design, much as we did in Section 8.1 for metals. Before illustrating the materials selection process for composites by way of a case study, it is useful to elaborate upon some of the mechanical design criteria that are specific to composites.

Unlike ductile metals, composite laminates containing fiber-reinforced thermosetting polymers do not exhibit gross ductile yielding. However, they do not behave as classic brittle materials, either. Under a static tensile load, many of these laminates show nonlinear characteristics attributed to sequential ply failures. One of the difficulties, then, in designing with laminar composites is to determine whether the failure of the first ply constitutes material failure, termed *first-ply failure* (FPF), or if ultimate failure of the composite constitutes failure. In many laminar composites, ultimate failure occurs soon after first ply failure, so that an FPF design approach is justified, as illustrated for two common laminar composites in Table 8.9 (see Section 5.4.3 for information on the notations used for laminar composites). In fact, the FPF approach is used for many aerospace and aircraft applications.

The behavior of a fiber-reinforced composite laminate in a fatigue load application is also quite different from that of metals. In a metal, nearly 80–90% of its fatigue life is spent in the formation of a critical crack. Generally, the fatigue crack in a metal is not detectable by present-day nondestructive testing techniques until it reaches the critical length. Once the fatigue crack attains the critical length, it propagates rapidly through the structure, resulting in catastrophic failure. In many polymeric composites, fatigue damage may appear at multiple locations in the first few hundred to a

Table 8.9   First Ply Failure (FPF) and Ultimate Tensile Properties of Two Common Laminar Composites

Material	Laminate	First-Ply Failure			Ultimate Failure		
		Stress MPa (ksi)	Strain (%)	Modulus, GPa (Msi)	Stress, MPa (ksi)	Strain (%)	Modulus, GPa (Msi)
S glass-epoxy	$[0_2/\pm 45]_S$	345.5 (50.1)	1.34	25.5 (3.7)	618.0 (89.6)	2.75	19.3 (2.8)
	$[0/90]_S$	89.7 (13.0)	0.38	23.4 (3.4)	547.6 (79.4)	2.75	19.3 (2.8)
HTS carbon carbon-epoxy	$[0_2/\pm 45]_S$	591.1 (85.7)	0.72	82.1 (11.9)	600.0 (87.0)	0.83	82.8 (12.0)
	$[0/90]_S$	353.1 (51.2)	0.45	78.6 (11.4)	549.0 (79.6)	0.72	72.4 (10.5)

Source: P. K. Mallick, *Fiber-Reinforced Composites*. Copyright © 1988 by Marcel Dekker, Inc.

thousand cycles. Some of these damages, such as surface craze marks, fiber splitting, and edge delaminations, may also be visible in the early stages of fatigue life. Unlike metals, the propagation and/or further accumulation of damage in a fiber-reinforced composite takes place in a progressive manner, resulting in a gradual loss of stiffness of the structure. Thus, the laminar composite continues to carry the load without catastrophic failure.

The result of these observations is that the basic design philosophy for fiber-reinforced composites, while similar to that for metals and other materials of construction as outlined earlier, is still evolving. Much of the design data in the areas of combined stresses, cumulative fatigue, and environmental damage, for example, are not available. Furthermore, there is little agreement on what constitutes a design failure and how to predict it. Industry-wide standards for material specifications, quality control, test methods, and failure analysis have not yet been developed for composites. For all of these reasons, the development of fiber-reinforced composite parts often relies upon empirical approaches and requires extensive prototyping and testing.

8.4.1   Problem Statement and Design Criteria

Design a driveshaft for an automobile. Specifically, there are five design criteria that must be met:

1. The shape constraint is that the materials must be able to be fabricated into the length and diameter requirements of the driveshaft, which are variable, depending on the size of the vehicle (see Table 8.10).

2. The static torque, critical rotational speed, and twist rate of a typical driveshaft must be met, which are variable, depending on the size of the vehicle (see Table 8.10).

Table 8.10   Design Specifications for a Typical Automotive Driveshaft

	Compact Cars	Medium-Sized Cars	Light Trucks	Medium-Sized Trucks	Heavy-Duty Trucks
Length (m)	1.14	1.40	1.91	1.91	2.54
Diameter (mm)	70	75	95	100	114
Static torque (kN · m)	1.70	2.83	3.39	8.14	10.85
Critical speed (r min^{-1})	12000	7200	6000	3000	5400
Twist rate (°kNm^{-1} m^{-1})	7.5	5.0	3.0	2.0	0.8
Maximum service temperature (°C)	120	120	130	110	90
Minimum service temperature (°C)	−40	−40	−40	−40	−40

Source: G. Lewis, *Selection of Engineering Materials*. Copyright © 1990 by Prentice-Hill, Inc.

3. The material must meet the maximum and minimum service temperatures of the driveshaft, which are typically −40 to 130°C.
4. The driveshaft must resist environmental degradation—for example, humidity, engine oil, road salt.
5. The driveshaft must be resistant to impact.

The specifications for criteria 1–3 are summarized in Table 8.10.

8.4.2   Problem Analysis

Let us first decide on the size of the vehicle we will design for, since some of the design specifications depend on the type of automobile. A medium-sized automobile has a nice combination of design specifications, so we will design for that. The second thing to recognize is that the first two design criteria, geometry and torsional mechanical properties, are the most important and are inextricably linked. So, we will divide our analysis into the geometry/torsional performance issues and the remaining issues, namely, environmental stability, use temperature, and impact resistance.

8.4.2.1   Geometry and Torsional Properties of Candidate Materials. From
a mechanics standpoint, a driveshaft can be approximated as a thin, rotating tube. The *first critical frequency*, N_{cr}, in units of rev/min (rpm), of a simply supported thin rotating tube is given by

$$N_{cr} = \frac{10.6\pi d_m}{L^2}\left(\frac{E_x}{\rho}\right)^{1/2} \tag{8.22}$$

where d_m is the mean diameter of the shaft, L is the shaft length, E_x is the axial elastic modulus (cf. Section 5.4.3.1) of the shaft material and ρ is its density. Rearrangement of Eq. (8.22) shows that L is proportional to $(E_x/\rho)^{0.25}$ and d_m is proportional to $(E_x/\rho)^{0.25}$. Obviously, the specific modulus is critical in this analysis. In addition, we

Figure 8.14   Torsional moment on a twisting shaft.

expect that specific shear strength and specific shear modulus will also be important given that the main mechanical stresses will arise from torsional loading.

Torsional stresses on the driveshaft are evaluated by determining the *applied torsional moment* (torque), T, as illustrated in Figure 8.14. Normally, the torsional moment could be simply related to the yield stress and shear modulus of the material. However, for thin-walled cylinders, the primary mode of failure is by *torsional buckling*. Torsional buckling determines the minimum wall thickness of the driveshaft for a given diameter. In general, it can be shown that critical minimum wall thickness to prevent torsional buckling in the driveshaft, t_{cr}, is given by

$$t_{cr} = \frac{2T}{K(TFM)\pi d_m^2} \tag{8.23}$$

where T is the torsional moment defined above, d_m is again the mean shaft diameter, K is a constant, and TFM is the *thickness figure of merit*, which is proportional to a collection of material properties:

$$TFM = \frac{E_x^{3/8} E_y^{5/8}}{(1 - v_{xy} v_{yx})^{5/8}} \tag{8.24}$$

Here, E_x and E_y are the axial and transverse elastic modulus of the driveshaft material, and v_{xy} and v_{yx} are the axial and transverse Poisson's ratios, respectively (see Section 5.4.3.1 for more information).

8.4.2.2   Secondary Properties of Candidate Materials. In terms of maximum and minimum use temperatures, we must simply ensure that our final candidate materials meet the use temperature requirements listed in Table 8.10. This design specification is not particularly restrictive, so we will save it for last. The most common measure of impact resistance is using an Izod impact test (cf. Figure 5.79). A similar test is called the *Charpy test*, a schematic diagram for which is shown in Figure 8.15. We will find relevant impact resistance data for our candidate materials, if available, and ensure that this design criterion is met as well.

As for environmental resistance, there exists a design chart that is somewhat useful for this case study, but, more importantly, may be of use in other designs. The compatibility of various materials in six common environments is shown in Figure 8.16. The suitability of a material for each of the six environments improves as you move from the center of the chart outward. In this case, resistance to organic solvents is of primary importance. We see that all ceramics and glasses, all alloys, and some polymers such as poly(tetrafluoroethylene), PTFE, will provide excellent resistance. Composites will provide good resistance, which may be satisfactory for our application.

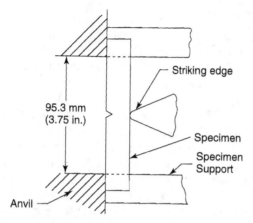

Figure 8.15 Schematic illustration of Charpy impact test. Reprinted, by permission, from P. K. Mallick, *Fiber-Reinforced Composites*, p. 249. Copyright © 1988 by Marcel Dekker, Inc.

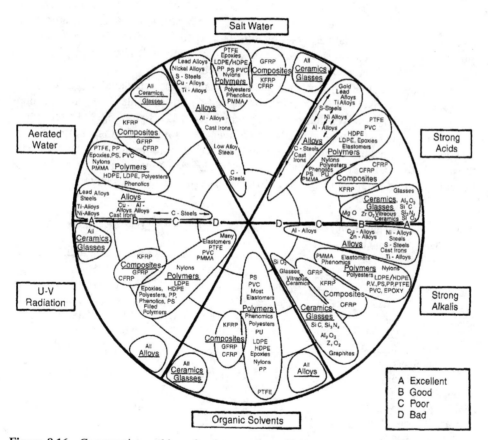

Figure 8.16 Comparative ranking of resistance of materials to attack by six common environments. Reprinted, by permission, from M. F. Ashby, *Materials Selection in Mechanical Design*, 2nd ed., p. 62. Copyright © 1999 by Michael F. Ashby.

8.4.3   Materials Selection

A materials selection chart of specific elastic modulus versus specific shear modulus would be of great help in this application. Unfortunately, one is not readily available, and to construct one without the aid of appropriate software would be too time-consuming. A similar plot of specific modulus versus specific strength will suffice for our initial materials selection guide. Such a plot has already been presented in Figure 5.57. A quick perusal of this figure for materials with both high specific modulus and high specific strength indicates that ceramics would be an excellent choice for this application, but we will not consider these here due to their inherent brittleness, which poses a distinct safety threat. However, research into ceramic driveshafts continues, and the consideration of these materials in the design process may soon be commonplace. Next to ceramics, the reinforced composites and alloys such as steel and aluminum alloys are the most likely candidates. Let us select three fiber-reinforced composites and two alloys for our comparison: ultrahigh-modulus graphite-fiber-reinforced epoxy resin; chopped-glass sheet molding compound; Kevlar® 49 fiber-reinforced epoxy resin; high strength steel; and 7075-T6 aluminum alloy.

The relevant properties of these materials for the torsional–mechanical analysis are listed in Table 8.11. On the basis of specific elastic modulus and specific shear modulus, the best materials are the graphite-fiber-reinforced epoxy resin, followed by either of the alloys, then the Kevlar® fiber-reinforced epoxy. The chopped glass sheet molding compound is obviously not a good choice.

The critical length of the driveshaft and *TFM* can then be calculated using Eq. (8.22) and (8.24), respectively with the values of d_m and N_{cr} from Table 8.10 for a medium-sized automobile. These results are presented in Table 8.12. Both of the fiber-reinforced epoxy composites have a decided advantage over the alloys due to their longer critical lengths. Since most medium-size automobile driveshafts are around 1 meter in length, the use of composites ensures that a one-piece design can be used. The use of either of the alloys would require a more costly two-piece design. The alloys have an advantage

Table 8.11   Relevant Properties of Candidate Materials for the Driveshaft of a Medium-Sized Automobile

Material	E_{xx} (GPa)	E_{yy} (GPa)	ν_{xy}	ν_{yx}[a]	G_{xy}	ρ (kg m^3)	E_{xx}/ρ (10^6 m)	G_{xy}/ρ (10^6 m)
High-strength steel	200	200	0.30	0.300	76	7833	2.60	0.99
7075-T6 aluminum alloy	70	70	0.34	0.340	27	2796	2.55	0.98
Ultrahigh-modulus graphite-fiber-reinforced epoxy resin[b]	303	7	0.25	0.006	6	1688	18.30	0.36
Chopped-glass sheet molding compound[b]	24	24	0.30	0.300	10	1933	1.27	0.53
Kevlar 49 fiber-reinforced epoxy resin[b]	86	5	0.33	0.019	2	1384	6.33	0.15

[a] $\nu_{yx} = \nu_{xy} E_{yy}/E_{xx}$.
[b] Volume fraction of fiber $= 0.65$.
Source: G. Lewis, *Selection of Engineering Materials*. Copyright © 1990 by Prentice-Hill, Inc.

Table 8.12   Critical Length and Thickness Figure of Merit for Driveshafts Made of Candidate Materials

Material	Critical Length, L_{cr} (m)	Thickness Figure of Merit (TFM)
High-strength steel	0.748	212
Al alloy	0.745	76
UHMGr/Ep	1.219	29
Kevlar 49/Ep	0.935	15

Source: G. Lewis, *Selection of Engineering Materials*. Copyright © 1990 by Prentice-Hill, Inc.

in terms of *TFM*, since they allow for much thinner-walled tubes to be used [recall from Eq. (8.23) and (8.24) that the critical thickness and TFM are inversely proportional].

In terms of the remaining criteria, we must simply ensure that the graphite-fiber-reinforced epoxy meets the minimum and maximum use temperature requirement, that it has sufficient impact resistance, and that it has acceptable environmental resistance. With regard to the latter, Figure 8.12 indicates that carbon-fiber-reinforced polymers (CFRP) have good to excellent resistance against organic solvents. Moreover, epoxies are some of the more resistant polymers, so we deem this criterion met. In terms of use temperature, we assume that this will be limited by the epoxy matrix, and not the carbon fibers. The minimum use temperature is not such an issue here due to the crosslinked nature of the epoxy resin. The upper use temperature of epoxies varies widely, depending primarily on the crosslink density and the type of hardener used. A rule of thumb is that the maximum use temperature is two-thirds of the cure temperature. In the worst case, epoxies can have heat deflection temperatures, where the material begins to deform, as low as 50°C. In the best case, the deflection temperatures can be as high as 300°C. So, depending upon the specific epoxy and field tests, the composite driveshaft may require a heat shield near the exhaust system. The Izod and Charpy impact energies of some common materials, including the five under consideration here, are listed in Table 8.13. In general, carbon-fiber-reinforced epoxies have lower impact energies than many metals. As a result, it may be necessary to coat the driveshaft with some energy-absorbing layer such as rubber to protect it from road debris. The exercise is not purely an academic one-driveshafts of filamentary composites have been in use for over a decade, such as in the Nissan 350Z.

Table 8.13 Impact Toughnesses of Various Materials

Material	Impact Energy, KJ/m^2 (ft-lb/in.^{2})	
	Charpy	Izod
S-glass-epoxy, $0°$, $v_f = 55\%$	734 (348)	—
Boron-epoxy, $0°$, $v_f = 55\%$	109–190 (51.5–90)	—
Kevlar 49-epoxy, $0°$, $v_f = 60\%$	317 (150)	158 (75)
AS carbon-epoxy, $0°$, $v_f = 60\%$	101 (48)	33 (15.5)
HMS carbon-epoxy, $0°$, $v_f = 60\%$	23 (11)	7.5 (3.6)
T-300 carbon-epoxy, $0°$, $v_f = 60\%$	132 (62.6)	67.3 (31.9)
4340 Steel ($R_c = 43$–46)	214 (102)	—
6061-T6 aluminum alloy	153 (72.5)	—
7075-T6 aluminum alloy	67 (31.7)	—

Source: P. K. Mallick, *Fiber-Reinforced Composites*. Copyright © 1988 by Marcel Dekker, Inc.

8.5 SELECTION OF MATERIALS AS TOOTH COATINGS

(Adapted from Lewis)

Enamel is a smooth, white, semitransparent material forming the outer, exposed part of teeth. It varies in thickness from 0.01 to about 2.5 mm, covering the dentin (cf. Section 1.5.2). Enamel is extremely hard and brittle, but not very strong. As a result, it can wear away over time. The main constituent inorganic phase (95 wt%) in enamel is calcium phosphate, in the form of hydroxyapatite, $Ca_{10}(OH)_2(PO_4)_6$ (cf. sect. 1.5.2). The organic phase (1 wt%) is made up of protein, carbohydrates, lipids, and other matter. The remaining phase (about 4 wt%) is water, mostly present as a shell surrounding the hydroxyapatite crystallites. The position of enamel relative to the other constituents in a human tooth was described in Chapter 1 (cf. Figure 1.91), and the mechanical properties of dentin and enamel were presented earlier in Table 5.14.

8.5.1 *Problem Statement and Design Criteria.* Propose a suitable replacement for enamel for use in coating a human tooth. The primary design criteria are:

1. High wear resistance and compressive strength.
2. Good insulation of the pulp against temperature extremes.

Table 8.14   Effect of Coupling Agents on Tensile Bond Strength of an Acrylic Resin on Teeth

Coupling Agent	Bond (Tensile) Strength (MPa)	
	Enamel	Dentin
None	1.67	0.00
NPG-GMA[a] (10%)	5.00	2.64
EDTA[b] + NPG-GMA	5.30	7.55
EDTA + NPG-GMA (after 20 days of immersion)	5.20	5.40

[a] N-(2-Hydroxy-3-methacryloxypropyl)-N-phenyl-lycine.
[b] Ethylene diamine tetraacetate, pH 7.
Source: G. Lewis, *Selection of Engineering Materials.* Copyright © 1990 by Prentice-Hill, Inc.

3. Aesthetic acceptability.
4. Excellent bonding to tooth enamel.
5. Ease of application (processibility).
6. Biocompatibility.

8.5.2 Problem Analysis

It is useful to divide the analysis into mechanical and nonmechanical properties. We will first consider the wear resistance and compressive strength considerations, then see if potential materials will meet the aesthetic, bonding, and biocompatibility criteria. It may be possible to incorporate secondary materials such as colorants, bonding agents, and compatiblizers that address these issues.

8.5.2.1 *Mechanical Properties of Candidate Materials.* The mechanical properties of enamel and dentin were presented earlier in Table 5.14. We will use these values as the basis for our material selection process. Of these properties, compressive strength is the most important. The candidate material should have a compressive strength at least that of enamel, which is about 384 MPa.

A property not listed in Table 5.14, but which is of paramount importance to this application, is wear rate. Wear rate was described briefly in Section 8.2.2, and those concepts apply here as well. As it applies to tooth enamel, abrasive wear occurs by fracture or chipping of the enamel, chemical erosion, which may arise due to acidic medications or drinks, dietary oxalate, or high oral hydrogen ion concentrations as a result of disease, physical erosion, or abrasion, which arises due to idiopathic mechanisms, dentrifices, toothbrushes, or abrasive diets. For example, the wear rate of enamel has been measured at about 10 μm/hour due to brushing with a toothbrush and toothpaste for 86,400 strokes [7].

8.5.2.2 *Secondary Properties of Candidate Materials.* The secondary criteria include a low thermal conductivity to reduce tooth sensitivity, color, adhesion, and

biocompatibility. In terms of thermal conductivity, the material should have a value as close to that of enamel as possible, which is 0.88 W/m · K. The thermal conductivity of dentin is approximately 0.59 W/m · K.

Ideally, the coating material will be transparent so that the natural color of the dentin will show through and eliminate the need for color-matching. If a composite material is to be used, this means that the refractive indices of both the matrix and reinforcement material should be matched. Alternatively, the material may be white or off-white. Regardless of the level of transparency, the material must be capable of resisting discoloration that arises from both (a) extrinsic sources such as food, drink, and tobacco and (b) intrinsic sources such as blood pigment staining, high-fluoride water, and drugs such as tetracycline.

Adhesive bonding can take place by one of two mechanisms. *Mechanical adhesion* is achieved by geometric effects that arise due to surface roughness and shrinkage. The other type of adhesion, *chemical adhesion*, is achieved by primary or secondary bonding (cf. Section 1.0.4). Most dental materials bond to tooth enamel by means of mechanical forces, the main vehicle being porosities for geometric hooking and rheological shrink-fitting. Roughness also plays a part by increasing the potential surface area of contact. Roughness is usually achieved by etching the tooth surface—for example, with phosphoric acid for 1–2 minutes. Bonding is enhanced by use of polyfunctional or difunctional coupling agents that chelate with dentinal calcium. Some of these coupling agents and their effect on the bond (tensile) strength of an acrylic resin are listed in Table 8.14. Bonding can also be enhanced by *mordants*, which are metal cations that are placed on the tooth surface to replace or supplement the calcium ions as sites for chelation. Calcifying agents, such as tricalcium phosphate, sodium fluoride, and sodium chloride, improve bonding, as do organic polyphosphonates, such as vinyl phosphonic acid and vinylbenzyl phosphonic acid.

Processability will be important insofar as the need for complex equipment will add to cost, and it may ultimately render the material useless for this type of application. An example is melting point. Even if all other factors were attractively addressed, a high melting point would require special melting equipment and may in fact be dangerous to the patient. There are, of course, alternative methods of delivery to melting, such as solvent evaporation, but the risks associated with those methods must also be evaluated.

Finally, since the material is to be used in the mouth, where it is in contact with the circulatory system, it must be nontoxic and biocompatible. Furthermore, there must not be any degradation products that are toxic, either as poisons or allergens. The biocompatibility of some materials was described briefly in Section 5.5, and processing techniques used to improve biocompatibility were described in Section 7.5.2.

8.5.3 Materials Selection

It is important to recognize that the unique combination of property requirements in this case study will complicate the materials selection process. In particular, the criterion of biocompatibility, while grouped with the other secondary criteria, is important enough to completely reject a candidate material if it is not met. As a result, we must further recognize that experimentation will be a critical part of this materials selection process and that substantial testing will be required before materials can be deemed acceptable or not.

Nonetheless, the wear-rate descriptions of Section 8.2.2 give us at least a starting point in the material selection process. We once again see that ceramics and refractory metals will provide the best (lowest) wear rates. However, processability must be carefully evaluated for these materials, particularly for the refractory metals due to their high melting points.

In terms of thermal conductivity, there are some generalizations that may prove helpful. Recall from Section 4.2.1.4 that thermal conduction can arise due to either electron or phonon conduction. A number of factors can influence the relative importance of these two conduction mechanisms, but for our purposes, structural order is probably the most important to consider. A highly ordered (i.e., highly crystalline material) will, in general, have a higher thermal conductivity than a disordered (i.e., amorphous) material. This trend is illustrated in Figure 8.17 where the thermal conductivities of some representative materials from three material classes are listed. Notice that, in general, polymers have the consistently lowest thermal conductivities. However, we know that their wear rates are much poorer than either ceramics or metals. We surmise from these data that ceramics may provide the most favorable combination of hardness, wear rate, and acceptably low thermal conductivity. The other issues of aesthetics, biocompatibility, bonding, and ease of application must be evaluated on a case-by-case basis.

Based upon the limited wear-rate and thermal conductivity data we have, then, it is not surprising that most commercial dental materials are ceramic-based. Some of the commercially available dental restorative materials are listed in Table 8.15. There have been a number of recent wear rate studies involving these, and other, dental materials [8–12], two of which are of particular importance to this case study. In the first study, Al-Hiyasat et al., [8] studied the wear rate of enamel against four dental

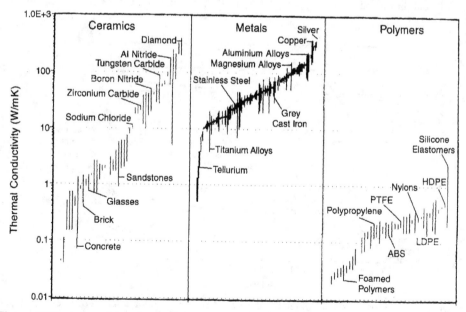

Figure 8.17 A materials selection bar chart showing ranges of thermal conductivity values for three of the material classes. Reprinted, by permission, from M. F. Ashby, *Materials Selection in Mechanical Design*, p. 33, 2nd ed. Copyright © 1999 by Michael F. Ashby.

Table 8.15   Some Commercially Available Dental Restorative Materials

Material	Type	Manufacturer
Z100	Composite	3M, St. Paul, MN
GIC	Glass ionomer ceramic	GC Corp., Tokyo, Japan
GIC, resin-modified	Polymer/glass ionomer ceramic	GC Corp., Tokyo, Japan
Duceram-LFC	Hydroxylated low fusing ceramic	Ducera Dental GmbH, Rosbach, Germany
Vitadur Alpha	Aluminous porcelain	Vita Zahnfabrik, Bad Säckingen, Germany
Vitadur Omega	Bonded-to-metal porcelain	Vita Zahnfabrik, Bad Säckingen, Germany
Cerc Vita Mark II	Machinable glass ceramic	Vita Zahnfabrik, Bad Säckingen, Germany
Dicor MGC	Glass ceramic	DeTrey/Dentsply
Type IV cast gold	71% gold casting alloy	Scientific Metal Co. Ltd., London, UK

ceramics (Vitadur Alpha, Vitadur Omega, Duceram-LFC, and Vita Mark II) and the gold casting alloy. In other words, instead of studying wear rates in a simulated tooth-brushing environment, the wear of real dental enamel as a result of friction against the other materials, as would be found during chewing or teeth-grinding, was studied. In the second study [9], the wear rate of three dental ceramics (Z100, GIC, and resin-modified GIC) as well as enamel was studied against an enamel surface in a similar apparatus, but at varying pH and with varying loads. The results of these studies under acidic conditions are summarized in Figure 8.18. While there was some variation of

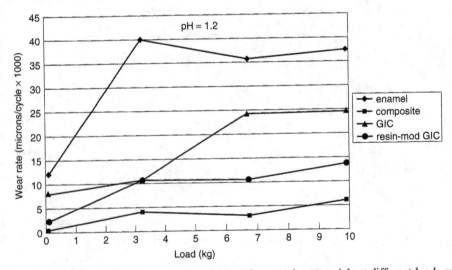

Figure 8.18   Wear rates for enamel and three dental restoration materials at different loads and pH 1.2. Reprinted from M. Shabanian and L. C. Richards, *In vitro* wear rates of materials under different loads and varying pH, *J. Prosth. Dent.*, **87**(6), 655. Copyright © 2002, with permission from Elsevier.

Cooperative Learning Exercise 8.5

(Extended Study, Spreadsheet Problem)

Due to the limited availability of test materials, the evaluation of biological materials often involves more statistical analysis to extract data than is typically used for the other materials classes. In this exercise, the use of statistics and spreadsheets will be illustrated in order to compare the experimental results of wear data for the candidate materials.

Work in groups of three. The raw data from reference 8 for the wear rate of enamel after 5000, 15,000, and 20,000 cycles in an experimental device against four dental restorative materials are presented in the table below.

	Mean Enamel Wear ± S.D. (mm). All data from 10 independent determinations.		
Material	5000 cycles	15,000 cycles	25,000 cycles
Alpha Porcelain	0.55 ± 0.10	0.78 ± 0.12	0.93 ± 0.15
Omega Porcelain	0.61 ± 0.13	0.83 ± 0.19	0.96 ± 0.20
Duceram-LFC	0.31 ± 0.06	0.45 ± 0.12	0.54 ± 0.15
Vita Mark II	0.41 ± 0.07	0.55 ± 0.13	0.65 ± 0.16
Gold	0.05 ± 0.02	0.07 ± 0.10	0.09 ± 0.03

Each person should select a data set for a given number of cycles in the experimental wear rate device—that is, pick a column of data. You are to perform the following analysis using a spreadsheet.

1. Create a made-up data set for each of the five values at your specified number of cycles. The only constraints are that the ten separate data points be positive values and that they produced an average value and standard deviation as given in the table. There is no one right answer to this problem.

2. Using a spreadsheet function or analysis tool, perform a Student's t-test to compare the mean values of each of the first four wear rates with that of gold (the last value in the column) at an agreed-upon confidence level—for example, 95% confidence level. If you have not yet learned how to do Student's t-test, check out any book on elementary statistics, and learn how to do a t-test, or use the Help menus in your spreadsheet package. The goal is to determine if the mean wear rates of the four ceramic materials are statistically different from the wear rate for gold. If done correctly, you will have performed four separate t-tests in this step.

3. Do one additional t-test to compare the mean wear rates of the Alpha and Omega Porcelain with each other for your number of cycles at the same confidence level as in the previous question.

Compare your answers with the other group members. Is there a change in the results with the number of cycles—that is, are the same number of wear rates different from gold at 15,000 cycles as for 5,000 cycles?

these results under neutral (pH = 7.0) and moderately acidic (pH = 3.3) conditions, the trends were similar.

The results of Figure 8.18 show that most commercially available dental restorative materials have wear rates that are lower (better) than human enamel. All of the materials listed in Table 8.15 have nominal colors equivalent to that of human teeth and are of acceptable biocompatibility. In particular, *glass ionomer ceramics* have become increasingly popular due to their favorable adhesion to dental tissues, fluoride release, and biocompatibility.

REFERENCES

Cited References

1. Ashby, M. F., *Materials Selection in Mechanical Design*, 2nd ed., Butterworth, Oxford, 1999.
2. Mueller, J. J., et al., Application of advanced materials and fabrication technology to let down valves for coal liquefaction systems, *Electrical Power Research Institute Report*, AF-305, January 1977.
3. Nielson, J. H., and A. Gilchrist, Erosion by a stream of solid particles, *Wear* **11**, 111–122, 123–143, (1968).
4. Finnie, I., and D. H. MacFadden, On the velocity dependence of the erosion of ductile metals by solid particles at low angles of incidence, *Wear* **48**, 181–190 (1978).
5. Evans, A. G., M. E. Gulden, and M. Rosenblatt, Impact damage in brittle materials in the elastic plastic range, *Proc. R. Soc.*, **A361**, 343–365 (1978).
6. Shetty, D. K., Erosive wear of advanced ceramics in coal-slurry streams, *Corrosion*, **38**(9), 500–509 (1982).
7. Rogers, B. J., Coating materials for teeth, Research Report 77–140, Lee Pharmaceuticals, South El Monte, CA, p. 5 (1977).
8. Al-Hiyasat, A. S., W. P. Saunders, S. W. Sharkey, G. McR. Smith, and W. H. Gilmour, Investigation of human enamel wear against four dental ceramics and gold, *J. Dentistry*, **26**, 487–495 (1998).
9. Shabanian, M., and L. C. Richards, *In vitro* wear rates of materials under different loads and varying pH, *J. Prosth. Dent.*, **87**(6), 650–656 (2002).
10. Burak, N., J. A. Kaidonis, L. C. Richards, and G. C. Townsend, Experimental studies of human dentine wear, *Arch. Oral Biol.*, **44**, 885–887 (1999).
11. Wendt, S. L., T. L. Ziemiecki, and K. F. Leinfelder, Proximal wear rates by tooth position of resin composite restorations, *J. Dentistry*, **24**(1–2), 33–39 (1996).
12. Pesun, I. J., A. K. Olson, J. S. Hodges, and G. C. Anderson, *in vivo* evaluation of the surface of posterior resin composite restorations: A pilot study, *J. Prosth. Dent.*, **84**(3), 353–359 (2000).

Materials Selection and Design — General

Budinski, K., *Engineering Materials, Properties and Selection*, Prentice-Hall, Englewood Cliffs, NJ, 1979.
Charles, J. A., F. A. A. Crane, and J. A. G. Furness, *Selection and Use of Engineering Materials*, 3rd ed., Butterworth, Oxford, 1987.
Dieter, G. E., *Engineering Design, A Materials and Processing Approach*, 2nd ed., McGraw-Hill, New York, 1991.

Ertas, A., and J. C. Jones, *The Engineering Design Process*, John Wiley & Sons, New York, 1993.

Farag, M. M., *Selection of Materials and Manufacturing Processes for Engineering Design*, Prentice-Hall, Englewood Cliffs, NJ, 1989.

French, M. J., *Conceptual Design for Engineers*, The Design Council, London and Springer, Berlin, 1985.

Howell, S. K., *Engineering Design and Problem Solving*, 2nd ed., Prentice-Hall, Upper Saddle River, NJ, 2002.

Lewis, G., *Selection of Engineering Materials*, Prentice-Hall, Englewood Cliffs, NJ, 1990.

Ullman, D. G., *The Mechanical Design Process*, McGraw-Hill, New York, 1992.

Materials Selection in Designing with Metals

ASM Metals Handbook, 10th ed., ASM International, Metals Park, OH, 1990.

Aluminum and Aluminum Alloys, A. L. Phillips, ed., American Welding Society, New York, 1967.

Materials Selection in Designing with Ceramics and Glasses

Wachtman, J. B., *Mechanical Properties of Ceramics*, John Wiley & Sons, New York, 1996.

Materials Selection in Designing with Polymers

Powell, P. C., and A. J. Ingen Housz, *Engineering with Polymers*, 2nd ed., Stanley Thornes, Cheltenham, UK, 1998.

Materials Selection in Designing with Composites

Mallick, P. K., *Fiber-Reinforced Composites: Materials, Manufacturing, and Design*, Marcel Dekker, New York, 1988.

Materials Selection in Designing with Biologics

Hill, D., *Design Engineering of Biomaterials for Medical Devices*, John Wiley & Sons, New York, 1998.

PROBLEMS

Level I

8.I.1 A component is presently made of brass. Use a modulus–density materials selection diagram, or look up the appropriate data, to suggest two metals which, in the same shape, would be stiffer.

8.I.2 List some materials that have a tensile modulus of $E > 200$ GPa and density $\rho < 2$ g/cm^3.

Level II

8.II.1 In Section 8.2, we saw how erosive wear can be different for brittle and ductile materials. In reality, most materials exhibit behavior that is a combination of brittle and ductile wear. A simple, empirical model has been developed to estimate the erosive wear, assuming that (a) the normal component of the velocity vector (y component in Figure 8.12) creates cracks from which fatigue failure or brittle failure initiates and (b) the tangential component (x component) of the velocity vector causes failure by plastic deformation. The relation expresses the total volume worn away per mass of abrasive particle, V/m, in terms of these two velocity components:

$$\frac{V}{m} = \frac{0.5(v_0 \sin\alpha - v_L)}{U_f} + \frac{0.5(v_0^2 \cos^2\alpha - v_p^2)}{U_s} \qquad (P8.1)$$

where U_f is the amount of energy needed to erode a unit volume of material by the fatigue mechanism, and U_s is the specific energy need to remove a unit volume of the surface material by shear deformation. The velocities v_L and v_p are the limiting velocity at which no fatigue failure takes place and the residual horizontal component of the particle velocity after impact, respectively.

Use the data below for the erosive wear for alumina particles impacting on a graphite-fiber-reinforced epoxy resin composite material to determine the four parameters U_f, U_s, v_L, and v_p in Eq. (P8.1). You may need to look up some additional data and make appropriate assumptions to solve this problem. You may also find a spreadsheet helpful for solving the four equations with four unknowns.

Impingement Angle, α (°)	Erosion Weight Loss (g)[a]
10	50×10^{-4}
20	75×10^{-4}
25	170×10^{-4}
40	130×10^{-4}

[a] Per gram of alumina impacted at a velocity of 91.4 m s^{-1}.
Source: N. P. Suh and A. P. L. Turner, *Elements of the Mechanical Behavior of Solids*. Washington, D.C.: Scripta Book Co., 1975, p. 577.

Level III

8.III.1 Collect timely information on the development of ceramic driveshafts, and perform an analysis similar to that used in Section 8.4 for composite driveshafts using the candidate ceramics you have found. How do your ceramic materials compare to those recommended in Section 8.4? If these materials are currently not used in production-level vehicles, propose some specific developments that must take place before they can be used in commercial driveshafts.

Energy Values for Single Bonds

Bond	Energy(kJ/mol)[a]	Bond	Energy(kJ/mol)[a]	Bond	Energy(kJ/mol)[a]
H–H	436	P–H	320	Si–Cl	359
C–C	348	As–H	245	Si–Br	289
Si–Si	177	O–H	463	Si–I	213
Ge–Ge	157	S–H	339	Ge–Cl	408
Sn–Sn	143	Se–H	277	N–F	270
N–N	161	Te–H	241	N–Cl	200
P–P	215	H–F	563	P–Cl	331
As–As	134	H–Cl	432	P–Br	274
Sb–Sb	126	H–Br	366	P–I	215
Bi–Bi	105	H–I	299	As–F	466
O–O	139	C–Si	290	As–Cl	288
S–S	213	C–N	292	As–Br	236
Se–Se	184	C–O	351	As–I	174
Te–Te	138	C–S	259	O–F	185
F–F	153	C–F	441	O–Cl	203
Cl–Cl	243	C–Cl	328	S–Cl	250
Br–Br	193	C–Br	276	S–Br	212
I–I	151	C–I	240	Cl–F	254
C–H	413	Si–O	369	Br–Cl	219
Si–H	295	Si–S	227	I–Cl	210
N–H	391	Si–F	541	I–Br	178

Source: Pauling, L., *The Nature of the Chemical Bond*, 3rd ed., Cornell University Press, New York, 1960.
[a] Divide values by 4.184 to obtain units of kcal/mol.

An Introduction to Materials Engineering and Science: For Chemical and Materials Engineers,
by Brian S. Mitchell
ISBN 0-471-43623-2 Copyright © 2004 John Wiley & Sons, Inc.

Structure of Some Common Polymers

Chemical Name	Mer Chemical Structure
Epoxy (diglycidyl ether of bisphenol A, DGEPA)	
Melamine-formaldehyde (melamine)	
Phenol-formaldehyde (phenolic)	
Polyacrylonitrile (PAN)	

An Introduction to Materials Engineering and Science: For Chemical and Materials Engineers,
by Brian S. Mitchell
ISBN 0-471-43623-2 Copyright © 2004 John Wiley & Sons, Inc.

Chemical Name	Mer Chemical Structure
Polyamide-imide (PAI)	
Polybutadiene	
Polybutylene terephthalate (PBT)	
Polycarbonate (PC)	
Polychloroprene	
Polychlorotrifluoroethylene	
Polydimethyl siloxane (silicone rubber)	
Polyetheretherketone (PEEK)	
Polyethylene (PE)	
Polyethylene terephthalate (PET)	

(continued)

Chemical Name	Mer Chemical Structure
Polyhexamethylene adipamide (nylon 6,6)	
Polyimide	
Polyisobutylene	
cis-Polyisoprene (natural rubber)	
Polymethyl methacrylate (PMMA)	
Polyphenylene oxide (PPO)	
Polyphenylene sulfide (PPS)	
Polyparaphenylene terephthalamide (aramid)	
Polypropylene (PP)	

Chemical Name	Mer Chemical Structure
Polystyrene (PS)	
Polytetrafluoroethylene (PTFE)	
Polyvinyl acetate (PVAc)	
Polyvinyl alcohol (PVA)	
Polyvinyl chloride (PVC)	
Polyvinyl fluoride (PVF)	
Polyvinylidene chloride (PVDC)	
Polyvinylidene fluoride (PVDF)	

Composition of Common Alloys

FERROUS ALLOYS

Standard Stainless and Heat-Resisting Steels

Type	UNS#	Composition, %, Balance Fe. Compositions are maximum unless noted.								
		C	Mn	P	S	Si	Cr	Ni	Mo	Other
201	S20100	0.15	5.5–7.5	0.06	0.03	1.00	16–18	3.5–5.5	—	N 0.25
202	S20200	0.15	7.5–10.0	0.06	0.03	1.00	17–19	4.0–6.0	—	N 0.25
205	S20500	0.12–0.25	14–15.5	0.06	0.03	1.00	16.5–18	1–1.75	—	N 0.32–0.4
301	S30100	0.15	2.0	0.045	0.03	1.00	16–18	6–8	—	—
302	S30200	0.15	2.0	0.045	0.03	1.00	17–19	8–10	—	—
302B	S30215	0.15	2.0	0.045	0.03	2.0–3.0	17–19	8–10	—	—
303	S30300	0.15	2.0	0.2	0.15 min	1.00	17–19	8–10	0.6 (opt.)	—
303 Se	S30323	0.15	2.0	0.2	0.06	1.00	17–19	8–10	—	Se 0.15 min
304	S30400	0.08	2.0	0.045	0.03	1.00	18–20	8–10.5	—	—
304L	S30403	0.03	2.0	0.045	0.03	1.00	18–20	8–12	—	—
304N	S30451	0.08	2.0	0.045	0.03	1.00	18–20	8–10.5	—	N 0.1–0.16
305	S30500	0.12	2.0	0.045	0.03	1.00	17–19	10.5–13	—	—
308	S30800	0.08	2.0	0.045	0.03	1.00	19–21	10–12	—	—
309	S30900	0.20	2.0	0.045	0.03	1.00	22–24	12–15	—	—
309S	S30908	0.08	2.0	0.045	0.03	1.00	22–24	12–15	—	—

An Introduction to Materials Engineering and Science: For Chemical and Materials Engineers,
by Brian S. Mitchell
ISBN 0-471-43623-2 Copyright © 2004 John Wiley & Sons, Inc.

		Composition, %, Balance Fe. Compositions are maximum unless noted.								
310	S31000	0.25	2.0	0.045	0.03	1.5	24–26	19–22	—	—
310S	S31008	0.08	2.0	0.045	0.03	1.5	24–26	19–22	—	—
314	S31400	0.25	2.0	0.045	0.03	1.5–3.0	23–26	19–22	—	—
316	S31600	0.08	2.0	0.045	0.03	1.00	16–18	10–14	2.0–3.0	—
316F	S31620	0.08	2.0	0.2	0.1 min	1.00	16–18	10–14	1.75–2.5	—
316L	S31603	0.03	2.0	0.045	0.03	1.00	16–18	10–14	2.0–3.0	—
316N	S31651	0.08	2.0	0.045	0.03	1.00	16–18	10–14	2.0–3.0	N 0.1–0.16
317	S31700	0.08	2.0	0.045	0.03	1.00	18–20	11–15	3.0–4.0	—
317L	S31703	0.03	2.0	0.045	0.03	1.00	18–20	11–15	3.0–4.0	—
321	S31200	0.08	2.0	0.045	0.03	1.00	17–19	9–12	—	Ti 5 × C min
329	S32900	0.10	2.0	0.04	0.03	1.00	25–30	3–6	1.0–2.0	—
330	S33000	0.08	2.0	0.04	0.03	0.75–1.5	17–20	34–37	—	—
347	S34700	0.08	2.0	0.045	0.03	1.00	17–19	9–13	—	Cb + Ta 10× C min
348	S34800	0.08	2.0	0.045	0.03	1.00	17–19	9–13	—	Cb + Ta 10× C min; Ta 0.10 max, Co 0.20 max
384	S38400	0.08	2.0	0.045	0.03	1.00	15–17	17–19	—	—
403	S40300	0.15	1.0	0.04	0.03	0.5	11.5–13	—	—	—
405	S40500	0.08	1.0	0.04	0.03	1.00	11.5–14.5	—	—	Al 0.1–0.3
409	S40900	0.08	1.0	0.045	0.045	1.00	10.5–11.75	—	—	Ti 6 × C min; 0.75 max
410	S41000	0.15	1.0	0.04	0.03	1.00	11.5–13.5	—	—	—
414	S41400	0.15	1.0	0.04	0.03	1.00	11.5–13.5	1.25–2.5	—	—
416	S41600	0.15	1.25	0.06	0.15 min	1.00	12–14	—	0.6 (opt.)	—
416 Se	S41623	0.15	1.25	0.06	0.06	1.00	12–14	—	—	Se 0.15 min
420	S41000	>0.15	1.0	0.04	0.03	1.00	12–14	—	—	—

(continued)

		Composition, %, Balance Fe. Compositions are maximum unless noted.								
Type	UNS#	C	Mn	P	S	Si	Cr	Ni	Mo	Other
420F	S42020	>0.15	1.25	0.06	0.15 min	1.00	12–14	—	0.6 (opt.)	—
422	S42200	0.2–0.25	1.0	0.025	0.025	0.75	11–13	0.5–1.0	0.75–1.25	V 0.15–0.30; W 0.75–1.25
429	S42900	0.12	1.0	0.04	0.03	1.00	14–16	—	—	—
430	S43000	0.12	1.0	0.04	0.03	1.00	16–18	—	—	—
430F	S43020	0.12	1.25	0.06	0.15 min	1.00	16–18	—	0.6 (opt.)	—
430FSe	S43023	0.12	1.25	0.06	0.06	1.00	16–18	—	—	Se 0.15 min
431	S43100	0.20	1.0	0.04	0.03	1.00	15–17	1.25–2.5	—	—
434	S43400	0.12	1.0	0.04	0.03	1.00	16–18	—	0.75–1.25	—
436	S43600	0.12	1.0	0.04	0.03	1.00	16–18	—	0.75–1.25	Cb + Ta 5 × C min, 0.70 max
440A	S44002	0.6–0.75	1.0	0.04	0.03	1.00	16–18	—	0.75	—
440B	S44003	0.75–0.95	1.0	0.04	0.03	1.00	16–18	—	0.75	—
440C	S44004	0.95–1.2	1.0	0.04	0.03	1.00	16–18	—	0.75	—
442	S44200	0.2	1.0	0.04	0.03	1.00	18–23	—	—	—
446	S44600	0.2	1.5	0.04	0.03	1.00	23–27	—	—	N 0.25
501	S50100	>0.1	1.0	0.04	0.03	1.00	4–6	—	0.4–0.65	—
502	S50200	0.1	1.0	0.04	0.03	1.00	4–6	—	0.4–0.65	—
503	S50300	0.15	1.0	0.04	0.04	1.00	6–8	—	0.45–0.65	—
504	S50400	0.15	1.0	0.04	0.04	1.00	8–10	—	0.9–1.1	—

Source: Engineering Properties of Steel, P. D. Harvey, ed., ASM, 1982.

IRON–NICKEL ALLOYS

	Composition, %, Balance Iron					
Type	Ni	Co	Cr	W	Al	Cu
Invar	36	—	—	—	—	—
Superinvar	31	5	—	—	—	—

	Composition, %, Balance Iron					
Type	Ni	Co	Cr	W	Al	Cu
Elinvar	32	—	5	2	—	—
Dumet	42	—	—	—	—	—
Kovar	29	17	—	—	—	—
Pllatinite	46	—	—	—	—	—
Permalloy	78.5	—	—	—	—	—
Hipernik	50	—	—	—	—	—
Perminvar	45	25	—	—	—	—
Alnico I	20	5	—	—	12	—
Alnico II	17	12.5	—	—	10	6
Alnico III	25	—	—	—	12	—
Alnico IV	28	5	—	—	12	—

CARBON STEELS

		Composition, %, Balance Iron		
AISI-SAE Designation	UNS #	C	Mn	S
1015	G10150	0.12–0.18	0.30–0.60	—
1020	G10200	0.17–0.23	0.30–0.60	—
1022	G10220	0.17–0.23	0.70–1.00	—
1030	G10900	0.27–0.34	0.60–0.90	—
1040	G10400	0.36–0.44	0.60–0.90	—
1050	G10500	0.47–0.55	0.60–0.90	—
1060	G10600	0.55–0.66	0.60–0.90	—
1080	G10800	0.74–0.88	0.60–0.90	—

(continued)

		Composition, %, Balance Iron		
AISI-SAE Designation	UNS #	C	Mn	S
1095	G10950	0.90–1.04	0.30–0.50	—
1118	G11180	0.14–0.20	1.30–1.60	0.08–0.13
1137	G11370	0.32–0.39	1.35–1.65	0.08–0.13
1141	G11410	0.37–0.45	1.35–1.65	0.08–0.13
1144	G11440	0.40–0.48	1.35–1.65	0.24–0.33

Source: *Materials Handbook*, G. S. Brady and H. R. Clauser, editors, McGraw-Hill, New York, 1991.

ALLOY STEELS

		Composition, %, Balance Iron							
AISI-SAE Designation	UNS #	C	Mn	P (max)	S (max)	Si	Cr	Ni	Mo
1330	G13300	0.28–0.33	1.60–1.90	0.035	0.040	0.15–0.30	—	—	—
1340	G13400	0.38–0.43	1.60–1.90	0.035	0.040	0.15–0.30	—	—	—
3140	—	0.38–0.43	0.70–0.90	0.040	0.040	0.20–0.35	0.55–0.75	1.10–1.40	—
4037	G40370	0.35–0.40	0.70–0.90	0.035	0.040	0.15–0.30	0.20–0.30	—	—
4042	G40420	0.40–0.45	0.70–0.90	0.035	0.040	0.15–0.30	0.80–1.10	—	0.15–0.25
4130	G41300	0.28–0.33	0.40–0.60	0.035	0.040	0.15–0.30	0.80–1.10	—	0.15–0.25
4140	G41400	0.38–0.43	0.75–1.00	0.035	0.040	0.15–0.30	0.80–1.10	—	0.15–0.25
4150	G41500	0.48–0.53	0.75–1.00	0.035	0.040	0.15–0.30	0.80–1.10	—	0.20–0.30
4320	G43200	0.17–0.22	0.45–0.65	0.035	0.040	0.15–0.30	0.40–0.60	1.65–2.00	0.20–0.30
4340	G43400	0.38–0.43	0.60–0.80	0.035	0.040	0.15–0.30	0.70–0.90	1.65–2.00	0.20–0.30
4620	G46200	0.17–0.22	0.45–0.65	0.035	0.040	0.15–0.30	1.65–2.00	—	0.20–0.30
4820	G48200	0.18–0.23	0.50–0.70	0.035	0.040	0.15–0.30	3.25–3.75	—	0.20–0.30
5046	G50460	0.43–0.48	0.75–1.00	0.035	0.040	0.15–0.30	0.20–0.35	—	—
50B46	G50461	0.44–0.49	0.75–1.00	0.035	0.040	0.15–0.30	0.20–0.35	—	—
5060	G50600	0.56–0.64	0.75–1.00	0.035	0.040	0.15–0.30	0.40–0.60	—	—
50B60	G50461	0.56–0.64	0.75–1.00	0.035	0.040	0.15–0.30	0.40–0.60	—	—

		Composition, %, Balance Iron							
AISI-SAE Designation	UNS #	C	Mn	P (max)	S (max)	Si	Cr	Ni	Mo
5130	G51300	0.28–0.33	0.70–0.90	0.035	0.040	0.15–0.30	0.80–1.10	—	—
5140	G51400	0.38–0.43	0.70–0.90	0.035	0.040	0.15–0.30	0.70–0.90	—	—
5150	G51500	0.48–0.53	0.70–0.90	0.035	0.040	0.15–0.30	0.70–0.90	—	—
5160	G51600	0.56–0.64	0.75–1.00	0.035	0.040	0.15–0.30	0.70–0.90	—	—
51B60	G51601	0.56–0.64	0.75–1.00	0.035	0.040	0.15–0.30	0.70–0.90	—	—
6150	G61500	0.48–0.53	0.70–0.90	0.035	0.040	0.15–0.30	0.8–1.10	—	—
81B45	G81451	0.43–0.48	0.75–1.00	0.035	0.040	0.15–0.30	0.35–0.55	0.20–0.40	0.08–0.15
8620	G86200	0.18–0.23	0.70–0.90	0.035	0.040	0.15–0.30	0.40–0.60	0.40–0.70	0.15–0.25
8630	G86300	0.28–0.33	0.70–0.90	0.035	0.040	0.15–0.30	0.40–0.60	0.40–0.70	0.15–0.25
8640	G86400	0.38–0.43	0.75–1.00	0.035	0.040	0.15–0.30	0.40–0.60	0.40–0.70	0.15–0.25
86B45	G86451	0.43–0.48	0.75–1.00	0.035	0.040	0.15–0.30	0.40–0.60	0.40–0.70	0.15–0.25
8650	G86500	0.48–0.53	0.75–1.00	0.035	0.040	0.15–0.30	0.40–0.60	0.40–0.70	0.15–0.25
8660	G86600	0.56–0.64	0.75–1.00	0.035	0.040	0.15–0.30	0.40–0.60	0.40–0.70	0.15–0.25
8740	G87400	0.38–0.43	0.75–1.00	0.035	0.040	0.15–0.30	0.40–0.60	0.40–0.70	0.20–0.30
9255	G92550	0.51–0.59	0.70–0.95	0.035	0.040	1.8–2.2	—	—	—
9260	G92600	0.56–0.64	0.75–1.00	0.035	0.040	1.8–2.2	—	—	—
9310	G93106	0.08–0.13	0.45–0.65	0.025	0.025	0.15–0.30	1.00–1.40	3.00–3.50	0.08–0.15
94B30	G94301	0.28–0.33	0.75–1.00	0.035	0.040	0.15–0.30	0.30–0.50	0.30–0.60	0.08–0.15

Source: *Materials Handbook*, G. S. Brady and H. R. Clauser, editors, McGraw-Hill, NY, 1991.

GRAY CAST IRONS

	Composition, %, Balance Fe			
UNS #	Mn	Si	P	S
F10004	0.50–0.80	2.3–2.8	0.15	0.15
F10005	0.60–0.90	2.0–2.4	0.12	0.15
F10009	0.60–0.90	1.6–2.1	0.12	0.12

(*continued*)

	Composition, %, Balance Fe			
F10006	0.60–0.90	1.9–2.3	0.10	0.16
F10007	0.60–0.90	1.8–2.2	0.08	0.16
F10010	0.60–0.90	1.3–1.8	0.08	0.12
F10011	0.60–0.90	1.3–1.8	0.08	0.12
F10008	0.70–1.00	1.8–2.1	0.07	0.16
F10012	0.60–0.90	1.95–2.4	0.07	0.12

Source: Materials Handbook, G. S. Brady and H. R. Clauser, editors, McGraw-Hill, New York, 1991.

ALUMINUM ALLOYS

Wrought Alloys									
Composition, %, Balance Aluminum									
Type	Si	Cu	Mn	Mg	Cr	Zn	Ni	Ti	Other
1100	—	0.12	—	—	—	—	—		—
2011	—	5.5	—	—	—	—	—		0.4 Bi; 0.4 Pb
2014	0.8	4.4	0.8	0.5	—	—	—		—
2024	—	4.4	0.6	1.5	—	—	—		—
2219	—	6.3	0.3	—	—	—	—	0.06	0.10 V; 0.18 Zr
2319	—	6.3	0.3	—	—	0.18	—	0.15	0.10 V
2618	0.18	2.3	—	1.6	—	—	1.0	0.07	1.1 Fe
3003	—	—	0.12	1.2	—	—	—	—	—
3004	—	—	1.2	1.0	—	—	—	—	—
3105	—	—	0.55	0.5	—	—	—	—	—
4032	12.2	0.9	—	1.0	—	—	0.9	—	—
4043	5.2	—	—	—	—	—	—	—	—

Wrought Alloys								
5005	—	—	—	0.8	—	—	—	—	—
5050	—	—	—	1.4	—	—	—	—	—
5052	—	—	—	2.5	0.25	—	—	—	—
5056	—	—	0.12	5.0	0.12	—	—	—	—
5083	—	—	0.7	4.4	0.15	—	—	—	—
5086	—	—	0.4	4.0	0.15	—	—	—	—
5154	—	—	—	3.5	0.25	—	—	—	—
5182	—	—	0.35	4.5	—	—	—	—	—
5252	—	—	—	2.5	—	—	—	—	—
5254	—	—	—	3.5	0.25	—	—	—	—
5356	—	—	0.12	5.0	0.12	—		0.13	—
5454	—	—	0.8	2.7	0.12			—	—
5456	—	—	0.8	5.1	0.12			—	—
5457	—	—	0.3	1.0	—	—	—	—	—
5652	—	—	—	2.5	0.25	—	—	—	—
5657	—	—	—	0.8	—	—	—	—	—
6005	0.8	—	—	0.5	—	—	—	—	—
6009	0.8	0.35	0.5	0.6	—	—	—	—	—
6010	1.0	0.35	-.5	0.8	—	—	—	—	—
6061	0.6	0.28	—	1.0	0.2	—	—	—	—
6063	0.4	—	—	0.7	—	—	—	—	—
6066	1.4	1.0	0.8	1.1	—	—	—	—	—
6070	1.4	0.28	0.7	0.8	—	—	—	—	—
6101	0.5	—	—	0.6	—	—	—	—	—
6151	0.9	—	—	0.6	0.25	—	—	—	—

(*continued*)

Wrought Alloys									
6201	0.7	—	—	0.8	—	—	—	—	—
6205	0.8	—	0.1	0.5	0.1	—	—	—	0.1 Zr
7049	—	1.5	—	2.5	0.15	7.6	—	—	—
7075	—	1.6	—	2.5	0.23	5.6	—	—	—
Cast Alloys									
201	—	4.6	0.35	0.35	—	—	—	0.25	0.7 Ag
206	0.10 max	4.6	0.35	0.25	—	—	—	0.22	0.15 Fe max
A206	0.05 max	4.6	0.35	0.25	—	—	—	0.22	0.10 Fe max
208	3.0	4.0	—	—	—	—	—	—	
242	—	4.0	—	1.5	—	—	2.0	—	
295	0.8	4.5	—	—	—	—	—	—	
296	2.5	4.5	—	—	—	—	—	—	
308	5.5	4.5	—	—	—	—	—	—	
319	6.0	3.5	—	—	—	—	—	—	
336	12.0	1.0	—	1.0	—	—	2.5	—	
354	9.0	1.8	—	0.50	—	—	—	—	
355	5.0	1.2	0.50 max	0.50	—	0.35 max	—	—	0.6 Fe max
C355	5.0	1.2	0.10 max	0.50	—	0.10 max	—	—	0.2 Fe max
356	7.0	0.25 max	0.35 max	0.32	—	0.35 max	—	—	0.6 Fe max
A356	7.0	0.25 max	0.10 max	0.35	—	0.10 max	—	—	0.2 Fe max
357	7.0	—	—	0.50	—	—	—	—	—
A357	7.0	—	—	0.6	—	—	—	0.15	0.005 Be

Wrought Alloys									
359	9.0	—	—	0.6	—	—	—	—	—
360	9.5	—	—	0.5	—	—	—	—	2.0 Fe max
A360	9.5	—	—	0.5	—	—	—	—	1.3 Fe max
380	8.5	3.5	—	—	—	—	—	—	2.0 Fe max
A380	8.5	3.5	—	—	—	—	—	—	1.3 Fe max
383	10.5	2.5	—	—	—	—	—	—	—
384	11.2	3.8	—	—	—	3.0 max	—	—	—
A384	11.2	3.8	—	—	—	1.0 max	—	—	—
390	17.0	4.5	—	0.6	—	1.3 max	—	—	—
A390	17.0	4.5	—	0.6	—	0.5 max	—	—	—
413	12.0	—	—	—	—	—	—	—	2.0 Fe max
A413	12.0	—	—	—	—	—	—	—	1.3 Fe max
443	5.2	0.6 max	—	—	—	—	—	—	—
A443	5.2	0.30 max	—	—	—	—	—	—	—
B443	5.2	0.15 max	—	—	—	—	—	—	—
C443	5.2	0.6 max max	—	—	—	—	—	—	2.0 Fe max
514	—	—	—	4.0	—	—	—	—	—

(continued)

Wrought Alloys									
518	—	—	—	8.0	—	—	—	—	—
520	—	—	—	10.0	—	—	—	—	—
535	—	—	0.18	6.8	—	—	—	0.18	—
A535	—	—	0.18	7.0	—	—	—	—	—
B535	—	—	—	7.0	—	—	—	0.18	—
712	—	—	—	0.6	0.5	5.8	—	0.2	—
713	—	0.7	—	0.35	—	7.5	—	—	—
771	—	—	—	0.9	0.13	7.0	—	0.15	—
850	—	1.0	—	—	—	—	1.0	—	6.2 Sn

Source: *Engineering Metallurgy*, Committee on Metallurgy, Pitman Publishing, 1957.

COPPER ALLOYS

Alloy	UNS #	Composition, %, Balance Copper						
		Ni	Zn	Sn	Al	Be	Pb	Other
Brass, naval		—	39.25	0.75	—	—	—	—
Muntz metal		—	40.0	—	—	—	—	—
Bronze, aluminum		—	—	—	8.0	—	—	—
Beryllium copper	C17200	—	—	—	—	1.9	—	Co or Ni
Brass, free-cutting	C36000	—	35.5	—	—	—	2.5	—
Nickel silver, 18%		18	17	—	—	—	—	—
Nickel silver, 13%		12.5	20.0	2.0	—	—	9.0	—
Constantan		45	—	—	—	—	—	—
Brass, red		—	5	5	—	—	5	—
Bronze, silicon		—	4	—	1	—	—	Fe 2.0; Mn 1.0

Bronze, tin		—	4	8	—	—	—	—
Cupronickel	C71500	30	—	—	—	—	—	—

TITANIUM ALLOYS

		Composition, %, Balance Titanium					
Type	UNS #	Al	Sn	Mo	V	Fe	Cu
Ti–5Al–2.5Sn	R54520	5	2.5				
Ti–8Al–1 Mo–1 V	R54810	8		1	1		
Ti–6Al–4 V	R56400	6			4		
Ti–6Al–6 V–2Sn	R56620	6	2		6		0.75
Ti–10 V–2Fe–3Al	—	3			10	2	

Source: *ASM Handbook*, American Society for Metals, Metals Park, OH, 1985.

SILVER ALLOYS

	Composition, %, Balance Silver							
Type	Au	Cu	Zn	Pt	Pb	Hg	Sn	Pd
Sterling silver	—	7.5	—	—	—	—	—	—
Coin silver (to 1966)	—	10	—	—	—	—	—	—
Brazing alloy	—	28	—	—	—	—	—	—
10-karat green gold	41.7	9	0.3	—	—	—	—	—
Dental amalgam	—	2	0.5	—	—	52	12.5	—
Gold solder	50	17.5	2	—	—	—	—	—
White gold	60	10	1	4	—	—	—	10
14-karat yellow gold	58.3	29.7	2	—	—	—	—	—
Soft solder	—	—	—	—	97.5	—	—	—

Source: *Practical Handbook of Materials Science*

MAGNESIUM ALLOYS

ASTM #	UNS #	Composition, %, Balance Magnesium					
		Al	Zn	Mn	Th	Zr	Rare Earths
AZ80A	M11800	8.5	0.5	0.12	—	—	—
HK31A	M13310	—	—	—	3.0	0.6	—
ZK60A	M16600	—	5.5	—	—	0.45	—
AM60A	M10600	6.0	—	0.13	—	—	—
EZ33A	M12330	—	2.7	—	—	0.6	3.3
AZ91A	M11910	9.0	—	0.13	—	0.7	—

Source: *ASM Handbook*, American Society for Metals, Metals Park, OH, 1985.

Surface and Interfacial Energies

SURFACE TENSIONS

Material	Temperature (°C)	Surface Tension (ergs/cm^2)
Ag[1]	1000	920
Al$_2$O$_3$[1]	2080	700
Al$_2$O$_3$[2]	2320	630
BaO[2]	2196	520
B$_2$O$_3$[1]	900	80
Bi$_2$O$_3$[2]	1098	213
BeO[2]	2843	415
CaO[2]	2860	670
CoO[2]	2078	550
Cu[1]	1120	1270
Cr$_2$O$_3$[2]	2573	812
FeO[1]	1420	585
FeO[2]	1641	545
GeO$_2$[2]	1389	250
H$_2$O[1]	25	72

(*continued*)

An Introduction to Materials Engineering and Science: For Chemical and Materials Engineers,
by Brian S. Mitchell
ISBN 0-471-43623-2 Copyright © 2004 John Wiley & Sons, Inc.

Material	Temperature (°C)	Surface Tension (ergs/cm^2)
K_2O^2	1400	156
$La_2O_3{}^2$	2573	560
MgO^2	3073	660
MnO^2	2058	630
$MoO_3{}^2$	1068	70
Na_2O^2	1400	308
$NaCl^1$	801	114
$Na(SO_4)^1$	884	196
$NaPO_3{}^1$	620	209
$Nb_2O_5{}^2$	1773	279
Pb^1	350	442
PbO^2	900	132
Pt^1	1770	1865
$P_2O_5{}^2$	836	60
$SiO_2{}^2$	1993	307
$Sm_2O_3{}^2$	2593	815
Sodium silicate^1	1000	250
$Ta_2O_5{}^2$	2150	280
$TiO_2{}^2$	2143	380
$Ti_2O_3{}^2$	2090	584
$V_2O_5{}^2$	943	80
$WO_3{}^2$	1743	100

Source: (1) Kingery, W. D., H. K. Bowen and D. R. Uhlmann, *Introduction to Ceramics*, John Wiley and Sons, New York, 1976.
(2) Eustathopoulos, N., M. G. Nicholas and B. Drevet, *Wettability at High Temperatures*, Pergamon, Oxford, 1999.

SOLID-SURFACE ENERGIES

System	Temperature (°C)	Surface Energy (ergs/cm^2)
Ag[1]	1183	1140 ± 90
Al$_2$O$_3$[2]	1850	905
Au[1]	1273	1400 ± 50
BaF$_2$, N$_2(l)$[2]	77 K	300
CaF$_2$, N$_2(l)$[2]	77 K	500
CaCO$_3$, N$_2(l)$[2]	77 K	300
Co[1]	1678	2282 ± 300
Cu[1]	1243	1650 ± 100
γ-Fe[1]	1648	2150 ± 325
δ-Fe[1]	1723	2220 ± 250
In[1]	420	674 ± 74
LiF, N$_2(l)$[2]	77 K	400
Mica, vacuum[2]	298 K	4500
MgO, N$_2(l)$[2]	77 K	1500
MgO[2]	25	1000
NaCl, N$_2(l)$[2]	77 K	300
Nb[1]	2523	2100 ± 100
Ni[1]	1488	2385 ± 100
Pb[1]	590	620 ± 20
Si, N$_2(l)$[2]	77 K	1800
Sapphire, $(10\bar{1}1)$[2] plane	298 K	6000
Sapphire, $(11\bar{2}3)$[2] plane	293 K	24,000
Sn[1]	488	685
Ti[1]	1873	1700

(*continued*)

System	Temperature (°C)	Surface Energy (ergs/cm^2)
TiC^2	1100	1190
Zn^1	653	830
Polymers (from critical surface tension)^3		
Polytetrafluoroethylene		19
Poly(dimethyl siloxane)		24
Poly(vinylidine fluoride)		25
Poly(vinyl fluoride)		28
Polyethylene		31
Polystyrene		33
Poly(hydroxyethyl methacrylate)		37
Poly(vinyl alcohol)		37
Poly(methyl methacrylate)		39
Poly(vinyl chloride)		39
Polycaproamide(Nylon 6)		42
Poly(ethylene oxide)-diol		43
Polyethylene terephthalate		43
Polyacrylonitrile		50

Source: (1) Eustathopoulos, N., M. G. Nicholas and B. Drevet, *Wettability at High Temperatures*, Pergamon, Oxford, 1999.
(2) Kingery, W. D., H. K. Bowen and D. R. Uhlmann, *Introduction to Ceramics*, John Wiley and Sons, New York, 1976.
(3) Ratner, B. D., A.S. Hoffman, F. J. Schoen, and J. E. Lemons, *Biomaterials Science*, Academic Press, New York, 1996.

INTERFACIAL ENERGIES

System	Temperature ($^\circ$C)	Interfacial Energy (ergs/cm^2)
$Al_2O_3(s)$–silicate glaze (l)	1000	<700
$Al_2O_3(s)$–Pb(l)	400	1440
$Al_2O_3(s)$–Ag(l)	1000	1770
$Al_2O_3(s)$–Fe(l)	1570	2300
$Ag(s)$–$Na_2SiO_3(l)$	900	1040
$Cu(s)$–$Na_2SiO_3(l)$	900	1500
$Cu(s)$–$Cu_2S(l)$	1131	90
$MgO(s)$–Ag (l)	1300	850
$MgO(s)$–Fe (l)	1725	1600
SiO_2(glass)–sodium silicate (l)	1000	<25
SiO_2(glass)–Cu (l)	1120	1370
TiC (s)–Cu (l)	1200	1225

Source: Kingery, W. D., H. K. Bowen and D. R. Uhlmann, *Introduction to Ceramics*, John Wiley and Sons, New York, 1976.

Thermal Conductivities
of Selected Materials

All values in units of W/m · K. Divide by 1.73 to obtain values in Btu/hr/ft^2/°F/ft. Divide by 419 to obtain values in cal/cm · K · s. Consult references for values at additional temperatures.

ELEMENTS, METALS, AND ALLOYS

	−190°C	0°C	20°C	100°C	1000°C

Source: Geankoplis, C. J., *Transport Processes and Unit Operations*, Prentice-Hall, Englewood Cliffs, 1993.

	−190°C	0°C	20°C	100°C	1000°C
Aluminum		202		—	
Copper		388		377	
Steel		—		45	

Source: Kittel, C., *Introduction to Solid State Physics*, John Wiley and Sons, New York, 1957.

	−190°C	0°C	20°C	100°C	1000°C
Aluminum	418	226			
Cadmium (‖hex. axis)	92.1	83.7			
Cadmium (⊥ hex. axis)	113	105			
Copper	578	394			
Gold	—	306			
Iron	184	92.1			
Magnesium	188	172			
Nickel	113	83.7			
Silver	427	418			
Sodium	155	138			

Source: Gaskell, D. R., *Transport Phenomena in Materials Engineering*, MacMillan, New York, 1992.

	−190°C	0°C	20°C	100°C	1000°C
1% carbon steel			43		
1% chrome steel			62		
304 stainless steel (18% Cr, 8% Ni)			14		
60% Pt, 40% Rh			46		
Al Bronze (90% Cu, 10% Al)			49		
Aluminum			236		
Beryllium			218		
Boron			31.7		
Brass (70% Cu, 30% Zn)			110		

	−190°C	0°C	20°C	100°C	1000°C
Cadmium			104		
Cast iron			52		
Chromium			95		
Cobalt			104		
Constantan (55% Cu, 45% Ni)			22		
Copper			401		
Germanium			67		
Gold			318		
Iron			83		
Lead			36		
Lithium			79		
Magnesium			157		
Manganese			7.7		
Molybdenum			139		
Nichrome (80% Ni, 10% Cr)			12		
Nickel			94		
Platinum			72		
Rhenium			49		
Rhodium			151		
Silicon			168		
Silver			428		
Sodium			135		
Tin			68		
Titanium			22		
Tungsten			182		
Uranium			27		
Vanadium			31		
Wrought iron			59		
Zinc			122		
Zirconium			23		

CERAMICS

	−190°C	0°C	20°C	100°C	1000°C

Source: Kingery, W. D., H. K. Bowen and D. R. Uhlmann, *Introduction to Ceramics*, John Wiley and Sons, New York, 1976.

	−190°C	0°C	20°C	100°C	1000°C
Al_2O_3				30.2	6.28
BeO				220	20.5
Fire-clay refractory				1.13	1.55
Fused silica glass				2.00	2.51
Graphite				180	62.8
$MgAl_2O_4$				15.1	5.86
MgO				37.7	7.11
Mullite				5.86	3.77
Porcelain				1.67	1.88
Soda-lime-silica glass				1.67	—
ThO_2				10.5	2.93
TiC				25.1	5.86
TiC cermet				33.5	8.37

(continued)

	−190°C	0°C	20°C	100°C	1000°C
$UO_{2.0}$				10.0	3.35
ZrO_2 (stabilized)				1.97	2.30

Source: *Oxide Handbook*, G. V. Samsanor, editor, Plenum, New York, 1973.

	−190°C	0°C	20°C	100°C	1000°C
CaO				15.2	7.79
NiO				12.4	4.48
TiO_2				6.53	3.31
VO_2				9.80	3.41
ZrO_2				1.97	2.0

Source: Kittel, C., *Introduction to Solid State Physics*, John Wiley and Sons, New York, 1957.

	−190°C	0°C	20°C	100°C	1000°C
CaF_2					
KCl					
KF					
NaCl					

Source: Gaskell, D. R., *Transport Phenomena in Materials Engineering*, MacMillan, New York, 1992.

Aluminum oxide	40
Asbestos	0.11
Beryllium oxide	302
Bricks	
Chrome	2.2
Common	0.7
Fireclay	1.0
Magnesite	4.0
Masonry	0.66
Silica	1.1
Cement mortar	0.9
Clay earth	1.4
Coal	0.24
Concrete	1.0
Diatomaceous earth	1.3
Fused quartz	1.3
Granite	3.0
Gypsum plaster	0.5
Limestone	2.0
Magnesium oxide	53
Marble	2.7
Rock wool	0.04
Sand	0.3
Sandstone	2.8

Source: *Materials Science & Engineering Handbook*, J. Shackelford and W. Alexander, editors, CRC Press, Boca Raton, FL, 1992.

Aluminum nitride	30.2	—	—
Aluminum oxide	25.1	—	5.9–6.7
Beryllium oxide	15.9–19.7	—	—
Boron carbide	27.2–28.9	—	—
Boron nitride (∥ to *a* axis)	—	—	12.4
Boron nitride (∥ to *c* axis)	—	—	26.7
Calcium oxide	—	15.5	7.96
Chromium carbide	190	—	—
Chromium diboride	20.5–31.8	—	—
Chromium oxide	10.0–33.0	—	—
Cordierite	3.2	—	—

	−190°C	0°C	20°C	100°C	1000°C
Hafnium carbide			22.2	—	—
Hafnium diboride			6.29	—	—
Hafnium oxide			11.4	—	—
Magnesium oxide			40.6	—	—
Mullite			—	6.1	3.8
Nickel oxide			12.1	—	—
Silicon carbide			—	—	21.4
Silicon nitride			30.2	—	—
Sillimanite			—	1.8	—
Spinel			—	14.7	—
Tantalum carbide			22.2	—	—
Tantalum diboride			10.9	—	—
Thorium oxide			10.1	8.4	2.9
Titanium carbide			17.1–31.0	—	—
Titanium diboride			24.3–26.0	—	—
Titanium nitride			28.9	—	—
Titanium oxide			—	6.7	3.4
Tungsten carbide			84.2	—	—
Uranium dioxide			—	10.5	3.4
Zircon			—	6.1	—
Zirconium carbide			20.5	—	—
Zirconium diboride			23.0–24.3	—	—
Zirconium oxide			—	2.1	—

GLASSES

	−190°C	0°C	20°C	100°C	1000°C

Source: Gaskell, D. R., *Transport Phenomena in Materials Engineering*, MacMillan, New York, 1992.

Glass fiber			0.035		
Glass wool			0.038		
Pyroceram			4.1		
Window glass			0.84		

POLYMERS

	−190°C	0°C	20°C	100°C	1000°C

Source: Geankoplis, C. J., *Transport Processes and Unit Operations*, Prentice-Hall, Englewood Cliffs, 1992.

Hard Rubber			0.15		

Source: *Materials Science & Engineering Handbook*, J. Shackelford and W. Alexander, editors, CRC Press, Boca Raton, FL, 1992.

ABS resins (molded, extruded)					
Heat resistant			0.21–0.35		
High impact			0.21–0.28		
Low temperature impact			0.13–0.24		
Medium impact			0.13–0.31		

(*continued*)

	−190°C	0°C	20°C	100°C	1000°C
Very high impact			0.02–0.24		
Acrylics (cast, molded, extruded)			0.21		
Alkyds (molded)			0.35–1.04		
Allyl diglycol carbonate			2.51		
Cellulose acetate (molded, extruded)			0.17–0.33		
Cellulose acetate butyrate (molded, extruded)			0.17–0.33		
Cellulose acetate propionate (molded, extruded)			0.17–0.33		
Chlorinated polyether			1.57		
Chlorinated polyvinyl chloride			1.64		
Epoxies (cast, molded)			0.17–0.86		
Fluorinated ethylene propylene (FEP)			0.21		
Nylon 6			2.08–2.92		
Nylon 6–6			2.94		
Nylon 6–10			2.60		
Phenylene oxides			1.9–3.1		
Polyacetal			0.22–0.28		
Polyarylsulfone			1.9		
Polycarbonate			0.19		
Polyester, thermoplastic			0.62–0.95		
Polyester, thermoset			0.17–0.21		
Polyethylene (all densities)			0.33		
Polyimides			6.58–11.7		
Polyphenylene sulfide			3.46		
Polypropylene			2.1–2.35		
Polypropylene, high impact			2.98		
Polystyrene			0.04–0.16		
Polytetrafluoroethylene (PTFE)			0.24		
Polytrifluorochloroethylene (PTFCE)			0.25		
Polyvinyl chloride			0.12–0.17		
Polyvinylidene fluoride (PVDF)			0.24		

Source: Gaskell, D. R., *Transport Phenomena in Materials Engineering*, MacMillan, New York, 1992.

	−190°C	0°C	20°C	100°C	1000°C
Foam rubber			0.030		
Hard rubber			0.16		
Polystyrene insulation			0.025		
Teflon			0.35		

COMPOSITES

	−190°C	0°C	20°C	100°C	1000°C

Source: Gaskell, D. R., *Transport Phenomena in Materials Engineering*, MacMillan, New York, 1992.

Plywood			0.12		

Source: *Materials Science & Engineering Handbook*, J. Shackelford and W. Alexander, editors, CRC Press, Boca Raton, FL, 1992.

Alkyd, glass-filled			0.35–0.52		
Polycarbonate, 40% glass-fiber-filled			0.22		
High-strength epoxy laminate			4.1		
Melamine, glass-fiber-filled			0.48		
Melamine, cellulose-filled			0.29–0.35		
Nylon 6, 30% glass-fiber-filled			2.92–5.66		

	−190°C	0°C	20°C	100°C	1000°C
Nylon 6-6, glass-fiber-filled			2.60–5.71		
Nylon 6–10, 30% glass-fiber-filled			6.06		
Polyester, high-strength glass-fiber-filled			2.28–2.91		
Polyimide, glass-filled			6.21		
Polyphenylene sulfide, 40% glass-fiber-filled			3.46		
Polystyrene, 30% glass-fiber-filled			0.20		
Polyurea, cellulose-filled			0.29–0.42		
Phenolic, wood-flour-filled			0.17–0.52		
Phenolic, glass-fiber-filled			0.35		

BIOLOGICS

	−190°C	0°C	20°C	100°C	1000°C

Source: Gaskell, D. R., *Transport Phenomena in Materials Engineering*, MacMillan, New York, 1992.

Human skin			0.37		
Balsa wood			0.55		
Cypress wood			0.097		
Fir wood			0.11		
Maple/Oak wood			0.17		
White pine wood			0.11		
Yellow pine wood			0.15		
Wool			0.038		

Source: Johnson, A. T., *Biological Process Engineering*, John Wiley and Sons, New York, 1999.

Animal skin			0.50		
Beeswax			0.40		
Bone			0.533		
Cat blood			23.0		
Dental amalgam			0.59		
Enamel			0.82		
Fleece			0.14–0.21		
Animal muscle			0.43–0.50		
Human blood			0.507		
Human skin			0.21–0.63		
Human fat			0.21–0.33		
Human muscle			0.41–0.50		
Kidney or liver			0.498		
Porcelain			1.00		
Pork fat			0.187		
Seal blubber			0.190		
Oyster shell			1.95–2.27		
Wool			0.036		
Cork			0.045		
Cotton			0.061		
Cypress wood			0.097		
Fir wood			0.17		
Leaves			0.24–0.50		
Animal coats			0.03–0.15		
Paper			0.13		

Diffusivities in Selected Systems

METALS AND ALLOYS

Source: Geankoplis, C. J., *Transport Processes and Unit Operations*, 3rd ed. Prentice-Hall, 1993.

Diffusing Species	System	$D_0, m^2/s$
H_2	Fe	2.59×10^{-13}
Al	Cu	1.3×10^{-34}

CERAMICS AND GLASSES

Diffusing Ion	System	D_0 cm^2/s	E_a kJ/mol (kcal/mol)	Comment
O^{2-}	Al_2O_3	1.9×10^3	36.3 (152)	Single crystal, $>1600°C$
O^{2-}	Al_2O_3	2.0×10^{-1}	26.3 (110)	Polycrystal, $>1450°C$
O^{2-}	Al_2O_3	6.3×10^{-8}	13.8 (57.6)	Polycrystal, $<1600°C$
O^{2-}	MgO	2.5×10^{-6}	14.9 (62.4)	$1300-1750°C$
O^{2-}	SiO	1.5×10^{-2}	17.0 (71.2)	Vitreous, $925-1225°C$
O^{2-}	ZrO_2 (Ca-stabilized)	1.0×10^{-2}	6.7 (28.1)	15 mol% CaO, $700-1100°C$
Al^{3+}	Al_2O_3	1.0×10^{-2}	27.2 (114)	Polycrystal, $1670-1905°C$
Ca^{2+}	ZrO_2 (Ca-stabilized)	2.8×10^0 (?)	26.1 (109)	16 mol% CaO
Mg^{2+}	MgO	2.3×10^{-1}	18.8 (78.7)	$1400-1600°C$
Mg^{2+}	$MgAl_2O_4$	2.0×10^2	18.6 (78)	–
Zr^{4+}	ZrO_2 (Ca-stabilized)	1.97	26.1 (109)	16 mol% CaO

*Adapted from McCauley, R. A., *Corrosion of Ceramics*, Marcel Dekker, New York, 1995.

An Introduction to Materials Engineering and Science: For Chemical and Materials Engineers, by Brian S. Mitchell
ISBN 0-471-43623-2 Copyright © 2004 John Wiley & Sons, Inc.

Source: Geankoplis, C. J., *Transport Processes and Unit Operations*, 3rd ed. Prentice-Hall, 1993.

Diffusing Species	System	D_0, m^2/s
He	SiO$_2$	$2.4–5.5 \times 10^{-14}$

POLYMERS

Source: Geankoplis, C. J., *Transport Processes and Unit Operations*, 3rd ed. Prentice-Hall, 1993.

Diffusing Species	System	D_0, m^2/s
O$_2$	Vulcanized rubber	2.1×10^{-10}
N$_2$	Vulcanized rubber	1.5×10^{-10}
CO$_2$	Vulcanized rubber	1.1×10^{-10}
H$_2$	Vulcanized Rubber	8.5×10^{-10}

BIOLOGICS

Source: Johnson, A. T., *Biological Process Engineering*, John Wiley and Sons, New York, 1999.

Diffusing Species	System	D_0, m^2/s
H$_2$O	Human skin	1.2×10^{-13}

Mechanical Properties of Selected Materials

METALS AND ALLOYS

Material; Form or Treatment	Modulus of Elasticity (GPa)	Yield Strength (MPa)	Tensile Strength (MPa)	Poisson's Ratio
Wrought Aluminum Alloys; see Appendix 3 for Composition				
1100		34–150	90–165	
2011		295–310	380–405	
2014		97–415	185–485	
2024		76–395	185–495	
2219		76–395	170–475	
2618		370	440	
3003		42–185	110–200	
3004		69–250	180–285	
3105		55–195	115–215	
4032		315	380	
4043		69–270	145–285	
5005		41–195	125–200	

An Introduction to Materials Engineering and Science: For Chemical and Materials Engineers,
by Brian S. Mitchell
ISBN 0-471-43623-2 Copyright © 2004 John Wiley & Sons, Inc.

Material; Form or Treatment	Modulus of Elasticity (GPa)	Yield Strength (MPa)	Tensile Strength (MPa)	Poisson's Ratio
5050		55–200	145–220	
5052		90–255	195–290	
5056		150–405	290–435	
5083		145–285	290–345	
5086		115–255	260–325	
5154		115–270	240–330	
5182		140–395	275–420	
5252		170–240	235–285	
5254		115–270	240–330	
5454		115–310	250–370	
5456		160–255	310–350	
5457		48–185	130–205	
5652		90–255	195–290	
5657		140–165	160–195	
6005		105–240	170–260	
6009		130–325	235–345	
6010		170	255	
6061		55–275	125–310	
6063		48–270	90–290	
6066		83–360	150–395	
6070		69–350	145–380	
6101		76	97	
6151		195	220	

(*continued*)

Material; Form or Treatment	Modulus of Elasticity (GPa)	Yield Strength (MPa)	Tensile Strength (MPa)	Poisson's Ratio
6201		300–310	330	
6205		140–290	260–310	
7075		105–505	230–570	
Cast Aluminum Alloys; see Appendix 3 for Compositions				
201		215–415	365–485	
206		345	435	
A206		345	435	
208		97	145	
242		125–290	185–325	
295		110–220	220–285	
296		130–180	255–275	
308		110	195	
319		125–185	185–280	
336		195–295	250–325	
354		285	380	
355		160–280	195–310	
356		140–210	175–265	
357		290	360	
A357		290	360	
359		255–290	330–345	
360		170	325	
A360		165	320	
380		165	330	

Material; Form or Treatment	Modulus of Elasticity (GPa)	Yield Strength (MPa)	Tensile Strength (MPa)	Poisson's Ratio
383		150	310	
384		165	330	
A384		165	330	
390		240–260	280–330	
A390		180–310	180–310	
413		140	300	
A413		130	290	
443		55	130	
B443		62	159	
C443		110	228	
514		85	170	
518		190	310	
520		180	330	
535		140	275	
712		170	240	
713		150	210–220	
771		275	345	
850		75	160	
Magnesium Alloys; see Appendix 3 for Compositions				
M11800		250	340	
M13310		200	255	
M16600		285	350	
M10600		130	220	

(continued)

Material; Form or Treatment	Modulus of Elasticity (GPa)	Yield Strength (MPa)	Tensile Strength (MPa)	Poisson's Ratio
M12330		110	160	
M11910		150	230	
Titanium Alloys; see Appendix 3 for Compositions				
R54520		807	862	
R54810		951	1000	
R56400		924	993	
R56620		1000	1069	
Ti–10V–2Fe–3Al		1200	1276	
Copper Alloys; see Appendix 3 for Compositions				
Brass, naval		152–276	386–448	
Muntz metal		138	372	
Bronze, aluminum		172–448	483–724	
Beryllium copper		221–717	483–1310	
Brass, free-cutting		303	483	
Nickel silver, 18%		172–483	400–724	
Nickel silver, 13%		124	241	
Constantan		207–448	414–586	
Brass, red		117	241	
Bronze, silicon				
Bronze, tin				
Cupronickel				
Commercially Pure Metals and Elements				
C, natural diamond	700–1200			0.10–0.30

Material; Form or Treatment	Modulus of Elasticity (GPa)	Yield Strength (MPa)	Tensile Strength (MPa)	Poisson's Ratio
C, synthetic diamond	800–925			0.20
C, extruded graphite	11			
C, isostatically molded graphite	11.7			
C, PAN carbon fiber	230–400			
Titanium	107	170	240	0.34
Tantalum	185	165	205	0.35
Tungsten	400	760	960	0.28
Molybdenum	320	500	630	0.32
Gold	77	205	130–220	0.42
Silver	74		170–296	0.37
Platinum	171	< 13.8	125–240	0.39
Tin	44.3	11		0.33
Zinc	104.5		134–186	0.25
Lead	18			0.40
Zirconium	94			
Nickel (cold drawn)	214			0.30
Copper	110			0.36
Aluminum	69			0.33

CERAMICS AND GLASSES

Material; Form or Treatment	Modulus of Elasticity (GPa)	Flectural Strength (MPa)	Compressive Strength (MPa)	Poisson's Ratio	Density g/cm^2
Al$_2$O$_3$	343–392	294–392	274–343	0.32 [1]	3.98
Glass, E-glass fiber	72.5				
BeO	392	147–196		0.34–0.38 [1]	3.02
CeO$_2$	294			0.27–0.31 [3]	7.13
Cr$_2$O$_3$					5.21
MgO	196–294	157–274	490–588	0.36 [1]	3.58
SiO$_2$, cristobalite					2.32
SiO$_2$, quartz	98		1960		2.65
SiO$_2$, silica glass	69	49–98	686–1862		2.20
TiO$_2$, rutile	98–196	69–167	274–823	0.28 [3]	4.24
ZrO$_2$, stabilized	147–196	176–784	980–2940	0.23–0.32 [3]	6.27
ZrO$_2$, monoclinic	245	176–784	980–2940		5.56
ZrO$_2$, high-strength	196	980–1470			5.7–6.1
Al$_6$Si$_2$O$_{13}$, mullite	49–147	108–186	392–588	0.238 [3]	3.16
MgAl$_2$O$_4$, spinel	255	147–167	1667	0.294 [3]	3.58
Mg$_2$Al$_4$Si$_5$O$_{20}$, cordierite	147	118	343–666	0.17–0.26 [3]	2.0–2.5
Sic, α	392–588	441–784	588–4116	0.16–0.17 [2]	3.22
TiC	294–392		745–1352	0.187–0.189 [3]	4.94
B$_4$C				0.207 [3]	2.52
WC	686	343–833	2646–3528	0.24 [3]	15.6
BN, hexagonal	49–78	49–78	69–98		2.27
Si$_3$N$_4$	294–392	490–980	490–784	0.22–0.30 [2]	3.17
GaAs, <100> orientation				0.30 [2]	
Glass, borosilicate (Pyrex)				0.20 [2]	

Material; Form or Treatment	Modulus of Elasticity (GPa)	Flectural Strength (MPa)	Compressive Strength (MPa)	Poisson's Ratio	Density g/cm^2
Glass, soda-line				0.23 [2]	
MoSi$_2$				0.158–0.172 [3]	

Source: Somiya; [1] *Oxide Handbook*; [2] *Callister*; [3] *Materials Handbook*

POLYMERS

Material; Form or Treatment	Modulus of Elasticity (GPa)	Yield Strength (MPa)	Tensile Strength (MPa)	Poisson's Ratio
Aramid, Kevlar 49 fiber	131	—	3600–4100	
Epoxy	2.41	—	27.6–90.0	
Nylon 66	1.59–3.79	55.1–82.8	94.5	0.39
Polybutylene terephthalate	1.93–3.00	56.6–60.0	56.6–60.0	
Polycarbonate	2.38	62.1	62.8–72.4	0.36
Polyester (thermoset)	2.06–4.41	—	41.4–89.7	
Rubber, nitrile	0.0034	—	6.9–24.1	
Rubber, styrene butadiene	0.002–0.010	—	12.4–20.7	
Rubber, silicone		—	10.3	
Phenol–formaldehyde	2.76–4.83	—	34.5–62.1	
Polyetheretherketone	1.10	91	70.3–103	
Polyethylene, low density	0.172–0.282	9.0–14.5	8.3–31.4	
Polyethylene, high density	1.08	26.2–33.1	22.1–31.0	
Polyethylene, ultrahigh molecular weight	0,69	21.4–27.6	38.6–48.3	

(*continued*)

Material; Form or Treatment	Modulus of Elasticity (GPa)	Yield Strength (MPa)	Tensile Strength (MPa)	Poisson's Ratio
Polyethylene terephthalate	2.76–4.14	59.3	48.3–72.4	
Polymethyl methacrylate	2.24–3.24	53.8–73.1	48.3–72.4	
Polypropylene	1.14–1.55	31.0–37.2	31.0–41.4	
Polystyrene	2.28–3.28	—	35.9–51.7	0.33
Polytetrafluoroethylene	0.40–0.55	—	20.7–34.5	0.46
Polyvinyl chloride	2.41–4.14	40.7–44.8	40.7–51.7	0.38

COMPOSITES

Material; Form or Treatment	Modulus of Elasticity (GPa)	Tensile Strength (MPa)	Density (g/cm^3)
Nylon 66/30% milled carbon fiber	8.76	109	1.30
Nylon 66/30% 0.25″ carbon fiber	10.07	241	1.28
Nylon 66/25% milled carbon fiber, 20% 0.25″ glass	11.79	114	1.42
SMC/33% glass fiber	11.7	103	1.67
SMC/35% carbon fiber	26.9	103	1.66
SMC/24% glass fiber, 26% carbon fiber	24	117	1.65
Epoxy/PAN-carbon fiber fabric (8-harness satin)	689	621	
Epoxy/PAN-carbon fiber fabric (plain weave)	66	552	

Material; Form or Treatment	Modulus of Elasticity (GPa)	Tensile Strength (MPa)	Density (g/cm^3)
Epoxy/pitch-carbon fiber fabric (8-harness satin)	40	379	
Epoxy/pitch-carbon fiber fabric (5-harness satin)	90–96.5	331–345	.
Aluminum(220)/35% PAN-carbon fiber (unidirectional)		483 (longitudinal)	
Aluminum(6061)/35% PAN-carbon fiber (cross-ply)		221 (longitudinal) 103 (transverse)	
Aluminum/boron fiber (unidirectional)	0.22	1207 (longitudinal) 10 (transverse)	
Aluminum/boron fiber (cross-ply, 0/90)	0.14	52 (longitudinal) 52 (transverse)	
Glass/short-fiber-reinforced	50–150		
Glass/continuous-fiber-reinforced	1600		

Source: Schwartz, *Matrix/Reinforcement*

BIOLOGICS

Material; Form or Treatment	Modulus of Elasticity (MPa)	Tensile Strength (MPa)	Poisson's Ratio	Density (g/cm^3)	
Arterial wall	1.0	0.5–1.72			Silver
Hyaline cartilage	0.4–19	1.3–18			Silver
Skin	6–40	2.5–16			Silver
Tendon/ligament	65–2500	30–300			Silver
Cortical bone	16,000–20,000	30–211			Silver
Cancellous bone	4600–15,000	51–93			Silver
Hydroxyapatite	19,000	600			Silver
Dentin, mineralized	14,700		0.31	2.14–2.18	O'Brien
Dentin, demineralized	260	29,600			O'Brien
Enamel	84,000–130,000	10.3	0.33	2.95–2.97	O'Brien

Source: Silver, *Biomaterials Science and Biocompatibility*

Electrical Conductivity of Selected Materials

Resistivities and conductivities are the numerical inverses of one another, unless both are listed, indicating two separate sources for the information. Both are highly temperature sensitive. Values listed here are nominally at room temperature. Consult Section 6.1 for further information on compositional and temperature dependences.

METALS AND ALLOYS

All elemental metals are commercially pure unless otherwise indicated.

Material	Conductivity, mho/m	Resistivity, ohm-m	Source
Silver	6.3×10^7		[1]
Copper	5.85×10^7		[1]
Gold	4.25×10^7		[1]
Aluminum	3.45×10^7		[1]
Al–1.2% Mn alloy	2.95×10^7		[1]
Sodium	2.1×10^7		[1]
Zinc		62.0×10^{-7}	[2]
Zirconium, grade 702		3.97×10^{-7}	[2]
Solder (60Sn–40Pb)		1.50×10^{-7}	[2]
Tin		1.11×10^{-7}	[2]
Lead		2.06×10^{-7}	[2]

(*continued*)

An Introduction to Materials Engineering and Science: For Chemical and Materials Engineers,
by Brian S. Mitchell
ISBN 0-471-43623-2 Copyright © 2004 John Wiley & Sons, Inc.

Material	Conductivity, mho/m	Resistivity, ohm-m	Source
Stainless steel, 301	0.14×10^7		[1]
Nichrome (80% Ni, 20% Cr)	0.093×10^7		[1]
Carbon, graphite	10^5		[1]
Germanium	2.2		[1]
Silicon (intrinsic)	4.3×10^{-4}		[1]
Carbon, diamond (natural)	$<10^{-14}$		[1]
Molybdenum		5.2×10^{-8}	[2]
Tantalum		13.5×10^{-8}	[2]
Tungsten		5.3×10^{-8}	[2]
Platinum		10.60×10^{-8}	[2]
Titanium		4.2×10^{-7}–5.2×10^{-7}	[2]
Titanium alloy Ti–6Al–4 V		17.1×10^{-7}	[2]
Titanium alloy Ti–5Al–2.5Sn		15.7×10^{-7}	[2]
Carbon, diamond (synthetic)		1.5×10^{-2}	[2]
Steel alloy, A36, 1040, 1020		1.6×10^{-7}	[2]
Steel alloy, 4140		2.2×10^{-7}	[2]
Steel alloy, 4340		2.48×10^{-7}	[2]
Stainless steel, 304		7.2×10^{-7}	[2]
Stainless steel, 316		7.4×10^{-7}	[2]
Stainless steel, 405, 440A		6.0×10^{-7}	[2]
Stainless steel, 17-7PH		8.3×10^{-7}	[2]
Iron, gray, G1800		15×10^{-7}	[2]
Iron, gray, G3000		9.5×10^{-7}	[2]
Iron, gray, G4000		8.5×10^{-7}	[2]
Iron, 60-40-18		5.5×10^{-7}	[2]
Iron, 80-55-06, 120-90-02		6.2×10^{-7}	[2]

Material	Conductivity, mho/m	Resistivity, ohm-m	Source
Aluminum alloy, 1100		2.9×10^{-8}	[2]
Aluminum alloy, 2024		3.4×10^{-8}	[2]
Aluminum alloy, 6061		3.7×10^{-8}	[2]
Aluminum alloy, 7075		5.22×10^{-8}	[2]
Aluminum alloy, 356.0		4.42×10^{-8}	[2]
Copper alloy, C11000		1.72×10^{-8}	[2]
Copper alloy, C172000 (beryllium–copper)		5.7×10^{-8}–1.15×10^{-7}	[2]
Copper alloy, C26000		6.2×10^{-8}	[2]
Copper alloy, C36000		6.6×10^{-8}	[2]
Copper alloy, C71500 (copper–30% nickel)		37.5×10^{-8}	[2]
Copper alloy, C93200		14.4×10^{-8}	[2]
Magnesium alloy, AZ31B		9.2×10^{-8}	[2]
Magnesium alloy, AZ91D		17×10^{-8}	[2]
Nickel 200		0.95×10^{-7}	[2]
Inconel 625		12.9×10^{-7}	[2]
Monel 400		5.47×10^{-7}	[2]
Invar		8.2×10^{-7}	[2]
Haynes alloy 25		8.9×10^{-7}	[2]
Kovar		4.9×10^{-7}	[2]
Mercury	1.04×10^6		[3]

CERAMICS AND GLASSES

Material	Conductivity, mho/m	Resistivity, ohm-m	Source
Al_2O_3	10^{-10}–10^{-12}		[1]
$Al_6Si_2O_{13}$ (mullite)		$>10^{12}$	[4]

(continued)

Material	Conductivity, mho/m	Resistivity, ohm-m	Source
AlN		$2 \times 10^9 - 10^{11}$	[4]
B_4C		$3 \times 10^{-3} - 8 \times 10^{-3}$	[4]
BeO	$10^{-12} - 10^{-15}$		[1]
BN		1.7×10^{11}	[4]
CrB_2		21×10^{-8}	[4]
E-glass, fiber		4×10^{14}	[2]
GaAs (intrinsic)		10^6	[2]
Glass, soda-lime		$10^{10} - 10^{11}$	[2]
Glass, window	$<10^{-10}$		[1]
Glass, silica	$<10^{-16}$		[1]
Glass-ceramic (Pyroceram)		2×10^{14}	[2]
HfB_2		10^{-7}	[4]
HfC		6×10^{-7}	[4]
InSb	17×10^3		[3]
$Mg_2Al_4Si_5O_{18}$ (Cordierite)		10^{12}	[4]
MgO		1.3×10^{13}	[4]
Mica	$10^{-11} - 10^{-15}$		[1]
$MoSi_2$		21.5×10^{-8}	[4]
PbS	38.4		[3]
Si_3N_4		$>10^{12}$	[2]
SiC (highly structurally-dependent)		$1.0 - 10^{10}$	[4]
SiO_2, fused		$>10^{18}$	[2]
TaB_2		68×10^{-8}	[4]
TaC		25×10^{-8}	[4]
TiB_2		$6.6 \times 10^{-8} - 28.4 \times 10^{-8}$	[4]
TiC		$0.3 \times 10^{-2} - 0.8 \times 10^{-2}$	[4]

Material	Conductivity, mho/m	Resistivity, ohm-m	Source
TiC	0.17×10^7		[1]
TiN		$11 \times 10^{-8} - 13 \times 10^{-7}$	[4]
TiO$_2$		$>10^{11}$	[5]
WC		5×10^{-7}	[5]
WSi$_2$		$33.4 \times 10^{-8} - 54.9 \times 10^{-8}$	[4]
ZrB$_2$		9.2×10^{-8}	[4]
ZrC		$61 \times 10^{-8} - 64 \times 10^{-8}$	[4]
ZrO$_2$, 3%Y$_2$O$_3$		10^{10}	[2]
ZrO2, monoclinic		$>10^{11}$	[5]

POLYMERS

Material	Conductivity, mho/m	Resistivity, ohm-m	Source
Polymethylmethacrylate	$<10^{-12}$		[1]
Polyethylene	$<10^{-14}$		[1]
Polystyrene	$<10^{-14}$		[1]
Polytetrafluoroethylene	$<10^{-16}$		[1]
Phenol-formaldehyde	$10^{-7} - 10^{-11}$		[1]
Polyethylene, low and high density		$10^{15} - 5 \times 10^{16}$	[2]
Polyethylene, ultrahigh molecular weight		$>5 \times 10^{14}$	[2]
Polypropylene		$>10^{14}$	[2]
Rubber, nitrile		3.5×10^8	[2]
Rubber, styrene butadiene		6×10^{11}	[2]
Rubber, silicone		10^{13}	[2]
Epoxy		$10^{10} - 10^{13}$	[2]
Polyamide (Nylon 66)		$10^{12} - 10^{13}$	[2]

(continued)

Material	Conductivity, mho/m	Resistivity, ohm-m	Source
Polybutylene terephthalate		4×10^{14}	[2]
Polycarbonate		2×10^{14}	[2]
Polyester		10^{13}	[2]
Polyetheretherketone		6×10^{14}	[2]
Polyethylene terephthalate		10^{12}	[2]
Polyvinyl chloride		$>10^{14}$	[2]
Cellulose acetate		$10^8 - 10^{13}$	[4]
Polyimide		4×10^{13}	[4]
Polyarylsulfone		$3.2 \times 10^{14} - 7.71 \times 10^{14}$	[4]

COMPOSITES

Material	Conductivity, mho/m	Resistivity, ohm-m	Source
Wood		$10^{14} - 10^{16}$	[2]
Concrete, dry		10^9	[2]
Polyimide, glass-fiber-reinforced		9.2×10^{13}	[4]
Nylon, glass-fiber-reinforced		$2.6 \times 10^{13} - 5.5 \times 10^{13}$	[4]
Polyester, glass-fiber-reinforced		$3.2 \times 10^{14} - 3.3 \times 10^{14}$	[4]
Phenylene oxide, glass-fiber-reinforced		10^{15}	[4]
Polyphenylene sulfide, 40% glass-fiber-reinforced		4.5×10^{12}	[4]
Polystyrene, 30% glass-fiber-reinforced		3.6×10^{14}	[4]
Polyacetal, 20% glass-fiber-reinforced		5×10^{12}	[4]

BIOLOGICS

Material	Conductivity, mho/cm	Resistivity, ohm-m	Source
Cortical bone, dry	10^{-9}		[6]
Cortical bone, wet (10% moisture)	10^{-3}	$3-4 \times 10^3$	[6]
Tissues, soft		$5-15$	[6]

[1] Ralls, K. M., T. H. Courtney and J. Wolff, *Introduction to Materials Science and Engineering*, John Wiley and Sons, New York, 1976.

[2] Callister, W., *Materials Science and Engineering. An Introduction*, 5th Ed. John Wiley & Sons, New York, 2000.

[3] Shackelford, J. F., *Introduction to Materials Science for Engineers*, 5th Ed., McGraw-Hill, New York, 2000.

[4] *Materials Science & Engineering Handbook*, J. Shackelford and W. Alexander, editors, CRL Press, Boca Raton, FL, 1992.

[5] Somiya, S., *Advanced Technical Ceramics*, Academic Press, Oxford, 1984.

[6] Brighton, C. T., J. Black and S. R. Pollack, *Electrical Properties of Bone and Cartilage*, Grune and Stratton, New York, 1979.

Refractive Index of Selected Materials

CERAMICS, GLASSES, AND IONIC SOLIDS

Material	Average Refractive Index	Source
Al_2O_3 (corundum, sapphire)	1.76	[1]
$Al_6Si_2O_{13}$ (mullite)	1.64	[1]
As_2S_3	2.66	[6]
BaF_2	1.48	[8]
$BaTiO_3$	2.40	[6]
C (diamond)	2.417	[2]
$CaCO_3$ (calcite)	1.65	[6]
CaF_2	1.43	[2]
CaO (lime)	1.84	[3]
CdTe	2.74	[6]
CuO	2.84	[3]
Glass, flint	1.65	[2]
Glass, soda–lime–silica	1.51–1.52	[1]
Glass from orthoclase	1.51	[1]
Glass, silica	1.458	[1]
Glass from albite	1.49	[1]
Glass, borosilicate (Pyrex)	1.47	[1]
$H_2O(l)$	1.33	[2]
$H_2O(s)$	1.30	[2]
$KAlSi_3O_8$ (orthoclase)	1.525	[1]
KBr	1.56	[8]
KCl	1.49	[2]
KI	1.673	[8]
LiF	1.392	[6]
$MgAl_2O_4$ (spinel)	1.72	[1]

An Introduction to Materials Engineering and Science: For Chemical and Materials Engineers, by Brian S. Mitchell
ISBN 0-471-43623-2 Copyright © 2004 John Wiley & Sons, Inc.

Material	Average Refractive Index	Source
MgF_2	1.38	[8]
MgO (periclase)	1.74	[1]
MnO (manganosite)	1.22	[3]
$NaAlSi_3O_8$ (albite)	1.529	[1]
NaCl	1.54	[2]
NaF	1.326	[6]
PbO (litharge)	2.67	[3]
PbS (galena)	3.91	[3]
Si	3.29	[6]
SiC	2.68	[3]
SiO_2 (quartz)	1.55	[1]
SrO	1.87	[2]
$SrTiO_3$	2.49	[6]
TiO_2 (rutile)	2.68	[6]
TiO_2 (anatase)	2.55	[3]
Y_2O_3	1.92	[6]
$ZiSiO_4$ (zircon)	1.95	[6]
ZnO	2.00	[3]
ZnSe	2.62	[6]
ZrO_2	2.1–2.2	[3]

POLYMERS

Material	Refractive Index	Source
Cellulosics (cellulose acetate)	1.46–1.50	[1]
Epoxies	1.55–1.60	[1]
Phenol-formaldehyde	1.47–1.50	[1]
Polyamides; e.g., Nylon 66	1.53	[1]
Polyarylsulfone	1.651	[4]
Polybutadiene/polystyrene copolymer	1.53	[1]
Polycarbonate	1.586	[4]
Polychloroprene	1.55–1.56	[1]
Polyester	1.52–1.57	[7]
Polyethylene, low density	1.51	[1]
Polyethylene terephthalate	1.64	[7]
Polyethylene, high density	1.545	[1]
Polyisoprene	1.52	[1]

(*continued*)

Material	Refractive Index	Source
Polymethylmethacrylate	1.49	[5]
Polypropylene	1.47	[1]
Polystyrene	1.59	[1]
Polytetrafluoroethylene	1.35–1.38	[1]
Polyvinyl chloride	1.54–1.55	[1]
Polyvinylidene fluoride	1.42	[4]
Urethanes	1.5–1.6	[1]

[1] Shackelford, J. F., *Introduction to Materials Science for Engineers*, 5th Ed., McGraw-Hill, New York, 2000.

[2] Ralls, K. M., T. H. Courtney and J. Wolff, *Introduction to Materials Science and Engineering*, John Wiley and Sons, New York, 1976.

[3] *Oxide Handbook*, G. V. Samsanor, Editor, Plenum, New York, 1973.

[4] *Materials Science & Engineering Handbook*, J. Shackelford and W. Alexander, editors, CRL Press, Boca Raton, FL, 1992.

[5] Callister, W., *Materials Science and Engineering. An Introduction*, 5th Ed. John Wiley & Sons, New York, 2000.

[6] Kingery, W. D., H. K. Bowen and D. R. Uhlmann, *Ceramics*, John Wiley and Sons, New York, 1976.

[7] Billmeyer, F. W., *Textbook of Polymer Science*, 3rd Edition John Wiley and Sons, New York, 1984.

[8] Fox, M., *Optical Properties of Solids*, Oxford University Press, New York, 2001.

Answers to Selected Problems

Chapter 1

1.I.2 63.2% for TiO_2, 1.0% for InSb.

1.I.5 (c) [010].

1.I.6 (a) No, it is a thermoset. (b) Yes, it is a thermoplastic.

1.I.12 6, 2 (actually, 3), 8 (actually 6), 6, 4, 6. Discrepancies are due to a combination of uncertainty in the estimation of ionic radii and bond directionality due to partially covalent character.

1.I.16 Gln $\rightarrow$ Pro, the others are structurally similar.

1.II.4 FCC: (111) 3, (200) 4, (220) 8, (311) 11, (222) 12; BCC (110) 2, (200) 4, (211) 6, (220) 8, (310) 10.

1.II.6 The molecular weight of the PMMA repeat unit is $M_0 = 5(12) + 2(16) + 8(1) = 100$ g/mol. The number-average molecular weight can be calculated from the polydispersity index

$$\overline{M}_n = \frac{\overline{M}_w}{2.2} = \frac{295000}{2.2} = 134,091,$$

which can then be used to calculate the number-average degree of polymerization

$$\overline{x}_n = \frac{\overline{M}_n}{M_0} = \frac{134,091}{100} = 1340$$

Chapter 2

2.I.3 $F = 2 - 2 + 1 = 1$; As a practical matter, we may retain this two-phase microstructure upon heating or cooling. But such a temperature change exhausts the freedom of the system and must be accompanied by changes in composition.

2.I.4 880 g α, 120 g Fe_3C

2.I.5 (a) $m_L = 1$ kg, $m_\alpha = 0$ kg; (b) $m_L = 667$ g, $m_\alpha = 333$ g; (c) $m_L = 0$ kg, $m_\alpha = 1$ kg.

An Introduction to Materials Engineering and Science: For Chemical and Materials Engineers, by Brian S. Mitchell
ISBN 0-471-43623-2 Copyright © 2004 John Wiley & Sons, Inc.

Chapter 3

3.I.1 $\Delta V = +0.153$ V.

3.I.4 $r = 4.42 \times 10^{-3}$ min^{-1}.

Chapter 4

4.I.1 4.74×10^{-3} kg-m^{-1}-s^{-1}.

4.I.2 5.8×10^{-3} Pa-s.

4.I.4   4.7; 9.0.

4.I.6 $J = 2.4 \times 10^{-9}$ kg/m^2-s.

4.II.1   2.83 W/m-K; 2839 W/m^2; 62.1°C.

4.II.2 $J_H = 2.32 \times 10^{-7}$ kg/m^2-s.

4.III.3 $\ln \mu = -2.49 + 15004/(T - 253)$.

4.III.4 $v_0 = 7.12 \times 10^{-13}(\Delta P/L) + 2.13 \times 10^{-17}(\Delta P/L)^{1.88}$.

4.III.5

$$Q_{tot} = Q_A + Q_B + Q_C = \left(\frac{k_A A_A}{\Delta x} + \frac{k_B A_B}{\Delta x} + \frac{k_C A_C}{\Delta x} \right)(T_1 - T_2)$$

$$= \left(\frac{1}{R_A} + \frac{1}{R_B} + \frac{1}{R_C} \right)(T_1 - T_2)$$

Chapter 5

5.I.1   108 GPa.

5.I.2 $\tau_{cr} = 20.2$ MPa.

5.I.3 310×10^3 MPa.

5.I.6   5.74 mm.

5.I.7   (a) 252 kJ/mol; (b) 1.75×10^{-5}% per hour.

5.I.8   57 MPa.

5.II.2 $E = 280$ GPa.

5.III.1   For sample GBC50 (a) 3 mm; (b) 0.28; (c) m = 389 kg.

5.III.2   For GBC50, $a = 0.2$ mm.

Chapter 6

6.I.1   0.16 ohm.

6.I.2   0.016 in. (0.04 cm).

6.I.3   (a) 10^5 mho/cm; (b) 6×10^4 mho/cm.

6.I.4 (a) 5.15×10^{22} atoms/cc; b) 3.68×10^{15} carrier electrons/cc.

6.I.5 (a) 1.4×10^7 silicon atoms per conduction electron (also 1.4×10^7 silicon atom per phosphorus atom); (b) $(655 \text{ Å})^3$.

6.I.6 (a) 1600 cm^2/vol-s; (b) Al, In, Ga.

6.I.7 4.33×10^{-3} m^2/V-s.

6.I.8 $H = 150$ A/m.

6.I.10 $R = 0.041$ and $T = 0.92$, assuming no absorption since glass is transparent.

6.I.11 $n_{\text{fused silica}} = 1.46$, $n_{\text{dense flint}} = 1.65$.

6.II.1 (a) $\sigma = 61.4$ (ohm-m)$^{-1}$; (b) $n = p = 1.16 \times 10^{21}$ m^{-3}.

6.II.4 18 μm.

Chapter 7

7.I.4 600 s^{-1}.

7.III.1 $n = 0.33$; Rabinowitsch factor $= (3n + 1)/4n = 1.5$.

Chapter 8

8.I.1 Metals that are generally stiffer than brass include steels, nickel alloys, molybdenum alloys, and tungsten alloys. All but steel are generally expensive. Molybdenum and tungsten alloys have high melting points.

8.I.2 Materials with a tensile modulus $E > 200$ GPa and density $\rho < 2$ Mg/m^3 include beryllium alloys and unaxial carbon-fiber-reinforced polymers. Beryllium alloys are both expensive and toxic.

An Introduction to Materials Engineering and Science: For Chemical and Materials Engineers,
by Brian S. Mitchell
ISBN 0-471-43623-2 Copyright © 2004 John Wiley & Sons, Inc.

Name	Symbol	Atomic Number	Atomic Weight
thallium	Tl	81	204.3833 (2)
thorium	Th	90	232.0381 (1)
thulium	Tm	69	168.93421 (2)
tin	Sn	50	118.71 (7)
titanium	Ti	22	47.867 (1)
tungsten	W	74	183.84 (1)
ununbium	Uub	112	[277]
ununnilium	Uun	110	[269]
unununium	Uuu	111	[272]
uranium	U	92	238.0289 (1)
vanadium	V	23	50.9415 (1)
xenon	Xe	54	131.29 (2)
ytterbium	Yb	70	173.04 (3)
yttrium	Y	39	88.90585 (2)
zinc	Zn	30	65.39 (2)
zirconium	Zr	40	91.224 (2)

The number in parentheses following the atomic weight value gives the uncertainty in the last digit. An entry in [square brackets] indicates the mass number of the longest-lived isotope of an element that has no stable isotopes and for which a standard atomic weight cannot be defined because of wide variability in isotopic composition (or complete absence) in nature.

Frequently Used Physical Constants

Quantity	Symbol	Value	Unit
UNIVERSAL			
standard acceleration of gravity	g	9.80665	$m\ s^{-2}$
speed of light in vacuum	$c,\ c_0$	299 792 458	$m\ s^{-1}$
magnetic permeability	μ_0	$4\pi \times 10^{-7}$	$N\ A^{-2}$
electric permittivity	ε_0	$8.854187817\ldots \times 10^{-12}$	$F\ m^{-1}$
Planck constant	h	$6.62606876(52) \times 10^{-34}$	$J\ s$
elementary charge	e	$1.602176462(63) \times 10^{-19}$	C
Bohr magneton	μ_B	$9.27400899(37) \times 10^{-24}$	$A \cdot m^2$
PHYSICO-CHEMICAL			
Avogadro constant	N_A	$6.02214199(47) \times 10^{23}$	mol^{-1}
Faraday constant	F	96 485.3415(39)	$C\ mol^{-1}$
molar gas constant	R	8.314 472(15)	$J\ mol^{-1}K^{-1}$
Boltzmann constant R/N_A	k_B	$1.380 6503(24) \times 10^{-23}$	$J\ K^{-1}$

Source: Peter J. Mohr and Barry N. Taylor, CODATA Recommended Values of the Fundamental Physical Constants: 1998, *Journal of Physical and Chemical Reference Data*, Vol. 28, No. 6, 1999 and *Reviews of Modern Physics*, Vol. 72, No. 2, 2000.

Conversion Factors

Name	To Convert From	To	Multiply By	Divide By
Acceleration	$ft \cdot s^{-2}$	$m \cdot s^{-2}$	0.3048	3.2810
Density	$lb \cdot ft^{-3}$	$kg \cdot m^{-3}$	16.02	6.243×10^{-2}
Energy	BTU	J	1055	9.478×10^{-4}
Energy	cal	J	4.1859	0.2389
Energy	erg	J	1.000×10^{-7}	1.000×10^{7}
Energy	ev	J	1.602×10^{-19}	6.242×10^{18}
Energy	ft*lbf	J	1.3557	0.7376
Force	dyne	N	1.000×10^{-5}	1.000×10^{5}
Force	lbf	N	4.4484	0.2248
Heat capacity	$BTU \cdot lb^{-1} \cdot {}^{\circ}F^{-1}$	$J \cdot kg^{-1} \cdot {}^{\circ}C^{-1}$	4188	2.388×10^{-4}
Heat transfer coefficient	$BTU \cdot hr^{-1} \; ft^{-2} \cdot {}^{\circ}F^{-1}$	$W \cdot m^{-2} \cdot {}^{\circ}C^{-1}$	5.6786	0.1761
Length	ft	m	0.3048	3.2810
Mass	amu	kg	1.661×10^{-27}	6.022×10^{26}
Mass	lb	kg	0.4535	2.2050
Power	$BTU \cdot hr^{-1}$	W	0.2931	3.4120
Power	hp	W	745.71	1.341×10^{-3}
Pressure	bar	Pa	1.000×10^{5}	1.000×10^{-5}
Pressure	$dyne \cdot cm^{-2}$	Pa	0.1000	10.0000
Pressure	in Hg	Pa	3377	2.961×10^{-4}
Pressure	in. water	Pa	248.82	4.019×10^{-3}
Pressure	$lbf \cdot in^{-2}$ (psi)	Pa	6897	1.450×10^{-4}
Pressure	mbar	Pa	100.00	1.000×10^{-2}
Pressure	mm Hg	Pa	133.3	7.501×10^{-3}
Pressure	std atm	Pa	1.013×10^{5}	9.869×10^{-6}
Specific heat	$BTU \cdot lb^{-1} \cdot {}^{\circ}F^{-1}$	$J \cdot kg^{-1} \cdot {}^{\circ}C^{-1}$	4186	2.389×10^{-4}
Specific heat	$cal \cdot g^{-1} \cdot {}^{\circ}C^{-1}$	$J \cdot kg^{-1} \cdot {}^{\circ}C^{-1}$	4186	2.389×10^{-4}
Thermal conductivity	$BTU \cdot hr^{-1} \cdot ft^{-1} \cdot {}^{\circ}F^{-1}$	$W \cdot m^{-1} \cdot {}^{\circ}C^{-1}$	1.7307	0.5778
Thermal conductivity	$BTU \cdot in \cdot hr^{-1} \cdot ft^{-2} \cdot {}^{\circ}F^{-1}$	$W \cdot m^{-1} \cdot {}^{\circ}C^{-1}$	0.1442	6.9340
Thermal conductivity	$cal \cdot cm^{-1} \cdot s^{-1} \cdot {}^{\circ}C^{-1}$	$W \cdot m^{-1} \cdot {}^{\circ}C^{-1}$	418.60	2.389×10^{-3}
Thermal conductivity	$cal \cdot ft^{-1} \cdot hr^{-1} \cdot {}^{\circ}F^{-1}$	$W \cdot m^{-1} \cdot {}^{\circ}C^{-1}$	6.867×10^{-3}	145.62
Torque	ft*lbf	N*m	1.3557	0.7376
Viscosity—absolute	centipoise	$N \cdot s \cdot m^{-2}$	1.000×10^{-3}	1000
Viscosity—absolute	$g \cdot cm^{-1} \cdot s^{-1}$	$N \cdot s \cdot m^{-2}$	0.1000	10
Viscosity—absolute	$lbf \cdot ft^{-1} \cdot s^{-1}$	$N \cdot s \cdot m^{-2}$	47.87	2.089×10^{-2}
Viscosity—absolute	$lb \cdot ft^{-1} \cdot s^{-1}$	$N \cdot s \cdot m^{-2}$	1.4881	0.6720
Volume	U.S. gallons	m^{3}	3.785×10^{-3}	264.20

Printed in the USA/Agawam, MA
September 19, 2017